Evolution Equations

Clay Mathematics Proceedings
Volume 17

Evolution Equations

Clay Mathematics Institute Summer School
Evolution Equations
Eidgenössische Technische Hochschule, Zürich, Switzerland
June 23–July 18, 2008

David Ellwood
Igor Rodnianski
Gigliola Staffilani
Jared Wunsch
Editors

American Mathematical Society
Clay Mathematics Institute

2010 *Mathematics Subject Classification.* Primary 35L05, 35L70, 35P25, 35Q41, 35Q55, 35Q76, 58J40, 58J47, 58J50, 83C57, 35-XX, 42-XX, 53-XX, 58-XX, 83-XX.

Library of Congress Cataloging-in-Publication Data

Clay Mathematics Institute. Summer School (2008 : Zürich, Switzerland)
Evolution equations : Clay Mathematics Institute Summer School, June 23–July 18, 2008, Eidgenössische Technische Hochschule, Zürich, Switzerland / David Ellwood, Igor Rodnianski, Gigliola Staffilani, Jared Wunsch, editors.
pages cm – (Clay mathematics proceedings ; volume 17)
Includes bibliographical references.
ISBN 978-0-8218-6861-4 (alk. paper)
1. Evolution equations. 2. Wave equation. I. Ellwood, D. (David), 1966– editor of compilation. II. Rodnianski, Igor, 1972– editor of compilation. III. Staffilani, Gigliola, 1966– editor of compilation. IV. Wunsch, Jared, 1971– editor of compilation. V. Title.

QC20.7.E88C53 2008
515′.353–dc23

2013002427

Published by the American Mathematical Society, Providence, RI,
for the Clay Mathematics Institute, Cambridge, MA.
Printed in the United States of America.

∞ The paper used in this book is acid-free and falls within the guidelines
established to ensure permanence and durability.
Visit the AMS home page at http://www.ams.org/
Visit the Clay Mathematics Institute home page at http://www.claymath.org/

10 9 8 7 6 5 4 3 2 1 18 17 16 15 14 13

Contents

Preface

This volume contains the lecture notes, rather loosely construed, from a summer school held at ETH in Zürich from June 23 to July 18, 2008. The school was hosted by both the mathematics and the physics departments at ETH, and the organizers would like to thank those departments for their hospitality, and many people, especially Gian Michele Graf, for invaluable organizational assistance. We are also grateful to Jörg Fröhlich and Horst Knörrer at ETH for scientific guidance and to Marcela Krämer from ETH and Amanda Battese and Candace Bott from CMI for their practical help. The dedication of the students in the school contributed inestimably to the breadth and accuracy of these notes, as did the help of a team of anonymous referees. Finally we would like to thank Vida Salahi for her work and dedication in managing the editorial process of this volume, as well as Naomi Kraker for dealing with the final stages of the project.

While we intended from the beginning to emphasize the unity of techniques and outlooks in the broad subject of evolution equations, procrastination and distraction prevented us from coordinating as extensively as we would have liked the content of the main courses. It therefore came as a very happy surprise, on arriving and starting the courses, that the subject of PDE seemed to enforce its unity on us, rather than vice-versa. In the first three weeks of the school, various common threads appeared, some of them anticipated and some not. The role of energy estimates, via commutator and multiplier arguments, had always been envisioned as one of the technical focuses of the school. The appearance of symmetry and approximate symmetry arguments, and of scaling arguments therefore came as no great surprise, arising throughout the main courses. Virial and Morawetz estimates formed the backbone of much of the beginning of the Wunsch-Mazzeo and Staffilani-Raphael courses. The general framework of extending local well-posedness to global via appropriate conserved quantities of course arose essentially in both the Staffilani-Raphaël course on the nonlinear Schrödinger equation and in the Rodnianski-Dafermos course, which focused more on hyperbolic equations arising from Lagrangian field theories. In both the Rodnianski-Dafermos course and that of Wunsch-Mazzeo, a good deal of Riemannian, pseudo-Riemannian, and symplectic geometry was shared, some expressly and some implicitly.

Other common themes shared by the main courses and various of the mini-courses included: The role of mixed long-distance, long-time asymptotics, leading to estimates for the wave operator along and orthogonal to the null cone and to radiation fields and the Lax-Phillips transform; nonlinear evolution equations as many-particle limits of linear many-body quantum-mechanical problems; analysis of blowup regimes through appropriate rescalings and both variational and dynamical techniques. Other central topics were: critical equations and blowup versus scattering; scattering itself, construed in terms of wave operators or in time-dependent formulation; parametrices in position space, in Fourier space, and in phase space and their different uses; local well-posedness via induction on energy; and concentration/compactness arguments.

"Evolution equations" is an area too rich in diverse phenomenology to ever be a single coherent subject, but we hope that this volume illuminates some of the major threads woven through it.

David Ellwood
Igor Rodnianski
Gigliola Staffilani
Jared Wunsch

October 2012

Clay Mathematics Proceedings
Volume **17**, 2013

Microlocal Analysis and Evolution Equations: Lecture Notes from 2008 CMI/ETH Summer School
April 25, 2013

Jared Wunsch

Contents

1. Introduction

The point of these notes, and the lectures from which they came, is not to provide a rigorous and complete introduction to microlocal analysis—many good ones now exist—but rather to give a quick and impressionistic feel for how the subject is used in practice. In particular, the philosophy is to crudely axiomatize the machinery of pseudodifferential and Fourier integral operators, and then to see what problems this enables us to solve. The primary emphasis is on application of commutator methods to yield microlocal energy estimates, and on simple parametrix constructions in the framework of the calculus of Fourier integral operators; the rigorous justification of the computations is kept as much as possible inside a black box. By contrast, the author has found that lecture courses focusing on a careful development of the inner workings of this black box can (at least when he is the lecturer) too easily bog down in technicality, leaving the students with no notion of why one might suffer through such agonies. The ideal education, of course, includes both approaches...

A wide range of more comprehensive and careful treatments of this subject are now available. Among those that the reader might want to consult for supplementary reading are [**17**], [**7**], [**22**], [**24**], [**26**], [**2**], [**28**], [**16**] (with the last three focusing

2010 *Mathematics Subject Classification.* Primary 35L05, 35P25, 35Q41, 58J40, 58J47, 58J50.

on the "semi-classical" point of view, which is not covered here). Hörmander's treatise [**11**], [**12**], [**13**], [**14**] remains the definitive reference on many aspects of the subject.

Some familiarity with the theory of distributions (or a willingness to pick it up) is a prerequisite for reading these notes, and fine treatments of this material include [**11**] and [**6**]. (Additionally, an appendix sets out the notation and most basic concepts in Fourier analysis and distribution theory.)

Much of the hard technical work in what follows has been shifted onto the reader, in the form of exercises. Doing at least some of them is essential to following the exposition. The exercises that are marked with a "star" are in general harder or longer than those without, in some cases requiring ideas not developed here.

The author has many debts to acknowledge in the preparation of these notes. The students at the CMI/ETH summer school were the ideal audience, and provided helpful suggestions on the exposition, as well as turning up numerous errors and inconsistencies in the notes (although many more surely remain). Discussions with Andrew Hassell, Michael Taylor, András Vasy, and Maciej Zworski were very valuable in the preparation of these lectures and notes. Rohan Kadakia kindly corrected a number of errrors in the final version of the manuscript. Finally, the author wishes to gratefully acknowledge Richard Melrose, who taught him most of what he knows of this subject: a strong influence of Melrose's own excellent lecture notes [**17**] can surely be detected here.

The author would like to thank the Clay Mathematics Institute and ETH for their sponsorship of the summer school, and MSRI for its hospitality in Fall 2008, while the notes were being revised. The author also acknowledges partial support from NSF grant DMS-0700318.

2. Prequel: energy methods and commutators

This section is supposed to be like the part of an action movie before the opening credits: a few explosions and a car chase to get you in the right frame of mind, to be followed by a more careful exposition of plot.

2.1. The Schrödinger equation on $\mathbb{R}^n$. Let us consider a solution ψ to the *Schrödinger equation* on $\mathbb{R} \times \mathbb{R}^n$:

$$i^{-1}\partial_t\psi - \nabla^2\psi = 0. \tag{2.1}$$

The complex-valued "wavefunction" ψ is supposed to describe the time-evolution of a free quantum particle (in rather unphysical units). We'll use the notation $\Delta = -\nabla^2$ (note the sign: it makes the operator positive, but is a bit non-standard).

Consider, for any self-adjoint operator A, the quantity

$$\langle A\psi, \psi\rangle$$

where $\langle\cdot,\cdot\rangle$ is the sesquilinear L^2-inner product on $\mathbb{R}^n$. In the usual interpretation of QM, this is the expectation value of the "observable" A. Since $\partial_t\psi = i\nabla^2\psi = -i\Delta\psi$, we can easily find the time-evolution of the expectation of A :

$$\partial_t\langle A\psi, \psi\rangle = \langle \partial_t(A)\psi, \psi\rangle + \langle A(-i\Delta)\psi, \psi\rangle + \langle A\psi, (-i\Delta)\psi\rangle.$$

Now, using the self-adjointness of Δ and the sesquilinearity, we may rewrite this as

$$\partial_t\langle A\psi, \psi\rangle = \langle \partial_t(A)\psi, \psi\rangle + i\langle [\Delta, A]\psi, \psi\rangle \tag{2.2}$$

where $[S, T]$ denotes the *commutator* $ST - TS$ of two operators (and $\partial_t(A)$ represents the derivative of the operator itself, which may have time-dependence). Note that this computation is a bit bogus in that it's a formal manipulation that we've done without regard to whether the quantities involved make sense, or whether the formal integration by parts (i.e. the use of the self-adjointness of Δ) was justified. For now, let's just keep in mind that this makes sense for sufficiently "nice" solutions, and postpone the technicalities.

If you want to learn things about $\psi(t, x)$, you might try to use (2.2) with a judicious choice of A. For instance, setting $A = \mathrm{Id}$ shows that the L^2-norm of $\psi(t, \cdot)$ is conserved. Additionally, choosing $A = \Delta^k$ shows that the H^k norm is conserved (see the appendix for a definition of this norm). In both these examples, we are using the fact that $[\Delta, A] = 0$.

A more interesting example might be the following: set $A = \partial_r$, the radial derivative. We may write the Laplace operator on $\mathbb{R}^n$ in polar coordinates as

$$\Delta = -\partial_r^2 - \frac{n-1}{r}\partial_r + \frac{\Delta_\theta}{r^2}$$

where Δ_θ is the Laplacian on S^{n-1}; thus we compute

$$[\Delta, \partial_r] = 2\frac{\Delta_\theta}{r^3} - \frac{(n-1)}{r^2}\partial_r.$$

EXERCISE 2.1. Do this computation! (Be aware that ∂_r is not a differential operator with smooth coefficients.)

This is kind of a funny looking operator. Note that Δ is self-adjoint, and ∂_r wants to be anti-self-adjoint, but isn't quite. In fact, it makes more sense to replace ∂_r by

$$A = (1/2)(\partial_r - \partial_r^*) = \partial_r + \frac{n-1}{2r},$$

which corrects ∂_r by a lower-order term to be anti-self-adjoint.

EXERCISE 2.2. Show that

$$\partial_r^* = -\partial_r - \frac{n-1}{r}.$$

Trying again, we get by dint of a little work:

$$[\Delta, \partial_r + \frac{n-1}{2r}] = \frac{2\Delta_\theta}{r^3} + \frac{(n-1)(n-3)}{2r^3}, \tag{2.3}$$

provided n, the dimension, is at least 4.

EXERCISE 2.3. Derive (2.3), where you should think of both sides as operators from Schwartz functions to tempered distributions (see the appendix for definitions). What happens if $n = 3$? If $n = 2$? Be very careful about differentiating negative powers of r in the context of distribution theory...

Why do we like (2.3)? Well, it has the very lovely feature that both summands on the RHS are *positive operators.* Let's plug this into (2.2) and integrate on a finite time interval:

$$\begin{aligned} i^{-1}\langle A\psi, \psi\rangle\Big|_0^T &= \int_0^T \left\langle \frac{2\Delta_\theta}{r^3}\psi, \psi \right\rangle + \left\langle \frac{(n-1)(n-3)}{2r^3}\psi, \psi \right\rangle dt \\ &= \int_0^T 2\left\| r^{-1/2}\nabla\psi \right\|^2 dt + \frac{(n-1)(n-3)}{2}\left\| r^{-3/2}\psi \right\|^2 dt, \end{aligned}$$

where $\not\nabla$ represents the (correctly scaled) angular gradient: $\not\nabla = r^{-1}\nabla_\theta$, where ∇_θ denotes the gradient on S^{n-1}.

Now, we're going to turn the way we use this estimate on its head, relative to what we did with conservation of L^2 and H^k norms: the *left*-hand-side can be estimated by a constant times the $H^{1/2}$ norm of the initial data. This should be at least plausible for the derivative term, since morally, half a derivative can be dumped on each copy of u, but is complicated by the fact that ∂_r is not a differential operator on $\mathbb{R}^n$ with smooth coefficients. The following (somewhat lengthy) pair of exercises goes somewhat far afield from the main thrust of these notes, but is necessary to justify our $H^{1/2}$ estimate.

In the sequel, we employ the useful notation $f \lesssim g$ to indicate that $f \le Cg$ for some $C \in \mathbb{R}^+$; when f and g are Banach norms of some function, C is always supposed to be independent of the function.

EXERCISE* 2.4.

(1) Verify that for $u \in \mathcal{S}(\mathbb{R}^n)$ with $n \ge 3$, $|\langle \partial_r u, u\rangle| \lesssim \|u\|^2_{H^{1/2}}$.
HINT: Use the fact that
$$\partial_r = \sum |x|^{-1} x^j \partial_{x^j}.$$
Check that $x/|x|$ is a bounded multiplier on both L^2 and H^1, and hence, by interpolation and duality, on $H^{-1/2}$. An efficient treatment of the interpolation methods you will need can be found in [**25**]. You will probably also need to use *Hardy's inequality* (see Exercise 2.5).

(2) Likewise, show that the $\langle r^{-1}u, u\rangle$ term is bounded by a multiple of $\|u\|^2_{H^{1/2}}$ (again, use Exercise 2.5).

EXERCISE 2.5. Prove *Hardy's inequality*: if $u \in H^1(\mathbb{R}^n)$ with $n \ge 3$, then
$$\frac{(n-2)^2}{4} \int \frac{|u|^2}{r^2}\, dx \le \int |\nabla u|^2\, dx.$$
HINT: In polar coordinates, we have for $u \in \mathcal{S}(\mathbb{R}^n)$
$$\int \frac{|u|^2}{r^2}\, dx = \int_{S^{n-1}} \int_0^\infty |u|^2 r^{n-3}\, dr\, d\theta.$$
Integrate by parts in the r integral, and apply Cauchy-Schwarz.

So we obtain, finally, the *Morawetz inequality*: if $\psi_0 \in H^{1/2}(\mathbb{R}^n)$, with $n \ge 4$ then

$$2\int_0^T \left\| r^{-1/2} \not\nabla \psi \right\|^2 dt + \frac{(n-1)(n-3)}{2} \int_0^T \left\| r^{-3/2}\psi \right\|^2 dt \lesssim \|\psi_0\|^2_{H^{1/2}}. \tag{2.4}$$

Now remember that we've been working rather formally, and there's no guarantee that either of the terms on the LHS is finite a priori. But the RHS is finite, so *since both terms on the LHS are positive, both must be finite, provided* $\psi_0 \in H^{1/2}$. (This is a dangerously sloppy way of reasoning—see the exercises below.) So we get, at one stroke two nice pieces of information: if $\psi_0 \in H^{1/2}$, we obtain the finiteness of both terms on the left.

Let's try and understand these. The term
$$\int_0^T \left\| r^{-3/2}\psi \right\|^2 dt$$

gives us a weighted estimate, which we can write as

$$\psi \in r^{3/2} L^2([0,T]; L^2(\mathbb{R}^n)) \tag{2.5}$$

for any T, or, more briefly, as

$$\psi \in r^{3/2} L^2_{\mathrm{loc}} L^2. \tag{2.6}$$

(The right side of (2.5) denotes the Hilbert space of functions that are of the form $r^{3/2}$ times an element of the space of L^2 functions on $[0,T]$ with values in the Hilbert space $L^2(\mathbb{R}^n)$; note that whenever we use the condensed notation (2.6), the Hilbert space for the time variables will precede that for the spatial variables.) So ψ can't "bunch up" too much at the origin. Incidentally, our whole setup was translation invariant, so in fact we can conclude

$$\psi \in |x - x_0|^{3/2} L^2_{\mathrm{loc}} L^2$$

for *any* $x_0 \in \mathbb{R}^n$, and ψ can't bunch up too much anywhere at all.

How about the other term? One interesting thing we can do is the following: Choose x_0, x_1 in $\mathbb{R}^n$, and let X be a smooth vector field with support disjoint from the line $\overline{x_0 x_1}$. Then we may write X in the form

$$\mathsf{X} = \mathsf{X}_0 + \mathsf{X}_1$$

with X_i smooth, and $\mathsf{X}_i \perp (x - x_i)$ for $i = 0, 1$; in other words, we split X into angular vector fields with respect to the origin of coordinates placed at x_0 and x_1 respectively. Moreover, we can arrange that the coefficients of X_i be bounded in terms of the coefficients of X (provided we bound the support uniformly away from $\overline{x_0 x_1}$). Thus, we can estimate for any such vector field X and any $u \in \mathcal{C}_c^\infty(\mathbb{R}^n)$

$$\int |\mathsf{X}u|^2 \, dx \lesssim \int \left| |x - x_0|^{-1/2} \not\nabla_0 u \right|^2 dx + \int \left| |x - x_1|^{-1/2} \not\nabla_1 u \right|^2 dx$$

where $\not\nabla_i$ is the angular gradient with respect to the origin of coordinates at x_i. Since for a solution of the Schrödinger equation, (2.4) tells us that the time integral of each of these latter terms is bounded by the squared $H^{1/2}$ norm of the initial data, we can assemble these estimates with the choices $\mathsf{X} = \chi \partial_{x^j}$ for any $\chi \in \mathcal{C}_c^\infty(\mathbb{R}^n)$ to obtain

$$\int_0^T \|\chi \nabla \psi\|^2 \, dt \lesssim \|\psi_0\|^2_{H^{1/2}}.$$

In more compact notation, we have shown that

$$\psi_0 \in H^{1/2} \Longrightarrow \psi \in L^2_{\mathrm{loc}} H^1_{\mathrm{loc}}.$$

This is called the *local smoothing estimate.* It says that on average in time, the solution is locally half a derivative smoother than the initial data was; one consequence is that in fact, with initial data in $H^{1/2}$, the solution is in H^1 in space at *almost every time.*

EXERCISE 2.6. Work out the Morawetz estimate in dimension 3. (This is in many ways the nicest case.) Note that our techniques yield no estimate in dimension 2, however.

In fact, if all we care about is the local smoothing estimate (and this is frequently the case) there is an easier commutator argument that we can employ to get just that estimate. Let $f(r)$ be a function on $\mathbb{R}^+$ that equals 0 for $r < 1$, is increasing, and equals 1 for $r \geq 2$. Set $A = f(r)\partial_r$ and employ (2.2) just as we did before. The commutant $f(r)\partial_r$ (as opposed to just ∂_r) has the virtue of actually being a smooth vector field on $\mathbb{R}^n$. So we can write

$$[\Delta, f(r)\partial_r] = -2f'(r)\partial_r^2 + 2r^{-3}f(r)\Delta_\theta + R$$

where R is a first order operator with coefficients in $\mathcal{C}_c^\infty(\mathbb{R}^n)$. As we didn't bother to make our commutant anti-self-adjoint, we might like to fix things up now by rewriting

$$[\Delta, f(r)\partial_r] = -2\partial_r^* f'(r)\partial_r + 2r^{-3}f(r)\Delta_\theta + R'$$

where R' is of the same type as R. Note that both main terms on the right are now nonnegative operators, and also that the term containing ∂_r^* is not, appearances to the contrary, singular at the origin, owing to the vanishing of f' there. Thus we obtain, by another use of (2.2),

$$\int_0^T \left\|\sqrt{f'(r)}\partial_r\psi\right\|^2 dt + \int_0^T \left\|\sqrt{f(r)}r^{-1/2}\not\nabla\psi\right\|^2 dt \lesssim \int_0^T |\langle R'\psi,\psi\rangle|\, dt + |\langle f(r)\partial_r\psi,\psi\rangle||_0^T. \tag{2.7}$$

Now the first term on the RHS is bounded by a multiple of $\|\psi_0\|^2_{H^{1/2}}$ (as R' is first order with coefficients in $\mathcal{C}_c^\infty(\mathbb{R}^n)$); the second term is likewise (since f is bounded with compactly supported derivative, and zero near the origin). This gives us an estimate of the desired form, valid on any compact subset of $\operatorname{supp} f \cap \operatorname{supp} f'$, which can be translated to contain any point.

EXERCISE 2.7. This exercise is on giving some rigorous underpinnings to some of the formal estimates above. It also gets you thinking about the alternative, Fourier-theoretic, picture of how might think about solutions to the Schrödinger equation.[1]

(1) Using the Fourier transform,[2] show that if $\psi_0 \in L^2(\mathbb{R}^n)$, there exists a unique solution $\psi(t,x)$ to (2.1) with $\psi(0,x) = \psi_0$.
(2) As long as you're at it, use the Fourier transform to derive the explicit form of the solution: show that

$$\psi(t,x) = \psi_0 * K_t$$

where K_t is the "Schrödinger kernel;" give an explicit formula for K_t.
(3) Use your explicit formula for K_t to show that if $\psi_0 \in L^1$ then $\psi(T,x) \in L^\infty(\mathbb{R}^n)$ for any $T \neq 0$.
(4) Show using the first part, i.e. by thinking about the solution operator as a Fourier multiplier, that if $\psi_0 \in H^s$ then $\psi(t,x) \in L^\infty(\mathbb{R}_t; H^s)$, hence give another proof that H^s regularity is conserved.
(5) Likewise, show that the Schrödinger evolution in $\mathbb{R}^n$ takes Schwartz functions to Schwartz functions.
(6) *Rigorously* justify the Morawetz inequality if $\psi_0 \in \mathcal{S}(\mathbb{R}^n)$. Then use a density argument to rigorously justify it for $\psi_0 \in H^{1/2}(\mathbb{R}^n)$.

---

[1] If you want to work hard, you might try to derive the local smoothing estimate from the explicit form of the Schrödinger kernel derived below. It's not so easy!

[2] See the appendix for a very brief review of the Fourier transform acting on tempered distributions and L^2-based Sobolev spaces.

2.2. The Schrödinger equation with a metric. Now let us change our problem a bit. Say we are on an n-dimensional manifold, or even just on $\mathbb{R}^n$ endowed with a complete non-Euclidean Riemannian metric g. There is a canonical choice for the Laplace operator in this setting:

$$\Delta = d^* d$$

where d takes functions to one-forms, and the adjoint is with respect to L^2 inner products on both (which of course also involve the volume form associated to the Riemannian metric). This yields, in coordinates,

$$\Delta = -\frac{1}{\sqrt{g}} \partial_{x^i} g^{ij} \sqrt{g} \partial_{x^j}, \tag{2.8}$$

where $\sum_{i,j=1}^n g^{ij} \partial_{x^i} \otimes \partial_{x^j}$ is the dual metric on forms (hence g^{ij} is the inverse matrix to g_{ij}) and g denotes $\det(g_{ij})$.

EXERCISE 2.8. Check this computation!

EXERCISE 2.9. Write the Euclidean metric on $\mathbb{R}^3$ in spherical coordinates, and use (2.8) to compute the Laplacian in spherical coordinates.

We can now consider the Schrödinger equation with the Euclidean Laplacian replaced by this new "Laplace-Beltrami" operator. By standard results in the spectral theory of self-adjoint operators,[3] there is still a solution in $L^\infty(\mathbb{R}; L^2)$ given any L^2 initial data—this generalizes our Fourier transform computation in Exercise 2.7—but its form and its properties are much harder to read off.

Computing commutators with this operator is a little trickier than in the Euclidean case, but certainly feasible; you might certainly try computing $[\Delta, \partial_r + (n - 1)/(2r)]$ where r is the distance from some fixed point.

EXERCISE 2.10. Write out the Laplace operator in Riemannian polar coordinates, and compute $[\Delta, \partial_r + (n-1)/(2r)]$ near $r = 0$.

But what happens when we get beyond the injectivity radius? Of course, the r variable doesn't make any sense any more. Moreover, if we try to think of ∂_r as the operator of differentiating "along geodesics emanating from the origin" then at a conjugate point to 0, we have the problem that we're somehow supposed to be be simultaneously differentiating in two different directions. One fix for this problem is to employ the calculus of pseudodifferential operators, which permits us to construct operators that behave differently depending on what direction we're looking in: we can make operators that separate out the different geodesics passing through the conjugate point, and do different things along them.

2.3. The wave equation. Let

$$\Box u \equiv (\partial_t^2 + \Delta) u = 0$$

denote the wave equation on $\mathbb{R} \times \mathbb{R}^n$ (recall that $\Delta = -\sum \partial_{x^i}^2$). For simplicity of notation, *let us consider only real-valued solutions in this section.*

[3]The operator Δ is manifestly formally self-adjoint, but in fact turns out to be essentially self-adjoint on $\mathcal{C}_c^\infty(X)$ for X any complete manifold.

The usual route to thinking about the energy of a solution to the wave equation is as follows. We consider the integral

$$0 = \int_0^T \langle \Box u, \partial_t u\rangle \, dt \tag{2.9}$$

where $\langle \cdot, \cdot \rangle$ is the inner product on $L^2(\mathbb{R}^n)$. Then integrating by parts in t and in x gives the conservation of

$$\|\partial_t u\|^2 + \|\nabla u\|^2.$$

We can recast this formally as a commutator argument, if we like, by considering the commutator with the indicator function of an interval:

$$0 = \int_{\mathbb{R}} \langle [\Box, 1_{[0,T]}(t)\partial_t] u, u\rangle \, dt.$$

The integral vanishes, at least formally, by self-adjointness of $\Box$—it is in fact a better idea to think of this whole thing as an inner product on $\mathbb{R}^{n+1}$:

$$\big\langle [\Box, 1_{[0,T]}(t)]\partial_t u, u\big\rangle_{\mathbb{R}^{n+1}}.$$

Having gone this far, we might like to replace the indicator function with something smooth, to give a better justification for this formal integration by parts; let $\chi(t)$ be a smooth approximator to the indicator function with $\chi' = \phi_1 - \phi_2$ with ϕ_1 and ϕ_2 nonnegative bump functions supported respectively in $(-\epsilon, \epsilon)$ and $(T-\epsilon, T+\epsilon)$, with $\phi_2(\cdot) = \phi_1(\cdot - T)$ Let $A = \chi(t)\partial_t + \partial_t \chi(t)$. Then we have

$$[\Box, A] = 2\partial_t \chi' \partial_t + \partial_t^2 \chi' + \chi' \partial_t^2,$$

and by (formal) anti-self-adjointness of ∂_t (and the fact that u is assumed real),

$$\begin{aligned} 0 = \langle [\Box, A]u, u\rangle_{\mathbb{R}^{n+1}} &= -2\langle \chi' \partial_t u, \partial_t u\rangle_{\mathbb{R}^{n+1}} + 2\big\langle \chi' u, \partial_t^2 u\big\rangle_{\mathbb{R}^{n+1}} \\ &= -2\langle \chi' \partial_t u, \partial_t u\rangle_{\mathbb{R}^{n+1}} + 2\big\langle \chi' u, \nabla^2 u\big\rangle_{\mathbb{R}^{n+1}} \\ &= -2\langle \chi' \partial_t u, \partial_t u\rangle_{\mathbb{R}^{n+1}} - 2\langle \chi' \nabla, \nabla u\rangle_{\mathbb{R}^{n+1}} \\ &= -2\int_{\mathbb{R}^{n+1}} \phi_1(t)(|u_t|^2 + |\nabla u|^2)\, dt\, dx \\ &\quad + 2\int_{\mathbb{R}^{n+1}} \phi_2(t)(|u_t|^2 + |\nabla u|^2)\, dt\, dx. \end{aligned}$$

Thus, the energy on the time interval $[T-\epsilon, T+\epsilon]$ (modulated by the cutoff ϕ_2) is the same as that in the time interval $[-\epsilon, \epsilon]$ (modulated by ϕ_1).

We can get fancier, of course. Finite propagation speed is usually proved by considering the variant of (2.9)

$$\int_{-T_2}^{-T_1} \int_{|x|^2 \le t^2} \Box u \, \partial_t u \, dx\, dt,$$

with $0 < T_1 < T_2$. Integrating by parts gives negative boundary terms, and we find that the energy in

$$\{t = -T_1, |x|^2 \le T_1^2\}$$

is bounded by that in

$$\{t = -T_2, |x|^2 \le T_2^2\}.$$

Hence if the solution has zero Cauchy data (i.e. value, time-derivative) on the latter surface, it also has zero Cauchy data on the former.

EXERCISE 2.11. Go through this argument to show finite propagation speed.

Making this argument into a commutator argument is messier, but still possible:

EXERCISE* 2.12. Write a positive commutator version of the proof of finite propagation speed, using smooth cutoffs instead of integrations by parts. (An account of energy estimates with smooth temporal cutoffs, in the general setting of Lorentzian manifolds, can be found in [**27**, Section 3].)

There is of course also a Morawetz estimate for the wave equation! (Indeed, this was what Morawetz originally proved.)

EXERCISE* 2.13. Derive (part of) the Morawetz estimate: Let u solve

$$\Box u = 0, (u, \partial_t u)|_{t=0} = (f, g)$$

on $\mathbb{R}^n$, with $n \geq 4$. Show that

$$\left\| r^{-3/2} u \right\|_{L^2_{\mathrm{loc}}(\mathbb{R}^{n+1})} \lesssim \|f\|_{H^1}^2 + \|g\|_{L^2}^2;$$

this is analogous to the weight part of the Morawetz estimate we derived for the Schrödinger equation. There is in fact no need for the local L^2 norm—the global spacetime estimate works too: prove this estimate, and use it to draw a conclusion about the long-time decay of a solution to the wave equation with Cauchy data in $\mathcal{C}_c^\infty(\mathbb{R}^n) \oplus \mathcal{C}_c^\infty(\mathbb{R}^n)$.

HINT: consider $\langle [\Box, \chi(t)(\partial_r + (n-1)/(2r))]u, u\rangle_{\mathbb{R}^{n+1}}$.

3. The pseudodifferential calculus

Recall that we hoped to describe a class of operators enriching the differential operators that would, among other things, enable us to deal properly with the local smoothing estimate on manifolds, where conjugate points caused our commutator arguments with ordinary differential operators to break down. One solution to this problem turns out to lie in the calculus of pseudodifferential operators.

3.1. Differential operators. What kind of a creature is a pseudodifferential operator? Well, first let's think more seriously about *differential* operators. A linear differential operator of order m is something of the form

$$P = \sum_{|\alpha| \leq m} a_\alpha(x) D^\alpha \tag{3.1}$$

where $D_j = i^{-1}(\partial/\partial x^j)$ and we employ "multiindex notation:"

$$D^\alpha = D_1^{\alpha_1} \dots D_n^{\alpha_n},$$

$$|\alpha| = \sum \alpha_j.$$

We will always take our coefficients to be smooth:

$$a_\alpha \in \mathcal{C}^\infty(\mathbb{R}^n).$$

We let

$$\mathrm{Diff}^m(\mathbb{R}^n)$$

denote the collection of all differential operators of order m on $\mathbb{R}^n$ (and will later employ the analogous notation on a manifold).

If $P \in \mathrm{Diff}^m(\mathbb{R}^n)$ is given by (3.1), we can associate with P a function by formally turning differentiation in x^j into a formal variable ξ_j with $(\xi_1, \dots, \xi_n) \in \mathbb{R}^n$:

$$p(x,\xi) = \sum a_\alpha(x)\xi^\alpha.$$

This is called the "total (left-) symbol" of P; of course, knowing p is equivalent to knowing P. Note that $p(x,\xi)$ is a rather special kind of a function on $\mathbb{R}^{2n}$: it is actually polynomial in the ξ variables with smooth coefficients. Let us write $p = \sigma_{\mathrm{tot}}(P)$.

Note that

$$\sigma_{\mathrm{tot}} : P \mapsto p$$

is *not* a ring homomorpism: we have

$$PQ = \sum_{\alpha,\beta} p_\alpha(x) D^\alpha q_\beta(x) D^\beta,$$

and if we expand out this product to be of the form

$$\sum_\gamma c_\gamma(x) D^\gamma,$$

then the coefficients c_γ will involve all kinds of derivatives of the q_β's. This is a pain, but on the other hand life would be pretty boring if the ring of differential operators were commutative.

If we make do with less, though, composition of operators doesn't look so bad. We let $\sigma_m(P)$, the *principal symbol* of P, just be the symbol of the top-order parts of P :

$$\sigma_m(P) = \sum_{|\alpha|=m} a_\alpha(x)\xi^\alpha.$$

Note that $\sigma_m(P)$ is a *homogeneous* degree-m polynomial in ξ, i.e., a polynomial such that $\sigma_m(P)(x,\lambda\xi) = \lambda^m \sigma_m(P)(x,\xi)$ for $\lambda \in \mathbb{R}$. As a result, we can reconstruct it from its value at $|\xi| = 1$, and it makes sense for many purposes to just consider it as a (rather special) smooth function on $\mathbb{R}^n \times S^{n-1}$. It turns out to make more invariant sense to regard the principal symbol as a homogeneous polynomial on $T^*\mathbb{R}^n$, so that once we have scaled away the action of $\mathbb{R}^+$, we may regard it as a function on $S^*\mathbb{R}^n$, the *unit* cotangent bundle of $\mathbb{R}^n$, which is simply defined as $T^*\mathbb{R}^n/\mathbb{R}^+$ (or identified with the bundle of unit covectors in, say, the Euclidean metric). To clarify when we are talking about the symbol on $S^*\mathbb{R}^n$, we define[4]

$$\hat{\sigma}_m(P) = \sigma_m(P)|_{|\xi|=1} \in \mathcal{C}^\infty(S^*\mathbb{R}^n).$$

Now it *is* the case that the principal symbol is a homomorphism:

PROPOSITION 3.1. *For P, Q differential operators of order m resp. m',*

$$\sigma_{m+m'}(PQ) = \sigma_m(P)\sigma_{m'}(Q).$$

(and likewise with $\hat{\sigma}$).

EXERCISE 3.1. Verify this!

Moreover, the principal symbol has another lovely property that the total symbol lacks: it behaves well under change of variables. If $y = \phi(x)$ is a change of

---

[4]The reader is warned that this notation is not a standard one.

variables, with ϕ a diffeomorphism, and if P is a differential operator in the x variables, we can of course define a pushforward of P by

$$(\phi_* P)f = P(\phi^* f)$$

Then in particular,

$$\phi_*(D_{x^j}) = \sum_k \frac{\partial y^k}{\partial x^j} D_{y^k},$$

hence

$$\phi_*(D_x^\alpha) = D_{x^1}^{\alpha_1} \dots D_{x^n}^{\alpha_n} = \left(\sum_{k_1=1}^n \frac{\partial y^{k_1}}{\partial x^1} D_{y^{k_1}} \right)^{\alpha_1} \dots \left(\sum_{k_n=1}^n \frac{\partial y^{k_n}}{\partial x^n} D_{y^{k_n}} \right)^{\alpha_n};$$

when we again try to write this in our usual form, as a sum of coefficients times derivatives, we end up with a hideous mess involving high derivatives of the diffeomorphism ϕ. *But,* if we restrict ourselves to dealing with principal symbols alone, the expression simplifies in both form and (especially) interpretation:

PROPOSITION 3.2. *If P is a differential operator given by* (3.1), *and $y = \phi(x)$, then*

$$\sigma_m(\phi_* P)(y, \eta) = \sum_{|\alpha|=m} a_\alpha(\phi^{-1}(y)) \left(\sum_{k_1=1}^n \frac{\partial y^{k_1}}{\partial x^1} \eta_{k_1} \right)^{\alpha_1} \dots \left(\sum_{k_n=1}^n \frac{\partial y^{k_n}}{\partial x^n} \eta_{k_n} \right)^{\alpha_n}$$

where η are the new variables "dual" to the y variables.

This corresponds exactly to the behavior of a function defined on the cotangent bundle: if ϕ is a diffeomorphism from $\mathbb{R}^n_x$ to $\mathbb{R}^n_y$, then it induces a map $\Phi = \phi^* : T^*\mathbb{R}^n_y \to T^*\mathbb{R}^n_x$, and

$$\sigma_m(\phi_* P) = \Phi^*(\sigma_m(P)).$$

EXERCISE 3.2. Prove the proposition, and verify this interpretation of it.

Notwithstanding its poor properties, it is nonetheless a useful fact that the map

$$\sigma_{\text{tot}} : P \mapsto p$$

is one-to-one and onto polynomials with smooth coefficients; it therefore has an inverse, which we shall denote

$$\text{Op}_\ell : p \mapsto P,$$

taking functions on $T^*\mathbb{R}^n$ that happen to be polynomial in the fiber variables to differential operators on $\mathbb{R}^n$. Op_ℓ is called a "quantization" map.[5] You may wonder about the ℓ in the subscript: it stands for "left," and has to do with the fact that we chose to write differential operators in the form (3.1) instead of as

$$P = \sum_{|\alpha| \le m} D^\alpha a_\alpha(x),$$

with the coefficients on the right. This would have changed the definition of σ_{tot} and hence of its inverse.

Note that $\text{Op}_\ell(x^j) = x^j$ (i.e. the operation of multiplication by x^j) while $\text{Op}_\ell(\xi_j) = D_j$.

Why not, you might ask, try to extend this quantization map to a more general class of functions on $T^*\mathbb{R}^n$? This is indeed how we obtain the calculus of pseudodifferential operators. The tricky point to keep in mind, however, is that for most

[5]It is far from unique, as will become readily apparent.

purposes, it is asking too much to deal with the quantizations of all possible functions on $T^*\mathbb{R}^n$, so we'll deal only with a class of functions that are somewhat akin to polynomials in the fiber variables.

3.2. Quantum mechanics. One reason why you might care about the existence of a quantization map, and give it such a suggestive name, lies in the foundations of quantum mechanics.

It is helpful to think about $T^*\mathbb{R}^n$ as being a classical *phase space,* with the x variables (in the base) being "position" and the ξ variables (the fiber variables) as "momenta" in the various directions. The general notion of *classical mechanics* (in its Hamiltonian formulation) is as follows: The state of a particle is a point in the phase space $T^*\mathbb{R}^n$, and moves along some curve in $T^*\mathbb{R}^n$ as time evolves; an *observable* $p(x,\xi)$ is a function on the phase space that we may evaluate at the state (x,ξ) of our particle to give a number (the observation). By contrast, a *quantum* particle is described by a complex-valued *function* $\psi(x)$ on $\mathbb{R}^n$, and a quantum observable is a self-adjoint *operator* P acting on functions on $\mathbb{R}^n$. Doing the same measurement repeatedly on identically prepared quantum states is not guaranteed to produce the same number each time, but at least we can talk about the *expected value* of the observation, and it's simply

$$\langle P\psi, \psi\rangle_{L^2(\mathbb{R}^n)}.$$

In the early development of quantum mechanics, physicists sought a way to transform the classical world into the quantum world, i.e. of taking functions on $T^*\mathbb{R}^n$ to operators on[6] $L^2(\mathbb{R}^n)$. This is, loosely speaking, the process of "quantization."

We now turn to the question of describing the dynamics in the quantum and classical worlds. To describe how the point in phase space corresponding to a classical particle in Hamiltonian mechanics evolves in time, we use the notion of the "Poisson bracket" of two observables. In coordinates, we can explicitly define

$$\{f,g\} \equiv \sum \frac{\partial f}{\partial \xi_j}\frac{\partial g}{\partial x^j} - \frac{\partial f}{\partial x^j}\frac{\partial g}{\partial \xi_j}$$

(this in fact makes invariant sense on any symplectic manifold). The map $g \mapsto \{f,g\}$ defines a vector field[7] (the Hamilton vector field) associated to f :

$$\mathsf{H}_f = \sum \frac{\partial f}{\partial \xi_j}\frac{\partial}{\partial x^j} - \frac{\partial f}{\partial x^j}\frac{\partial}{\partial \xi_j}.$$

The classical time-evolution is along the flow generated by the Hamilton vector field associated to the *energy function* of our system, i.e. the flow along H_h for some given $h \in \mathcal{C}^\infty(T^*\mathbb{R}^n)$. By contrast, the wavefunction for a quantum particle evolves in time according to the Schrödinger equation (2.1), with $-\nabla^2$ in general replaced by a self-adjoint "Hamiltonian operator" H whose principal symbol is the energy function h.[8] By a mild generalization of (2.2), the time derivative of the

[6]Well, they are not necessarily going to be defined on all of L^2; the technical subtleties of unbounded self-adjoint operators will mostly not concern us here, however.

[7]We use the geometers' convention of identifying a vector and the directional derivative along it.

[8]For honest physical applications, one really ought to introduce the semi-classical point of view here, carrying Planck's constant along as a small parameter and using an associated notion of principal symbol.

expectation of an observable A is related to the commutator

$$[H, A].$$

One of the essential features of quantum mechanics is that

$$\sigma_{m+m'}([H, A]) = i\{\sigma_m(H), \sigma_{m'}(A)\},$$

so that the time-evolution of the quantum observable A is related to the classical evolution of its symbol along the Hamilton flow; this is the "correspondence principle" between classical and quantum mechanics.[9]

3.3. Quantization. How might we construct a quantization map extending the usual quantization on fiber-polynomials?

Let $\mathcal{F}$ denote the Fourier transform (see Appendix for details). Then we may write, on $\mathbb{R}^n$,

$$\begin{aligned}(D_{x^j}\psi)(x) = \mathcal{F}^{-1}\xi_j\mathcal{F}u &= (2\pi)^{-n}\int e^{ix\cdot\xi}\xi_j \int e^{-iy\cdot\xi}\psi(y)\,dy\,d\xi \\ &= \frac{1}{2\pi}\iint \xi_j e^{i(x-y)\cdot\xi}\psi(y)\,dy\,d\xi.\end{aligned}$$

Likewise, since $\mathcal{F}^{-1}\mathcal{F} = I$, we of course have

$$(x^j\psi)(x) = (2\pi)^{-n}\iint x^j e^{i(x-y)\cdot\xi}\psi(y)\,dy\,d\xi.$$

Going a bit further, we see that at least for a fiber polynomial $a(x,\xi) = \sum a_\alpha(x)\xi^\alpha$ we have

$$(3.2)\quad (\mathrm{Op}_\ell(a)\psi)(x) = \sum a_\alpha(x)D^\alpha\psi(x) = (2\pi)^{-n}\iint a(x,\xi)e^{i(x-y)\cdot\xi}\psi(y)\,dy\,d\xi;$$

stripping away the function ψ, we can also simply write the Schwartz kernel (see Appendix) of the operator $\mathrm{Op}_\ell(a)$ as

$$\kappa\big(\mathrm{Op}_\ell(a)\big) = (2\pi)^{-n}\int a(x,\xi)e^{i(x-y)\cdot\xi}\,d\xi.$$

(Making sense of the integrals written above is not entirely trivial: Given $\psi \in \mathcal{S}(\mathbb{R}^n)$, we can make sense of the ξ integral in (3.2), which looks (potentially) divergent, by observing that

$$(1+|\xi|^2)^{-k}(1+\Delta_y)^k e^{i(x-y)\cdot\xi} = e^{i(x-y)\cdot\xi}$$

for all $k \in \mathbb{N}$; repeatedly integrating by parts in y then moves the derivatives onto ψ. This method brings down an arbitrary negative power of $(1+|\xi|^2)$ at the cost of differentiating ψ, thus making the ξ integral convergent.[10] Similar arguments yield continuity of $\mathrm{Op}_\ell(a)$ as a map $\mathcal{S}(\mathbb{R}^n) \to \mathcal{S}(\mathbb{R}^n)$, hence we can extend to let $\mathrm{Op}_\ell(a)$ act on $\psi \in \mathcal{S}'$ by duality. For more details, cf. [**17**].)

EXERCISE* 3.3. Verify the vague assertions in the parenthetical remark above. You may wish to consult, for example, the beginning of [**10**].

---

[9]In the semi-classical setting, the correspondence principle tells that we can in a sense recover CM from QM in the limit when Planck's constant tends to zero. What we have in this setting is a correspondence principle that works at high energies, i.e. in doing computations with high-frequency waves.

[10]This kind of integration by parts argument is ubiquitous in the subject, and somewhat scanted in these notes, relative to its true importance.

This of course suggests that we use (3.2) as the *definition* of $\mathrm{Op}_\ell(a)$ for more general observables ("symbols") a. And we do. In $\mathbb{R}^n$, we set

$$(\mathrm{Op}_\ell(a)\psi)(x) = \frac{1}{(2\pi)^n}\int a(x,\xi)e^{i(x-y)\cdot\xi}\psi(y)\,dy\,d\xi. \tag{3.3}$$

We can define the pseudodifferential operators on $\mathbb{R}^n$ to be just the range of this quantization map on some reasonable set of symbols a, to be discussed below.

On a Riemannian manifold, we can make similar constructions global by cutting off near the diagonal and using the exponential map and its inverse. The pseudodifferential operators are those whose Schwartz kernels[11] near the diagonal look like (3.3) in local coordinates, and that away from the diagonal are allowed to be arbitrary functions in $\mathcal{C}^\infty(X\times X)$. If the manifold is noncompact, we will often assume further that operators are *properly supported*, i.e. that both left- and right-projection give proper maps from the support of the Schwartz kernel to X.

3.4. The pseudodifferential calculus.

DEFINITION 3.3. A function a on $T^*\mathbb{R}^n$ is a *classical symbol* of order m if

- $a \in \mathcal{C}^\infty(T^*\mathbb{R}^n)$
- On $|\xi|>1$, we have
$$a(x,\xi) = |\xi|^m \tilde{a}(x,\hat{\xi},|\xi|^{-1}),$$
where $\tilde{a}$ is a smooth function on $\mathbb{R}^n_x \times S^{n-1}_{\hat{\xi}} \times \mathbb{R}^+$, and
$$\hat{\xi} = \frac{\xi}{|\xi|} \in S^{n-1}.$$

We then write $a \in S^m_{\mathrm{cl}}(T^*\mathbb{R}^n)$.

It is convenient to introduce the notation
$$\langle\xi\rangle = (1+|\xi|^2)^{1/2},$$
so that $\langle\xi\rangle$ behaves like $|\xi|$ near infinity, but is smooth and nonvanishing at 0. A fancy way of saying that a is a classical symbol of order m is thus to simply say that a is equal to $\langle\xi\rangle^m$ times *a smooth function on the fiberwise radial compactification of* $T^*\mathbb{R}^n$, denoted $\overline{T}^*\mathbb{R}^n$. This compactification is defined as follows: We can diffeomorphically identify $\mathbb{R}^n_\xi$ with the interior of the unit ball by first mapping it to the upper hemisphere of $S^n \subset \mathbb{R}^{n+1}$ by mapping

$$\xi \mapsto \left(\frac{\xi}{\langle\xi\rangle}, \frac{1}{\langle\xi\rangle}\right) \tag{3.4}$$

and identifying this latter space with the interior of the ball. Then $1/\langle\xi\rangle$ becomes a *boundary defining function,* i.e. one that cuts out the boundary nondegenerately as its zero-set; $1/|\xi|$ is also a valid boundary defining function near the boundary of the ball, i.e. away from its singularity.

A very important consequence is that we can write a Taylor series for a near $|\xi|^{-1}=0$ (the "sphere at infinity") to obtain

$$a(x,\xi) \sim \sum_{j=0}^\infty a_{m-j}(x,\hat{\xi})|\xi|^{m-j}, \quad \text{with } a_{m-j} \in \mathcal{C}^\infty(\mathbb{R}^n \times S^{n-1}),$$

[11]For some remarks on the Schwartz kernel theorem, see the Appendix.

and where the tilde denotes an "asymptotic expansion"—truncating the expansion at the $|\xi|^{m-N}$ term gives an error that is $O(|\xi|^{m-N-1})$.[12]

If X is a Riemannian manifold, we may define $S^m_{\mathrm{cl}}(T^*X)$ in the same fashion, insisting that these conditions hold in local coordinates.[13]

(For later use, we will also want symbols in a more general geometric setting: if E is a vector bundle we define

$$S^m_{\mathrm{cl}}(E)$$

to consist of smooth functions having an asymptotic expansion, as above, in the fiber variables. Often, we will be concerned with trivial examples like $E = \mathbb{R}^n_x \times \mathbb{R}^k_\xi$, where we will usually use Greek letters to distinguish the fiber variables.)

The classical symbols are the functions that we will "quantize" into operators using the definition (3.3). As with fiber-polynomials, the symbol that we quantize to make a given operator will transform in a complicated manner under change of variables, but the *top order* part of the symbol, $a_m(x, \hat{\xi}) \in \mathcal{C}^\infty(S^*\mathbb{R}^n)$, will transform invariantly.

EXERCISE 3.4. We say that a function $a \in \mathcal{C}^\infty(T^*X)$ is a *Kohn-Nirenberg symbol* of order m on T^*X (and write $a \in S^m_{\mathrm{KN}}(T^*X)$) if for all α, β,

$$\sup \langle\xi\rangle^{|\beta|-m} |\partial_x^\alpha \partial_\xi^\beta a| = C_{\alpha,\beta} < \infty. \tag{3.5}$$

Check that $S^m_{\mathrm{cl},c}(T^*\mathbb{R}^n) \subset S^m_{\mathrm{KN}}(T^*\mathbb{R}^n)$, where the extra subscript c denotes compact support in the base variables. Find examples of Kohn-Nirenberg symbols compactly supported in x that are not classical symbols.[14]

In the interests of full disclosure, it should be pointed out that it is the Kohn-Nirenberg symbols, rather than the classical ones defined above, that are conventionally used in the definition of the pseudodifferential calculus.

At this point, as discussed in the previous section, we are in a position to "define" the pseudodifferential calculus as sketched at the end of the previous section: it consists of operators whose Schwartz kernels near the diagonal look like the quantizations of classical symbols, and away from the diagonal are smooth. While our quantization procedure so far has been restricted to $\mathbb{R}^n$, the theory is in fact cleanest on compact manifolds, so *we shall state the properties of the calculus only for X a compact n-manifold.*[15] Most of the properties continue to hold on noncompact manifolds provided we are a little more careful either to control the behavior of the symbols at infinity, or if we restrict ourselves to "properly supported" operators, where the projections to each factor of the support of the Schwartz kernels give proper maps. We will therefore not shy away from pseudodifferential operators on $\mathbb{R}^n$, for instance, even though they are technically a bit distinct; indeed we will only use them in situations where we could in fact localize, and work on a large torus instead.

---

[12]This does not, of course, mean that the series has to converge, or, if it converges, that it has to converge to a : we never said a had to be *analytic* in $|\xi|^{-1}$, after all.

[13]One should of course check that the conditions for being a classical symbol are in fact coordinate invariant.

[14]Note that most authors use S^m to denote S^m_{KN}.

[15]Some remarks about the noncompact case will be found in the explanatory notes that follow.

Instead of trying to make a definition of the calculus and read off its properties, we shall simply try to axiomatize these objects:

THE SPACE OF PSEUDODIFFERENTIAL OPERATORS $\Psi^*(X)$ ON A COMPACT MANIFOLD X ENJOYS THE FOLLOWING PROPERTIES. (Note that this enumeration is followed by further commentary.)

(I) (Algebra property) $\Psi^m(X)$ is a vector space for each $m \in \mathbb{R}$. If $A \in \Psi^m(X)$ and $B \in \Psi^{m'}(X)$ then $AB \in \Psi^{m+m'}(X)$. Also, $A^* \in \Psi^m(X)$. Composition of operators is associative and distributive. The identity operator is in $\Psi^0(X)$.

(II) (Characterization of smoothing operators) We let

$$\Psi^{-\infty}(X) = \bigcap_m \Psi^m(X);$$

the operators in $\Psi^{-\infty}(X)$ are exactly those whose Schwartz kernels are $\mathcal{C}^\infty$ functions on $X \times X$, and can also be characterized by the property that they map distributions to smooth functions on X.

(III) (Principal symbol homomorphism) There is family of linear "principal symbol maps" $\hat{\sigma}_m : \Psi^m(X) \to \mathcal{C}^\infty(S^*X)$ such that if $A \in \Psi^m(X)$ and $B \in \Psi^{m'}(X)$,

$$\hat{\sigma}_{m+m'}(AB) = \hat{\sigma}_m(A)\hat{\sigma}_{m'}(B)$$

and

$$\hat{\sigma}_m(A^*) = \overline{\hat{\sigma}_m(A)}$$

We think of the principal symbol either as a function on the unit cosphere bundle S^*X or as a homogeneous function of degree m on T^*X, depending on the context, and we let $\sigma_m(A)$ denote the latter.

(IV) (Symbol exact sequence) There is a short exact sequence

$$0 \to \Psi^{m-1}(X) \to \Psi^m(X) \xrightarrow{\hat{\sigma}_m} \mathcal{C}^\infty(S^*X) \to 0,$$

hence the principal symbol of order m is 0 if and only if an operator is of order $m-1$.

(V) There is a linear "quantization map" $\mathrm{Op} : S^m_{\mathrm{cl}}(T^*X) \to \Psi^m(X)$ such that if $a \sim \sum_{j=0}^\infty a_{m-j}(x, \hat{\xi})|\xi|^{m-j} \in S^m_{\mathrm{cl}}(T^*X)$ then

$$\hat{\sigma}_m(\mathrm{Op}(a)) = a_m(x, \hat{\xi}).$$

The map Op is onto, modulo $\Psi^{-\infty}(X)$.

(VI) (Symbol of commutator) If $A \in \Psi^m(X)$, $B \in \Psi^{m'}(X)$ then[16] $[A, B] \in \Psi^{m+m'-1}(X)$, and we have

$$\sigma_{m+m'}([A, B]) = i\{\sigma_m(a), \sigma_{m'}(b)\}.$$

(VII) (L^2-boundedness, compactness) If $A = \mathrm{Op}(a) \in \Psi^0(X)$ then $A : L^2(X) \to L^2(X)$ is bounded, with a bound depending on finitely many constants $C_{\alpha,\beta}$ in (3.5). Moreoever, if $A \in \Psi^m(X)$, then

$$A \in \mathcal{L}(H^s(X), H^{s-m}(X)) \text{ for all } s \in \mathbb{R}.$$

---

[16]That the order is $m + m' - 1$ follows from Properties (III), (IV).

Note in particular that A maps $\mathcal{C}^\infty(X) \to \mathcal{C}^\infty(X)$. As a further consequence, note that operators of negative order are compact operators on $L^2(X)$.

(VIII) (Asymptotic summation) Given $A_j \in \Psi^{m-j}(X)$, with $j \in \mathbb{N}$, there exists $A \in \Psi^m(X)$ such that

$$A \sim \sum_j A_j,$$

which means that

$$A - \sum_{j=0}^{N} A_j \in \Psi^{m-N-1}(X)$$

for each N.

(IX) (Microsupport) Let $A = \mathrm{Op}(a) + R$, $R \in \Psi^{-\infty}(X)$. The set of $(x_0, \hat{\xi}_0) \in S^*X$ such that $a(x,\xi) = O(|\xi|^{-\infty})$ for $x, \hat{\xi}$ in some neighborhood of $(x_0, \hat{\xi}_0)$ is well-defined, independent of our choice of quantization map. Its complement is called the *microsupport* of A, and is denoted $\mathrm{WF}'\,A$. We moreover have

$$\mathrm{WF}'\,AB \subseteq \mathrm{WF}'\,A \cap \mathrm{WF}'\,B, \quad \mathrm{WF}'(A+B) \subseteq \mathrm{WF}'\,A \cup \mathrm{WF}'\,B,$$
$$\mathrm{WF}'\,A^* = \mathrm{WF}'\,A.$$

The condition $\mathrm{WF}'\,A = \emptyset$ is equivalent to $A \in \Psi^{-\infty}(X)$.

COMMENTARY:

(I) If we begin by defining our operators on $\mathbb{R}^n$ by the formula (3.3), with $a \in S^m_{\mathrm{cl}}(T^*\mathbb{R}^n)$, it is quite nontrivial to verify that the composition of two such operators is of the same type; likewise for adjoints. Much of the work that we are omitting in developing the calculus goes into verifying this property.

(II) On a non-compact manifold, it is only among, say, properly supported operators that elements of $\Psi^{-\infty}(X)$ are characterized by mapping distributions to smooth functions.

(III) Note that there is no sensible, *invariant*, way to associate, to an operator A, a "total symbol" a such that $A = \mathrm{Op}(a)$. As we saw before, a putative "total symbol" even for differential operators would be catastrophically bad under change of variables. Moreover, as we also saw for differential operators, it's a little hard to see what the total symbol of the composition is. This principal symbol map is a compromise that turns out to be extremely useful, especially when coupled with the asymptotic summation property, in making iterative arguments.

(IV) A good way to think of this is that $\hat{\sigma}_m$ is just the obstruction to an operator in $\Psi^m(X)$ being of order $m-1$.

(V) The map Op is far from unique. Even on $\mathbb{R}^n$, for instance, we can use Op_ℓ as defined by (3.2) but we could also use the "Weyl" quantization

$$(\mathrm{Op}_W(a)\psi)(x) = (2\pi)^{-n} \iint a((x+y)/2, \xi) e^{i(x-y)\cdot\xi} \psi(y)\, dy\, d\xi$$

or the "right" quantization

$$(\mathrm{Op}_r(a)\psi)(x) = (2\pi)^{-n} \iint a(y, \xi) e^{i(x-y)\cdot\xi} \psi(y)\, dy\, d\xi$$

or any of the obvious interpolating choices. On a manifold the choices to be made are even more striking. One convenient choice that works globally on a manifold is what might be called "Riemann-Weyl" quantization: Fix a Riemannian metric g. Given $a \in S^m_{\mathrm{cl}}(T^*X)$, define the Schwartz kernel of an operator A by

$$\kappa(A)(x,y) = (2\pi)^{-n} \int \chi(x,y) a(m(x,y),\xi) e^{i(\exp_y^{-1}(x),\xi)}\, dg_\xi;$$

here χ is a cutoff localizing near the diagonal and in particular, within the injectivity radius; $m(x,y)$ denotes the midpoint of the shortest geodesic between x, y, exp denotes the exponential map, and the round brackets denote the pairing of vectors and covectors. The "Weyl" in the name refers to the evaluation of a at $m(x,y)$ as opposed to x or y (which give rise to corresponding "left" and "right" quantizations respectively—also acceptable choices). The "Riemann" of course refers to our use of a choice of metric.

We will often only employ a single simple consequence of the existence of a quantization map: given $a_m \in \mathcal{C}^\infty(S^*X)$ and $m \in \mathbb{R}$, there exists $A \in \Psi^m(X)$ with principal symbol a_m and with $\mathrm{WF}'\, A = \operatorname{supp} a_m$.

(VI) A priori of course $AB - BA \in \Psi^{m+m'}(X)$; however the principal symbol vanishes, by the commutativity of $\mathcal{C}^\infty(S^*X)$. Hence the need for a lower-order term, which is subtler, and noncommutative. That the Poisson bracket is well-defined independent of coordinates reflects the fact that T^*X is naturally a symplectic manifold, and the Poisson bracket is well-defined on such a manifold (see §4.1 below).

EXERCISE 3.5. Check (by actually performing a change of coordinates) that if $f, g \in \mathcal{C}^\infty(T^*X)$, then $\{f, g\}$ is well-defined, independent of coordinates.

This property is the one which ties classical dynamics to quantum evolution, as the discussion in §3.2 shows.

(VII) Remarkably, the mapping property is one that can be derived from the other properties of the calculus purely algebraically, with the only analytic input being boundedness of operators in $\Psi^{-\infty}(X)$. This is the famous Hörmander "square-root" argument—see [**10**], as well as Exercise 3.12 below.

On noncompact manifolds, restricting our attention to properly supported operators gives boundedness $L^2 \to L^2_{\mathrm{loc}}$.

The compactness of negative order operators of course follows from boundedness, together with Rellich's lemma, but is worth emphasizing; we can regard $\hat{\sigma}_0$ as the "obstruction to compactness" in general. On noncompact manifolds, this compactness property fails quite badly, resulting in much interesting mathematics.

(VIII) This follows from our ability to do the corresponding "asymptotic summation" of total symbols, which in turn is precisely "Borel's Lemma," which tells us that any sequence of coefficients are the Taylor coefficients of a $\mathcal{C}^\infty$ function; here we are applying the result to smooth functions on the radial compactification of T^*X, and the Taylor series is in the variable $\sigma = |\xi|^{-1}$, at $\sigma = 0$.

(IX) Since the total symbol is not well-defined, it is not so obvious that the microsupport is well-defined; verifying this requires checking how the total symbol transforms under change of coordinates; likewise, we may verify that the (highly non-invariant) formula for the total symbol of the composition respects microsupports to give information about $\mathrm{WF}' AB$.

3.5. Some consequences. If you believe that there exists a calculus of operators with the properties enumerated above, well, then you believe quite a lot! For instance:

THEOREM 3.4. *Let $P \in \Psi^m(X)$ with $\hat{\sigma}_m(P)$ nowhere vanishing on S^*X. Then there exists $Q \in \Psi^{-m}(X)$ such that*

$$QP - I, PQ - I \in \Psi^{-\infty}(X).$$

In other words, P has an approximate inverse ("parametrix") which succeeds in inverting it modulo smoothing operators.

An operator P with nonvanishing principal symbol is said to be *elliptic.* Note that this theorem gives us, via the Sobolev estimates of (VII), the usual elliptic regularity estimates. In particular, we can deduce

$$Pu \in \mathcal{C}^\infty(X) \Longrightarrow u \in \mathcal{C}^\infty(X).$$

EXERCISE 3.6. Prove this.

PROOF. Let $q_{-m} = (1/\hat{\sigma}_m(P))$; let $Q_{-m} \in \Psi^{-m}(X)$ have principal symbol q_{-m}. (Such an operator exists by the exactness of the short exact symbol sequence.) Then by (III),

$$\hat{\sigma}_0(PQ_{-m}) = 1,$$

hence by (IV),[17]

$$PQ_{-m} - I = R_{-1} \in \Psi^{-1}(X).$$

Now we try to correct for this "error term:" pick $Q_{-m-1} \in \Psi^{-m-1}(X)$ with

$$\hat{\sigma}_{-m-1}(Q_{-m-1}) = -\hat{\sigma}_{-1}(R_{-1})/\hat{\sigma}_m(P).$$

Then we have

$$P(Q_{-m} + Q_{-m-1}) - I = R_{-2} \in \Psi^{-2}(X).$$

Continuing iteratively, we get a series of $Q_j \in \Psi^{-m-j}$ such that

$$P(Q_{-m} + \cdots + Q_{-m-N}) - I \in \Psi^{-N-1}(X).$$

Using (VIII), pick

$$Q \sim \sum_{j=-m}^{-\infty} Q_j.$$

This gives the desired parametrix:

EXERCISE 3.7.

(1) Check that $PQ - I \in \Psi^{-\infty}(X)$.

[17]The identity operator has principal symbol equal to 1, since the symbol map is a homomorphism.

(2) Check that $QP - I \in \Psi^{-\infty}(X)$. (HINT: First check that a left parametrix exists; you may find it helpful to take adjoints. Then check that the left parametrix must agree with the right parametrix.)

□

EXERCISE 3.8. Show that an elliptic pseudodifferential operator on a compact manifold is Fredholm. (HINT: You can show, for instance, that the kernel is finite dimensional by observing that the existence of a parametrix implies that the identity operator on the kernel is equal to a smoothing operator, which is compact.)

EXERCISE* 3.9.

(1) Let X be a compact manifold. Show that if $P \in \Psi^m(X)$ is elliptic, and has an actual inverse operator P^{-1} as a map from smooth functions to smooth functions, then $P^{-1} \in \Psi^{-m}(X)$. (HINT: Show that the parametrix differs from the inverse by an operator in $\Psi^{-\infty}(X)$—remember that an operator is in $\Psi^{-\infty}(X)$ if and only if it maps distributions to smooth functions.)
(2) More generally, show that if $P \in \Psi^m(X)$ is elliptic, then there exists a generalized inverse of P, inverting P on its range, mapping to the orthocomplement of the kernel, and annihilating the orthocomplement of the range, that lies in $\Psi^{-m}(X)$.

EXERCISE* 3.10. Let X be compact, and P an elliptic operator on X, as above, with positive order. Using the spectral theorem for compact, self-adjoint operators, show that if $P^* = P$, then there is an orthornormal basis for $L^2(X)$ of eigenfunctions of P, with eigenvalues tending to $+\infty$. Show that the eigenfunctions are in $\mathcal{C}^\infty(X)$. (HINT: show that there exists a basis of such eigenfunctions for the generalized inverse Q and then see what you can say about P.)

EXERCISE 3.11. Let X be compact.

(1) Show that the principal symbol of Δ, the Laplace-Beltrami operator on a compact Riemannian manifold, is just
$$|\xi|_g^2 \equiv \sum g^{ij}(x)\xi_i\xi_j,$$
the metric induced on the cotangent bundle.
(2) Using the previous exercise, conclude that there exists an orthonormal basis for $L^2(X)$ of eigenfunctions of Δ, with eigenvalues tending toward $+\infty$.

EXERCISE 3.12. Work out the Hörmander "square root trick" on a compact manifold X as follows.

(1) Show that if $P \in \Psi^0(X)$ is self-adjoint, with positive principal symbol, then P has an approximate square root, i.e. there exists $Q \in \Psi^0(X)$ such that $Q^* = Q$ and $P - Q^2 \in \Psi^{-\infty}(X)$. (HINT: Use an iterative construction, as in the proof of existence of elliptic parametrices.)
(2) Show that operators in $\Psi^{-\infty}(X)$ are L^2-bounded.
(3) Show that an operator $A \in \Psi^0(X)$ is L^2-bounded. (HINT: Take an approximate square root of $\lambda I - A^*A$ for $\lambda \gg 0$.)

As usual, let Δ denote the Laplacian on a compact manifold. By Exercise 3.12, there exists an operator $A \in \Psi^1(X)$ such that $A^2 = \Delta + R$, with $R \in \Psi^{-\infty}(X)$. By abstract methods of spectral theory, we know that $\sqrt{\Delta}$ exists as an unbounded operator on $L^2(X)$. (This is a very simple use of the functional calculus: merely take $\sqrt{\Delta}$ to act by multiplication by λ_j on each ϕ_j, where (ϕ_j, λ_j^2) are the eigenfunctions and eigenvalues of the Laplacian, from Exercise 3.11.) In fact, we can improve this argument to obtain:

PROPOSITION 3.5.

$$\sqrt{\Delta} \in \Psi^1(X).$$

Indeed, it follows from a theorem of Seeley that all complex powers of a self-adjoint, elliptic pseudodifferential operator[18] on a compact manifold are pseudodifferential operators.

All proofs of the proposition seem to introduce an auxiliary parameter in some way, and the following (taken directly from [**24**, Chapter XII, §1]) seems one of the simplest. An alternative approach, using the theory of elliptic boundary problems, is sketched in [**26**, pp.32-33, Exercises 4–6].

PROOF. Let A be the self-adjoint parametrix constructed in Exercise 3.12, so that

$$A^2 - \Delta = R \in \Psi^{-\infty}(X).$$

By taking a parametrix for the square root of A, in turn, we obtain

$$A = B^2 + R'$$

with $B \in \Psi^{1/2}(X)$ and $R' \in \Psi^{-\infty}$, both self-adjoint; then pairing with a test function ϕ shows that

$$\langle A\phi, \phi\rangle \geq \langle R'\phi, \phi\rangle \geq -C\|u\|^2$$

for some $C \in \mathbb{R}$. Thus, A can only have finitely many nonpositive eigenvalues (since it has a compact generalized inverse) hence its eigenvalues can accumulate only at $+\infty$). So we may alter A by the smoothing operator projecting off of these eigenspaces, and maintain

$$A^2 - \Delta = R \in \Psi^{-\infty}(X)$$

(with a different R, of course) while now ensuring that A is positive.

Now we may write, using the spectral theorem,

$$(\Delta')^{-1/2} = \frac{1}{2\pi i}\int_\Gamma z^{-1/2}((\Delta') - z)^{-1}\, dz$$

where Γ is a contour encircling the positive real axis counterclockwise, and given by $\operatorname{Im} z = \operatorname{Re} z$ for z sufficiently large, and Δ' is given by Δ minus the projection onto constants (hence has no zero eigenvalue). (The integral converges in norm, as self-adjointness of Δ' yields

$$\|((\Delta') - z)^{-1}\|_{L^2\to L^2} \lesssim |\operatorname{Im} z|^{-1}.)$$

Likewise, since $A^2 = \Delta' + R$ (with R yet another smoothing operator) we may write

$$A^{-1} = \frac{1}{2\pi i}\int_\Gamma z^{-1/2}((\Delta') + R - z)^{-1}\, dz$$

[18]Seeley's theorem is better yet: self-adjointness is unnecessary.

Hence

$$(\Delta')^{-1/2} - A^{-1} = \frac{1}{2\pi i}\int_\Gamma z^{-1/2}\big[((\Delta')-z)^{-1} - ((\Delta')+R-z)^{-1}\big]\,dz$$
$$= \frac{1}{2\pi i}\int_\Gamma z^{-1/2}((\Delta')-z)^{-1}R((\Delta')+R-z)^{-1}\,dz.$$

Now the integrand, $z^{-1/2}((\Delta')-z)^{-1}R((\Delta')+R-z)^{-1}$, is for each z a smoothing operator, and decays fast enough that when applied to any $u \in \mathcal{D}'(X)$, the integral converges to an element of $\mathcal{C}^\infty(X)$ (in particular, the integral converges in $\mathcal{C}^0(X)$, even after application of Δ^k on the left, for any k). Hence

$$(\Delta')^{-1/2} - A^{-1} = E \in \Psi^{-\infty}(X);$$

thus we also obtain

$$(\Delta')^{1/2} = (A^{-1}+E)^{-1} \in \Psi^1(X);$$

as $(\Delta')^{1/2}$ differs from $\Delta^{1/2}$ by the smoothing operator of projection onto constants, this shows that

$$\Delta^{1/2} \in \Psi^1(X). \quad \square$$

4. Wavefront set

If $P \in \Psi^m(X)$ and $(x_0, \xi_0) \in S^*X$, we say P is elliptic at (x_0,ξ_0) if $\hat\sigma_m(P)(x_0,\xi_0) \neq 0$. Of course if P is elliptic at each point in S^*X, it is elliptic in the sense defined above. We let

$$\operatorname{ell}(P) = \{(x,\xi) : P \text{ is elliptic at } (x,\xi)\},$$

and let

$$\Sigma_P = S^*X \backslash \operatorname{ell}(X);$$

Σ_P is known as the *characteristic set* of P.

EXERCISE 4.1.
(1) Show that $\operatorname{ell} P \subseteq \operatorname{WF}' P$.
(2) If P is a differential operator of order m of the form $\sum a_\alpha(x) D^\alpha$ then show that $\operatorname{WF}' P = \pi^*(\bigcup \operatorname{supp} a_\alpha)$, while $\operatorname{ell} P$ may be smaller.

The following "partition of unity" result, and variants on it, will frequently be useful in discussing microsupports. It yields an operator that is microlocally the identity on a compact set, and microsupported close to it.

LEMMA 4.1. *Given $K \subset U \subset S^*X$ with K compact, U open, there exists a self-adjoint operator $B \in \Psi^0(X)$ with*

$$\operatorname{WF}'(\operatorname{Id} - B) \cap K = \emptyset, \ \operatorname{WF}' B \subset U.$$

EXERCISE 4.2. Prove the lemma. (HINT: You might wish to try constructing B in the form

$$\operatorname{Op}(\psi\sigma_{\text{tot}}(\operatorname{Id}))$$

where $\sigma_{\text{tot}}(\operatorname{Id})$ is the total symbol of the identity (which is simply 1 for all the usual quantizations on $\mathbb{R}^n$) and ψ is a cutoff function equal to 1 on K and supported in U. Then make B self-adjoint.)

THEOREM 4.2. *If $P \in \Psi^m(X)$ is elliptic at (x_0,ξ_0), there exists a* microlocal elliptic parametrix *$Q \in \Psi^{-m}(X)$ such that*

$$(x_0,\xi_0) \notin \operatorname{WF}'(PQ-I) \cup \operatorname{WF}'(QP-I).$$

In other words, you should think of Q as inverting P *microlocally near* (x_0, ξ_0).

EXERCISE 4.3. Prove the theorem. (HINT: If B is a microlocal partition of unity as in Lemma 4.1, microsupported sufficiently close to (x_0, ξ_0) and microlocally the identity in a smaller neighborhood, then show

$$W = BP + \lambda \operatorname{Op}(\langle \xi \rangle^m)(\operatorname{Id} - B)$$

is globally elliptic provided $\lambda \in \mathbb{C}$ is chosen appropriately. Now, using the existence of an elliptic parametrix for W, prove the theorem.)

Let u be a distribution on a manifold X. We define the *wavefront set* of u as follows.

DEFINITION 4.3. The wavefront set of u,

$$\operatorname{WF} u \subseteq S^*X,$$

is given by

$$(x_0, \xi_0) \notin \operatorname{WF} u$$

if and only if there exists $P \in \Psi^0(X)$, elliptic at (x_0, ξ_0), such that

$$Pu \in \mathcal{C}^\infty.$$

EXERCISE 4.4. Show that the choice of $\Psi^0(X)$ in this definition is immaterial, and that we get the same definition of $\operatorname{WF} u$ if we require $P \in \Psi^m(X)$ instead.

Note that the wavefront set is, from its definition, a closed set. Instead of viewing $\operatorname{WF} u$ as a subset of S^*X, we also, on occasion, think of $\operatorname{WF} u$ as a *conic* subset of $T^*X \backslash o$, with o denoting the zero section; a conic set in a vector bundle is just one that is invariant under the $\mathbb{R}^+$ action on the fibers.

An important variant is as follows: we say that

$$(x_0, \xi_0) \notin \operatorname{WF}^m u$$

if and only if there exists $P \in \Psi^m(X)$, elliptic at (x_0, ξ_0) such that

$$Pu \in L^2(X).$$

PROPOSITION 4.4. $\operatorname{WF} u = \emptyset$ *if and only if* $u \in \mathcal{C}^\infty(X)$; $\operatorname{WF}^m u = \emptyset$ *if and only if* $u \in H^m_{loc}(X)$.

The wavefront set serves the purpose of measuring not just where, but also in what (co-)direction, a distribution fails to be in $\mathcal{C}^\infty(X)$ (or H^m in the case of the indexed version). It is instructive to think about testing for such regularity, at least on $\mathbb{R}^n$, by localizing and Fourier transforming. Given $(x_0, \hat{\xi}_0) \in S^*\mathbb{R}^n$, let $\phi \in \mathcal{C}_c^\infty(\mathbb{R}^n)$ be nonzero at x_0; let $\gamma \in \mathcal{C}^\infty(\mathbb{R}^n)$ be given by

$$\gamma(\xi) = \psi\Big(\Big|\frac{\xi}{|\xi|} - \hat{\xi}_0\Big|\Big)\chi(|\xi|)$$

where ψ is a cutoff function supported near $x = 0$ and $\chi(t) \in \mathcal{C}^\infty(\mathbb{R})$ is equal to 0 for $t < 1$ and 1 for $t > 2$. Think of γ as a cutoff in a cone of directions near ξ_0, but modified to be smooth at the origin. (We will use such a construction frequently, and refer in future to a function such as γ as a "conic cutoff near direction $\hat{\xi}_0$.".)

Now note that $\phi(x)\gamma(\xi)$ is a symbol of order zero, and

$$\operatorname{Op}_\ell(\phi(x)\gamma(\xi))^* = \operatorname{Op}_r(\phi(x)\gamma(\xi))u = (2\pi)^{-n}\mathcal{F}^{-1}\gamma(\xi)\mathcal{F}(\phi u). \tag{4.1}$$

By definition, if $\mathrm{Op}_\ell(\phi(x)\gamma(\xi))^* u \in \mathcal{C}^\infty$, then $(x_0, \xi_0) \notin \mathrm{WF}\, u$. Note that since ϕu has compact support, we automatically have $\mathcal{F}(\phi u) \in \mathcal{C}^\infty$, hence $\mathcal{F}^{-1}\gamma\mathcal{F}(\phi u)$ is rapidly decreasing. Since $\mathcal{F}$ is an isomorphism from $\mathcal{S}(\mathbb{R}^n)$ to itself, we see that it in fact suffices to have

$$\gamma\mathcal{F}(\phi u) \in \mathcal{S}(\mathbb{R}^n)$$

to be able to conclude that $(x_0, \xi_0) \notin \mathrm{WF}\, u$. Conversely, one can check that the class of operators of the form

$$\mathrm{Op}_\ell(\phi(x)\gamma(\xi))^*$$

is rich enough that this in fact amounts to a *characterization* of wavefront set:

PROPOSITION 4.5. *We have* $(x_0, \xi_0) \notin \mathrm{WF}\, u$ *if and only if there exist* ϕ, γ *as above with*

$$\gamma\mathcal{F}(\phi u) \in \mathcal{S}(\mathbb{R}^n).$$

EXERCISE 4.5. Prove the Proposition. (HINT: If $A \in \Psi^0(\mathbb{R}^n)$ is elliptic at (x_0, ξ_0) and $Au \in \mathcal{C}^\infty(\mathbb{R}^n)$, construct $B = \mathrm{Op}_\ell(\phi(x)\gamma(\xi))^*$ as above so that $\mathrm{WF}'\, B$ is contained in the set where A is elliptic. Hence there is a microlocal parametrix Q such that $B(QA - I) \in \Psi^{-\infty}(X)$.)

Note that if u is smooth near x_0, then we have $\phi u \in \mathcal{C}_c^\infty(\mathbb{R}^n)$ for appropriately chosen ϕ, hence there is no wavefront set in the fiber over x_0.

If, by contrast, u is not smooth in any neighborhood of x_0, then we of course do not have $\mathcal{F}(\phi u) \in \mathcal{S}$, although it is in $\mathcal{C}^\infty$; the wavefront set includes the directions in which it fails to be rapidly decaying.

Thus, we can easily see that in fact the projection to the base variables of $\mathrm{WF}\, u$ is the *singular support* of u, i.e. the points which have no neighborhood in which the distribution u is a $\mathcal{C}^\infty$ function.

EXERCISE 4.6. Let $\Omega \subset \mathbb{R}^n$ be a domain with smooth boundary. Show that $\mathrm{WF}\, 1_\Omega = SN^*(\partial\Omega)$, the spherical normal bundle of the boundary. (HINT: You may want to use the fact that the definition of $\mathrm{WF}\, u$ is coordinate-invariant.)

We have a result constraining the wavefront set of a solution to a PDE or, more generally, a pseudodifferential equation, directly following from the definition:

THEOREM 4.6. *If* $Pu \in \mathcal{C}^\infty(X)$, *then* $\mathrm{WF}\, u \subseteq \Sigma_P$.

PROOF. By definition, $Pu \in \mathcal{C}^\infty(X)$ means that $\mathrm{WF}\, u \cap \mathrm{ell}\, P = \emptyset$. □

THEOREM 4.7. *If* $P \in \Psi^*(X)$, $\mathrm{WF}\, Pu \subseteq \mathrm{WF}\, u \cap \mathrm{WF}'\, P$.

EXERCISE 4.7. Prove this, using microlocal elliptic parametrices for the inclusion in $\mathrm{WF}\, u$.

The property of pseudodifferential operators that $\mathrm{WF}\, Pu \subseteq \mathrm{WF}\, u$ is called "microlocality:" the operators are not "local," in that they do move *supports* of distributions around, but they don't move *singularities*, even in the refined sense of wavefront set.

We shall also need related results on Sobolev based wavefront sets in what follows:

PROPOSITION 4.8. *If* $P \in \Psi^m(X)$, $\mathrm{WF}^{k-m}\, Pu \subseteq \mathrm{WF}^k\, u \cap \mathrm{WF}'\, P$ *for all* $k \in \mathbb{R}$.

COROLLARY 4.9. *Let $P \in \Psi^m(X)$. If*

$$\mathrm{WF}' P \cap \mathrm{WF}^m u = \emptyset$$

then

$$Pu \in L^2(X).$$

EXERCISE 4.8. Prove the proposition (again using a microlocal elliptic parametrix) and the corollary.

We will have occasion to use the following relationship between ordinary and Sobolev-based wavefront sets:

PROPOSITION 4.10.

$$\mathrm{WF}\, u = \overline{\bigcup_k \mathrm{WF}^k u}.$$

EXERCISE 4.9. Prove the proposition.

EXERCISE 4.10. Let $\Box$ denote the wave operator,

$$\Box u = D_t^2 u - \Delta u$$

on $M = \mathbb{R} \times X$ with X a Riemannian manifold. Show if $\Box u = 0$ then the wavefront set of u is a subset of the "wave cone" $\{\tau^2 = |\xi|_g^2\}$ where τ is the dual variable to t and ξ to x in $T^*(M)$.

EXERCISE 4.11.

(1) Let $k < n$, and let $\iota : \mathbb{R}^k \to \mathbb{R}^n$ denote the inclusion map.
Show that there is a continuous *restriction map* on compactly supported distributions with no wavefront set conormal to $\mathbb{R}^k$:

$$\iota^* : \{u \in \mathcal{E}'(\mathbb{R}^n) : \mathrm{WF}\, u \cap SN^*(\mathbb{R}^k) = \emptyset\} \to \mathcal{E}'(\mathbb{R}^k).$$

HINT: Show that it suffices to consider u supported in a small neighborhood of a single point in $\mathbb{R}^k$. Then take the Fourier transform of u and try to integrate in the conormal variables to obtain the Fourier transform of the restriction.

(2) Show that, with the notation of the previous part,

$$\mathrm{WF}\, \iota^* u \subseteq \iota_*(\mathrm{WF}\, u)$$

where $\iota_* : T^*_{\mathbb{R}^k}\mathbb{R}^n \to T^*\mathbb{R}^k$ is the naturally defined projection map.

(3) Show that both the previous parts make sense, and are valid, for restriction to an embedded submanifold Y of a manifold X.

(4) Show that if u is a distribution on $\mathbb{R}^k_x$ and v is a distribution on $\mathbb{R}^l_y$ then $w = u(x)v(y)$ is a distribution on $\mathbb{R}^{k+l}$ and

$$\mathrm{WF}\, w \subseteq \big[(\operatorname{supp} u, 0) \times \mathrm{WF}\, v\big] \cup \big[\mathrm{WF}\, u \times (\operatorname{supp} v, 0)\big] \cup \mathrm{WF}\, u \times \mathrm{WF}\, v.$$

(HINT: Localize and Fourier transform, as in (4.1).)

You might wonder: given P, can the wavefront set of a solution to $Pu = 0$ be any closed subset of Σ? The answer is no, there are, in general, further constraints. To talk about them effectively, we should digress briefly back into geometry.

4.1. Hamilton flows. We now amplify the discussion §3.2 of Hamiltonian mechanics and symplectic geometry, generalizing it to a broader geometric context.

Let N be a symplectic manifold, that is to say, one endowed with a closed, nondegenerate[19] two-form. (Our prime example is $N = T^*X$, endowed with the form $\sum d\xi_j \wedge dx^j$; by Darboux's theorem, every symplectic manifold in fact locally looks like this.)

Given a real-valued function $a \in \mathcal{C}^\infty(N)$, we can make a *Hamilton vector field* from a as follows: by nondegeneracy, there is a unique vector field H_a such that $\iota_{\mathsf{H}_a}\omega \equiv \omega(\cdot, \mathsf{H}_a) = da$.

EXERCISE 4.12. Check that in local coordinates in T^*X,

$$\mathsf{H}_a = \sum_{j=1}^n \frac{\partial a}{\partial \xi_j}\partial_{x^j} - \frac{\partial a}{\partial x^j}\partial_{\xi_j}.$$

Thus, for any smooth function b, we may define the Poisson bracket

$$\{a, b\} = \mathsf{H}_a(b)$$

EXERCISE 4.13. Check that the Poisson bracket is antisymmetric.

It is easy to verify that the flow along H_a preserves both the symplectic form and the function a : we have from Cartan's formula (and since ω is closed):

$$\mathcal{L}_{\mathsf{H}_a}(\omega) = d\iota_{\mathsf{H}_a}\omega = d(da) = 0;$$

also,

$$\mathsf{H}_a(a) = da(\mathsf{H}_a) = \omega(\mathsf{H}_a, \mathsf{H}_a) = 0.$$

The integral curves of the vector field H_a are called the *bicharacteristics* of a and those lying inside $\Sigma_a = \{a = 0\}$ are called *null bicharacteristics*.

EXERCISE* 4.14.

(1) Show that the bicharacteristics of $|\xi|_g = (\sigma_2(\Delta))^{1/2}$ project to X to be geodesics. The flow along the Hamilton vector field of $|\xi|_g$ is known as *geodesic flow*.
(2) Show that the null bicharacteristics of $\sigma_2(\Box)$ are lifts to $T^*(\mathbb{R} \times X)$ of geodesics of X, traversed both forward and backward at unit speed.

Recall that the setting of symplectic manifolds is exactly that of Hamiltonian mechanics: given such a manifold, we can regard it as the phase space for a particle; specifying a function (the "energy" or "Hamiltonian") gives a vector field, and the flow along this vector field is supposed to describe the time-evolution of our particle in the phase space.

EXERCISE 4.15. Check that the phase space evolution of the harmonic oscillator Hamiltonian, $(1/2)(\xi^2 + x^2)$ on $T^*\mathbb{R}$, agrees with what you learned in physics class long ago.

---

[19]Nondegeneracy of ω means that contraction with ω is an isomorphism from T_pN to T_p^*N at each point.

4.2. Propagation of singularities.

THEOREM 4.11 (Hörmander). *Let $Pu \in \mathcal{C}^\infty(X)$, with $P \in \Psi^m(X)$ an operator with real principal symbol. Then* $\operatorname{WF} u$ *is a union of maximally extended null bicharacteristics of $\hat{\sigma}_m(P)$ in S^*X.*

We should slightly clarify the usage here: to make sense of these null bicharacteristics, we should actually take the Hamilton vector field of the homogeneous version of the symbol, $\sigma_m(P)$; this is a homogeneous vector field, and its integral curves thus have well-defined projections onto S^*X. If the Hamilton vector field should be "radial" at some point $q \in T^*X$, i.e. coincide with a multiple of the vector field $\xi \cdot \partial_\xi$ there, then the projection of the integral curve through q is just a single point in S^*X, and the theorem gives no further information about wavefront set at that point.

For $P = \Box$, the theorem says that the wavefront set lies in the "light cone," and propagates forward and backward at unit speed along geodesics. If we take the fundamental solution to the wave equation[20] $u = \sin(t\sqrt{\Delta}/\sqrt{\Delta})\delta_p$, it is not hard to compute that in fact for small, nonzero time,[21]

$$\operatorname{WF} u \subseteq N^*\{d(\cdot, p) = |t|\} \equiv \mathcal{L};$$

This is a generalization of Huygens's Principle, which tells us that in $\mathbb{R} \times \mathbb{R}^n$, for n odd, the support of the fundamental solution is on this expanding sphere (but which is a highly unstable property). Note that $\mathcal{L}$ is in fact the bicharacteristic flowout of all covectors in Σ projecting to $N^*(\{p\})$ at $t = 0$, and under this interpretation, $\mathcal{L} \subset T^*(\mathbb{R} \times X)$ makes sense for *all* times, not just for short time, regardless of the metric geometry. We shall return to and amplify this point of view in §9.

EXERCISE 4.16.

(1) Suppose that $\Box u = 0$ on $\mathbb{R} \times \mathbb{R}^n$ and $u(t, x) \in \mathcal{C}^\infty$ for $(t, x) \in (-\epsilon, \epsilon) \times B(0, 1)$ for some $\epsilon > 0$. Show, using Theorem 4.11, that $u \in \mathcal{C}^\infty$ on $\{|x| < 1 - |t|\}$. Can you show this more directly using the energy methods described in §2.3?

(2) Suppose that $\Box u = 0$ on $\mathbb{R} \times \mathbb{R}^n$ and $u(t, x) \in \mathcal{C}^\infty$ for $(t, x) \in (-\epsilon, \epsilon) \times (B(0, 1) \backslash B(0, 1/2))$ for some $\epsilon > 0$. Show, using the theorem, that $u \in \mathcal{C}^\infty$ in $\{|x| < 1 - |t|\} \cap \{|t| \in (3/4, 1)\}$

PROOF. [22] Note that we already know that $\operatorname{WF} u \subseteq \Sigma_P$ by Theorem 4.6, hence what remains to be proved is the flow-invariance.

Let $q \in \Sigma_P \subset S^*X$. By homogeneity of $\sigma_m(P)$, we can write the Hamilton vector field in T^*X in a neighborhood of q as

$$\mathsf{H}_p = |\xi|^{m-1}(\mathsf{V} + h\mathsf{R}), \tag{4.2}$$

where R denotes the radial vector field $\xi \cdot \partial_\xi$, h is a function on S^*X, and V is the pullback under quotient of a vector field on S^*X itself, i.e. V is homogeneous of

[20] This is the spectral-theoretic way of writing the solution with initial value 0 and initial time-derivative δ_p.

[21] Well, I am cheating a bit here, as we haven't stated any results allowing us to relate the wavefront set of Cauchy data for the wavefront set of the solution to the equation. To understand how to do this, you should read [**17**].

[22] This proof is very close to those employed by Melrose in [**17**] and [**18**].

degree zero with no radial component, hence of the form $\sum_j f_j(x,\hat\xi)\partial_{\hat\xi_j} + g_j(x,\hat\xi)\partial_{x^j}$. Note that if a is homogeneous of degree l then

$$\mathsf{R}a = la. \tag{4.3}$$

(Exercise: Verify these consequences of homogeneity.)

By the comments above, we may take $\mathsf{V} \neq 0$ near q; otherwise the theorem is void. Thus, without loss of generality, we may employ a coordinate system $\alpha_1, \dots, \alpha_{2n-1}$ for S^*X in which

$$\mathsf{V} = \partial_{\alpha_1}, \tag{4.4}$$

hence using $\alpha, |\xi|$ as coordinates in T^*X,

$$\mathsf{H}_p = \partial_{\alpha_1} + h\mathsf{R};$$

we may shift coordinates so that $\alpha(q) = 0$. We split the α variables into α_1 and $\alpha' = (\alpha_2, \dots, \alpha_{n-1})$.

Since $\mathrm{WF}\, u$ is closed, it suffices to prove the following: if $q \notin \mathrm{WF}\, u$ then $\Phi_t(q) \notin \mathrm{WF}\, u$ for $t \in [-1,1]$, where Φ_t denotes the flow generated by V.[23] (This will show that the intersection of $\mathrm{WF}\, u$ with the bicharacteristic through q is both open and closed, hence is the whole thing.)

We can make separate arguments for $t \in [0,1]$ and $t \in [-1,0]$, and will do so (in fact, we will leave one case to the reader).

For simplicity, let us take $Pu = 0$; we leave the case of an inhomogeneous equation for the reader (it introduces extra terms, but no serious changes will in fact be necessary in the proof).

Since $\mathrm{WF}\, u$ is closed, our assumption that $q \notin \mathrm{WF}\, u$ tells us that there is in fact a 2δ-neighborhood of 0 in the α coordinates that is disjoint from $\mathrm{WF}\, u$; we are trying to extend this regularity along the rest of the set $(\alpha_1, \alpha') \in [0,1] \times 0$. We proceed as follows: let

$$s_0 = \sup\{s : \mathrm{WF}^s\, u \cap \{(\alpha_1, \alpha') \in [0,1] \times B(0,\delta)\} = \emptyset\}. \tag{4.5}$$

Pick any $s < s_0$. We will show that in fact

$$\mathrm{WF}^{s+1/2}\, u \cap \{(\alpha_1, \alpha') \in [0,1] \times B(0,\delta)\} = \emptyset, \tag{4.6}$$

thus establishing that $s_0 = \infty$, which is the desired result (by Proposition 4.10). One can regard this strategy as iteratively obtaining more and more regularity for u along the bicharacteristic (i.e. the idea is that we start by knowing *some* possibly very bad regularity, and we step by step conclude that we can improve upon this regularity, half a derivative at a time). More colloquially, the idea is that the "energy," as measured by testing the distribution u by pseudodifferential operators, should be comparable at different points along the bicharacteristic curve.

Now we prove the estimates that yield (4.6) via commutator methods. Let $\phi(s)$ be a cutoff function with

$$\begin{aligned} &\phi(t) > 0 \text{ on } (-1,1), \\ &\operatorname{supp}\phi = [-1,1] \end{aligned} \tag{4.7}$$

[23]Of course, we are assuming here that the interval $[-1,1]$ remains in our coordinate neighborhood; rescale the coordinates if necessary to make this so.

Let $\phi_\delta(s) = \phi(\delta^{-1}s)$; arrange that $\sqrt{\phi} \in \mathcal{C}^\infty$. Let χ be a cutoff function equal to 1 on $(0,1)$ and with $\chi' = \psi_1 - \psi_2$, with ψ_1 supported on $(-\delta, \delta)$ and ψ_2 on $(1-\delta, 1+\delta)$; we will further assume that $\sqrt{\chi}, \sqrt{\psi_i} \in \mathcal{C}^\infty$.

EXERCISE 4.17. Verify that cutoffs with these properties exist.

In our coordinate system for S^*X, let

$$\hat{a} = \phi_\delta(|\alpha'|)\chi(\alpha_1)e^{-\lambda\alpha_1} \in \mathcal{C}^\infty(S^*X),$$

with $\lambda \gg 0$ to be chosen presently. Passing to the corresponding function on $a \in \mathcal{C}^\infty(T^*X)$ that is homogeneous of degree $2s - m + 2$, we have

$$\begin{aligned}(4.8)\quad \mathsf{H}_p(a) = |\xi|^{2s+1}\big(- \lambda\phi_\delta(|\alpha'|)\chi(\alpha_1)e^{-\lambda\alpha_1} \\ + \phi_\delta(|\alpha'|)(\psi_1 - \psi_2)e^{-\lambda\alpha_1} + h(\alpha)(2s - m + 2)a\big)\end{aligned}$$

with h given by (4.2). Since a 2δ coordinate neighborhood of the origin was assumed absent from $\mathrm{WF}\, u$, we have in particular ensured that $\operatorname{supp} \phi_\delta(|\alpha'|)\psi_1(\alpha_1)$ is contained in $(\mathrm{WF}\, u)^c$. We also have $\operatorname{supp} \hat{a} \subset (\mathrm{WF}^s\, u)^c$ by (4.5), since $s < s_0$.

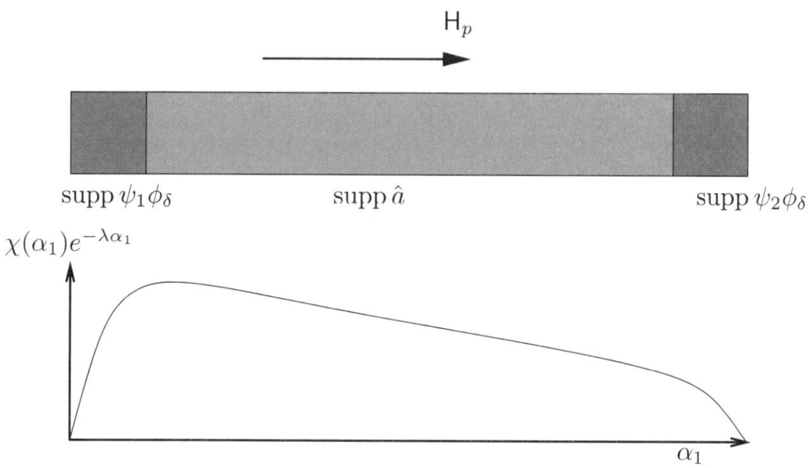

FIGURE 1. The support of the commutant and its value along the line $\alpha' = 0$. The support of the term $\psi_1(\alpha_1)\phi_\delta(|\alpha'|)$ is arranged to be contained in the complement of $\mathrm{WF}\, u$, while the support of the whole of a is arranged to be in the complement of $\mathrm{WF}^s\, u$.

Let $A \in \Psi^{2s-m+2}(X)$ be given by the quantization of a.[24] Since $\sigma_m(P)$ is real by assumption, we have $P^* - P \in \Psi^{m-1}(X)$. (Exercise: Check this!) Thus the "commutator" $P^*A - AP$, which is a priori of order $2s+2$, has vanishing principal symbol of order $2s+2$, hence it in fact lies in $\Psi^{2s+1}(X)$, and we may write

$$(P^*A - AP) = [P, A] + (P^* - P)A,$$

[24]I.e., really A is given by cutting off a near $\xi = 0$ to give a smooth total symbol and quantizing that.

with

$$(4.9)\quad \begin{aligned} i\sigma_{2s+1}([P,A]+(P^*-P)A) &= \mathsf{H}_p(a)+\sigma_{m-1}(P^*-P)a \\ &= -\lambda\phi_\delta(|\alpha'|)\chi(\alpha_1)e^{-\lambda\alpha_1}|\xi|^{2s+1}+\phi_\delta(|\alpha'|)(\psi_1-\psi_2)e^{-\lambda\alpha_1}|\xi|^{2s+1} \\ &\qquad + (i\sigma_{m-1}(P^*-P)+h(\alpha)(2s-m+2))a, \end{aligned}$$

by (4.2),(4.3), and (4.4). If $\lambda \gg 0$ is chosen sufficiently large, we may absorb the third term into the first, and write the RHS of (4.9) as

$$-f(\alpha)\phi_\delta(|\alpha'|)\chi(\alpha_1)+\phi_\delta(|\alpha'|)(\psi_1-\psi_2)e^{-\lambda\alpha_1}$$

with $f > 0$ on the support of $\phi_\delta\chi$.

Let $B \in \Psi^{(2s+1)/2}(X)$ be obtained by quantization of

$$|\xi|^{s+1/2}(f(\alpha)\phi_\delta(|\alpha'|)\chi(\alpha_1))^{1/2};$$

and let $C_i \in \Psi^{(2s+1)/2}(X)$ be obtained by quantization of

$$|\xi|^{s+1/2}(\phi_\delta(|\alpha'|)\psi_i(\alpha_1))^{1/2}e^{-\lambda\alpha_1/2}.$$

Then by the symbol calculus, i.e. by Properties III, IV of the calculus of pseudodifferential operators,

$$(4.10)\qquad i(P^*A-AP)=i(P^*-P)A+i[P,A]=-B^*B+C_1^*C_1-C_2^*C_2+R$$

with $R \in \Psi^{2s}(X)$, hence of lower order than the other terms; moreover we have $\mathrm{WF}' R \subset \operatorname{supp}\hat{a}$.

Now we "pair" both sides of (4.10) with our solution u. We have

$$i\langle (P^*A-AP)u,u\rangle = \langle (-B^*B+C_1^*C_1-C_2^*C_2+R)u,u\rangle;$$

as we are taking $Pu=0$, the LHS vanishes.[25] We thus have, rearranging this equation,

$$(4.11)\qquad \|Bu\|^2+\|C_2u\|^2=\|C_1u\|^2+\langle Ru,u\rangle.$$

I claim that the RHS is finite: Recall that R lies in $\Psi^{2s}(X)$. Let Λ be an operator of order s, elliptic on $\mathrm{WF}' R$ and with $\mathrm{WF}'\Lambda$ contained in the complement of $\mathrm{WF}^s u$.

EXERCISE 4.18. Show that such a Λ exists.

Thus, letting Υ be a microlocal parametrix for Λ on $\mathrm{WF}' R$, we have

$$\mathrm{WF}' R \cap \mathrm{WF}'(\mathrm{Id}-\Lambda\Upsilon)=\emptyset,$$

hence

$$R-\Lambda\Upsilon R = E \in \Psi^{-\infty}(X).$$

Thus,

$$|\langle Ru,u\rangle| \le |\langle \Upsilon Ru,\Lambda^* u\rangle|+|\langle Eu,u\rangle| < \infty$$

by Corollary 4.9 since $\mathrm{WF}'\Upsilon R\cup \mathrm{WF}'\Lambda^* \subset (\mathrm{WF}^s u)^c$ (and since E is smoothing). Returning to (4.11), we also note that the term $\|C_1u\|^2$ is finite by our assumptions on the location of $\mathrm{WF}^{s+1/2} u$ (and another use of Corollary (4.9)). Thus,

$$\|Bu\| < \infty,$$

and consequently,

$$\mathrm{WF}^{s+1/2} u \cap \operatorname{ell} B = \emptyset,$$

which was the desired estimate. □

[25] In the case of an inhomogeneous equation, it is of course here that extra terms arise.

EXERCISE 4.19. Now see how the argument should be modified to yield absence of $\mathrm{WF}^{s+1/2}\, u$ on

$$\{\alpha' \in [-1,0], \alpha' = 0\}.$$

One cheap alternative to going through the whole proof might be to notice that we also have $(-P)u \in \mathcal{C}^\infty$, and that $\mathsf{H}_{-p} = -\mathsf{H}_p$; thus, the "forward propagation" that we have just proved should yield backward propagation along H_p as well.

THE FINE PRINT: Now, having done all that, note that it was a cheat. In particular, we didn't know a priori that we could apply any of the operators that we used to u and obtain an L^2 function, let alone justify the formal integrations by parts used to move adjoints across the pairings. Therefore, to make the above argument rigorous, we need to modify it with an *approximation argument.* This is similar to the situation in Exercise 2.7, except in that case, we had a natural way of obtaining smooth solutions to the equation which approximated the desired one: we could replace our initial data ψ_0 for the Schrödinger equation by, for instance, $e^{-\epsilon\Delta}\psi_0$; the solution at later time is then just $e^{-\epsilon\Delta}\psi$, and we can consider the limit $\epsilon \downarrow 0$. In the general case to which this theorem applies, though, we do not have any convenient families of smoothing operators commuting with P. So we instead take the tack of smoothing our *operators* rather than the solution u. We should manufacture a family of smoothing operators G_ϵ that strongly approach the identity as $\epsilon \downarrow 0$, and replace A by AG_ϵ everywhere it appears above. If we do this sensibly, then the analogs of the estimates proved above yield the desired estimates in the $\epsilon \downarrow 0$ limit. Of course, we need to know how G_ϵ passes through commutators, etc., so the right thing to do is to take the G_ϵ themselves to be pseudodifferential approximations of the identity, something like

$$G_\epsilon = \mathrm{Op}_\ell(\varphi(\epsilon|\xi|))$$

on $\mathbb{R}^n$, with $\varphi \in \mathcal{C}_c^\infty(\mathbb{R})$ a cutoff equal to 1 near 0. We content ourselves with referring the interested reader to [**18**] for the analogous development in the "scattering calculus" including details of the approximation argument.

EXERCISE 4.20.

(1) Show the following variant of Theorem 4.11: if $P \in \Psi^m(X)$ is an operator with real principal symbol, and $Pu \in \mathcal{C}^\infty(X)$, show that $\mathrm{WF}^k\, u$ is a union of maximally extended bicharacteristics of P for each $k \in \mathbb{R}$. (Hint: the proof is a subset of the proof of Theorem 4.11.)
(2) Show the following inhomogeneous variant of Theorem 4.11: if $P \in \Psi^m(X)$ is an operator with real principal symbol, and $Pu = f$, show that $\mathrm{WF}\, u \backslash \mathrm{WF}\, f$ is a union of maximally extended bicharacteristics of P.

EXERCISE 4.21.

(1) What does Theorem 4.11 tell us about solutions to the Schrödinger equation? (Hint: not much.)
(2) Nonetheless: let $\psi(t,x)$ be a solution to the Schrödinger equation on $\mathbb{R} \times X$ with (X,g) a Riemannian manifold; suppose that $\psi(0,x) = \psi_0 \in H^{1/2}(X)$. Define a set $S_1 \subset S^*X$ by

$$q \notin S_1 \iff \text{there exists } A \in \Psi^1(X),\ q \in \operatorname{ell}(A), \qquad \text{such that } \int_0^1 \|A\psi\|^2\, dt < \infty.$$

(In other words, S_1 is a kind of wavefront set measuring where in the phase space S^*X we have $\psi \in L^2([0,1]; H^1(X))$—cf. Exercise 2.7.)

Show that S_1 is invariant under the geodesic flow on S^*X. (See Exercise 4.14 for the definition of geodesic flow.)

(Hint: use (2.2) with A an appropriately chosen pseudodifferential operator of order zero, constructed much like the ones used in proving Theorem 4.11.)

Reflect on the following interpretation: "propagation of L^2H^1 regularity for the Schrödinger equation occurs at infinite speed along geodesics."

5. Traces

It turns out to be of considerable interest in spectral geometry to consider the *traces* of operators manufactured from Δ, the Laplace-Beltrami operator on a compact[26] Riemannian manifold. The famous question posed by Kac [**15**], "Can one hear the shape of a drum," has a natural extension to this context: Recall from Exercise 3.11 that there exists an orthonormal basis ϕ_j of eigenfunctions of Δ with eigenvalues $\lambda_j^2 \to +\infty$; what, one wonders, can one recover of the geometry of a Riemannian manifold from the sequence of frequencies λ_j? Using PDE methods to understand traces of functions of the Laplacian has led to a better understanding of these inverse spectral problems.

Recall from Proposition 3.5 that $\sqrt{\Delta}$ is a first-order pseudodifferential operator on X. It is a slightly inconvenient fact that while $\sqrt{\Delta} \in \Psi^1(X)$, $\sqrt{\Delta} \notin \Psi^1(\mathbb{R} \times X)$: its Schwartz kernel is easily seen to be singular away from the diagonal. But this turns out be be of little practical importance for our considerations here: it is close enough!

Let us now consider the operator

$$U(t) = e^{-it\sqrt{\Delta}} \tag{5.1}$$

which can be defined by the functional calculus to act as the scalar operator $e^{-it\lambda_j}$ on each ϕ_j. $U(t)$ is unitary, and indeed is the solution operator to the Cauchy problem for the equation

$$(\partial_t + i\sqrt{\Delta})u = 0; \tag{5.2}$$

that is to say, if $u = U(t)f$, we have

$$(\partial_t + i\sqrt{\Delta})u = 0, \quad \text{and } u(0,x) = f(x).$$

Equation (5.2) is easily seen to be very closely related to the wave equation: if u solves (5.2) then applying $\partial_t - i\sqrt{\Delta}$, we see that u also satisfies the wave equation. Of course, (5.2) only requires a single Cauchy datum, unlike the wave equation, so the trade-off is that the Cauchy data of u as a solution to $\Box u = 0$ are constrained: we have

$$u(0,x) = f(x), \quad \partial_t u(0,x) = -i\sqrt{\Delta} f.$$

The real and imaginary parts of the operator $U(t)$ are exactly the solution operators to the (more usual) Cauchy problem for the wave equation with $u(0,x) = f(x), \partial_t u(0,x) = 0$ and with $u(0,x) = 0, \partial_t u(0,x) = -i\sqrt{\Delta} f(x)$ respectively.

---

[26]We especially emphasize that X denotes a *compact* manifold throughout this section.

Why is the operator $U(t)$ of interest? Well, suppose that we are interested in the sequence of λ_j's. It makes sense to combine these numbers into a generating function, and certainly one option would be to take the exponential sum[27]

$$\sum_j e^{-it\lambda_j}$$

This is, at least formally, nothing but the trace of the operator $U(t)$. One of the principal virtues of this generating function is that if we let $N(\lambda)$ denote the "counting function"

$$N(\lambda) = \#\{\lambda_j \le \lambda\},$$

then we have

$$N'(\lambda) = \sum_j \delta(\lambda - \lambda_j),$$

hence

$$\sum e^{-it\lambda_j} = (2\pi)^{n/2}\mathcal{F}_{\lambda\to t}(N'(\lambda))(t).$$

This is all a bit optimistic, as $U(t)$ is easily seen to be not of trace class—for example at $t = 0$ it is the identity. So we should try and think of $\operatorname{Tr} U(t)$ as a *distribution.* We do know that for any test function $\varphi(t) \in \mathcal{S}(\mathbb{R})$ and any $f \in L^2(X)$,

$$\begin{aligned} \int \varphi(t)U(t)f\,dt &= \int (1+D_t^2)^{-k}(1+D_t^2)^k(\varphi(t))U(t)f\,dt \\ &= \int (1+D_t^2)^k(\varphi(t))(1+D_t^2)^{-k}U(t)f\,dt \\ &= \int (1+D_t^2)^k(\varphi(t))(1+\Delta)^{-k}U(t)f\,dt, \end{aligned} \tag{5.3}$$

since $D_t^2 U = \Delta U$. Here we can, if we like, consider $(1+\Delta)^{-k}$ to be defined by the functional calculus; it is in fact pseudodifferential, of order $-2k$. We easily obtain (using either point of view) the estimate:

$$(1+\Delta)^{-k}U(t) : L^2(X) \to H^{2k}(X);$$

hence, for $k \gg 0$, the operator $(1+\Delta)^{-k}U(t)$ is of trace class.

EXERCISE 5.1. Prove that this operator is of trace class for $k \gg 0$. (HINT: One easy route is to think about first choosing k large enough that the Schwartz kernel is continuous, hence the operator is Hilbert-Schmidt; then you can take k even larger to get a trace-class operator, by factoring into a product of two Hilbert-Schmidt operators (see Appendix).)

Equation (5.3) thus establishes that

$$\operatorname{Tr} U(t) : \varphi \mapsto \operatorname{Tr} \int \varphi(t)U(t)\,dt$$

makes sense as a distribution on $\mathbb{R}$. We can thus write

$$\operatorname{Tr} U(t) = (2\pi)^{n/2}\mathcal{F}(N')(t). \tag{5.4}$$

[27]This choice of generating function, corresponding to taking the wave trace, is of course one choice among many. Some other approaches include taking the trace of the complex powers of the Laplacian or the heat trace. The idea of using (at least some version of) the wave trace originates with Levitan and Avakumovič.

where both sides are defined as distributions. Our next goal is to try to understand the left side of this equality through PDE methods.

EXERCISE 5.2. Show that if the Schwartz kernel $K(x, y)$ of a bounded, normal operator T on $L^2(X)$ is in $\mathcal{C}^k(X)$ for sufficiently large k, then T is of trace-class and

$$\operatorname{Tr} T = \int K(x, x)\, dg(x).$$

(HINT: Check that K is trace-class as in the previous exercise. Then apply the spectral theorem for compact normal operators, and use the basis of eigenfunctions of K when computing the trace. The crucial thing to check is that if φ_j are the eigenfunctions, then

$$\sum \varphi_j(x)\overline{\varphi_j(y)} = \delta_\Delta,$$

the delta-distribution at the diagonal, since this is nothing but a spectral resolution of the identity operator.)

As a consequence of Exercise 5.2, we can compute the distribution $\operatorname{Tr} U(t)$ in another way if we can compute the Schwartz kernel of $U(t)$. Indeed, knowing even rather crude things about $U(t)$ can give us some useful information here.

THEOREM 5.1. *Let Φ_t be the geodesic flow, i.e. the flow generated by the Hamilton vector field of $|\xi|_g \equiv (\sum g^{ij}\xi_i\xi_j)^{1/2}$. Then*

$$\operatorname{WF} U(t)f = \Phi_t(\operatorname{WF} f).$$

We begin with a lemma:

LEMMA 5.2. *Let $(\partial_t + i\sqrt{\Delta})u = 0$. Then*

$$(x_0, \xi_0) \in \operatorname{WF} u|_{t=t_0}$$

if and only if

$$(t = t_0, \tau = -|\xi_0|, x_0, \xi_0) \in \operatorname{WF} u.$$

PROOF. [28] Suppose $q = (x_0, \xi_0) \in \operatorname{WF} u|_{t=t_0}$. Since $\tilde{q} = (t = t_0, \tau = -|\xi_0|, x_0, \xi_0)$ is the only vector in $\Sigma_{\partial_t + i\sqrt{\Delta}}$ that projects to (x_0, ξ_0), it must lie in the wavefront set of u by Exercise 4.11.

The converse is harder. Suppose $q \notin \operatorname{WF} u|_{t=t_0}$. Let $v = H(t - t_0)u$, with H denoting the Heaviside function. Then

$$(\partial_t + i\sqrt{\Delta})v = \delta(t - t_0)u(t_0, x) \equiv f.$$

and v vanishes identically for $t < t_0$. By the last part of Exercise 4.11,

$$\tilde{q} \notin \operatorname{WF} f,$$

hence (since $\operatorname{WF} f$ only lies over $t = t_0$) certainly no points along the bicharacteristic through $\tilde{q}$ lie in $\operatorname{WF} f$. Moreover, no points along this bicharacteristic lie in $\operatorname{WF} v$ for $t < t_0$ (since v is in fact zero there). Hence by the version of the propagation of singularities in the second part of Exercise 4.20, this bicharacteristic is absent from $\operatorname{WF} u$. In particular, $\tilde{q} \notin \operatorname{WF} u$. □

[28] I am grateful to András Vasy for showing me this proof.

Theorem 5.1 now follows directly[29] from the lemma and Theorem 4.11.

We now require a result on microlocal partitions of unity somewhat generalizing Lemma 4.1:

EXERCISE 5.3. Let ρ_j, $j = 1, \dots, N$ be a smooth partition of unity for S^*X. Show that there exists $A_j \in \Psi^0(X)$ with $\mathrm{WF}' A_j = \operatorname{supp} \rho_j$, $\hat{\sigma}_0(A_j) = \rho_j$, $A_j^* = A_j$, and

$$\sum_{j=1}^N A_j^2 = \mathrm{Id} - R,$$

with $R \in \Psi^{-\infty}(X)$.

For a distribution u, let $\operatorname{singsupp} u$ (the "singular support" of u) be the projection of its wavefront set, i.e. the complement of the largest open set on which it is in $\mathcal{C}^\infty$.

THEOREM 5.3.

$$\operatorname{singsupp} \operatorname{Tr} U(t) \subseteq \{0\} \cup \{\textit{lengths of closed geodesics on } X\}.$$

This theorem is due to Chazarain and to Duistermaat-Guillemin.

We begin with the following dynamical result:

LEMMA 5.4. *Let L not be the length of any closed geodesic. Then there exists $\epsilon > 0$ and a cover U_i of S^*X by open sets such that for $t \in (L - \epsilon, L + \epsilon)$, there exists no geodesic with start- and endpoints both contained in the same U_i.*

EXERCISE 5.4.

(1) Prove the lemma. (HINT: The cosphere bundle is compact.)
(2) As long as you're at it, show that 0 is an isolated point in the set of lengths of closed geodesics ("length spectrum"), and that the length spectrum is a closed set.

We now prove Theorem 5.3.

PROOF. Let L not be the length of any closed geodesic on X. Let U_j be a cover of S^*X as given by Lemma 5.4. Let ρ_j be a partition of unity subordinate to U_j and let A_j be a microlocal partition of unity as in Exercise 5.3. Then, calculating with distributions on $\mathbb{R}^1$, we have

$$\begin{aligned} \operatorname{Tr} U(t) &= \sum_j \operatorname{Tr} A_j^2 U(t) + \operatorname{Tr} RU(t) \\ &= \sum_j \operatorname{Tr} A_j U(t) A_j + \operatorname{Tr} RU(t) \end{aligned}$$

and, more generally,

$$D_t^{2m} \operatorname{Tr} U(t) = \sum_j \operatorname{Tr} A_j \Delta^m U(t) A_j + \operatorname{Tr} R \Delta^m U(t).$$

[29] Here is one of the places where we should worry about the fact that $\sqrt{\Delta}$ is not a pseudodifferential operator on $\mathbb{R} \times X$. This problem is seen not to affect the proof of Hörmander's theorem if we note that composing $\sqrt{\Delta}$ with a pseudodifferential operator that is microsupported in a neighborhood of the characteristic set $\{|\tau| = -|\xi|_g\}$ yields an operator that *is* pseudodifferential, and that the symbol calculus extends to such compositions. (The author confesses that this is not entirely a trivial matter.)

Let u be a distribution on X; then $\mathrm{WF}\, A_j u \subseteq \mathrm{WF}'\, A_j \subset U_j$. Thus Theorem 5.1 gives

$$\mathrm{WF}\, \Delta^m U(t) A_j u \subseteq \Phi_t(U_j).$$

But by construction, this set is disjoint from U_j and hence from $\mathrm{WF}'\, A_j$. Hence for any m,[30]

$$A_j \Delta^m U(t) A_j \in L^\infty([L-\epsilon, L+\epsilon]; \Psi^{-\infty}(X));$$

consequently,

$$D_t^{2m} \operatorname{Tr} U(t) \in L^\infty([L-\epsilon, L+\epsilon]). \quad \square$$

EXERCISE 5.5. Show that in the special case of $X = S^1$, Theorem 5.3 can be deduced from the Poisson summation formula. For this reason it is often referred to as the *Poisson relation.*

One is tempted to conclude from (5.4) and Theorem 5.3 that one can "hear" the lengths of closed geodesics on a manifold, since the right side of (5.4) is determined by the spectrum, and the left side seems to be a distribution from whose singularities we can read off the lengths of closed geodesics. The trouble with this approach is that we do not know with any certainty from Theorem 5.3 that the putative singularities in $\operatorname{Tr} U(t)$ at lengths of closed geodesics are actually there: perhaps the distribution is, after all, miraculously smooth. Thus, proving actual inverse spectral results requires somewhat more care, as we shall see. To this end, we will begin studying the operator $U(t)$ more constructively in the following section.

6. A parametrix for the wave operator

In order to learn more about the wave trace, we will have to bite the bullet and construct an approximation ("parametrix") for the fundamental solution to the wave equation on a manifold. The approach will have a similar iterative flavor to the technique we used to construct an approximate inverse for an elliptic operator, but we have now left the comfortable world of pseudodifferential operators: the parametrix we construct is going to be something rather different. Exactly what, and how to systematize the kinds of calculation we do here, will be discussed later on.

As this construction will be local, we will work in a single coordinate patch, which we identify with $\mathbb{R}^n$; for the sake of exposition, we omit the coordinate maps and partitions of unity necessary to glue this construction into a Riemannian manifold.

Consider once again the "half-wave equation"[31]

$$(D_t + \sqrt{\Delta})u = 0 \tag{6.1}$$

on $\mathbb{R}^n$, where Δ is the Laplace-Beltrami operator with respect to a metric g. Our goal is to find a distribution u approximately solving (6.1) with initial data

$$u(0, x, y) = \delta(x - y)$$

for any $y \in \mathbb{R}^n$. Recall that if we let U denote the exact solution to (6.1) with initial data $\delta(x-y)$ then U can also be interpreted as (the Schwartz kernel of) the

[30]We technically have to work just a little to obtain the uniformity in time: observe that $A_j \Delta^m U(t) A_j$ are a continuous (or even smooth) *family* of smoothing operators. We have been avoiding the topological issues necessary to easily dispose of such matters, however.

[31]Remember that $D_t = i^{-1}\partial_t$.

"solution operator" mapping initial data f to the solution $e^{-it\sqrt{\Delta}}f$ with that initial data, evaluated at time t; this is why we denote it U, as we did above, and why we will often think of our parametrix $u(t,x,y)$ as a family in t of integral kernels of operators on $\mathbb{R}^n$.

We do not expect $U(t,x,y)$ or our parametrix for it to be the Schwartz kernel of a pseudodifferential operator, as it moves wavefront set around, by Theorem 4.11; recall that pseudodifferential operators are *microlocal*, which is to say they don't do that. But we will try and construct our parametrix $u(t,x,y)$ as something of *roughly* the same form, which is to say as an oscillatory integral

$$u(t,x,y) = \int a(t,x,\eta)e^{i\Phi}\,d\eta$$

where the main difference is that the "phase function" $\Phi = \Phi(t,x,y,\eta)$ will be something a good deal more interesting than $(x-y)\cdot\eta$; indeed, this phase function is where all the geometry of the problem turns out to reside.

First, let's write our initial data as an oscillatory integral:

$$\delta(x-y) = (2\pi)^{-n}\int e^{i(x-y)\cdot\eta}\,d\eta.$$

Let us now try, as an Ansatz, modifying the phase as it varies in t,x by setting

$$u(t,x,y) = (2\pi)^{-n}\int a(t,x,\eta)e^{i(\phi(t,x,\eta)-y\cdot\eta)}\,d\eta; \tag{6.2}$$

then if $\phi(0,x,\eta) = x\cdot\eta$ and $a(0,x,\eta) = 1$, we recover our initial data; moreover, if ϕ were to remain unchanged as t varied we would have nothing but a family of pseudodifferential operators. Let us assume that a is a classical symbol of order 0 in η, so that we have an asymptotic expansion

$$a \sim a_0 + |\eta|^{-1}a_{-1} + |\eta|^{-2}a_{-2} + \dots, \quad a_j = a_j(t,x,\hat{\eta}).$$

Let us further assume that ϕ is homogeneous in η of degree 1, hence matches the homogeneity[32] of $x\cdot\eta$.

Now if u solves the half-wave equation, it solves the wave equation, hence we have

$$\Box u = 0;$$

As we seek an approximate solution, we will instead accept

$$\Box u \in \mathcal{C}^\infty((-\epsilon,\epsilon)_t \times \mathbb{R}^n).$$

Our strategy is to plug (6.2) into this equation and see what is forced upon us. To this end, note that if we have an expression

$$v = (2\pi)^{-n}\int b(t,x,y,\eta)e^{i(\phi(t,x,\eta)-y\cdot\eta)}\,d\eta; \tag{6.3}$$

where b is a symbol of order $-\infty$, then v lies in $\mathcal{C}^\infty$, as the integral converges absolutely, together with all its t,x,y derivatives. So terms of this form will be acceptable errors.

Applying $\Box$ to (6.2), we group terms according to their order in η. The "worst case" terms involve factors of η^2, and can only be produced by second-order terms

[32]That is is then likely to be singular at $\eta = 0$ will not in fact concern us, as it will turn out that we may as well assume that a vanishes near $\eta = 0$.

in $\square$, with all derivatives falling on the exponential term. Since the second-order terms in Δ are just

$$\sum g^{ij}(x) D_i D_j,$$

we can write the term this produces from the phase as $|d_x\phi|_g^2$ or, equivalently, $|\nabla_x\phi|_g^2$. Thus, the equation that we need to solve to make the η^2 terms vanish is just

$$(\partial_t\phi)^2 - |\nabla_x\phi|_g^2 = 0. \tag{6.4}$$

Recall that we further want our phase to agree with the standard pseudodifferential one at time zero, i.e. we want

$$\phi(0,x,\eta) = x\cdot\eta. \tag{6.5}$$

Combining this information with (6.4) we easily see that we in particular have

$$(\partial_t\phi|_{t=0})^2 = |\eta|_g^2,$$

and we need to make an arbitrary choice of sign in solving this to get the initial time-derivative: we will choose[33]

$$\partial_t\phi|_{t=0} = -|\eta|_g. \tag{6.6}$$

If our metric is the Euclidean metric, we can easily solve (6.4), (6.5), and (6.6) by setting

$$\phi(t,x,\eta) = x\cdot\eta - t|\eta|.$$

More generally, the construction of a phase satisfying (6.4),(6.5) and (6.6) is the classic construction of Hamilton-Jacobi theory, and is sketched in the following exercise.

EXERCISE 6.1.

(1) Show that equation (6.4) is equivalent to the statement that for each η, the graph of $d_{t,x}\phi(t,x,\eta)$ is contained in the set

$$\Lambda = \{\tau^2 - |\xi|_g^2 = 0\} \subset T^*(\mathbb{R}_t \times \mathbb{R}^n_x)$$

(where the variables τ and ξ are the canonical dual variables to t and x respectively). The condition (6.5) implies

$$d_x\phi(t,x,\eta)|_{t=0} = \eta\cdot dx.$$

Equation (6.6) gives further

$$d_{t,x}\phi(t,x,\eta)|_{t=0} = -|\eta|\,dt + \eta\cdot dx; \tag{6.7}$$

accordingly, for fixed η, let

$$G_0 = \{t=0, x\in\mathbb{R}^n, \tau = -|\eta|, \xi = \eta\} \subset T^*(\mathbb{R}\times\mathbb{R}^n).$$

(2) Let H denote the Hamilton vector field of $\tau^2 - |\xi|_g^2$. Show that flow along H preserves Λ and that H is transverse to G_0.

[33] We will use this solution for reasons that will become apparent presently—it is the right one to solve (5.2) and not merely the wave equation.

(3) Show that there is a solution to (6.4),(6.7) for $t \in (-\epsilon, \epsilon)$ where the graph of $d_{t,x}\phi$ is given by flowing out the set G_0 under H. (Among other things, you need to check that the resulting smooth manifold is indeed the graph of the differential of a function.) Show that this solution can be integrated to give a solution to (6.4),(6.5).

Employing the phase ϕ constructed in Exercise 6.1, we have now solved away the homogeneous degree-two (in η) terms in the application of $\Box$ to our parametrix. We thus move on to the degree-one terms, which are as follows:

$$2D_t\phi D_t a_0 - 2\langle D_x\phi, D_x\rangle_g a_0 + r_1(t,x,y,\eta) \tag{6.8}$$

where r_1 is a homogeneous function of degree 1 *independent of* a_0, i.e. determined completely by ϕ. Given that ϕ solves the eikonal equation, we can rewrite (6.8) by factoring out $|\nabla_x\phi|$ and noting that our sign choice $\partial_t\phi = -|\nabla_x\phi|$ must persist away from $t = 0$ (for a short time, anyway). In this way we obtain

$$2\partial_t a_0 + 2\Big\langle \frac{\nabla_x\phi}{|\nabla_x\phi|_g}, \partial_x\Big\rangle_g a_0 - \tilde{r}_1 = 0,$$

with $\tilde{r}_1$ homogeneous of degree 0. This is a *transport equation* that we would like to solve, with the initial condition $a_0(0,x,y,\eta) = 1$ (the symbol of the identity operator). We can easily see that a solution exists with the desired initial condition $a_0(0,y,\eta) = 1$, as, letting

$$\mathsf{H} = 2\partial_t + 2\Big\langle \frac{\nabla_x\phi}{|\nabla_x\phi|_g}, \partial_x\Big\rangle_g$$

we see that H is a nonvanishing vector field, transverse to $t = 0$, hence we may solve

$$\mathsf{H}a_0 = \tilde{r}_1, \quad a_0|_{t=0} = 1$$

by standard ODE methods.

Now we consider degree-zero terms in η. We find that they are of the form

$$2D_t\phi D_t a_{-1} - 2\langle D_x\phi, D_x\rangle_g a_{-1} + r_0(t,x,y,\eta)$$

where r_0 only depends on a_0 and ϕ (i.e. not on a_{-1}). Thus, we may use the same procedure as above to find a_{-1} with initial value zero, making the degree-zero term vanish. (Note that the vector field H along which we need to flow remains the same as in the previous step.)

We continue in this manner, solving successive transport equations along the flow of H so as to drive down the order in η of the error term. Finally we Borel sum the resulting symbols, obtaining a symbol

$$a(t,x,\eta) \in S^0_{\mathrm{cl}}(\mathbb{R}^{2n}_{x,y} \times \mathbb{R}^n_\eta)$$

such that

$$a(0,x,\eta) = 1,$$

and

$$\begin{aligned} \Box u = \Box\Big((2\pi)^{-n}\int a(t,x,\eta)e^{i(\phi(t,x,\eta)-y\cdot\eta)}\,d\eta\Big) \\ = (2\pi)^{-n}\int b(t,x,y,\eta)e^{i(\phi(t,x,\eta)-y\cdot\eta)}\,d\eta \in \mathcal{C}^\infty((-\epsilon,\epsilon)\times X), \end{aligned} \tag{6.9}$$

since $b \in S^{-\infty}$.

Now we need to check that (6.9) implies that in fact u differs by a smooth term from the actual solution. We will show soon (in the next section) that our choice of the phase implies that[34] $\operatorname{WF} u \subset \{\tau < 0\}$. Hence, using this fact, we have

$$(\partial_t - i\sqrt{\Delta})(\partial_t + i\sqrt{\Delta})u = f \in \mathcal{C}^\infty. \tag{6.10}$$

Now $\partial_t - i\sqrt{\Delta}$ is elliptic on $\tau < 0$, so, letting Q denote a microlocal elliptic parametrix, we have

$$Q(\partial_t - i\sqrt{\Delta}) = I + E$$

with $\operatorname{WF}' E \cap \operatorname{WF} u = \emptyset$. Thus, applying Q to both sides of (6.10), we have

$$(\partial_t + i\sqrt{\Delta})u \in \mathcal{C}^\infty.$$

Also, as we have arranged that $a(0, x, \eta) = 1$, we have got our initial data exactly right: $u(0, x, y) = \delta(x - y)$. Letting U denote the actual solution operator to (5.2), we thus find

$$(\partial_t + i\sqrt{\Delta})(u - U) \in \mathcal{C}^\infty, \quad u(0, x, y) - U(0, x, y) = 0;$$

hence by global energy estimates[35] we have

$$u - U \in \mathcal{C}^\infty((-\epsilon, \epsilon) \times \mathbb{R}^n).$$

7. The wave trace

Our treatment of this material (and, in part, that of the previous section) closely follows the treatment in [**7**], which is in turn based on work of Hörmander [**9**].

Recall that, if $N(\lambda) = \#\{\lambda_j \le \lambda\}$ and $U(t)$ is given by (5.1), then

$$\operatorname{Tr} U(t) = (2\pi)^{n/2}\mathcal{F}(N'(\lambda)). \tag{7.1}$$

Thus, the singularities of $\operatorname{Tr} U(t)$ are related to the growth of $N(\lambda)$. We think that $\operatorname{Tr} U(t)$ should have singularities at zero, together with lengths of closed geodesics; since $U(0)$ is the identity (which has a very divergent trace), the singularity at $t = 0$, at least, seems certain to appear. We will thus spend some time discussing this singularity of the wave trace and its consequences for spectral geometry.

What is the form of the singularity of $\operatorname{Tr} U(t)$ at $t = 0$? Our parametrix from the previous section was

$$u(t, x, y) = (2\pi)^{-n} \int a(t, x, \eta) e^{i(\phi(t,x,\eta) - y\cdot\eta)}\, d\eta,$$

where $\phi(t, x, \eta) = x \cdot \eta - t|\eta|_{g(x)} + O(t^2)$, and $a(t, x, \eta) = 1 + O(t)$. Thus,

$$u(t, x, x) = (2\pi)^{-n} \int a(t, x, \eta) e^{i(-t|\eta|_{g(x)} + O(t^2|\eta|))}\, d\eta, \tag{7.2}$$

where we have used the homogeneity of the phase in writing the error term as $O(t^2|\eta|)$.

---

[34] This can also be verified directly, with localization, Fourier transform, and elbow grease.

[35] We can either use the estimates developed in §2.3, adapted to this variable coefficient setting, and with a power of the Laplacian applied to the solution (in order to gain derivatives); or we can apply Theorem 4.11, which is overkill.

Formally, we would now like to conclude that the singularity at $t = 0$ is approximately that of

$$u(t, x, x) = (2\pi)^{-n} \int e^{-it|\eta|_{g(x)}}\, d\eta$$

so that integrating in x would give, if all goes well,

$$\begin{aligned} \operatorname{Tr} U(t) &\sim \int u(t, x, x)\, dx \\ &\sim (2\pi)^{-n} \iint e^{-it|\eta|_g}\, d\eta\, dx \\ &= (2\pi)^{-n} \iiint_{\sigma>0, |\theta|=1} e^{-it\sigma|\theta|_g} \sigma^{n-1}\, d\sigma\, d\theta\, dx \\ &= (2\pi)^{-n/2} \iint \mathcal{F}(\sigma^{n-1} H(\sigma))(t|\theta|_g)\, d\theta\, dx, \end{aligned} \tag{7.3}$$

with H denoting the Heaviside function. (Recall that the notation $f \sim g$ means that $(f/g) \to 1$, in this case as $t \to 0$.) If we crudely try to solve (7.1) for $N'(\lambda)$ by applying an inverse Fourier transform to $\operatorname{Tr} U(t)$ and pretending that the singularity of $\operatorname{Tr} U(t)$ at $t = 0$ is all that matters, we find, formally, that (7.3) yields

$$\begin{aligned} N'(\lambda) &\sim (2\pi)^{-n/2} \mathcal{F}^{-1}_{t\to\lambda} \operatorname{Tr} U(t) \\ &\sim (2\pi)^{-n} \iint_{|\theta|=1} |\theta|_g^{-1} \Big(\frac{\lambda}{|\theta|_g}\Big)^{n-1} d\theta\, dx \\ &= (2\pi)^{-n} \lambda^{n-1} \iint_{|\theta|=1} |\theta|_g^{-n}\, d\theta\, dx. \end{aligned}$$

Integrating would formally yield

$$\begin{aligned} N(\lambda) &\sim (2\pi)^{-n} \frac{\lambda^n}{n} \iint_{|\theta|=1} |\theta|_g^{-n}\, d\theta\, dx \\ &= (2\pi)^{-n} \lambda^n \iiint_{|\theta|=1, \rho\in(0,1)} |\theta|_g^{-n} \rho^{n-1}\, d\rho\, d\theta\, dx \\ &= (2\pi)^{-n} \lambda^n \iiint_{|\sigma\theta|_g<1} \sigma^{n-1}\, d\sigma d\theta\, dx, \end{aligned}$$

where we have, in the last line, set $\sigma = \rho/|\theta|_g$, with the result that definition of the region of integration now involves the metric. This last quantity can easily be seen to be simply the volume in phase space of the set $|\xi|_g < 1$, otherwise known as the unit ball bundle.[36] Thus, we obtain *formally*

$$N(\lambda) \sim (2\pi)^{-n} \lambda^n \operatorname{Vol}(B^*X) = (2\pi)^{-n} \operatorname{Vol}(\{|\xi|_g < \lambda\}).$$

This is all nonsense, of course, for several different reasons. First, we were very imprecise about dropping higher order terms in t in computing the asymptotics of the trace as $t \to 0$. Furthermore, we formally computed with N' as if it were a smooth function, but of course N' is quite singular (a sum of delta distributions). Moreover, and potentially most seriously, there are in general infinitely many singularities in $\operatorname{Tr} U(t)$ that might be contributing to the asymptotic behavior of its Fourier transform: we have been concerning ourselves only with the one near $t = 0$.

---

[36]Recall that on a symplectic manifold (N^{2n}, ω) we have a naturally defined volume form ω^n, and it is this volume that we are integrating over the unit ball here.

However: the above argument does give the right leading order asymptotics, the so-called "Weyl Law." What follows is (the outline of) a rigorous version of the above argument.

To begin, we need a cutoff function to localize us near the singularity at $t = 0$, where our parametrix is valid.

EXERCISE 7.1. Show that there exists $\rho \in \mathcal{S}(\mathbb{R})$ with $\hat{\rho}$ compactly supported, $\hat{\rho}(0) = 1$, $\hat{\rho}(t) = \hat{\rho}(-t)$, $\rho(\lambda) > 0$ for all λ, and $\hat{\rho}$ supported in an arbitrarily small neighborhood of 0. (HINT: Start with a smooth, compactly supported $\hat{\rho}$; convolve with its complex conjugate, and scale.)

We now consider

$$\begin{aligned}\mathcal{F}^{-1}_{t\to\lambda}(\hat{\rho}(t)\operatorname{Tr} u(t))\\ &= (2\pi)^{-n-1/2}\iiint \hat{\rho}(t)a(t,x,\eta)e^{i(t(\lambda-|\eta|_g)+O(t^2|\eta|))}\,dx\,d\eta\,dt\\ &= (2\pi)^{-n-1/2}\iiint \hat{\rho}(t)a(t,x,\lambda\sigma\theta)e^{it\lambda(1-\sigma+O(t^2\sigma))}(\lambda\sigma)^{n-1}\,dx\,d\sigma\,d\theta\,dt;\end{aligned}$$

here we have used the change of variables $\eta = \lambda\sigma\theta$ with $|\theta| = 1$. We now employ the *method of stationary phase* to estimate the asymptotics of the integral in t, σ. If $\hat{\rho}$ is chosen supported sufficiently close to the origin, then the unique stationary point on the support of the amplitude is at $\sigma = 1$, $t = 0$; we thus obtain a complete asymptotic expansion in λ beginning with the terms

$$A\lambda^{n-1} + O(\lambda^{n-2})$$

where

$$A = n(2\pi)^{-n}\operatorname{Vol}(B^*X).$$

EXERCISE* 7.2. Do this stationary phase computation. If you don't know about the method of stationary phase, this is your chance to learn it, e.g. from [**11**].

Thus, since $u - U \in \mathcal{C}^\infty((-\epsilon,\epsilon)\times\mathbb{R}^n)$, (7.1) yields

PROPOSITION 7.1.

$$(\rho * N')(\lambda) \sim A\lambda^{n-1} + O(\lambda^{n-2}).$$

We now try to make a "Tauberian" argument to extract the desired asymptotics of $N(\lambda)$ from this estimate.

LEMMA 7.2.

$$N(\lambda+1) - N(\lambda) = O(\lambda^{n-1}).$$

PROOF. By Proposition 7.1 and since $N'(\lambda) = \sum \delta(\lambda - \lambda_j)$, we have

$$\sum \rho(\lambda-\lambda_j) \sim A\lambda^{n-1} + O(\lambda^{n-2});$$

thus, by positivity of $\rho(\lambda)$,

$$(\inf_{[-1,1]}\rho)\,(\#\{\lambda_j : \lambda - 1 < \lambda_j < \lambda+1\}) \leq \sum \rho(\lambda-\lambda_j) = O(\lambda^{n-1}),$$

and the estimate follows as the infimum is strictly positive. □

This yields at least a crude estimate:

COROLLARY 7.3.

$$N(\lambda) = O(\lambda^n).$$

A more technically useful result is:

COROLLARY 7.4.

$$N(\lambda - \tau) - N(\lambda) \lesssim \langle\tau\rangle^n \langle\lambda\rangle^{n-1}.$$

EXERCISE 7.3. Prove the corollaries. (For the latter, begin with the intermediate estimate $\langle\tau\rangle\langle|\lambda| + |\tau|\rangle^{n-1}$.)

Now we work harder.

EXERCISE 7.4. Show that we can antidifferentiate the convolution to get

$$\int_{-\infty}^{\lambda} (\rho * N')(\mu)\, d\mu = (\rho * N)(\lambda).$$

As a result, we of course have

$$(\rho * N)(\lambda) = A\lambda^n/n + O(\lambda^{n-1}) = B\lambda^n + O(\lambda^{n-1})$$

where $B = A/n = (2\pi)^{-n}\operatorname{Vol}(B^*X)$.

Thus, since $\int \rho(\mu)\, d\mu = 1$,

$$\begin{aligned} N(\lambda) &= (N * \rho)(\lambda) - \int (N(\lambda - \mu) - N(\lambda))\rho(\mu)\, d\mu \\ &= B\lambda^n + O(\lambda^{n-1}) - \int O(\langle\mu\rangle^n \langle\lambda\rangle^{n-1})\rho(\mu)\, d\mu \\ &= B\lambda^n + O(\lambda^{n-1}), \end{aligned}$$

where we have used Corollary 7.4 in the penultimate equality. We record what we have now obtained as a theorem, better known as Weyl's law with remainder term. This form of the remainder term is sharp, and not so easy to obtain by other means.

THEOREM 7.5.

$$N(\lambda) = (2\pi)^{-n}\operatorname{Vol}(B^*X)\lambda^n + O(\lambda^{n-1}).$$

As noted above, it is perhaps suggestive to view the main term as the volume of the sublevel set in phase space $\{(x, \xi) : \sigma(\Delta)(x, \xi) \le \lambda^2\}$. Weyl's law is one of the most beautiful instances of the quantum-classical correspondence, in which we can deduce something about a quantum quantity (the counting function for eigenvalues, also known as energy levels) in terms of a classical quantity, in this case the volume of a region of phase space.

EXERCISE* 7.5. Show that the error term in Weyl's law is sharp on spheres.

8. Lagrangian distributions

The form of the parametrix that we used for the wave equation turns out to be a special case of a very general and powerful class of distributions, known as *Lagrangian distributions*, introduced by Hörmander. Here we will give a very sketchy introduction to the general theory of Lagrangian distributions, and see both how it systematizes and extends our parametrix construction for the wave equation and how (in principle, at least) it can be made to yield the Duistermaat-Guillemin trace formula, which gives us an explicit description of the singularities of the wave trace.

We begin with a special case of the theory.

8.1. Conormal distributions. Let X be a smooth manifold of dimension n and let Y be a submanifold of codimension k. The *conormal distributions* with respect to Y are a special class of distributions having wavefront set[37] in the conormal bundle of Y, N^*Y. Let us suppose that Y is locally cut out by defining functions $\rho_1, \dots, \rho_k \in \mathcal{C}^\infty(X)$, i.e. that (at least locally), $\{\rho_1 = \dots = \rho_k = 0\} = Y$, and $d\rho_1, \dots, d\rho_k$ are linearly independent on Y. Then we may (locally) extend the ρ_j's to a complete coordinate system

$$(x_1, \dots x_k, y_1, \dots y_{n-k})$$

with

$$x_1 = \rho_1, \dots, x_k = \rho_k,$$

so that $Y = \{x = 0\}$. In these coordinates, how might we write down some distributions with wavefront set lying only in N^*Y? Well, we can try to make things that are singular in the x variables at $x = 0$, with the y's behaving like smooth parameters. How do we create singularities at $x = 0$? One very nice answer is in the following:

LEMMA 8.1. *Let $a(\xi) \in S^m_{cl}(\mathbb{R}^k_\xi)$ for some m. Then* $\operatorname{WF} \mathcal{F}^{-1}(a) \subseteq N^*(\{0\})$.

PROOF. Writing

$$\mathcal{F}^{-1}(a)(x) = (2\pi)^{-k/2} \int a(\xi) e^{i\xi\cdot x}\, d\xi,$$

we first note that

$$\mathcal{F}^{-1}(a)(x) \in H^{-m-k/2-\epsilon}(\mathbb{R}^k)$$

for any $a \in S^m_{\mathrm{cl}}$ and for all $\epsilon > 0$. Moreover for all j,

$$\begin{aligned}(x^i D_{x^j})\mathcal{F}^{-1}(a)(x) &= (2\pi)^{-k/2} \int a(\xi)(x^i D_{x^j}) e^{i\xi\cdot x}\, dx \\ &= (2\pi)^{-k/2} \int x^i \xi_j a(\xi) e^{i\xi\cdot x}\, d\xi \\ &= (2\pi)^{-k/2} \int \xi_j a(\xi) D_{\xi_i} e^{i\xi\cdot x}\, d\xi \\ &= -(2\pi)^{-k/2} \int D_{\xi_i}(\xi_j a(\xi)) e^{i\xi\cdot x}\, d\xi,\end{aligned}$$

where we have integrated by parts in the final line. Note that if $a \in S^m_{\mathrm{cl}}$ then $D_{\xi_i}(\xi_j a(\xi)) \in S^m_{\mathrm{cl}}$ too (cf. Exercise 3.4). Thus we also have

$$(x^i D_{x^j})\mathcal{F}^{-1}(a)(x) \in H^{-m-k/2-\epsilon}(\mathbb{R}^k).$$

Iterating this argument gives

$$(x_{i_1} D_{x_{j_1}}) \dots (x_{i_l} D_{x_{j_l}})\mathcal{F}^{-1}(a)(x) \in H^{-m-k/2-\epsilon}(\mathbb{R}^k) \tag{8.1}$$

for all choices of indices and all $l \in \mathbb{N}$. Thus $\mathcal{F}^{-1}a$ is smooth[38] away from $x = 0$. □

[37]Recall that we have defined the wavefront set to lie in S^*X but it is often convenient to regard it as a *conic* subset of $T^*X\backslash o$, with o denoting the zero-section.

[38]We are of course proving more than the lemma states here: (8.1) gives a more precise "conormality" estimate that is valid uniformly across the origin.

By the same token, we have more generally,

PROPOSITION 8.2. *Let $\rho_1, \ldots, \rho_k$ be (local) defining functions for $Y \subset X$ and let*

$$a \in S_{cl}^{m+(n-2k)/4}(\mathbb{R}^n_x \times \mathbb{R}^k_\xi) \tag{8.2}$$

be compactly supported in x. Then

$$u(x) = (2\pi)^{-(n+2k)/4} \int_{\mathbb{R}^k} a(x, \theta) e^{i(\rho_1\theta_1 + \cdots + \rho_k\theta_k)}\, d\theta \tag{8.3}$$

has wavefront set contained in N^*Y. Moreover there exists $s \in \mathbb{R}$ such that if $V_1, \ldots V_l$ are vector fields tangent to Y, then

$$V_1 \ldots V_l u \in H^s.$$

EXERCISE 8.1. Prove the proposition. You will probably find it helpful to change to a coordinate system $(x_1, \ldots, x_k, y_1, \ldots, y_{n-k})$ in which $x_1, \ldots, x_k = \rho_1, \ldots, \rho_k$. Note that in this coordinate system, any vector field tangent to Y can be written

$$\sum a_{ij}(x,y) x^i \partial_{x^j} + \sum b_j(x,y) \partial_{y^j}.$$

What values of s, the Sobolev exponent in the proposition, are allowable?

DEFINITION 8.3. A distribution $u \in \mathcal{D}'(X)$ is a *conormal distribution* with respect to Y, of order m, if it can (locally) be written in the form (8.3) with symbol as in (8.2).

While it may appear that the definition of conormal distributions depends on the choice of the defining functions ρ_j, this is in fact not the case. The rather peculiar-looking convention on the orders of distributions is not supposed to make much sense just yet.

Note that examples of conormal distributions include $\delta(x) \in \mathbb{R}^n$ (conormal with respect to the origin), and more generally, delta distributions along submanifolds. Also quite pertinent is the example of pseudodifferential operators: if $A = \mathrm{Op}_\ell(a) \in \Psi^m(X)$ then the Schwartz kernel of A is a conormal distribution with respect to the diagonal in $X \times X$, of order m. (This goes at least some of the way to explaining the convention on orders.) Indeed, we could (at some pedagogical cost) simply have introduced conormal distributions and then used the notion to define the Schwartz kernels of pseudodifferential operators in the first place.

8.2. Lagrangian distributions. We now introduce a powerful generalization of conormal distributions, the class of *Lagrangian distributions.*[39] We begin by introducing some underlying geometric notions.

An important notion from symplectic geometry is that of a *Lagrangian submanifold* $\mathcal{L}$ of a symplectic manifold N^{2n}. This is a submanifold of dimension n on which the symplectic form vanishes. We can always find local coordinates in which the symplectic form is given by $\omega = \sum dx^i \wedge dy^i$ and $\mathcal{L} = \{y = 0\}$, so there are no interesting local invariants of Lagrangian manifolds.

A *conic Lagrangian manifold* in T^*X is a Lagrangian submanifold of $T^*X \backslash o$ that is invariant under the $\mathbb{R}^+$ action on the fibers. (Here, o denotes the zero-section.)

[39] These were first studied by Hörmander [**10**].

Among the most important examples of conic Lagrangians are the following: let $Y \subset X$ be any submanifold; then $N^*Y \subset T^*X$ is a conic Lagrangian.

EXERCISE 8.2. Verify this.

The trick to defining Lagrangian distributions is to figure out how to associate a *phase function* ϕ with a conic Lagrangian $\mathcal{L}$ in T^*X.

DEFINITION 8.4. A nondegenerate phase function is a smooth function $\phi(x, \theta)$, locally defined on a coordinate neighborhood of $X \times \mathbb{R}^k$, such that ϕ is homogeneous of degree 1 in θ and such that the differentials $d(\partial\phi/\partial\theta_j)$ are linearly independent on the set

$$C = \left\{(x, \theta) : \frac{\partial \phi}{\partial \theta_j} = 0 \text{ for all } j = 1, \dots, k\right\}.$$

The phase function is said to locally *parametrize* the conic Lagrangian $\mathcal{L}$ if

$$C \ni (x, \theta) \mapsto (x, d_x\phi)$$

is a local diffeomorphism from C to $\mathcal{L}$.

EXERCISE 8.3.

(1) Show that, in the notation of the definition above, C is automatically a manifold, and the map $C \ni (x, \theta) \mapsto (x, d_x\phi)$ is automatically a local diffeomorphism from C to its image, which is a conic Lagrangian.
(2) Show that if ρ_j are definining functions for $Y \subset X$ then

$$\phi = \sum \rho_j \theta_j$$

is a nondegenerate parametrization of N^*Y.
(3) What Lagrangian is parametrized by the phase function used in our parametrix for the half-wave operator in the Euclidean case, given by

$$\phi(t, x, y, \theta) = (x - y) \cdot \theta - t|\theta|?$$

It turns out that every conic Lagrangian manifold has a local parametrization; the trouble is, in fact, that it has lots of them.

DEFINITION 8.5. A *Lagrangian distribution* of order m with respect to the Lagrangian $\mathcal{L}$ as one that is given, locally near any point in X, by a finite sum of oscillatory integrals of the form

$$(2\pi)^{-(n+2k)/4} \int_{\mathbb{R}^k} a(x, \theta) e^{i\phi(x,\theta)}\, d\theta$$

where

$$a \in S^{m+(n-2k)/4}_{\text{cl}}(\mathbb{R}^n_x \times \mathbb{R}^k_\theta)$$

and where ϕ is a nondegenerate phase function parametrizing $\mathcal{L}$. Let $I^m(X, \mathcal{L})$ denote the space of all Lagrangian distributions on X with respect to $\mathcal{L}$ of order m.

Note that the connection between k, the number of phase variables, and the geometry of $\mathcal{L}$ is not obvious; indeed, it turns out that we have some choice in how many phase variables to use. As there are many different ways to parametrize a given conic Lagrangian manifold, one tricky aspect of the theory of Lagrangian distributions is necessarily the proof that using different parametrizations (possibly involving different numbers of phase variables) gives us the same class of distributions.

The analogue of the *iterated regularity* property of conormal distributions, i.e. our ability to repeatedly differentiate along vector fields tangent to Y, turns out to be as follows:

PROPOSITION 8.6. *Let $u \in I^m(X, \mathcal{L})$. There exists s such that for any $l \in \mathbb{N}$ and for any $A_1, \dots, A_l \in \Psi^1(X)$ with $\sigma_1(A_j)|_{\mathcal{L}} = 0$, we have*

$$A_1 \dots A_l u \in H^s(X).$$

Of course, once this holds for one s, it holds for all smaller values; the precise range of possible values of s is related to the order m of the Lagrangian distribution; we will not pursue this relationship here, however. This iterated regularity property of Lagrangian distributions completely characterizes them if we use "Kohn-Nirenberg" symbols (as in Exercise 3.4) instead of "classical" ones (see [**14**]).

8.3. Fourier integral operators. Fourier integral operators ("FIO's") quantize classical *maps* from a phase space to itself just as pseudodifferential operators quantize classical *observables* (i.e. functions on the phase space). The maps from phase space to itself that we may quantize in this manner are the *symplectomorphisms*, exactly the class of transformations of phase space that arise in classical mechanics. We recall that a symplectomorphism between symplectic manifolds is a diffeomorphism that preserves the symplectic form. We further define a *homogeneous symplectomorphism* from T^*X to T^*X to be one that is homogeneous in the fiber variables, i.e. commutes with the $\mathbb{R}^+$ action on the fibers.

An important class of homogeneous symplectomorphisms is those obtained as follows:

EXERCISE 8.4. Show that the time-1 flowout of the Hamilton vector field of a homogeneous function of degree 1 on T^*X is a homogeneous symplectomorphism.

Given a homogeneous symplectomorphism $\Phi : T^*X \to T^*X$, consider its *graph* $\Gamma_\Phi \subset (T^*X\backslash o) \times (T^*X\backslash o)$. Since Φ is a symplectomorphism, we have

$$\iota^*\pi_L^*\omega = \iota^*\pi_R^*\omega,$$

where ι is inclusion of Γ_Φ in $(T^*X\backslash o) \times (T^*X\backslash o)$, and $\pi_\bullet$ are the left and right projections. If we alter Γ_Φ slightly, forming

$$\Gamma'_\Phi = \{(x_1, \xi_1, x_2, \xi_2) : (x_1, \xi_1, x_2, -\xi_2) \in \Gamma_\Phi\},$$

and let ι' denote the inclusion of this manifold, then we find that a sign is flipped, and

$$(\iota')^*\pi_L^*\omega + (\iota')^*\pi_R^*\omega = 0;$$

since $\Omega = (\pi_L^*\omega + \pi_R^*\omega)$ is just the symplectic form on

$$T^*(X \times X) = T^*X \times T^*X,$$

we thus find that Γ'_Φ is *Lagrangian* in $T^*(X \times X)$. In fact, it is easily to verify that given a diffeomorphism Φ, Γ'_Φ is Lagrangian if and only if Φ is a symplectomorphism.

EXERCISE 8.5. Check this.

Now we simply define the class of Fourier integral operators (of order m) associated with the symplectomorphism Φ of X to be those operators from smooth functions to distributions whose Schwartz kernels lie in the Lagrangian distributions

$$I^m(X \times X, \Gamma'_\Phi).$$

It would be nice if this class of operators turned out to have good properties such as behaving well under composition, as pseudodifferential operators certainly do. We note right off the bat that these operators *include* pseudodifferential operators, as well as a number of other, familiar examples:

(1) $\Psi^m(X) = I^m(X \times X, \Gamma'_{\mathrm{Id}})$.

(2) In $\mathbb{R}^n$, fix α and let $Tf(x) = f(x - \alpha)$ Then T has Schwartz kernel

$$\delta(x - x' - \alpha)$$

which is clearly conormal of order zero at $x - x' - \alpha = 0$. Note that this is certainly *not* a pseudodifferential operator, as it moves wavefront around; indeed, it is associated with the symplectomorphism $\Phi(x, \xi) = (x + \alpha, \xi)$, and it it no coincidence that

$$\mathrm{WF}\, Tf = \Phi(\mathrm{WF}\, f).$$

(3) As a generalization of the previous example, note that if $\phi : X \to X$ is a diffeomorphism, then we may set

$$Tf(x) = f(\phi(x));$$

this is a FIO associated to the homogeneous symplectomorphism

$$\Phi(x, \xi) = (\phi^{-1}(x), \phi^*_{\phi^{-1}(x)}(\xi))$$

induced by ϕ on T^*X.

EXERCISE 8.6. Work out this last example carefully.

Now it turns out to be helpful to actually consider a broader class of FIO's than we have described so far. Instead of just using Lagrangian submanifolds of $T^*(X \times X)$ given by $\Gamma' = \Gamma'_\Phi$ where Φ is a symplectomorphism, we just require that Γ' be a reasonable Lagrangian (and we allow operators between different manifolds while we are at it):

DEFINITION 8.7. Let X, Y be two manifolds (not necessarily of the same dimension). A *homogeneous canonical relation* from T^*Y to T^*X is a homogeneous submanifold Γ of $(T^*X\backslash o) \times (T^*Y\backslash o)$, closed in $T^*(X \times Y)\backslash o$ such that

$$\Gamma' \equiv \{(x, \xi, y, \eta) : (x, \xi, y, -\eta) \in \Gamma\}$$

is Lagrangian in $T^*(X \times Y)$.

We can view Γ as giving a multivalued generalization of a symplectomorphism, with

$$\Gamma(y, \eta) \equiv \{(x, \xi) : (x, \xi, y, \eta) \in \Gamma\}.$$

and, more generally, if $S \subset T^*Y$ is conic,

$$\Gamma(S) \equiv \{(x, \xi) : \text{ there exists } (y, \eta) \in S, \text{ with } (x, \xi, y, \eta) \in \Gamma\}. \tag{8.4}$$

DEFINITION 8.8. A Fourier integral operator of order m associated to a homogeneous canonical relation Γ is an operator from $\mathcal{C}_c^\infty(Y)$ to $\mathcal{D}'(X)$ with Schwartz kernel in

$$I^m(X \times Y, \Gamma').$$

EXERCISE 8.7. Show that a homogeneous canonical relation Γ is associated to a symplectomorphism if and only if its projections onto both factors T^*X and T^*Y are diffeomorphisms.

EXERCISE 8.8.

(1) Let $Y \subset X$ be a submanifold. Show that the operation of restriction of a smooth function on X to Y is an FIO.
(2) Endow X with a metric, and consider the volume form dg_Y on Y arising from the restriction of this metric; show that the map taking a function f on Y to the distribution $\phi \mapsto \int_Y \phi|_Y(y)f(y)dg_Y$ is an FIO. (Think of it as just multiplying f by the delta-distribution along Y, which makes sense if we choose a metric.) What is the relationship between the restriction FIO and this one, which you might think of as an extension map?

In the special case that Γ is a canonical relation that is locally the graph of a symplectomorphism, we say it is a *local canonical graph.*

We now briefly enumerate the properties of the FIO calculus, somewhat in parallel with our discussion of pseudodifferential operators. These theorems are considerably deeper, however. In preparation for our discussion of composition, suppose that

$$\Gamma_1 \subset T^*X\backslash o \times T^*Y\backslash o,$$
$$\Gamma_2 \subset T^*Y\backslash o \times T^*Z\backslash o$$

are homogeneous canonical relations. We say that Γ_1 and Γ_2 are *transverse* if the manifolds

$$\Gamma_1 \times \Gamma_2 \text{ and } T^*X \times \Delta_{T^*Y} \times T^*Z$$

intersect transversely in $T^*X \times T^*Y \times T^*Y \times T^*Z$; here Δ_{T^*Y} denotes the diagonal submanifold.

EXERCISE 8.9. Show that if either Γ_1 or Γ_2 is the graph of a symplectomorphism, then Γ_1 and Γ_2 are transverse.

In what follows, we will as usual assume for simplicity that all manifolds are compact.[40] In the following list of properties, some are special to FIO's, that is to say, Lagrangian distributions on product manifolds, viewed as operators; others are more generally properties of Lagrangian distributions per se, hence their statements do not necessarily involve products of manifolds. In the interests of brevity, we focus on the deeper properties, and omit trivialities such as associativity of composition. Note also that for brevity we will systematically confuse operators with their Schwartz kernels.

(I) (Algebra property) If $S \in I^m(X \times Y, \Gamma_1')$ and $T \in I^{m'}(Y \times Z, \Gamma_2')$ and Γ_1 and Γ_2 are transverse, then

$$S \circ T \in I^{m+m'}(X \times Z, (\Gamma_1 \circ \Gamma_2)'),$$

[40] In the absence of this assumption, we need as usual to add various hypotheses of properness.

where

$$\begin{aligned}(8.5)\quad \Gamma_1\circ\Gamma_2=\{(x,\xi,z,\zeta):(x,\xi,y,\eta)\in\Gamma_1 \\ \text{and } (y,\eta,z,\zeta)\in\Gamma_2 \text{ for some } (y,\eta)\}.\end{aligned}$$

Moreover,

$$S^*\in I^m(Y\times X,(\Gamma^{-1})')$$

where Γ^{-1} is obtained from Γ by switching factors.

(II) (Characterization of smoothing operators) The distributions in $I^{-\infty}(X,\mathcal{L})$ are exactly those in $\mathcal{C}^\infty(X)$; composition of an operator $S\in I^m(X\times Y,\Gamma')$ on either side with a smoothing operator (i.e. one with smooth Schwartz kernel) yields a smoothing operator.

(III) (Principal symbol homomorphism) There is family of linear "principal symbol maps"

$$\sigma_m: I^m(X,\mathcal{L})\to\frac{S^{m+(\dim X)/4}_{\mathrm{cl}}(\mathcal{L};L)}{S^{m-1+(\dim X)/4}_{\mathrm{cl}}(\mathcal{L};L)}. \tag{8.6}$$

Here L is a certain canonically defined line bundle on $\mathcal{L}$ (see the commentary below), and $S^m_{\mathrm{cl}}(\mathcal{L};L)$ denotes L-valued symbols. We may identify the quotient space in (8.6) with

$$\mathcal{C}^\infty(S^*\mathcal{L};L),$$

and we call the resulting map $\hat{\sigma}_m$ instead. If S, T, are as in (I), with canonical relations Γ_1,Γ_2 intersecting transversely,

$$\sigma_{m+m'}(ST)=\sigma_m(S)\sigma_{m'}(T)$$

and

$$\sigma_m(A^*)=s^*\overline{\sigma_m(A)},$$

where s is the map interchanging the two factors. The product of the symbols, at $(x,\xi,z,\zeta)\in\Gamma_1\circ\Gamma_2$, is defined as

$$\sigma_m(S)(x,\xi,y,\eta)\cdot\sigma_{m'}(T)(y,\eta,z,\zeta)$$

evaluated at (the unique) (y,η) such that $(x,\xi,y,\eta)\in\Gamma_1$, $(y,\eta,z,\zeta)\in\Gamma_2$.

(IV) (Symbol exact sequence) There is a short exact sequence

$$0\to I^{m-1}(X,\mathcal{L})\to I^m(X,\mathcal{L})\overset{\hat{\sigma}_m}{\to}\mathcal{C}^\infty(S^*\mathcal{L};L)\to 0.$$

Hence the symbol is 0 if and only if an operator is of lower order.

(V) Given $\mathcal{L}$, there is a linear "quantization map"

$$\mathrm{Op}: S^{m+(\dim X)/4}_{\mathrm{cl}}(\mathcal{L};L)\to I^m(X,\mathcal{L})$$

such that if

$$a\sim\sum_{j=0}^\infty a_{m+(\dim X)/4-j}(x,\hat{\xi})|\xi|^{m+(\dim X)/4-j}\in S^{m+(\dim X)/4}_{\mathrm{cl}}(\mathcal{L};L)$$

then

$$\sigma_m(\mathrm{Op}(a))=a_{m+(\dim X)/4}(x,\hat{\xi}).$$

The map Op is onto, modulo $\mathcal{C}^\infty(X)$.

(VI) (Product with vanishing principal symbol) If $P \in \mathrm{Diff}^m(X)$ is self-adjoint and $u \in I^{m'}(X, \mathcal{L})$, with $\mathcal{L} \subset \Sigma_P \equiv \{\sigma_m(P) = 0\}$, then

$$Pu \in I^{m+m'-1}(X, \mathcal{L})$$

and

$$\sigma_{m+m'-1}(Pu) = i^{-1}\mathsf{H}_p(\sigma_{m'}(u)),$$

with H_p denoting the Hamilton vector field.

(VII) (L^2-boundedness, compactness) If $T \in I^m(X \times Y, \Gamma)$ is associated to a local canonical graph, then

$$T \in \mathcal{L}(H^s(Y), H^{s-m}(X)) \text{ for all } s \in \mathbb{R}.$$

Negative-order operators of this type acting on $L^2(X)$ are thus compact.

(VIII) (Asymptotic summation) Given $u_j \in I^{m-j}(X, \mathcal{L})$, with $j \in \mathbb{N}$, there exists $u \in I^m(X, \mathcal{L})$ such that

$$u \sim \sum_j u_j,$$

which means that

$$u - \sum_{j=0}^{N} u_j \in I^{m-N-1}(X, \mathcal{L})$$

for each N.

(IX) (Microsupport) The microsupport of $T \in I^m(X \times Y, \Gamma')$ is well defined as the largest conic subset $\tilde{\Gamma} \subset \Gamma$ on which the symbol is $O(|\xi|^{-\infty})$. We have

$$\mathrm{WF}\, Tu \subseteq \tilde{\Gamma}(\mathrm{WF}\, u)$$

for any distribution u on Y, where the action of $\tilde{\Gamma}$ on $\mathrm{WF}\, u$ is given by (8.4). Furthermore,

$$\mathrm{WF}'(S \circ T) \subseteq \mathrm{WF}'\, S \circ \mathrm{WF}'\, T.$$

COMMENTARY:

(I) This is a major result. Since FIO's include pseudodifferential operators, this includes the composition property for pseudodifferential operators as a special case. Another special case, when Z a point, yields the statement that an FIO applied to a Lagrangian distribution on the manifold Y with respect to the Lagrangian $\mathcal{L} \subset T^*Y$ is a Lagrangian distribution associated to $\Gamma(\mathcal{L})$, where Γ is the canonical relation of the FIO and $\Gamma(\mathcal{L})$ is defined by (8.4).

One remarkable corollary of this result is as follows: As will be discussed below, what our parametrix construction in §6 really showed was that for t sufficiently small, and fixed, we have

$$e^{-it\sqrt{\Delta}} \in I^0(X \times X, \mathcal{L}_t)$$

where $\mathcal{L}_t$ is the backwards geodesic flowout, for time t, in the left factor of $N^*\Delta$, of the conormal bundle to the diagonal in $T^*(X \times X)$.

EXERCISE* 8.10. Verify this assertion! (Try this now, but fear not: we will discuss this example further in §9 and you can try again then.)

Now $e^{-it\sqrt{\Delta}}$ is a one-parameter group and so the composition property for FIO's allows us to conclude that in fact $e^{-it\sqrt{\Delta}}$ is an FIO for *all* times t, associated to the same flowout described above. The interesting subtlety is that while $\mathcal{L}_t$ is an inward- or outward-pointing conormal bundle for small positive resp. negative time (i.e. in the regime where our parametrix construction worked directly), for t exceeding the injectivity radius, it ceases to be a conormal bundle, while remaining a smooth Lagrangian manifold in $T^*(X \times X)$.

(III) Modulo bundle factors, the principal symbol is defined as follows: if $u \in I^m(X, \mathcal{L})$ is given by

$$u = (2\pi)^{-(n+2k)/4} \int_{\mathbb{R}^k} a(x,\theta) e^{i\phi(x,\theta)}\, d\theta,$$

then $\sigma_m(u)$ is defined by first restricting $a(x,\theta)$ to the manifold

$$C = \{(x,\theta) : d_\theta \phi = 0\};$$

as ϕ is a nondegenerate phase function, this manifold is locally diffeomorphic (via a homogeneous diffeomorphism) to $\mathcal{L}$, hence we may identify $a|_C$ with a function on $\mathcal{L}$; transferring this function to $\mathcal{L}$ via the local diffeomorphism and taking the top-order homogeneous term in the asymptotic expansion gives the principal symbol.

Much has been swept under the rug here—for a proper discussion, see, e.g., [**10**]. In particular, the line bundle L contains not just the density factors that we have been studiously ignoring—the Schwartz kernel of an operator from functions to functions on X is actually a "right-density" on $X \times X$, i.e. a section of the pullback of the bundle $|\Omega^n(X)|$ in the right factor—but also the celebrated "Keller-Maslov index," which is related to the indeterminacy in choosing the phase function parametrizing the Lagrangian. We will not enter into a serious discussion of these issues here. We have also omitted discussion of the geometry of composing canonical relations, and the fact that transverse canonical relations compose to give a new canonical relation, with a unique point y, η such that $(x, \xi, y, \eta) \in \Gamma_1$, $(y, \eta, z, \zeta) \in \Gamma_2$ whenever $(x, \xi, z, \zeta) \in \Gamma_1 \circ \Gamma_2$.

(VI) There is a more general version of this statement valid for any $P \in \Psi^m(X)$ characteristic on $\mathcal{L}$, but it involves the notion of subprincipal symbol, which requires some explanation; see [**5**, §5.2–5.3]. Moreover, if we are a little more honest about making this computation work invariantly, so that the symbol has a density factor in it (one factor in the line bundle L,) then we should really write

$$\sigma_{m+m'-1}(Pu) = i^{-1} \mathcal{L}_{\mathsf{H}_p} \sigma_{m'}(u),$$

where $\mathcal{L}_Z$ denotes the Lie derivative along the vector field Z.

(VII) This is fairly easy to prove, as if T of order m is associated to a symplectomorphism from Y to X, it is easy to check from the previous properties that T^*T is an FIO associated with the canonical relation given by the identity map, and hence

$$T^*T \in \Psi^{2m}(Y),$$

and we may invoke boundedness results for the pseudodifferential calculus. In cases when T is not associated to a local canonical graph, this argument fails badly (i.e. interestingly), and the optimal mapping properties are a subject of ongoing research.

Finally, as with the pseudodifferential calculus, we may define a notion of ellipticity for FIO's, and the above properties imply that (microlocal) parametrices exist for the inverses of elliptic operators associated to symplectomorphisms.

9. The wave trace, redux

Let us briefly revisit our construction of the parametrix for the half-wave equation in the light of the FIO calculus. Here is what we did, in hindsight: we sought a distribution

$$u \in I^m(\mathbb{R} \times X \times X, \mathcal{L})$$

for some Lagrangian $\mathcal{L}$, and some order m, with

$$u(0, x, y) = \delta(x - y)$$

such that

$$(D_t + \sqrt{\Delta_x})u \in I^{-\infty}((-\epsilon, \epsilon) \times X \times X, \mathcal{L}) = \mathcal{C}^\infty((-\epsilon, \epsilon) \times X \times X).$$

We begin by sorting out what m, the order of u, should be. Since

$$u|_{t=0} = \delta(x - y) = (2\pi)^{-n} \int_{\mathbb{R}^n} e^{i(x-y)\cdot\theta}\, d\theta,$$

we were led us to a solution that for t small was of the form

$$\int_{\mathbb{R}^n} a(t, x, y, \theta) e^{i\Phi(t,x,y,\theta)}\, d\theta$$

with a a symbol of order zero such that $a(0, x, y, \theta) = 1$, and Φ a nondegenerate phase function such that $\Phi(0, x, y, \theta) = (x - y) \cdot \theta$. This was certainly the rough form of our earlier Ansatz; it should now be regarded as a Lagrangian distribution, of course. Since $\dim(\mathbb{R} \times X \times X) = 2n+1$ and we have n phase variables $\theta_1, \ldots, \theta_n$, the convention on orders of FIO's leads to $m = -1/4$.

Now we address the following question: what Lagrangian $\mathcal{L}$ ought we to choose? Since

$$\Box_{t,x} \in \mathrm{Diff}^2(\mathbb{R} \times X \times X) \subset \Psi^2(\mathbb{R} \times X \times X),$$

we a priori would have

$$\Box u \in I^{7/4}(\mathbb{R} \times X \times X, \mathcal{L});$$

as we would like smoothness of $\Box u$, we ought to start by making the principal symbol of $\Box u$ vanish. The symbol of $\Box$ vanishes only on

$$\Sigma_\Box = \{\tau^2 = |\xi|_g^2\}$$

hence the easiest way to ensure vanishing of the principal symbol is simply to arrange that

$$\mathcal{L} \subset \Sigma_\Box. \tag{9.1}$$

Now, recall that our initial conditions were to be

$$u(0, x, y) = \delta(x - y),$$

where we may view this as a Lagrangian distribution on $X \times X$ with respect to $N^*\Delta$, the conormal to the diagonal:

$$N^*\Delta = \{(x, y, \xi, \eta) : x = y, \xi = -\eta\}.$$

It is not difficult to check that the requirement that $u|_{t=0}$ gives this lower-dimensional Lagrangian[41] together with the requirement (9.1) that $\mathcal{L}$ should lie in the characteristic set implies that $\mathcal{L} \cap \{t = 0\}$ should just consist of points in $\Sigma_\Box$ *projecting* to points in $N^*\Delta$, i.e. that we should in fact have

$$\mathcal{L} \cap \{t = 0\} = \{(t = 0, \tau = -|\eta|_g, x = y, \xi = -\eta)\} \subset T^*(\mathbb{R} \times X \times X).$$

Here we have chosen the sign $\tau = -|\eta|_g$ in view of our real interest, which is in solving

$$(D_t + \sqrt{\Delta})u = 0$$

rather than the full wave equation;[42] we have thus kept $\mathcal{L}$ inside the characteristic set of $D_t + \sqrt{\Delta}$, which is one of the two components of $\Sigma_\Box$.

Let $\mathcal{L}_0$ now denote $\mathcal{L} \cap \{t = 0\}$. The set $\mathcal{L}_0$ is a manifold on which the symplectic form vanishes (an "isotropic" manifold), of dimension one less than half the dimension of $T^*(\mathbb{R} \times X \times X)$. (Exercise: Check this! Most of the work is done already, as $N^*(\Delta)$ is Lagrangian in $T^*(X \times X)$.)

We now proceed as follows to find a Lagrangian (necessarily one dimensional larger) containing $\mathcal{L}_0$: let $\mathsf{H} = \mathsf{H}_\Box$ denote the Hamilton vector field of the symbol of the wave operator, in the variables (t, x, τ, ξ). (I.e., take the Hamilton vector field of $\Box_{(t,x)}$ on the cotangent bundle of $\mathbb{R} \times X \times X$—nothing interesting happens in y, η.) By construction, $\mathcal{L}_0 \subset \Sigma_\Box$; we now *define* $\mathcal{L}$ to be the union of integral curves of H passing through points in $\mathcal{L}_0$. More concretely, these are all backwards unit-speed parametrized geodesics beginning at $(x = y, \xi = -\eta)$, where (x, ξ) evolves along the geodesic flow, and (y, η) are fixed. (Meanwhile, t is evolving at unit speed, and τ is constrained by the requirement that we are in the characteristic set so that $\tau = -|\xi|_g$.) The manifold $\mathcal{L}$ stays inside $\Sigma_\Box$ (indeed, inside the component that is $\Sigma_{D_t+\sqrt{\Delta}}$) since H is tangent to this manifold; moreover, $\mathcal{L}$ is automatically Lagrangian since ω vanishes on $\mathcal{L}_0$ and $\sigma_2(\Box)$ does as well, so that for $\mathsf{Y} \in T\mathcal{L}_0$, we further have

$$\omega(\mathsf{Y}, \mathsf{H}) = (d(\sigma_2(\Box)), \mathsf{Y}) = \mathsf{Y}\sigma_2(\Box) = 0.$$

This gives vanishing of ω on the tangent space to $\mathcal{L}$ at points along $t = 0$; to conclude it more generally, just recall that the flow generated by a Hamilton vector field is a family of symplectomorphisms.

EXERCISE 9.1. Check that $\mathcal{L}$ is in fact the *only* connected conic Lagrangian manifold passing through $\mathcal{L}_0$ and lying in $\Sigma_\Box$. (HINT: Observe that H is in fact the *unique* vector at each point along $\mathcal{L}_0$ that has the property $\omega(\mathsf{Y}, \mathsf{H}) = 0$ for all $\mathsf{Y} \in T\mathcal{L}_0$.)

Thus, to recapitulate, if we obtain $\mathcal{L}$ by flowing out $\mathcal{L}_0$ (the lift of the conormal bundle of the diagonal to the characteristic set of $D_t + \sqrt{\Delta}$) along H, the Hamilton vector field of $\Box$, we produce a Lagrangian on which $\Box$ is characteristic.

[41] We really ought to think a bit about restriction of Lagrangian distributions here: this is best done by regarding the restriction operator itself as an FIO (cf. Exercise 8.8). We shall omit further discussion of this point, but remark that it should at least seem plausible that the Lagrangian manifold associated to the restriction is the projection (i.e. pullback under inclusion), of the Lagrangian in the ambient space—cf. Exercise 4.11.

[42] We have chosen to emphasize this distinction only at this critical juncture only because as it is in some respects more pleasant to deal with $\Box$ than with the half-wave operator when possible.

EXERCISE 9.2. Show that the phase function $\phi(t,x,\eta)-y\cdot\eta$ that we constructed explicitly in §6 does indeed parametrize

$$\mathcal{L} = \{(t,\tau,x,\xi,y,-\eta) : \tau = -|\xi|_g,\ (x,\xi) = \Phi_t(y,\eta)\}$$

(with Φ_t denoting geodesic flow, i.e. the flow generated by the Hamilton vector field of $|\xi|_g$) over $|t| \ll 1$.

Compare our solution to the eikonal equation using Hamilton-Jacobi theory in Exercise 6.1 to what we have done here.

We now remark that while our parametrization of the Lagrangian in §6 worked only for small t, the definition given here of $\mathcal{L} \subset T^*(\mathbb{R}\times X\times X)$ makes sense *globally* in t, not merely for short time. When t is small and positive and y fixed, the projection of $\mathcal{L}$ to (x,ξ) is just the inward-pointing conormal bundle to an expanding geodesic sphere centered at y; when t exceeds the injectivity radius of X, $\mathcal{L}$ ceases to be a conormal bundle, but remains a well-behaved smooth Lagrangian.

Let us now return from our lengthy digression on the construction of $\mathcal{L}$ to recall what it gets us. Solving the eikonal equation, i.e. choosing $\mathcal{L}$, has reduced our error term by one order, and we have achieved

$$\Box u \in I^{3/4}(\mathbb{R}\times X\times X,\mathcal{L});$$

to proceed further, we invoke Property (VI) of FIO's, to compute

$$\sigma_{3/4}(\Box u) = i^{-1}\mathsf{H}\sigma_{-1/4}(u);$$

setting this equal to zero yields our first transport equation, and it is solved by simply insisting that $\sigma_{-1/4}(u)$ be constant along the flow, hence equal to 1, its value at $t=0$ (which was dictated by our δ-function initial data).

Now we have achieved $\Box u = r_{-1/4} \in I^{-1/4}$ Adding an element $u_{-5/4}$ of $I^{-5/4}(\mathbb{R}\times X\times X,\mathcal{L})$ to solve this error away and again applying (VI) yields the transport equation

$$i^{-1}\mathsf{H}(\sigma_{-5/4}(u_{-5/4})) = -\sigma_{-1/4}(r_{-1/4}),$$

which we may solve as before. Continuing in this manner and asymptotically summing the resulting terms, we have our parametrix $u \in I^{-1/4}(\mathbb{R}\times X\times X,\mathcal{L})$.

Now we describe, *very roughly*, how to use the FIO calculus to compute the singularities of $\operatorname{Tr} U(t)$ at lengths of closed geodesics.

Let T denote the operator $\mathcal{C}^\infty(\mathbb{R}\times X\times X)\to\mathcal{C}^\infty(\mathbb{R})$ given by[43]

$$T : f(t,x,y)\mapsto \int_X f(t,x,x)\,dx.$$

Thus, $\operatorname{Tr} U = T(U)$, and we seek to identify this composition as a Lagrangian distribution on $\mathbb{R}^1$; such a distribution is thus conormal to some set of points; as we saw above (and will see again below) these points may only be the lengths of closed geodesics, together with 0.

The Schwartz kernel of T is the distribution

$$\delta(t-t')\delta(x-y)$$

[43]It is here that our omission of density factors becomes most serious: T should really act on *densities* defined along the diagonal, so that the integral over X is well-defined. Fortunately, U itself should be a *right*-density (i.e. a section of the density bundle lifted from the right factor); restricted to the diagonal, this yields a density of the desired type.

on $\mathbb{R} \times \mathbb{R} \times X \times X$; it is thus conormal to $t = t', x = y$, i.e. is a Lagrangian distribution with respect to the Lagrangian

$$\{t = t', x = y, \tau = -\tau', \xi = -\eta\}.$$

Noting that if we reshuffle the factors into $(\mathbb{R} \times X) \times (\mathbb{R} \times X)$, the distribution $\delta(t - t')\delta(x - y)$ becomes the kernel of the identity operator, we can easily see that the order of this Lagrangian distribution is 0. Thus,

$$T \in I^0(\mathbb{R} \times \mathbb{R} \times X \times X, \Gamma')$$

where the relation $\Gamma : T^*(\mathbb{R} \times X \times X) \to T^*\mathbb{R}$ maps as follows:

$$\Gamma(t, \tau, x, \xi, y, \eta) = \begin{cases} \emptyset, \text{ if } (x, \xi) \neq (y, -\eta) \\ (t, \tau), \text{ if } (x, \xi) = (y, -\eta). \end{cases}$$

Let $\mathcal{L}$ be the Lagrangian for our parametrix u constructed above. If an interval about $L \in \mathbb{R}$ contains no lengths of closed geodesics, then we see that no points in $\mathcal{L}$ lie over $\{(x, \xi) = (y, -\eta)\}$ for t near L, hence $\Gamma(\mathcal{L})$ has no points over this interval, i.e. the composition Tu is smooth in this interval. *This gives another proof of the Poisson relation,* Theorem 5.3.

If, by contrast, there is a closed geodesic of length L, then

$$\{(L, \tau) : \tau < 0\} \in \Gamma(\mathcal{L}).$$

Note that in effect we get a contribution from every (x, ξ) lying along the geodesic, and that in particular, the fiber over (L, τ) of the projection on the left factor

$$\left(T^*\mathbb{R} \times \Delta_{T^*(\mathbb{R} \times X \times X) \times T^*(\mathbb{R} \times X \times X)}\right) \cap (\Gamma \times \mathcal{L}) \to T^*\mathbb{R}$$

(giving the composition $\Gamma(\mathcal{L})$) consists of at least a whole geodesic of length L, rather than a single point. Thus, the composition of these canonical relations *is not transverse* and the machinery described thus far does not apply. In [**3**], Duistermaat-Guillemin remedied this deficiency by constructing a theory of composition of FIO's with canonical relations intersecting *cleanly.*

DEFINITION 9.1. Two manifolds X, Y intersect *cleanly* if $X \cap Y$ is a manifold with $T(X \cap Y) = TX \cap TY$ at points of intersection.

For instance, pairs of coordinate axes intersect cleanly but not transversely in $\mathbb{R}^n$. In general, in the notation of Property (I), if the intersection of the product of canonical relations $\Gamma_1 \times \Gamma_2$ with the partial diagonal $T^*X \times \Delta \times T^*Z$ is clean, we define the *excess*, e, to be the dimension of the fiber of the projection from this intersection to $T^*X \times T^*Z$; this is zero in the case of transversality. Duistermaat-Guillemin show:

$$S \circ T \in I^{m+m'+e/2}(X \times Z, (\Gamma_1 \circ \Gamma_2)')$$

i.e. composition goes as before, but with a change in order. In addition the symbol of the product is obtained by *integrating* the product of the symbols over the e-dimensional fiber of the projection in what turns out to be an invariant way.

Let us now assume that there are finitely many closed geodesics of length L, and that they are *nondegenerate* in the following sense. For each closed bicharacteristic (i.e. lift to S^*X of a closed geodesic) $\gamma \subset S^*X$, pick a point $p \in \gamma$ and let $Z \subset S^*X$ be a small patch of a hypersurface through p transverse to γ. Shrinking Z as necessary, we can consider the map $P_\gamma : Z \to Z$ taking a point to its first intersection with Z under the bicharacteristic flow on S^*X. This is called a *Poincaré map.* Since

$P_\gamma(p) = p$, we can consider $dP_\gamma : T_pZ \to T_pZ$. We say that the closed geodesic is *nondegenerate* if $\mathrm{Id} - dP_\gamma$ is invertible. Note that this condition is independent of our choices of p and Z, as are the eigenvalues of $\mathrm{Id} - dP_\gamma$.

The following is due to Duistermaat-Guillemin [**3**]:

THEOREM 9.2. *Assume that all closed geodesics of length L on X are nondegenerate. Then*

$$\lim_{t\to L} (t - L) \operatorname{Tr} U(t) = \sum_{\gamma \text{ of length } L} \frac{L}{2\pi} i^{\sigma_\gamma} \left|\mathrm{Id} - dP_\gamma\right|^{-1/2},$$

where P_γ is the Poincaré map corresponding to the geodesic γ, and σ_γ is the number of conjugate points along the geodesic.

A proof of this theorem requires understanding the symbol of the clean composition Tu (where u is our parametrix for the half-wave equation). This lies beyond the scope of these notes. We merely note that we are in the setting of clean composition with excess 1, hence locally near $t = L$,

$$Tu \in I^{0-1/4+1/2}(\mathbb{R}, \{t = L, \tau < 0\}).$$

This Lagrangian is easily seen to be parametrized, locally near $t = L$, by the phase function with one fiber variable[44]

$$\phi(t, \theta) = \begin{cases} (t - L)\theta, \theta < 0, \\ 0, \theta \geq 0; \end{cases}$$

hence we may write

$$Tu = (2\pi)^{-3/4} \int_0^\infty a(t, \theta) e^{-i(t-L)\theta}\, d\theta,$$

where $a \in S^0(\mathbb{R} \times \mathbb{R})$ has an asymptotic expansion $a \sim a_0 + |\theta|^{-1} a_{-1} + \dots$. Our task is to find the leading-order behavior of Tu, and this is of course dictated by its principal symbol. To top order, a is given by the constant function $a_0(L, 1)$, hence Tu is (to leading order) a universal constant times $a_0(L, 1)$ times the Fourier transform of the Heaviside function, evaluated at $t - L$. Thus, the limit in the statement of the theorem is, up to a constant factor, just the value of $a_0(L, 1)$. The whole problem, then, is to compute the principal symbol of this clean composition, and we refer the interested reader to [**3**] for the (rather tricky) computation.[45]

10. A global calculus of pseudodifferential operators

10.1. The scattering calculus on $\mathbb{R}^n$. We now return to some of the problems discussed in §2, involving operators on noncompact manifolds. Recall that the Morawetz estimate on $\mathbb{R}^n$, for instance, hinged upon a *global* commutator argument, involving the commutator of the Laplacian with $(1/2)(D_r + D_r^*)$ on $\mathbb{R}^n$. Generalizing this estimate to noncompact manifolds will require some understanding of differential and pseudodifferential operators that is uniform near infinity.

[44]This phase function should of course be modified to make it smooth across $\theta = 0$, but making this modification will only add a term in $\mathcal{C}^\infty(\mathbb{R})$ to the Lagrangian distribution we write down.

[45]We note that the factor i^{σ_γ} is the contribution of the (in)famous Keller-Maslov index, and is in many ways the subtlest part of the answer.

Recall that thus far, we have focused on the calculus of pseudodifferential operators on compact manifolds; in discussing operators on $\mathbb{R}^n$, we have avoided as far as possible any discussion of asymptotic behavior at spatial infinity. Thus, our next step is to discuss a calculus of operators—initially just on $\mathbb{R}^n$—that involves sensible bounds near infinity.

Thus, let us consider pseudodifferential symbols defined on all of $T^*\mathbb{R}^n$ with no restrictions on the support in the base variables, with asymptotic expansions in *both* the base and fiber variables, both separately and jointly. To this end, note that changing to variables $|x|^{-1}$, $\hat{x}$, $|\xi|^{-1}$, and $\hat{\xi}$ amounts to *compactifying* the base and fiber variables of $T^*\mathbb{R}^n$ radially, to make the space $B^n_x \times B^n_\xi$, with B^n denoting the closed unit ball. (Recall that we defined a radial compactification map in (3.4), and that while $\langle \xi \rangle^{-1}$ and $\langle x \rangle^{-1}$ are what we should really use as defining functions for the spheres at infinity, $|\xi|^{-1}$ and $|x|^{-1}$ are acceptable substitutes as long as we stay away from the origin in the corresponding variables.) The space $B^n \times B^n$ is a *manifold with codimension-two corners,* i.e. a manifold locally modelled on $[0,1) \times [0,1) \times \mathbb{R}^{2n-2}$; its boundary is the union of the two smooth hypersurfaces $S^{n-1}_x \times B^n_\xi$ and $B^n_x \times S^{n-1}_\xi$. In our local coordinates, $|x|^{-1}$ and $|\xi|^{-1}$ are the defining functions for the two boundary hypersurfaces, i.e. the variables locally in $[0,1)$, while a choice of $n-1$ of each of the $\hat{x}$ and $\hat{\xi}$ variables gives the remaining $\mathbb{R}^{n-2}$.

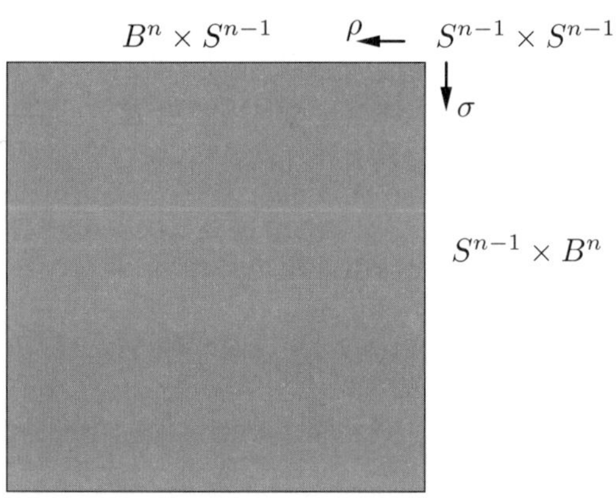

FIGURE 2. The manifold with corners $B^n \times B^n$ in the case $n = 1$. At the top (and bottom) are the boundary faces from $B^n \times S^{n-1}$ arising from the compactification of the second factor—this is "fiber infinity." At left (and right) are the faces from $S^{n-1} \times B^n$, arising from compactification of the first factor—this is "spatial infinity." The corner(s) at which these faces meet is $S^{n-1} \times S^{n-1}$. The functions $\rho = |x|^{-1}$ and $\sigma = |\xi|^{-1}$ can be locally taken as defining functions for the spatial infinity resp. fiber infinity boundary faces. The disconnectedness of $B^n \times S^{n-1}$ and $S^{n-1} \times B^n$ is of course a feature unique to dimension one.

We now let[46]

$$S^{m,l}_{\mathrm{sc}}(T^*\mathbb{R}^n)$$

denote the space of $a \in \mathcal{C}^\infty(T^*\mathbb{R}^n)$ such that[47]

$$\langle \xi \rangle^{-m} \langle x \rangle^{-l} a \in \mathcal{C}^\infty(B^n \times B^n). \tag{10.1}$$

This condition gives asymptotic expansions (i.e., Taylor series) in various regimes:

$$\begin{aligned} a(x,\xi) &\sim \sum |\xi|^{m-j} a_{\bullet,j}(x,\hat{\xi}), \text{ as } \xi \to \infty, \; x \in U \Subset \mathbb{R}^n \cong (B^n)^\circ \\ a(x,\xi) &\sim \sum |x|^{l-i} a_{i,\bullet}(\hat{x},\xi), \text{ as } x \to \infty, \; \xi \in V \Subset \mathbb{R}^n \cong (B^n)^\circ \\ a(x,\xi) &\sim \sum |x|^{l-i} |\xi|^{m-j} a_{ij}(\hat{x},\hat{\xi}), \text{ as } x, \xi \to \infty. \end{aligned} \tag{10.2}$$

Finally, let

$$\Psi^{m,l}_{\mathrm{sc}}(\mathbb{R}^n)$$

denote the space consisting of the (left) quantizations of these symbols. The "sc" stands for "scattering."[48]

This is an algebra of pseudodifferential operators, containing all ordinary pseudodifferential operators on $\mathbb{R}^n$ with compactly supported Schwartz kernels. The algebra of scattering pseudodifferential operators enjoys all the good properties of our usual algebra, plus some more that derive from its good behavior at infinity. We can compose operators to get new operators, and if $A \in \Psi^{m,l}_{\mathrm{sc}}(\mathbb{R}^n)$, $B \in \Psi^{m',l'}_{\mathrm{sc}}(\mathbb{R}^n)$, we have $AB \in \Psi^{m+m',l+l'}_{\mathrm{sc}}(\mathbb{R}^n)$. Likewise, adjoints preserve orders. What is novel here, however, is the principal symbol map.

As the symbols defined by (10.1) are those that, up to overall factors, are smooth functions on $B^n \times B^n$, we can define the *principal symbol* of order m, l of the operator $\mathrm{Op}(a)$ as

$$\hat{\sigma}_{m,l}(A) = \langle \xi \rangle^{-m} \langle x \rangle^{-l} a|_{\partial(B^n \times B^n)};$$

this can be further split into pieces corresponding to the restrictions to the two boundary hypersurfaces:

$$\hat{\sigma}_{m,l}(A) = (\hat{\sigma}^{\xi}_{m,l}(A), \hat{\sigma}^{x}_{m,l}(A))$$

where

$$\hat{\sigma}^{\xi}_{m,l}(A)(x,\hat{\xi}) \in \mathcal{C}^\infty(B^n \times S^{n-1})$$

is nothing but the ordinary principal symbol, rescaled by a power of $\langle x \rangle$, and

$$\hat{\sigma}^{x}_{m,l}(A)(\hat{x},\xi) \in \mathcal{C}^\infty(S^{n-1} \times B^n)$$

is the novel piece of the symbol, measuring the behavior of the operator at spatial infinity. Note that these two pieces of the principal symbol are not independent:

---

[46]This space should really be called $S^{m,l}_{\mathrm{cl,sc}}$, with the cl once again indicating "classicality" (as opposed to Kohn-Nirenberg type of estimates alone). We omit the cl so as not to clutter up the notation.

[47]We are abusing notation here by ignoring the diffeomorphism of radial compactification, thus identifying $\mathcal{C}^\infty(B^n \times B^n)$ directly with a space of functions on $\mathbb{R}^n \times \mathbb{R}^n$.

[48]This is a space of operators considered by many authors; as we are following roughly the treatment of Melrose [**18**], we have adopted his notation for the space. Note, however, that we have reversed the sign from his convention for the order l.

they must agree at the *corner*, $S^{n-1} \times S^{n-1}$. We may also choose to think of the principal symbol as

$$\sigma^{m,l}(A) \in S^{m,l}_{\mathrm{sc}}(T^*\mathbb{R}^n)/S^{m-1,l-1}_{\mathrm{sc}}(T^*\mathbb{R}^n),$$

and we will often confuse the symbol with its equivalence class; this is usually less confusing than keeping track of the rescaling factor $\langle\xi\rangle^m\langle x\rangle^l$.

The principal symbol short exact sequence thus reads:

$$0 \to \Psi^{m-1,l-l}_{\mathrm{sc}}(\mathbb{R}^n) \to \Psi^{m,l}(\mathbb{R}^n) \overset{\hat{\sigma}_{m,l}}{\to} \mathcal{C}^\infty(\partial(B^n \times B^n)) \to 0.$$

Thus, vanishing of this symbol yields improvement in both orders at once; correspondingly, vanishing of one part of the symbol gives improvement in just one order:

$$0 \to \Psi^{m-1,l}_{\mathrm{sc}}(\mathbb{R}^n) \to \Psi^{m,l}(\mathbb{R}^n) \overset{\hat{\sigma}^\xi_{m,l}}{\to} \mathcal{C}^\infty(B^n \times S^{n-1}) \to 0,$$

$$0 \to \Psi^{m,l-1}_{\mathrm{sc}}(\mathbb{R}^n) \to \Psi^{m,l}(\mathbb{R}^n) \overset{\hat{\sigma}^x_{m,l}}{\to} \mathcal{C}^\infty(S^{n-1} \times B^n) \to 0.$$

The symbol of the product of two scattering operators is indeed the product of the symbols,[49] as (equivalence classes of) smooth functions on $\partial(B^n \times B^n)$.

The symbol of the commutator of two scattering operators (which is of lower order than the product in both filtrations) is, as one might suspect, given by i times the Poisson bracket of the symbols.

The residual calculus is particularly nice in this setting: instead of merely consisting of smoothing operators, it consists of operators that are "Schwartzing"—they create decay as well as smoothness:

$$R \in \Psi^{-\infty,-\infty}_{\mathrm{sc}}(\mathbb{R}^n) \iff R : \mathcal{S}'(\mathbb{R}^n) \to \mathcal{S}(\mathbb{R}^n).$$

One problem with using the ordinary calculus for global matters is that we can only conclude compactness of operators of negative order for compactly supported operators. Here, we have a much more precise result:

PROPOSITION 10.1. *An operator in $\Psi^{0,0}_{sc}(\mathbb{R}^n)$ is bounded on $L^2(\mathbb{R}^n)$; an operator of order (m,l) with $m, l < 0$ is compact on $L^2(\mathbb{R}^n)$.*

Associated to the expanded notion of symbol, there is are associated notions of ellipticity (nonvanishing of the principal symbol) and of WF$'$ (lack of infinite order vanishing of the total symbol). We have an associated family of Sobolev spaces:

$$u \in H^{m,l}_{\mathrm{sc}}(\mathbb{R}^n) \iff \forall A \in \Psi^{m,l}_{\mathrm{sc}}(\mathbb{R}^n),\ Au \in L^2(\mathbb{R}^n).$$

Operators in the calculus act on this scale of Sobolev spaces in the obvious way. Since smoothing operators are "Schwartzing," it is not hard to see that

$$H^{-\infty,-\infty}_{\mathrm{sc}}(\mathbb{R}^n) = \mathcal{S}(\mathbb{R}^n).$$

(We will return to an explicit description of these Sobolev spaces shortly.)

There is also an associated wavefront set:

$$\mathrm{WF}_{\mathrm{sc}}\, u \subset \partial(B^n \times B^n)$$

[49]It is exactly this innocuous statement, which the reader might think routine, that separates the scattering calculus from many other choices of pseudodifferential calculus on noncompact manifolds: typically the "symbol at infinity" (here $\hat{\sigma}^x_{m,l}(\hat{x},\xi)$) will compose under operator composition in a more complex, noncommutative way.

is defined by

$$p \notin \mathrm{WF}_{\mathrm{sc}}\, u \Longleftrightarrow \text{ there exists } A \in \Psi^{0,0}_{\mathrm{sc}}(\mathbb{R}^n), \text{ elliptic at } p, \text{ with } Au \in \mathcal{S}.$$

In $(B^n_x)^\circ \times S^{n-1}_\xi \subset \partial(B^n \times B^n)$, (i.e., in the usual cotangent bundle of $\mathbb{R}^n$) this definition just coincides with ordinary wavefront set; but "at infinity," i.e. in $S^{n-1}_x \times B^n_\xi$, it measures something new. To see what, let us consider some examples.

EXAMPLE 10.2.

(1) *Constant coefficient vector fields on* $\mathbb{R}^n$: If $v \in \mathbb{R}^n$ and $P = i^{-1} v \cdot \nabla$, then, we can write

$$P = \mathrm{Op}_\ell(v \cdot \xi);$$

the principal symbol is thus

$$\sigma_{1,0}(P) = v \cdot \xi$$

(2) Likewise, the symbol of the Euclidean Laplacian Δ is $\sigma_{2,0}(\Delta) = |\xi|^2$. Note that the Laplacian is *not elliptic* in the scattering calculus, as its principal symbol vanishes at $\xi = 0$ on the boundary face $S^{n-1}_x \times B^n_\xi$. This should come as no suprise, as Δ has nullspace in $\mathcal{S}'(\mathbb{R}^n)$ (given by harmonic polynomials) that does not lie in L^2, hence is not consistent with elliptic regularity in the scattering calculus sense: if Q is elliptic in the scattering calculus,

$$Qu \in \mathcal{S}(\mathbb{R}^n) \Longrightarrow u \in \mathcal{S}(\mathbb{R}^n).$$

On the other hand, consider $\mathrm{Id} + \Delta$. We have $\mathrm{Id} \in \Psi^{0,0}_{\mathrm{sc}}(\mathbb{R}^n)$, hence adding it certainly does not alter the "ordinary" part of the symbol, living on $(B^n)^\circ \times S^{n-1}$. But it *does* affect the symbol in $S^{n-1} \times B^n$: we have

$$\sigma_{2,0}(\mathrm{Id} + \Delta) = 1 + |\xi|^2;$$

$\mathrm{Id} + \Delta$ *is* an elliptic operator in the scattering calculus, and of course it is the case that $(\mathrm{Id} + \Delta)u \in \mathcal{S}(\mathbb{R}^n)$ implies that u is likewise Schwartz.

(3) If we vary the metric from the Euclidean metric to some other metric g, we may or may not obtain a scattering differential operator; for example, if g were periodic, we certainly would not, as the total symbol of Δ would clearly lack an asymptotic expansion as $|x| \to \infty$. Suppose, however, that we may write in spherical coordinates on $\mathbb{R}^n$

$$g = dr^2 + r^2 \sum h_{ij}(r^{-1}, \theta) d\theta^i d\theta^j \quad \text{for } r > R_0 \gg 0.$$

where h_{ij} is a smooth function of its arguments, and

$$h_{ij}(0, \theta) d\theta^i d\theta^j$$

is the standard metric on the "sphere at infinity." We will call such a metric *asymptotically Euclidean.* Then the corresponding Laplace operator is in the scattering calculus.

EXERCISE 10.1. Check that this operator does lie in the scattering calculus.

Let Δ denote the Laplacian with respect to an asymptotically Euclidean metric. Then

$$(\mathrm{Id} + \Delta)^{-1} \in \Psi^{-2,0}_{\mathrm{sc}}(\mathbb{R}^n).$$

(4) $\langle x\rangle^2(\mathrm{Id}+\Delta) \in \Psi_{\mathrm{sc}}^{2,2}(\mathbb{R}^n)$ and has symbol $\langle x\rangle^2(1+|\xi|^2)$. This is globally elliptic.

By the last example, we find that

$$u \in H_{\mathrm{sc}}^{2,2}(\mathbb{R}^n) \Longleftrightarrow \langle x\rangle^2(\mathrm{Id}+\Delta)u \in L^2(\mathbb{R}^n);$$

interpolation and duality arguments allow us to conclude more generally that the scattering Sobolev spaces coincide with the usual weighted Sobolev spaces:

$$H_{\mathrm{sc}}^{m,l}(\mathbb{R}^n) = \langle x\rangle^{-l} H^m(\mathbb{R}^n).$$

We now turn to some examples illustrating the scattering wavefront set. Consider the plane wave

$$u(x) = e^{i\alpha\cdot x}.$$

We have

$$(D_{x^j} - \alpha_j)u = 0 \text{ for all } j = 1, \dots, n.$$

The symbol of the operator $D_{x^j} - \alpha_j$ is $\xi_j - \alpha_j$, hence the intersection of the characteristic sets of these operators is just the points in $S^{n-1} \times B^n$ where $\xi = \alpha$. As a consequence, we have

$$\mathrm{WF}_{\mathrm{sc}}(e^{i\alpha\cdot x}) \subseteq \{(\hat{x}, \xi) \in S^{n-1} \times \mathbb{R}^n : \xi = \alpha\}$$

(here we are as usual identifying $(B^n)^\circ \cong \mathbb{R}^n$). In fact this containment turns out to be equality, as we see by the following characterization of scattering wavefront set.

PROPOSITION 10.3. *Let $p = (\hat{x}_0, \xi_0) \in S^{n-1} \times \mathbb{R}^n$. We have*

$$p \notin \mathrm{WF}_{sc}\, u$$

if and only if there exist cutoff functions $\phi \in \mathcal{C}_c^\infty(\mathbb{R}^n)$ nonzero at ξ_0 and $\gamma \in \mathcal{C}^\infty(\mathbb{R}^n)$ nonzero in a conic neighborhood of the direction $\hat{x}_0$ such that

$$\phi\mathcal{F}(\gamma u) \in \mathcal{S}(\mathbb{R}^n).$$

This is of course closely analogous to the characterization of ordinary wavefront set in Proposition 4.5, and is proved in an analogous manner. Note that if u is a Schwartz function in a set of the form

$$\{\Big|\frac{x}{|x|} - \hat{x}_0\Big| < \epsilon, |x| > R_0\}$$

for any $\epsilon > 0$, $R_0 \gg 0$, then there is no scattering wavefront set at points of the form $(\hat{x}_0, \xi)$ for any $\xi \in \mathbb{R}^n$. Thus, this new piece of the wavefront set measures the asymptotics of u in different directions toward spatial infinity: $\hat{x}_0$ provides the direction, while the value of ξ_0 records oscillatory behavior of a specific frequency.

There is also, of course, a similar characterization of $\mathrm{WF}_{\mathrm{sc}}\, u$ inside $S^{n-1} \times S^{n-1}$. We leave this as an exercise for the reader.

10.2. Applications of the scattering calculus. As an example of how we might use the scattering calculus to obtain global results on manifolds, let us return to the *local smoothing estimate* from §2.1. Recall that if ψ satisfies the Schrödinger equation (2.1) on $\mathbb{R}^n$ with initial data $\psi_0 \in H^{1/2}$, this estimate (or, at least, one version of it) tells us that

$$\psi \in L^2_{\mathrm{loc}}(\mathbb{R}_t; H^1_{\mathrm{loc}}(\mathbb{R}^n)), \tag{10.3}$$

hence the solution is (locally) half a derivative smoother than the data, on average. How might we obtain this estimate on a manifold, with Δ replaced by the Laplace-Beltrami operator (which we also denote Δ)? For a start, note that (10.3) fails badly on compact manifolds; in particular, recall that since $[\Delta, \Delta^s] = 0$ for all $s \in \mathbb{R}$, the H^s norms are conserved under the evolution, hence if $\psi_0 \notin H^s$, with $s > 1/2$, then we certainly do not have[50] $\psi \in L^2_{\text{loc}}(\mathbb{R}_t; H^s)$. So if we seek a broader geometric context for this estimate, we had better try noncompact manifolds.

Recall that we initially obtained the estimate by a commutator argument with the Morawetz commutant

$$\partial_r + \frac{n-1}{2r},$$

which actually gave more information; we noted that we could, instead, have used a simpler commutant $f(r)D_r$, with $f(r) = 0$ near $r = 0$, nondecreasing, and equal to 1 for $r \geq 2$ (say): this gives a commutator with a term

$$\chi'(r)D_r^2$$

which, when paired with ψ and integrated in time, tests for H^1 regularity in an annular neighborhood of the origin (which could have been translated to be anywhere); other terms in the commutator are positive also, modulo estimable error terms, and we thus obtain the local smoothing estimate. Generalizing this is tricky, as the positivity of the symbol of the term

$$i[\Delta, D_r]$$

on $\mathbb{R}^n$ is delicate: the symbol of this commutator is given by the Poisson bracket

$$\{|\xi|^2, \xi \cdot \hat{x}\} = 2\xi \cdot \partial_x(\xi \cdot \hat{x}) = \frac{2}{|x|}(|\xi|^2 - (\xi \cdot \hat{x})^2)$$

which is nonnegative but does actually vanish at $\xi \parallel x$, i.e. in radial directions. If we perturb the Euclidean metric a bit, and replace $|\xi|^2$ with $|\xi|_g^2$, the symbol of the Laplace-Beltrami operator, but leave the inner product $\langle \xi, x\rangle = \sum \xi_j x^j$, then this computation fails to give positivity. So we have to be more careful. We might try to adapt $\sum \xi_j x^j$ to the new metric instead, but this is problematic, as it doesn't really make much invariant sense. Moreover, it seems even more problematic upon interpretation: what positivity of $\{|\xi|_g^2, a\}$ *means* is just that a is increasing along the bicharacteristic flow of $|\xi|_g^2$, i.e. is increasing along (the lifts to the cosphere bundle of) geodesics. This is clearly impossible if there are any closed (i.e., periodic) geodesics, or indeed if there are geodesics that remain in a compact set for all time, hence our difficulty in obtaining an estimate on compact manifolds.

EXERCISE 10.2. Suppose that a geodesic γ remains in a compact subset of $\mathbb{R}^n$ (equipped with a non-Euclidean metric) for all $t > 0$. Let $p = (\gamma(0), (\gamma'(0))^*) \in T^*\mathbb{R}^n$ (with $*$ denoting dual under the metric). Show that there cannot exist a smooth $a \in \mathcal{C}^\infty(T^*\mathbb{R}^n)$ with $\{|\xi|_g^2, a\} \geq \epsilon > 0$ and $a(p) \neq 0$.

[50]Note that this argument fails on $\mathbb{R}^n$ exactly because of the distinction between local and global Sobolev regularity: there is nothing preventing a solution on $\mathbb{R}^n$ with initial data in $H^{1/2}$ from being locally H^1—or even smooth on arbitrarily large compact sets—in return for having nasty behavior near infinity.

DEFINITION 10.4. Let g be an asymptotically Euclidean metric on $\mathbb{R}^n$, and let γ be a geodesic. We say that γ is *not trapped forward/backward* if

$$\lim_{t\to\pm\infty} |\gamma(t)| = \infty.$$

We say that γ is *trapped* if it is trapped both forward *and* backward. We also use the same notation for the bicharacteristic projecting to γ. Moreover, we say that a point in $S^*\mathbb{R}^n$ along a non-(forward/backward)-trapped geodesic is itself non-(forward/backward)-trapped.

It is a theorem of Doi [**4**] that the local smoothing estimate (10.3) *cannot hold* near a trapped geodesic. (The total failure of (10.3) on compact manifolds should make this plausible, but it turns out to be considerably more delicate to show that it fails even if the only trapping is, for instance, a single, highly unstable, closed geodesic.) As a result we will require some strong geometric hypotheses in in order to find a general context in which (10.3) holds.

The following is a result of Craig-Kappeler-Strauss [**1**]:

THEOREM 10.5. *Consider ψ a solution to the Schrödinger equation on asymptotically Euclidean space, with $\psi_0 \in H^{1/2}(\mathbb{R}^n)$. The estimate* (10.3) *holds microlocally at any (x_0, ξ_0) that lies on a nontrapped bicharacteristic, i.e. for any $A \in \Psi^1(\mathbb{R}^n)$ compactly supported and microsupported sufficiently near to (x_0, ξ_0), we have for any $T > 0$,*[51]

$$\int_0^T \|A\psi\|^2 \, dt \lesssim \|\psi_0\|^2_{H^{1/2}}.$$

PROOF. We will prove the theorem by using a commutator argument in the scattering calculus. To begin, we recall from Exercise 4.21 that the set along which microlocal $L^2_{\mathrm{loc}}H^1$ regularity holds is invariant under the geodesic flow. Hence it suffices just to obtain regularity of this form *somewhere* along the geodesic γ. The convenient place to do this is out near infinity.

In order to make a commutator argument, note that it is very useful to have a quantity that behaves monotonically along the flow. We refer to points in $T^*\mathbb{R}^n$ near infinity (i.e. for $|x| \gg 0$) as *incoming* if $\hat{\xi} \cdot \hat{x} < 0$ and *outgoing* if $\hat{\xi} \cdot \hat{x} > 0$ (this corresponds to moving toward or away from the origin, respectively, under asymptotically Euclidean geodesic flow). Heuristically, under the classical evolution, points move from being incoming to being outgoing. More precisely, we observe that the Hamilton vector field of $p \equiv \sigma_{2,0}(\Delta)$ is given by

$$\mathsf{H}_p = -\sum \xi_i \xi_j \frac{\partial g^{ij}(x)}{\partial x^k} \partial_{\xi_k} + 2 \sum \xi_i g^{ij}(x) \partial_{x^j}.$$

Recalling that g^{ij} has an asymptotic expansion with leading term given by the identity metric, we can write this as

$$\mathsf{H}_p = 2\xi \cdot \partial_x + O(|x|^{-1}|\xi|)\partial_x + O(|x|^{-1}|\xi|^2)\partial_\xi \tag{10.4}$$

(where in fact the whole vector field is homogeneous of degree 1 in ξ).

EXERCISE 10.3. Verify (10.4).

---

[51] More generally, we can replace the Sobolev exponents $1/2$ and 1 by s and $s + 1/2$ respectively; in particular, L^2 initial data gives an $L^2H^{1/2}$ estimate.

Thus,

$$\mathsf{H}_p(\hat{\xi}\cdot\hat{x}) = \frac{|\xi|}{|x|}\left(1-(\hat{\xi}\cdot\hat{x})^2\right) + O(|\xi||x|^{-1}).$$

This is thus positive, as long as $\hat{\xi}\cdot\hat{x}$ is away from ± 1, and $|x|$ is large,[52] i.e., as long as we stay away from precisely incoming or outgoing points. Thus, we manufacture a scattering symbol for a commutant that has increase owing to the increase in "outgoingness:" Let $\chi(s)$ denote a smooth function that equals 0 for $s < 1/4$ and 1 for $s > 1/2$, with χ' a square of a smooth function, nonzero in the interior of its support. Let $\chi_\delta(s) = \chi(\delta s)$. We choose

$$a(x,\xi) = |\xi|_g \chi(-\hat{\xi}\cdot\hat{x})\chi_\delta(|x|)\chi(|\xi|_g).$$

Thus a is supported at *incoming* points at which $|x| \geq 1/(4\delta) \gg 0$; the first χ factor localizes near incoming points, and the factor of χ_δ keeps $|x|$ large. (The factor $\chi(|\xi|_g)$ simply cuts off near the origin in ξ to yield a smooth symbol.) Under the flow on the support of a, x tends to decrease and we become more outgoing, so the tendency is the *leave* the support of a along the flow. This is the essential point in the following:

EXERCISE 10.4. Check that $a \in S^{1,0}_{\mathrm{sc}}(T^*\mathbb{R}^n)$ and that if δ is chosen sufficiently small, we may write

$$\mathsf{H}_p a = -b^2 - c^2$$

where

(1) $b \in S^{1,-1/2}_{\mathrm{sc}}(T^*\mathbb{R}^n)$ is supported in $\operatorname{supp}\chi'(-\hat{\xi}\cdot\hat{x})\chi_\delta(|x|)$
(2) $c \in S^{1,-1/2}_{\mathrm{sc}}(T^*\mathbb{R}^n)$ is supported in $\operatorname{supp}\chi(-\hat{\xi}\cdot\hat{x})\chi_\delta'(|x|)$ and nonzero on the interior of that set.

(Note that $|\xi|_g$ is annihilated by H_p, so the terms containing $|\xi|_g$ simply do not contribute.)

Now let $A \in \Psi^{1,0}_{\mathrm{sc}}(\mathbb{R}^n)$ have principal symbol a. Then we have

$$i[\Delta, A] = -B^*B - C^*C + R$$

with $B = \mathrm{Op}(b)$, $C = \mathrm{Op}(c) \in \Psi^{1,-1/2}_{\mathrm{sc}}(\mathbb{R}^n)$, and $R \in \Psi^{1,-2}_{\mathrm{sc}}(\mathbb{R}^n)$.

Hence,

$$\int_0^T \|C\psi\|^2\,dt \leq \left|\langle A\psi,\psi\rangle\big|_0^T\right| + \left|\int_0^T \langle R\psi,\psi\rangle\,dt\right|.$$

As $\langle A\psi,\psi\rangle$ is bounded by the $L^\infty H^{1/2}$ norm of ψ and hence by $\|\psi_0\|^2_{H^{1/2}}$, and the R term likewise,[53] we obtain

$$\int_0^T \|C\psi\|^2\,dt \lesssim \|\psi_0\|^2_{H^{1/2}}. \tag{10.5}$$

---

[52] Largeness of ξ plays no role because of homogeneity of the Hamilton vector field of the principal symbol of Δ.

[53] In fact, the R term is considerably better than necessary for this step, as it has weight -2 rather than just 0 (which would be all we need to obtain the estimate). The astute reader may thus recognize that we are far from using the full power of the scattering calculus here. A proof of the global estimate in Exercise 10.6 requires a more serious use of the symbol calculus, however, as do the estimates which are the focus of [**1**], which show that microlocal decay of the initial data yields higher regularity of the solution along bicharacteristics.

EXERCISE* 10.5. Show that for any $R_0 > 0$, there exists $\delta > 0$ sufficiently small that if $(x_0, \xi_0) \in T^*\mathbb{R}^n \cap \{|x| < R_0\}$ lies along a non-backward trapped bicharacteristic, some point on that bicharacteristic with $t \ll 0$ lies in $\operatorname{ell} C$, with $C = \operatorname{Op}(c)$ constructed as above.

Thus, rays starting close to the origin that pass through $|x| \sim \delta^{-1}$ for $t \ll 0$ are incoming when they do so. This is an exercise in ODE. You might begin by showing that if a backward bicharacteristic starting in $\{|x| < R_0\}$ passes through the hypersurface $|x| = R'$ with $R' \gg 0$, then it must have $\hat{\xi} \cdot \hat{x} < 0$ there, and that $\hat{\xi} \cdot \hat{x}$ will keep decreasing thereafter along the backward flow.

Given a non-backward-trapped point $q \in S^*\mathbb{R}^n$, Exercise 10.5 tells us that we may construct a commutant A as above so that the commutator term C is elliptic somewhere along the bicharacteristic through q. Equation 10.5 tells us that we have the desired L^2H^1 estimate on $\operatorname{ell} C$, and the flow-invariance from Exercise 4.21 yields the same conclusion at q. Thus, we have proved the desired result at non-backward-trapped points. It remains to consider non-forward-trapped points.

Suppose, then, that $q = (x_0, \xi_0) \in T^*\mathbb{R}^n$ is non-forward-trapped; then note that $q' = (x_0, -\xi_0)$ is non-backward-trapped. Consider then the function $\overline{\psi}$: if

$$(D_t + \Delta)\psi = 0$$

then

$$(-D_t + \Delta)\overline{\psi} = 0,$$

i.e.

$$\tilde{\psi}(t, x) = \overline{\psi}(T - t, x)$$

again solves the Schrödinger equation. Of course, by unitarity,

$$\left\|\tilde{\psi}(0, x)\right\|_{H^{1/2}} = \|\psi_0\|_{H^{1/2}}.$$

Since q' is non-backward trapped, we thus find that there exists $C \in \Psi_{\mathrm{sc}}^{1,-1/2}(\mathbb{R}^n)$, elliptic at q', with

$$\int_0^T \left\|C\tilde{\psi}\right\|^2 dt \lesssim \|\tilde{\psi}(0, x)\|_{H^{1/2}}^2 = \|\psi_0\|_{H^{1/2}}^2;$$

on the other hand,

$$\begin{aligned}\left\|C\tilde{\psi}(t, \cdot)\right\|^2 &= \left\|C\overline{\psi}(T - t, \cdot)\right\|^2 \\ &= \left\|\overline{C}\psi(T - t, \cdot)\right\|^2,\end{aligned}$$

where

$$C = \operatorname{Op}_\ell(c(x, \xi)), \text{ and } \overline{C} = \operatorname{Op}_\ell(\overline{c}(x, -\xi));$$

thus, $\overline{C}$ tests for regularity at q, and we have obtained the desired estimate at q. □

COROLLARY 10.6. *On an asymptotically Euclidean space with no trapped geodesics, the local smoothing estimate holds everywhere.*

EXERCISE* 10.6. *(Global (weighted) smoothing.)* Show that if there are no trapped geodesics, and $\psi_0 \in L^2$, we have

$$\int_0^T \left\|\langle x\rangle^{-1/2-\epsilon}\psi\right\|_{H^{1/2}}^2 dt \lesssim \|\psi_0\|_{L^2}^2$$

for every $\epsilon > 0$. (This is a bit involved; a solution can be found, e.g., in Appendix II of [**8**].)

10.3. The scattering calculus on manifolds. We can generalize the description of the scattering calculus to manifolds quite easily, following the prescription of Melrose [**18**]. Let X be a compact manifold with boundary. We will, in practice, think of the interior, X°, as a noncompact manifold (with a complete metric) that just happens to come pre-equipped with a compactification to X. Our motivating example will be $X = B^n$, where X° is then diffeomorphically identified with $\mathbb{R}^n$ via the radial compactification map. Recall that on $\mathbb{R}^n$, radially compactified to the ball, we used coordinates near S^{n-1}, the "boundary at infinity," given by

$$\rho = \frac{1}{|x|}, \ \theta = \frac{x}{|x|},$$

where in fact ρ together with an appropriate choice of $n-1$ of the θ's furnish local coordinates near a point. In these coordinates, what do constant coefficient vector fields on $\mathbb{R}^n$ look like? We have

$$\partial_{x^j} = \rho\partial_{\theta^j} - \rho\sum \theta^k\theta^j\partial_{\theta^k} - \rho^2\theta^j\partial_\rho.$$

Recall moreover that functions in $\mathcal{C}^\infty(B^n)$ correspond exactly, under radial (un)compactification, to symbols of order zero on $\mathbb{R}^n$. So in fact it is easy to check more generally that vector fields on $\mathbb{R}^n$ with zero-symbol coefficients correspond exactly to vector fields on B^n that, near S^{n-1}, take the form

$$a(\rho,\theta)\rho^2\partial_\rho + \sum b_j(\rho,\theta)\rho\partial_{\theta^j},$$

with $a, b_j \in \mathcal{C}^\infty(B^n)$.

We generalize this notion as follows. Given our manifold X, let $\rho \in \mathcal{C}^\infty(X)$ denote a *boundary defining function*, i.e.

$$\rho \geq 0 \text{ on X}, \ \rho^{-1}(0) = \partial X, \ d\rho \neq 0 \text{ on } \partial X.$$

Let θ^j be local coordinates on ∂X. We define *scattering vector fields on* X to be those that can be written locally, near ∂X, in the form

$$a(\rho,\theta)\rho^2\partial_\rho + \sum b_j(\rho,\theta)\rho\partial_{\theta^j},$$

with $a, b_j \in \mathcal{C}^\infty(X)$. Let

$$\mathcal{V}_{\mathrm{sc}}(X) = \{\text{scattering vector fields on } X\}$$

EXERCISE 10.7.

(1) Show that $\mathcal{V}_{\mathrm{sc}}(X)$ is well-defined, independent of the choices of ρ, θ.
(2) Let $\mathcal{V}_{\mathrm{b}}(X)$ denote the space of smooth vector fields on X tangent to ∂X. Show that
$$\mathcal{V}_{\mathrm{sc}}(X) = \rho\mathcal{V}_{\mathrm{b}}(X)$$
(3) Show that both $\mathcal{V}_{\mathrm{sc}}(X)$ and $\mathcal{V}_{\mathrm{b}}(X)$ are Lie algebras.

As we can locally describe the elements of $\mathcal{V}_{\mathrm{sc}}(X)$ as the $\mathcal{C}^\infty$-span of n vector fields, $\mathcal{V}_{\mathrm{sc}}(X)$ is itself the space of sections of a *vector bundle*, denoted

$$^{\mathrm{sc}}TX.$$

There is also of course a dual bundle, denoted

$$^{\mathrm{sc}}T^*X,$$

whose sections are the $\mathcal{C}^\infty$-span of the one-forms

$$\frac{d\rho}{\rho^2}, \frac{d\theta^j}{\rho}.$$

Over X°, we may of course canonically identify ${}^{\mathrm{sc}}T^*X$ with T^*X, and the canonical one-form on the latter pulls back to give a canonical one-form

$$\xi\frac{d\rho}{\rho^2} + \eta\cdot\frac{d\theta}{\rho} \tag{10.6}$$

defining coordinates ξ, η on the fibers of ${}^{\mathrm{sc}}T^*X$.

The scattering calculus on $\mathbb{R}^n$ is concocted to contain scattering vector fields:

EXERCISE 10.8. Show that $\Psi^{1,0}_{\mathrm{sc}}(\mathbb{R}^n) \supset \mathcal{V}_{\mathrm{sc}}(B^n)$.

We can, following Melrose, define the scattering calculus more generally as follows. Let ${}^{\mathrm{sc}}\overline{T}^*X$ denote the *fiber-compactification* of the bundle ${}^{\mathrm{sc}}T^*X$, i.e. we are radially compactifying each fiber to a ball, just as we did globally in compactifying $T^*\mathbb{R}^n$ to $B^n \times B^n$, only this time, the base is already compact. Now let

$$S^{m,l}_{\mathrm{sc}}({}^{\mathrm{sc}}T^*X) = \sigma^{-m}\rho^{-l}\mathcal{C}^\infty({}^{\mathrm{sc}}\overline{T}^*X),$$

where σ is a boundary defining function for the fibers. We can (by dint of some work!) quantize these "total" symbols to a space of operators, denoted

$$\Psi^{m,l}_{\mathrm{sc}}(X).$$

(Note that in the case $X = B^n$, we recover what we were previously writing as $\Psi^{m,l}_{\mathrm{sc}}(\mathbb{R}^n)$; the latter usage, with $\mathbb{R}^n$ instead of the more correct B^n, was an abuse of the usual notation.) The principal symbol of a scattering operator is, in this invariant picture, a smooth function on $\partial({}^{\mathrm{sc}}\overline{T}^*X)$; or equivalently, an equivalence class of smooth functions on ${}^{\mathrm{sc}}\overline{T}^*X$; or, in the partially uncompactified picture, an equivalence class of smooth symbols on ${}^{\mathrm{sc}}T^*X$. (It is this last point of view that we shall mostly adopt.) In the coordinates defined by the canonical one-form (10.6), we have

$$\sigma_{1,0}(\rho^2 D_\rho) = \xi, \ \sigma_{1,0}(\rho D_{\theta^j}) = \eta_j. \tag{10.7}$$

Recall that the Euclidean metric may be written in polar coordinates as

$$d(\rho^{-1})^2 + (\rho^{-1})^2 h(\theta, d\theta)$$

with h denoting the standard metric on S^{n-1}. We can generalize this to define a *scattering metric* as one on a manifold with boundary X that can be written in the form

$$\frac{d\rho^2}{\rho^4} + \frac{h(\rho,\theta,d\theta)}{\rho^2}$$

locally near ∂X, with ρ a boundary defining function, and h now a smooth *family* in ρ of metrics on ∂X.[54]

[54]The usual definition, as in [**18**], is a little more general, allowing $d\rho$ terms in h; however, it was shown by Joshi-Sá Barreto that these terms can always be eliminated by appropriate choice of coordinates.

EXERCISE 10.9.

(1) Show that if g is a scattering metric on X, then the Laplace operator with respect to g can be written
$$\Delta = (\rho^2 D_\rho)^2 + O(\rho^3) D_\rho + \rho^2 \Delta_\theta$$
where Δ_θ is the family of Laplacians on ∂X associated to the family of metrics $h(r, \theta, d\theta)$.

(2) Show that for $\lambda \in \mathbb{C}$,
$$\sigma_{2,0}(\Delta - \lambda^2) = \xi^2 + |\eta|_h^2 - \lambda^2.$$
(Note that this entails noticing that you can drop the $O(\rho^3)D_\rho$ terms for different reasons at the the two different boundary faces of ${}^{\mathrm{sc}}\overline{T}^* X$. The term $-\lambda^2$ is of course only relevant at the $\rho = 0$ face; it does not contribute to the part of the symbol at fiber infinity, as it is a lower-order term there.)

As a consequence of Exercise 10.9, note as before that for $\lambda \in \mathbb{R}$, the Helmholz operator $\Delta - \lambda^2$ is not elliptic in the scattering sense: there are points in ${}^{\mathrm{sc}}T^*_{\partial X} X$ where $\xi^2 + |\eta|_h^2 = \lambda^2$.

We now turn to scattering wavefront set $\mathrm{WF}_{\mathrm{sc}}$, which can, as one might expect, be defined in the usual manner as a subset of
$$\partial({}^{\mathrm{sc}}\overline{T}^* X),$$
hence is a subset of boundary faces at fiber infinity and at spatial infinity (i.e., over ∂X). The scattering wavefront set is the obstruction to a distribution lying in $\dot{\mathcal{C}}^\infty(X)$, where $\dot{\mathcal{C}}^\infty(X)$ denotes the set of smooth functions on X decaying to infinite order at ∂X. This space is the analogue of the space of Schwartz functions in our compactified picture:

EXERCISE 10.10. Show that pullback under the radial compactification map sends $\dot{\mathcal{C}}^\infty(B^n)$ to $\mathcal{S}(\mathbb{R}^n)$.

By (10.7), it is not hard to see that
$$(\rho^2 D_\rho - \alpha)u = 0 \Longrightarrow \mathrm{WF}_{\mathrm{sc}}\, u \subset \{\rho = 0, \xi = \alpha\},$$
$$(\rho D_{\theta^j} - \beta)u = 0 \Longrightarrow \mathrm{WF}_{\mathrm{sc}}\, u \subset \{\rho = 0, \eta_j = \beta\}.$$
The following variant provides a useful family of examples (and can be proved with only a little more thought): if $a(\rho, \theta)$ and $\phi(\rho, \theta) \in \mathcal{C}^\infty(X)$, then[55]
$$\mathrm{WF}_{\mathrm{sc}}\left(a(\rho,\theta) e^{i\phi(\rho,\theta)/\rho}\right) = \{(\rho = 0, \theta, d(\phi(\rho,\theta)/\rho) : (0,\theta) \in \text{ess-supp}\, a\},$$
where $\text{ess-supp}\, a \subseteq \partial X$ denotes the "essential support" of a, i.e. the points near which a is not $O(\rho^\infty)$.

Of course, if
$$(\Delta - \lambda^2) u = f \in \dot{\mathcal{C}}^\infty(X), \tag{10.8}$$
then we have, by microlocal elliptic regularity,
$$\mathrm{WF}_{\mathrm{sc}}\, u \subset \{\rho = 0,\ \xi^2 + |\eta|_h^2 = \lambda^2\}.$$

[55] The distribution $ae^{i\phi}$ used here is a simple example of a *Legendrian distribution*. The class of Legendrian distributions on manifolds with boundary, introduced by Melrose-Zworski [**19**], stands in the same relationship to Lagrangian distributions as scattering wavefront set does to ordinary wavefront set.

In fact, there is a propagation of singularities theorem for scattering operators of real principal type that further constrains the scattering wavefront set of a solution to (10.8): it must be invariant under the (appropriately rescaled) Hamilton vector field of the symbol of $\Delta - \lambda^2$.

EXERCISE* 10.11. Let $\omega = d(\xi\, d\rho/\rho^2 + \eta \cdot d\theta/\rho)$ and let

$$p = \xi^2 + |\eta|_h^2 - \lambda^2;$$

show that up to an overall scaling factor, the Hamilton vector field of p with respect to the symplectic form ω is, on the face, $\rho = 0$ just

$$\mathsf{H}_p = 2\xi\eta \cdot \partial_\eta - 2|\eta|_{h_0}^2 \partial_\xi + \mathsf{H}_{h_0}$$

where $h_0 = h|_{\rho=0}$, and H_{h_0} is the Hamilton vector field of h_0, i.e. (twice) geodesic flow on ∂X.

Show that maximally extended bicharacteristics of H_p project to the θ variables to be geodesics of length π. (Hint: reparametrize the flow.)

(For a careful treatment of the material in this exercise and indeed in this section, see [**18**].)

Appendix

We give an extremely sketchy account of some background material on Fourier transforms, distribution theory, and Sobolev spaces. For further details, see, for instance, [**25**] or [**11**].

Let $\mathcal{S}(\mathbb{R}^n)$, the *Schwartz space,* denote the space

$$\{\phi \in \mathcal{C}^\infty(\mathbb{R}^n) : \sup \left|x^\alpha \partial_x^\beta \phi\right| < \infty\ \forall \alpha, \beta\},$$

topologized by the seminorms given by the suprema. The dual space to $\mathcal{S}(\mathbb{R}^n)$, denoted $\mathcal{S}'(\mathbb{R}^n)$, is the space of tempered distributions.

For $\phi \in \mathcal{S}(\mathbb{R}^n)$, let

$$\mathcal{F}\phi(\xi) = (2\pi)^{-n/2} \int \phi(x) e^{-i\xi\cdot x}\, dx.$$

Then $\mathcal{F}\phi \in \mathcal{S}(\mathbb{R}^n)$, too; indeed, $\mathcal{F} : \mathcal{S}(\mathbb{R}^n) \to \mathcal{S}(\mathbb{R}^n)$ is an isomorphism, and its inverse is closely related:

$$\mathcal{F}^{-1}\psi(x) = (2\pi)^{-n/2} \int \psi(\xi) e^{+i\xi\cdot x}\, dx.$$

We can, by duality, then define $\mathcal{F}$ on tempered distributions.

Let $\mathcal{E}'(\mathbb{R}^n)$ denote the space of compactly supported distributions on $\mathbb{R}^n$. When X is a compact manifold without boundary, we let $\mathcal{D}'(X)$ denote the dual space of $\mathcal{C}^\infty(X)$.

We define the (L^2-based) Sobolev spaces by

$$H^s(\mathbb{R}^n) = \{u \in \mathcal{S}'(\mathbb{R}^n) : \langle\xi\rangle^s \mathcal{F}u(\xi) \in L^2(\mathbb{R}^n)\},$$

where $\langle\xi\rangle = (1 + |\xi|^2)^{1/2}$. If s is a positive integer, this definition coincides exactly with the space of L^2 functions having s distributional derivatives also lying in L^2. We note that the operation of multiplication by a Schwartz function is a bounded map on each H^s; this is most easily proved by interpolation arguments similar to (but easier than) those alluded to in Exercise 2.4—cf. [**25**].

Throughout these notes we will take for granted the *Schwartz kernel theorem*, not so much as a result to be quoted but as a world-view. Recall that this result says *any* continuous linear operator

$$\mathcal{S}(\mathbb{R}^n) \to \mathcal{S}'(\mathbb{R}^n)$$

is of the form

$$u \mapsto \int k(x,y)u(y)\,dy$$

for a unique $k \in \mathcal{S}'(\mathbb{R}^n \times \mathbb{R}^n)$; a corresponding result also holds on all the manifolds that we will consider. We thus consistently take the liberty of confusing operators with their Schwartz kernels, although we let $\kappa(A)$ denote the Schwartz kernel of the operator A when we wish to emphasize the difference.

Some results relating Schwartz kernels to traces are important for our discussion of the wave trace. Recall that an operator T on a separable Hilbert space is called *Hilbert-Schmidt* if

$$\sum_j \|Te_j\|^2 < \infty$$

where $\{e_j\}$ is any orthonormal basis. In the special case when our Hilbert space is $L^2(X)$ with X a manifold, the condition to be Hilbert-Schmidt turns out to be easy to verify in terms of the Schwartz kernel: T is Hilbert-Schmidt if and only if $\kappa(T)$, its Schwartz kernel,[56] lies in $L^2(X \times X)$.

A *trace-class* operator is one such that

$$\sum_{i,j} |\langle Te_i, f_j\rangle| < \infty$$

for every pair of orthormal bases $\{e_i\}, \{f_j\}$. It turns out to be the case that an operator T is trace-class if and only if it can be written

$$T = PQ$$

with P, Q Hilbert-Schmidt. The *trace* of a trace-class operator is given by

$$\sum_i \langle Te_i, e_i\rangle$$

over an orthonormal basis: this turns out to be well-defined. We refer the reader to **[20]** for further discussion of trace-class and Hilbert-Schmidt operators.

References

[1] Craig, W., Kappeler, T., Strauss, W. *Microlocal dispersive smoothing for the Schrödinger equation,* Comm. Pure Appl. Math. 48 (1995), no. 8, 769–860.

[2] Dimassi, Mouez; Sjöstrand, Johannes, *Spectral asymptotics in the semi-classical limit,* London Mathematical Society Lecture Note Series, 268. Cambridge University Press, Cambridge, 1999.

[3] Duistermaat, J. J.; Guillemin, V. W. *The spectrum of positive elliptic operators and periodic bicharacteristics,* Invent. Math. 29 (1975), no. 1, 39–79.

[4] Doi, S.-I. *Smoothing effects of Schrödinger evolution groups on Riemannian manifolds,* Duke Math. J. 82 (1996), no. 3, 679–706.

[5] Duistermaat, J. J.; Hörmander, L. *Fourier integral operators. II,* Acta Math. 128 (1972), no. 3-4, 183–269.

[56] It is probably best to think of X as a *Riemannian* manifold here, so that the Schwartz kernel is a function, which we can integrate against test functions via the metric density, and likewise integrate the kernel.

[6] Friedlander, F. G., *Introduction to the theory of distributions* Second edition. With additional material by M. Joshi. Cambridge University Press, Cambridge, 1998.

[7] Grigis, A. and Sjöstrand, J., *Microlocal analysis for differential operators. An introduction.* London Mathematical Society Lecture Note Series, 196. Cambridge University Press, Cambridge, 1994.

[8] *A Strichartz inequality for the Schrödinger equation on non-trapping asymptotically conic manifolds* (with Andrew Hassell and Terence Tao), Comm. PDE., **30** (2005), 157–205.

[9] Hörmander, L., *The spectral function of an elliptic operator,* Acta Math. 121 (1968), 193–218.

[10] L. Hörmander, *Fourier Integral Operators I,* Acta Math. 127 (1971), 79–183.

[11] Hörmander, L. *The analysis of linear partial differential operators. I. Distribution theory and Fourier analysis.* Second edition. Grundlehren der Mathematischen Wissenschaften, 256. Springer-Verlag, Berlin, 1990.

[12] Hörmander, L. *The analysis of linear partial differential operators. II. Differential operators with constant coefficients.* Grundlehren der Mathematischen Wissenschaften, 257. Springer-Verlag, Berlin, 1983.

[13] Hörmander, L. *The analysis of linear partial differential operators. III. Pseudodifferential operators.* Grundlehren der Mathematischen Wissenschaften, 274. Springer-Verlag, Berlin, 1985.

[14] Hörmander, L. *The analysis of linear partial differential operators. IV. Fourier integral operators.* Grundlehren der Mathematischen Wissenschaften, 275. Springer-Verlag, Berlin, 1985.

[15] Kac, M. *Can one hear the shape of a drum?* Amer. Math. Monthly 73 1966 no. 4, part II, 1–23.

[16] Martinez, André, *An introduction to semiclassical and microlocal analysis,* Universitext. Springer-Verlag, New York, 2002.

[17] R. Melrose *Lecture notes on microlocal analysis,* available at `www-math.mit.edu/~rbm/Lecture_Notes.html`

[18] R. B. Melrose, *Spectral and scattering theory for the Laplacian on asymptotically Euclidian spaces*, Spectral and scattering theory (Sanda, 1992), Dekker, New York, 1994, pp. 85–130.

[19] R. B. Melrose and M. Zworski, *Scattering metrics and geodesic flow at infinity,* Invent. Math. **124** (1996), no. 1-3, 389–436.

[20] Reed, Michael and Simon, Barry, *Methods of modern mathematical physics I: Functional analysis* Second edition, Academic Press, Inc., New York, 1980.

[21] Seeley, R. T., *Complex powers of an elliptic operator,* 1967 Singular Integrals (Proc. Sympos. Pure Math., Chicago, Ill., 1966) 288–307 Amer. Math. Soc., Providence, R.I.

[22] Shubin, M. A., *Pseudodifferential operators and spectral theory,* Second edition. Springer-Verlag, Berlin, 2001.

[23] Stein, E. M. *Singular integrals and differentiability properties of functions* Princeton Mathematical Series, No. 30 Princeton University Press, Princeton, N.J. 1970.

[24] Taylor, M. E. *Pseudodifferential operators,* Princeton Mathematical Series, 34. Princeton University Press, Princeton, N.J., 1981.

[25] Taylor, M. E., *Partial differential equations. I. Basic theory* Applied Mathematical Sciences, 115. Springer-Verlag, New York, 1996.

[26] Taylor, M. E. *Partial differential equations. II. Qualitative studies of linear equations* Applied Mathematical Sciences, 116. Springer-Verlag, New York, 1996.

[27] A. Vasy, *The wave equation on asymptotically Anti-de Sitter spaces,* Anal. PDE, to appear.

[28] M. Zworski, *Semiclassical analysis*, AMS Graduate Studies in Mathematics, American Mathematical Society, Providence, 2012.

Department of Mathematics, Northwestern University, Evanston Illinois 60208
E-mail address: `jwunsch@math.northwestern.edu`

Clay Mathematics Proceedings
Volume **17**, 2013

Some Global Aspects of Linear Wave Equations

Dean Baskin and Rafe Mazzeo

ABSTRACT. This paper surveys a few aspects of the global theory of wave equations. This material is structured around the contents of a minicourse given by the second author during the CMI/ETH Summer School on evolution equations during the Summer of 2008.

1. Introduction

The week-long minicourse on which this brief survey paper is based came after a vigorous, detailed and outstanding series of lectures by Jared Wunsch on the applications of microlocal analysis to the study of linear wave equations. Both lecture series took place at the Clay Mathematics Institute Summer School at ETH Zürich in 2008. The goal of this minicourse was to describe a few topics which involve global aspects of wave theory, relying at least to some extent on the microlocal underpinnings from Wunsch's lectures. The first of these topics is an account of some striking consequences that can be derived from the finite propagation speed property. While this had been applied in various interesting ways before, the systematic development of this principle appears in the very influential paper of Cheeger, Gromov and Taylor [**CGT82**]. We recall how this property, applied to solutions of the wave equation associated to a Laplace-type operator, can be used to obtain estimates for solutions of various related operators. We present only one application of this, which is a lovely argument due to Gilles Carron which estimates the off-diagonal decay profile of the Green function for generalized Laplace-type operators on globally symmetric spaces of noncompact type. This result had caught the lecturer's eye in the months before this Clay meeting and nicely illustrates the unexpected power of the finite propagation speed method. Following this, the remainder of the lectures reviewed several different approaches to scattering theory and described a few of the relationships between these. The primary goal, however, was to introduce the Friedlander radiation fields and explain how they give a concrete realization of the Lax–Phillips translation representation. We follow suit here, recalling the outlines of a few of the numerous successful approaches to scattering theory and culminating in a discussion of these radiation fields.

This paper attempts to give some feel for what was presented in these lectures. The reader should be warned that the topics covered here are in many places old-fashioned and we omit any mention of many of the most important recent advances and trends in scattering theory. The material here is meant to indicate a few things that can be accomplished, often with not very sophisticated machinery by modern standards. We typically make very restrictive assumptions in order to convey the main essence of the ideas. We give references for further reading interspersed inter alia, but do not make any claim to a comprehensive bibliography.

2010 *Mathematics Subject Classification.* Primary 35L05, 35P25, 58J50, 53C35.

The material assembled here is based on the notes of the first-named author; the lecturer (and second author) is extremely grateful to him not only for this careful recording of the lectures, but also for his enthusiasm during the lectures and his very substantial assistance in writing this paper. We did discuss at some point, but later abandoned, the possibility of writing a much more exhaustive treatment of some of the topics here, particularly the theory of radiation fields. That will unfortunately have to wait for another day and other authors. We hope that this survey accomplishes what the original lectures also attempted, which is to whet the reader's curiosity to learn more about this subject. Needless to say, wave theory is an immense subject and we mention here only a very small set of possible topics.

Throughout this paper we focus on properties of solutions, and of the solution operator, for the wave operator

$$\Box_V = D_t^2 - L, \quad \text{where} \quad L = \nabla^*\nabla + V \tag{1.1}$$

acting on sections of some bundle E over a Riemannian manifold (M, g), where ∇ is the covariant derivative of some connection on E and V is a (self-adjoint) potential of order 0, which can either be scalar or an endomorphism of E. For simplicity we typically assume that V is smooth and compactly supported, although neither of these properties is present in almost any of the interesting physical or geometric applications. Furthermore, we often discuss only the scalar Laplacian and its perturbations, although the extension of all results below to this slightly more general framework is usually just notational. Finally, here and below we write $D = \frac{1}{i}\partial$.

As noted above, we take advantage of the luxury of being able to refer back to the excellent lecture notes by Jared Wunsch [**Wun08**] covering his longer mini-course. Those notes provide a nice introduction for many central themes and results in the subject, including the existence of solutions of the equation $\Box u = f$ with vanishing Cauchy data, or of $\Box u = 0$ with prescribed nonzero Cauchy data, along a noncharacteristic hypersurface, the positive commutator method leading to Hörmander's renowned theorem on propagation of singularities of solutions, the finite propagation speed property, and much else besides. Using this as a blanket resource, we can dive right into the material at hand.

There are now many terrific monographs concerning the local and global aspects of wave equations. Michael Taylor's three-volume series [**Tay11**] belongs high on this list; it contains an amazing amount of information about many different topics. Other recent monographs with a particular focus on hyperbolic equations include those by Alinhac [**Ali09**], Lax [**Lax06**], Rauch [**Rau12**]; we mention also the new book by Zworski on semiclassical analysis [**Zwo12**].

The first part of this survey, in § 2, focuses on the finite propagation speed property for solutions of the wave equation. After sketching a proof of this property in § 2.1, we state some key facts about the Cheeger–Gromov–Taylor theory in § 2.2, which leads to the discussion in § 2.3 of Carron's application of these ideas to estimate certain geometric operators on globally symmetric spaces of noncompact type. The second part, § 3, presents a few different perspectives in scattering theory. We begin in § 3.1 with some topics in stationary scattering theory, then move on in § 3.2 to several formulations of time-dependent scattering theory: progressing wave solutions, Møller wave operators, Lax–Phillips theory, and the theory of Friedlander radiation fields.

The authors are very grateful to the Clay Foundation for making this Summer School possible – it was a lot of fun and the large attendance and enthusiasm of the participants was amazing. We also appreciate the forbearance by the editors of this volume for their (relative) tolerance for the length of time between the original lectures and when this paper was finally written. Both authors are very grateful to many people for teaching us about many of the topics here. We thank, in particular, Gilles Carron, Richard Melrose, Gunther Uhlmann, Andras Vasy and Jared Wunsch. Gilles Carron and Andras Vasy also gave some helpful remarks on this paper.

D.B. is supported by NSF Postdoctoral Fellowship DMS-1103436; R.M. is supported by the NSF grant DMS-1105050.

2. Finite propagation speed and its consequences

Although [**Wun08**] contains a proof of the basic finite propagation speed property for the operator $\Box_V$, we begin by recalling this familiar argument very briefly. We then show how using the functional calculus one can write the Schwartz kernels of various functions of the elliptic operator L in terms of the Schwartz kernel of the wave operator. This leads directly to the important Cheeger–Gromov–Taylor theory which uses finite propagation speed to obtain interesting estimates for these Schwartz kernels. We illustrate this with an outline of Carron's estimates for the resolvent and heat kernel of generalized Laplacians on symmetric spaces of noncompact type.

2.1. Finite propagation speed. The fundamental identity behind finite propagation speed is the observation that for any sufficiently regular function u,

$$\operatorname{div}_x(u_t \nabla u) = u_t \Box_0 u + \frac{1}{2}\partial_t(u_t^2 + |\nabla u|^2). \tag{2.1}$$

We suppose that the space on which we are doing calculations has a global time function t and moreover, splits as $\mathbb{R} \times M$, with a static Lorentzian metric $-dt^2 + h$, where (M, h) is a Riemannian manifold. A hypersurface $Y \subset \mathbb{R} \times M$ is called spacelike if its unit normal ν (with respect to this Lorentzian metric) satisfies $\nu \cdot \nu < 0$. Suppose that $\Omega \subset \mathbb{R} \times M$ is a domain bounded by two spacelike hypersurfaces, $\partial\Omega = Y_1 \cup Y_2$, which meet transversely along a codimension two submanifold, and that u is a solution of the homogeneous wave equation, $\Box_0 u = 0$. Integrate (2.1) over Ω. The left side is transformed using the divergence theorem; the first term on the right vanishes while the second term is also transformed to a boundary integral. If $\nu_j = (\nu_{t,j}, \nu_{x,j})$ is the upward-pointing unit normal to Y_j, decomposed into its vertical (t) and horizontal (x) components, then we obtain

$$\int_{Y_1} (|u_t|^2 + |\nabla u|^2)|\nu_{t,1}| - 2u_t \cdot \partial_{\nu_1} u|\nu_{x,1}|$$
$$= \int_{Y_2} (|u_t|^2 + |\nabla u|^2)|\nu_{t,2}| - 2u_t \cdot \partial_{\nu_2} u|\nu_{x,2}|.$$

Since ν_j is timelike, the integrand on each side is bounded from below by $c(|u_t|^2 + |\nabla u|^2)$ for some $c > 0$ which depends on Y_j. We conclude that if $u_t = \nabla u = 0$ on Y_1, then these same quantities must also vanish on Y_2. Finally, if Ω is foliated by spacelike hypersurfaces, then the vanishing of $(u_t, \nabla u)$ on the bottom (spacelike) boundary of Ω can be propagated throughout this entire region, and hence if u vanishes at Y_1, then $u \equiv 0$ in Ω.

If we consider wave operators with terms of order 0 or 1, then this calculation can be adapted to show that if Ω is foliated by spacelike hypersurfaces Z_s, $0 \le s \le 1$, then the integral over Z_s of $|u_t|^2 + |\nabla u|^2$ satisfies a differential inequality, and the fact that it vanishes when $s = 0$ implies that it vanishes for all $s \le 1$.

To interpret this calculation, we observe that there many natural domains Ω which can be foliated by spacelike hypersurfaces in this way. Indeed, suppose that $p = (t_1, x_1)$ is any point, and $\mathcal{D}^-_{t_1,x_1}$ denotes the (backward) domain of dependence of this point, i.e. the set of points in $\mathbb{R} \times M$ which can be reached by timelike paths traveling backward in t and emanating from (t_1, x_1). Let Y be one of the level sets $\{t = t_0\}$ where $t_0 < t_1$. Then the region $\Omega = \{(t,x) \in \mathcal{D}^-_{t_1,x_1} : t \ge t_0\}$ can be shown to have a spacelike foliation by submanifolds Y_s which all intersect along the submanifold $\{(t,x) \in \mathcal{D}^-_{t_1,x_1} : t = t_0\}$. Thus any homogeneous solution of $\Box_V u = 0$ which vanishes along with its normal derivative along $\{t = t_0\}$ vanishes throughout this Ω. This implies that if the Cauchy data of u at t_0 is supported in some subset K, then the Cauchy data of u at $t_1 = t_0 + \tau$, where $\tau > 0$, is supported in the subset $K_\tau = \{(t_1, x) : \mathrm{dist}_g(x, K) \le \tau\}$, which is precisely what is meant by saying that the support of a solution propagates with speed 1. For more general variable coefficient hyperbolic equations, the speed of propagation may be variable but is still finite.

2.2. Cheeger–Gromov–Taylor theory. Consider the fundamental solution for the problem

$$\Box_V u = 0, \quad u|_{t=0} = \phi, \ \partial_t u|_{t=0} = 0.$$

It is customary to write this solution operator as $\cos(t\sqrt{L})$, so that the solution $u(t,x)$ is equal to $\cos(t\sqrt{L})\phi$. We assume for simplicity that L has no negative eigenvalues so that $\left\|\cos(t\sqrt{L})\right\| \le 1$. This is an instance of the functional calculus for self-adjoint operators, which are defined in purely abstract terms using the spectral theorem and can be used to describe solution operators for various equations involving L. There are many interesting examples, including prominently the resolvent and heat operator

$$R_L(\lambda) := (L - \lambda^2)^{-1} \qquad \text{and} \qquad e^{-tL}.$$

The abstract definitions of these operators (i.e. defined using the spectral theorem) are all well and good, but in order to use them one usually wishes to know much more about their mapping properties. For example, a priori, using only these abstract definitions, we only know how one of these functions of L acts on L^2 functions, but not on other function spaces. The goal then is to obtain a more concrete understanding of the Schwartz kernels of any one of these operators. Of course, there is a lot of theory devoted to doing just this. Thus the classical theory of pseudodifferential operators gives a nice picture of the resolvent for λ varying in a compact region in $\mathbb{C}$ disjoint from the spectrum, while the theory of semiclassical pseudodifferential operators provides a means to understand this family of operators as λ tends to infinity in various directions in the complex plane. Similarly, the well-known heat-kernel parametrix construction, cf. [**BGV92**], gives a way to understand the asymptotic behavior of the Schwartz kernel of the solution operator for the heat equation in various regimes of the space $\mathbb{R}^+ \times M \times M$. These theories and constructions give very precise information, but are often very intricate, and furthermore, it is often hard to use these ideas directly to say anything interesting about global behavior of these Schwartz kernels. The idea in [**CGT82**] is that one can extract, often in a rather simple way, some very useful global behavior of these kernels using mainly the finite propagation speed property of $\cos(t\sqrt{L})$ and some other simple properties, such as the fact that the norm of $\cos(t\sqrt{L})$ as a bounded operator on L^2 never exceeds 1.

To explain this, suppose that $f(s)$ is a smooth, even function on $\mathbb{R}$ which decays sufficiently rapidly so that the following manipulations are justified. Assuming $L \geq 0$ for simplicity, we define $f(\sqrt{L})$ using the spectral theorem, but at the same time we can spectrally synthesize this function of L directly from the wave kernel:

$$f(\sqrt{L}) = \frac{1}{2\pi} \int_{-\infty}^{\infty} \hat{f}(s) \cos(s\sqrt{L})\, ds.$$

The simple but crucial observation is that this is not just an identity about abstract self-adjoint operators, but also calculates the Schwartz kernel of $f(\sqrt{L})$ in terms of the Schwartz kernel of the wave operator.

The following discussion is drawn from the paper [**CGT82**]. Suppose that f has the property that its Fourier transform $\hat{f}(s)$ is integrable, along with a certain number of its derivatives, on $\mathbb{R} \setminus (-\epsilon, \epsilon)$ for any $\epsilon > 0$. The first key result is that under such a hypothesis, if $u \in L^2$ has support in a ball $B_r(y)$, then for $R > r$,

$$||f(\sqrt{L})u||_{L^2(M \setminus B_R(y))} \leq \pi^{-1} ||u||_{L^2} \int_{R-r}^{\infty} |\hat{f}(s)|\, ds.$$

The proof is very simple. We know that $\cos(s\sqrt{L})u$ has support in $B_{r+|s|}(y)$, so that

$$||f(\sqrt{L})u|| \leq \frac{1}{\pi} \left|\left| \int_{R-r}^{\infty} \hat{f}(s) \cos(s\sqrt{L})u\, ds \right|\right| \leq \frac{1}{\pi} ||u|| \int_{R-r}^{\infty} |\hat{f}(s)|\, ds.$$

A very similar argument gives bounds for $||L^p f(\sqrt{L}) L^q u||$ depending on the integral of some higher derivatives of $\hat{f}(s)$ over the same half-line. The particularly useful aspect of this is that the integrals of $|\partial_s^\ell \hat{f}|$ which appear on the right in these estimates start at $R - r$ rather than at 0, and hence if these functions decay at some rate, then the right sides of these inequalities exhibit the corresponding decay. Assuming we are on a space with appropriate local uniformity of the metric (or coefficients of L), then we can deduce from this some off-diagonal pointwise estimates for the Schwartz kernel $f(\sqrt{L})(z, w)$. By off-diagonal we mean that the estimates are valid in any region where $\operatorname{dist}(z, w) \gg 0$. One reason for assuming this local uniformity for L is that these arguments require bounds on the injectivity radius and volumes of geodesic balls, for example, in order to pass from L^2 to pointwise estimates.

2.3. Carron's theorem. This subsection provides a concrete example of how this all works. We describe some of the main features in the paper [**Car10**] of Carron which uses the ideas above to derive fairly accurate pointwise bounds on the off-diagonal decay of the resolvent kernel and heat-kernel for Laplace-type operators on symmetric spaces of noncompact type.

In order to describe this we must first explain at least a small amount about the geometry of these spaces. This is recounted elsewhere in much greater detail; the classic reference is [**Hel84**], but we refer (self-servingly) to [**MV05**] for an analyst's point of view of this geometry.

Symmetric spaces are distinguished amongst general Riemannian manifolds by the richness of their isometry groups. Their defining property is that the geodesic reflection around any point $(exp_p(v) \mapsto \exp_p(-v))$ extends to a global isometry; Cartan's classic characterization is that any such space is necessarily of the form G/K, where G is a semisimple Lie group and $K \subset G$ is a maximal compact subgroup, endowed with an invariant metric. Because of this, almost all of the basic structure theory can be reduced to algebra and hence described quite explicitly. We shall focus on one particular realization stemming from the polar decomposition $G = KAK$, where K is as above and A is a maximal connected abelian subgroup. For a symmetric space X of noncompact type, this subgroup A is isomorphic (and isometric) to a copy of $\mathbb{R}^k$ for some k, where the positive integer k is called the rank of X. Using this polar decomposition, we identify $G/K \cong KA$. The map $\Phi : K \times A \to X$, $\Phi(k,a) = ka$, is surjective, but far from injective.

It is best to think of the simplest special case, the real hyperbolic space $\mathbb{H}^n$; here $A \cong \mathbb{R}$ and $K = \mathrm{SO}(n)$. The image of the origin $0 \in A$ via Φ is a single point $o \in X$, and this point is fixed by the entire (left) action of K. The space X is the union of geodesic lines through o which all intersect pairwise only at this point. The group K acts transitively on this space of geodesic lines through o with stabilizer $\mathrm{SO}(n-1)$. Note that there are elements of K which take a geodesic to itself but reverses its orientation; this means that we get a less redundant 'parametrization' by restricting Φ to $K \times \mathbb{R}^+$. Geometrically, we have the familiar picture of $\mathbb{H}^n$ as $\mathbb{R}^+ \times S^{n-1}$ with the warped product metric $dr^2 + \sinh^2 r \, d\theta^2$.

For a general symmetric space X of rank $k > 1$, this picture generalizes as follows. The space X is the union of the various images of A by elements $k \in K$, and all of these images intersect at o, though $kA \cap k'A$ often consists of a larger subspace. These translates of A by $k \in K$ should be thought of as the radial directions in X. Another important piece of structure is the existence of a finite set of linear functionals $\Lambda = \{\alpha_j\}$ on A called the roots. These divide into positive and negative roots, $\Lambda = \Lambda^+ \cup \Lambda^-$, and the positive roots determine a (closed) sector $V \subset A$ by $V = \{\alpha_j \geq 0 \; \forall \; \alpha_j \in \Lambda^+\}$. This sector V is the analogue of the half-line in $\mathbb{H}^n$, and the restriction of Φ to $K \times V$ is still surjective, and if $K' \subset K$ denotes the isotropy group at a generic point, then we can regard X as being the product $V \times (K/K')$ with certain submanifolds of K/K' collapsed along various boundary faces of V. In terms of this data, we can finally write down the multiply-warped product metric

$$g = da^2 + \sum_j \sinh^2 \alpha_j \, dn_j^2,$$

where the sum is over positive roots, da^2 is the Euclidean metric on A and dn_j^2 is a metric on a certain subbundle of the tangent bundle of K/K' corresponding to the root α_j.

For simplicity here we just discuss the scalar Laplacian Δ on X and, following the theme of this section, consider the problem of estimating the Schwartz kernels of $f(\sqrt{-\Delta})$ for suitable functions f. Because Δ commutes with all isometries on X, the Schwartz kernel $K_f(x,x')$ of this operator depends effectively on a smaller number of variables. Given any pair of distinct points $x, x' \in X$, choose an isometry φ of X so that $\varphi(x') = o$ and $\varphi(x)$ lies in some particular copy of A. If we ask that $\varphi(x) = a \in V \subset A$, then φ is almost uniquely determined. We thus have that $K_f(x, x') = K_f(\varphi(x), o) = K_f(a, 0)$. In other words, K_f is really only a function of the k Euclidean variables $a = (a_1, \ldots, a_k)$. In particular, when the rank of X is 1, then K_f reduces to a function of one variable $r \geq 0$.

This reduction points to the difficulty of studying functions of the Laplacian on symmetric spaces of rank greater than 1. Indeed, while the resolvent kernel $R(\lambda; x, x')$ on a space of rank 1 depends only on $\mathrm{dist}(x, x')$ and hence can be analyzed completely by ODE methods, the same is not true when the rank of X is larger. Similarly, even in the rank 1 setting, the heat kernel $H(t, x, x')$ depends on two variables, t and $\mathrm{dist}(x, x')$, but unlike the Euclidean case, there is no extra homogeneity which reduces this further to a function of one variable. Thus the problem which Carron's theorem answers is how to give good estimates on these reduced functions, $R(\lambda, a)$ and $H(t, a)$, where $a \in A$ is the 'relative position' between $x, x' \in X$.

THEOREM 2.2 (Carron [**Car10**]). *Let X be a symmetric space of noncompact type and rank k and consider the Schwartz kernels $R(\lambda, a)$ and $H(t, a)$ of the resolvent $(-\Delta - \lambda_0 - \lambda^2)^{-1}$ and heat operator $e^{t\Delta}$, written in reduced form as above. The number λ_0 here is the bottom of the spectrum of $-\Delta$; it may be calculated explicitly. Then*

$$|R(\lambda, a)| \leq Ce^{-\rho(a) - \mathrm{Re}(\lambda)\ \mathrm{dist}(a, o)}$$

and

$$|H(t, a)| \leq Ce^{-\lambda_0 t - \rho(a) - \mathrm{dist}^2(a, 0)/4t}\, \Phi_t(a).$$

The function $\Phi_t(a)$ is a somewhat messy but quite explicit and understandable function which is a rational function of a and certain powers of t. The linear functional ρ on A is half the sum of the 'restricted' positive roots; this is a standard object which appears frequently.

It is known that the upper bounds given here are sharp in the sense that there are lower bounds that differ just by the constant multiple for these same kernels. We refer also to the papers [**AJ99**] and [**LM10**] sharper bounds obtained by different and more complicated methods.

The proofs of these estimates are clever but not very long, and in the remainder of this section we give a few of the ideas which go into them.

The first step is that if $a \in A$ is arbitrary and $\epsilon \in (0, 1)$, then we can estimate from above and below the volume of the set $KB(a, \epsilon)$, where $B(a, \epsilon)$ is a ball of radius ϵ in A centered around a. This can be done because we have very good information on the Jacobian determinant for the coordinate change implicit in some natural coordinatizations induced by $K \times A \to X$.

Let us first study the resolvent. Fix $a \in A$ such that $\mathrm{dist}(a, o) \geq 2$. We shall obtain a pointwise estimate for $|R(\lambda, a)|$ in $B(o, 1)$ starting from L^2 estimates in this same ball of functions of the form $u = R(\lambda)\sigma$, where σ varies over all L^2 functions in the "annular shell" $D := KB(a, 1)$ which vanish outside D. Thus,

$$u = R(\lambda, \cdot)\sigma = \int_0^\infty \frac{e^{-\lambda\xi}}{\lambda} \cos(\xi\sqrt{-\Delta - \lambda_0})\sigma\, d\xi.$$

Using that $||\cos(\xi\sqrt{-\Delta - \lambda_0})||_{L^2\to L^2} = 1$ as well as finite propagation speed, because of the support properties of σ, we obtain

$$||u||_{L^2(B(o,1))} \leq \int_{\mathrm{dist}(a,o)-2}^{\infty} \frac{e^{-\mathrm{Re}(\lambda)\xi}}{|\lambda|} ||\sigma||_{L^2} \leq \frac{1}{|\lambda|^2} e^{-\mathrm{Re}(\lambda)(\mathrm{dist}(a,o)-2)} ||\sigma||_{L^2}.$$

From here, using local elliptic estimates, we obtain that

$$|u(o)| \leq Ce^{-\mathrm{Re}(\lambda)\,\mathrm{dist}(a,o)} ||\sigma||_{L^2(D)}.$$

In other words, this estimates the norm of the mapping T defined by $L^2(D) \ni \sigma \mapsto R(\lambda)\sigma|_o$, whence (using the $L^\infty \to L^\infty$ norm of TT^*),

$$\int_D |R(\lambda, x, o)|^2 \, dx \leq Ce^{-2\mathrm{Re}(\lambda)\,\mathrm{dist}(a,o)}. \tag{2.3}$$

We next wish to find a similar estimate where the integral on the left is only over some ball $B(ka, 1/4) \subset D$ rather than the entire annular region D. More specifically we assert that

$$\mathrm{Vol}(B(ka, 1/4)) \int_{B(ka,1/4)} |R(\lambda, x, o)|^2 \, dx \leq Ce^{-2\mathrm{Re}(\lambda)\,\mathrm{dist}(a,o)}.$$

This must hold, since if it were to fail for every $B(ka, 1/4)$, then the sum over all such balls would lead to a violation of (2.3).

Finally, noting that the volume of this ball is approximately $e^{2\rho(a)}$, and applying the same local elliptic estimates as before to estimate the value at a point in terms of a local L^2 norm, we conclude that

$$|R(\lambda, a, o)| \leq Ce^{-\mathrm{Re}(\lambda)\,\mathrm{dist}(a,o) - \rho(a)}.$$

This is the desired off-diagonal decay estimate.

The corresponding argument to estimate the off-diagonal behavior of the heat kernel proceeds in a very similar way, substituting local parabolic estimates for local elliptic estimates. We refer to [**Car10**] for details.

It is worth remarking that there are other very effective ways to establish so-called Gaussian bounds for heat kernels under rather general circumstances. We mention in particular the beautiful theory developed by Grigor'yan and Saloff-Coste, see [**SC02**], [**Gri09**]. These techniques work in far more general circumstances, and depend on quite different underlying principles. However, one point of interest in Carron's work is that he is able to obtain the correct 'subexponential' factor $\Phi_t(a)$ in the estimate of $|H(t, a)|$, which might be impossible using those more general approaches.

3. Scattering theory

For the second and longer part of this survey, we turn to an entirely different aspect of the global theory of wave equations and discuss some approaches to mathematical scattering theory. This classical subject has deep physical origins, and has received numerous mathematical formulations. While these approaches are mostly equivalent, the correspondences between them are not always obvious. In the following pages we first review one point of view on stationary scattering theory, then turn to some perspectives on the corresponding time-dependent theory. This is all done with a distinctly PDE (rather than, say, operator-theoretic) focus. We conclude with a discussion of a more abstract functional analytic setup of scattering theory due to Lax and Phillips centered around the notion of a translation representation and explain how the theory of radiation fields developed by Friedlander provides a concrete realization of the translation representation.

There are numerous settings in which to introduce any of these topics, including scattering by potentials, which is the study of Schrödinger operators $-\Delta + V$ on $\mathbb{R}^n$, or scattering by obstacles, which studies these same operators but on exterior domains $\mathbb{R}^n \setminus \mathcal{O}$ with some elliptic boundary condition at $\partial\mathcal{O}$. There are also significant differences between the cases n odd and n even in each of these theories. Finally, it is also natural to consider these same problems on manifolds which are asymptotically Euclidean or asymptotically conic at infinity (or indeed, have some other type of asymptotically regular geometry, e.g. asymptotically hyperbolic). Each setting requires different sets of techniques, and in order to make this exposition as simple as possible, we focus on the combination of hypotheses where everything works out most simply. Namely, we study the scattering theory associated to $L = -\Delta + V$ on an *odd*-dimensional Euclidean space $\mathbb{R}^n$, with the strong assumption that $V \in \mathcal{C}_0^\infty$. We describe the structure theory for solutions of the Helmholtz equation $(L - \lambda^2)u = 0$, and for $\Box_V := \Box + V = D_t^2 - L$, the time-dependent wave equation, and give some indication how objects in these respective settings correspond to one another.

There are very many excellent references to each part of what we discuss (and much that is closely related that we do not discuss), so we relegate almost all of the technicalities to those sources. We mention in particular [**RS78**, Vol. IV], [**Tay11**, Ch. 9], [**Per83**], [**Yaf10**] and [**Mel95**]. The material on radiation fields is spread over several papers, starting from the original work by F.G. Friedlander [**Fri80**]. There is a forthcoming and detailed survey of this subject by Melrose and Wang [**MW**], to which the discussion here is intended to be an introduction.

3.1. Stationary scattering theory. The stationary formulation of scattering theory concerns the elliptic operator $L-\lambda^2$, where here and below, $L = -\Delta+V$, with $V \in \mathcal{C}_0^\infty$ (and real-valued!). It is obvious that L is bounded below, i.e.

$$\int_{\mathbb{R}^n} (Lu)\overline{u}\, dV \geq -C \int_{\mathbb{R}^n} |u|^2\, dV$$

for all $u \in \mathcal{C}_0^\infty(\mathbb{R}^n)$, and with little more work one can also prove that it has a unique self-adjoint extension as an unbounded operator on $L^2(\mathbb{R}^n)$. Indeed, this is yet another consequence of the finite speed of propagation, see [**Che73**]. Its spectrum is contained in a half-line $[-C, \infty)$; the positive ray $[0, \infty)$ comprises the entire continuous spectrum, and there are a finite number of L^2 eigenvalues in the $[-C, 0)$. If we allow V to be less regular, simple examples show that this negative interval may contain an infinite sequence of such eigenvalues converging to 0; the basic example of this is when $V(x) = -1/|x|$, which is the potential for the Schrödinger operator modeling the hydrogen atom.

Assume initially that λ lies in the lower half-plane $\Im\lambda < 0$. Provided that $\lambda \neq -i|\lambda_j|$ corresponding to any of the negative eigenvalues $-\lambda_j^2 < 0$, the operator $L - \lambda^2$ has an L^2 bounded inverse,

$$R_V(\lambda) = (L - \lambda^2)^{-1}.$$

This is called the resolvent and is a meromorphic family of bounded operators on L^2 with poles in the lower half-plane at the points $-i|\lambda_j|$; these are all simple since L is self-adjoint.

The first issue is to show that the continuous spectrum ($\lambda^2 \in [0,\infty)$) is absolutely continuous, or in other words, that the singular continuous part of the spectrum is empty. More specifically, we must find an L-invariant orthogonal splitting $L^2(\mathbb{R}^n) = \mathcal{H}_{\mathrm{pp}} \oplus \mathcal{H}_{\mathrm{ac}}$, so that the restriction of L to $\mathcal{H}_{\mathrm{pp}}$ is discrete, while the restriction of L to $\mathcal{H}_{\mathrm{ac}}$ is absolutely continuous. It is a classical theorem due to Friedrichs that in this setting any L^2 eigenvalue of L is strictly negative. The proof consists of showing that if any such eigenvalue is positive, then the corresponding eigenfunction must vanish outside a compact set, which violates standard unique continuation theorems. (This uses that V is compactly supported – if V only decays rapidly then the argument is a bit more intricate.) By the general spectral theorem, the absolute continuity of $L|_{\mathcal{H}_{\mathrm{ac}}}$ is equivalent to the existence of a unitary isomorphism $U : \mathcal{H}_{\mathrm{ac}} \longrightarrow L^2(\mathbb{R}; Y)$, where Y is an auxiliary Hilbert space, so that the self-adjoint operator $U \circ L \circ U^{-1}$ on $L^2(\mathbb{R}; Y)$ is multiplication by the coordinate function $t \in \mathbb{R}$. One of the goals of scattering theory is to exhibit this unitary isomorphism explicitly, which is done using the Møller wave operators, see below. A closely related goal is to understand the structure of generalized (non-L^2) solutions to the equation $(L - \lambda^2)u = 0$, $\lambda^2 > 0$. The key tool for all these questions is the resolvent $R_V(\lambda)$, introduced above.

Let us first consider the free Laplacian $L_0 = -\Delta$ on $\mathbb{R}^n$. When $\lambda \in \mathbb{R} \setminus \{0\}$, the nullspace $\mathcal{E}(\lambda)$ of the operator $-\Delta - \lambda^2$ (acting on tempered distributions) contains the plane wave solutions $e^{i\lambda z\cdot\omega}$ for any $\omega \in \mathbf{S}^{n-1}$. Any linear combination of these plane waves also lies in $\mathcal{E}(\lambda)$, and indeed, general superpositions of these plane wave solutions span all of $\mathcal{E}(\lambda)$. We explain this more carefully. For any $g \in \mathcal{C}^\infty(\mathbf{S}^{n-1})$, define

$$u(z) = \int_{\mathbf{S}^{n-1}} e^{i\lambda z\cdot\omega} g(\omega)\, d\omega.$$

This is a solution of $(-\Delta - \lambda^2)u = 0$, and the most general (polynomially bounded) element of $\mathcal{E}(\lambda)$ can always be obtained from this same representation but allowing g to be a distribution. The "smooth" elements of $\mathcal{E}(\lambda)$ are those where g is smooth.

We can look at this a different way. Note that since $\omega \mapsto z \cdot \omega$ is a Morse function on $\mathbf{S}^{n-1}$, and has critical points $\omega = \pm z/|z|$, the stationary phase lemma shows that (assuming g is smooth), the integral expression for u has an asymptotic expansion of the form

$$u(z) \sim e^{i\lambda|z|}|z|^{-\frac{n-1}{2}} \sum_{j=0}^{\infty} |z|^{-j} a_{+,j}(\theta) + e^{-i\lambda|z|}|z|^{-\frac{n-1}{2}} \sum_{j=0}^{\infty} |z|^{-j} a_{-,j}(\theta). \tag{3.1}$$

Here $z = |z|\theta$, $\theta \in \mathbf{S}^{n-1}$ are polar coordinates on $\mathbb{R}^n$. As part of this, one obtains that up to a multiple of 2π, $a_{\pm,0} = i^{\mp(n-1)/2} g(\pm\theta)$. Closely related is the assertion that any $u \in \mathcal{E}(\lambda)$ has an expansion of this same form and moreover, fixing any $a_{+,0} \in \mathcal{D}'(S^{n-1})$, there is a unique $u \in \mathcal{E}(\lambda)$ with this distribution as its leading coefficient. It is reasonable to regard the operator $\mathcal{P} : a_{+,0} \mapsto u$ as solving a Dirichlet problem at infinity for $-\Delta - \lambda^2$, and hence we call $\mathcal{P}$ the *Poisson operator*.

The free *scattering operator* at energy λ is the map $\mathcal{S}_0(\lambda)$ sending the function $a_{+,0}$ to $a_{-,0}$. Using the explicit representation above, we see that in this free setting, $\mathcal{S}_0(\lambda)a(\theta) = i^{n-1}a(-\theta)$; it is just a constant multiple of the antipodal map.

Proceeding slightly further with the free problem, suppose that $\operatorname{Im}\lambda < 0$. Using the Fourier transform, one can determine the inverse of $-\Delta - \lambda^2$ (as an operator on Schwartz functions) via

$$R_0(\lambda)f = (-\Delta - \lambda^2)^{-1} f = (2\pi)^{-n} \int_{\mathbb{R}^n} e^{iz\cdot\zeta}(|\zeta|^2 - \lambda^2)^{-1}\hat{f}(\zeta)\, d\zeta.$$

When n is odd, this has a particularly simple form: there is a simple polynomial $p_n(\alpha)$ of degree $(n-1)/2$ such that the Schwartz kernel of $R_0(\lambda)$ can be written as

$$|z - z'|^{2-n} p_n(\lambda|z - z'|)e^{-i\lambda|z-z'|}. \tag{3.2}$$

(In particular, $p_3(\alpha)$ is simply a constant.) There is a related but slightly more complicated formula when n is even. This explicit expression shows that as a function of λ, $R_0(\lambda)$ continues holomorphically from the lower "physical" half-plane $\{\Im\lambda < 0\}$ to the entire complex plane when $n \geq 3$ is odd. When $n = 1$, this continuation has a simple pole at $\lambda = 0$, and when n is even, there is a similar continuation but to the infinitely sheeted logarithmic Riemann surface branched at the origin. To make sense of this, one can say that this Schwartz kernel continues as a holomorphic function taking values in distributions; an alternate and equivalent sense is to regard the continuation taking values in the space of bounded operators $L^2_c \to L^2_{\mathrm{loc}}$, (this domain space consists of compactly supported L^2 functions). From (3.1) and stationary phase, one proves that if $f \in \mathcal{C}^\infty_c(\mathbb{R}^n)$, then

$$R_0(\lambda)f = e^{-i\lambda|z|}|z|^{-\frac{n-1}{2}} w,$$

where w is a smooth function on the radial compactification of $\mathbb{R}^n$. This last assertion about smoothness on the compactification is simply a concise way of stating that w has an asymptotic expansion

$$w \sim \sum_{j=0}^{\infty} w_j(\theta)|z|^{-j}.$$

Let us now pass to the analogous considerations for the operator L. Some versions of all the structural results about solutions remain true. These are typically proved by a perturbative argument, which means that one no longer has explicit formulæ. The starting point is the Lippmann–Schwinger formula, which gives a relationship between $R_0(\lambda)$ and $R_V(\lambda)$ in the region in the λ-plane where they both make sense. This states that

$$R_V(\lambda) = R_0(\lambda)\left(I + VR_0(\lambda)\right)^{-1} = \left(I + R_0(\lambda)V\right)^{-1} R_0(\lambda).$$

The issue is to prove that the inverses of $I+VR_0(\lambda)$ and $I+R_0(\lambda)V$ make sense, and to do this one observes that $VR_0(\lambda)$ and $R_0(\lambda)V$ are compact operators (between suitable function spaces), so that one can invoke the analytic Fredholm theorem to obtain that these inverses, and hence $R_V(\lambda)$ itself, are meromorphic on the region where $R_0(\lambda)$ is holomorphic (hence on $\mathbb{C}$ when n is odd and greater than 1).

The argument sketched earlier that L has no L^2 eigenvalues embedded in the continuous spectrum implies that $R_V(\lambda)$ has no poles on the real axis. (The argument for regularity at $\lambda = 0$ requires slightly more care.) On the other hand, the negative eigenvalues λ_j of L correspond to poles of $R_V(\lambda)$ at $-i|\lambda_j|$. The new and perhaps unexpected phenomenon is that $R_V(\lambda)$ may have poles in the upper half-plane (and indeed, this always occurs if V is nontrivial). These poles are known as the resonances of L, and their location and distribution has been the target of much research.

Let $\mathcal{E}_V(\lambda)$ denote the nullspace of $L - \lambda^2$ (say in $\mathcal{S}'(\mathbb{R}^n)$). Just as in the free case, this space may be generated using "distorted" plane waves. These are defined as follows. For any $\omega \in \mathbf{S}^{n-1}$ and $\lambda \in \mathbb{R} \setminus \{0\}$, there is a function $W_{\lambda,\omega}$ which is smooth on the radial compactification of $\mathbb{R}^n$ so that

$$\phi_{\lambda,\omega}(z) = e^{i\lambda z\cdot\omega} + e^{-i\lambda|z|}|z|^{-\frac{n-1}{2}} W_{\lambda,\omega}$$

lies in $\mathcal{E}_V(\lambda)$. Note that the second term here is simply $R_0(\lambda)(-Ve^{i\lambda z\cdot\omega})$. Superpositions of these can be used as before to generate all elements of $\mathcal{E}_V(\lambda)$. Indeed, if $g \in \mathcal{C}^\infty(\mathbf{S}^{n-1})$, then the general "smooth" element of $\mathcal{E}_V(\lambda)$ can be written as

$$u(z) = \int_{\mathbf{S}^{n-1}} \phi_{\lambda,\omega} g(\omega)\, d\omega.$$

Using stationary phase as before, this integral has an asymptotic expansion of exactly the same form as (3.1). The leading coefficient $a_{+,0}(\theta)$ is again just (a multiple of) g, but now the other leading coefficient $a_{-,0}(\theta)$ is not simply the reflection $g(-\theta)$, but rather a sum of this reflection plus an extra term which is an integral over S^n involving both g and V. The scattering operator $\mathcal{S}_V(\lambda)$, which sends $a_{+,0} \mapsto a_{-,0}$, is again unitary, and is the sum of the antipodal operator and another term which has a smooth Schwartz kernel. The map $\mathcal{P}_V(\lambda)$ which sends $a_{+,0}$ to u is again called the Poisson operator.

The results and definitions above continue to hold in suitably modified form not only for obstacle scattering, but also in the rather general setting of asymptotically Euclidean or asymptotically conic manifolds (these are called *scattering manifolds* [**Mel94**] by Melrose). For more on this as well as many further details about everything discussed above, we refer to the book of Melrose [**Mel95**]; see also [**Mel94**] and [**MZ96**].

3.2. Time-dependent scattering. We now turn our attention to the time-dependent formulation of scattering theory, and its relationship with stationary scattering. This time-dependent theory involves the study of "large time" properties of solutions of the wave equation. The connection with the stationary approach is via the Fourier transform in time; indeed, this Fourier transform carries $L - D_t^2$ to $L - \lambda^2$, and asymptotic properties of as $|t| \to \infty$ correspond to 'local in λ' properties of the latter operator. For the wave equation associated to $L = -\Delta + V$, where V is compactly supported, the intuitive picture is that one sends in a wave for times $t \ll 0$ from some direction at infinity and then observes what happens as this wave interacts with the potential and then scatters into a sum of plane waves as $t \nearrow +\infty$. Amongst the many good sources for this material, we refer to the books of Friedlander [**Fri75**], Lax [**Lax06**], Lax–Phillips [**LP89**], Taylor [**Tay11**] and Melrose [**Mel95**].

3.2.1. *Progressing wave solutions.* We begin by describing the special class of progressing wave solutions for wave operators. The calculations here go back to the dawn of microlocal analysis and can be regarded as the nexus of many constructions and ideas in that field. This construction is quite geometric and it is most naturally phrased in terms of the wave operator on a general Lorentzian metric g. The special case of a static metric $g = -\,dt^2 + h$ on the product of $\mathbb{R}$ with a Riemannian manifold (M, h) is of particular interest, and we discuss at the end how this specializes for the particular operator $\Box_V = \Box + V$ on Minkowski space. For more details, we refer the reader to the book of Friedlander [**Fri75**].

Thus let (X, g) be a Lorentzian manifold and consider $\Box_g + V$, where $V \in \mathcal{C}_c^\infty(X)$. We look for solutions u to $(\Box_g + V)u = 0$ which have the form

$$u = \varphi\, \alpha(\Gamma),$$

where φ is smooth, α is a distribution on $\mathbb{R}$ which models the 'wave form' of the solution, and Γ is a function on X with nowhere vanishing gradient which we call the phase function. To be concrete, we typically let $\alpha = \delta$, the Dirac delta function, or $\alpha = x_+^k$ for some $k \geq 0$, but the key feature we require of α is that it behave like a homogeneous function in the sense that its successive derivatives and integrals are progressively more or less smooth than α itself. Of course, it is usually impossible to choose solutions of $(\Box_g + V)u = 0$ which have this precise form, but the goal is to add increasingly higher order correction terms of a similar form involving the integrals of α so that, in the end, this initial expression is the first term in some asymptotic expansion of an exact solution.

The first step is to calculate

$$\begin{aligned}(\Box_g + V)\, u &= \frac{1}{\sqrt{|g|}}\partial_i \left(g^{ij}\sqrt{|g|}\partial_j u\right) + Vu \\ &= \alpha''(\Gamma) g\left(\nabla\Gamma, \nabla\Gamma\right)\varphi + \alpha'(\Gamma)\left(2g\left(\nabla\Gamma, \nabla\varphi\right) + \varphi\Box_g\Gamma\right) \\ &\qquad + \alpha(\Gamma)\left(\Box\varphi + V\varphi\right).\end{aligned}$$

As indicated above, assume that α_k is a sequence of distributions on $\mathbb{R}$ such that $\alpha_k = \alpha'_{k+1}$. (Again, refer to the basic example $\alpha_0 = \delta$, $\alpha_{k+1} = \frac{1}{k!}x_+^k$.) Let us now assume that

$$u \sim \sum_{k\geq 0} u_k = \sum_{k \geq 0} \varphi_k(t, z)\alpha_k(\Gamma). \tag{3.3}$$

We apply the calculation above and group together the terms of the same order (where the order of α_k is k and each derivative lowers the order by 1).

Grouping terms of the same order, we attempt to choose φ_k so that each term vanishes. The only term of order -2 is $\varphi_0\alpha_0''(\Gamma)g\left(\nabla\Gamma, \nabla\Gamma\right)$, so the first requirement is that

$$g\left(\nabla\Gamma, \nabla\Gamma\right) = 0.$$

This is known as the *eikonal equation* and states that $\nabla\Gamma$ is a *null-vector* for the metric g. This is a global nonlinear Hamilton–Jacobi equation for Γ. In the special case $X = \mathbb{R} \times M$, $g = -\, dt^2 + h$, the eikonal equation can be written as

$$\left(\partial_t\Gamma\right)^2 = \left|\nabla_h\Gamma\right|^2;$$

if we write $\Gamma = t - S$, where S is a function on M, then

$$\left|\nabla_h S\right|^2 = 1.$$

It is straightforward to see that the level sets $S = \text{const}$ are at constant distance from one another, so in general, $S(x) = \text{dist}_h\left(x, Z\right)$ where Z is some fixed level set of S. Even in the more general Lorentzian setting, the function Γ incorporates a lot of the distance geometry of g.

In any case, fix a solution Γ of the eikonal equation. We have now arranged that the term of order -2 vanishes. In fact, for any k, the term containing $g(\nabla\Gamma, \nabla\Gamma)$

vanishes, and so the equations for the higher coefficients simplify to *transport equations*. In particular, the term of order -1 reduces to

$$\alpha_0'(\Gamma)\left(2g\left(\nabla\Gamma, \nabla\varphi_0\right) + \varphi_0\Box\Gamma\right).$$

Since $\nabla\Gamma$ is nowhere vanishing, this is a linear ODE for φ_0 along the integral curves of $\nabla\Gamma$, which means that given any initial conditions for φ_0 on the characteristic surface $\Gamma =$ constant we may solve this equation locally.

The term of order $k-1$ yields an inhomogeneous transport equation for φ_k in terms of $\Gamma, \varphi_0, \ldots, \varphi_k$. We solve this transport equation with vanishing initial data and proceed inductively to choose all φ_k.

It is possible to asymptotically sum the series (3.3). This means that we can choose a function v with the property that

$$v - \sum_{k=0}^{N} \varphi_k \alpha_k(\Gamma)$$

is as smooth as the next term in the series, $\varphi_{N+1}\alpha_{N+1}(\Gamma)$. By construction, $(\Box_g + V)v = f \in \mathcal{C}^\infty(X)$. We must now invoke a theorem guaranteeing the existence of a smooth solution w for the initial value problem $(\Box_g + V)w = f$ with vanishing Cauchy data vanishes, where f is smooth. Given this, then $u = v - w$ is a solution of the original equation and the expansion we have calculated determines the singularity profile of u. Note that these singularities of u occur precisely along the union of level sets $\Gamma = c$ where one (and hence every) α_k is singular at c.

For the special case where $g = -dt^2 + dx^2$ on Minkowski space, fix $\omega \in \mathbf{S}^{n-1}$ and consider the equation

$$\left(\partial_t^2 - \Delta_z + V\right) u = 0, \qquad u = \delta(t - z\cdot\omega) \text{ when } t \ll 0.$$

The eikonal equation $|\nabla\Gamma|_g^2 = 0$ has solution $\Gamma(t, z) = t - z\cdot\omega$. This gives a global solution of the wave equation for all t when $V \equiv 0$. However, by the propagation of singularities theorem, the wave front set of the solution u for the perturbed problem with this initial data in the distant past agrees with that of this exact free solution. Hence it makes sense to look for a solution of the perturbed problem of the form

$$u \sim \delta(t - z\cdot\omega) + \sum_{k\geq 0} \varphi_k(t, z) x_+^k(t - z\cdot\omega),$$

for some choice of smooth functions φ_k. This fits exactly into the scheme above (and was, of course, the setting for the original version of these calculations). The first transport equation is

$$2\left(\partial_t - \omega\cdot\nabla_z\right)\varphi_0 = 0,$$

which means that φ_0 is a function of $t = z\cdot\omega$ and z; its Cauchy data is defined on the hypersurface $t = z\cdot\omega$, and the equation dictates that it must be constant along the lines parallel to ω.

Once we have determined $\varphi_0, \ldots, \varphi_k$, then the $(k+1)^{\text{st}}$ transport equation is

$$2(k+1)\left(\partial_t - \omega\cdot\nabla_z\right)\varphi_{k+1} = -\left(\Box + V\right)\varphi_k,$$

which we solve with vanishing initial data. Carrying this procedure out for all k determines the Taylor series of u along the hypersurface $\{t = z\cdot\omega\}$. As described earlier, we can use the Borel Lemma to choose an asymptotic sum v for this series, so that $(\Box + V)\,v = f$ is smooth and v satisfies the correct "initial condition" for

$t \ll 0$. We can then find a *smooth* correction term w which solves away this error term. Thus $u = v - w$ is an exact solution

The calculations here were historical precursors to the more elaborate but ultimately very similar ones which come up in the construction of Fourier integral operators. Indeed, solving the eikonal equation for Γ is the direct analogue of solving the eikonal equation for the phase of an FIO. For potential scattering, keeping track of the parametric dependence on ω fixes the phase; the solutions of the transport equations are the coefficients in the expansion of the amplitude, and these correspond to the terms in the expansion for the symbol of the FIO.

3.2.2. *Møller wave operators.* We now turn to another perspective on time-dependent scattering, which is through the definition of the so-called Møller wave operators. This can be regarded as a formalization of the discussion above; there we described how to calculate the profile of the solution obtained by "sending in" a delta function along a particular direction. Our goal now is to put this information together into a map which compares the long-time evolution with respect to the perturbed equation against that for the free equation.

Let us suppose now that $g = -dt^2 + h$ is a static Lorentzian metric. For any ($\mathcal{C}_c^\infty$) potential V, define the wave evolution operator

$$U_V(t) : C_c^\infty(\mathbb{R}^n) \times C_c^\infty(\mathbb{R}^n) \to C_c^\infty(\mathbb{R}^n) \times C_c^\infty(\mathbb{R}^n),$$

where, if u solves the Cauchy problem

$$(\Box + V)\, u = 0, \quad (u, \partial_t u)\,|_{t=0} = (\phi, \psi),$$

then $U_V(t_0)(\phi, \psi) = (u, \partial_t u)|_{t=t_0}$. The free wave evolution operator $U_0(t)$ is defined analogously using solutions for $\Box u = 0$ instead. Uniqueness of solutions of these Cauchy problems implies that U_V and U_0 are groups, i.e. $U_*(t)^{-1} = U_*(-t)$ and $U_*(t+s) = U_*(t)U_*(s)$ for $* = 0$ or V.

Now define the *Møller wave operators* $W_\pm$ by

$$W_\pm(\phi, \psi) = \lim_{t\to\pm\infty} U_V(-t)U_0(t)(\phi, \psi),$$

when the limit exists. This limit is meant to be taken in the sense of strong operator convergence. If we define the *energy space*

$$H_E = \left\{ (\phi, \psi) : \int \psi^2 + |\nabla_z \phi|^2 \, dV < \infty \right\},$$

then $W_\pm$ extends by continuity to all of H_E. It can be proved that if certain local measurements of this energy decay appropriately, then $-\Delta + V$ has no L^2 eigenvalues and this extension is an isomorphism of H_E to itself. If $-\Delta + V$ does have L^2 eigenvalues, then $\mathcal{H}_{\mathrm{pp}}$ determines a finite dimensional subspace in H_E and $W_\pm$ is an isomorphism from H_E onto the orthogonal complement of $\mathcal{H}_{\mathrm{pp}}$, which we denote $H_E^\perp$.

Since $U_0(t)$ and $U_V(t)$ are unitary, the wave operators $W_\pm$ are characterized by the property that

$$\|U_V(t)W_\pm(\phi, \psi) - U_0(t)(\phi, \psi)\|_{H_E} \to 0 \text{ as } t \to \pm\infty$$

for all $(\phi, \psi) \in H_E^\perp$. Now define the *scattering operator*

$$\mathcal{S} = W_+^{-1} W_-;$$

this is an isomorphism of $H_E^\perp$. It describes the relationship between the asymptotic free wave emerging as $t \nearrow +\infty$ for a solution of the perturbed equation $(\Box + V)u = 0$ in terms of the incoming free wave for $t \ll 0$.

These operators lead directly to the unitary isomorphism mentioned earlier which intertwines L (or rather, its restriction to $\mathcal{H}_{\mathrm{ac}}$), with a simple multiplication operator. In other words, the existence and properties of the wave operators and scattering matrix proves that the singular continuous spectrum of L is empty.

There are many other settings where one can define analogues of the Møller wave and scattering operators. Classically this is done for exterior domains, and more recently on asymptotically Euclidean or conic manifolds (where the structure of the scattering matrix is quite intriguing, see [**MZ96**]), as well as other geometric settings such as asymptotically hyperbolic manifolds, etc. There is also a parallel and vigorous line of research concerning the possibility of defining the analogues of wave and scattering operators for various classes of nonlinear evolution equations.

3.2.3. *Lax–Phillips theory and radiation fields.* In this final section we present yet another approach to scattering theory. This is the more abstract approach developed by Lax and Phillips [**LP89**], which has played an influential paradigmatic role. Directly following this we describe the theory pioneered by Friedlander [**Fri80**] on what he called the radiation fields associated to solutions of a linear wave equation. These describe certain asymptotic information about waves, and beyond their purely analytic appeal, they also provide a beautiful realization of the Lax–Phillips theory. These radiation fields have received quite a lot of attention in recent years, and the theory has been extended to various nonlinear settings as well. There is a forthcoming and much more detailed survey specifically about radiation fields [**MW**] to which we direct the reader.

Throughout this section we fix a Hilbert space $\mathcal{H}$ and a unitary semigroup $U(t)$ which acts on it. The specific application we have in mind is that $\mathcal{H}$ is the space H_E of finite energy initial data for the wave equation on $\mathbb{R}^n$ with n odd and $U(t)$ is the wave evolution operator. More precisely, let $\mathcal{H}_0$ be the completion of the space $\mathcal{C}_c^\infty(\mathbb{R}^n) \times \mathcal{C}_c^\infty(\mathbb{R}^n)$ with respect to the norm

$$\|(\phi, \psi)\|_{\mathcal{H}_0}^2 = \int_{\mathbb{R}^n} \left(|\nabla \phi(z)|^2 + |\psi(z)|^2 \right) \, dz;$$

then, for $(\phi, \psi) \in \mathcal{H}_0$, let $U_0(t)(\phi, \psi)$ be as defined in the previous section. The unitarity of U_0 corresponds to conservation of energy for solutions of this wave equation.

Return now to the general formulation.

DEFINITION 3.4. *A closed subspace $\mathcal{D} \subset \mathcal{H}$ is called* outgoing, *respectively* incoming, *if*

(i) $U(t)\mathcal{D} \subset \mathcal{D}$ *for* $t > 0$, *respectively* $t < 0$,
(ii) $\bigcap_{t \in \mathbb{R}} U(t)\mathcal{D} = \{0\}$, *and*
(iii) $\overline{\bigcup_{t \in \mathbb{R}} U(t)\mathcal{D}} = \mathcal{H}$.

In the example above, the space $\mathcal{D}_+$ consists of the pairs $(\phi, \psi) \in \mathcal{H}_0$ for which the solution $u(t, z)$ vanishes for $|z| \leq t$ when $t \geq 0$. Continuous dependence of solutions of the wave equation on initial data shows that $\mathcal{D}_+$ is a closed subspace. The first and second properties follow from the observation that if $(\phi, \psi) \in \mathcal{H}_0$, then by finite propagation speed, the solution of the wave equation with initial data $U(s)(\phi, \psi)$ vanishes for $|z| \leq t + s$.

The third property is more subtle. For the unperturbed wave equation in odd dimensions, it is a consequence of Huygens' principle; in even dimensions, one may prove it using local energy decay, but it can also be proved fairly explicitly via the Radon transform. We say more about this later.

The fundamental result of Lax–Phillips theory is the existence of a translation representation:

THEOREM 3.5 ([**LP89**, Chapter II, Theorem 3.1]). *Let $U(t)$ be a group of unitary operators on $\mathcal{H}$, and $\mathcal{D}$ an outgoing subspace with respect to $U(t)$. Then there exists a Hilbert space $\mathcal{K}$ and an isometric isomorphism*

$$\Phi : \mathcal{H} \to L^2\left((-\infty,\infty);\mathcal{K}\right)$$

such that $\Phi(\mathcal{D}) = L^2\left((0,\infty);\mathcal{K}\right)$ and $\Phi \circ U(t) = T_t \circ \Phi$, where $(T(t)f)(s) = f(s-t)$ is the standard translation action of $\mathbb{R}$ on $L^2(\mathbb{R};\mathcal{K})$. The isomorphism Φ is unique up to an isomorphism of $\mathcal{K}$.

The isomorphism given here is called an *outgoing translation representation* of $U(t)$. There is an essentially identical result giving an isomorphism Φ' which maps an incoming subspace $\mathcal{D}_-$ to $L^2\left((-\infty,0);\mathcal{K}\right)$ and intertwines $U(t)$ with $T(t)$. This is called an *incoming translation representation.* The auxiliary Hilbert space $\mathcal{K}$ may be taken to be the same as for the outgoing translation representation, but of course the map Φ' is different than Φ.

Returning again to the unperturbed wave equation in $\mathbb{R}^n$, n odd, there is an explicit way to obtain the translation representations using the Radon transform.

DEFINITION 3.6. *For any $f \in \mathcal{C}_c^\infty(\mathbb{R}^n)$, define the* Radon transform

$$(Rf)(s,\theta) = \int_{\langle z,\theta\rangle = s} f(z)\, d\sigma(z),$$

where $d\sigma(z)$ is surface measure on the hyperplane $\langle z,\theta\rangle = s$. Clearly $Rf \in \mathcal{C}_c^\infty(\mathbb{R}\times \mathbf{S}^{n-1})$.

A key property of the Radon transform for our purposes is that it is invertible and in fact the inversion formula is quite explicit:

$$f(z) = \frac{1}{2\left(2\pi\right)^{n-1}} \int_{\mathbf{S}^{n-1}} \left(\left|D_s\right|^{n-1} Rf\right)(z\cdot\theta,\theta)\, d\theta,$$

where $|D_s|$ is defined by conjugating multiplication by $|\sigma|$ with respect to the Fourier transform. A remarkable fact, which can be proved by direct computation, is that R intertwines the Laplacians on $\mathbb{R}^n$ and $\mathbb{R}$,

$$R\Delta f = \partial_s^2 Rf.$$

We now define the *Lax–Phillips transform*: for n odd, and $(\phi,\psi) \in \mathcal{C}_c^\infty(\mathbb{R}^n) \times \mathcal{C}_c^\infty(\mathbb{R}^n)$, let

$$\mathrm{LP}(\phi,\psi)(s,\theta) = \frac{1}{(2\pi)^{(n-1)/2}} \left(D_s^{(n+1)/2}\left(R\phi\right)(s,\theta) - D_s^{(n-1)/2}\left(R\psi\right)(s,\theta)\right).$$

THEOREM 3.7 (See [**Mel95**, Section 3.4]). *For n odd, the Lax–Phillips transform* LP *extends to a unitary isomorphism*

$$\mathrm{LP} : \mathcal{H}_0 \to L^2\left(\mathbb{R}; L^2(\mathbf{S}^{n-1})\right),$$

and is a translation representation,

$$(\mathrm{LP}\, U_0(t)(\phi,\psi))\,(s,\theta) = (T_t\, \mathrm{LP}(\phi,\psi))\,(s,\theta) = (\mathrm{LP}(\phi,\psi))\,(s-t,\theta).$$

One consequence of Theorem 3.7 is that $\mathcal{H}_0$ splits as an orthogonal direct sum of the incoming and outgoing subspaces:

$$\mathcal{H}_0 = \mathcal{D}_+ \oplus \mathcal{D}_-. \tag{3.8}$$

In particular, in this special case, the outgoing and incoming isomorphisms Φ and Φ' are equal.

Now consider the wave equation with potential. As before, assume that n is odd and $V \in \mathcal{C}_c^\infty(\mathbb{R}^n)$ is real-valued. Choose R so that $\operatorname{supp} V \subseteq B(0,R)$. Let $U(t)$ be the group associated to the Cauchy problem

$$\Box u + Vu = 0, \qquad (u, \partial_t u)|_{t=0} = (\phi,\psi), \tag{3.9}$$

i.e. $U(t)(\phi,\psi) = (u(t), \partial_t u(t))$. Since V does not depend on t, there is a conserved energy,

$$\|(u(t),\partial_t u(t))\|_E^2 = \int_{\mathbb{R}^n} \left(|\partial_t u(t,z)|^2 + |\nabla u(t,z)|^2 + V(z)\, |u(t,z)|^2 \right) dz. \tag{3.10}$$

The Hilbert space $\mathcal{H}$ is the set of pairs (ϕ,ψ) for which this energy is finite. It is not hard to see, using the Sobolev inequality, that $\mathcal{H}$ and $\mathcal{H}_0$ consist of the same pairs of elements, although the norm is different. The energy extends to the bilinear pairing on $\mathcal{H}$:

$$\left\langle \begin{pmatrix} \phi_1 \\ \psi_1 \end{pmatrix}, \begin{pmatrix} \phi_2 \\ \psi_2 \end{pmatrix} \right\rangle = \int_{\mathbb{R}^n} \left(\nabla\phi_1 \cdot \nabla\overline{\phi_2} + V(z)\phi_1\overline{\phi_2} + \psi_1\overline{\psi_2} \right) dz. \tag{3.11}$$

Consider now the operator

$$A = \begin{pmatrix} 0 & 1 \\ \Delta - V & 0 \end{pmatrix};$$

this is anti-symmetric with respect to the pairing (3.11). The wave group $U(t)$ can be regarded instead as the solution operator for the system

$$\partial_t \begin{pmatrix} u_0 \\ u_1 \end{pmatrix} = A \begin{pmatrix} u_0 \\ u_1 \end{pmatrix}, \qquad \begin{pmatrix} u_0 \\ u_1 \end{pmatrix}\bigg|_{t=0} = \begin{pmatrix} \phi \\ \psi \end{pmatrix}.$$

We now make a simplifying assumption that $L = -\Delta + V$ has no L^2 eigenvalues, or equivalently, that A has no such eigenvalues. Without this assumption, the results below require a projection off the finite dimensional space $\mathcal{H}_{\mathrm{pp}}$. We refer to [**LP66**] for more details about how to proceed without this assumption. The advantage of this assumption is that now the energy (3.10) is positive definite.

For this perturbed problem, we define the incoming and outgoing subspaces $\mathcal{D}_{\pm,R} \subset \mathcal{H}$ to consist of those elements (ϕ,ψ) so that $U_0(t)(\phi,\psi)$ vanishes in $|z| \le t + R$ for $t \ge 0$, respectively $|z| \le -t + R$ for $t \le 0$. Thus, in terms of the free incoming and outgoing subspaces, $\mathcal{D}_{\pm,R} = U_0(\pm R)\mathcal{D}_\pm$. The verification that these satisfy all the correct properties relies on the following

LEMMA 3.12. *If* $\mathbf{f} = (\phi,\psi) \in \mathcal{D}_{+,R}$*, then* $U_0(t)\mathbf{f} = U(t)\mathbf{f}$ *for* $t > 0$*; the analogous statement holds for* $\mathcal{D}_{-,R}$ *when* $t < 0$*.*

We now use this to show that $\mathcal{D}_{+,R}$ is an outgoing subspace for $U(t)$ on $\mathcal{H}$. Indeed, by this lemma, the first two properties follow from the corresponding properties of $\mathcal{D}_+$. For the third property, suppose we know that for any compact subset $K \subset \mathbb{R}^n$ and any solution u of (3.9), we have

$$\lim_{t\to\infty} \|u(t)\|_{E,K}^2 := \lim_{t\to\infty} \int_K \left(|\partial_t u(t,z)|^2 + |\nabla u(t,z)|^2 + V(z)|u(t,z)|^2 \right) dz = 0.$$

This is called local energy decay, and is known to be true in many circumstances. Now consider the initial data $\mathbf{f} = (\phi, \psi) \in \mathcal{H}$ with $\mathbf{f} \perp \bigcup U(t)\mathcal{D}_{+,R}$ with respect to the pairing (3.11). Thus $U(t)\mathbf{f} \perp \mathcal{D}_{+,R}$ for any t, and in particular, $U(t)\mathbf{f} \perp \mathcal{D}_{+,R}$ with respect to the standard pairing on $\dot{H}^1 \times L^2$. This shows that $U_0(-R)U(t)\mathbf{f} \perp \mathcal{D}_+$ with respect to the standard pairing, and hence $U_0(-R)U(t)\mathbf{f} \in \mathcal{D}_-$ and $U_0(-2R)U(t)\mathbf{f} \in \mathcal{D}_{-,R}$.

Consider now $v(s,z) = U(s)U_0(-2R)U(t)\mathbf{f}$. By Lemma 3.12, $v(s,z)$ agrees with $U_0(s)U_0(-2R)U(t)\mathbf{f}$ for $s < 0$ and thus vanishes for $|z| \le -s + R$ for $s < 0$.

Now we bring in the local energy decay. This implies that for any $\epsilon > 0$, if t is sufficiently large then $\|U(t)\mathbf{f}\|_{E,B(5R)} < \epsilon$. For such t, finite propagation speed implies both

$$\|U_0(-2R)U(t)\mathbf{f}\|_{E,B(3R)} < \epsilon, \quad \text{and} \quad \|U(-2R)U(t)\mathbf{f}\|_{E,B(3R)} < \epsilon.$$

Because the two equations and the initial data agree outside $B(R)$, using finite propagation speed again, we get that $U_0(-2R)U(t)\mathbf{f} = U(-2R)U(t)\mathbf{f}$ for $|z| > 3R$ and hence

$$\|U_0(-2R)U(t)\mathbf{f} - U(-2R)U(t)\mathbf{f}\|_E < 2\epsilon.$$

Because $U(t)$ is unitary with respect to (3.11), applying $U(2R-t)$ to the difference shows that

$$\|U(2R-t)U_0(-2R)U(t)\mathbf{f} - \mathbf{f}\|_E < 2\epsilon.$$

Finally, since t is large, $2R - t < 0$ and so $U(2R-t)U_0(-2R)U(t)\mathbf{f} = U_0(2R-t)U_0(-2R)U(t)\mathbf{f}$ by Lemma 3.12. This shows that in fact

$$\|U_0(-t)U(t)\mathbf{f} - \mathbf{f}\|_E < 2\epsilon.$$

Because $U_0(-R)U(t)\mathbf{f} \in \mathcal{D}_-$, the first term here is an element of $\mathcal{D}_{-,t-R}$ and thus vanishes for $|z| \le t - R$. Taking t even larger gives

$$\|\mathbf{f}\|_E < 2\epsilon,$$

and therefore $\mathbf{f} = 0$. This establishes the third property.

Theorem 3.5 asserts the existence of incoming and outgoing translation representations for the incoming and outgoing subspaces $\mathcal{D}_{-,R}$ and $\mathcal{D}_{+,R}$. We shall give a a concrete realization of these using the so-called radiation fields.

Our next goal is to show that a particular quantitative rate of local energy decay implies that the local energy actually decays exponentially.

THEOREM 3.13 (See [**LP89**, Chapter V, Theorem 3.2]). *Suppose that for each compact subset $K \subset \mathbb{R}^n$ there is a function $c_K(t)$ which tends to 0 as $t \to \infty$, such that if the Cauchy data $u(0)$ have support in K, then*

$$\|u(t)\|_{E,K}^2 \le c_K(t) \|u(0)\|_E^2 . \tag{3.14}$$

Then there are positive constants C and α depending on K such that if $u(0)$ is supported in K, then

$$\|u(t)\|_{E,K} \leq Ce^{-\alpha t} \|u(0)\|_E \tag{3.15}$$

for all $t > 0$.

The proof uses the compactness properties of the Lax–Phillips semigroup $Z(t)$, which we introduce now. If $P_{\pm,R}$ are the orthogonal projections onto the orthocomplements of $\mathcal{D}_{\pm,R}$, then $Z(t)$ is given for $t \geq 0$ by

$$Z(t) = P_{+,R} U(t) P_{-,R}.$$

The local energy decay hypothesis in the theorem statement implies that, for t large enough, $Z(t)$ has norm bounded by $1/2$, and repeated application of $Z(t)$ leads to the exponential decay.

We are now in a position to introduce the radiation field of a solution u to the perturbed wave equation. The idea is to identify initial data for u with a normalized limit of the solution along outgoing (or incoming) light rays. As before, we start with the definition of these radiation fields for the unperturbed operator.

Suppose that u solves $\Box_0 u = 0$ with initial data (ϕ, ψ). Introduce coordinates $s = t - |z|$ and $x = |z|^{-1}$; these parametrize the family of outgoing light rays and the position along them. Now define the auxiliary function

$$v : \mathbb{R}_s \times (0,\infty)_x \times \mathbf{S}^{n-1}_\theta \to \mathbb{R}, \qquad v(s,x,\theta) = x^{-\frac{n-1}{2}} u\left(s + \frac{1}{x}, \frac{1}{x}\theta\right).$$

Here $\frac{1}{x}\theta$ is simply z in polar coordinates. Finite speed of propagation implies that v vanishes for $s \ll 0$, and so has a smooth extension across $x = 0$ for $s \ll 0$. Since $x^2 g_M$ is nondegenerate at $x = 0$, v satisfies a hyperbolic equation that is also nondegenerate across $x = 0$ and so v extends smoothly across $x = 0$. We then define the forward radiation field operator $\mathcal{R}_+$ by

$$\mathcal{R}_+(\phi,\psi)(s,\theta) = \partial_s v(s,0,\theta).$$

The derivative of v is included here to make

$$\mathcal{R}_+ : \mathcal{H}_0 \to L^2(\mathbb{R} \times \mathbf{S}^{n-1})$$

an isometric isomorphism. Furthermore, the Minkowski metric is static, so $\mathcal{R}_+$ intertwines wave evolution and translation in s:

$$\mathcal{R}_+ U_0(T)(\phi,\psi)(s,\theta) = \mathcal{R}_+(\phi,\psi)(s - T, \theta).$$

Now observe that if $\mathbf{f} \in \mathcal{D}_+$, then $\mathcal{R}_+\mathbf{f}$ vanishes when $s \geq 0$. This follows from the unitarity of the radiation field operator, and the fact that the inverse image of those functions in $L^2(\mathbb{R} \times \mathbf{S}^{n-1})$ supported in the nonpositive half-cylinder form an outgoing subspace $\widetilde{\mathcal{D}}_+$:

$$\widetilde{\mathcal{D}}_+ = \left\{\mathcal{R}_+^{-1} f : f(s,\theta) = 0 \text{ for } s > 0\right\}.$$

Indeed, this is a closed subspace; the first and second properties follow directly from the fact that $\mathcal{R}_+$ is a translation representation, while the third property follows from the surjectivity of $\mathcal{R}_+$. One may also define $\widetilde{\mathcal{D}}_-$ via the backward radiation field $\mathcal{R}_-$; this encodes information from solutions in the limit as $t \searrow -\infty$. For the free wave equation, $\mathcal{R}_+$ has an explicit expression in terms of the Radon transform, and this can be used to show that $\mathcal{H}_0 = \widetilde{\mathcal{D}}_+ \oplus \widetilde{\mathcal{D}}_-$.

For the perturbed equation the forward and backward radiation fields, $\mathcal{L}_\pm$, are defined in the same way. We can also define the scattering operator $\mathcal{A}$ using the radiation fields by

$$\mathcal{A} = \mathcal{L}_+ \mathcal{L}_-^{-1}.$$

Thus $\mathcal{A}$ maps data at past null infinity into data at future null infinity. The relationship to the scattering operator $\mathcal{S}$ introduced in Section 3.2.2 is that

$$\mathcal{S} = \mathcal{R}_+^{-1} \mathcal{L}_+ \mathcal{L}_-^{-1} \mathcal{R}_- = \mathcal{R}_+^{-1} \mathcal{A} \mathcal{R}_-.$$

The conjugation of $\mathcal{A}$ by the Fourier transform in s corresponds to the *scattering matrix* employed by Melrose in [**Mel94**].

The radiation field exists and is a unitary operator in a variety of geometric settings. On asymptotically Euclidean spaces, this is due to Friedlander [**Fri80, Fri01**] and Sá Barreto [**SB03, SB08**]; on asymptotically real and complex hyperbolic manifolds it was proved by Sá Barreto [**SB05**], and Guillarmou and Sá Barreto [**GSB08**], respectively. In the asymptotically Euclidean and real hyperbolic cases, Sá Barreto and Wunsch [**SBW05**] prove that it is a Fourier integral operator with canonical relation given by the sojourn relation, a close relative of the Busemann function in each of these geometric settings. The radiation field has also been defined in certain nonlinear and non-static situations. In particular, the first author and Sá Barreto show [**BSB12**] that it exists and is norm-preserving for certain semilinear wave equations in $\mathbb{R}^3$, while Wang [**Wan10, Wan11**] defined the radiation field for the Einstein equations on perturbations of Minkowski space when the spatial dimension is at least 4. Forthcoming work of the first author, Vasy, and Wunsch [**BVW**] analyzes the $s \to \infty$ asymptotics of the radiation field on (typically non-static) perturbations of Minkowski space.

References

[AJ99] J.-P. Anker and L. Ji. Heat kernel and Green function estimates on noncompact symmetric spaces. *Geom. Funct. Anal.*, 9(6):1035–1091, 1999.

[Ali09] Serge Alinhac. *Hyperbolic partial differential equations.* Universitext. Springer, Dordrecht, 2009.

[BGV92] Nicole Berline, Ezra Getzler, and Michèle Vergne. *Heat kernels and Dirac operators*, volume 298 of *Grundlehren der Mathematischen Wissenschaften [Fundamental Principles of Mathematical Sciences].* Springer–Verlag, Berlin, 1992.

[BSB12] D. Baskin and A. Sá Barreto. Radiation fields for semilinear wave equations. Preprint, arXiv:1208.2743, August 2012.

[BVW] D. Baskin, A. Vasy, and J. Wunsch. Asymptotics of radiation fields on asymptotically Minkowski spaces. In preparation.

[Car10] Gilles Carron. Estimées des noyaux de Green et de la chaleur sur les espaces symétriques. *Anal. PDE*, 3(2):197–205, 2010.

[CGT82] Jeff Cheeger, Mikhail Gromov, and Michael Taylor. Finite propagation speed, kernel estimates for functions of the Laplace operator, and the geometry of complete Riemannian manifolds. *J. Differential Geom.*, 17(1):15–53, 1982.

[Che73] Paul R. Chernoff. Essential self-adjointness of powers of generators of hyperbolic equations. *J. Functional Analysis*, 12:401–414, 1973.

[Fri75] F. G. Friedlander. *The wave equation on a curved space-time.* Cambridge University Press, Cambridge, 1975. Cambridge Monographs on Mathematical Physics, No. 2.

[Fri80] F. G. Friedlander. Radiation fields and hyperbolic scattering theory. *Math. Proc. Cambridge Philos. Soc.*, 88(3):483–515, 1980.

[Fri01] F. G. Friedlander. Notes on the wave equation on asymptotically Euclidean manifolds. *J. Funct. Anal.*, 184(1):1–18, 2001.

[Gri09] Alexander Grigor'yan. *Heat kernel and analysis on manifolds*, volume 47 of *AMS/IP Studies in Advanced Mathematics*. American Mathematical Society, Providence, RI, 2009.

[GSB08] Colin Guillarmou and Antônio Sá Barreto. Scattering and inverse scattering on ACH manifolds. *J. Reine Angew. Math.*, 622:1–55, 2008.

[Hel84] Sigurdur Helgason. *Groups and geometric analysis*, volume 113 of *Pure and Applied Mathematics*. Academic Press Inc., Orlando, FL, 1984. Integral geometry, invariant differential operators, and spherical functions.

[Lax06] Peter D. Lax. *Hyperbolic partial differential equations*, volume 14 of *Courant Lecture Notes in Mathematics*. New York University Courant Institute of Mathematical Sciences, New York, 2006. With an appendix by Cathleen S. Morawetz.

[LM10] N. Lohoué and S. Mehdi. Estimates for the heat kernel on differential forms on Riemannian symmetric spaces and applications. *Asian J. Math.*, 14(4):529–580, 2010.

[LP66] Peter D. Lax and Ralph S. Phillips. Analytic properties of the Schrödinger scattering matrix. In *Perturbation Theory and its Applications in Quantum Mechanics (Proc. Adv. Sem. Math. Res. Center, U. S. Army, Theoret. Chem. Inst., Univ. of Wisconsin, Madison, Wis., 1965)*, pages 243–253. Wiley, New York, 1966.

[LP89] Peter D. Lax and Ralph S. Phillips. *Scattering theory*, volume 26 of *Pure and Applied Mathematics*. Academic Press Inc., Boston, MA, second edition, 1989. With appendices by Cathleen S. Morawetz and Georg Schmidt.

[Mel94] Richard B. Melrose. Spectral and scattering theory for the Laplacian on asymptotically Euclidian spaces. In *Spectral and scattering theory (Sanda, 1992)*, volume 161 of *Lecture Notes in Pure and Appl. Math.*, pages 85–130. Dekker, New York, 1994.

[Mel95] Richard B. Melrose. *Geometric scattering theory*. Stanford Lectures. Cambridge University Press, Cambridge, 1995.

[MV05] Rafe Mazzeo and András Vasy. Analytic continuation of the resolvent of the Laplacian on symmetric spaces of noncompact type. *J. Funct. Anal.*, 228(2):311–368, 2005.

[MW] R.B. Melrose and F. Wang. Radon transforms and radiation fields. To appear.

[MZ96] Richard Melrose and Maciej Zworski. Scattering metrics and geodesic flow at infinity. *Invent. Math.*, 124(1-3):389–436, 1996.

[Per83] Peter A. Perry. *Scattering theory by the Enss method*, volume 1 of *Mathematical Reports*. Harwood Academic Publishers, Chur, 1983. Edited by B. Simon.

[Rau12] Jeffrey Rauch. *Hyperbolic partial differential equations and geometric optics*, volume 133 of *Graduate Studies in Mathematics*. American Mathematical Society, Providence, RI, 2012.

[RS78] Michael Reed and Barry Simon. *Methods of modern mathematical physics. IV. Analysis of operators*. Academic Press [Harcourt Brace Jovanovich Publishers], New York, 1978.

[SB03] Antônio Sá Barreto. Radiation fields on asymptotically Euclidean manifolds. *Comm. Partial Differential Equations*, 28(9-10):1661–1673, 2003.

[SB05] Antônio Sá Barreto. Radiation fields, scattering, and inverse scattering on asymptotically hyperbolic manifolds. *Duke Math. J.*, 129(3):407–480, 2005.

[SB08] Antônio Sá Barreto. A support theorem for the radiation fields on asymptotically Euclidean manifolds. *Math. Res. Lett.*, 15(5):973–991, 2008.

[SBW05] Antônio Sá Barreto and Jared Wunsch. The radiation field is a Fourier integral operator. *Ann. Inst. Fourier (Grenoble)*, 55(1):213–227, 2005.

[SC02] Laurent Saloff-Coste. *Aspects of Sobolev-type inequalities*, volume 289 of *London Mathematical Society Lecture Note Series*. Cambridge University Press, Cambridge, 2002.

[Tay11] Michael E. Taylor. *Partial differential equations. I - III*, volumes 115–117 of *Applied Mathematical Sciences*, 2nd edition. Springer–Verlag, New York, 2011.

[Wan10] Fang Wang. *Radiation Field for Einstein Vacuum Equations*. PhD thesis, Massachusetts Institute of Technology, Aug 2010.

[Wan11] Fang Wang. Radiation field for Einstein vacuum equations with spacial dimension $n \geq 4$. In preparation, 2011.

[Wun08] Jared Wunsch. Microlocal analysis and evolution equations. Clay Mathematics Institute Proceedings, Summer School on Evolution Equations, Zürich 2008.

[Yaf10] D. R. Yafaev. *Mathematical scattering theory*, volume 158 of *Mathematical Surveys and Monographs*. American Mathematical Society, Providence, RI, 2010. Analytic theory.

[Zwo12] Maciej Zworski. *Semiclassical analysis*, volume 138 of *Graduate Studies in Mathematics*. American Mathematical Society, Providence, RI, 2012.

Department of Mathematics, Northwestern University, Evanston, Illinois 60208
E-mail address: dbaskin@math.northwestern.edu

Department of Mathematics, Stanford University, Stanford, California 94305
E-mail address: mazzeo@math.stanford.edu

Clay Mathematics Proceedings
Volume **17**, 2013

Lectures on Black Holes and Linear Waves

Mihalis Dafermos and Igor Rodnianski

ABSTRACT. These lecture notes, based on a course given at the Zürich Clay Summer School (June 23–July 18 2008), review our current mathematical understanding of the global behaviour of waves on black hole exterior backgrounds. Interest in this problem stems from its relationship to the non-linear stability of the black hole spacetimes themselves as solutions to the Einstein equations, one of the central open problems of general relativity. After an introductory discussion of the Schwarzschild geometry and the black hole concept, the classical theorem of Kay and Wald on the boundedness of scalar waves on the exterior region of Schwarzschild is reviewed. The original proof is presented, followed by a new more robust proof of a stronger boundedness statement. The problem of decay of scalar waves on Schwarzschild is then addressed, and a theorem proving quantitative decay is stated and its proof sketched. This decay statement is carefully contrasted with the type of statements derived heuristically in the physics literature for the asymptotic tails of individual spherical harmonics. Following this, our recent proof of the boundedness of solutions to the wave equation on axisymmetric stationary backgrounds (including slowly-rotating Kerr and Kerr-Newman) is reviewed and a new decay result for slowly-rotating Kerr spacetimes is stated and proved. This last result was announced at the summer school and appears in print here for the first time. A discussion of the analogue of these problems for spacetimes with a positive cosmological constant $\Lambda > 0$ follows. Finally, a general framework is given for capturing the red-shift effect for non-extremal black holes. This unifies and extends some of the analysis of the previous sections. The notes end with a collection of open problems. [This version has an Addendum reviewing subsequent developments up to December 2011.]

CONTENTS

2010 *Mathematics Subject Classification.* Primary 83C57, 83C05, 35L05.

1. Introduction: General relativity and evolution

Black holes are one of the fundamental predictions of general relativity. At the same time, they are one of its least understood (and most often misunderstood) aspects. These lectures intend to introduce the black hole concept and the analysis of waves on black hole backgrounds $(\mathcal{M}, g)$ by means of the example of the scalar wave equation

$$\Box_g \psi = 0. \tag{1}$$

We do not assume the reader is familiar with general relativity, only basic analysis and differential geometry. In this introductory section, we briefly describe general relativity in outline form, taking from the beginning the evolutionary point of view which puts the Cauchy problem for the *Einstein equations*–the system of nonlinear partial differential equations (see (2) below) governing the theory–at the centre. The problem (1) can be viewed as a poor man's linearisation for the Einstein equations. Study of (1) is then intimately related to the problem of the dynamic stability of the black hole spacetimes $(\mathcal{M}, g)$ themselves. Thus, one should view the subject of these lectures as intimately connected to the very tenability of the black hole concept in the theory.

1.1. General relativity and the Einstein equations. General relativity postulates a 4-dimensional Lorentzian manifold $(\mathcal{M}, g)$–*space-time*–which is to satisfy the *Einstein equations*

$$R_{\mu\nu} - \frac{1}{2} g_{\mu\nu} R = 8\pi T_{\mu\nu}. \tag{2}$$

Here, $R_{\mu\nu}$, R denote the *Ricci* and *scalar* curvature of g, respectively, and $T_{\mu\nu}$ denotes a symmetric 2-tensor on $\mathcal{M}$ termed the *stress-energy-momentum tensor* of matter. (Necessary background on Lorentzian geometry to understand the above notation is given in Appendix A.) The equations (2) in of themselves do not close, but must be coupled to "matter equations" satisfied by a collection $\{\Psi_i\}$ of matter fields defined on $\mathcal{M}$, together with a constitutive relation determining $T_{\mu\nu}$ from $\{g, \Psi_i\}$. These equations and relations are stipulated by the relevant continuum field theory (electromagnetism, fluid dynamics, etc.) describing the matter. The formulation of general relativity represents the culmination of the classical field-theoretic world-view where physics is governed by a closed system of partial differential equations.

Einstein was led to the system (2) in 1915, after a 7-year struggle to incorporate gravity into his earlier principle of relativity. In the field-theoretic formulation of the "Newtonian" theory, gravity was described by the *Newtonian potential* ϕ satisfying the *Poisson equation*

$$\triangle\phi = 4\pi\mu, \tag{3}$$

where μ denotes the *mass-density* of matter. It is truly remarkable that the constraints of consistency were so rigid that incorporating gravitation required finally a complete reworking of the principle of relativity, leading to a theory where Newtonian gravity, special relativity and Euclidean geometry each emerge as limiting aspects of one dynamic geometrical structure–the Lorentzian metric–naturally living on a 4-dimensional spacetime continuum. A second remarkable aspect of general relativity is that, in contrast to its Newtonian predecessor, the theory is non-trivial even in the absence of matter. In that case, we set $T_{\mu\nu} = 0$ and the system (2) takes the form

$$R_{\mu\nu} = 0. \tag{4}$$

The equations (4) are known as the *Einstein vacuum equations.* Whereas (3) is a linear elliptic equation, (4) can be seen to form a closed system of non-linear (but *quasilinear*) wave equations. Essentially all of the characteristic features of the dynamics of the Einstein equations are already present in the study of the vacuum equations (4).

1.2. Special solutions: Minkowski, Schwarzschild, Kerr. To understand a theory like general relativity where the fundamental equations (4) are non-linear, the first goal often is to identify and study important *explicit solutions*, i.e., solutions which can be written in closed form.[1] Much of the early history of general relativity centred around the discovery and interpretation of such solutions. The simplest explicit solution to the Einstein vacuum equations (4) is *Minkowski space* $\mathbb{R}^{3+1}$. The next simplest solution of (4) is the so-called *Schwarzschild solution*, written down [**139**] already in 1916. This is in fact a one-parameter family of solutions $(\mathcal{M}, g_M)$, the parameter M identified with *mass.* See (5) below for the metric form. The Schwarzschild family lives as a subfamily in a larger two-parameter family of explicit solutions $(\mathcal{M}, g_{M,a})$ known as the *Kerr solutions*, discussed in Section 5.1. These were discovered only much later [**99**] (1963).

When the Schwarzschild solution was first written down in local coordinates, the necessary concepts to understand its geometry had not yet been developed. It took nearly 50 years from the time when Schwarzschild was first discovered for its global geometry to be sufficiently well understood so as to be given a suitable name: Schwarzschild and Kerr were examples of what came to be known as *black hole* spacetimes[2]. The Schwarzschild solution also illustrates another feature of the Einstein equations, namely, the presence of singularities.

We will spend Section 2 telling the story of the emergence of the black hole notion and sorting out what the distinct notions of "black hole" and "singularity" mean. For the purpose of the present introductory section, let us take the notion of "black hole" as a "black box" and make some general remarks on the role of explicit solutions, whatever might be their properties. These remarks are relevant for any physical theory governed by an evolution equation.

[1] The traditional terminology in general relativity for such solutions is *exact solutions.*

[2] This name is due to John Wheeler.

1.3. Dynamics and the stability problem. Explicit solutions are indeed suggestive as to how general solutions behave, but only if they are appropriately "stable". In general relativity, this notion can in turn only be understood after the problem of dynamics for (4) has been formulated, that is to say, the *Cauchy problem*.

In contrast to other non-linear field theories arising in physics, in the case of general relativity, even formulating the Cauchy problem requires addressing several conceptual issues (e.g. in what sense is (4) hyperbolic?), and these took a long time to be correctly sorted out. Important advances in this process include the identification of the harmonic gauge by de Donder [**70**], the existence and uniqueness theorems for general quasilinear wave equations in the 1930's based on work of Friedrichs, Schauder, Sobolev, Petrovsky, Leray and others, and Leray's notion of global hyperbolicity [**112**]. The well-posedness of the appropriate Cauchy problem for the vacuum equations (4) was finally formulated and proven in celebrated work of Choquet-Bruhat [**33**] (1952) and Choquet-Bruhat–Geroch [**35**] (1969). See Appendix B for a concise survey of these developments and the precise statement of the existence and uniqueness theorems and some comments on their proof.

In retrospect, much of the confusion in early discussions of the Schwarzschild solution can be traced to the lack of a dynamic framework to understand the theory. It is only in the context of the language provided by [**35**] that one can then formulate the dynamical stability problem and examine the relevance of various explicit solutions.

The stability of Minkowski space was first proven in the monumental work of Christodoulou and Klainerman [**51**]. See Appendix B.5 for a formulation of this result. The dynamical stability of the Kerr family as a family of solutions to the Cauchy problem for the Einstein equations, even restricted to parameter values near Schwarzschild, i.e. $|a| \ll M$,[3] is yet to be understood and poses an important challenge for the mathematical study of general relativity in the coming years. See Section 5.6 for a formulation of this problem. In fact, even the most basic linear properties of waves (e.g. solutions of (1)) on Kerr spacetime backgrounds (or more generally, backgrounds near Kerr) have only recently been understood. In view of the wave-like features of the Einstein equations (4) (see in particular Appendix B.4), this latter problem should be thought of as a prerequisite for understanding the non-linear stability problem.

1.4. Outline of the lectures. The above linear problem will be the main topic of these lectures: We shall here develop from the beginning the study of the linear homogeneous wave equation (1) on fixed black hole spacetime backgrounds $(\mathcal{M}, g)$. We have already referred in passing to the content of some of the later sections. Let us give here a complete outline: Section 2 will introduce the black hole concept and the Schwarzschild geometry in the wider context of open problems in general relativity. Section 3 will concern the basic boundedness properties for solutions ψ of (1) on Schwarzschild exterior backgrounds. Section 4 will concern quantitative decay properties for ψ. Section 5 will move on to spacetimes $(\mathcal{M}, g)$ "near" Schwarzschild, including slowly rotating Kerr, discussing boundedness and decay properties for solutions to (1) on such $(\mathcal{M}, g)$, and ending in Section 5.6 with a formulation of the non-linear stability problem for Kerr, the open problem which

[3]Note that without symmetry assumptions one cannot study the stability problem for Schwarzschild per se. Only the larger Kerr family can be stable.

in some sense provides the central motivation for these notes. Section 6 will consider the analogues of these problems in spacetimes with a positive cosmological constant Λ, Section 7 will give a multiplier-type estimate valid for general non-degenerate Killing horizons which quantifies the classical red-shift effect. The importance of the red-shift effect as a stabilising mechanism for the analysis of waves on black hole backgrounds will be a common theme throughout these lectures. The notes end with a collection of open problems in Section 8.

The proof of Theorem 5.2 of Section 5 as well as all results of Section 7 appear in print in these notes for the first time. The discussion of Section 3.3 as well as the proof of Theorem 4.1 have also been streamlined in comparison with previous presentations. We have given a guide to background literature in Sections 3.4, 4.4, 5.5 and 6.3.

We have tried to strike a balance in these notes between making the discussion self-contained and providing the necessary background to appreciate the place of the problem (1) in the context of the current state of the art of the Cauchy problem for the Einstein equations (2) or (4) and the main open problems and conjectures which will guide this subject in the future. Our solution has been to use the history of the Schwarzschild solution as a starting point in Section 2 for a number of digressions into the study of gravitational collapse, singularities, and the weak and strong cosmic censorship conjectures, deferring, however, formal development of various important notions relating to Lorentzian geometry and the well-posedness of the Einstein equations to a series of Appendices. We have already referred to these appendices in the text. The informal nature of Section 2 should make it clear that the discussion is not intended as a proper survey, but merely to expose the reader to important open problems in the field and point to some references for further study. The impatient reader is encouraged to move quickly through Section 2 at a first reading. The problem (1) is itself rather self-contained, requiring only basic analysis and differential geometry, together with a good understanding of the black hole spacetimes, in particular, their so-called causal geometry. The discussion of Section 2 should be more than enough for the latter, although the reader may want to supplement this with a more general discussion, for instance [**55**].

These notes accompanied a series of lectures at a summer school on "Evolution Equations" organised by the Clay Mathematics Institute, June–July 2008. The centrality of the evolutionary point of view in general relativity is often absent from textbook discussions. (See however the recent [**133**].) We hope that these notes contribute to the point of view that puts general relativity at the centre of modern developments in partial differential equations of evolution.

2. The Schwarzschild metric and black holes

Practically all concepts in the development of general relativity and much of its history can be told from the point of view of the Schwarzschild solution. We now readily associate this solution with the black hole concept. It is important to remember, however, that the Schwarzschild solution was first discovered in a thoroughly classical astrophysical setting: it was to represent the vacuum region outside a star. The black hole interpretation–though in some sense inevitable–historically only emerged much later.

The most efficient way to present the Schwarzschild solution is to begin at the onset with Kruskal's maximal extension as a point of departure. Instead, we shall take advantage of the informal nature of the present notes to attempt a more conversational and "historical" presentation of the Schwarzschild metric and its interpretation.[4] Although certainly not the quickest route, this approach has the advantage of highlighting the themes which have become so important in the subject–in particular, singularities, black holes and their event horizons–with the excitement of their step-by-step unravelling from their origin in a model for the simplest of general relativistic stars. The Schwarzschild solution will naturally lead to discussions of the Oppenheimer-Snyder collapse model, the cosmic censorship conjectures, trapped surfaces and Penrose's incompleteness theorems, and recent work of Christodoulou on trapped surface formation in vacuum collapse, and we elaborate on these topics in Sections 2.6–2.8. (The discussion in these three last sections was not included in the lectures, however, and is not necessary for understanding the rest of the notes.)

2.1. Schwarzschild's stars. The most basic self-gravitating objects are stars. In the most primitive stellar models, dating from the 19th century, stars are modelled by a self-gravitating fluid surrounded by vacuum. Moreover, to a first approximation, classically stars are *spherically symmetric* and *static*.

It should not be surprising then that early research on the Einstein equations (2) would address the question of the existence and structure of general relativistic stars in the new theory. In view of our above discussion, the most basic problem is to understand spherically symmetric, static metrics, represented in coordinates (t, r, θ, ϕ), such that the spacetime has two regions: In the region $r \leq R_0$–the interior of the star–the metric should solve a suitable Einstein-matter system (2) with appropriate matter, and in the region $r \geq R_0$–the exterior of the star–the spacetime should be vacuum, i.e. the metric should solve (4).

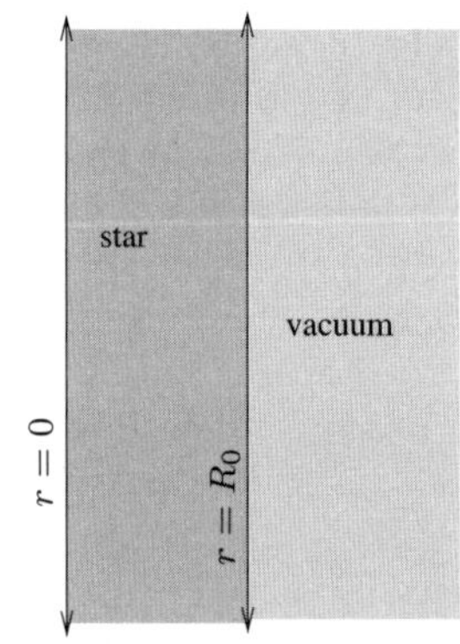

This is the problem first addressed by Schwarzschild [**139, 140**], already in 1916. Schwarzschild considered the vacuum region first [**139**] and arrived[5] at the

[4]This in no way should be considered as a true attempt at the history of the solution, simply a pedagogical approach to its study. See for example [**76**].

[5]As is often the case, the actual history is more complicated. Schwarzschild based his work on an earlier version of Einstein's theory which, while obtaining the correct vacuum equations, imposed a condition on admissible coordinate systems which would in fact exclude the coordinates of (5). Thus he had to use a rescaled r as a coordinate. Once this condition was removed from the theory, there is no reason not to take r itself as the coordinate. It is in this sense that these coordinates can reasonably be called "Schwarzschild coordinates".

one-parameter family of solutions:

$$g = -\left(1 - \frac{2M}{r}\right) dt^2 + \left(1 - \frac{2M}{r}\right)^{-1} dr^2 + r^2(d\theta^2 + \sin^2\theta\, d\phi^2). \tag{5}$$

Every student of this subject should explicitly check that this solves (4) (**Exercise**).

In [**140**], Schwarzschild found interior metrics for the darker shaded region $r \leq R_0$ above. In this region, matter is described by a perfect fluid. We shall not write down explicitly such metrics here, as this would require a long digression into fluids, their equations of state, etc. See [**44**]. Suffice it to say here that the existence of such solutions required that one take the constant M positive, and the value R_0 marking the boundary of the star always satisfied $R_0 > 2M$. The constant M could then be identified with the total mass of the star as measured by considering the orbits of far-away test particles.[6] In fact, for most reasonable matter models, static solutions of the type described above only exist under a stronger restriction on R_0 (namely $R_0 \geq 9M/4$) now known as the *Buchdahl inequality*. See [**14, 2, 97**].

The restriction on R_0 necessary for the existence of Schwarzschild's stars appears quite fortuitous: It is manifest from the form (5) that the components of g are singular if the (t, r, θ, ϕ) coordinate system for the vacuum region is extended to $r = 2M$. But a natural (if perhaps seemingly of only academic interest) question arises, namely, what happens if one does away completely with the star and tries simply to consider the expression (5) for all values of r? This at first glance would appear to be the problem of understanding the gravitational field of a "point particle" with the particle removed.[7]

For much of the history of general relativity, the degeneration of the metric functions at $r = 2M$, when written in these coordinates, was understood as meaning that the gravitational field should be considered singular there. This was the famous Schwarzschild "singularity".[8] Since "singularities" were considered "bad" by most pioneers of the theory, various arguments were concocted to show that the behaviour of g where $r = 2M$ is to be thought of as "pathological", "unstable", "unphysical" and thus, the solution should not be considered there. The constraint on R_0 related to the Buchdahl inequality seemed to give support to this point of view. See also [**75**].

With the benefit of hindsight, we now know that the interpretation of the previous paragraph is incorrect, on essentially every level: neither is $r = 2M$ a singularity, nor are singularities–which do in fact occur!–necessarily to be discarded! Nor is it true that non-existence of static stars renders the behaviour at $r = 2M$–whatever it is–"unstable" or "unphysical"; on the contrary, it was an early hint of gravitational collapse! Let us put aside this hindsight for now and try to discover for ourselves the geometry and "true" singularities hidden in (5), as well as the correct framework for identifying "physical" solutions. In so doing, we are retracing in part the steps of early pioneers who studied these issues without the benefit of the global geometric framework we now have at our disposal. All the notions referred to above will reveal themselves in the next subsections.

---

[6]Test particles in general relativity follow timelike geodesics of the spacetime metric. **Exercise**: Explain the statement claimed about far-away test particles. See also Appendix B.2.3.

[7]Hence the title of [**139**].

[8]Let the reader keep in mind that there is a good reason for the quotation marks here and for those that follow.

2.2. Extensions beyond the horizon. The fact that the behaviour of the metric at $r = 2M$ is not singular, but simply akin to the well-known breakdown of the coordinates (5) at $\theta = 0, \pi$ (this latter breakdown having never confused anyone...), is actually quite easy to see, and there is no better way to appreciate this than by doing the actual calculations. Let us see how to proceed.

First of all, before even attempting a change of coordinates, the following is already suggestive: Consider say a future-directed[9] ingoing radial null geodesic. The image of such a null ray is in fact depicted below:

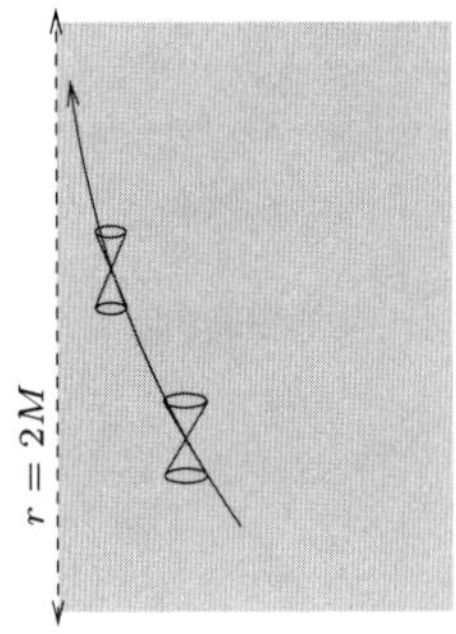

One can compute that this has finite affine length to the future, i.e. these null geodesics are *future-incomplete*, while scalar curvature invariants remain bounded as $s \to \infty$. It is an amusing **exercise** to put oneself in this point of view and carry out the above computations in these coordinates.

Of course, as such the above doesn't show anything.[10] But it turns out that indeed the metric can be extended to be defined on a "bigger" manifold. One defines a new coordinate

$$t^* = t + 2M \log(r - 2M).$$

This metric then takes the form

$$g = -\left(1 - \frac{2M}{r}\right)(dt^*)^2 + \frac{4M}{r}dt^*\, dr + \left(1 + \frac{2M}{r}\right)dr^2 + r^2 d\sigma_{\mathbb{S}^2} \tag{6}$$

on $r > 2M$. Note that $\frac{\partial}{\partial t^*} = \frac{\partial}{\partial t}$, each interpreted in its respective coordinate system. But now (6) can clearly be defined in the region $r > 0$, $-\infty < t^* < \infty$, and, by explicit computation or better, by analytic continuation, the metric (6) must satisfy (4) for all $r > 0$.

Transformations similar to the above were already known to Eddington and Lemaitre [**111**] in the early 1930's. Nonetheless, from the point of view of that time, it was difficult to interpret their significance. The formalisation of the manifold concept and associated language had not yet become common knowledge to physicists (or most mathematicians for that matter), and in any case, there was no selection principle as to what should the underlying manifold $\mathcal{M}$ be on which a solution g to (4) should live, or, to put it another way, the domain of g in (4) is not specified a priori by the theory. So, even if the solutions (6) exist, how do we know that they are "physical"?

---

[9]We time-orient the metric by ∂_t. See Appendix A.

[10]Consider for instance a cone with the vertex removed...

This problem can in fact only be clarified in the context of the *Cauchy problem* for (2) coupled to appropriate matter. Once the Cauchy problem for (4) is formulated correctly, then one can assign a *unique* spacetime to an appropriate notion of *initial data set*. This is the *maximal development* of Appendix B. It is only the initial data set, and the matter model, which can be judged for "physicality". One cannot throw away the resulting maximal development just because one does not like its properties!

From this point of view, the question of whether the extension (6) was "physical" was resolved in 1939 by Oppenheimer and Snyder [**125**]. Specifically, they showed that the extension (6) for $t \geq 0$ arose as a subset of the solution to the Einstein equations coupled to a reasonable (to a first approximation at least) matter model, evolving from physically plausible initial data. With hindsight, the notion of black hole was born in that paper.

Had history proceeded differently, we could base our further discussion on [**125**]. Unfortunately, the model [**125**] was ahead of its time. As mentioned in the introduction, the proper language to formulate the Cauchy problem in general only came in 1969 [**35**]. The interpretation of explicit solutions remained the main route to understanding the theory. We will follow thus this route to the black hole concept–via the geometric study of so-called maximally extended Schwarzschild–even though this spacetime is not to be regarded as "physical". It was through the study of this spacetime that the relevant notions were first understood and the important Penrose diagrammatic notation was developed. We shall return to [**125**] only in Section 2.5.3.

2.3. The maximal extension of Synge and Kruskal. Let us for now avoid the question of what the underlying manifold "should" be, a question whose answer requires physical input (see paragraphs above), and simply ask the purely mathematical question of how big the underlying manifold "can" be. This leads to the notion of a "maximally extended" solution. In the case of Schwarzschild, this will be a spacetime which, although not to be taken as a model for anything *per se*, can serve as a reference for the formulation of all important concepts in the subject.

To motivate this notion of "maximally extended" solution, let us examine our first extension a little more closely. The light cones can be drawn as follows:

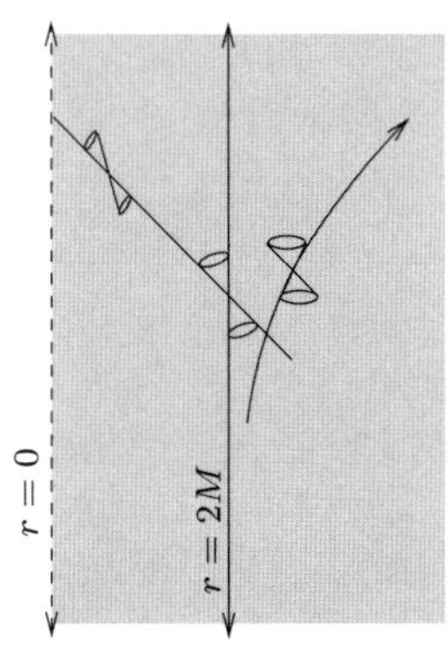

Let us look say at null geodesics. One can see (**Exercise**) that future directed null geodesics either approach $r = 0$ or are future-complete. In the former case, scalar invariants of the curvature blow up in the limit as the affine parameter approaches its supremum (**Exercise**). The spacetime is thus "singular" in this sense. It thus follows from the above properties that the above spacetime is *future null geodesically incomplete*, but also *future null geodesically inextendible* as a C^2 Lorentzian metric, i.e. there does not exist a larger 4-dimensional Lorentzian manifold with C^2 metric such that the spacetime above embeds isometrically into the larger one such that a future null geodesic passes into the extension.

On the other hand, one can see that past-directed null geodesics are not all complete, yet no curvature quantity blows up along them (**Exercise**). Again, this suggests that something may still be missing!

Synge was the first to consider these issues systematically and construct "maximal extensions" of the original Schwarzschild metric in a paper [**146**] of 1950. A more concise approach to such a construction was given in a celebrated 1960 paper [**107**] of Kruskal. Indeed, let $\mathcal{M}$ be the manifold with differentiable structure given by $\mathcal{U} \times \mathbb{S}^2$ where $\mathcal{U}$ is the open subset $T^2 - R^2 < 1$ of the (T, R)-plane. Consider the metric g

$$g = \frac{32M^3}{r} e^{-r/2M}(-dT^2 + dR^2) + r^2 d\sigma_{\mathbb{S}}^2$$

where r is defined implicitly by

$$T^2 - R^2 = \left(1 - \frac{r}{2M}\right) e^{r/2M}.$$

The region $\mathcal{U}$ is depicted below:

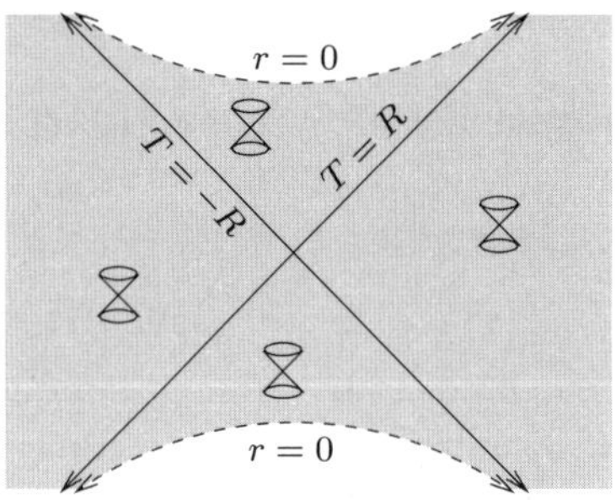

This is a spherically symmetric 4-dimensional Lorentzian manifold satisfying (4) such that the original Schwarzschild metric is isometric to the region $R > |T|$ (where t is given by $\tanh\left(\frac{t}{4M}\right) = T/R$), and our previous partial extension is isometric to the region $T > -R$ (**Exercise**). It can be shown now (**Exercise**) that $(\mathcal{M}, g)$ is *inextendible as a C^2 (in fact C^0) Lorentzian manifold*, that is to say, if

$$i : (\mathcal{M}, g) \to (\widetilde{\mathcal{M}}, \tilde{g})$$

is an isometric embedding, where $(\widetilde{M}, \tilde{g})$ is a C^2 (in fact C^0) 4-dimensional Lorentzian manifold, then necessarily $i(\mathcal{M}) = \widetilde{\mathcal{M}}$.

The above property defines the sense in which our spacetime is "maximally" extended, and thus, $(\mathcal{M}, g)$ is called sometimes *maximally-extended Schwarzschild*. In later sections, we will often just call it "the Schwarzschild solution".

Note that the form of the metric is such that the light cones are as depicted. Thus, one can read off much of the causal structure by sight.

It may come as a surprise that in maximally-extended Schwarzschild, there are *two* regions which are isometric to the original $r > 2M$ Schwarzschild region.

Alternatively, a Cauchy surface[11] will have topology $\mathbb{S}^2 \times \mathbb{R}$ with *two* asymptotically flat ends. This suggests that this spacetime is not to be taken as a physical model. We will discuss this later on. For now, let us simply try to understand better the global geometry of the metric.

2.4. The Penrose diagram of Schwarzschild. There is an even more useful way to represent the above spacetime. First, let us define null coordinates $U = T - R$, $V = T + R$. These coordinates have infinite range. We may rescale them by $u = u(U)$, $v = v(V)$ to have finite range. (Note the freedom in the choice of u and v!) The domain of (u, v) coordinates, when represented in the plane where the axes are at 45 and 135 degrees with the horizontal, is known as a *Penrose diagram* of Schwarzschild. Such a Penrose diagram is depicted below[12]:

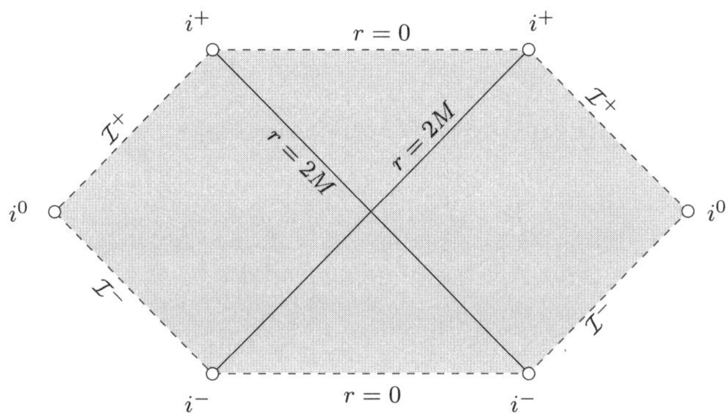

In more geometric language, one says that a Penrose diagram corresponds to the image of a bounded conformal map

$$\mathcal{M}/\mathrm{SO}(3) = \mathcal{Q} \to \mathbb{R}^{1+1},$$

where one makes the identification $v = t+x$, $u = t-x$ where (t, x) are now the standard coordinates $\mathbb{R}^{1+1}$ represented in the standard way on the plane. We further assume that the map preserves the time orientation, where Minkowski space is oriented by ∂_t. (In our application, this is a fancy way of saying that $u'(U), v'(V) > 0$). It follows that the map preserves the causal structure of $\mathcal{Q}$. In particular, we can "read off" the radial null geodesics of $\mathcal{M}$ from the depiction.

Now we may turn to the boundary induced by the causal embedding. We define $\mathcal{I}^\pm$ to be the boundary components as depicted.[13] These are characterized geometrically as follows: $\mathcal{I}^+$ are limit points of future-directed null rays in $\mathcal{Q}$ along which $r \to \infty$. Similarly, $\mathcal{I}^-$ are limit points of past-directed null rays for which $r \to \infty$. We call $\mathcal{I}^+$ *future null infinity* and $\mathcal{I}^-$ *past null infinity*. The remaining boundary components i^0 and $i^\pm$ depicted are often given the names *spacelike* infinity and *future (past) timelike* infinity, respectively.

In the physical application, it is important to remember that asymptotically flat[14] spacetimes like our $(\mathcal{M}, g)$ are not meant to represent the whole universe[15],

[11] See Appendix A.

[12] How can (u, v) be chosen so that the $r = 0$ boundaries are horizontal lines? (**Exercise**)

[13] Our convention is that open endpoint circles are not contained in the intervals they bound, and dotted lines are not contained in the regions they bound, whereas solid lines are.

[14] See Appendix B.2.3 for a definition.

[15] The study of that problem is what is known as "cosmology". See Section 6.

but rather, the gravitational field in the vicinity of an isolated self-gravitating system. $\mathcal{I}^+$ is an idealisation of far away observers who can receive radiation from the system. In this sense, "we"–as astrophysical observers of stellar collapse, say–are located at $\mathcal{I}^+$. The ambient causal structure of $\mathbb{R}^{1+1}$ allows us to talk about $J^-(p) \cap \mathcal{Q}$ for $p \in \mathcal{I}^+$[16] and this will lead us to the black hole concept. Therein lies the use of the Penrose diagram representation.

The systematic use of the conformal point of view to represent the global geometry of spacetimes is one of the many great contributions of Penrose to general relativity. These representations can be traced back to the well-known "spacetime diagrams" of special relativity, promoted especially by Synge [**147**]. The "formal" use of Penrose diagrams in the sense above goes back to Carter [**28**], in whose hands these diagrams became a powerful tool for determining the global structure of all classical black hole spacetimes. It is hard to overemphasise how important it is for the student of this subject to become comfortable with these representations.

2.5. The black hole concept. With Penrose diagram notation, we may now explain the black hole concept.

2.5.1. *The definitions for Schwarzschild.* First an important remark: In Schwarzschild, the boundary component $\mathcal{I}^+$ enjoys a limiting affine completeness. More specifically, normalising a sequence of ingoing radial null vectors by parallel transport along an outgoing geodesic meeting $\mathcal{I}^+$, the affine length of the null geodesics generated by these vectors, parametrized by their parallel transport (restricted to $J^-(\mathcal{I}^+)$), tends to infinity:

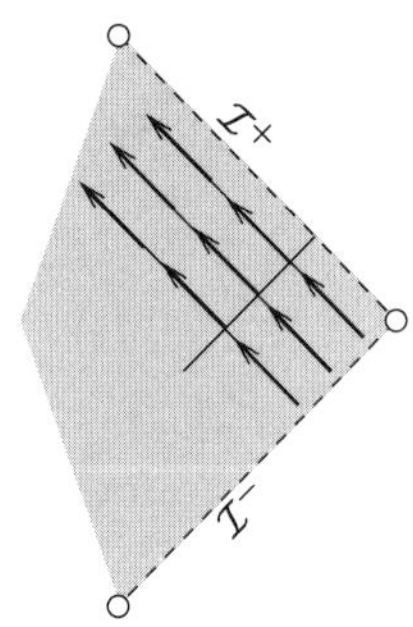

This has the interpretation that far-away observers in the radiation zone can observe for all time. (This is in some sense related to the presence of timelike geodesics near infinity of infinite length, but the completeness is best formulated with respect to $\mathcal{I}^+$.) A similar statement clearly holds for $\mathcal{I}^-$.

Given this completeness property, we define now the *black hole* region to be $\mathcal{Q} \setminus J^-(\mathcal{I}^+)$, and the *white hole* region to be $\mathcal{Q} \setminus J^+(\mathcal{I}^-)$. Thus, the black hole corresponds to those points of spacetime which cannot "send signals" to future null infinity, or, in the physical interpretation, to far-away observers who (in view of the completeness property!) nonetheless can observe radiation for infinite time.

The future boundary of $J^-(\mathcal{I}^+)$ in $\mathcal{Q}$ (alternatively characterized as the past boundary of the black hole region) is a null hypersurface known as the *future event horizon*, and is denoted by $\mathcal{H}^+$. Exchanging past and future, we obtain the *past event horizon* $\mathcal{H}^-$. In maximal Schwarzschild, $\{r = 2M\} = \mathcal{H}^+ \cup \mathcal{H}^-$. The subset $J^-(\mathcal{I}^+) \cap J^+(\mathcal{I}^-)$ is known as the *domain of outer communications*.

[16]Refer to Appendix A for $J^\pm$.

2.5.2. *Minkowski space.* Note that in the case of Minkowski space, $\mathcal{Q} = \mathbb{R}^{3+1}/\mathrm{SO}(3)$ is a manifold with boundary since the SO(3) action has a locus of fixed points, the centre of symmetry. A Penrose diagram of Minkowski space is easily seen to be:

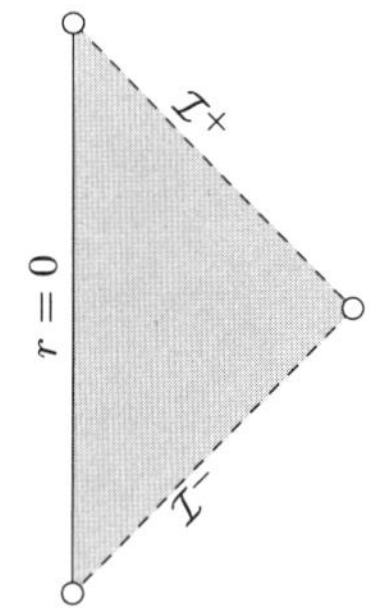

Here $\mathcal{I}^+$ and $\mathcal{I}^-$ are characterized as before, and enjoy the same completeness property as in Schwarzschild. One reads off immediately that $J^-(\mathcal{I}^+) \cap \mathcal{Q} = \mathcal{Q}$, i.e. $\mathbb{R}^{3+1}$ does not contain a black hole under the above definitions.

2.5.3. *Oppenheimer-Snyder.* Having now the notation of Penrose diagrams, we can concisely describe the geometry of the Oppenheimer-Snyder solutions referred to earlier, without giving explicit forms of the metric. Like Schwarzschild's original picture of the gravitational field of a spherically symmetric star, these solutions involve a region $r \leq R_0$ solving (2) and $r \geq R_0$ satisfying (4). The matter is described now by a pressureless fluid which is initially assumed homogeneous in addition to being spherically symmetric. The assumption of staticity is however dropped, and for appropriate initial conditions, it follows that $R_0(t^*) \to 0$ with respect to a suitable time coordinate t^*. (In fact, the Einstein equations can be reduced to an o.d.e. for $R_0(t^*)$.) We say that the star "collapses".[17] A Penrose diagram of such a solution (to the future of a Cauchy hypersurface) can be seen to be of the form:

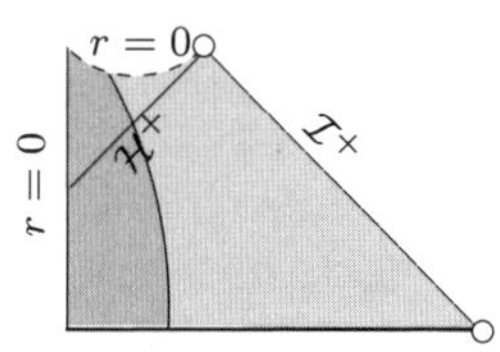

The lighter shaded region is isometric to a subset of maximal Schwarzschild, in fact a subset of the original extension of Section 2.2. In particular, the completeness property of $\mathcal{I}^+$ holds, and as before, we identify the black hole region to be $\mathcal{Q} \setminus J^-(\mathcal{I}^+)$.

In contrast to maximal Schwarzschild, where the initial configuration is unphysical (the Cauchy surface has two ends and topology $\mathbb{R} \times \mathbb{S}^2$), here the initial configuration is entirely plausible: the Cauchy surface is topologically $\mathbb{R}^3$, and its geometry is not far from Euclidean space. The Oppenheimer-Snyder model [**125**] should be viewed as the most basic black hole solution arising from physically plausible regular initial data.[18]

[17]Note that $R_0(t^*) \to 0$ does not mean that the star collapses to "a point", merely that the spheres which foliate the interior of the star shrink to 0 area. The limiting singular boundary is a spacelike hypersurface as depicted.

[18] Note however the end of Section 2.6.2.

It is traditional in general relativity to "think" Oppenheimer-Snyder but "write" maximally-extended Schwarzschild. In particular, one often imports terminology like "collapse" in discussing Schwarzschild, and one often reformulates our definitions replacing $\mathcal{I}^+$ with one of its connected components, that is to say, we will often write $J^-(\mathcal{I}^+) \cap J^+(\mathcal{I}^+)$ meaning $J^-(\mathcal{I}^+_A) \cap J^+(\mathcal{I}^-_A)$, etc. In any case, the precise relation between the two solutions should be clear from the above discussion. In view of Cauchy stability results [**91**], sufficiently general theorems about the Cauchy problem on maximal Schwarzschild lead immediately to such results on Oppenheimer-Snyder. (See for instance the exercise in Section 3.2.6.) One should always keep this relation in mind.

2.5.4. *General definitions?* The above definition of black hole for the Schwarzschild metric should be thought of as a blueprint for how to define the notion of black hole region in general. That is to say, to define the black hole region, one needs

(1) some notion of future null infinity $\mathcal{I}^+$,
(2) a way of identifying $J^-(\mathcal{I}^+)$, and
(3) some characterization of the "completeness" of $\mathcal{I}^+$.[19]

If $\mathcal{I}^+$ is indeed complete, we can define the black hole region as

"the complement in $\mathcal{M}$ of $J^-(\mathcal{I}^+)$".

For spherically symmetric spacetimes arising as solutions of the Cauchy problem for (2), one can show that there always exists a Penrose diagram, and thus, a definition can be formalised along precisely these lines (see [**60**]). For spacetimes without symmetry, however, even defining the relevant asymptotic structure so that this structure is compatible with the theorems one is to prove is a main part of the problem. This has been accomplished definitively only in the case of perturbations of Minkowski space. In particular, Christodoulou and Klainerman [**51**] have shown that spacetimes arising from perturbations of Minkowski initial data have a complete $\mathcal{I}^+$ in a well defined sense, whose past can be identified and is indeed the whole spacetime. See Appendix B.5. That is to say, small perturbations of Minkowski space cannot form black holes.

2.6. Birkhoff's theorem. Formal Penrose diagrams are a powerful tool for understanding the global causal structure of spherically symmetric spacetimes. Unfortunately, however, it turns out that the study of spherically symmetric *vacuum* spacetimes is not that rich. In fact, the Schwarzschild family parametrizes all spherically symmetric vacuum spacetimes in a sense to be explained in this section.

2.6.1. *Schwarzschild for $M < 0$.* Before stating the theorem, recall that in discussing Schwarzschild we have previously restricted to parameter value $M > 0$. For the uniqueness statement, we must enlarge the family to include all parameter values.

If we set $M = 0$ in (5), we of course obtain Minkowski space in spherical polar coordinates. A suitable maximal extension is Minkowski space as we know it, represented by the Penrose diagram of Section 2.5.2.

---

[19]The characterization of completeness can be formulated for general asymptotically flat vacuum space times using the results of [**51**]. This formulation is due to Christodoulou [**47**]. Previous attempts to formalise these notions rested on "asymptotic simplicity" and "weak asymptotic simplicity". See [**91**]. Although the qualitative picture suggested by these notions appears plausible, the detailed asymptotic behaviour of solutions to the Einstein equations turns out to be much more subtle, and Christodoulou has proven [**48**] that these notions cannot capture even the simplest generic physically interesting systems.

On the other hand, we may also take $M < 0$ in (5). This is so-called *negative mass Schwarzschild*. The metric element (5) for such M is now regular for all $r > 0$. The limiting singular behaviour of the metric at $r = 0$ is in fact essential, i.e. one can show that along inextendible incomplete geodesics the curvature blows up. Thus, one immediately arrives at a maximally extended solution which can be seen to have Penrose diagram:

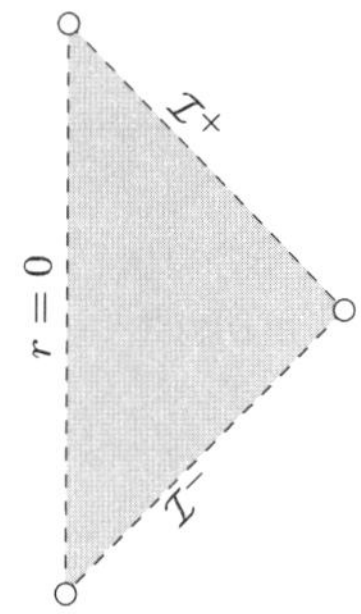

Note that in contrast to the case of $\mathbb{R}^{3+1}$, the boundary $r = 0$ is here depicted by a dotted line denoting (according to our conventions) that it is not part of $\mathcal{Q}$!

2.6.2. *Naked singularities and weak cosmic censorship.* The above spacetime is interpreted as having a "naked singularity". The traditional way of describing this in the physics literature is to remark that the "singularity" $\mathcal{B} = \{r = 0\}$ is "visible" to $\mathcal{I}^+$, i.e., $J^-(\mathcal{I}^+) \cap \mathcal{B} \neq \emptyset$. From the point of view of the Cauchy problem, however, this characterization is meaningless because the above maximal extension is not globally hyperbolic, i.e. it is not uniquely characterized by an appropriate notion of initial data.[20] From the point of view of the Cauchy problem, one must not consider maximal extensions but the *maximal Cauchy development of initial data*, which by definition is globally hyperbolic (see Theorem B.4 of Appendix B). Considering an inextendible spacelike hypersurface Σ as a Cauchy surface, the maximal Cauchy development of Σ would be the darker shaded region depicted below:

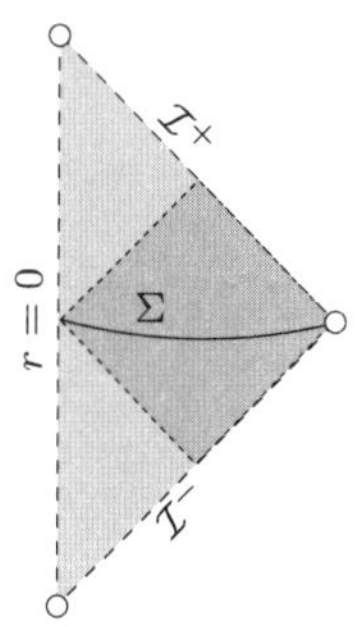

The proper characterization of "having a naked singularity", from the point of view of the darker shaded spacetime, is that its $\mathcal{I}^+$ is incomplete. Of course, this example does not say anything about the dynamic formation of naked singularities, because the inital data hypersurface Σ is already in some sense "singular", for instance, it is geodesically incomplete, and the curvature blows up along incomplete geodesics. The dynamic formation of a naked singularity from regular, complete initial data

[20]See Appendix A for the definition of global hyperbolicity.

would be pictured by:

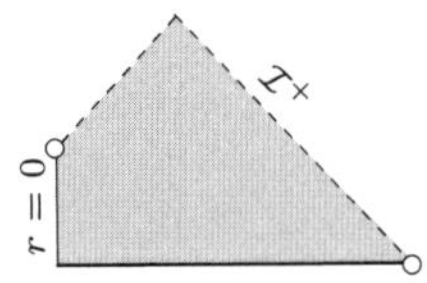

where we are to understand also in the above that $\mathcal{I}^+$ is incomplete. The conjecture that for generic asymptotically flat[21] initial data for "reasonable" Einstein-matter systems, the maximal Cauchy development "possesses a complete $\mathcal{I}^+$" is known as *weak cosmic censorship.*[22]

In light of the above conjecture, the story of the Oppenheimer-Snyder solution and its role in the emergence of the black hole concept does have an interesting epilogue. Recall that in the Oppenheimer-Snyder solutions, the region $r \leq R_0$, in addition to being spherically symmetric, is homogeneous. It turns out that by considering spherically symmetric initial data for which the "star" is no longer homogeneous, Christodoulou has proven that one can arrive at spacetimes for which "naked singularities" form [**39**] with Penrose diagram as above and with $\mathcal{I}^+$ incomplete. Moreover, it is shown in [**39**] that this occurs for *an open subset* of initial data within spherical symmetry, with respect to a suitable topology on the set of spherically symmetric initial data. Thus, weak cosmic censorship is violated in this model, at least if the conjecture is restricted to spherically symmetric data.

The fact that in the Oppenheimer-Snyder solutions black holes formed appears thus to be a rather fortuitous accident! Nonetheless, we should note that the failure of weak cosmic censorship in this context is believed to be due to the inappropriateness of the pressureless model, not as indicative of actual phenomena. Hence, the restriction on the matter model to be "reasonable" in the formulation of the conjecture. In a remarkable series of papers, Christodoulou [**45, 47**] has shown weak cosmic censorship to be true for the Einstein-scalar field system under spherical symmetry. On the other hand, he has also shown [**43**] that the assumption of genericity is still necessary by explicitly constructing solutions of this system with incomplete $\mathcal{I}^+$ and Penrose diagram as depicted above.[23]

2.6.3. *Birkhoff's theorem.* Let us understand now by "Schwarzschild solution with parameter M" (where $M \in \mathbb{R}$) the maximally extended Schwarzschild metrics described above.

We have the so-called *Birkhoff's theorem*:

THEOREM 2.1. *Let $(\mathcal{M}, g)$ be a spherically symmetric solution to the vacuum equations* (4). *Then it is locally isometric to a Schwarzschild solution with parameter M, for some $M \in \mathbb{R}$.*

In particular, spherically symmetric solutions to (4) possess an additional Killing field not in the Lie algebra so(3). (**Exercise**: Prove Theorem 2.1. Formulate and prove a global version of the result.)

[21]See Appendix B.2.3 for a formulation of this notion. Note that asymptotically flat data are in particular complete.

[22]This conjecture is originally due to Penrose [**127**]. The present formulation is taken from Christodoulou [**47**].

[23]The discovery [**43**] of these naked singularities led to the discovery of so-called critical collapse phenomena [**37**] which has since become a popular topic of investigation [**87**].

2.6.4. *Higher dimensions.* In $3+1$ dimensions, spherical symmetry is the only symmetry assumption compatible with asymptotic flatness (see Appendix B.2.3), such that moreover the symmetry group acts transitively on 2-dimensional orbits. Thus, Birkhoff's theorem means that vacuum gravitational collapse cannot be studied in a $1+1$ dimensional setting by imposing symmetry. The simplest models for dynamic gravitational collapse thus necessarily involve matter, as in the Oppenheimer-Snyder model [**125**] or the Einstein-scalar field system studied by Christodoulou [**41, 45**]

Moving, however, to $4+1$ dimensions, asymptotically flat manifolds can admit a more general $SU(2)$ symmetry acting transitively on 3-dimensional group orbits. The Einstein vacuum equations (4) under this symmetry admit 2 dynamical degrees of freedom and can be written as a nonlinear system on a $1+1$-dimensional Lorentzian quotient $\mathcal{Q} = \mathcal{M}/SU(2)$, where the dynamical degrees of freedom of the metric are reflected by two nonlinear scalar fields on $\mathcal{Q}$. This symmetry–known as "Triaxial Bianchi IX"–was first identified by Bizon, Chmaj and Schmidt [**16, 17**] who derived the equations on $\mathcal{Q}$ and studied the resulting system numerically. The symmetry includes spherical symmetry as a special case, and thus, is admitted in particular by $4+1$-dimensional Schwarzschild[24]. The nonlinear stability of the Schwarzschild family as solutions of the vacuum equations (4) can then be studied–within the class of Triaxial Bianchi IX initial data–as a $1+1$ dimensional problem. Asymptotic stability for the Schwarzschild spacetime in this setting has been recently shown in the thesis of Holzegel [**93, 62, 94**], adapting vector field multiplier estimates similar to Section 4 to a situation where the metric is not known a priori. The construction of the relevant mutipliers is then quite subtle, as they must be normalised "from the future" in a bootstrap setting. The thesis [**93**] is a good reference for understanding the relation of the linear theory to the non-linear black hole stability problem. See also Open problem 13 in Section 8.6.

2.7. Geodesic incompleteness and "singularities". Is the picture of gravitational collapse as exhibited by Schwarzschild (or better, Oppenheimer-Snyder) stable? This question is behind the later chapters in the notes, where essentially the considerations hope to be part of a future understanding of the stability of the exterior region up to the event horizon, i.e. the closure of the past of null infinity to the future of a Cauchy surface. (See Section 5.6 for a formulation of this open problem.) What is remarkable, however, is that there is a feature of Schwarzschild which can easily be shown to be "stable", without understanding the p.d.e. aspects of (2): its geodesic incompleteness.

2.7.1. *Trapped surfaces.* First a definition: Let $(\mathcal{M}, g)$ be a time-oriented Lorentzian manifold, and S a closed spacelike 2-surface. For any point $p \in S$, we may define two null mean curvatures $\mathrm{tr}\chi$ and $\mathrm{tr}\bar{\chi}$, corresponding to the two future-directed null vectors $n(x)$, $\bar{n}(x)$, where n, $\bar{n}$ are normal to S at x. We say that S is *trapped* if $\mathrm{tr}\chi < 0$, $\mathrm{tr}\bar{\chi} < 0$.

Exercise: Show that points $p \in \mathcal{Q} \setminus \mathrm{clos}(J^-(\mathcal{I}^+))$ correspond to trapped surfaces of $\mathcal{M}$. Can there be other trapped surfaces? (Refer also for instance to [**12**].)

[24]**Exercise**: Work out explicitly the higher dimensional analogue of the Schwarzschild solution for all dimensions.

2.7.2. *Penrose's incompleteness theorem.*

THEOREM 2.2. *(Penrose 1965* [**126**]*) Let* $(\mathcal{M}, g)$ *be globally hyperbolic*[25] *with non-compact Cauchy surface* Σ*, where* g *is a* C^2 *metric, and let*

$$R_{\mu\nu}V^\mu V^\nu \geq 0 \tag{7}$$

for all null vectors V*. Then if* $\mathcal{M}$ *contains a closed trapped two-surface* S*, it follows that* $(\mathcal{M}, g)$ *is future causally geodesically incomplete.*

This is the celebrated *Penrose incompleteness theorem.*

Note that solutions of the Einstein vacuum equations (4) satisfy (7). (Inequality (7), known as the null convergence condition, is also satisfied for solutions to the Einstein equations (2) coupled to most plausible matter models, specifically, if the energy momentum tensor $T_{\mu\nu}$ satisfies $T_{\mu\nu}V^\mu V^\nu \geq 0$ for all null V^μ.) On the other hand, by definition, the unique solution to the Cauchy problem (the so-called *maximal Cauchy development of initial data*) is globally hyperbolic (see Appendix B.3). Thus, the theorem applies to the maximal development of (say) asymptotically flat (see Appendix B.2.3) vacuum initial data containing a trapped surface. Note finally that by Cauchy stability [**91**], the presence of a trapped surface in $\mathcal{M}$ is clearly "stable" to perturbation of initial data.

From the point of view of gravitational collapse, it is more appropriate to define a slightly different notion of trapped. We restrict to $S \subset \Sigma$ a Cauchy surface such that S bounds a disc in Σ. We then can define a unique outward null vector field n along S, and we say that S is trapped if $\mathrm{tr}\chi < 0$ and antitrapped[26] if $\mathrm{tr}\underline{\chi} > 0$, where $\mathrm{tr}\underline{\chi}$ denotes the mean curvature with respect to a conjugate "inward" null vector field. The analogue of Penrose's incompleteness theorem holds under this definition. One may also prove the interesting result that antitrapped surface cannot form if they are not present initially. See [**49**].

Note finally that there are related incompleteness statements due to Penrose and Hawking [**91**] relevant in cosmological (see Section 6) settings.

2.7.3. *"Singularities" and strong cosmic censorship.* Following [**49**], we have called Theorem 2.2 an "incompleteness theorem" and not a "singularity theorem". This is of course an issue of semantics, but let us further discuss this point briefly as it may serve to clarify various issues. The term "singularity" has had a tortuous history in the context of general relativity. As we have seen, its first appearance was to describe something that turned out not to be a singularity at all–the "Schwarzschild singularity". It was later realised that behaviour which could indeed reasonably be described by the word "singularity" did in fact occur in solutions, as exemplified by the $r = 0$ singular "boundary" of Schwarzschild towards which curvature scalars blow up. The presence of this singular behaviour "coincides" in Schwarzschild with the fact that the spacetime is future causally geodesically incomplete–in fact, the curvature blows up along all incomplete causal geodesics. In view of the fact that it is the incompleteness property which can be inferred from Theorem 2.2, it was tempting to redefine "singularity" as geodesic incompleteness (see [**91**]) and to call Theorem 2.2 a "singularity theorem".

This is of course a perfectly valid point of view. But is it correct then to associate the incompleteness of Theorem 2.2 to "singularity" in the sense of "breakdown" of the metric? Breakdown of the metric is most easily understood with

[25] See Appendix A.

[26] Note that there exist other conventions in the literature for this terminology. See [**12**].

curvature blowup as above, but more generally, it is captured by the notion of "inextendibility" of the Lorentzian manifold in some regularity class. We have already remarked that maximally-extended Schwarzschild is inextendible in the strongest of senses, i.e. as a C^0 Lorentzian metric. It turns out, however, that the statement of Theorem 2.2, even when applied to the maximal development of complete initial data for (4), is compatible with the solution being extendible as a C^∞ Lorentzian metric such that *every* incomplete causal geodesic of the original spacetime enter the extension! This is in fact what happens in the case of Kerr initial data. (See Section 5.1 for a discussion of the Kerr metric.) The reason that the existence of such extensions does not contradict the "maximality" of the "maximal development" is that these extensions fail to be globally hyperbolic, while the "maximal development" is "maximal" in the class of globally hyperbolic spacetimes (see Theorem B.4 of Appendix B). In the context of Kerr initial data, Theorem 2.2 is thus not saying that breakdown of the metric occurs, merely that globally hyperbolicity breaks down, and thus further extensions cease to be predictable from initial data.[27]

A similar phenomenon is exhibited by the Reissner-Nordström solution of the Einstein-Maxwell equations [**91**], which, unlike Kerr, is spherically symmetric and thus admits a Penrose diagram representation:

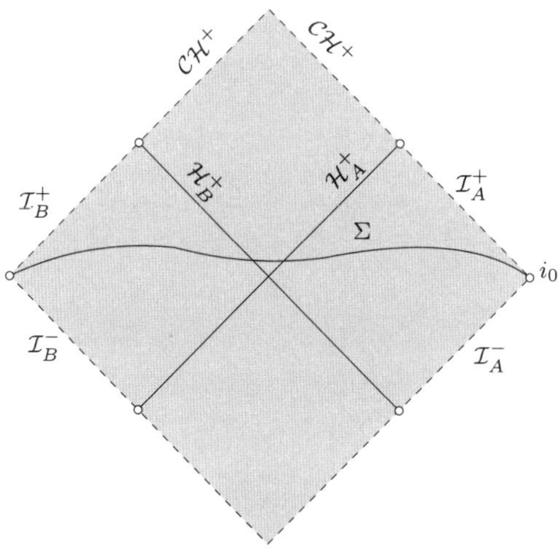

What is drawn above is the maximal development of Σ. The spacetime is future causally geodesically incomplete, but can be extended smoothly to a $(\tilde{\mathcal{M}}, \tilde{g})$ such that all inextendible geodesics leave the original spacetime. The boundary of $(\mathcal{M}, g)$ in the extension corresponds to $\mathcal{CH}^+$ above. Such boundaries are known as *Cauchy horizons.*

The *strong cosmic censorship conjecture* says that the maximal development of *generic* asymptotically flat initial data for the vacuum Einstein equations is

[27]Further confusion can arise from the fact that "maximal extensions" of Kerr constructed with the help of analyticity are still geodesically incomplete and inextendible, in particular, with the curvature blowing up along all incomplete causal geodesics. Thus, one often talks of the "singularities" of Kerr, referring to the ideal singular boundaries one can attach to such extensions. One must remember, however, that these extensions are of no relevance from the point of view of the Cauchy problem, and in any case, their singular behaviour in principle has nothing to do with Theorem 2.2.

inextendible as a suitably regular Lorentzian metric.[28] One can view this conjecture as saying that whenever one has geodesic incompleteness, it is due to breakdown of the metric in the sense discussed above. (In view of the above comments, for this conjecture to be true, the behaviour of the Kerr metric described above would have to be *unstable* to perturbation.[29]) Thus, if by the term "singularity" one wants to suggest "breakdown of the metric", it is only a positive resolution of the strong cosmic censorship conjecture that would in particular (generically) make Theorem 2.2 into a true "singularity theorem".

2.8. Christodoulou's work on trapped surface formation in vacuum. These notes would not be complete without a brief discussion of the recent breakthrough by Christodoulou [**53**] on the understanding of trapped surface formation for the vacuum.

The story begins with Christodoulou's earlier [**41**], where a condition is given ensuring that trapped surfaces form for spherically symmetric solutions of the Einstein-scalar field system. The condition is that the difference in so-called Hawking mass m of two concentric spheres on an outgoing null hypersurface be sufficiently large with respect to the difference in area radius r of the spheres. This is a surprising result as it shows that trapped surface formation can arise from initial conditions which are as close to dispersed as possible, in the sense that the supremum of the quantity $2m/r$ can be taken arbitrarily small initially.

The results of [**41**] lead immediately (see for instance [**61**]) to the existence of smooth spherically symmetric solutions of the Einstein-scalar field system with Penrose diagram

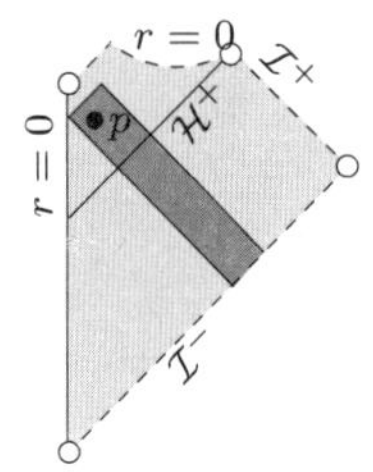

where the point p depicted corresponds to a trapped surface, and the spacetime is past geodesically complete with a complete past null infinity, whose future is the entire spacetime, i.e., the spacetime contains no white holes.[30] Thus, black hole formation can arise from spacetimes with a complete regular past.[31]

---

[28]As with weak cosmic censorship, the original formulation of this conjecture is due to Penrose [**128**]. The formulation given here is from [**47**]. Related formulations are given in [**54, 118**]. One can also pose the conjecture for compact initial data, and for various Einstein-matter systems. It should be emphasised that "strong cosmic censorship" does not imply "weak cosmic censorship". For instance, one can imagine a spacetime with Penrose diagram as in the last diagram of Section 2.6.2, with incomplete $\mathcal{I}^+$, but still inextendible across the null "boundary" emerging from the centre.

[29]Note that the instability concerns a region "far inside" the black hole interior. The black hole exterior is expected to be stable (as in the formulation of Section 5.6), hence these notes. See [**58, 59**] for the resolution of a spherically symmetric version of this problem, where the role of the Kerr metric is played by Reissner-Nordström metrics.

[30]The triangle "under" the darker shaded region can in fact be taken to be Minkowski.

[31]The singular boundary in general consists of a possibly empty null component emanating from the regular centre, and a spacelike component where $r = 0$ in the limit and across which the spacetime is inextendible as a C^0 Lorentzian metric. (This boundary could "bite off" the top corner of the darker shaded rectangle.) The null component arising from the centre can be

In [**53**], Christodoulou constructs vacuum solutions by prescribing a characteristic initial value problem with data on (what will be) $\mathcal{I}^-$. This $\mathcal{I}^-$ is taken to be past complete, and in fact, the data is taken to be trivial to the past of a sphere on $\mathcal{I}^-$. Thus, the development will include a region where the metric is Minkowski, corresponding precisely to the lower lighter shaded triangle above. It is shown that–as long as the incoming energy per unit solid angle in all directions[32] is sufficiently large in a strip of $\mathcal{I}^-$ right after the trivial part, where sufficiently large is taken in comparison with the affine length of the generators of $\mathcal{I}^-$–a trapped surface arises in the domain of development of the data restricted to the past of this strip. Comparing with the spherically symmetric picture above, this trapped surface would arise precisely as before in the analogue of the darker shaded region depicted.

In contrast to the spherically symmetric case, where given the lower triangle, existence of the solution in the darker shaded region (at least as far as trapped surface formation) follows immediately, for vacuum collapse, showing the existence of a sufficiently "big" spacetime is a major difficulty. For this, the results of [**53**] exploit a hierarchy in the Einstein equations (4) in the context of what is there called the "short pulse method". This method may have many other applications for nonlinear problems.

One could in principle hope to extend [**53**] to show the formation of black hole spacetimes in the sense described previously. For this, one must first extend the initial data suitably, for instance so that $\mathcal{I}^-$ is complete. If the resulting spacetime can be shown to possess a complete future null infinity $\mathcal{I}^+$, then, since the trapped surface shown to form can be proven (using the methods of the proof of Theorem 2.2) *not* to be in the past of null infinity, the spacetime will indeed contain a black hole region.[33] Of course, resolution of this problem would appear comparable in difficulty to the stability problem for the Kerr family (see the formulation of Section 5.6).

3. The wave equation on Schwarzschild I: uniform boundedness

In the remainder of these lectures, we will concern ourselves solely with linear wave equations on black hole backgrounds, specifically, the scalar linear homogeneous wave equation (1). As explained in the introduction, the study of the solutions to such equations is motivated by the stability problem for the black hole spacetimes themselves as solutions to (4). The equation (1) can be viewed as a poor man's linearisation of (4), neglecting tensorial structure. Other linear problems with a much closer relationship to the study of the Einstein equations will be discussed in Section 8.

3.1. Preliminaries.

Let $(\mathcal{M}, g)$ denote (maximally-extended) Schwarzschild with parameter $M > 0$. Let Σ be an arbitrary *Cauchy surface*, that is to say,

shown to be empty generically after passing to a slightly less regular class of solutions, for which well-posedness still holds. See Christodoulou's proof of the cosmic censorship conjectures [**45**] for the Einstein-scalar field system.

[32] This is defined in terms of the shear of $\mathcal{I}^-$.

[33] In spherical symmetry, the completeness of null infinity follows immediately once a single trapped surface has formed, for the Einstein equations coupled to a wide class of matter models. See for instance [**60**]. For vacuum collapse, Christodoulou has formulated a statement on trapped surface formation that would imply weak cosmic censorship. See [**47**].

a hypersurface with the property that every inextendible causal geodesic in $\mathcal{M}$ intersects Σ precisely once. (See Appendix A.)

PROPOSITION 3.1.1. *If $\uppsi \in H^2_{\rm loc}(\Sigma)$, $\uppsi' \in H^1_{\rm loc}(\Sigma)$, then there is a unique ψ with $\psi|_{\mathcal{S}} \in H^2_{\rm loc}(\mathcal{S})$, $n_{\mathcal{S}}\psi|_{\mathcal{S}} \in H^1_{\rm loc}(\mathcal{S})$, for all spacelike $\mathcal{S} \subset \mathcal{M}$, satisfying*

$$\Box_g \psi = 0, \qquad \psi|_\Sigma = \uppsi, \qquad n_\Sigma \psi|_\Sigma = \uppsi',$$

where n_Σ denotes the future unit normal of Σ. For $m \geq 1$, if $\uppsi \in H^{m+1}_{\rm loc}$, $\uppsi' \in H^m_{\rm loc}$, then $\psi|_{\mathcal{S}} \in H^{m+1}_{\rm loc}(\mathcal{S})$, $n_{\mathcal{S}}\psi|_{\mathcal{S}} \in H^m_{\rm loc}(\mathcal{S})$. Moreover, if $\uppsi_1, \uppsi'_1$, and $\uppsi_2, \uppsi'_2$ are as above and $\uppsi_1 = \uppsi_2$, $\uppsi'_1 = \uppsi'_2$ in an open set $\mathcal{U} \subset \Sigma$, then $\psi_1 = \psi_2$ in $\mathcal{M} \setminus J^\pm(\Sigma \setminus \mathrm{clos}(\mathcal{U}))$.

We will be interested in understanding the behaviour of ψ in the exterior of the black hole and white hole regions, up to and including the horizons. It is enough of course to understand the behaviour in the region

$$\mathcal{D} \doteq \mathrm{clos}\left(J^-(\mathcal{I}^+_A) \cap J^+(\mathcal{I}^-_A)\right) \cap \mathcal{Q}$$

where $\mathcal{I}^\pm_A$ denote a pair of connected components of $\mathcal{I}^\pm$, respectively, with a common limit point.[34]

Moreover, it suffices (**Exercise**: Why?) to assume that $\Sigma \cap \mathcal{H}^- = \emptyset$, and that we are interested in the behaviour in $J^-(\mathcal{I}^+) \cap J^+(\Sigma)$. Note that in this case, by the domain of dependence property of the above proposition, we have that the solution in this region is determined by $\uppsi|_{\mathcal{D}\cap\Sigma}$, $\uppsi'|_{\mathcal{D}\cap\Sigma}$. In the case where Σ itself is spherically symmetric, then its projection to $\mathcal{Q}$ will look like:

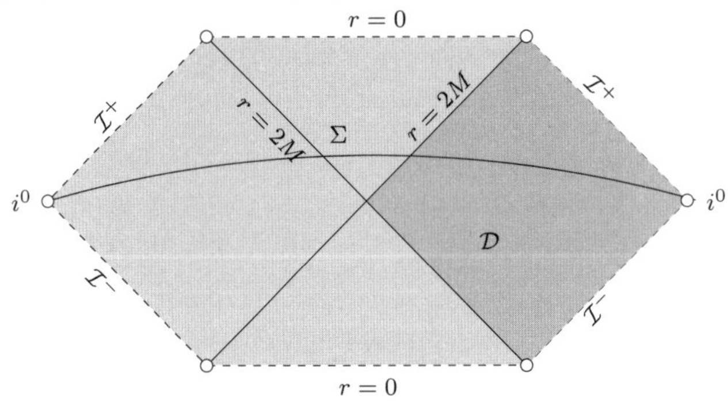

If Σ is not itself spherically symmetric, then its projection to $\mathcal{Q}$ will in general have open interior. Nonetheless, we shall always depict Σ as above.

3.2. The Kay–Wald boundedness theorem. The most basic problem is to obtain uniform boundedness for ψ. This is resolved in the celebrated:

THEOREM 3.1. *Let ψ, $\uppsi$, $\uppsi'$ be as in Proposition 3.1.1, with $\uppsi \in H^{m+1}_{\rm loc}(\Sigma)$, $\uppsi' \in H^m_{\rm loc}(\Sigma)$ for a sufficiently high m, and such that $\uppsi$, $\uppsi'$ decay suitably at i^0. Then there is a constant D depending on $\uppsi$, $\uppsi'$ such that*

$$|\psi| \leq D$$

in $\mathcal{D}$.

[34]We will sometimes be sloppy with distinguishing between $\pi^{-1}(p)$ and p, where $\pi : \mathcal{M} \to \mathcal{Q}$ denotes the natural projection, distinguishing $J^-(p)$ and $J^-(p) \cap \mathcal{Q}$, etc. The context should make clear what is meant.

The proof of this theorem is due to Wald [**151**] and Kay–Wald [**98**]. The "easy part" of the proof (Section 3.2.3) is a classic application of vector field commutators and multipliers, together with elliptic estimates and the Sobolev inequality. The main difficulties arise at the horizon, and these are overcome by what is essentially a clever trick. In this section, we will go through the original argument, as it is a nice introduction to vector field multiplier and commutator techniques, as well as to the geometry of Schwarzschild. We will then point out (Section 3.2.7) various disadvantages of the method of proof. Afterwards, we give a new proof that in fact achieves a stronger result (Theorem 3.2). As we shall see, the techniques of this proof will be essential for future applications.

3.2.1. *The Killing fields of Schwarzschild.* Recall the symmetries of $(\mathcal{M}, g)$: $(\mathcal{M}, g)$ is spherically symmetric, i.e. there is a basis of Killing vectors $\{\Omega_i\}_{i=1}^3$ spanning the Lie algebra so(3). These are sometimes known as *angular momentum operators*. In addition, there is another Killing field T (equal to ∂_t in the coordinates (5)) which is hypersurface orthogonal and future directed timelike near i^0. This Killing field is in fact timelike everywhere in $J^-(\mathcal{I}^+) \cap J^+(\mathcal{I}^-)$, becoming null and tangent to the horizon, vanishing at $\mathcal{H}^+ \cap \mathcal{H}^-$. We say that the Schwarzschild metric in $J^-(\mathcal{I}^+) \cap J^+(\mathcal{I}^-)$ is *static*. T is spacelike in the black hole and white hole regions.

Note that whereas in Minkowski space $\mathbb{R}^{3+1}$, the Killing fields at any point span the tangent space, this is no longer the case for Schwarzschild. We shall return to this point later.

3.2.2. *The current J^T and its energy estimate.* Let φ_t denote the 1-parameter group of diffeomorphisms generated by the Killing field T. Define $\Sigma_\tau = \varphi_t(\Sigma \cap \mathcal{D})$. We have that $\{\Sigma_\tau\}_{\tau \geq 0}$ defines a spacelike foliation of

$$\mathcal{R} \doteq \cup_{\tau \geq 0} \Sigma_\tau.$$

Define

$$\mathcal{H}^+(0, \tau) \doteq \mathcal{H}^+ \cap J^+(\Sigma_0) \cap J^-(\Sigma_\tau),$$

and

$$\mathcal{R}(0, \tau) \doteq \cup_{0 \leq \bar{\tau} \leq \tau} \Sigma_{\bar{\tau}}.$$

Let n^μ_Σ denote the future directed unit normal of Σ, and let $n^\mu_{\mathcal{H}}$ define a null generator of $\mathcal{H}^+$, and give $\mathcal{H}^+$ the associated volume form.[35]

Let $J^T_\mu(\psi)$ denote the energy current defined by applying the vector field T as a multiplier, i.e.

$$J^T_\mu(\psi) = \mathbf{T}_{\mu\nu}(\psi) T^\nu = (\partial_\mu \psi \partial_\nu \psi - \frac{1}{2} g_{\mu\nu} \partial^\alpha \psi \partial_\alpha \psi) T^\nu$$

with its associated current $K^T(\psi)$,

$$K^T(\psi) = {}^T\pi^{\mu\nu} \mathbf{T}_{\mu\nu}(\psi) = \nabla^\mu J^T_\mu(\psi),$$

where $\mathbf{T}_{\mu\nu}$ denotes the standard energy momentum tensor of ψ (see Appendix D). Since T is Killing, and $\nabla^\mu \mathbf{T}_{\mu\nu} = 0$, it follows that $K^T(\psi) = 0$, and the divergence

[35] Recall that for null surfaces, the definition of a volume form relies on the choice of a normal. All integrals in what follow will always be with respect to the natural volume form, and in the case of a null hypersurface, with respect to the volume form related to the given choice of normal. See Appendix C.

theorem (See Appendix C) applied to J^T_μ in the region $\mathcal{R}(0,\tau)$ yields

$$\int_{\Sigma_\tau} J^T_\mu(\psi)n^\mu_{\Sigma_\tau} + \int_{\mathcal{H}^+(0,\tau)} J^T_\mu(\psi)n^\mu_{\mathcal{H}} = \int_{\Sigma_0} J^T_\mu(\psi)n^\mu_{\Sigma_0}. \tag{8}$$

See

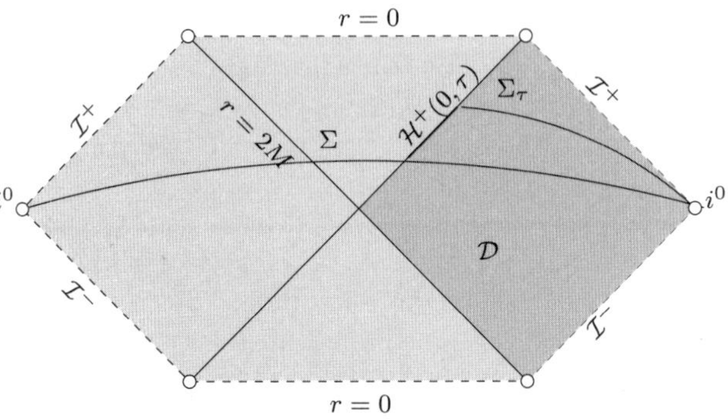

Since T is future-directed causal in $\mathcal{D}$, we have

$$J^T_\mu(\psi)n^\mu_\Sigma \geq 0, \qquad J^T_\mu(\psi)n^\mu_{\mathcal{H}} \geq 0. \tag{9}$$

Let us fix an $r_0 > 2M$. It follows from (8), (9) that

$$\int_{\Sigma_\tau\cap\{r\geq r_0\}} J^T_\mu(\psi)n^\mu_{\Sigma_\tau} \leq \int_{\Sigma_0} J^T_\mu(\psi)n^\mu_{\Sigma_0}.$$

As long as $-g(T, n_{\Sigma_0}) \leq B$ for some constant B,[36] we have

$$B(r_0,\Sigma)((\partial_t\psi)^2 + (\partial_r\psi)^2 + |\nabla\!\!\!/\psi|^2) \geq J^T_\mu(\psi)n^\mu \geq b(r_0,\Sigma)((\partial_t\psi)^2 + (\partial_r\psi)^2 + |\nabla\!\!\!/\psi|^2).$$

Here, $|\nabla\!\!\!/\psi|^2$ denotes the induced norm on the group orbits of the $SO(3)$ action, with $\nabla\!\!\!/$ the gradient of the induced metric on the group orbits. We thus have

$$\int_{\Sigma_\tau\cap\{r\geq r_0\}} (\partial_t\psi)^2 + (\partial_r\psi)^2 + |\nabla\!\!\!/\psi|^2 \leq B(r_0,\Sigma)\int_{\Sigma_0} J^T_\mu(\psi)n^\mu_{\Sigma_0}.$$

3.2.3. *T as a commutator and pointwise estimates away from the horizon.* We may now commute the equation with T (See Appendix E), i.e., since $[\Box_g, T] = 0$, if $\Box_g\psi = 0$ then $\Box_g(T\psi) = 0$. We thus obtain an estimate

$$\int_{\Sigma_\tau\cap\{r\geq r_0\}} (\partial_t^2\psi)^2 + (\partial_r\partial_t\psi)^2 + |\nabla\!\!\!/\partial_t\psi|^2 \leq B(r_0,\Sigma)\int_{\Sigma_0} J^T_\mu(T\psi)n^\mu_{\Sigma_0}. \tag{10}$$

Exercise: By elliptic estimates and a Sobolev estimate show that if $\psi(x) \to 0$ as $x \to i^0$, then (10) implies that for $r \geq r_0$,

$$|\psi|^2 \leq B(r_0,\Sigma)\left(\int_{\Sigma_0} J^T_\mu(\psi)n^\mu_{\Sigma_0} + \int_{\Sigma_0} J^T_\mu(T\psi)n^\mu_{\Sigma_0}\right), \tag{11}$$

for solutions ψ of $\Box_g\psi = 0$.

The right hand side of (11) is finite under the assumptions of Theorem 3.1, for $m = 1$. Thus, proving the estimate of Theorem 3.1 away from the horizon poses no difficulty. The difficulty of Theorem 3.1 is obtaining estimates which hold up to the horizon.

[36] For definiteness, one could choose Σ to be a surface of constant t^* defined in Section 2.2, or alternatively, require that it be of constant t for large r.

Remark: The above argument via elliptic estimates clearly also holds for Minkowski space. But in that case, there is an alternative "easier" argument, namely, to commute with all translations.[37] We see thus already that the lack of Killing fields in Schwarzschild makes things more difficult. We shall again return to this point later.

3.2.4. *Degeneration at the horizon.* As one takes $r_0 \to 2M$, the constant $B(r_0, \Sigma)$ provided by the estimate (11) blows up. This is precisely because T becomes null on $\mathcal{H}^+$ and thus its control over derivatives of ψ degenerates. Thus, one cannot prove uniform boundedness holding up to the horizon by the above.

Let us examine more carefully this degeneration on various hypersurfaces.

On Σ_τ, we have only

$$J^T_\mu(\psi)n^\mu_{\Sigma_\tau} \geq B(\Sigma_\tau)((\partial_{t^*}\psi)^2 + (1-2M/r)(\partial_r\psi)^2 + |\nabla\!\!\!/\psi|^2). \tag{12}$$

We see the degeneration in the presence of the factor $(1 - 2M/r)$. Note that (**Exercise**) $1 - 2M/r$ vanishes to first order on $\mathcal{H}^+ \setminus \mathcal{H}^-$. Alternatively, one can examine the flux on the horizon $\mathcal{H}^+$ itself. For definiteness, let us choose $n_{\mathcal{H}^+} = T$ in $\mathcal{R} \cap \mathcal{H}^+$. We have

$$J^T_\mu(\psi)T^\mu = (T\psi)^2. \tag{13}$$

Comparing with the analogous computation on a null cone in Minkowski space, one sees that a term $|\nabla\!\!\!/\psi|^2$ is "missing".

Are estimates of the terms (12), (13) enough to control ψ? It is a good idea to play with these estimates on your own, allowing yourself to commute the equation with T and Ω_i to obtain higher order estimates. **Exercise**: Why does this not lead to an estimate as in (11)?

It turns out that there is a way around this problem and the degeneration on the horizon is suggestive. For suppose there existed a $\tilde{\psi}$ such that

$$\Box_g\tilde{\psi} = 0, \qquad T\tilde{\psi} = \psi. \tag{14}$$

Let us see immediately how one can obtain estimates on the horizon itself. For this, we note that

$$J^T_\mu(\tilde{\psi})T^\mu + J^T_\mu(\psi)T^\mu = \psi^2 + (T\psi)^2.$$

Commuting now with the whole Lie algebra of isometries, we obtain

$$\begin{aligned} J^T_\mu(\tilde{\psi})T^\mu + J^T_\mu(\psi)T^\mu + \sum_i J^T_\mu(\Omega_i\tilde{\psi})T^\mu + J^T_\mu(T\psi)T^\mu \cdots \\ = \psi^2 + (T\psi)^2 + \sum_i(\Omega_i\psi)^2 + (T^2\psi)^2 + \cdots . \end{aligned}$$

Clearly, by a Sobolev estimate applied on the horizon, together with the estimate

$$\int_{\mathcal{H}^+\cap\mathcal{R}} J^T_\mu(\Gamma^{(\alpha)}\tilde{\psi})n^\mu_{\mathcal{H}} \leq \int_{\Sigma_0} J^T_\mu(\Gamma^{(\alpha)}\tilde{\psi})n^\mu_{\Sigma_0}$$

for $\Gamma = T, \Omega_i$ (here (α) denotes a multi-index of arbitrary order), we would obtain

$$|\psi|^2 \leq B \sum_{\Gamma=T,\Omega_i} \sum_{|(\alpha)|\leq 2} \int_{\Sigma_0} J^T_\mu(\Gamma^{(\alpha)}\tilde{\psi})n^\mu_{\Sigma_0} \tag{15}$$

on $\mathcal{H}^+ \cap \mathcal{R}$.

---

[37]Easier, but not necessarily better...

It turns out that the estimate (15) can be extended to points not on the horizon by considering $t = c$ surfaces. Note that these hypersurfaces all meet at $\mathcal{H}^+ \cap \mathcal{H}^-$. It is an informative calculation to examine the nature of the degeneration of estimates on such hypersurfaces because it is of a double nature, since, in addition to T becoming null, the limit of (subsets of) these spacelike hypersurfaces approaches the null horizon $\mathcal{H}^+$. We leave the details as an **exercise**.

3.2.5. *Inverting an elliptic operator.* So can a $\tilde{\psi}$ satisfying (14) actually be constructed? We have

PROPOSITION 3.2.1. *Suppose m is sufficiently high, ψ, ψ' decay suitably at i^0, and $\psi|_{\mathcal{H}^+ \cap \mathcal{H}^-} = 0$, $\Xi\psi|_{\mathcal{H}^+ \cap \mathcal{H}^-} = 0$ for some spherically symmetric timelike vector field Ξ defined along $\mathcal{H}^+ \cap \mathcal{H}^-$. Then there exists a $\tilde{\psi}$ satisfying $\Box_g \tilde{\psi} = 0$ with $T\tilde{\psi} = \psi$ in $\mathcal{D}$, and moreover, the right hand side of* (15) *is finite.*

Formally, one sees that on $t = c$ say, if we let $\bar{g}$ denote the induced Riemannian metric, and if we impose initial data

$$T\tilde{\psi}|_{t=c} = \psi,$$

$$\tilde{\psi}|_{t=c} = A^{-1}T\psi,$$

where $A = \triangle_{(1-2M/r)^{-1}\bar{g}} + (2M/r^2)(1-2M/r)\partial_r$, and let $\tilde{\psi}$ solve the wave equation with this data, then

$$T\tilde{\psi} = \psi$$

as desired.

So to use the above, it suffices to ask whether the initial data for $\tilde{\psi}$ above can be constructed and have sufficient regularity so as for the right hand side of (15) to be defined. To impose the first condition, since $T = 0$ along $\mathcal{H}^+ \cap \mathcal{H}^-$, one must have that ψ vanish there to some order. For the second condition, note first that the metric $(1 - 2M/r)^{-1}\bar{g}$ has an asymptotically hyperbolic end and an asymptotically flat end. Thus, to construct $A^{-1}T\psi$ suitably well-behaved[38], one must have that $T\psi$ decays appropriately towards the ends. We leave to the reader the task of verifying that the assumptions of the Proposition are sufficient.

3.2.6. *The discrete isometry.* Proposition 3.2.1, together with estimates (15) and (11), yield the proof of Theorem 3.1 in the special case that the conditions of Proposition 3.2.1 happen to be satisfied. In the original paper of Wald [**151**], one took Σ_0 to coincide with $t = 0$ and restricted to data ψ, ψ' which were supported in a compact region not containing $\mathcal{H}^+ \cap \mathcal{H}^-$. Clearly, however, this is a deficiency, as general solutions will be supported in $\mathcal{H}^+ \cap \mathcal{H}^-$. (See also the last exercise below.)

It turns out, however, that one can overcome the restriction on the support by the following trick: Note that the previous proposition produces a $\tilde{\psi}$ such that $T\tilde{\psi} = \psi$ on all of $\mathcal{D}$. We only require however that $T\tilde{\psi} = \psi$ on $\mathcal{R}$. The idea is to define a new $\bar{\psi}$, $\bar{\psi}'$ on Σ, such that $\bar{\psi} = \psi$, $\bar{\psi}' = \psi'$ on Σ_0 and, denoting by $\bar{\psi}$ the solution to the Cauchy problem with the new data, $\bar{\psi}|_{\mathcal{H}^+ \cap \mathcal{H}^-} = 0$, $\Xi\bar{\psi}|_{\mathcal{H}^+ \cap \mathcal{H}^-} = 0$. By the previous proposition and the domain of dependence property of Proposition 3.1.1, we will have indeed constructed a $\tilde{\psi}$ with $T\tilde{\psi} = \psi$ in $\mathcal{R}$ for which the right hand side of (15) is finite.

Remark that Schwarzschild admits a discrete symmetry generated by the map $R \to -R$ in the Kruskal R-coordinate defined in section 2.3. Define $\bar{\psi}$, $\bar{\psi}'$ so that $\bar{\psi}(R, \cdot) = -\bar{\psi}(-R, \cdot)$, $\bar{\psi}'(R, \cdot) = -\bar{\psi}'(-R, \cdot)$.

[38]so that we may apply to this quantity the arguments of Section 3.2.4.

PROPOSITION 3.2.2. *Under the above assumptions, it follows that*

$$\bar{\psi}(R,\cdot) = -\bar{\psi}(-R,\cdot).$$

The proof of the above is left as an exercise in preservation of symmetry for solutions of the wave equation. It follows immediately that

$$\bar{\psi}|_{\mathcal{H}^+\cap\mathcal{H}^-} = 0$$

and that

$$\partial_U\bar{\psi} = -\partial_V\bar{\psi},$$

and thus $(\partial_U + \partial_V)\bar{\psi} = 0$. Here U and V are the bounded null coordinates of Section 2.3. In view of the above remarks and Proposition 3.2.1 with $\Xi = \partial_U + \partial_V$, we have shown the full statement of Theorem 3.1.

Exercise: Work out explicit regularity assumptions and quantitative dependence on initial data in Theorem 3.1, describing in particular decay assumptions necessary at i^0.

Exercise: Prove the analogue of Theorem 3.1 on the Oppenheimer-Snyder spacetime discussed previously. *Hint: One need not know the explicit form of the metric, the statement given about the Penrose diagram suffices.* Convince yourself that the original restricted version of Theorem 3.1 due to Wald [**151**], where the support of ψ is restricted near $\mathcal{H}^+\cap\mathcal{H}^-$, is not sufficient to yield this result.

3.2.7. *Remarks.* The clever proof described above successfully obtains pointwise boundedness for ψ up to the horizon $\mathcal{H}^+$. Does this really close the book, however, on the boundedness question? From various points of view, it may be desirable to go further.

(1) Even though one obtains the "correct" pointwise result, one does not obtain boundedness at the horizon for the energy measured by a local observer, that is to say, bounds for

$$\int_{\Sigma_\tau} J^{n_{\Sigma_\tau}}_\mu(\psi)n^\mu_{\Sigma_\tau}.$$

This indicates that it would be difficult to use this result even for the simplest non-linear problems.

(2) One does not obtain boundedness for transverse derivatives to the horizon, i.e. in (t^*, r) coordinates, $\partial_r\psi$, $\partial_r^2\psi$, etc. (**Exercise**: Why not?)

(3) The dependence on initial data is somewhat unnatural. (**Exercise**: Work out explicitly what it is.)

As far as the method of proof is concerned, there are additional shortcomings when the proof is viewed from the standpoint of possible future generalisations:

(4) To obtain control at the horizon, one must commute (see (15)) with all angular momentum operators Ω_i. Thus the spherical symmetry of Schwarzschild is used in a fundamental way.

(5) The exact staticity is fundamental for the construction of $\tilde{\psi}$. It is not clear how to generalise this argument in the case say where T is not hypersurface orthogonal and Killing but one assumes merely that its deformation tensor ${}^T\pi_{\mu\nu}$ decays. This would be the situation in a bootstrap setting of a non-linear stability problem.

(6) The construction of $\bar{\psi}$ requires the discrete isometry of Schwarzschild, which again, cannot be expected to be stable.

3.3. The red-shift and a new proof of boundedness. We give in this section a new proof of boundedness which overcomes the shortcomings outlined above. In essence, the previous proof limited itself by relying solely on Killing fields as multipliers and commutators. It turns out that there is an important physical aspect of Schwarzschild which can be captured by other vector-field multipliers and commutators which are not however Killing. This is related to the celebrated *red-shift effect.*

3.3.1. *The classical red-shift.* The *red-shift effect* is one of the most celebrated aspects of black holes. It is classically described as follows: Suppose two observers, A and B are such that A crosses the event horizon and B does not. If A emits a signal at constant frequency as he measures it, then the frequency at which it is received by B is "shifted to the red".

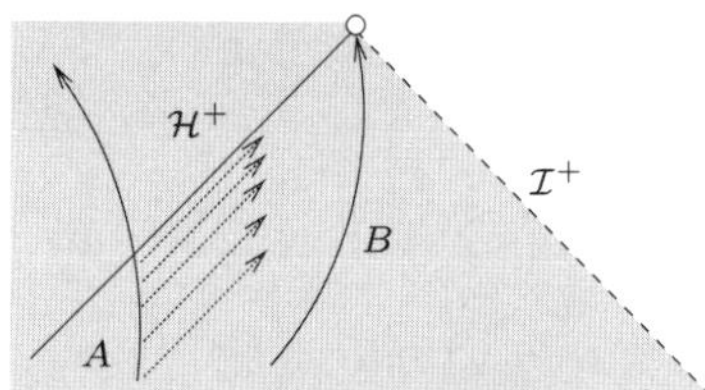

The consequences of this for the appearance of a collapsing star to far-away observers were first explored in the seminal paper of Oppenheimer-Snyder [**125**] referred to at length in Section 2. For a nice discussion, see also the classic textbook [**117**].

The red-shift effect as described above is a global one, and essentially depends only on the fact that the proper time of B is infinite whereas the proper time of A before crossing $\mathcal{H}^+$ is finite. In the case of the Schwarzschild black hole, there is a "local" version of this red-shift: If B also crosses the event horizon but at advanced time later than A:

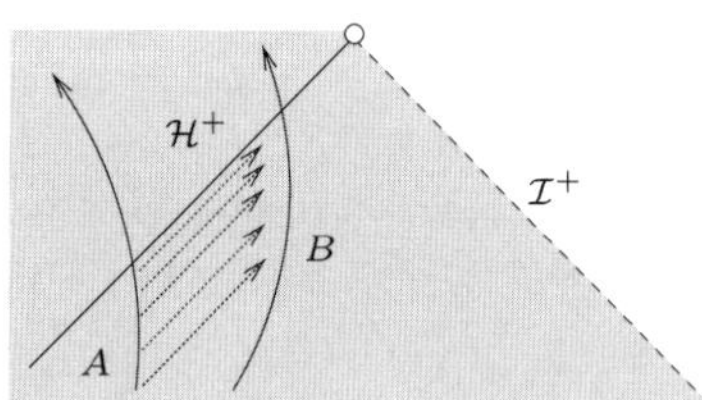

then the frequency at which B receives at his horizon crossing time is shifted to the red by a factor depending exponentially on the advanced time difference of the crossing points of A and B.

The exponential factor is determined by the so-called *surface gravity*, a quantity that can in fact be defined for all so-called Killing horizons. This localised red-shift effect depends only on the positivity of this quantity. We shall understand this more general situation in Section 7. Let us for now simply explore how we can "capture" the red-shift effect in the Schwarzschild geometry.

3.3.2. *The vector fields N, Y, and $\hat{Y}$.* It turns out that a "vector field multiplier" version of this localised red-shift effect is captured by the following

PROPOSITION 3.3.1. *There exists a φ_t-invariant smooth future-directed timelike vector field N on $\mathcal{R}$ and a positive constant $b > 0$ such that*

$$K^N(\psi) \geq bJ^N_\mu(\psi)N^\mu$$

on $\mathcal{H}^+$.

(See Appendix D for the J^N, K^N notation.)

PROOF. Note first that since T is tangent to $\mathcal{H}^+$, it follows that given any $\sigma < \infty$, there clearly exists a vector field Y on $\mathcal{R}$ such that

(1) Y is φ_t invariant and spherically symmetric.
(2) Y is future-directed null on $\mathcal{H}^+$ and transverse to $\mathcal{H}^+$, say $g(T, Y) = -2$.
(3) On $\mathcal{H}^+$,

$$\nabla_Y Y = -\sigma\,(Y + T). \tag{16}$$

Since T is tangent to $\mathcal{H}^+$, along which Y is null, we have

$$g(\nabla_T Y, Y) = 0. \tag{17}$$

From properties 1 and 2, and the form of the Schwarzschild metric, one computes (**Exercise**)

$$g(\nabla_T Y, T) \doteq 2\kappa > 0 \tag{18}$$

on $\mathcal{H}^+$. Defining a local frame E_1, E_2 for the SO(3) orbits, we note

$$g(\nabla_{E_i} Y, Y) = \frac{1}{2} E_i g(Y, Y) = 0,$$

$$g(\nabla_{E_1} Y, E_2) = -g(Y, \nabla_{E_1} E_2) = -g(Y, \nabla_{E_2} E_1) = g(\nabla_{E_2} Y, E_1).$$

Writing thus

$$\nabla_T Y = -\kappa Y + a^1\, E_1 + a^2\, E_2 \tag{19}$$

$$\nabla_Y Y = -\sigma\, T - \sigma\, Y \tag{20}$$

$$\nabla_{E_1} Y = h^1_1\, E_1 + h^2_1\, E_2 - \frac{1}{2} a^1\, Y \tag{21}$$

$$\nabla_{E_2} Y = h^1_2\, E_1 + h^2_2\, E_2 - \frac{1}{2} a^2\, Y \tag{22}$$

with ($h^2_1 = h^1_2$), we now compute

$$\begin{aligned} K^Y \;\; = \;\; & \frac{1}{2}\left(\mathbf{T}(Y,Y)\kappa + \mathbf{T}(T,Y)\sigma + \mathbf{T}(T,T)\sigma\right) \\ & - \frac{1}{2}(\mathbf{T}(E_1, Y)a^1 + \mathbf{T}(E_2, Y)a^2) \\ & + \mathbf{T}(E_1, E_1)h^1_1 + \mathbf{T}(E_2, E_2)h^1_2 + \mathbf{T}(E_1, E_2)(h^2_1 + h^1_2) \end{aligned}$$

where we denote the energy momentum tensor by $\mathbf{T}$, to prevent confusion with T. (Note that, in view of the fact that $\mathcal{Q}$ imbeds as a totally geodesic submanifold of $\mathcal{M}$, we have in fact $a^1 = a^2 = 0$. This is of no importance in our computations, however.) It follows immediately in view again of the algebraic properties of $\mathbf{T}$, that

$$\begin{aligned} K^Y \;\; \geq \;\; & \frac{1}{2}\kappa\, \mathbf{T}(Y, Y) + \frac{1}{4}\sigma\, \mathbf{T}(T, Y + T) \\ & - c\mathbf{T}(T, Y + T) - c\sqrt{\mathbf{T}(T, Y + T)\mathbf{T}(Y, Y)} \end{aligned}$$

where c is independent of the choice of σ. It follows that choosing σ large enough, we have

$$K^Y \geq b\, J^{T+Y}_\mu (T+Y)^\mu.$$

So just set $N = T + Y$, noting that $K^N = K^T + K^Y = K^Y$. □

The computation (18) represents a well known property of stationary black holes holes and the constant κ is the so-called *surface gravity*. (See [**148**].) Note that since Y is φ_t-invariant and T is Killing, we have

$$g(\nabla_T Y, T) = g(\nabla_Y T, T) = -g(\nabla_T T, Y)$$

on $\mathcal{H}^+$. On the other hand

$$g(\nabla_T T, E_i) = -g(\nabla_{E_i} T, T) = 0,$$

since T is null on $\mathcal{H}^+$. Thus, κ is alternatively characterized by

$$\nabla_T T = \kappa T$$

on $\mathcal{H}^+$. We will elaborate on this in Section 7, where a generalisation of Proposition 3.3.1 will be presented.

Exercise: Relate the strength of the red-shift with the constant κ, for the case where observers A and B both cross the horizon, but B at advanced time v later than A.

If one desires an explicit form of the vector field, then one can argue as follows: Define first the vector field $\hat{Y}$ by

$$\hat{Y} = \frac{1}{1-2M/r}\partial_u. \tag{23}$$

(See Appendix F.) Note that this vector field satisfies $g(\nabla_{\hat{Y}}\hat{Y}, T) = 0$. Define

$$Y = (1 + \delta_1(r-2M))\hat{Y} + \delta_2(r-2M)T.$$

It suffices to choose δ_1, δ_2 appropriately.

The behaviour of N away from the horizon is of course irrelevant in the above proposition. It will be useful for us to have the following:

COROLLARY 3.1. *Let Σ be as before. There exists a φ_t-invariant smooth future-directed timelike vector field N on $\mathcal{R}$, constants $b > 0$, $B > 0$, and two values $2M < r_0 < r_1 < \infty$ such that*

(1) $K^N \geq b\, J^N_\mu n^\mu_\Sigma$ *for* $r \leq r_0$,
(2) $N = T$ *for* $r \geq r_1$,
(3) $|K^N| \leq B J^T_\mu n^\mu_\Sigma$, *and* $J^N_\mu n^\mu_\Sigma \sim J^T n^\mu_\Sigma$ *for* $r_0 \leq r \leq r_1$.

3.3.3. *N as a multiplier.* Recall the definition of $\mathcal{R}(0,\tau)$. Applying the energy identity with the current J^N in this region, we obtain

$$\begin{aligned}\int_{\Sigma_\tau} J^N_\mu n^\mu_\Sigma + \int_{\mathcal{H}^+(0,\tau)} J^N_\mu n^\mu_{\mathcal{H}} + \int_{\{r\leq r_0\}\cap\mathcal{R}(0,\tau)} K^N \\ = \int_{\{r_0\leq r\leq r_1\}\cap\mathcal{R}(0,\tau)} (-K^N) + \int_{\Sigma_0} J^N_\mu n^\mu_\Sigma.\end{aligned} \tag{24}$$

The reason for writing the above identity in this form will become apparent in what follows. Note that since N is timelike at $\mathcal{H}^+$, we see all the "usual terms" in the flux integrals, i.e.

$$J^N_\mu n^\mu_{\mathcal{H}} \sim (\partial_{t^*}\psi)^2 + |\nabla\!\!\!/\psi|^2,$$

and

$$J^N_\mu n^\mu_{\Sigma_\tau} \sim (\partial_{t^*}\psi)^2 + (\partial_r\psi)^2 + |\nabla\!\!\!/\psi|^2.$$

The constants in the $\sim$ depend as usual on the choice of the original Σ_0 and the precise choice of N.

Now the identity (24) also holds where Σ_0 is replaced by $\Sigma_{\tau'}$, $\mathcal{H}^+(0,\tau)$ is replaced by $\mathcal{H}^+(\tau',\tau)$, and $\mathcal{R}(0,\tau)$ is replaced by $\mathcal{R}(\tau',\tau)$, for an arbitrary $0 \le \tau' \le \tau$.

We may add to both sides of (24) an arbitrary multiple of the spacetime integral $\int_{\{r\ge r_0\}\cap\mathcal{R}(\tau',\tau)} J^T_\mu n^\mu_\Sigma$. In view of the fact that

$$\int_{\{r\ge r'\}\cap\mathcal{R}(\tau',\tau)} J^N_\mu n^\mu_\Sigma \sim \int_{\tau'}^{\tau}\left(\int_{\{r\ge r'\}\cap\Sigma_{\bar\tau}} J^N_\mu n^\mu_\Sigma\right) d\bar\tau$$

for any $r' \ge 2M$ (where $\sim$ depends on Σ_0, N), from the inequalities shown and property 3 of Corollary 3.1 we obtain

$$\int_{\Sigma_\tau} J^N_\mu n^\mu_\Sigma + b\int_{\tau'}^{\tau}\left(\int_{\Sigma_{\bar\tau}} J^N_\mu n^\mu_\Sigma\right) d\bar\tau \le B\int_{\tau'}^{\tau}\left(\int_{\Sigma_{\bar\tau}} J^T_\mu n^\mu_\Sigma\right) d\bar\tau + \int_{\Sigma_{\tau'}} J^N_\mu n^\mu_\Sigma.$$

On the other hand, in view of our previous (8), (9), we have

$$\int_{\tau'}^{\tau}\left(\int_{\Sigma_{\bar\tau}} J^T_\mu n^\mu_\Sigma\right) d\bar\tau \le (\tau-\tau')\int_{\Sigma_0} J^T_\mu n^\mu_\Sigma. \tag{25}$$

Setting

$$f(\tau) = \int_{\Sigma_\tau} J^N_\mu n^\mu_\Sigma$$

we have that

$$f(\tau) + b\int_{\tau'}^{\tau} f(\bar\tau)d\bar\tau \le BD(\tau-\tau') + f(\tau') \tag{26}$$

for all $\tau \ge \tau' \ge 0$, from which it follows (**Exercise**) that $f \le B(D + f(0))$. (We use the inequality with $D = \int_{\Sigma_0} J^T_\mu n^\mu_{\Sigma_0}$.) In view of the trivial inequality

$$\int_{\Sigma_0} J^T_\mu n^\mu_{\Sigma_0} \le B\int_{\Sigma_0} J^N_\mu n^\mu_{\Sigma_0},$$

we obtain

$$\int_{\Sigma_\tau} J^N_\mu n^\mu_{\Sigma_\tau} \le B\int_{\Sigma_0} J^N_\mu n^\mu_{\Sigma_0}. \tag{27}$$

We have obtained a "local observer's" energy estimate. This addresses point 1 of Section 3.2.7.

3.3.4. $\hat{Y}$ *as a commutator.* It turns out (**Exercise**) that from (27), one could obtain pointwise bounds as before on ψ by commuting with angular momentum operators Ω_i. No construction of $\tilde\psi$, $\bar\psi$, etc., would be necessary, and this would thus address points 3, 5, 6 of Section 3.2.7.

Commuting with Ω_i clearly would not address however point 4. Moreover, it would not address point 2. **Exercise**: Why not?

It turns out that one can resolve this problem by applying N not only as a multiplier, but also as a commutator. The calculations are slightly easier if we more simply commute with $\hat{Y}$ defined in (23).

PROPOSITION 3.3.2. *Let ψ satisfy $\Box_g\psi = 0$. Then we may write*

$$\Box_g(\hat{Y}\psi) = \left(\frac{2}{r} - \frac{2M}{r^2}\right)\hat{Y}(\hat{Y}(\psi)) - \frac{4}{r}(\hat{Y}(T\psi)) + P_1\psi \tag{28}$$

where P_1 is the first order operator $P_1\psi \doteq \frac{2}{r^2}(T\psi - \hat{Y}\psi)$.

This is proven easily with the help of Appendix E. As we shall see, the sign of the first term on the right hand side of (28) is important. We will interpret this computation geometrically in terms of the sign of the surface gravity in Theorem 7.2 of Section 7.

Let us first note that our boundedness result gives us in particular

$$\int_{\{r\le r_0\}\cap\mathcal{R}(0,\tau)} K^N(\psi) \le BD\tau \tag{29}$$

where D comes from initial data. (**Exercise**: Why?) Commute now the wave equation with T and apply the multiplier N. See Appendix E. One obtains in particular an estimate for

$$\int_{\{r\le r_0\}\cap\mathcal{R}(0,\tau)} (\hat{Y}T\psi)^2 \le B\int_{\{r\le r_0\}\cap\mathcal{R}(0,\tau)} K^N(T\psi) \le BD\tau, \tag{30}$$

where again D refers to a quantity coming from initial data. Commuting now the wave equation with $\hat{Y}$ and applying the multiplier N, one obtains an energy identity of the form

$$\begin{aligned}
&\int_{\Sigma_\tau} J^N_\mu(\hat{Y}\psi)n^\mu_\Sigma + \int_{\mathcal{H}^+(0,\tau)} J^N_\mu(\hat{Y}\psi)n^\mu_{\mathcal{H}} + \int_{\{r\le r_0\}\cap\mathcal{R}(0,\tau)} K^N(\hat{Y}\psi)\\
&= \int_{\{r_0\le r\le r_1\}\cap\mathcal{R}(0,\tau)} (-K^N(\hat{Y}(\psi))\\
&\quad + \int_{\{r\le r_0\}\cap\mathcal{R}(0,\tau)} \mathcal{E}^N(\hat{Y}\psi) + \int_{\{r\ge r_0\}\cap\mathcal{R}(0,\tau)} \mathcal{E}^N(\hat{Y}\psi)\\
&\quad + \int_{\Sigma_0} J^N_\mu(\hat{Y}\psi)n^\mu_\Sigma,
\end{aligned}$$

where $J^N(\hat{Y}\psi)$, $K^N(\hat{Y}\psi)$ are defined by (123), (124), respectively, with $\hat{Y}\psi$ replacing ψ, and

$$\begin{aligned}
\mathcal{E}^N(\hat{Y}\psi) &= -(N\hat{Y}\psi)\left(\frac{2}{r}\hat{Y}(\hat{Y}(\psi)) - \frac{4}{r}(\hat{Y}(T\psi)) + P_1\psi\right)\\
&= -\frac{2}{r}(\hat{Y}(\hat{Y}(\psi)))^2\\
&\quad - \frac{2}{r}((N-\hat{Y})\hat{Y}\psi)(\hat{Y}\hat{Y}\psi) + \frac{4}{r}(N\hat{Y}\psi)(\hat{Y}(T\psi))\\
&\quad - (N\hat{Y}\psi)P_1\psi.
\end{aligned}$$

The first term on the right hand side has a good sign! Applying Cauchy-Schwarz and the fact that $N - \hat{Y} = T$ on $\mathcal{H}^+$, it follows that choosing r_0 accordingly, one obtains that the second two terms can be bounded in $r \le r_0$ by

$$\epsilon K^N(\hat{Y}\psi) + \epsilon^{-1}(\hat{Y}T\psi)^2$$

whereas the last term can be bounded by

$$\epsilon K^N(\hat{Y}\psi) + \epsilon^{-1}K^N(\psi).$$

In view of (29) and (30), one obtains

$$\int_{\{r\le r_0\}\cap\mathcal{R}(0,\tau)} \mathcal{E}^N(\hat{Y}\psi) \le \epsilon \int_{\{r\le r_0\}\cap\mathcal{R}(0,\tau)} K^N(\hat{Y}\psi) + B\epsilon^{-1}D\tau.$$

Exercise: Show how from this one can arrive again at an inequality (26).

Commuting repeatedly with T, $\hat{Y}$, the above scheme plus elliptic estimates yield natural H^m estimates for all m. Pointwise estimates for all derivatives then follow by a standard Sobolev estimate.

3.3.5. *The statement of the boundedness theorem.* We obtain finally

THEOREM 3.2. *Let Σ be a Cauchy hypersurface for Schwarzschild such that $\Sigma\cap\mathcal{H}^-=\emptyset$, let $\Sigma_0=\mathcal{D}\cap\Sigma$, let Σ_τ denote the translation of Σ_0, let n_{Σ_τ} denote the future normal of Σ_τ, and let $\mathcal{R}=\cup_{\tau\ge 0}\Sigma_\tau$. Assume $-g(n_{\Sigma_0},T)$ is uniformly bounded. Then there exists a constant C depending only on Σ_0 such that the following holds. Let ψ, $\uppsi$, $\uppsi'$ be as in Proposition* 3.1.1, *with $\uppsi\in H^{k+1}_{\rm loc}(\Sigma)$, $\uppsi'\in H^k_{\rm loc}(\Sigma)$, and*

$$\int_{\Sigma_0} J^T_\mu(T^m\psi)n^\mu_{\Sigma_0} < \infty$$

for $0\le m\le k$. Then

$$|\nabla^{\Sigma_\tau}\psi|_{H^k(\Sigma_\tau)} + |n\psi|_{H^k(\Sigma_\tau)} \le C\left(|\nabla^{\Sigma_0}\uppsi|_{H^k(\Sigma_0)} + |\uppsi'|_{H^k(\Sigma_0)}\right).$$

If $k\ge 1$, then we have

$$\sum_{0\le m\le k-1}\ \sum_{m_1+m_2=m,m_i\ge 0} |(\nabla^\Sigma)^{m_1}n^{m_2}\psi| \le C\left(\lim_{x\to i^0}|\uppsi| + |\nabla^{\Sigma(0)}\uppsi|_{H^k(\Sigma_0)} + |\uppsi'|_{H^k(\Sigma_0)}\right)$$

in $\mathcal{R}$.

Note that $(\nabla^\Sigma)^{m_1}n^{m_2}\psi$ denotes an m_1-tensor on the Riemannian manifold Σ_τ, and $|\cdot|$ on the left hand side of the last inequality above just denotes the induced norm on such tensors.

3.4. Comments and further reading. The first discussion of the wave equation on Schwarzschild is perhaps the work of Regge and Wheeler [**131**], but the true mathematical study of this problem was initiated by Wald [**151**], who proved Theorem 3.1 under the assumption that ψ vanished in a neighbourhood of $\mathcal{H}^+\cap\mathcal{H}^-$. The full statement of Theorem 3.1 and the proof presented in Section 3.2 is due to Kay and Wald [**98**]. The present notes owe a lot to the geometric view point emphasised in the works [**151, 98**].

Use of the vector field Y as a multiplier was first introduced in our [**65**], and its use is central in [**66**] and [**67**]. In particular, the property formalised by Proposition 3.3.1 was discovered there. It appears that this may be key to a stable understanding of black hole event horizons. See Section 3.5 below, as well as Section 7, for a generalisation of Proposition 3.3.1.

It is interesting to note that in [**66, 67**], Y had always been used in conjunction with vector fields X of the type to be discussed in the next section (which require a more delicate global construction) as well as T. This meant that one always had to obtain *more* than boundedness (i.e. decay!) in order to obtain the proper boundedness result at the horizon. Consequently, one had to use many aspects of the structure of Schwarzschild, particularly, the trapping to be discussed in later lectures. The argument given above, where boundedness is obtained using only N and T as multipliers is presented for the first time in a self-contained fashion in these lectures. The argument can be read off, however, from the more general argument of [**68**] concerning perturbations of Schwarzschild including Kerr. The use of $\hat{Y}$ as a commutator to estimate higher order quantities also originates in [**68**]. The geometry behind this computation is further discussed in Section 7.

Note that the use of Y together with T is of course equivalent to the use of N and T. We have chosen to give a name to the vector field $N = T + Y$ merely for convenience. Timelike vector fields are more convenient when perturbing...

Another remark on the use of $\hat{Y}$ as a commutator: Enlarging the choice of commutators has proven very important in previous work on the global analysis of the wave equation. In a seminal paper, Klainerman [**100**] showed improved decay for the wave equation on Minkowski space in the interior region by commutation with scaling and Lorentz boosts. This was a key step for further developments for long time existence for quasilinear wave equations [**101**].

The distinct role of multipliers and commutators and the geometric considerations which enter into their construction is beautifully elaborated by Christodoulou [**52**].

3.5. Perturbing? Can the proof of Theorem 3.2 be adapted to hold for spacetimes "near" Schwarzschild? To answer this, one must first decide what one means by the notion of "near". Perhaps the simplest class of perturbed metrics would be those that retain the same differentiable structure of $\mathcal{R}$, retain $\mathcal{H}^+$ as a null hypersurface, and retain the Killing field T. One infers (without computation!) that the statement of Proposition 3.3.1 and thus Corollary 3.1 is stable to such perturbations of the metric. Therein lies the power of that Proposition and of the multiplier N. (In fact, see Section 7.) Unfortunately, one easily sees that our argument proving Theorem 3.2 is still unstable, even in the class of perturbations just described. The reason is the following: Our argument relies essentially on an a priori estimate for $\int_{\Sigma_\tau} J^T_\mu n^\mu$ (see (25)), which requires T to be non-spacelike in $\mathcal{R}$. When one perturbs, T will in general become spacelike in a region of $\mathcal{R}$. (As we shall see in Section 5.1, this happens in particular in the case of Kerr. The region where T is spacelike is known as the ergoregion.)

There is a sense in which the above is the only obstruction to perturbing the above argument, i.e. one can solve the following

Exercise: Fix the differentiable structure of $\mathcal{R}$ and the vector field T. Let g be a metric sufficiently close to Schwarzschild such that $\mathcal{H}^+$ is null, and suppose T is Killing and non-spacelike in $\mathcal{R}$, and T is null on $\mathcal{H}^+$. Then Theorem 3.2 applies. (In fact, one need not assume that T is non-spacelike in $\mathcal{R}$, only that T is null on the horizon.) See also Section 7.

Exercise: Now do the above where T is not assumed to be Killing, but ${}^T\pi_{\mu\nu}$ is assumed to decay suitably. What precise assumptions must one impose?

This discussion may suggest that there is in fact no stable boundedness argument, that is to say, a "stable argument" would of necessity need to prove more than boundedness, i.e. decay. We shall see later that there is a sense in which this is true and a sense in which it is not! But before exploring this, let us understand how one can go beyond boundedness and prove quantitative decay for waves on Schwarzschild itself. It is quantitative decay after all that we must understand if we are to understand nonlinear problems.

4. The wave equation on Schwarzschild II: quantitative decay rates

Quantitative decay rates are central for our understanding of non-linear problems. To discuss energy decay for solutions ψ of $\Box_g\psi = 0$ on Schwarzschild, one must consider a different foliation. Let $\tilde{\Sigma}_0$ be a spacelike hypersurface terminating on null infinity and define $\tilde{\Sigma}_\tau$ (for $\tau \geq 0$) by future translation.

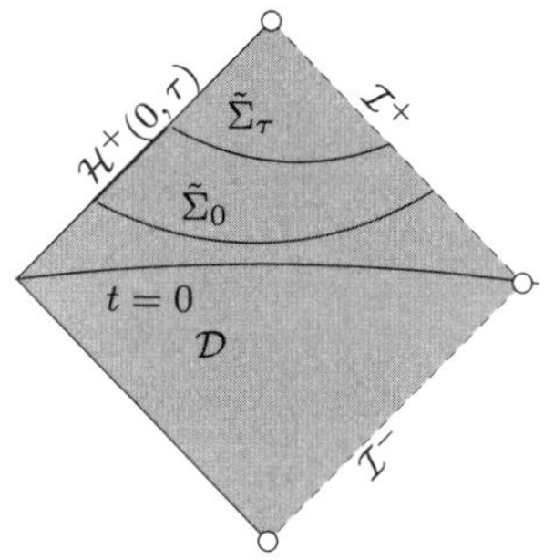

The main result of this section is the following

THEOREM 4.1. *There exists a constant C depending only on $\tilde{\Sigma}_0$ such that the following holds. Let $\psi \in H^4_{\rm loc}$, $\psi' \in H^3_{\rm loc}$, and suppose $\lim_{x\to i^0}\psi = 0$ and*

$$E_1 = \sum_{|(\alpha)|\leq 3}\sum_{\Gamma=\{\Omega_i\}}\int_{t=0} r^2 J^{n_0}_\mu(\Gamma^{(\alpha)}\psi)n_0^\mu < \infty$$

where n_0 denotes the unit normal of the hypersurface $\{t=0\}$. Then

$$\int_{\tilde{\Sigma}_\tau} J^N_\mu(\psi)n^\mu_{\tilde{\Sigma}_\tau} \leq CE_1\tau^{-2}, \tag{31}$$

where N is the vector field of Section 3.3.2. Now let $\psi \in H^7_{\rm loc}$, $\psi' \in H^6_{\rm loc}$, $\lim_{x\to i^0}\psi = 0$, and suppose

$$E_2 = \sum_{|(\alpha)|\leq 6}\sum_{\Gamma=\{\Omega_i\}}\int_{t=0} r^2 J^{n_0}_\mu(\Gamma^{(\alpha)}\psi)n_0^\mu < \infty.$$

Then

$$\sup_{\tilde{\Sigma}_\tau}\sqrt{r}|\psi| \leq C\sqrt{E_2}\tau^{-1}, \qquad \sup_{\tilde{\Sigma}_\tau} r|\psi| \leq C\sqrt{E_2}\tau^{-1/2}. \tag{32}$$

The fact that (31) "loses derivatives" is a fundamental aspect of this problem related to the trapping phenomenon, to be discussed in what follows, although the precise number of derivatives lost above is wasteful. Indeed, there are several aspects in which the above results can be improved. See Proposition 4.2.1 and the exercise of Section 4.3.

We can also express the pointwise decay in terms of advanced and retarded null coordinates u and v. Defining[39] $v = 2(t + r^*) = 2(t + r + 2M\log(r-2M))$, $u = 2(t - r^*) = 2(t - r - 2M\log(r-2M))$, it follows in particular from (32) that

$$|\psi| \leq CE_2(|v|+1)^{-1}, \qquad |r\psi| \leq C(r_0)E_2(\max\{u,1\})^{-\frac{1}{2}}, \tag{33}$$

where the first inequality applies in $\mathcal{D}\cap{\rm clos}(\{t\geq 0\})$, whereas the second applies only in $\mathcal{D}\cap\{t\geq 0\}\cap\{r\geq r_0\}$, with $C(r_0)\to\infty$ as $r_0\to 2M$. See also Appendix F.

[39] The strange convention on the factor of 2 is chosen simply to agree with [65].

Note that, as in Minkowski space, the first inequality of (33) is sharp as a *uniform* decay rate in v.

4.1. A spacetime integral estimate. The zero'th step in the proof of Theorem 4.1 is an estimate for a spacetime integral whose integrand should control the quantity

$$\chi J^N_\mu(\psi) n^\mu_{\tilde{\Sigma}_\tau} \tag{34}$$

where χ is a φ_t-invariant weight function such that χ degenerates only at infinity. Estimates of the spacetime integral (34) have their origin in the classical virial theorem, which in Minkowski space essentially arises from applying the energy identity to the current J^V with $V = \frac{\partial}{\partial r}$.

Naively, one might expect to be able to obtain an estimate of the form say

$$\int_{\tilde{\mathcal{R}}(0,\tau)} \chi J^N_\mu(\psi) n^\mu_{\tilde{\Sigma}_\tau} \leq B \int_{\tilde{\Sigma}_0} J^N_\mu n^\mu_{\tilde{\Sigma}_0}, \tag{35}$$

for such a χ. It turns out that there is a well known high-frequency obstruction for the existence of an estimate of the form (35) arising from *trapped null geodesics.* This problem has been long studied in the context of the wave equation in Minkowski space outside of an obstacle, where the analogue of trapped null geodesics are straight lines which reflect off the obstacle's boundary in such a way so as to remain in a compact subset of space. In Schwarzschild, one can easily infer from a continuity argument the existence of a family of null geodesics with i^+ as a limit point.[40] But in view of the integrability of geodesic flow, one can in fact understand all such geodesics explicitly.

Exercise: Show that the hypersurface $r = 3M$ is spanned by null geodesics. Show that from every point in $\mathcal{R}$, there is a codimension-one subset of future directed null directions whose corresponding geodesics approach $r = 3M$, and all other null geodesics either cross $\mathcal{H}^+$ or meet $\mathcal{I}^+$.

The timelike hypersurface $r = 3M$ is traditionally called the *photon sphere.* Let us first see how one can capture this high frequency obstruction.

4.1.1. *A multiplier X for high angular frequencies.* We look for a multiplier with the property that the spacetime integral it generates is positive definite. Since in Minkowski space, this is provided by the vector field ∂_r, we will look for simple generalisations. Calculations are slightly easier when one considers ∂_{r^*} associated to Regge-Wheeler coordinates (r^*, t). See Appendix F.2 for the definition of this coordinate system.[41] For $X = f(r^*)\partial_{r^*}$, where f is a general function, we obtain the formula

$$K^X = \frac{f'}{1-2M/r}(\partial_{r^*}\psi)^2 + \frac{f}{r}\left(1 - \frac{3M}{r}\right)|\nabla\!\!\!\!/\,\psi|^2 - \frac{1}{4}\left(2f' + 4\frac{r-2M}{r^2}f\right)\nabla^\alpha\psi\nabla_\alpha\psi.$$

Here f' denotes $\frac{df}{dr^*}$. We can now define a "modified" current

$$J^{X,w}_\mu = J^X_\mu(\psi) + \frac{1}{8}w\partial_\mu(\psi^2) - \frac{1}{8}(\partial_\mu w)\psi^2$$

[40] This can be thought of as a very weak notion of what it would mean for a null geodesic to be trapped from the point of view of decay results with respect to the foliation $\tilde{\Sigma}_\tau$.

[41] Remember, when considering coordinate vector fields, one has to specify the entire coordinate system. When considering ∂_r, it is here to be understood that we are using Schwarzschild coordinates, and when considering ∂_{r^*}, it is to be understood that we are using Regge-Wheeler coordinates. The precise choice of the angular coordinates is of course irrelevant.

associated to the vector field X and the function w. Let

$$K^{X,w} = \nabla^\mu J_\mu^{X,w}.$$

Choosing

$$w = f' + 2\frac{r-2M}{r^2}f + \frac{\delta(r-2M)}{r^5}\left(1-\frac{3M}{r}\right)f,$$

we have

$$\begin{aligned} K^{X,w} &= \left(\frac{f'}{1-2M/r} - \frac{\delta f}{2r^4}\left(1-\frac{3M}{r}\right)\right)(\partial_{r^*}\psi)^2 \\ &\quad + \frac{f}{r}\left(1-\frac{3M}{r}\right)\left(\left(1-\frac{\delta(r-2M)}{2r^4}\right)|\nabla\!\!\!/\psi|^2 + \frac{\delta}{2r^3}(\partial_t\psi)^2\right) \\ &\quad - \left(\frac{1}{8}\Box_g\left(2f' + 4\frac{r-2M}{r^2}f + 2\frac{\delta(r-2M)}{r^5}\left(1-\frac{3M}{r}\right)f\right)\right)\psi^2. \end{aligned}$$

Recall that in view of the spherical symmetry of $\mathcal{M}$, we may decompose

$$\psi = \sum_{\ell\geq 0, |m|\leq \ell} \psi_{\ell,m}(r,t)Y_{m,\ell}(\theta,\phi)$$

where $Y_{m,\ell}$ are the so-called *spherical harmonics*, each summand satisfies again the wave equation, and the convergence is in L^2 of the SO(3) orbits.

Let us assume that $\psi_{\ell,m} = 0$ for spherical harmonic number $\ell \leq L$ for some L to be determined. We look for $K^{X,w}$ such that $\int_{\mathbb{S}^2} K^{X,w} \geq 0$, but also $\int_{\mathbb{S}^2} |J_\mu^{X,w}n^\mu| \leq B\int_{\mathbb{S}^2} J_\mu^N n^\mu$. Here $\int_{\mathbb{S}^2}$ denotes integration over group orbits of the SO(3) action. For such ψ, in view of the resulting inequality

$$\frac{L(L+1)}{r^2}\int_{\mathbb{S}^2}\psi^2 \leq \int_{\mathbb{S}^2}|\nabla\!\!\!/\psi|^2,$$

it follows that taking L sufficiently large and $0<\delta<1$ sufficiently small so that $1-\frac{\delta(1-2M/r)}{2r^3} \geq \frac{1}{2}$, it clearly suffices to construct an f with the following properties:

(1) $|f| \leq B$,
(2) $f' \geq B(1-2M/r)r^{-4}$,
(3) $f(r=3M) = 0$,
(4) $-\frac{1}{8}\Box_g\left(2f' + 4\frac{r-2M}{r^2}f + 2\frac{\delta(r-2M)}{r^5}\left(1-\frac{3M}{r}\right)f\right)(r=3M) > 0$,
(5) $\frac{1}{8}\Box_g\left(2f' + 4\frac{r-2M}{r^2}f + 2\frac{\delta(r-2M)}{r^5}\left(1-\frac{3M}{r}\right)f\right) \leq \tilde{B}r^{-3}$

for some constants B, $\tilde{B}$. **Exercise**. Show that one can construct such a function.

Note the significance of the photon sphere!

4.1.2. *A multiplier X for all frequencies.* Constructing a multiplier for all spherical harmonics, so as to capture in addition "low frequency" effects, is more tricky. It turns out, however, that one can actually define a current which does not require spherical harmonic decomposition at all. The current is of the form:

$$\begin{aligned} J_\mu^{\mathbf{X}}(\psi) &= eJ_\mu^N(\psi) + J_\mu^{X^a}(\psi) + \sum_i J_\mu^{X^b,w^b}(\Omega_i\psi) \\ &\quad - \frac{1}{2}\frac{r(f^b)'}{f^b(r-2M)}\left(\frac{r-2M}{r^2} - \frac{(r^*-\alpha-\alpha^{1/2})}{\alpha^2+(r^*-\alpha-\alpha^{1/2})^2}\right)X_\mu^b\psi^2. \end{aligned}$$

Here, N is as in Section 3.3.2, $X^a = f^a\partial_{r^*}$, $X^b = f^b\partial_{r^*}$, the warped current $J^{X,w}$ is defined as in Section 4.1.1,

$$f^a = -\frac{C_a}{\alpha r^2} + \frac{c_a}{r^3},$$

$$f^b = \frac{1}{\alpha}\left(\tan^{-1}\frac{r^* - \alpha - \alpha^{1/2}}{\alpha} - \tan^{-1}(-1-\alpha^{-1/2})\right),$$

$$w^b = \frac{1}{8}\left((f^b)' + 2\frac{r-2M}{r^2}f^b\right),$$

and e, C_a, c_a, α are positive parameters which must be chosen accordingly. With these choices, one can show (after some computation) that the divergence $K^{\mathbf{X}} = \nabla^\mu J^{\mathbf{X}}_\mu$ controls in particular

$$\int_{\mathbb{S}^2} K^{\mathbf{X}}(\psi) \geq b\chi \int_{\mathbb{S}^2} J^N_\mu(\psi)n^\mu, \tag{36}$$

where χ is non-vanishing but decays (polynomially) as $r \to \infty$. Note that in view of the normalisation (125) of the r^* coordinate, $X^b = 0$ precisely at $r = 3M$. The left hand side of the inequality (36) controls also second order derivatives which degenerate however at $r = 3M$. We have dropped these terms. It is actually useful for applications that the $J^{X^a}(\psi)$ part of the current is not "modified" by a function w^a, and thus ψ itself does not occur in the boundary terms. That is to say

$$|J^{\mathbf{X}}_\mu(\psi)n^\mu| \leq B\left(J^N_\mu(\psi)n^\mu + \sum_{i=1}^{3} J^N_\mu(\Omega_i\psi)n^\mu\right). \tag{37}$$

On the event horizon $\mathcal{H}^+$, we have a better one-sided bound

$$-J^{\mathbf{X}}_\mu(\psi)n^\mu_{\mathcal{H}^+} \leq B\left(J^T_\mu(\psi)n^\mu_{\mathcal{H}^+} + \sum_{i=1}^{3} J^T_\mu(\Omega_i\psi)n^\mu_{\mathcal{H}^+}\right). \tag{38}$$

For details of the construction, see [**67**].

In view of (36), (37) and (38), together with the previous boundedness result Theorem 3.2, one obtains in particular the estimate

$$\int_{\tilde{\mathcal{R}}(\tau',\tau)} \chi J^N_\nu(\psi)n^\nu_{\tilde{\Sigma}} \leq B\int_{\tilde{\Sigma}(\tau')}\left(J^N_\mu(\psi) + \sum_{i=1}^{3} J^N_\mu(\Omega_i\psi)\right)n^\mu_{\tilde{\Sigma}_\tau}, \tag{39}$$

for some nonvanishing φ_t-invariant function χ which decays polynomially as $r \to \infty$.

On the other hand, considering the current $J^{\mathbf{X}}_\mu(P_{\leq L}\psi) + J^{X,w}_\mu((I - P_{\leq L})\psi)$, where $J^{X,w}_\mu$ is the current of Section 4.1.1 and $P_{\leq L}\psi$ denotes the projection to the space spanned by spherical harmonics with $\ell \leq L$, we obtain the estimate

$$\int_{\tilde{\mathcal{R}}(\tau',\tau)} \chi h J^N_\nu(\psi)n^\nu_{\tilde{\Sigma}} \leq B\int_{\tilde{\Sigma}(\tau')} J^N_\mu(\psi)n^\mu_{\tilde{\Sigma}_\tau}, \tag{40}$$

where h is any smooth nonnegative function $0 \leq h \leq 1$ vanishing at $r = 3M$, and B depends also on the choice of function h.

4.2. The Morawetz conformal Z multiplier and energy decay. How does the estimate (39) assist us to prove decay?

Recall that energy decay can be proven in Minkowski space with the help of the so-called *Morawetz current.* Let

$$Z = u^2\partial_u + v^2\partial_v \tag{41}$$

and define

$$J^{Z,w}_\mu(\psi) = J^Z_\mu(\psi) + \frac{tr^*(1-2M/r)}{2r}\psi\partial_\mu\psi - \frac{r^*(1-2M/r)}{4r}\psi^2\partial_\mu t.$$

(Here (u,v), (r^*,t) are the coordinate systems of Appendix F.) Setting $M=0$, this corresponds precisely to the current introduced by Morawetz [**119**] on Minkowski space.

It is a good **exercise** to show that (for $M>0$!) the coefficients of this current are C^0 but not C^1 across $\mathcal{H}^+\cup\mathcal{H}^-$.

To understand how one hopes to use this current, let us recall the situation in Minkowski space. There, the significance of (41) arises since it is a conformal Killing field. Setting $M=0$, $r^*=r$ in the above one obtains[42]

$$\int_{t=\tau} J^{Z,w}_\mu n^\mu \geq 0, \tag{42}$$

$$K^{Z,w} = 0. \tag{43}$$

The inequality (42) remains true in the Schwarzschild case and one can obtain exactly as before

$$\int_{t=\tau} J^{Z,w}_\mu n^\mu \geq b\int_{t=\tau} u^2(\partial_u\psi)^2 + v^2(\partial_v\psi)^2 + \left(1-\frac{2M}{r}\right)(u^2+v^2)|\nabla\!\!\!/\psi|^2. \tag{44}$$

(In fact, we have dropped positive 0'th order terms from the right hand side of (44), which will be useful for us later on in Section 4.3.) Note that away from the horizon, we have that

$$\int_{t=\tau} J^{Z,w}_\mu n^\mu \geq b(r_0,R)\tau^2\int_{\{t=\tau\}\cap\{r_0\leq r\leq R\}} J^N_\mu n^\mu. \tag{45}$$

Thus, if the left hand side of (45) could be shown to be bounded, this would prove the first statement of Theorem 4.1 where $\tilde\Sigma_\tau$ is replaced however with $\{t=\tau\}\cap\{r_0\leq r\leq R\}$.

In the case of Minkowski space, the boundedness of the left hand side of (44) follows immediately by (43) and the energy identity

$$\int_{t=\tau} J^{Z,w}_\mu + \int_{0\leq t\leq\tau} K^{Z,w} = \int_{t=0} J^{Z,w}_\mu \tag{46}$$

as long as the data are suitably regular and decay so as for the right hand side to be bounded. For Schwarzschild, one cannot expect (43) to hold, and this is why we have introduced the X-related currents.

First the good news: There exist constants $r_0<R$ such that

$$K^{Z,w}\geq 0$$

[42]The reason for introducing the 0'th order terms is because the wave equation is not conformally invariant. It is remarkable that one can nonetheless obtain positive definite boundary terms, although a slightly unsettling feature is that this positivity property (42) requires looking specifically at constant $t=\tau$ surfaces and integrating.

for $r \leq r_0$, and in fact

$$K^{Z,w} \geq b\frac{t}{r^3}\psi^2 \tag{47}$$

for $r \geq R$ and some constant b. These terms have the "right sign" in the energy identity (46). In $\{r_0 \leq r \leq R\}$, however, the best we can do is

$$-K^{Z,w} \leq B\,t\,(|\nabla\!\!\!/\psi|^2 + |\psi|^2).$$

This is the bad news, although, in view of the presence of trapping, it is to be expected. Using also (47), we may estimate

$$\begin{aligned}\int_{0\leq t\leq\tau} -K^{Z,w} &\leq B\int_{\{0\leq t\leq\tau\}\cap\{r_0\leq r\leq R\}} t\,J^N_\mu n^\mu \\ &\leq B\tau\int_{\{0\leq t\leq\tau\}\cap\{r_0\leq r\leq R\}} J^N_\mu n^\mu.\end{aligned} \tag{48}$$

In view of the fact that the first integral on the right hand side of (48) is bounded by (39), and the weight τ^2 in (45), applying the energy identity of the current $J^{Z,w}$ in the region $0 \leq t \leq \tau$, we obtain immediately a preliminary version of the first statement of the Theorem 4.1, but with τ^2 replaced by τ, and the hypersurfaces $\tilde{\Sigma}_\tau$ replaced by $\{t = \tau\} \cap \{r' \leq r \leq R'\}$ for some constants r', R', but where B depends on these constants. (Note the geometry of this region. All $\{t = \text{constant}\}$ hypersurfaces have common boundary $\mathcal{H}^+\cap\mathcal{H}^-$. **Exercise**: Justify the integration by parts (46), in view of the fact that Z and w are only C^0 at $\mathcal{H}^+\cup\mathcal{H}^-$.)

Using the current J^T and an easy geometric argument, it is not difficult to replace the hypersurfaces $\{t = \tau\} \cap \{r' \leq r \leq R'\}$ above with $\tilde{\Sigma}_\tau \cap \{r \geq r'\}$,[43] obtaining

$$\int_{\tilde{\Sigma}_\tau\cap\{r\geq r'\}} J^N_\mu(\psi)n^\mu \leq B\tau^{-1}\left(\int_{t=0} J^{Z,w}_\mu(\psi)n^\mu + \int_{\tilde{\Sigma}_0} J^N_\mu(\psi)n^\mu + \sum_{i=1}^3 J^N_\mu(\Omega_i\psi)n^\mu\right). \tag{49}$$

To obtain decay for the nondegenerate energy near the horizon, note that by the pigeonhole principle in view of the boundedness of the left hand side of (39) and what has just been proven, there exists (**exercise**) a dyadic sequence $\tilde{\Sigma}_{\tau_i}$ for which the first statement of Theorem 4.1 holds, with τ^{-2} replaced by τ_i^{-1}. Finally, by Theorem 3.2, we immediately (**exercise**: why?) remove the restriction to the dyadic sequence.

We have thus obtained

$$\int_{\tilde{\Sigma}_\tau} J^N_\mu(\psi)n^\mu \leq B\tau^{-1}\left(\int_{t=0} J^{Z,w}_\mu(\psi)n^\mu + \int_{\tilde{\Sigma}_0} J^N_\mu(\psi)n^\mu + \sum_{i=1}^3 J^N_\mu(\Omega_i\psi)n^\mu\right). \tag{50}$$

The statement (50) loses one power of τ in comparison with the first statement of Theorem 3.2. How do we obtain the full result? First of all, note that, commuting once again with Ω_j, it follows that (50) holds for ψ replaced with $\Omega_j\psi$. Now we may partition $\tilde{\mathcal{R}}(0,\tau)$ dyadically into subregions $\tilde{\mathcal{R}}(\tau_i,\tau_{i+1})$ and revisit the X-estimate

[43] Hint: Use (44) to estimate the energy on $\{t = t_0\} \cap J^+(\tilde{\Sigma}_\tau)$ with weights in τ. Send $t_0 \to \infty$ and estimate backwards to $\tilde{\Sigma}_\tau$ using conservation of the J^T flux.

(39) on each such region. In view of (50) applied to both ψ and $\Omega_j\psi$, the estimate (39) gives

$$\int_{\tilde{\mathcal{R}}(\tau_i,\tau_{i+1})} \chi J^N_\nu(\psi) n^\nu_{\tilde{\Sigma}} \leq BD\tau_i^{-1}, \tag{51}$$

where D is a quantity coming from data. Summing over i, this gives us that

$$\int_{\mathcal{R}(0,\tau)} t\,\chi J^N_\nu(\psi) n^\nu_{\tilde{\Sigma}} \leq BD(1+\log|\tau+1|).$$

This estimates in particular the first term on the right hand side of the first inequality of (48). Applying this inequality, we obtain as before (49), but with $\tau^{-2}(1+\log|\tau+1|)$ replacing τ. Using (51) and a pigeonhole principle, one improves this to (50), with $\tau^{-2}(1+\log|\tau+1|)$ now replacing τ. Iterating this argument again one removes the log (**exercise**).

Note that this loss of derivatives in (31) simply arises from the loss in (39). If Ω_i could be replaced by Ω_i^ϵ in (39), then the loss would be 3ϵ. The latter refinement can in fact be deduced from the original (31) using in addition work of Blue-Soffer [**21**]. Running the argument of this section with the ϵ-loss version of (31), we obtain now

PROPOSITION 4.2.1. *For any $\epsilon > 0$, statement* (31) *holds with* 3 *replaced by* ϵ *in the definition of E_1 and C replaced by C_ϵ.*

4.3. Pointwise decay. To derive pointwise decay for ψ itself, we should remember that we have in fact dropped a good 0'th order term from the estimate (44). In particular, we have also

$$\int_{t=\tau} J^{Z,w}_\mu(\psi) n^\mu \geq b \int_{\{t=\tau\}\cap\{r\geq r_0\}} (\tau^2 r^{-2}+1)\psi^2.$$

From this and the previously derived bounds, pointwise decay can be shown easily by applying Ω_i as commutators and Sobolev estimates. See [**65**] for details.

Exercise: Derive pointwise decay for all derivatives of ψ, including transverse derivatives to the horizon of any order, by commuting in addition with $\hat{Y}$ as in the proof of Theorem 3.2.

4.4. Comments and further reading.

4.4.1. *The X-estimate.* The origin of the use of vector field multipliers of the type X (as in Section 4.1) for proving decay for solutions of the wave equation goes back to Morawetz. (These identities are generalisations of the classical virial identity, which has itself a long and complicated history.) In the context of Schwarzschild black holes, the first results in the direction of such estimates were in Laba and Soffer [**110**] for a certain "Schrödinger" equation (related to the Schwarzschild t-function), and, for the wave equation, in Blue and Soffer [**19**]. These results were incomplete (see [**20**]), however, and the first estimate of this type was actually obtained in our [**65**], motivated by the original calculations of [**19, 110**]. This estimate required decomposition of ψ into individual spherical harmonics ψ_ℓ, and choosing the current $J^{X,w}$ separately for each ψ_ℓ. A slightly different approach to this estimate is provided by [**20**]. A somewhat simpler choice of current $J^{X,w}$ which provides an estimate for all sufficiently high spherical harmonics was first presented by Alinhac [**1**]. Our Section 4.1.1 is similar in spirit. The first estimate not requiring a spherical harmonic decomposition was obtained in [**67**]. This is the current of Section 4.1.2. The problem of reducing the loss of derivatives in (39) has been addressed in Blue-Soffer [**21**].[44] The results of [**21**] in fact also apply to the Reissner-Nordström metric.

A slightly different construction of a current as in Section 4.1.2 has been given by Marzuola and collaborators [**116**]. This current does not require commuting with Ω_i. In their subsequent [**115**], the considerations of [**116**] are combined with ideas from [**65, 67**] to obtain an estimate which does not degenerate on the horizon: One includes a piece of the current J^N of Section 3.3.2 and exploits Proposition 3.1.

4.4.2. *The Z-estimate.* The use of vector-field multipliers of the type Z also goes back to celebrated work of Morawetz, in the context of the wave equation outside convex obstacles [**119**]. The geometric interpretation of this estimate arose later, and the use of Z adapted to the causal geometry of a non-trivial metric first appears perhaps in the proof of stability of Minkowski space [**51**]. The decay result Theorem 4.1 was obtained in our [**65**]. A result yielding similar decay away from the horizon (but weaker decay along the horizon) was proven independently in a nice paper of Blue and Sterbenz [**22**]. Both [**22**] and [**65**] make use of a current based on the vector field Z. In [**22**], the error term analogous to $K^{Z,w}$ of Section 4.2 was controlled with the help of an auxiliary collection of multipliers with linear weights in t, chosen at the level of each spherical harmonic, whereas in [**65**], these error terms are controlled directly from (39) by a dyadic iteration scheme similar to the one we have given here in Section 4.2. The paper [**22**] does not obtain estimates for the non-degenerate energy flux (31); moreover, a slower pointwise decay rate near the horizon is achieved in comparison to Theorem 4.1. Motivated by [**65**], the authors of [**22**] have since given a different argument [**23**] to obtain just the pointwise estimate (32) on the horizon, exploiting the "good" term in $K^{Z,w}$ near the horizon. The proof of Theorem 4.1 presented in Section 4.2 is a slightly modified version of the scheme in [**65**], avoiding spherical harmonic decompositions (for obtaining (39)) by using in particular the result of [**67**].

4.4.3. *Other results.* Statement (32) of Theorem 4.1 has been generalised to the Maxwell case by Blue [**18**]. In fact, the Maxwell case is much "cleaner", as the current J^Z need not be modified by a function w, and its flux is pointwise positive through any spacelike hypersurface. The considerations near the horizon follow [**23**] and thus the analogue of (31) is not in fact obtained, only decay for the degenerate flux of J^T. Nevertheless, the non-degenerate (31) for Maxwell can be proven following the methods of this section, using in particular currents associated to the vector field Y (**Exercise**).

To our knowledge, the above discussion exhausts the quantitative pointwise and energy decay-type statements which are known for general solutions of the wave equation on Schwarzschild.[45] The best previously known results on general solutions of the wave equation were non-quantitative decay type statements which we briefly mention. A pointwise decay without a rate was first proven in the thesis of Twainy [**149**]. Scattering and asymptotic completeness statements for the wave, Klein-Gordon, Maxwell and Dirac equations have been obtained by [**72, 73, 5, 4, 122**]. These type of statements are typically insensitive to the amount of trapping. See the related discussion of Section 4.6, where the statement of

[44]A related refinement, where h of (40) is replaced by a function vanishing logarithmically at $3M$, follows from [**115**] referred to below.

[45]For fixed spherical harmonic $\ell = 0$, there is also the quantitative result of [**63**], to be mentioned in Section 4.6.

Theorem 4.1 is compared to non-quantitative statements heuristically derived in the physics literature.

4.5. Perturbing? Use of the J^N current "stabilises" the proof of Theorem 4.1 with respect to considerations near the horizon. There is, however, a sense in which the above argument is still fundamentally attached to Schwarzschild. The approach taken to derive the multiplier estimate (36) depends on the structure of the trapping set, in particular, the fact that trapped null geodesics approach a codimension-1 subset of spacetime, the photon sphere. Overcoming the restrictiveness of this approach is the fundamental remaining difficulty in extending these techniques to Kerr, as will be accomplished in Section 5.3. Precise implications of this fact for multiplier estimates are discussed further in [**1**].

4.6. Aside: Quantitative vs. non-quantitative results and the heuristic tradition. The study of wave equations on Schwarzschild has a long history in the physics literature, beginning with the pioneering Regge and Wheeler [**131**]. These studies have all been associated with showing "stability".

A seminal paper is that of Price [**130**]. There, insightful heuristic arguments were put forth deriving the asymptotic tail of each spherical harmonic ψ_ℓ evolving from compactly supported initial data, suggesting that for $r > 2M$,

$$\psi_\ell(r,t) \sim C_\ell t^{-(3+2\ell)}. \tag{52}$$

These arguments were later extended by Gundlach et al [**88**] to suggest

$$\psi_\ell|_{\mathcal{H}^+} \sim C_\ell v^{-(3+2\ell)}, \qquad r\psi_\ell|_{\mathcal{I}^+} \sim \bar{C}_\ell u^{-(2+\ell)}. \tag{53}$$

Another approach to these heuristics via the analytic continuation of the Green's function was followed by [**31**]. The latter approach in principle could perhaps be turned into a rigorous proof, at least for solutions not supported on $\mathcal{H}^+ \cap \mathcal{H}^-$. See [**114, 106**] for just (52) for the $\ell = 0$ case.

Statements of the form (52) are interesting because, if proven, they would give the fine structure of the tail of the solution. However, it is important to realise that statements like (52) in of themselves would not give quantitative bounds for the size of the solution at all later times in terms of initial data. In fact, the above heuristics do not even suggest what the best such quantitative result would be, they only give a heuristic *lower* bound on the best possible quantitative decay rate in a theorem like Theorem 4.1.

Let us elaborate on this further. For fixed spherical harmonic, by compactness a statement of the form (52) would immediately yield some bound

$$|\psi_\ell|(r,t) \le D(r,\psi_\ell)t^{-3}, \tag{54}$$

for some constant D depending on r *and on the solution itself.* It is not clear, however, what the sharp such quantitative inequality of the form (54) is supposed to be when the constant is to depend on a natural quantity associated to data. It is the latter, however, which is important for the nonlinear stability problem.

There is a setting in which a quantitative version of (54) has indeed been obtained: The results of [**63**] (which apply to the nonlinear problem where the scalar field is coupled to the Einstein equation, but which can be specialised to the decoupled case of the $\ell = 0$ harmonic on Schwarzschild) prove in particular that

$$|n_{\Sigma_\tau}\psi_0| + |\psi_0| \le C_\epsilon D(\psi,\psi')\tau^{-3+\epsilon}, \qquad |r\psi_0| \le CD(\psi,\psi')\tau^{-2} \tag{55}$$

where C_ϵ depends only on ϵ, and $D(\psi, \psi')$ is a quantity depending only on the initial J^T energy and a pointwise weighted C^1 norm. In view of the relation between τ, u, and v, (55) includes also decay on the horizon and null infinity as in the heuristically derived (53). The fact that the power 3 indeed appears in both in the quantitative (55) and in (54) may be in part accidental. See also [**15**].

For general solutions, i.e. for the sum over spherical harmonics, the situation is even worse. In fact, a statement like (52) a priori gives no information whatsoever of any sort, even of the non-quantitative kind. It is in principle compatible with $\limsup_{t\to\infty} \psi(r,t) = \infty$.[46] It is well known, moreover, that to understand quantitative decay rates for general solutions, one must quantify trapping. This is not, however, captured by the heuristics leading to (52), essentially because for fixed ℓ, the effects of trapping concern an intermediate time interval not reflected in the tail. It should thus not be surprising that these heuristics do not address the fundamental problem at hand.

Another direction for heuristic work has been the study of so-called *quasi-normal modes.* These are solutions with time dependence $e^{-i\omega t}$ for ω with negative imaginary part, and appropriate boundary conditions. These occur as poles of the analytic continuation of the resolvent of an associated elliptic problem, and in the scattering theory literature are typically known as *resonances.* Quasinormal modes are discussed in the nice survey article of Kokkotas and Schmidt [**104**]. Rigorous results on the distribution of resonances have been achieved in Bachelot–Motet-Bachelot [**7**] and Sá Barreto-Zworski [**135**]. The asymptotic distribution of the quasi-normal modes as $\ell \to \infty$ can be thought to reflect trapping. On the other hand, these modes do not reflect the "low-frequency" effects giving rise to tails. Thus, they too tell only part of the story. See, however, the case of Schwarzschild-de Sitter in Section 6.

Finally, we should mention Stewart [**144**]. This is to our knowledge the first clear discussion in the physics literature of the relevance of trapping on the Schwarzschild metric in this context and the difference between quantitative and non-quantitative decay rates. It is interesting to compare Section 3 of [**144**] with what has now been proven: Although the predictions of [**144**] do not quite match the situation in Schwarzschild (it is in particular incompatible with (52)), they apply well to the Schwarzschild-de Sitter case developed in Section 6.

The upshot of the present discussion is the following: Statements of the form (52), while interesting, may have little to do with the problem of non-linear stability of black holes, and are perhaps more interesting for the lower bounds that they suggest.[47] In fact, in view of their non-quantitative nature, these results are less relevant for the stability problem than the quantitative boundedness theorem of Kay and Wald. Even the statement of Section 3.2.3 cannot be derived as a corollary of the statement (52), nor would knowing (52) simplify in any way the proof of Section 3.2.3.

5. Perturbing Schwarzschild: Kerr and beyond

We now turn to the problem of perturbing the Schwarzschild metric and proving boundedness and decay for the wave equation on the backgrounds of such perturbed

[46]Of course, *given* the quantitative result of Theorem 3.2 and the statement (52), one could then infer that for each $r > 2M$, then $\lim_{t\to\infty} \phi(r,t) = 0$, without however a rate (**exercise**).

[47]See for instance the relevance of this in [**59**].

metrics. Let us recall our dilemma: The boundedness argument of Section 3 required that T remains causal everywhere in the exterior. In view of the comments of Section 3.5, this is clearly unstable. On the other hand, the decay argument of Section 4 requires understanding the trapped set and in particular, uses the fact that in Schwarzschild, a certain codimension-1 subset of spacetime–the photon sphere–plays a special role. Again, as discussed in Section 4.5, this special structure is unstable.

It turns out that nonetheless, these issues can be addressed and both boundedness (see Theorem 5.1) and decay (see Theorem 5.2) can be proven for the wave equation on suitable perturbations of Schwarzschild. As we shall see, the boundedness proof (See Section 5.2) turns out to be more robust and can be applied to a larger class of metrics–but it too requires some insight from the Schwarzschild decay argument! The decay proof (See Section 5.3) will require us to restrict to exactly Kerr spacetimes.

Without further delay, perhaps it is time to introduce the Kerr family...

5.1. The Kerr metric. The *Kerr metric* is a 2-parameter family of metrics first discovered **[99]** in 1963. The parameters are called *mass* M and specific angular momentum a, i.e. angular momentum per unit mass. In so-called Boyer-Lindquist local coordinates, the metric element takes the form:

$$-\left(1-\frac{2M}{r\left(1+\frac{a^2\cos^2\theta}{r^2}\right)}\right)dt^2+\frac{1+\frac{a^2\cos^2\theta}{r^2}}{1-\frac{2M}{r}+\frac{a^2}{r^2}}dr^2+r^2\left(1+\frac{a^2\cos^2\theta}{r^2}\right)d\theta^2$$
$$+r^2\left(1+\frac{a^2}{r^2}+\left(\frac{2M}{r}\right)\frac{a^2\sin^2\theta}{r^2\left(1+\frac{a^2\cos^2\theta}{r^2}\right)}\right)\sin^2\theta\, d\phi^2$$
$$-4M\frac{a\sin^2\theta}{r\left(1+\frac{a^2\cos^2\theta}{r^2}\right)}dt\, d\phi.$$

The vector fields ∂_t and ∂_ϕ are Killing. We say that the Kerr family is *stationary* and *axisymmetric*.[48] Traditionally, one denotes

$$\Delta = r^2 - 2Mr + a^2.$$

If $a = 0$, the Kerr metric clearly reduces to Schwarzschild (5).

Maximal extensions of the Kerr metric were first constructed by Carter **[29]**. For parameter range $0 \le |a| < M$, these maximal extensions have black hole regions and white hole regions bounded by future and past event horizons $\mathcal{H}^\pm$ meeting at a bifurcate sphere. The above coordinate system is defined in a domain of outer communications, and the horizon will correspond to the limit $r \to r_+$, where r_+ is the larger positive root of $\Delta = 0$, i.e.

$$r_+ = M + \sqrt{M^2 - a^2}.$$

Since the motivation of our study is the Cauchy problem for the Einstein equations, it is more natural to consider not maximal extensions, but maximal developments of complete initial data. (See Appendix B.) In the Schwarzschild case, the maximal development of initial data on a Cauchy surface Σ as described previously coincides with maximally-extended Schwarzschild. In Kerr, if we are to take an asymptotically flat (with two ends) hypersurface in a maximally extended

[48]There are various conventions on the meaning of the words "stationary" and "axisymmetric" depending on the context. Let us not worry about this here...

Kerr for parameter range $0 < |a| < M$, then its maximal development will have a smooth boundary in maximally-extended Kerr. This boundary is what is known as a *Cauchy horizon*. We have already discussed this phenomenon in Section 2.7.3 in the context of strong cosmic censorship. The maximally extended Kerr solutions are quite bizarre, in particular, they contain closed timelike curves. This is of no concern to us here, however. By definition, for us the term "Kerr metric $(\mathcal{M}, g_{M,a})$" will always denote the maximal development of a complete asymptotically flat hypersurface Σ, as above, with two ends. One can depict the Penrose-diagrammatic representation of a suitable two-dimensional timelike slice of this solution as below:

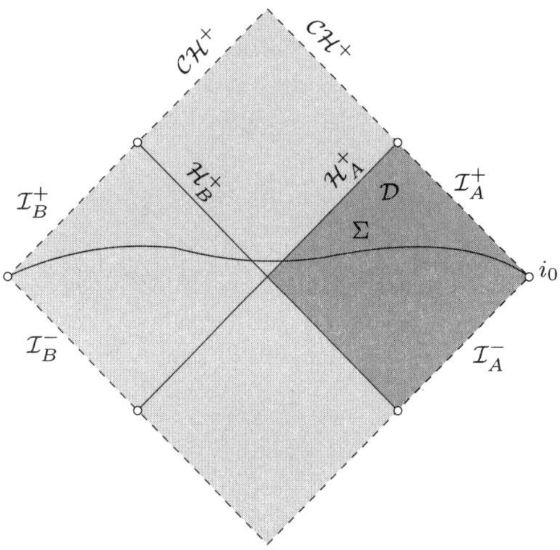

This depiction coincides with the standard Penrose diagram of the spherically symmetric Reissner-Nordström metric [**91, 148**].

With this convention in mind, we note that the dependence of $g_{M,a}$ on a is smooth in the range $0 \leq |a| < M$. In particular, Kerr solutions with small $|a| \ll M$ can be viewed as close to Schwarzschild.

One can see this explicitly in the subregion of interest to us by passing to a new system of coordinates. Define

$$t^* = t + \bar{t}(r)$$
$$\phi^* = \phi + \bar{\phi}(r)$$

where

$$\frac{d\bar{t}}{dr}(r) = (r^2 + a^2)/\Delta, \qquad \frac{d\bar{\phi}}{dr}(r) = a/\Delta.$$

(These coordinates are often known as *Kerr-star coordinates.*) These coordinates are regular across $\mathcal{H}^+ \setminus \mathcal{H}^-$.[49] We may finally define a coordinate $r_{\text{Schw}} = r_{\text{Schw}}(r, a)$ such that which takes $[r_+, \infty) \to [2M, \infty)$ with smooth dependence in a and such that $r_{\text{Schw}}(r, 0)$ is the identity map. In particular, if we define Σ_0 by $\mathcal{D} = \{t^* = 0\}$, and define $\mathcal{R} = \mathcal{D} \cap \{t^* \geq 0\}$, and fix r_{Schw}, t^*, ϕ^* Schwarzschild coordinates, then the metric functions of $g_{M,a}$ written in terms of these coordinates as defined previously depend smoothly on a for $0 \leq |a| < M$ in $\mathcal{R}$, and, for $a = 0$, reduce to the Schwarzschild metric form in (r, t^*, ϕ, θ) coordinates where t^* is defined from Schwarzschild t as above.

[49]Of course, one again needs two coordinate systems in view of the breakdown of spherical coordinates. We shall suppress this issue in the discussion that follows.

We note that $\partial_t = \partial_{t^*}$ in the intersection of the coordinate systems. We immediately note that ∂_t is spacelike on the horizon, except where $\theta = 0, \pi$, i.e. on the axis of symmetry. Note that we shall often abuse notation (as we just have done) and speak of ∂_t on the horizon or at $\theta = 0$, where of course the (r, t, θ, ϕ) coordinate system breaks down, and formally, this notation is meaningless.

In general, the part of the domain of outer communications plus horizon where ∂_t is spacelike is known as the *ergoregion*. It is bounded by a hypersurface known as the *ergosphere*. The ergosphere meets the horizon on the axis of symmetry $\theta = 0, \pi$.

The ergosphere allows for a particle "process", originally discovered by Penrose [**127**], for extracting energy out of a black hole. This came to be known as the *Penrose process*. In his thesis, Christodoulou [**38**] discovered the existence of a quantity–the so-called *irreducible mass* of the black hole–which he showed to be always nondecreasing in a Penrose process. The analogy between this quantity and entropy led later to a subject known as "black hole thermodynamics" [**8, 11**]. This is currently the subject of intense investigation from the point of view of high energy physics.

In the context of the study of $\Box_g\psi = 0$, we have already discussed in Section 5 the effect of the ergoregion: It is precisely the presence of the ergoregion that makes our previous proof of boundedness for Schwarzschild not immediately generalise for Kerr. Moreover, in contrast to the Schwarzschild case, there is no "easy result" that one can obtain away from the horizon analogous to Section 3.2.3. In fact, the problem of proving any sort of boundedness statement for general solutions to $\Box_g\psi = 0$ on Kerr had been open until very recently. We will describe in the next section our recent resolution [**68**] of this problem.

5.2. Boundedness for axisymmetric stationary black holes. We will derive a rather general boundedness theorem for a class of axisymmetric stationary black hole exteriors near Schwarzschild. The result (Theorem 5.1) will include slowly rotating Kerr solutions with parameters $|a| \ll M$.

We have already explained in what sense the Kerr metric is "close" to Schwarzschild in the region $\mathcal{R}$. Let us note that with respect to the coordinates r_{Schw}, t^*, ϕ^*, θ in $\mathcal{R}$, then ∂_{t^*} and ∂_{ϕ^*} are Killing for both the Schwarzschild and the Kerr metric. The class of metrics which will concern us here are metrics defined on $\mathcal{R}$ such that the metric functions are close to Schwarzschild in a suitable sense[50], and ∂_{t^*}, ∂_{ϕ^*} are Killing, where these are defined with respect to the ambient Schwarzschild coordinates.

There is however an additional geometric assumption we shall need, and this is motivated by a geometric property of the Kerr spacetime, to be described in the section that follows immediately.

5.2.1. *Killing fields on the horizon.* Let us here remark a geometric property of the Kerr spacetime itself which turns out to be of utmost importance in what follows: Let V denote a null generator of $\mathcal{H}^+$. Then

$$V \in \text{Span}\{\partial_{t^*}, \partial_{\phi^*}\}. \tag{56}$$

There is a deep reason why this is true. For stationary black holes with non-degenerate horizons, a celebrated argument of Hawking [**91**] retrieves a second Killing field in the direction of the null generator V. Thus, if ∂_{t^*} and ∂_{ϕ^*} span the complete set of Killing fields, then V must evidently be in their span.

[50]This requires moving to an auxiliary coordinate system. See [**68**].

In fact, choosing V accordingly we have

$$V = \partial_{t^*} + (a/2Mr_+)\partial_{\phi^*} \tag{57}$$

(For the Kerr solution, we have that there exists a timelike direction in the span of ∂_{t^*} and ∂_{ϕ^*} for all points outside the horizon. We shall not explicitly make reference to this property, although in view of Section 7, one can infer this property (**exercise**) for small perturbations of Schwarzschild of the type considered here, i.e., given any point p outside the horizon, there exists a Killing field V (depending on p) such that $V(p)$ is timelike.)

5.2.2. *The axisymmetric case.* From (57), it follows that there is a constant $\omega_0 > 0$, depending only on the parameters a and M, such that if

$$|\partial_{t^*}\psi|^2 \geq \omega_0 |\partial_{\phi^*}\psi|^2, \tag{58}$$

on $\mathcal{H}^+$, then the flux satisfies

$$J^T_\mu(\psi) n^\mu_{\mathcal{H}^+} \geq 0. \tag{59}$$

Note also that, for fixed M, we can take

$$\omega_0 \to 0, \qquad \text{as} \qquad a \to 0. \tag{60}$$

There is an immediate application of (58). Let us restrict for the moment to axisymmetric solutions, i.e. to ψ such that $\partial_\phi \psi = 0$. It follows that (58) trivially holds. As a result, our argument proving boundedness is stable, i.e. Theorem 3.2 holds for axisymmetric solutions of the wave equation on Kerr spacetimes with $|a| \ll M$. (See the exercise of Section 3.5.) In fact, the restriction on a can be be removed (**Exercise**, or go directly to Section 7).

Let us note that the above considerations make sense not only for Kerr but for the more general class of metrics on $\mathcal{R}$ close to Schwarzschild such that ∂_{t^*}, ∂_{ϕ^*} are Killing, $\mathcal{H}^+$ is null and (56) holds. In particular, (58) implies (59), where in (60), the condition $a \to 0$ is replaced by the condition that the metric is taken suitably close to Schwarzschild. **The discussion which follows will refer to metrics satisfying these assumptions.**[51] For simplicity, the reader can specialise the discussion below to the case of a Kerr metric with $|a| \ll M$.

5.2.3. *Superradiant and non-superradiant frequencies.* There is a more general setting where we can make use of (58). Let us suppose for the time being that we could take the Fourier transform $\hat{\psi}(\omega)$ of our solution ψ in t^* and then expand in azimuthal modes ψ_m, i.e. modes associated to the Killing vector field ∂_{ϕ^*}.

If we were to restrict ψ to the frequency range

$$|\omega|^2 \geq \omega_0 m^2, \tag{61}$$

then (58) and thus (59) holds after integrating along $\mathcal{H}^+$. In view of this, frequencies in the range (61) are known as *nonsuperradiant frequencies.* The frequency range

$$|\omega|^2 \leq \omega_0 m^2 \tag{62}$$

determines the so-called *superradiant frequencies.* In the physics literature, the main difficulty of this problem has traditionally been perceived to "lie" with these frequencies.

[51]They are summarised again in the formulation of Theorem 5.1.

Let us pretend for the time being that using the Fourier transform, we could indeed decompose

$$\psi = \psi_\sharp + \psi_\flat \tag{63}$$

where $\psi_\sharp$ is supported in (61), whereas $\psi_\flat$ is supported in (62).

In view of the discussion immediately above and the comments of Section 5.2.2, it is plausible to expect that one could indeed prove boundedness for $\psi_\sharp$ in the manner of the proof of Theorem 3.2. In particular, if one could localise the integrated version of (59) to arbitrary sufficiently large subsegments $\mathcal{H}(\tau', \tau'')$, one could obtain

$$\int_{\Sigma_\tau} J^{n_{\Sigma_\tau}}_\mu(\psi_\sharp) n^\mu_{\Sigma_\tau} \leq B \int_{\Sigma_0} J^{n_{\Sigma_0}}_\mu(\psi_\sharp) n^\mu_{\Sigma_0}. \tag{64}$$

This would leave $\psi_\flat$. Since this frequency range does not suggest a direct boundedness argument, it is natural to revisit the decay mechanism of Schwarzschild. We have already discussed (see Section 4.5) the instability of the decay argument; this instability arose from the structure of the set of trapped null geodesics. At the heuristic level, however, it is easy to see that, if one can take ω_0 sufficiently small, then solutions supported in (62) cannot be trapped. In particular, for $|a| \ll M$, **superradiant frequencies for $\Box_g\psi = 0$ on Kerr are not trapped**. This will be the fundamental observation allowing for the boundedness theorem. Let us see how this statement can be understood from the point of view of energy currents.

5.2.4. *A stable energy estimate for superradiant frequencies.* We continue here our heuristic point of view, where we assume a decomposition (63) where $\psi_\flat$ is supported in (62). In particular, one has an inequality

$$\int_{-\infty}^{\infty}\int_0^{2\pi} \omega_0^2(\partial_\phi\psi_\flat)^2\, d\phi^*\, dt^* \geq \int_{-\infty}^{\infty}\int_0^{2\pi} (\partial_t\psi_\flat)^2\, d\phi^*\, dt^* \tag{65}$$

for all (r, θ). We shall see below that (65) allows us easily to construct a suitable stable current for Schwarzschild.

It may actually be a worthwhile **exercise** for the reader to come up with a suitable current for themselves. The choice is actually quite flexible in comparison with the considerations of Section 4.1. Our choice (see [**68**]) is defined by

$$J^{\mathbf{X}} = eJ^N + J^{X_a} + J^{X_b, w_b} \tag{66}$$

where here, N is the vector field of Section 3.3.2, $X_a = f_a\partial_{r^*}$, with

$$\begin{aligned} f_a &= -r^{-4}(r_0)^4, \qquad \text{for } r \leq r_0 \\ f_a &= -1, \qquad \text{for } r_0 \leq r \leq R_1, \\ f_a &= -1 + \int_{R_1}^{r} \frac{d\tilde r}{4\tilde r} \qquad \text{for } R_1 \leq r \leq R_2, \\ f_a &= 0 \text{ for } r \geq R_2, \end{aligned}$$

$X_b = f_b\partial_{r^*}$ with

$$f_b = \chi(r^*)\pi^{-1}\int_0^{r^*} \frac{\alpha}{x^2 + \alpha^2}$$

and $\chi(r^*)$ is a smooth cutoff with $\chi = 0$ for $r^* \le 0$ and $\chi = 1$ for $r^* \ge 1$. Here r and r^* are Schwarzschild coordinates.[52] The function $w_\flat$ is given by

$$w_\flat = f_\flat' + \frac{2}{r}(1 - 2M/r)(1 - M/r)f_\flat.$$

The parameters e, α, r_0, R_1, R_2 must be chosen accordingly!

Restricting to the range (62), using (65), with some computation we would obtain

$$\int_{-\infty}^{\infty}\int_0^{2\pi} K^{\mathbf{X}}(\psi_\flat)\,d\phi^*\,dt^* \ge b\int_{-\infty}^{\infty}\int_0^{2\pi} \chi J_\mu^{n_\Sigma}(\psi_\flat)n_\Sigma^\mu\,d\phi^*\,dt^*, \tag{67}$$

for all (r,θ).

The above inequality can immediately be seen to be stable to small[53] axisymmetric, stationary perturbations of the Schwarzschild metric. That is to say, for such metrics, if $\psi_\flat$ is supported in (62) (where frequencies here are defined by Fourier transform in coordinates t^*, ϕ^*), then the inequality (67) holds as before. In particular, (67) holds for Kerr for small $|a| \ll M$.

How would (67) give boundedness for $\psi_\flat$? We need in fact to suppose something slightly stronger, namely that (67) holds localised to $\mathcal{R}(0,\tau)$. Consider the currents

$$J = J^N + e_2 J^{\mathbf{X}}, \qquad K = \nabla^\mu J_\mu,$$

where e_2 is a positive parameters, and J^N is the current of Section 3.3.2. Then, for metrics g close enough to Schwarzschild, and for e_2 sufficiently small, we would have from a localised (67) that

$$\int_{\mathcal{R}(0,\tau)} K(\psi_\flat) \ge 0,$$

$$\int_{\mathcal{H}(0,\tau)} J_\mu(\psi_\flat)n_{\mathcal{H}}^\mu \ge 0,$$

and thus

$$\int_{\Sigma_\tau} J_\mu(\psi_\flat)n_{\Sigma_\tau}^\mu \le \int_{\Sigma_0} J_\mu(\psi_\flat)n_{\Sigma_0}^\mu.$$

Moreover, for g sufficiently close to Schwarzschild and e_1, e_2 suitably defined, we also have (**exercise**)

$$\int_{\Sigma_\tau} J_\mu^{n_{\Sigma_\tau}}(\psi_\flat)n^\mu \le B\int_{\Sigma_\tau} J_\mu(\psi_\flat)n_{\Sigma_\tau}^\mu.$$

We thus would obtain

$$\int_{\Sigma_\tau} J_\mu^{n_{\Sigma_\tau}}(\psi_\flat)n^\mu \le B\int_{\Sigma_0} J_\mu^{n_{\Sigma_0}}(\psi_\flat)n^\mu. \tag{68}$$

Adding (68) and (64), we would obtain

$$\int_{\Sigma_\tau} J_\mu^{n_{\Sigma_\tau}}(\psi)n^\mu \le B\int_{\Sigma_0} J_\mu^{n_{\Sigma_0}}(\psi)n^\mu$$

[52]Since we are dealing now with general perturbations of Schwarzschild, we shall now use r for what we previously denoted by r_{Schw}. Note that in the special case that our metric is Kerr, this r is different from the Boyer-Lindquist r.

[53]Of course, in view of the degeneration towards i^0, it is important that smallness is understood in a weighted sense.

provided that we could also estimate say

$$\int_{\Sigma_0} J_\mu^{n_{\Sigma_0}}(\psi_\sharp)n^\mu \leq B \int_{\Sigma_0} J_\mu^{n_{\Sigma_0}}(\psi)n^\mu. \tag{69}$$

5.2.5. *Cutoff and decomposition.* Unfortunately, things are not so simple!

For one thing, to take the Fourier transform necessary to decompose in frequency, one would need to know a priori that $\psi(t^*, \cdot)$ is in $L^2(t^*)$. What we want to prove at this stage is much less. A priori, ψ can grow exponentially in t^*. In order to apply the above, one must cut off the solution appropriately in time.

This is achieved as follows. For definiteness, define Σ_0 to be $t^* = 0$, and Σ_τ as before. We will also need two auxiliary families of hypersurfaces defined as follows. (The motivation for considering these will be discussed in Section 5.2.6.) Let χ be a cutoff such that $\chi(x) = 0$ for $x \geq 0$ and $\chi = 1$ for $x \leq -1$, and define $t^\pm$ by

$$t^+ = t^* - \chi(-r+R)(1+r-R)^{1/2}$$

and

$$t^- = t^* + \chi(-r+R)(1+r-R)^{1/2}$$

where R is a large constant, which must be chosen appropriately. Let us define then

$$\Sigma^+(\tau) \doteq \{t^+ = \tau\}, \qquad \Sigma^-(\tau) \doteq \{t^- = \tau\}.$$

Finally, we define

$$\mathcal{R}(\tau_1, \tau_2) = \bigcup_{\tau_1 \leq \tau \leq \tau_2} \Sigma(\tau),$$

$$\mathcal{R}^+(\tau_1, \tau_2) = \bigcup_{\tau_1 \leq \tau \leq \tau_2} \Sigma^+(\tau),$$

$$\mathcal{R}^-(\tau_1, \tau_2) = \bigcup_{\tau_1 \leq \tau \leq \tau_2} \Sigma^-(\tau).$$

Let ξ now be a cutoff function such that $\xi = 1$ in $J^+(\Sigma_1^-) \cap J^-(\Sigma_{\tau-1}^+)$, and $\xi = 0$ in $J^+(\Sigma_\tau^+) \cap J^-(\Sigma_0^-)$. We may finally define

$$\psi_{\text{✂}} = \xi\psi.$$

The function $\psi_{\text{✂}}$ is a solution of the inhomogeneous equation

$$\Box_g \psi_{\text{✂}} = F, \qquad F = 2\nabla^\alpha \xi \, \nabla_\alpha \psi + \Box_g \xi \, \psi.$$

Note that F is supported in $\mathcal{R}^-(0,1) \cup \mathcal{R}^+(\tau - 1, \tau)$.

Another problem is that sharp cutoffs in frequency behave poorly under localisation. We thus do the following: Let ζ be a smooth cutoff supported in $[-2, 2]$ with the property that $\zeta = 1$ in $[-1, 1]$, and let $\omega_0 > 0$ be a parameter to be determined later. For an arbitrary Ψ of compact support in t^*, define

$$\Psi_\flat(t^*, \cdot) \doteq \sum_{m \neq 0} e^{im\phi^*} \int_{-\infty}^{\infty} \zeta((\omega_0 m)^{-1}\omega)\, \hat{\Psi}_m(\omega, \cdot)\, e^{i\omega t^*} d\omega,$$

$$\Psi_\sharp(t^*, \cdot) \doteq \Psi_0 + \sum_{m \neq 0} e^{im\phi^*} \int_{-\infty}^{\infty} \left(1 - \zeta((\omega_0 m)^{-1}\omega)\right) \hat{\Psi}_m(\omega, \cdot)\, e^{i\omega t^*} d\omega.$$

Note of course that $\Psi_\sharp + \Psi_\flat = \Psi$. We shall use the notation $\psi_\flat$ for $(\psi_{\text{✂}})_\flat$ and $\psi_\sharp$ for $(\psi_{\text{✂}})_\sharp$. Note that $\psi_\flat$, $\psi_\sharp$ satisfy

$$\Box_g \psi_\flat = F_\flat, \qquad \Box_g \psi_\sharp = F_\sharp. \tag{70}$$

5.2.6. *The bootstrap.* With $\psi_\flat$, $\psi_\sharp$ well defined, we now try to fill in the argument heuristically outlined before.

We wish to show the boundedness of

$$\mathbf{q} \doteq \sup_{0\le\tilde\tau\le\tau} \int_{\Sigma_{\tilde\tau}} J^N_\mu n^\mu. \tag{71}$$

We will argue by continuity in τ. We have already seen heuristically how to obtain a bound for $\mathbf{q}$ in Sections 5.2.3 and 5.2.4. When interpreted for the $\psi_\flat$, $\psi_\sharp$ defined above, these arguments produce error terms from:

- the inhomogeneous terms $F_\flat$, $F_\sharp$ from (70)
- the fact that we wish to localise estimates (59) and (65) to subregions $\mathcal{H}^+(\tau',\tau'')$ and $\mathcal{R}(\tau',\tau'')$ resepectively
- the fact that (69) is not exactly true.

These error terms can be controlled by $\mathbf{q}$ itself. For this, one studies carefully the time-decay of $F_\flat$, $F_\sharp$ away from the cutoff region $\mathcal{R}^-(0,1)\cup\mathcal{R}^+(\tau-1,\tau)$ using classical properties of the Fourier transform. An important subtlety arises from the presence of 0'th order terms in ψ, and it is here that the divergence of the region $\mathcal{R}^\pm$ from $\mathcal{R}(0,\tau)$ is exploited to exchange decay in τ and r.

To close the continuity argument, it is essential not only that the error terms be controlled by $\mathbf{q}$ itself, but that a small constant is retrieved, i.e. that the error terms are controlled by $\epsilon\mathbf{q}$, so that they can be absorbed. For this, use is made of the fact that for metrics in the allowed class sufficiently close to Schwarzschild (in the Kerr case, for $|a|\ll M$), one can control a priori the exponential growth rate of (71) to be small. See [**68**].

5.2.7. *Pointwise bounds.* Having proven the uniform boundedness of (71), one argues as in the proof of Theorem 3.2 to obtain higher order energy and pointwise bounds. In particular, the positivity property in the computation of Proposition 3.3.2 is stable. (It turns out that this positivity property persists in fact for much more general black hole spacetimes and there is in fact a geometric reason for this! See Chapter 7.)

5.2.8. *The boundedness theorem.* We have finally

THEOREM 5.1. *Let g be a metric defined on the differentiable manifold $\mathcal{R}$ with stratified boundary $\mathcal{H}^+\cup\Sigma_0$, and let T and $\Phi=\Omega_1$ be Schwarzschild Killing fields. Assume*

(1) *g is sufficiently close to Schwarzschild in an appropriate sense*
(2) *T and Φ are Killing with respect to g*
(3) *$\mathcal{H}^+$ is null with respect to g and T and Φ span the null generator of $\mathcal{H}^+$.*

Then the statement of Theorem 3.2 holds.

See [**68**] for the precise formulation of the closeness assumption 1.

COROLLARY 5.1. *The result applies to Kerr, and to the more general Kerr-Newman family (solving Einstein-Maxwell), for parameters $|a|\ll M$ (and also $|Q|\ll M$ in the Kerr-Newman case).*

Thus, we have quantitative pointwise and energy bounds for ψ and arbitrary derivatives on slowly rotating Kerr and Kerr-Newman exteriors.

5.3. Decay for Kerr. To obtain decay results analogous to Theorem 4.1, one needs to understand trapping. For general perturbations of Schwarzschild of the class considered in Theorem 5.1, it is not a priori clear what stability properties one can infer about the nature of the trapped set, and how these can be exploited. But for the Kerr family itself, the trapping structure can easily be understood, in view of the complete integrability of geodesic flow discovered by Carter [**29**]. The codimensionality of the trapped set persists, but in contrast to the Schwarzschild case where trapped null geodesics all approach the codimension-1 subset $r = 3M$ of spacetime, in Kerr, this codimensionality must be viewed in phase space.

5.3.1. *Separation.* There is a convenient way of doing phase space analysis in Kerr spacetimes, namely, as discovered by Carter [**30**], the wave equation can be separated. Walker and Penrose [**153**] later showed that both the complete integrability of geodesic flow and the separability of the wave equation have their fundamental origin in the presence of a *Killing tensor.*[54] In fact, as we shall see, in view of its intimate relation with the integrability of geodesic flow, Carter's separation of $\Box_g$ immediately captures the codimensionality of the trapped set.

The separation of the wave equation requires taking the Fourier transform, and then expanding into oblate spheroidal harmonics. As before, taking the Fourier transform requires cutting off in time. We shall here do the cutoff, however, in a somewhat different fashion.

Let Σ_τ be defined specifically as $t^* = \tau$. Given $\tau' < \tau$, define $\mathcal{R}(\tau', \tau)$ as before, and let ξ be a cutoff function as in Section 5.2.5, but with $\Sigma_{\tau'+1}$ replacing Σ_1^-, $\Sigma_{\tau'}$ replacing Σ_0^-, and Σ_τ replacing Σ_τ^+, $\Sigma_{\tau-1}$ replacing $\Sigma_{\tau-1}^+$. Define as before

$$\psi_{\text{✂}} = \xi\psi.$$

The function $\psi_{\text{✂}}$ is a solution of the inhomogeneous equation

$$\Box_g \psi_{\text{✂}} = F, \qquad F = 2\nabla^\alpha \xi \, \nabla_\alpha \psi + \Box_g \xi \, \psi.$$

Note that F is supported in $\mathcal{R}(\tau', \tau'+1) \cup \mathcal{R}(\tau-1, \tau)$.

Since $\psi_{\text{✂}}$ is compactly supported in t for each fixed $r > r_+$, we may consider its Fourier transform $\hat{\psi}_{\text{✂}} = \hat{\psi}_{\text{✂}}(\omega, \cdot)$. We may now decompose

$$\hat{\psi}_{\text{✂}}(\omega, \cdot) = \sum_{m,\ell} R^\omega_{m\ell}(r) S_{m\ell}(a\omega, \cos\theta) e^{im\phi},$$

$$\hat{F}(\omega, \cdot) = \sum_{m,\ell} F^\omega_{m\ell}(r) S_{m\ell}(a\omega, \cos\theta) e^{im\phi},$$

where $S_{m\ell}$ are the oblate spheroidal harmonics. For each $m \in \mathbb{Z}$, and fixed ω, these are a basis of eigenfunctions $S_{m\ell}$ satisfying

$$-\frac{1}{\sin\theta}\frac{d}{d\theta}\left(\sin\theta \frac{d}{d\theta} S_{m\ell}\right) + \frac{m^2}{\sin^2\theta} S_{m\ell} - a^2\omega^2 \cos^2\theta S_{m\ell} = \lambda_{m\ell} S_{m\ell},$$

and, in addition, satisfying the orthogonality conditions with respect to the θ variable,

$$\int_0^{2\pi} d\varphi \int_{-1}^{1} d(\cos\theta) e^{im\phi} S_{m\ell}(a\omega, \cos\theta)\, e^{-im'\phi} S_{m'\ell'}(a\omega, \cos\theta) = \delta_{mm'}\delta_{\ell\ell'}.$$

---

[54] See [**32, 108**] for recent higher-dimensional generalisations of these properties.

Here, the $\lambda_{m\ell}(\omega)$ are the eigenvalues associated with the harmonics $S_{m\ell}$. Each of the functions $R^\omega_{m\ell}(r)$ is a solution of the following problem

$$\Delta\frac{d}{dr}\left(\Delta\frac{R^\omega_{m\ell}}{dr}\right)+\left(a^2m^2+(r^2+a^2)^2\omega^2-4aMrm\omega-\Delta(\lambda_{m\ell}+a^2\omega^2)\right)R^\omega_{m\ell}$$
$$=\Delta\left((r^2+a^2\cos^2\theta)F\right)^\omega_{m\ell}.$$

Note that if $a=0$, we typically label $S_{m\ell}$ by $\ell\geq|m|$ such that

$$\lambda_{m\ell}(\omega)=\ell(\ell+1)/2.$$

With this choice, $S_{m\ell}$ coincides with the standard spherical harmonics $Y_{m\ell}$.

Given any $\omega_1>0$, $\lambda_1>0$ then we can choose a such that for $|\omega|\leq\omega_1$, $\lambda_{m\ell}\leq\lambda_1$, then

$$|\lambda_{m\ell}-\ell(\ell+1)/2|\leq\epsilon.$$

Rewriting the equation for the oblate spheroidal function

$$-\frac{1}{\sin\theta}\frac{d}{d\theta}\left(\sin\theta\frac{d}{d\theta}S_{m\ell}\right)+\frac{m^2}{\sin^2\theta}S_{m\ell}=\lambda_{m\ell}S_{m\ell}+a^2\omega^2\cos^2\theta S_{m\ell},$$

the smallest eigenvalue of the operator on the left hand side of the above equation is $m(m+1)$. This implies that

$$\lambda_{m\ell}\geq m(m+1)-a^2\omega^2. \tag{72}$$

This will be all that we require about $\lambda_{m\ell}$. For a more detailed analysis of $\lambda_{m\ell}$, see [**81**].

5.3.2. *Frequency decomposition.* Let ζ be a sharp cutoff function such that $\zeta=1$ for $|x|\leq1$ and $\zeta=0$ for $|x|>1$. Note that

$$\zeta^2=\zeta. \tag{73}$$

Let ω_1, λ_1 be (potentially large) constants to be determined, and λ_2 be a (potentially small) constant to be determined.

Let us define

$$\psi_\flat=\int_{-\infty}^{\infty}\zeta(\omega/\omega_1)\sum_{m,\ell:\lambda_{m\ell}(\omega)\leq\lambda_1}R^\omega_{m\ell}(r)S_{m\ell}(a\omega,\cos\theta)e^{im\phi}e^{i\omega t}d\omega,$$

$$\psi_{\text{ᵭ}}=\int_{-\infty}^{\infty}\zeta(\omega/\omega_1)\sum_{m,\ell:\lambda_{m\ell}(\omega)>\lambda_1}R^\omega_{m\ell}(r)S_{m\ell}(a\omega,\cos\theta)e^{im\phi}e^{i\omega t}d\omega,$$

$$\psi_\natural=\int_{-\infty}^{\infty}(1-\zeta(\omega/\omega_1))\sum_{m,\ell:\lambda_{m\ell}(\omega)\geq\lambda_2\omega^2}R^\omega_{m\ell}(r)S_{m\ell}(a\omega,\cos\theta)e^{im\phi}e^{i\omega t}d\omega,$$

$$\psi_\sharp=\int_{-\infty}^{\infty}(1-\zeta(\omega/\omega_1))\sum_{m,\ell:\lambda_{m\ell}(\omega)<\lambda_2\omega^2}R^\omega_{m\ell}(r)S_{m\ell}(a\omega,\cos\theta)e^{im\phi}e^{i\omega t}d\omega.$$

We have clearly

$$\psi_{✂}=\psi_\flat+\psi_{\text{ᵭ}}+\psi_\natural+\psi_\sharp.$$

For quick reference, we note:

- $\psi_\flat$ is supported in $|\omega|\leq\omega_1$, $\lambda_{m\ell}\leq\lambda_1$,
- $\psi_{\text{ᵭ}}$ is supported in $|\omega|\leq\omega_1$, $\lambda_{m\ell}>\lambda_1$,
- $\psi_\natural$ is supported in $|\omega|\geq\omega_1$, $\lambda_{m\ell}\geq\lambda_2\omega^2$ and
- $\psi_\sharp$ is supported in $|\omega|\geq\omega_1$, $\lambda_{m\ell}<\lambda_2\omega^2$.

5.3.3. *The trapped frequencies.* Trapping takes place in $\psi_\natural$. We show here how to construct a multiplier for this frequency range.

Defining a coordinate r^* by

$$\frac{dr^*}{dr} = \frac{r^2+a^2}{\Delta}$$

and setting

$$u(r) = (r^2+a^2)^{1/2} R^{\omega}_{m\ell}(r), \qquad H(r) = \frac{\Delta((r^2+a^2\cos^2\theta)F)^{\omega}_{m\ell}(r)}{(r^2+a^2)^{3/2}},$$

then u satisfies

$$\frac{d^2}{(dr^*)^2}u + (\omega^2 - V^{\omega}_{m\ell}(r))u = H$$

where

$$V^{\omega}_{m\ell}(r) = \frac{4Mram\omega - a^2m^2 + \Delta(\lambda_{m\ell}+\omega^2a^2)}{(r^2+a^2)^2} + \frac{\Delta(3r^2-4Mr+a^2)}{(r^2+a^2)^3} - \frac{3\Delta^2r^2}{(r^2+a^2)^4}.$$

Consider the following quantity

$$Q = f\left(\left|\frac{du}{dr^*}\right|^2 + (\omega^2 - V)|u|^2\right) + \frac{df}{dr^*}\mathrm{Re}\left(\frac{du}{dr^*}\bar{u}\right) - \frac{1}{2}\frac{d^2f}{dr^{*2}}|u|^2.$$

Then, with the notation $' = \frac{d}{dr^*}$,

$$Q' = 2f'|u'|^2 - fV'|u|^2 + \mathrm{Re}(2f\bar{H}u' + f'\bar{H}u) - \frac{1}{2}f'''|u|^2. \tag{74}$$

For $\psi_\natural$, we have

$$\lambda_{m\ell} + \omega^2a^2 \geq (\lambda_2 + a^2)\omega^2 \geq (\lambda_2+a^2)\omega_1^2. \tag{75}$$

We set

$$V_0 = (\lambda_{m\ell}+\omega^2a^2)\frac{r^2-2Mr}{(r^2+a^2)^2}$$

so that

$$V_1 = V - V_0 = \frac{4Mram\omega - a^2m^2 + a^2(\lambda_{m\ell}+\omega^2a^2)}{(r^2+a^2)^2} + \frac{\Delta(3r^2-4Mr+a^2)}{(r^2+a^2)^3} - \frac{3\Delta^2r^2}{(r^2+a^2)^4}.$$

Using (72), (75), we easily see that

$$\begin{aligned} r^3|V_1'| + \left|\left(\frac{(r^2+a^2)^4}{\Delta r^2}V_1'\right)'\right| &\leq C\Delta r^{-2}\left(|am\omega| + a^2(\lambda_{m\ell}+a^2\omega^2)+1\right) \\ &\leq \epsilon\Delta r^{-2}(\lambda_{m\ell}+a^2\omega^2), \end{aligned} \tag{76}$$

where ϵ can be made arbitrarily small, if ω_1 is chosen sufficiently large, and a is chosen $a<\epsilon$. On the other hand

$$\begin{aligned} V_0' &= 2\frac{\Delta}{(r^2+a^2)^4}(\lambda_{m\ell}+\omega^2a^2)\left((r-M)(r^2+a^2) - 2r(r^2-2Mr)\right) \\ &= -2\frac{\Delta r^2}{(r^2+a^2)^4}\left(\lambda_{m\ell}+\omega^2a^2\right)\left(r - 3M + a^2\frac{r-M}{r^2}\right). \end{aligned} \tag{77}$$

This computation implies that V_0' has a simple zero in the a^2 neighbourhood of $r = 3M$. Furthermore,

$$\left(\frac{(r^2+a^2)^4}{\Delta r^2}V_0'\right)' \leq -\Delta r^{-2}(\lambda_{m\ell}+\omega^2 a^2).$$

From the above and (76), it follows that for ω_1 sufficiently large and a sufficiently small, we have

$$\left(\frac{(r^2+a^2)^4}{\Delta r^2}V'\right)' \leq -\frac{1}{2}\Delta r^{-2}(\lambda_{m\ell}+\omega^2 a^2).$$

This alone implies that V' has *at most* a simple zero.

To show that V' indeed has a zero we examine the boundary values at r_+ and ∞. From (77) we see that

$$\frac{(r^2+a^2)^4}{\Delta r^2}V_0' \sim C(\lambda_{m\ell}+\omega^2 a^2)$$

for some positive constant C on the horizon and

$$\frac{(r^2+a^2)^4}{\Delta r^2}V_0' \sim -2r(\lambda_{m\ell}+\omega^2 a^2)$$

near $r = \infty$. On the other hand, from the inequality as applied to the first term on the right hand side of (76), it follows that

$$\left|\frac{(r^2+a^2)^4}{\Delta r^2}V_1'\right| \leq \epsilon r(\lambda_{m\ell}+\omega^2 a^2),$$

where ϵ can be chosen arbitrarily small if ω_1 is chosen sufficiently large and a sufficiently small. Thus, for suitable choice of ω_1, it follows that

$$\begin{aligned}\frac{(r^2+a^2)^4}{\Delta r^2}V'\Big|_{r_+} &= \frac{(r^2+a^2)^4}{\Delta r^2}(V_0'+V_1')\Big|_{r_+}\\ &> 0 > \frac{(r^2+a^2)^4}{\Delta r^2}(V_0'+V_1')\Big|_{\infty} = \frac{(r^2+a^2)^4}{\Delta r^2}V'\Big|_{\infty},\end{aligned}$$

and thus V' has a unique zero. Let us denote the r-value of this zero by $r^\omega_{m\ell}$.

We now choose f so that

(1) $f' \geq 0$,
(2) $f \leq 0$ for $r \leq r^\omega_{m\ell}$ and $f \geq 0$ for $r \geq r^\omega_{m\ell}$,
(3) $-fV' - \frac{1}{2}f''' \geq c$.

Property 3 can be verified by ensuring that $f'''(r^\omega_{m\ell}) < 0$ as well as requiring that $f''' < 0$ at the horizon. We may moreover normalise f to -1 on the horizon. Finally, we may assume that there exists an R such that for all $r \geq R$, f is of the form:

$$f = \tan^{-1}\frac{r^* - \alpha - \sqrt{\alpha}}{\alpha} - \tan^{-1}(-1-\alpha^{-1/2})$$

In particular, for $r \geq R$, the function f will not depend on ω, ℓ, m.

Note the similarity of this construction with that of Section 4.1.1, modulo the need for complete separation to centre the function f appropriately.

Integrating the identity (74) and using that $u \to 0$ as $r \to \infty$ we obtain that for any compact set K_1 in r^* and a certain compact set K_2 (which in particular does not contain $r = 3M$), there exists a positive constant $b > 0$ so that

$$b\int_{K_1}(|u'|^2+|u|^2)dr + b(\lambda_{m\ell}+\omega^2)\int_{K_2}|u|^2dr$$
$$\leq \left(|u'|^2+(\omega^2-V)|u|^2\right)(r_+) + \int_{-\infty}^{\infty}\mathrm{Re}(2f\bar{H}u'+f'\bar{H}u)\,dr^*.$$

On the horizon $r = r_+$, we have $u' = (i\omega + (iam/2Mr_+))u$ and

$$V(r_+) = \frac{4Mram\omega - a^2m^2}{(r_+^2+a^2)^2}.$$

Therefore, we obtain

$$b\int_{K_1}(|u'|^2+|u|^2)\,dr^* + b(\lambda_{m\ell}+\omega^2)\int_{K_2}|u|^2\,dr^*$$
$$\leq (\omega^2+\epsilon m^2)|u|^2(r_+) + \int_{-\infty}^{\infty}\mathrm{Re}(2f\bar{H}u'+f'\bar{H}u)\,dr^*. \tag{78}$$

We now wish to reinstate the dropped indices m, ℓ, ω, and sum over m, ℓ and integrate over ω. Note that by the orthogonality of the $S^{\omega}_{m\ell}$, it follows that for any functions α and β with coefficients defined by

$$\hat{\alpha}(\omega,\cdot) = \sum_{m,\ell}\alpha^{\omega}_{m\ell}(r)S_{m\ell}(a\omega,\cos\theta)e^{im\phi}, \qquad \hat{\beta}(\omega,\cdot) = \sum_{m,\ell}\beta^{\omega}_{m\ell}(r)S_{m\ell}(a\omega,\cos\theta)e^{im\phi},$$

we have

$$\int \alpha^2(t^*,r,\theta,\varphi)\sin\theta d\varphi\,d\theta\,dt = \int_{-\infty}^{\infty}\sum_{m,\ell}|\alpha^{\omega}_{m\ell}(r)|^2d\omega,$$

$$\int \alpha\cdot\beta\sin\theta d\varphi\,d\theta\,dt = \int_{-\infty}^{\infty}\sum_{m,\ell}\alpha^{\omega}_{m\ell}\cdot\bar{\beta}^{\omega}_{m\ell}d\omega.$$

Clearly, the summed and integrated left hand side of (78) bounds

$$b\int_{-\infty}^{\infty}dt^*\int_{K_1}\left((\partial_r\psi_{\natural})^2+\psi_{\natural}^2\right)dV_g + b\int_{K_2}\sum_i(\partial_i\psi_{\natural})^2\,dV_g.$$

Similarly, we read off immediately that the first term on the right hand side of (78) upon summation and integration yields precisely

$$\int_{\mathcal{H}_+}\left((T\psi_{\natural})^2+\epsilon(\partial_{\phi^*}\psi_{\natural})^2\right).$$

Note that we can bound

$$\begin{aligned}\int_{\mathcal{H}_+}\left((T\psi_{\natural})^2+\epsilon(\partial_{\phi^*}\psi_{\natural})^2\right) &\leq \int_{\mathcal{H}^+}((T\psi_{✂})^2+\epsilon(\partial_{\phi^*}\psi_{✂})^2)\\ &\leq B\int_{\Sigma_{\tau'}}J^N_{\mu}(\psi)n^{\mu}_{\Sigma}+\epsilon\int_{\mathcal{H}(\tau',\tau)}(\partial_{\phi^*}\psi)^2.\end{aligned}$$

(**Exercise**: Why?)

The "error term" of the right hand side of (78) is more tricky. To estimate the second summand of the integrand, note that

$$
\begin{aligned}
&\int_{|\omega|\geq\omega_1} \sum_{m,\ell:\lambda_{m\ell}(\omega)\geq\lambda_2\omega^2} (f')^\omega_{m\ell}(r)\bar{H}^\omega_{m\ell}(r)u^\omega_{m\ell}(r)d\omega \\
\leq\ & \int_{|\omega|\geq\omega_1} \sum_{m,\ell:\lambda_{m\ell}(\omega)\geq\lambda_2\omega^2} \delta^{-1}\Delta^{-1}(r^2+a^2)|(f')^\omega_{m\ell}H^\omega_{m\ell}|^2(r)+\delta\Delta(r^2+a^2)^{-1}|u^\omega_{m\ell}|^2d\omega \\
\leq\ & \int_{|\omega|\geq\omega_1} \sum_{m,\ell:\lambda_{m\ell}(\omega)\geq\lambda_2\omega^2} \delta^{-1}\Delta^{-1}(r^2+a^2)B|H^\omega_{m\ell}|^2(r)+\delta\Delta|R^\omega_{m\ell}|^2d\omega \\
\leq\ & \delta^{-1}B\Delta\int(F_\natural)^2\sin\theta\, d\phi\, d\theta\, dt+\delta\Delta\int(\psi_\natural)^2\sin\theta\, d\phi\, d\theta\, dt \\
\leq\ & \delta^{-1}B\Delta\int F^2\sin\theta\, d\phi\, d\theta\, dt+\delta\Delta\int\psi^2\sin\theta\, d\phi\, d\theta\, dt,
\end{aligned}
$$

where δ can be chosen arbitrarily. In particular, this estimate holds for $r\leq R$. For $r\geq R$, in view of the fact that f is independent of ω, m, ℓ, we have in fact

$$
\begin{aligned}
\int_{|\omega|\geq\omega_1} & \sum_{m,\ell:\lambda_{m\ell}(\omega)\geq\lambda_2\omega^2} (f')(r)\bar{H}^\omega_{m\ell}(r)u^\omega_{m\ell}(r)d\omega \\
&= f'(r)\int_{|\omega|\geq\omega_1} \sum_{m,\ell:\lambda_{m\ell}(\omega)\geq\lambda_2\omega^2} \bar{H}^\omega_{m\ell}(r)u^\omega_{m\ell}(r)d\omega \\
&= f'(r)(r^2+a^2)^{-1}\int((r^2+a^2\cos^2\theta)F)_\natural\psi_\natural\sin\theta\, d\phi\, d\theta\, dt \\
&= f'(r)(r^2+a^2)^{-1}\int((r^2+a^2\cos^2\theta)F)_\natural\psi_{\ding{34}}\sin\theta\, d\phi\, d\theta\, dt,
\end{aligned}
$$

where for the last line we have used (73). The first summand of the error integrand of (78) can be estimated similarly.

We thus obtain

$$
\begin{aligned}
b\int_{\mathcal{R}} & \chi\left((\partial_r\psi_\natural)^2+\psi_\natural^2\right)+b\int_{\mathcal{R}}\chi h J^N_\mu(\psi_\natural)N^\mu \\
\leq & B\int_{\Sigma_{\tau'}} J^N_\mu(\psi)n^\mu_\Sigma+\epsilon\int_{\mathcal{H}(\tau',\tau)}(\partial_\phi\psi)^2+\delta^{-1}B\int_{\mathcal{R}\cap\{r\leq R\}}F^2 \\
& +\delta\int_{\mathcal{R}\cap\{r\leq R\}}\psi^2+(\partial_r\psi)^2 \\
& +\int_{-\infty}^{\infty}dt^*\int_{r\geq R}\Big(2f(r^2+a^2)^{-1/2}((r^2+a^2\cos^2\theta)F)_\natural\partial_{r^*}((r^2+a^2)^{1/2}\psi_{\ding{34}}) \\
& \qquad +f'((r^2+a^2\cos^2\theta)F)_\natural\psi_{\ding{34}}\Big)\frac{\Delta}{r^2+a^2}\sin\theta\, d\phi^*\, d\theta\, dr^*,
\end{aligned}
\tag{79}
$$

where χ is a cutoff which degenerates at infinity and h is a function $0\leq h\leq 1$ which vanishes in a suitable neighbourhood of $r=3M$.

5.3.4. *The untrapped frequencies.* Given λ_2 sufficiently small and *any* choice of ω_1, λ_1, then, for a sufficiently small (where sufficiently small depends on these latter two constants), it follows that for $♪=\flat$, $\reflectbox{$\flat$}$, $\sharp$, we may produce currents of type $J_\mu^{\mathbf{X}_{\text{♪}}}$ as in Section 5.2.4 such that

$$b\int_{\mathcal{R}} \chi J_\mu^N(\psi_{\text{♪}})N^\mu + \tilde{\chi}\psi_{\text{♪}}^2 \leq \int_{\mathcal{R}} K^{\mathbf{X}_{\text{♪}}}(\psi_{\text{♪}})$$

for χ a suitable cutoff function degenerating at infinity, and $\tilde{\chi}$ a suitable cutoff function degenerating at infinity and vanishing in a neighbourhood of $\mathcal{H}^+$. These currents can in fact be chosen independently of a for such small a, and moreover, they can be chosen so that, defining

$$\mathcal{E}^{\mathbf{X}_{\text{♪}}} \doteq \nabla^\mu J_\mu^{\mathbf{X}_{\text{♪}}} - K^{\mathbf{X}_{\text{♪}}},$$

we have on the one hand

$$\int_{\mathcal{R}\cap\{r\geq R\}} \mathcal{E}^{\mathbf{X}_{\text{♪}}} = \int_{-\infty}^{\infty} dt^* \int_{r\geq R} \Big(2f(r^2+a^2)^{-1/2}((r^2+a^2\cos^2\theta)F)_{\text{♪}}\partial_{r^*}((r^2+a^2)^{1/2}\psi_{\text{✂}})$$
$$+f'((r^2+a^2\cos^2\theta)F)_{\text{♪}}\psi_{\text{✂}}\Big)\frac{\Delta}{r^2+a^2}\sin\theta\, d\phi^*\, d\theta\, dr^*$$

for the f of Section 5.3.3, and on the other hand, for the region $r \leq R$, we have

$$\int_{\mathcal{R}\cap\{r\leq R\}} \mathcal{E}^{\mathbf{X}_{\text{♪}}} \leq B\delta^{-1}\int_{\mathcal{R}\cap\{r\leq R\}} F^2 + B\delta\int_{\mathcal{R}\cap\{r\leq R\}} \psi_{\text{✂}}^2 + (\partial_r\psi_{\text{✂}})^2 + \chi J_\mu^N(\psi_{\text{✂}})n^\mu$$

where χ is supported near the horizon and away from a neighbourhood of $3M$.

Moreover, one can show as in Section 5.2.6 that

$$\begin{aligned}
-\int_{\mathcal{H}} J_\mu^{\mathbf{X}_{\text{♪}}}(\psi_{\text{♪}})n^\mu &\leq -\int_{\mathcal{H}} J_\mu^T(\psi_{\text{♪}})n^\mu \\
&\leq -\int_{\mathcal{H}} J_\mu^T(\psi_{\text{✂}})n^\mu \\
&\leq B\int_{\Sigma_{\tau'}} J_\mu^N(\psi)n^\mu.
\end{aligned}$$

(**Exercise**: Prove the last inequality.)

From the identity

$$\int_{\mathcal{H}^+} J_\mu^{\mathbf{X}_{\text{♪}}}(\psi_{\text{♪}})n_{\mathcal{H}}^\mu + \int_{\mathcal{R}} K^{\mathbf{X}_{\text{♪}}}(\psi_{\text{♪}}) = \int_{\mathcal{R}} \mathcal{E}^{\mathbf{X}_{\text{♪}}}(\psi_{\text{♪}})$$

and the above remarks, one obtains finally an estimate

$$\int_{\mathcal{R}} \chi(J^N_\mu(\psi_\flat) + J^N_\mu(\psi_\natural) + J^N_\mu(\psi_\sharp))n^\mu_{\Sigma_\tau}$$
$$\leq B\int_{\Sigma_{\tau'}} J^N_\mu(\psi)n^\mu + B\delta^{-1}\int_{\mathcal{R}\cap\{r\leq R\}} F^2$$
$$+ B\delta\int_{\mathcal{R}\cap\{r\leq R\}} \psi^2 + (\partial_r\psi)^2 + \chi J^N_\mu(\psi)N^\mu$$
$$+ \int_{-\infty}^{\infty} dt^* \int_{r\geq R}\left(2f(r^2+a^2)^{-1/2}\sum_{♪=\flat,\natural,\sharp}((r^2+a^2\cos^2\theta)F)_♪\partial_{r^*}((r^2+a^2)^{1/2}\psi_{✂})\right.$$
$$\left. +f'\sum_{♪=\flat,\natural,\sharp}((r^2+a^2\cos^2\theta))F)_♪\psi_{✂}\right)\frac{\Delta}{r^2+a^2}\sin\theta\, d\phi^*\, d\theta\, dr^*. \tag{80}$$

5.3.5. *The integrated decay estimates.* Now, we will add (79), (80) and the energy identity of $eJ^Y(\psi)$

$$\int_{\tilde{\Sigma}_\tau} J^N_\mu(\psi)n^\mu_{\tilde{\Sigma}_\tau} + \int_{\tilde{\mathcal{R}}(\tau',\tau)\cap\{r\leq r_0\}} eK^Y(\psi)$$
$$= -\int_{\mathcal{H}(\tau',\tau)} eJ^Y_\mu(\psi)n^\mu_{\mathcal{H}} + \int_{\tilde{\mathcal{R}}(\tau',\tau)\cap\{r_0\leq r\leq r_1\}} eK^Y(\psi) + \int_{\tilde{\Sigma}_{\tau'}} J^N_\mu(\psi)n^\mu_{\tilde{\Sigma}_{\tau'}} \tag{81}$$

for a small e with $\epsilon \ll e$, and where $r_0 < r_1 < 3M$ are as in Corollary 3.1, and r_1 is in the support of K_2.

In the resulting inequality, the left hand side bounds in particular

$$\int_{\mathcal{R}(\tau'+1,\tau-1)} \chi(hJ^N_\mu(\psi)N^\mu + (\partial_r\psi)^2) \tag{82}$$

where χ is a cutoff decaying at infinity, $\tilde{\chi}$ is a cutoff decaying at infinity and vanishing at $\mathcal{H}^+$ and h is a function with $0 \leq h \leq 1$ such that h vanishes precisely in a neighbourhood of $r = 3M$. (As $a \to 0$, this neighbourhood can be chosen smaller and smaller in the sense of the coordinate r.)

Let us examine the right hand side of the resulting inequality.

The second term of the first line of the right hand side of (79) is absorbed by the first term on the right hand side of (81) provided that $\epsilon \ll e$.

The third term of the first line of the right hand side of (79) and the second term of (80) are bounded by

$$B\delta^{-1}\int_{\Sigma_{\tau'}} J^N_\mu(\psi)n^\mu_{\Sigma_{\tau'}},$$

in view of Theorem 5.1.

The second line of the right side of (79) and the third term of (80) can be absorbed by (82), provided that δ is chosen suitably small, whereas the second term of the right hand side of (81) can be absorbed by (82), provided that e is sufficiently small.

The fourth terms of the right hand sides of (79) and (80) combine to yield

$$\int_{-\infty}^{\infty} dt^* \int_{r\geq R} \Big(2f(r^2+a^2)^{-1/2}(r^2+a^2\cos^2\theta)F\partial_{r^*}((r^2+a^2)^{1/2}\psi_{\text{✂}})$$
$$+f'(r^2+a^2\cos^2\theta)F\psi_{\text{✂}}\Big)\,\frac{\Delta}{r^2+a^2}\sin\theta\, d\phi^*\, d\theta\, dr^*.$$

Note where F is supported and how it decays. Using our boundedness Theorem 5.1, a Hardy inequality and integration by parts we may now bound this term by

$$B\int_{\Sigma_{\tau'}} J^N_\mu(\psi)n^\mu_{\Sigma_\tau}.$$

But the remaining terms on the right hand side of (79), (80) and (81) are also of this form! We thus obtain

PROPOSITION 5.3.1. *There exists a φ_t-invariant weight χ, degenerating only at i_0, and a second φ_t-invariant weight h, which vanishes on a neighbourhood of $r=3M$, and a constant $B>0$ such that the following estimates hold for all $\tau'\leq\tau$,*

$$\int_{\mathcal{R}(\tau',\tau)} \chi h J^N_\mu(\psi)N^\mu+\chi\psi^2 \leq B\int_{\Sigma_{\tau'}} J^N_\mu(\psi)n^\mu_{\Sigma_{\tau'}},$$

$$\int_{\mathcal{R}(\tau',\tau)} \chi J^N_\mu(\psi)N^\mu+\chi\psi^2 \leq B\int_{\Sigma_{\tau'}} (J^N_\mu(\psi)+J^N_\mu(T\psi))n^\mu_{\Sigma_{\tau'}},$$

for all solutions $\Box_g\psi=0$ on Kerr.

Similar estimates could be shown on regions $\tilde{\mathcal{R}}(\tau',\tau)$, $\tilde{\Sigma}'_\tau$, after having derived a priori suitable decay of ψ in r.[55]

5.3.6. *The Z-estimate.* To turn integrated decay as in Proposition 5.3.1 into decay of energy and pointwise decay, we must adapt the argument of Section 4.2.

Let V be a ϕ_t-invariant vector field such that $V=\partial_{t^*}$ for $r\geq r_++c_2$ and $V=\partial_{t^*}+(a/2Mr_+)\partial_{\phi^*}$ for $f\leq r_++c_1$ for some $c_1<c_2$, and such that V is timelike in $\mathcal{R}\setminus\mathcal{H}^+$. Note that V is Killing except in $r_++c_1\leq r\leq r_++c_2$. As $a\to 0$, we can construct such a V with c_2 arbitrarily small.

Now let us define u and v to be the Schwarzschild[56] coordinates

$$u=t-r^*_{\text{Schw}},$$
$$v=t+r^*_{\text{Schw}}.$$

With respect to the coordinates (u,v,ϕ^*,θ), defining $\underline{L}=\partial_u$, then $\underline{L}$ vanishes smoothly along the horizon. Define $\bar{L}=V-\underline{L}$. Finally, define the vector field

$$Z=u^2L+v^2\underline{L}.$$

Note that under these choices Z is null on $\mathcal{H}^+$. With w as before, the currents $J^{Z,w}$ together with J^N can be used to control the energy fluxes on Σ_τ with weights. Use of the energy identities of $J^{Z,w}$ and J^N leads to estimates of the form

$$\int_{\Sigma_\tau}\chi\psi^2+\int_{\Sigma_\tau\cap\{r\precsim\tau\}} J^N_\mu(\psi)n^\mu_{\tilde{\Sigma}_\tau}\leq B\,D\tau^{-2}+B\,\tau^{-2}\int_{\mathcal{R}(0,\tau)}\mathcal{E}, \tag{83}$$

[55]In the section that follows, we shall in fact localise the above estimate in a different way applying a cutoff function. The resulting 0'th order terms which arise can be controlled using the "good" 0'th order term in the boundary integrals of $J^{Z,w}$.

[56]Recall that we are considering both the Kerr and Schwarzschild metric on the fixed differentiable structure $\mathcal{R}$ as described in Section 5.1.

where χ is a cutoff function supported suitably, and where $\mathcal{E}$ is an error term arising from the part of $K^{Z,w}$ which has the "wrong" sign; D arises from data.

We may partition

$$\mathcal{E} = \mathcal{E}_1 + \mathcal{E}_2 + \mathcal{E}_3$$

where

- $\mathcal{E}_1$ is supported in some region $r_0 \le r \le R_0$,
- $\mathcal{E}_2$ is supported in $r \le r_0$ and
- $\mathcal{E}_3$ is supported in $r \ge R_0$.

Recall that $L+\underline{L}$ is Killing for $r \ge 2M + c_2$. It follows (**Exercise**) that choosing $c_2 < r_0$, there are no terms growing quadratically in t for $\mathcal{E}_1$, $\mathcal{E}_3$. Moreover, by our construction, Z depends smoothly on a away from the horizon. The behaviour near the horizon is more subtle as Z itself is not smooth! We shall return to this when discussing $\mathcal{E}_2$.

In view of our above remarks, we have that

$$\mathcal{E}_1 \le B\,t(J^N_\mu(\psi)N^\mu + \psi^2),$$

just like in the case of Schwarzschild. In view of Proposition 5.3.1, this leads to the following estimate: If $\hat{\psi} = \psi$ in $\mathcal{R}(\tau', \tau'') \cap \{r \le R_0\}$, where $\hat{\psi}$ solves again $\Box_g \hat{\psi} = 0$, then

$$\int_{\mathcal{R}(\tau',\tau'')} \mathcal{E}_1(\psi) = \int_{\mathcal{R}(\tau',\tau'')} \mathcal{E}_1(\hat{\psi}) \le B\tau' \int_{\Sigma_{\tau'}} (J^N_\mu(\hat{\psi}) + J^N_\mu(T\hat{\psi}))n^\mu_{\tilde{\Sigma}_{\tau'}}. \tag{84}$$

The introduction of $\hat{\psi}$ is related to our localisation procedure we shall carry out in what follows.

Recall that in the Schwarzschild case, for R_0 suitably chosen, there is no $\mathcal{E}_3$ term, as the term $K^{Z,w}$ has a good sign in that region. (See Section 4.2.) Examining the r-decay of error terms in the smooth dependence of Z in a, we obtain

$$\mathcal{E}_3 \le \epsilon\, t r^{-2} J^N_\mu(\psi) N^\mu$$

where ϵ can be made arbitrarily small if a is small. If $\tau'' - \tau' \sim \tau' \sim t$, this leads to an estimate

$$\begin{aligned}\int_{\mathcal{R}(\tau',\tau'')} \mathcal{E}_3(\psi) \quad &\le \quad \epsilon(\tau''-\tau')(\tau''+\tau') \int_{\Sigma_{\tau'}\cap\{r \lesssim \tau''-\tau'\}} J^N_\mu(\psi) n^\mu_{\Sigma_{\tau'}} \\ &\quad + \epsilon \log|\tau''-\tau'| \int_{\Sigma_{\tau'}} J^N_\mu(\psi) n^\mu_{\Sigma_{\tau'}}.\end{aligned} \tag{85}$$

In the region $r_+ + c_1 \le r \le r_+ + c_2$, then, choosing r_0 such that $\mathcal{E}_2$ is absent in Schwarzschild, we can argue without computation from the smooth dependence on a that

$$\mathcal{E}_2 \le \epsilon\, t^2 (J^N_\mu(\psi)N^\mu + \psi^2)$$

where ϵ can be made arbitrarily small by choosing a small. The necessity of a quadratically growing error term arises from the fact that $L + \underline{L}$ is not Killing in this region.[57]

As we have already mentioned, an important subtlety occurs near the horizon $\mathcal{H}^+$ where Z fails to be C^1. This means that $\mathcal{E}_2$ is not necessarily small in local

[57]Alternatively, one can keep $L + \underline{L}$ Killing at the expense of Z failing to be causal on the horizon. This would lead to errors of a similar nature.

coordinates, and one must understand how to bound the singular terms. It turns out that these singular terms have a structure:

PROPOSITION 5.3.2. *Let $\hat{V}$, $\hat{Y}$, E_1, E_2 extend V to a null frame in $r \le r_+ + c_1$. We have*

$$\mathcal{E}_2 \le \epsilon v |\log(r - r_+)|^p (\mathbf{T}(\hat{Y}, \hat{V}) + \mathbf{T}(\hat{V}, \hat{V})) + \epsilon v\, J^N_\mu(\psi) N^\mu.$$

PROOF. The warping function w can be chosen as in Schwarzschild near $\mathcal{H}^+$, and thus, the extra terms it generates are harmless. For the worst behaviour, it suffices to examine now K^Z itself. We must show that terms of the form:

$$|\log(r - r_+)|^p (T(\hat{Y}, \hat{Y}))$$

do not appear in the computation for K^Z.

The relevant property follows from examining the covariant derivative of Z with respect to the null frame:

$$\nabla_{\hat{V}} Z = 2u(\hat{V}u)\underline{L} + 2v\hat{V}(v)L + v^2\nabla_{\hat{V}}V - 4r^* v\nabla_{\hat{V}}\underline{L} + 4(r^*)^2\nabla_{\hat{V}}\underline{L},$$

$$\nabla_{\hat{Y}} Z = 2u(\hat{Y}u)\underline{L} + 2v(\hat{Y}v)L + v^2\nabla_{\hat{Y}}V - 4r^* v\nabla_{\hat{Y}}\underline{L} + 4(r^*)^2\nabla_{\hat{Y}}\underline{L},$$

$$\nabla_{E_1} Z = 2u(E_1 u)\underline{L} + 2v(E_1 v)L + v^2\nabla_{E_1}V - 4r^* v\nabla_{E_1}\underline{L} + 4(r^*)^2\nabla_{E_1}\underline{L},$$

$$\nabla_{E_2} Z = 2u(E_2 u)\underline{L} + 2v(E_2 v)L + v^2\nabla_{E_2}V - 4r^* v\nabla_{E_2}\underline{L} + 4(r^*)^2\nabla_{E_2}\underline{L}.$$

□

To estimate now $\mathcal{E}_2$, we first remark that with Proposition 5.3.1, we can obtain the following refinement of the red-shift multiplier construction of Corollary 3.1:

PROPOSITION 5.3.3. *If we weaken the requirement that N be smooth in Corollary 3.1 with the statement that N is C^0 at $\mathcal{H}^+$ and smooth away from $\mathcal{H}^+$, then given $p \ge 0$, we may construct an N as in Corollary 3.1 where property 1 is replaced by the stronger inequality:*

$$K^N(\psi) \ge b_p |\log(r - r_+)|^p (\mathbf{T}(\hat{Y}, \hat{V}) + \mathbf{T}(\hat{V}, \hat{V}))$$

for $r \le r_0$.

It now follows immediately from Proposition 5.3.1 that with ψ and $\hat{\psi}$ as before, we have

$$\int_{\mathcal{R}(\tau', \tau'')} \mathcal{E}_2(\psi) \le \epsilon(\tau')^2 \int_{\Sigma_{\tau'}} J^N_\mu(\hat{\psi}) n^\mu_{\Sigma_{\tau'}}. \tag{86}$$

To obtain energy decay from (83), (85), (84) and (86), we argue now by continuity. Introduce the bootstrap assumptions

$$\int_{\Sigma_\tau \cap \{r \lesssim \tau\}} J^N_\mu(\psi) N^\mu + \chi\psi^2 \le C\, D\tau^{-2+2\delta}, \tag{87}$$

$$\int_{\Sigma_\tau \cap \{r \lesssim \tau\}} J^N_\mu(T\psi) N^\mu \le C'\, D\tau^{-1+2\delta} \tag{88}$$

for a $\delta > 0$.

Now dyadically decompose the interval $[0,\tau]$ by $\tau_i < \tau_{i+1}$. Using (84) and the above, we obtain

$$\begin{aligned}
\int_{\mathcal{R}(0,\tau)} \mathcal{E}_1(\psi) &\leq \sum_i \int_{\mathcal{R}(\tau_i,\tau_{i+1})} \mathcal{E}_1(\psi) \\
&\leq \sum_i \tau_i \int_{\Sigma_{\tau_i}} (J^N_\mu(\hat\psi) + J^N_\mu(T\hat\psi))N^\mu \\
&\leq \sum_i \tau_i \int_{\Sigma_{\tau_i}\cap\{r\lesssim\tau_{i+1}-\tau_i\}} (J^N_\mu(\psi) + J^N_\mu(T\psi))N^\mu + \chi\psi^2 \\
&\leq \sum_i \tau_i(\tau_i^{-2+2\delta}CD + \tau_i^{-1+2\delta}C'D) \\
&\leq \delta^{-1}(CD\tau^{-1+2\delta} + C'D\tau^{2\delta}).
\end{aligned} \tag{89}$$

Here, $\hat\psi$ is constructed separately on each dyadic region $\mathcal{R}(\tau_i,\tau_{i+1})$ by throwing a cutoff on $\psi|_{\Sigma_{\tau_i}}$ equal to 1 in $r \lesssim \tau_{i+1}-\tau_i$ and vanishing in $\tau_{i+1}-\tau_i \lesssim r$, solving again the initial value problem in $\mathcal{R}(\tau_i,\tau_{i+1})$, and exploiting the domain of dependence property. See the original [**65**] for this localisation scheme. The parameters of the "dyadic" decomposition must be chosen accordingly for the constants to work out. Similarly, using (86) we obtain

$$\begin{aligned}
\int_{\mathcal{R}(0,\tau)} \mathcal{E}_2(\psi) &\leq \sum_i \int_{\mathcal{R}(\tau_i,\tau_{i+1})} \mathcal{E}_2(\psi) \\
&\leq \epsilon \sum_i \tau_i^2 \int_{\Sigma_{\tau_i}} J^N_\mu(\hat\psi)N^\mu \\
&\leq \epsilon \sum_i \tau_i^2 \int_{\Sigma_{\tau_i}\cap\{r\lesssim\tau_{i+1}-\tau_i\}} J^N_\mu(\psi)N^\mu + \chi\psi^2 \\
&\leq \epsilon \sum_i \tau_i^2\tau_i^{-2+2\delta}CD \\
&\leq \epsilon\delta^{-1}\tau^{2\delta}CD
\end{aligned} \tag{90}$$

and using (85)

$$\begin{aligned}
\int_{\mathcal{R}(0,\tau)} \mathcal{E}_3(\psi) &\leq \sum_i \int_{\mathcal{R}(\tau_i,\tau_{i+1})} \mathcal{E}_3(\psi) \\
&\leq \epsilon \sum_i \left(\tau_i^2 \int_{\Sigma_{\tau_i}} J^N_\mu(\psi)N^\mu + \int_{\Sigma_{\tau_i}} J^N_\mu(\psi)n^\mu_{\Sigma_{\tau_i}}\right) \\
&\leq \epsilon \sum_i (\tau_i^2\tau_i^{-2+2\delta}CD + D\log\tau') \\
&\leq \epsilon\delta^{-1}\tau^{2\delta}CD.
\end{aligned} \tag{91}$$

For $T\psi$ we obtain

$$\int_{\mathcal{R}(0,\tau)} \mathcal{E}_1(T\psi) \leq BD\tau, \tag{92}$$

$$\begin{aligned}\int_{\mathcal{R}(0,\tau)} \mathcal{E}_2(T\psi) &\leq \sum_i \int_{\mathcal{R}(\tau_i,\tau_{i+1})} \mathcal{E}_2(T\psi)\\ &\leq \epsilon \sum_i \tau_i^2 \int_{\Sigma_{\tau_i}} J^N_\mu(T\hat{\psi})N^\mu\\ &\leq \epsilon \sum_i \tau_i^2 \int_{\Sigma_{\tau_i}\cap\{r\lesssim\tau_{i+1}-\tau_i\}} J^N_\mu(T\psi)N^\mu + \chi(T\psi)^2\\ &\leq \epsilon \sum_i \tau_i^2(\tau_i^{-1+2\delta}C'D + \tau_i^{-2+2\delta}CD)\\ &\leq \epsilon\delta^{-1}\tau^{1+2\delta}C'D + \epsilon\delta^{-1}\tau^{2\delta}CD, \end{aligned} \tag{93}$$

$$\begin{aligned}\int_{\mathcal{R}(0,\tau)} \mathcal{E}_3(T\psi) &\leq \sum_i \int_{\mathcal{R}(\tau_i,\tau_{i+1})} \mathcal{E}_3(T\psi)\\ &\leq \epsilon \sum_i \left(\tau_i^2 \int_{\Sigma_{\tau_i}} J^N_\mu(T\psi)N^\mu + \int_{\Sigma_{\tau_i}} J^N_\mu(T\psi)n^\mu_{\Sigma_{\tau_i}}\right)\\ &\leq \epsilon \sum_i (\tau_i^2\tau_i^{-1+2\delta}C'D + D\log\tau_i)\\ &\leq \epsilon\delta^{-1}\tau^{1+2\delta}C'D. \end{aligned} \tag{94}$$

We use here the algebra of constants where $B\epsilon = \epsilon$. The constant D is a quantity coming from data. **Exercise**: What is D and why is (92) true?

For $\epsilon \ll \delta$ and C' sufficiently large, we see that from (83) applied to $T\psi$ in place of ψ, using (92), (93), we improve (88).

On the other hand choosing $C' \ll C$ and then τ sufficiently large, we have

$$\tau^{-2}\delta^{-1}(CD\tau^{-1+2\delta} + C'D\tau^{2\delta}) \leq \frac{1}{2}CD\tau^{-2+2\delta}$$

and thus, again for $\epsilon \ll \delta$, using (89), (90) we can improve (87) from (83).

Once one obtains (87), then decay can be extended to decay in $\tilde{\Sigma}_\tau$ by the argument of Section 4.2, by applying conservation of the J^T flux backwards.[58]

5.3.7. *Pointwise bounds.* In any region $r \leq R$, we may now obtain pointwise decay bounds simply by further commutation with T, N as in Section 3.3.4. To obtain the correct pointwise decay statement towards null infinity, one must also commute the equation with a basis Ω_i for the Lie algebra of the *Schwarzschild* metric, exploiting the r-weights of these vector fields. Defining $\tilde{\Omega}_i = \zeta(r)\Omega_i$, where ζ is a cutoff which vanishes for $r \leq R_0$, where $3M \ll R_0$, and, setting $\tilde{\psi} = \tilde{\Omega}\psi$, we have

$$\Box_g\tilde{\psi} = F_1\partial^2\psi + F_2\partial\psi$$

where $F_1 = O(r^{-2})$ and $F_2 = O(r^{-3})$. Having estimates already for ψ, $T\psi$, one can may apply the X and Z estimates as before for $\tilde{\psi}$, only, in view of the F_2 term, now one must exploit also the X-estimate in $D^+(\Sigma_{\tau_i}\cap\{r \lesssim \tau_{i+1}-\tau_i\})\cap J^-(\Sigma_{\tau_{i+1}})$. We leave this as an **exercise**.

[58]Note that in view of the fact that we argued by continuity to obtain (87), we could not obtain this extended decay through $\tilde{\Sigma}_\tau$ earlier. This is why we have localised as in [**65**], not as in Section 4.2.

5.3.8. *The decay theorem.* We have obtained thus

THEOREM 5.2. *Let $(\mathcal{M}, g)$ be Kerr for $|a| \ll M$, $\mathcal{D}$ be the closure of its domain of dependence, let Σ_0 be the surface $\mathcal{D} \cap \{t^* = 0\}$, let ψ, ψ' be initial data on Σ_0 such that $\psi \in H^s_{\mathrm{loc}}(\Sigma)$, $\psi' \in H^{s-1}_{\mathrm{loc}}(\Sigma)$ for $s \geq 1$, and $\lim_{x \to i^0} \psi = 0$, and let ψ be the corresponding unique solution of $\Box_g \psi = 0$. Let φ_τ denote the 1-parameter family of diffeomorphisms generated by T, let $\tilde{\Sigma}_0$ be an arbitrary spacelike hypersurface in $J^+(\Sigma_0 \setminus \mathcal{U})$ where $\mathcal{U}$ is an open neighbourhood of the asymptotically flat end*[59]*, and define $\tilde{\Sigma}_\tau = \varphi_\tau(\tilde{\Sigma}_0)$. Let $s \geq 3$ and assume*

$$E_1 \doteq \int_{\Sigma_0} r^2 (J^{n_0}_\mu(\psi) + J^{n_0}_\mu(T\psi) + J^{n_0}_\mu(TT\psi)) n^\mu_0 < \infty.$$

Then there exists a $\delta > 0$ depending on a (with $\delta \to 0$ as $a \to 0$) and a B depending only on $\tilde{\Sigma}_0$ such that

$$\int_{\tilde{\Sigma}_\tau} J^N(\psi) n^\mu_{\tilde{\Sigma}_\tau} \leq B E_1 \, \tau^{-2+2\delta}.$$

Now let $s \geq 5$ and assume

$$E_2 \doteq \sum_{|\alpha| \leq 2} \sum_{\Gamma = \{T, N, \Omega_i\}} \int_{\Sigma_0} r^2 (J^{n_0}_\mu(\Gamma^\alpha \psi) + J^{n_0}_\mu(\Gamma^\alpha T\psi) + J^{n_0}_\mu(\Gamma^\alpha TT\psi)) n^\mu_0 < \infty$$

where Ω_i are the Schwarzschild angular momentum operators. Then

$$\sup_{\tilde{\Sigma}_\tau} \sqrt{r} |\psi| \leq B \sqrt{E_2} \, \tau^{-1+\delta}, \qquad \sup_{\tilde{\Sigma}_\tau} r |\psi| \leq B \sqrt{E_2} \, \tau^{(-1+\delta)/2}.$$

One can obtain decay for arbitrary derivatives, including transversal derivatives to $\mathcal{H}^+$, using additional commutation by N. See [**69**].

5.4. Black hole uniqueness. In the context of the vacuum equations (4), the Kerr solution plays an important role not only because it is believed to be stable, but because it is believed to be the only stationary black hole solution.[60] This is the celebrated *no-hair "theorem"*. In the case of the Einstein-Maxwell equations, there is an analogous no-hair "theorem" stating uniqueness for Kerr-Newman. A general reference is [**92**].

Neither of these results is close to being a theorem in the generality which they are often stated. Reasonably definitive statements have only been proven in the much easier static case, and in the case where axisymmetry is assumed a priori and the horizon is assumed connected, i.e. that there is one black hole. Axisymmetry can be inferred from stationarity under various special assumptions, including the especially restrictive assumption of analyticity. See [**57**] for the latest on the analytic case, and [**96**] for new interesting results in the direction of removing the analyticity assumption in inferring axisymmetry from stationarity.

Nonetheless, the expectation that black hole uniqueness is true reasonably raises the question: why the interest in more general black holes, allowed in Theorem 5.1?

For a classical "astrophysical" motivation, note that black hole solutions can in principle exist in the presence of persistent atmospheres. Perhaps the simplest such constructions would involve solutions of the Einstein-Vlasov system, where matter

[59] This is just the assumption that $\tilde{\Sigma}_0$ "terminates" on null infinity

[60] A further extrapolation leads to the "belief" that all vacuum solutions eventually decompose into n Kerr solutions moving away from each other.

is described by a distribution function on phase space invariant under geodesic flow. These black hole spacetimes would in general not be Kerr even in their vacuum regions. Recent speculations in high energy physics yield other possible motivations: There are now a variety of "hairy black holes" solving Einstein-matter systems for non-classical matter, like Yang-Mills fields [**141**], and a large variety of vacuum black holes in higher dimensions [**77**], many of which are currently the topic of intense study.

There is, however, a second type of reason, which is relevant even when we restrict our attention to the vacuum equations (4) in dimension 4. The less information one must use about the spacetime to obtain quantitative control on fields, the better chance one has at obtaining a stability theorem. The essentially non-quantitative[61] aspect of our current limited understanding of black hole uniqueness should make it clear that these arguments probably will not have a place in a stability proof. Indeed, it would be an interesting problem to explore the possibility of obtaining a more quantitative version of uniqueness theorems (in a neighbourhood of Kerr) following ideas in this section.

5.5. Comments and further reading. Theorem 5.1 was proven in [**68**]. In particular, this provided the first global result of any kind for general solutions of the Cauchy problem on a (non-Schwarzschild) Kerr background. Theorem 5.2 was first announced at the Clay Summer School where these notes were lectured. Results in the direction of Proposition 5.3.1 are independently being studied in work in progress by Tataru-Tohaneanu[62] and Andersson-Blue[63].

The best previous results concerning Kerr had been obtained by Finster and collaborators in an important series of papers culminating in [**79**]. See also [**80**]. The methods of [**79**] are spectral theoretic, with many pretty applications of contour integration and o.d.e. techniques. The results of [**79**] do not apply to general solutions of the Cauchy problem, however, only to individual azimuthal modes, i.e. solutions ψ_m of fixed m. In addition, [**79**] imposes the restrictive assumption that $\mathcal{H}^+ \cap \mathcal{H}^-$ not be in the support of the modes. (Recall the discussion of Section 3.2.6.) Under these assumptions, the main result stated in [**79**] is that

$$\lim_{t\to\infty} \psi_m(r,t) = 0 \tag{95}$$

for any $r > r_+$. Note that the reason that (95) did not yield any statement concerning general solutions, i.e. the sum over m–not even a non-quantitative one–is that one did not have a quantitative boundedness statement as in Theorem 5.1. Moreover, one should mention that even for fixed m, the results of [**79**] are in principle compatible with the statement

$$\sup_{\mathcal{H}^+} \psi_m = \infty,$$

i.e. that the azimuthal modes blow up along the horizon. See the comments in Section 4.6. It is important to note, however, that the statement of [**79**] need not restrict to $|a| \ll M$, but concerns the entire subextremal range $|a| < M$. Thus, the statement (95) of [**79**] is currently the only known statement available in the

[61] As should be apparent by the role of analyticity or Carleman estimates.

[62] communication from Mihai Tohaneanu, a summer school participant who attended these lectures

[63] lecture of P. Blue, Mittag-Leffler, September 2008

literature concerning azimuthal modes on Kerr spacetimes for large but subextremal a.

There has also been interesting work on the Dirac equation [**78, 90**], for which superradiance does not occur, and the Klein-Gordon equation [**89**]. For the latter, see also Section 8.3.

5.6. The nonlinear stability problem for Kerr. We have motivated these notes with the nonlinear stability problem of Kerr. Let us give finally a rough formulation.

CONJECTURE 5.1. *Let* $(\Sigma, \bar{g}, K)$ *be a vacuum initial data set (see Appendix* B.2*) sufficiently close (in a weighted sense) to the initial data on Cauchy hypersurface in the Kerr solution* $(\mathcal{M}, g_{M,a})$ *for some parameters* $0 \leq |a| < M$. *Then the maximal vacuum development* $(\mathcal{M}, g)$ *possesses a complete null infinity* $\mathcal{I}^+$ *such that the metric restricted to* $J^-(\mathcal{I}^+)$ *approaches a Kerr solution* $(\mathcal{M}, g_{M_f,a_f})$ *in a uniform way (with respect to a foliation of the type* $\tilde{\Sigma}_\tau$ *of Section* 4*) with quantitative decay rates, where* M_f, a_f *are near* M, a *respectively.*

Let us make some remarks concerning the above statement. Under the assumptions of the above conjecture, $(\mathcal{M}, g)$ certainly contains a trapped surface S by Cauchy stability. By Penrose's incompleteness theorem (Theorem 2.2), this implies that $(\mathcal{M}, g)$ is future causally geodesically incomplete. By the methods of the proof of Theorem 2.2, it is easy to see that $S \cap J^-(\mathcal{I}^+) = \emptyset$. Thus, as soon as $\mathcal{I}^+$ is shown to be complete, it would follow that the spacetime has a black hole region in the sense of Section 2.5.4.[64]

In view of this, one can also formulate the problem where the initial data are assumed close to Kerr initial data on an incomplete subset of a Cauchy hypersurface with one asymptotically flat end and bounded by a trapped surface. This is in fact the physical problem[65], but in view of Cauchy stability, it is equivalent to the formulation we have given above. Note also the open problem described in the last paragraph of Section 2.8.

In the spherically symmetric analogue of this problem where the Einstein equations are coupled with matter, or the Bianchi-triaxial IX vacuum problem discussed in Section 2.6.4, the completeness of null infinity can be inferred easily without detailed understanding of the geometry [**60, 62**]. One can view this as an "orbital stability" statement. In this spherically symmetric case, the asymptotic stability can then be studied a posteriori, as in [**63, 94**]. This latter problem is much more difficult.

In the case of Conjecture 5.1, in contrast to the symmetric cases mentioned above, one does *not* expect to be able to show any weaker stability statement than the asymptotic stability with decay rates as stated. Note that it is only the Kerr family as a whole–*not* the Schwarzschild subfamily–which is expected to be asymptotically stable: Choosing $a = 0$ certainly does not imply that $a_f = 0$. On the other hand, if $|a| \ll M$, then by the formulation of the above conjecture, it would follow that $|a_f| \ll M_f$. It is with this in mind that we have considered the $|a| \ll M$ case in these lecture notes.

[64]Let us also remark the obvious fact that the above conjecture implies in particular that weak cosmic censorship holds in a neighbourhood of Kerr data.

[65]Cf. the comments on the relation between maximally-extended Schwarzschild and Oppenheimer-Snyder.

6. The cosmological constant Λ and Schwarzschild-de Sitter

Another interesting setting for the study of the stability problem are black holes within *cosmological spacetimes*. Cosmological spacetimes–as opposed to asymptotically flat spacetimes (See Appendix B.2.3), which model spacetime in the vicinity of an isolated self-gravitating system–are supposed to model the whole universe. The working hypothesis of classical cosmology is that the universe is approximately homogeneous and isotropic (sometimes known as the *Copernican principle* [**91**]). In the Newtonian theory, it was not possible to formulate a cosmological model satisfying this hypothesis.[66] One of the major successes of general relativity was that the theory allowed for such solutions, thus making cosmology into a mathematical science.

In the early years of mathematical cosmology, it was assumed that the universe should be static[67]. To allow for such static cosmological solutions, Einstein modified his equations (2) by adding a 0'th order term:

$$R_{\mu\nu} - \frac{1}{2}g_{\mu\nu}R + \Lambda g_{\mu\nu} = 8\pi T_{\mu\nu}. \tag{96}$$

Here Λ is a constant known as *the cosmological constant.* When coupled with a perfect fluid, this system admits a static, homogeneous, isotropic solution with $\Lambda > 0$ and topology $\mathbb{S}^3 \times \mathbb{R}$. This spacetime is sometimes called the *Einstein static universe.*

Cosmological solutions with various values of the parameter Λ were studied by Friedmann and Lemaitre, under the hypothesis of exact homogeneity and isotropy. Static solutions are in fact always unstable under perturbation of initial data. Typical homogeneous isotropic solutions expand or contract, or both, beginning and or ending in singular configurations. As with the early studies (referred to in Sections 2.2) illuminating the extensions of the Schwarzschild metric across the horizon, these were ahead of their time.[68] (See the forthcoming book [**123**] for a history of this fascinating early period in the history of mathematical cosmology.) These predictions were taken more seriously with Hubble's observational discovery of the expansion of the universe, and the subsequent evolutionary theories of matter, but the relevance of the solutions near where they are actually singular was taken seriously only after the incompleteness theorems of Penrose and Hawking–Penrose were proven (see Section 2.7).

We shall not go into a general discussion of cosmology here, nor tell the fascinating story of the ups and downs of Λ–from its adoption by Einstein to his subsequent well-known rejection of it, to its later "triumphant" return in current cosmological models, taking a very small positive value, the "explanation" of which is widely regarded as one of the outstanding puzzles of theoretical physics. Rather, let us pass directly to the object of our study here, one of the simplest examples of an inhomogeneous "cosmological" spacetime, where non-trivial small scale structure occurs in an ambient expanding cosmology. This is the Schwarzschild–de Sitter solution.

---

[66] It is possible, however, if one geometrically reinterprets the Newtonian theory and allows space to be–say–the torus. See [**132**]. These reinterpretations, of course, postdate the formulation of general relativity.

[67] much like in the early studies of asymptotically flat spacetimes discussed in Section 2.1

[68] In fact, the two are very closely related! The interior region of the Oppenheimer-Snyder collapsing star is precisely isometric to a region of a Friedmann universe. See [**117**].

6.1. The Schwarzschild-de Sitter geometry. Again, this metric was discovered in local coordinates early in the history of general relativity, independently by Kottler [**105**] and Weyl [**155**]. Fixing $\Lambda > 0$,[69] Schwarzschild-de Sitter is a one-parameter family of solutions of the form

$$-(1-2M/r-\Lambda r^2)dt^2+(1-2M/r-\Lambda r^2)^{-1}dr^2+r^2d\sigma_{\mathbb{S}^2}. \tag{97}$$

The black hole case is the case where $0 < M < \frac{1}{3\sqrt{\Lambda}}$. A maximally-extended solution (see [**28, 85**]) then has as Penrose diagram the infinitely repeating chain:

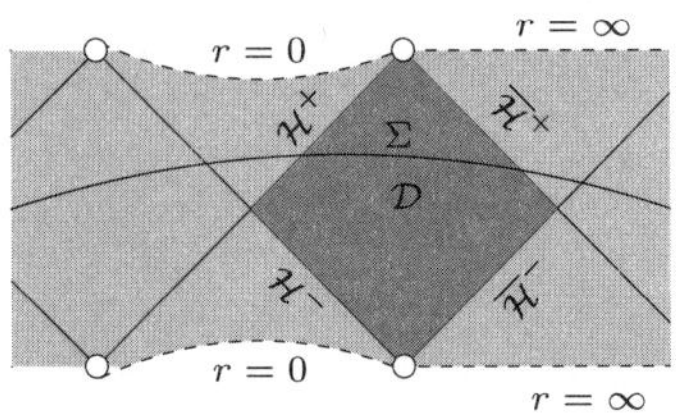

To construct "cosmological solutions" one often takes spatially compact quotients. (One can also glue such regions into other cosmological spacetimes. See [**56**]. For more on the geometry of this solution, see [**10**].)

6.2. Boundedness and decay. The region "analogous" to the region studied previously for Schwarzschild and Kerr is the darker shaded region $\mathcal{D}$ above. The horizon $\overline{\mathcal{H}}^+$ separates $\mathcal{D}$ from an "expanding" region where the spacetime is similar to the celebrated de Sitter space. If Σ is a Cauchy surface such that $\Sigma\cap\mathcal{H}^- = \Sigma\cap\overline{\mathcal{H}}^- = \emptyset$, then let us define $\Sigma_0 = \mathcal{D}\cap\Sigma$, and let us define Σ_τ to be the translates of Σ_0 by the flow φ_t generated by the Killing field T $(= \frac{\partial}{\partial t})$. Note that, in contrast to the Schwarzschild or Kerr case, Σ_0 is compact.

We have

THEOREM 6.1. *The statement of Theorem 3.2 holds for these spacetimes, where Σ, Σ_0, Σ_τ are as above, and $\lim_{x\to i^0}|\psi|$ is replaced by $\sup_{x\in\Sigma_0}|\psi|$.*

PROOF. The proof of the above theorem is as in the Schwarzschild case, except that in addition to the analogue of N, one must use a vector field $\bar{N}$ which plays the role of N near the "cosmological horizon" $\bar{\mathcal{H}}^+$. It is a good **exercise** for the reader to think about the properties required to construct such a $\bar{N}$. A general construction of such a vector field applicable to all non-extremal stationary black holes is done in Section 7. □

As for decay, we have

THEOREM 6.2. *For every $k\geq 0$, there exist constants C_k such that the following holds. Let $\psi\in H^{k+1}_{\rm loc}$, $\psi'\in H^k_{\rm loc}$, and define*

$$E_k \doteq \sum_{|(\alpha)|\leq k}\sum_{\Gamma=\{\Omega_i\}}\int_{\Sigma_0} J^{n_{\Sigma_\tau}}_\mu(\Gamma^\alpha\psi)n^\mu_{\Sigma_\tau}.$$

Then

$$\int_{\Sigma_\tau} J^{n_{\Sigma_\tau}}_\mu(\psi)n^\mu_{\Sigma_\tau} \leq C_kE_k\tau^{-k}. \tag{98}$$

[69]The expression (97) with $\Lambda<0$ defines *Schwarzschild–anti-de Sitter*. See Section 8.4.

For $k > 1$ we have

$$\sup_{\Sigma_\tau} |\psi - \psi_0| \leq C_k \sqrt{E_k} \tau^{\frac{-k+1}{2}}, \tag{99}$$

where ψ_0 denotes the 0'th spherical harmonic, for which we have for instance the estimate

$$\sup_{\Sigma_\tau} |\psi_0| \leq \sup_{x \in \Sigma_0} \psi_0 + C_0 \sqrt{E_0(\psi_0, \psi_0')}. \tag{100}$$

The proof of this theorem uses the vector fields T, Y and $\bar{Y}$ (alternatively N, $\bar{N}$), together with a version of X as multipliers, and requires commutation of the equation with Ω_i to quantify the loss caused by trapping. (Like Schwarzschild, the Schwarzschild-de Sitter metric has a photon sphere which is at $r = 3M$ for all values of Λ in the allowed range. See **[86]** for a discussion of the optical geometry of this metric and its importance for gravitational lensing.) An estimate analogous to (39) is obtained, but without the χ weight, in view of the compactness of Σ_0. The result of the Theorem follows essentially immediately, in view of Theorem 6.1 and a pigeonhole argument. No use need be made of a vector field of the type Z as in Section 4.2. Note that for $\psi =$ constant, $E_k = 0$, so removing the 0'th spherical harmonic in (99) is necessary. See **[66]** for details.

Note that if Ω_i can be replaced by Ω_i^ϵ in (39), then it follows that the loss in derivatives for energy decay at any polynomial rate k in (98) can be made arbitrarily small. If Ω_i could be replaced by $\log \Omega_i$, then what would one obtain? (**Exercise**)

It would be a nice **exercise** to commute with $\hat{Y}$ as in the proof of Theorem 6.1, to obtain pointwise decay for arbitrary derivatives of k. See the related exercise in Section 4.3 concerning improving the statement of Theorem 4.1.

6.3. Comments and further reading. Theorem 6.2 was proven in **[66]**. Independently, the problem of the wave equation on Schwarzschild-de Sitter has been considered in a nice paper of Bony-Häfner **[24]** using methods of scattering theory. In that setting, the presence of trapping is manifest by the appearance of resonances, that is to say, the poles of the analytic continuation of the resolvent.[70] The relevant estimates on the distribution of these necessary for the analysis of **[24]** had been obtained earlier by Sá Barreto and Zworski **[135]**.

In contrast to Theorem 6.2, the theorem of Bony-Häfner **[24]** makes the familiar restrictive assumption on the support of initial data discussed in Section 3.2.5. For these data, however, the results of **[24]** obtain better decay than Theorem 6.2 away from the horizon, namely exponential, at the cost of only an ϵ derivative. The decay results of **[24]** degenerate at the horizon, in particular, they do not retrieve even boundedness for ψ itself. However, using the result of **[24]** together with the analogue of the red-shift Y estimate as used in the proof of Theorem 6.2, one can prove exponential decay up to and including the horizon, i.e. exponential decay in the parameter τ (**Exercise**). This still requires, however, the restrictive hypothesis of **[24]** concerning the support of the data. It would be interesting to sort out whether the restrictive hypothesis can be removed from **[24]**, and whether this fast decay is stable to perturbation. There also appears to be interesting work in progress by Sá Barreto, Melrose and Vasy **[150]** on a related problem.

[70]In the physics literature, these are known as *quasi-normal modes*. See **[104]** for a nice survey, as well as the discussion in Section 4.6.

One should expect that the statement of Theorem 6.1 holds for the wave equation on axisymmetric stationary perturbations of Schwarzschild-de Sitter, in particular, slowly rotating Kerr-de Sitter, in analogy to Theorem 5.1.

Finally, we note that in many context, more natural than the wave equation is the conformally covariant wave equation $\Box_g\psi - \frac{1}{6}R\psi = 0$. For Schwarzschild-de Sitter, this is then a special case of Klein-Gordon (106) with $\mu > 0$. The analogue of Theorem 6.1 holds by virtue of Section 7.2. **Exercise**: Prove the analogue of Theorem 6.2 for this equation.

7. Epilogue: The red-shift effect for non-extremal black holes

We give in this section general assumptions for the existence of vector fields Y and N as in Section 3.3.2. As an application, we can obtain the boundedness result of Theorem 3.2 or Theorem 6.1 for all classical non-extremal black holes for general nonnegative cosmological constant $\Lambda \geq 0$. See [**91, 148, 28**] for discussions of these solutions.

7.1. A general construction of vector fields Y and N. Recall that a *Killing horizon* is a null hypersurface whose normal is Killing [**92, 148**]. Let $\mathcal{H}$ be a sufficiently regular Killing horizon with (future-directed) generator the Killing field V, which bounds a spacetime $\mathcal{D}$. Let φ_t^V denote the one-parameter family of transformations generated by V, assumed to be globally defined for all $t \geq 0$. Assume there exists a spatial hypersurface $\Sigma \subset \mathcal{D}$ transverse to V, such that $\Sigma \cap \mathcal{H} = S$ is a compact 2-surface. Consider the region

$$\mathcal{R}' = \cup_{t\geq 0}\varphi_t^V(\Sigma)$$

and assume that $\mathcal{R}' \cap \mathcal{D}$ is smoothly foliated by $\varphi_t(\Sigma)$.

Note that

$$\nabla_V V = \kappa V$$

for some function $\kappa : \mathcal{H} \to \mathbb{R}$.

THEOREM 7.1. *Let $\mathcal{H}$, $\mathcal{D}$, $\mathcal{R}'$, Σ, V, φ_t^V be as above. Suppose $\kappa > 0$. Then there exists a ϕ_t^V-invariant future-directed timelike vector field N on $\mathcal{R}'$ and a constant $b > 0$ such that*

$$K^N \geq b\, J_\mu^N N^\mu$$

in an open φ_t-invariant (for $t \geq 0$) subset $\tilde{\mathcal{U}} \subset \mathcal{R}'$ containing $\mathcal{H} \cap \mathcal{R}'$.

PROOF. Define Y on S so that Y is future directed null, say

$$g(Y, V) = -2, \tag{101}$$

and orthogonal to S. Moreover, extend Y off S so that

$$\nabla_Y Y = -\sigma(Y + V) \tag{102}$$

on S. Now push Y forward by φ_t^V to a vector field on $\mathcal{U}$. Note that all the above relations still hold on $\mathcal{H}$.

It is easy to see that the relations (19)–(22) hold as before, where E_1, E_2 are a local frame for $T_p\varphi_t^V(S)$. Now a^1, a^2 are not necessarily 0, hence our having included them in the original computation! We define as before

$$N = V + Y.$$

Note that it is the compactness of S which gives the uniformity of the choice of b in the statement of the theorem. □

We also have the following commutation theorem

THEOREM 7.2. *Under the assumptions of the above theorem, if ψ satisfies $\Box_g\psi = 0$, then for all $k \geq 1$.*

$$\Box_g(Y^k\psi) = \kappa_k Y^{k+1}\psi + \sum_{0\leq|m|\leq k+1,\, 0\leq m_4\leq k} c_m E_1^{m_1} E_2^{m_2} T^{m_3} Y^{m_4}\psi$$

on $\mathcal{H}^+$, where $\kappa_k > 0$.

PROOF. From (19)–(22), we deduce that relative to the null frame (on the horizon) V, Y, E_1, E_2 the deformation tensor ${}^Y\pi$ takes the form

$${}^Y\pi_{YY} = 2\sigma, \quad {}^Y\pi_{VV} = 2\kappa, \quad {}^Y\pi_{VY} = \sigma, \quad {}^Y\pi_{YE_i} = 0, \quad {}^Y\pi_{VE_i} = a^i, \quad {}^Y\pi_{E_iE_j} = h_i^j.$$

As a result the principal part of the commutator expression–the term $2\,{}^Y\pi^{\alpha\beta}D_\alpha D_\beta\psi$ can be written as follows

$$2\,{}^Y\pi^{\alpha\beta}\nabla_\alpha\nabla_\beta\psi = \kappa\nabla^2_{YY}\psi + \sigma(\nabla^2_{VV} + \nabla^2_{YV})\psi - a^i\nabla^2_{YE_i}\psi + 2h^i_j\nabla^2_{E_iE_j}\psi.$$

The result now follows by induction on k. □

7.2. Applications. The proposition applies in particular to sub-extremal Kerr and Kerr-Newman, as well as to both horizons of sub-extremal Kerr-de Sitter, Kerr-Newman-de Sitter, etc. Let us give the following general, albeit somewhat awkward statement:

THEOREM 7.3. *Let $(\mathcal{R}, g)$ be a manifold with stratified boundary $\mathcal{H}^+ \cup \Sigma$, such that $\mathcal{R}$ is globally hyperbolic with past boundary the Cauchy hypersurface Σ, where Σ and $\mathcal{H}$ are themselves manifolds with (common) boundary S. Assume*

$$\mathcal{H}^+ = \cup_{i=1}^n \mathcal{H}_i^+, \qquad S = \cup_{i=1}^n S_i,$$

where the unions are disjoint and each $\mathcal{H}_i^+$, S_i is connected. Assume each $\mathcal{H}_i^+$ satisfies the assumptions of Theorem 7.1 with future-directed Killing field V_i, some subset $\Sigma_i \subset \Sigma$, and cross section a connected component S_i of S. Let us assume there exists a Killing field T with future complete orbits, and φ_t is the one-parameter family of transformations generated by T. Let $\tilde{\mathcal{U}}_i$ be given by Theorem 7.1 and assume that there exists a $\mathcal{V}$ as above such that

$$\mathcal{R} = \varphi_t(\Sigma \setminus \mathcal{V}) \cup \cup_{i=1}^n \tilde{\mathcal{U}}_i.$$

and

$$-g((\varphi_t^{V_i})_* n_\Sigma, n_{\Sigma_\tau}) \leq B$$

where $\Sigma_\tau = \varphi_\tau(\Sigma)$, $\varphi_t^{V_i}$ represents the one-parameter family of transformations generated by V^i, and the last inequality is assumed for all values of t, τ where the left hand side can be defined. Finally, let ψ be a solution to the wave equation and assume that for any open neighbourhood $\mathcal{V}$ of S in Σ, there exists a positive constant $b_\mathcal{V} > 0$ such that

(103) $$J^T_\mu(T^k\psi)n^\mu_\Sigma \geq b_\mathcal{V} J^{n_\Sigma}_\mu(T^k\psi)n^\mu_\Sigma$$

in $\Sigma \setminus \mathcal{V}$ and

(104) $$T\psi = c_i V_i\psi$$

on $\mathcal{H}_i^+$. It follows that the first statement of Theorem 3.2 holds for ψ.

Assume in addition that Σ is compact or asymptotically flat, in the weak sense of the validity of a Sobolev estimate (11) *near infinity. Then the second statement of Theorem* 3.2 *holds for ψ.*

In the case where T is assumed timelike in $\mathcal{R} \setminus \mathcal{H}^+$, then (104) is automatic, whereas (103) holds if

$$-g(T,T) \geq -b_{\mathcal{V}}\, g(n_\mu, T)$$

in $\Sigma \setminus \mathcal{V}$. Thus we have

COROLLARY 7.1. *The above theorem applies to Reissner-Nordström, Reissner-Nordström-de Sitter, etc, for all subextremal range of parameters. Thus Theorem* 3.2 *holds for all such metrics.*[71]

On the other hand, (104), (103) can be easily seen to hold for *axisymmetric* solutions ψ_0 of $\Box_g \psi = 0$ on backgrounds in the Kerr family (see Section 5.2). We thus have

COROLLARY 7.2. *The statement of Theorem* 3.2 *holds for axisymmetric solutions ψ_0 of for Kerr-Newman and Kerr-Newman-de Sitter for the full subextremal range of parameters.*[72]

Let us also mention that the the theorems of this section apply to the Klein-Gordon equation $\Box_g \psi = \mu^2 \psi$, as well as to the Maxwell equations (**Exercise**).

8. Open problems

We end these notes with a discussion of open problems. Some of these are related to Conjecture 5.1, but all have independent interest.

8.1. The wave equation. The decay rates of Theorem 4.1 are sharp as uniform decay rates in v for any nontrivial class of initial data. On the other hand, it would be nice to obtain more decay in the interior, possibly under a stronger assumption on initial data.

OPEN PROBLEM 1. *Show that there exists a $\delta > 0$ such that* (31) *holds with τ replaced with $\tau^{-2(1+\delta)}$, for a suitable redefinition of E_1. Show the same thing for Kerr spacetimes with $|a| \ll M$.*

At the very least, it would be nice to obtain this result for the energy restricted to $\tilde{\Sigma}_\tau \cap \{r \leq R\}$.

Recall how the algebraic structure of the Kerr solution is used in a fundamental way in the proof of Theorem 5.2. On the other hand, one would think that the validity of the results should depend only on the robustness of the trapping structure. This suggests the following

OPEN PROBLEM 2. *Show the analogue of Theorem* 5.2 *for the wave equation on metrics close to Schwarzschild with as few as possible geometric assumptions on the metric.*

[71]In the $\Lambda = 0$ case this range is $M > 0$, $0 \leq |Q| < M$. **Exercise**: What is it for $\Lambda > 0$?

[72]In the $\Lambda = 0$ case this range is $M > 0$, $0 \leq |Q| < \sqrt{M^2 - a^2}$. **Exercise**: What is it for $\Lambda > 0$?

For instance, can Theorem 5.2 be proven under the assumptions of Theorem 5.1? Under even weaker assumptions?

Our results for Kerr require $|a| \ll M$. Of course, this is a "valid" assumption in the context of the nonlinear stability problem, in the sense that if this condition is assumed on the parameters of the initial reference Kerr solution, one expects it holds for the final Kerr solution. Nonetheless, one certainly would like a result for all cases. See the discussion in Section 5.5.

OPEN PROBLEM 3. *Show the analogue of Theorem* 5.2 *for Kerr solutions in the entire subextremal range* $0 \leq |a| < M$.

The extremal case $|a| = M$ may be quite different in view of the fact that Section 7 cannot apply:

OPEN PROBLEM 4. *Understand the behaviour of solutions to the wave equation on extremal Reissner-Nordström, extremal Schwarzschild-de Sitter, and extremal Kerr.*

Turning to the case of $\Lambda > 0$, we have already remarked that the analogue of Theorems 6.1 and 6.2 should certainly hold in the case of Kerr-de Sitter. In the case of both Schwarzschild-de Sitter and Kerr-de Sitter, another interesting problem is to understand the behaviour in the region $\mathcal{C} = J^+(\overline{\mathcal{H}_A^+}) \cap J^+(\overline{\mathcal{H}_B^+})$, where $\mathcal{H}_A^+$, $\mathcal{H}_B^+$ are two cosmological horizons meeting at a sphere:

OPEN PROBLEM 5. *Understand the behaviour of solutions to the wave equation in region* $\mathcal{C}$ *of Schwarzschild-de Sitter and Kerr-de Sitter, in particular, their behaviour along* $r = \infty$ *as* i^+ *is approached.*

Let us add that in the case of cosmological constant, in some contexts it is appropriate to replace $\Box_g$ with the conformally covariant wave operator $\Box_g - \frac{1}{6}R$. In view of the fact that R is constant, this is a special case of the Klein-Gordon equation discussed in Section 8.3 below.

8.2. Higher spin. The wave equation is a "poor man's" linearisation of the Einstein equations (4). The role of linearisation in the mathematical theory of nonlinear partial differential equations is of a different nature than that which one might imagine from the formal "perturbation" theory which one still encounters in the physics literature. Rather than linearising the equations, one considers the solution of the nonlinear equation from the point of view of a related linear equation that it itself satisfies.

In the case of the simplest nonlinear equations (say (107) discussed in Section 8.6 below), typically this means freezing the right hand side, i.e. treating it as a given inhomogeneous term. In the case of the Einstein equations, the proper analogue of this procedure is much more geometric. Specifically, it amounts to looking at the so called Bianchi equations

$$\nabla_{[\mu} R_{\nu\lambda]\rho\sigma} = 0, \tag{105}$$

which are already linear as equations for the curvature tensor when g is regarded as fixed. For more on this point of view, see [**51**]. The above equations for a field $S_{\lambda\mu\nu\rho}$ with the symmetries of the Riemann curvature tensor are in general known as the spin-2 equations. This motivates:

OPEN PROBLEM 6. *State and prove the spin-2 version of Theorems* 5.1 *or* 5.2 *(or Open problem* 1*) on Kerr metric backgrounds or more generally, metrics settling down to Kerr.*

In addition to [**51**], a good reference for these problems is [**50**], where this problem is resolved just for Minkowski space. In contrast to the case of Minkowski space, an additional difficulty in the above problem for the black hole setting arises from the presence of nontrivial stationary solutions provided by the curvature tensor of the solutions themselves. This will have to be accounted for in the statement of any decay theorem. From the "linearisation" point of view, the existence of stationary solutions is of course related to the fact that it is the 2-parameter Kerr family which is expected to be stable, not an individual solution.

8.3. The Klein-Gordon equation. Another important problem is the Klein-Gordon equation

$$\Box_g \psi = \mu\psi. \tag{106}$$

A large body of heuristic studies suggest the existence of a sequence of quasinormal modes (see Section 4.6) approaching the real axis from below in the Schwarzschild case. When the metric is perturbed to Kerr, it is thought that essentially this sequence "moves up" and produces exponentially growing solutions. See [**158, 71**]. This suggests

OPEN PROBLEM 7. *Construct an exponentially growing solution of* (106) *on Kerr, for arbitrarily small* $\mu > 0$ *and arbitrary small* a.

Interestingly, if one fixed m, then adapting the proof of Section 5.2, one can show that for $\mu > 0$ sufficiently small and a sufficiently small, depending on m, the statement of Theorem 5.1 holds for (106) for such Kerr's. This is consistent with the quasinormal mode picture, as one must take $m \to \infty$ for the modes to approach the real axis in Schwarzschild. This shows how misleading fixed-m results can be when compared to the actual physical problem.

8.4. Asymptotically anti-de Sitter spacetimes. In discussing the cosmological constant we have considered only the case $\Lambda > 0$. This is the case of current interest in cosmology. On the other hand, from the completely different viewpoint of high energy physics, there has been intense interest in the case $\Lambda < 0$. See [**84**].

The expression (97) for $\Lambda < 0$ defines a solution known as *Schwarzschild-anti-de Sitter*. A Penrose diagramme of this solution is given below.

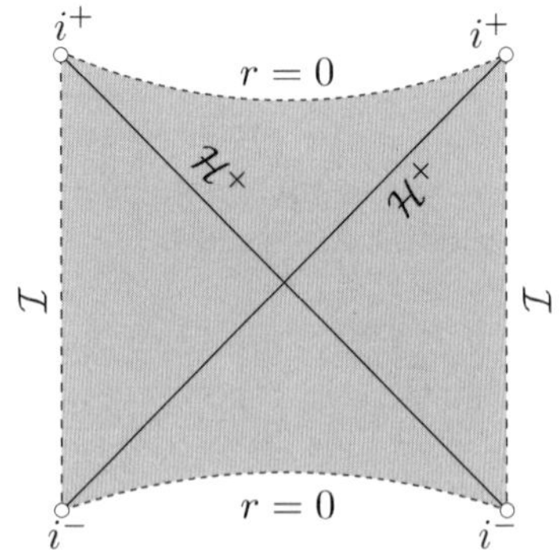

The timelike character of infinity means that this solution is not globally hyperbolic. As with Schwarzschild-de Sitter, Schwarzschild-anti-de Sitter can be viewed as a subfamily of a larger Kerr-anti de Sitter family, with similar properties.

Again, as with Schwarzschild-de Sitter, the role of the wave equation is in some contexts replaced by the conformally covariant wave equation. Note that this corresponds to (106) with a negative $\mu = 2\Lambda/3 < 0$.

Even in the case of anti-de Sitter space itself (set $M = 0$ in (97)), the question of the existence and uniqueness of dynamics is subtle in view of the timelike character of the ideal boundary $\mathcal{I}$. It turns out that dynamics are unique for (106) only if the $\mu \geq 5\Lambda/12$, whereas for the total energy to be nonnegative one must have $\mu \geq 3\Lambda/4$. Under our conventions, the conformally covariant wave equation lies between these values. See [**6, 26**].

OPEN PROBLEM 8. *For suitable ranges of* μ, *understand the boundedness and blow-up properties for solutions of* (106) *on Schwarzschild-anti de Sitter and Kerr-anti de Sitter.*

See [**109, 27**] for background.

8.5. Higher dimensions. All the black hole solutions described above have higher dimensional analogues. See [**77, 120**]. These are currently of great interest from the point of view of high energy physics.

OPEN PROBLEM 9. *Study all the problems of Sections* 8.1–8.4 *in dimension greater than* 4.

Higher dimensions also brings a wealth of explicit black hole solutions such that the topology of spatial sections of $\mathcal{H}^+$ is no longer spherical. In particular, in 5 spacetime dimensions there exist "black string" solutions, and much more interestingly, asymptotically flat "black ring" solutions with horizon topology $S^1 \times S^2$. See [**77**].

OPEN PROBLEM 10. *Investigate the dynamics of the wave equation* $\Box_g \psi = 0$ *and related equations on black ring backgrounds.*

8.6. Nonlinear problems. The eventual goal of this subject is to study the global dynamics of the Einstein equations (4) themselves and in particular, to resolve Conjecture 5.1.

It may be interesting, however, to first look at simpler non-linear equations on fixed black hole backgrounds and ask whether decay results of the type proven here are sufficient to show non-linear stability.

The simplest non-linear perturbation of the wave equation is

$$\Box_g \psi = V'(\psi) \tag{107}$$

where $V = V(x)$ is a potential function. Aspects of this problem on a Schwarzschild background have been studied by [**121, 64, 22, 115**].

OPEN PROBLEM 11. *Investigate the problem* (107) *on Kerr backgrounds.*

In particular, in view of the discussion of Section 8.3, one may be able to construct solutions of (107) with $V = \mu\psi^2 + |\psi|^p$, for $\mu > 0$ and for arbitrarily large p, arising from arbitrarily small, decaying initial data, which blow up in finite time. This would be quite interesting.

A nonlinear problem with a stronger relation to (4) is the wave map problem. Wave maps are maps $\Phi : \mathcal{M} \to \mathcal{N}$ where $\mathcal{M}$ is Lorentzian and $\mathcal{N}$ is Riemannian, which are critical points of the Lagrangian

$$\mathcal{L}(\Phi) = \int |d\Phi|^2_{g_N}$$

In local coordinates, the equations take the form

$$\Box_{g_M} \Phi^k = -\Gamma^k_{ij} g^{\alpha\beta}_M (\partial_\alpha \Phi^i \partial_\beta eta \Phi^j),$$

where Γ^k_{ij} denote the Christoffel symbols of g_N. See the lecture notes of Struwe [**145**] for a nice introduction.

OPEN PROBLEM 12. *Show global existence in the domain of outer communications for small data solutions of the wave map problem, for arbitrary target manifold $\mathcal{N}$, on Schwarzschild and Kerr backgrounds.*

All the above problems concern fixed black hole backgrounds. One of the essential difficulties in proving Conjecture 5.1 is dealing with a black hole background which is not known a priori, and whose geometry must thus be recovered in a bootstrap setting. It would be nice to have more tractable model problems which address this difficulty. One can arrive at such problems by passing to symmetry classes. The closest analogue to Conjecture 5.1 in such a context is perhaps provided by the results of Holzegel [**94**], which concern the dynamic stability of the 5-dimensional Schwarzschild as a solution of (4), restricted under Triaxial Bianchi IX symmetry. See Section 2.6.4. In the symmetric setting, one can perhaps attain more insight on the geometric difficulties by attempting a large-data problem. For instance

OPEN PROBLEM 13. *Show that the maximal development of asymptotically flat triaxial Bianchi IX vacuum initial data for the* 5*-dimensional vacuum equations containing a trapped surface settles down to Schwarzschild.*

The analogue of the above statement has in fact been proven for the Einstein-scalar field system under spherical symmetry [**40, 63**]. In the direction of the above, another interesting set of problems is provided by the Einstein-Maxwell-charged scalar field system under spherical symmetry. For both the charged-scalar field system and the Bianchi IX vacuum system, even more ambitious than Open problem 13 would be to study the strong and weak cosmic censorship conjectures, possibly unifying the analysis of [**45, 58, 59**]. Discussion of these open problems, however, is beyond the scope of the present notes.

9. Acknowledgements

These notes were the basis for a course at the Clay Summer School on Evolution Equations which took place at ETH, Zürich from June 23–July 18, 2008. A shorter version of this course was presented as a series of lectures at the Mittag-Leffler Institute in September 2008.

The authors thank ETH for hospitality while these notes were written, as well as the Clay Mathematics Institute. M. D. thanks in addition the Mittag-Leffler Institute in Stockholm. M. D. is supported in part by a grant from the European Research Council. I. R. is supported in part by NSF grant DMS-0702270.

10. Addendum: December 2011

It has been over 3 years since our July 2008 Clay Summer School Lectures in Zürich and the subsequent posting of these Lecture Notes shortly thereafter to the arxiv. The intervening period has witnessed remarkable progress concerning the study of waves on black holes, at a rate in no way foreseen by us. It is especially satisfying that so much of this progress has been accomplished by participants in the Clay Summer School (Aretakis, Baskin, Blue, Holzegel, Schlue, Smulevici, Tohaneanu) as well as by two of the other lecturers (Vasy, Wunsch)!

In submitting a final version of these notes for publication by the CMI, we wish to record, at least briefly, some of the highlights of these rapid subsequent developments–hence this Addendum. These exciting works have clarified issues, resolved fully or partially open problems, fulfilled prophesies, but also, modified (at least to some extent) various aspects of our point of view. For instance, were we to rewrite these notes, we would certainly replace Sections 4.2 *and* 5.3.6 *with an exposition of the results described in Section* 10.5 *below, which give what we believe to be a definitive approach to obtaining robust pointwise decay from integrated local energy decay. As another example, our discussion of finer polynomial tails in Section* 4.6 *would certainly be enhanced by an exposition of the results described in Section* 10.7 *below. We have resisted, however, the temptation to modify the original text with the benefit of this hindsight. Our lecture notes were not meant as a definitive treatment of the subject, but rather, as a snapshot of the field as it stood in the Summer of 2008. Moreover, these lecture notes double as an original research paper, giving for the first time a proof of integrated local energy decay on slowly rotating Kerr (Proposition* 5.3.1*), pointwise decay on Kerr (Theorem* 5.2*), and the general red-shift multiplier and commutation constructions (Theorems* 7.1, 7.2 *and* 7.3*) that have proven very useful in much subsequent work. We feel that in view of this double role, it is important to preserve the notes' original form for the historical record.*

We have thus confined all references to subsequent developments to this Addendum, leaving the rest of the text "as is", except for various typographical changes and corrections to minor errors in some formulas which could cause confusion. We thank particularly Stefanos Aretakis, Gustav Holzegel, Igor Khavkine, Jan Sbierski and Volker Schlue for their careful readings and for pointing out many such errors in the original version of these notes.

Mihalis Dafermos
Igor Rodnianski
Cambridge (UK and USA), December 2011

10.1. Two new approaches to dispersion on slowly-rotating Kerr $|a| \ll M$. Since the Clay Summer School, two additional approaches to integrated local energy decay for the wave equation on Kerr exteriors with $|a| \ll M$, originally proven as Proposition 5.2 of these notes, have been completed.

The first such additional approach is due to Tataru–Tohaneanu:

D. Tataru and M. Tohaneanu *A local energy estimate on Kerr black hole backgrounds* Int. Math. Res. Not. 2011, no. 2, 248–292 and was in fact posted to the arxiv in parallel with the arxiv version of these lecture notes. Recall from the discussion of Section 5.3 that the main difficulty for proving Proposition 5.2, after the stability issues at the horizon had been sorted out in [**65, 68**] using the red-shift, was capturing the obstruction posed by trapping in the high-frequency limit. In our own approach, as given for the first time in these lecture notes, this difficulty was resolved by using Carter's separation of the wave equation as a tool to frequency-localise the energy current constructions of Section 5.3. Tataru–Tohaneanu instead appeal to separation at the level of the equations of geodesic flow, but microlocalise according to the standard pseudodifferential calculus applied only in a neighbourhood of the Schwarzschild photon sphere with respect to an ambient Euclidean coordinate system. The method of red-shift commutation, introduced in our previous [**68**], is then applied so as to complete the argument, giving also an alternative proof of the pointwise boundedness statement of [**68**], when the latter is specialised to the exactly Kerr case.

The second new approach to Proposition 5.2 of these lecture notes, due to Andersson–Blue, and appearing in:

L. Andersson and P. Blue *Hidden symmetries and decay for the wave equation on the Kerr spacetime*, arXiv:0908.2265,

replaces the above two frequency localisation techniques by a third, which combines classical vector field multipliers with commutation by a second order differential operator constructed from the so-called Carter tensor. Carter's separation of the wave equation is in fact intimately connected with these operators and the relevant positivity computation can be directly translated to the formalism of Section 5.3. (In this language, one is choosing an f as in formula (74) with polynomial ω-dependence which has an interpretation as commutation by a differential operator.) The fact that the implicit frequency analysis is accomplished using only differential operators gives the Andersson–Blue argument many attractive features. The result is slightly weaker, however, than that given by our previous method (as well as that described in the paragraph above), as commutation gives rise to an estimate at the level of a weighted H^3 norm, rather than H^1 as in Proposition 5.2.

Let us add that we ourselves have given yet another proof of Proposition 5.2 in our

M. Dafermos and I. Rodnianski *Decay for solutions of the wave equation on Kerr exterior spacetimes I-II: The cases $|a| \ll M$ or axisymmetry* arXiv:1010.5132

For the high frequency domain, this proof follows closely the proof of these notes, but in the new argument, the full potential (no pun intended!) of the separation is exploited to construct novel low-frequency currents, which make the proof completely independent of both our previous decay work on Schwarzschild [**65**] and of our previous general boundedness theorem [**68**], both of which were used (albeit simply as a convenience) in the proof contained in Section 5.3. In particular, the above paper yields as a by-product *yet another proof* of local energy decay for the Schwarzschild case, completely self-contained, and having the additional advantage that, in providing a systematic approach to low frequencies, the proof gives a blue-print which can be applied to a wide variety of spacetimes. This has indeed proven useful for subsequent developments in the extremal case and in AdS (see Sections 10.3 and 10.10 below).

Moreover, the above new proof unifies the small $|a| \ll M$ case with the case of axisymmetric solutions in the full subextremal range $|a| < M$, where superradiance is absent and one can appeal to Theorem 7.3.

In contrast, as we shall see, the case of general, non-axisymmetric solutions in the full range $|a| < M$ required a new insight, which we turn to immediately the next section!

10.2. The full subextremal range $|a| < M$. (cf. Open problem 3)

The case $|a| \ll M$ is characterized by the fact that superradiance is a small parameter. This played a fundamental role in both the general boundedness result [**68**] which inaugurated the study of the wave equation on Kerr, as well as the subsequent decay results just described.

Let us briefly recall the role of the small parameter for both boundedness and decay:

In our original general boundedness result [**68**], the smallness of $|a|$ was exploited first to show that the difficulties of superradiance and trapping were "disjoint". Essentially this can be understood in physical space: The ergoregion is in a small neighbourhood of the horizon, while trapping is confined to a region near the Schwarzschild photon sphere; for $|a| \ll M$, these two regions are well separated. Using only the separation with respect to ω and m, this allowed one to construct a multiplier current with positive bulk term (and without degeneration) *for the superradiant frequencies* but bypass constructing such a current for the non-superradiant frequencies, relying instead on an independent boundedness argument. In this, the smallness of $|a|$ is exploited a second time so as to ensure the positivity of the boundary terms in the energy current applied to the superradiant frequencies–in effect, here one uses that the "strength" of superradiance can be taken as a small parameter.

In the decay problem for slowly rotating Kerr spacetimes, one does not need to handle separately the superradiant and non-superradiant frequencies, for essentially one applies to all frequencies the argument which above was applied only to the superradiant frequencies, at the expense however of now having to face the difficulty of capturing trapping. Nonetheless, the smallness of $|a|$ in its second manifestation described above, namely as allowing for the "strength" of superradiance to be taken as a small parameter, is exploited just as above, for the control of the boundary terms. This applies both to our approach and that of Tataru–Tohaneanu mentioned above. In the work of Andersson–Blue, a similar scheme is used again requiring small $|a|$ for control of the boundary terms, but with the use of N replaced by a vector field in the span of the Killing fields T and Φ.

In turning to the general subextremal range $|a| < M$, it is not too difficult to see (in the context of the frequency localisation given by Carter's separation) that currents generating a non-negative bulk term can still be constructed–here one is in particular implicitly exploiting the fact that the dynamics of geodesic flow near the set of trapped geodesics remain normally hyperbolic. These constructions, however, as such, do not allow one to control the boundary terms. For this, it turned out that one must return to the insight of the boundedness paper concerning the "disjointness" of the trapped and superradiant frequencies. Fortuitously, it turns out that **trapping and superradiance remain disjoint in the whole subextremal range.** In contrast to the $|a| \ll M$ case, however, this is not obvious at all from pure physical-space considerations as the ergoregion in general now contains trapped null geodesics. It is thus very much a phase space phenomenon. This disjointness can moreover be quantified, in particular, one can exploit the superradiant/non-superradiant decomposition in the multiplier constructions to ensure that the boundary terms are also controlled. One obtains thus the precise analogue of Proposition 5.2 for the whole subextremal range $|a| < M$. See Section 11 of

M. Dafermos and I. Rodnianski *The black hole stability problem for linear scalar perturbations* arXiv:1010.5137

In view also of the results to be discussed in Section 10.5 below, the above result is sufficient to obtain the full set of decay estimates in the whole subextremal range. Thus, with the above, the study of the scalar wave equation on the subextremal Kerr family is complete.

Let us conclude this discussion with a remark about the miraculous disjointness of the trapping and superradiance phenomena. A special case of this disjointness is the absence of trapped null geodesics which are orthogonal to ∂_t. This is related to the conditional pseudoconvexity property that had played a fundamental role in the Ionsecu–Klainerman approach to uniqueness of Kerr via unique continuation [**96**]. It would be of great interest to understand more conceptually the origin of this feature. See also the next section for a discussion of the extremal case $|a| = M$.

10.3. The extremal case $Q = M$ or $|a| = M$. (cf. Open problem 4)

The simplest example of an extremal black hole spacetime is extremal Reissner–Nordström with parameters $Q = M$. As we have discussed in Section 8.1, on such spacetimes the red-shift factor on the horizon vanishes. Thus, even a uniform boundedness result in the style of Theorem 7.3 is now non-trivial.

This problem was taken up by Aretakis in a series of papers

S. Aretakis *Stability and instability of extreme Reissner–Nordström black hole spacetimes for linear scalar perturbations I* Comm. Math. Phys. **307** (2011), 17–63

S. Aretakis *Stability and instability of extreme Reissner–Nordström black hole spacetimes for linear scalar perturbations II*, Ann. Henri Poincare **12** (2011), 1491–1538

which we shall describe briefly in what follows.

First, one sees easily that in the extremal case, there is no pure-vector field translation-invariant current satisfying $K^N \geq 0$ near the horizon, where N is timelike at $\mathcal{H}^+$. With a suitable modification term, this problem can be overcome, and the above series of papers indeed begins by constructing a current $J^{N,1}$ satisfying

$$K^{N,1} \geq 0 \tag{108}$$

near the horizon. It is however *not* possible to obtain

$$K^{N,1} \geq J^N_\mu N^\mu. \tag{109}$$

The bulk term associated to the $J^{N,1}$ energy identity thus must still degenerate at the horizon.

In view of the failure of (109) to hold, the above current is still not sufficient to be used together with just T, in the manner of the argument of Section 3.3, to obtain uniform boundedness up to and including the horizon.

Given, however, an analogue of an X-estimate, the nonnegativity property (108) near the horizon is then sufficient to retrieve the boundedness of the boundary terms $J^{N,1}$, and thus uniform boundedness of the non-degenerate energy. One sees that in the extremal case, the problem of boundedness for the non-degenerate energy is inextricably coupled with local energy decay.[73]

The next result of the above series of papers indeed obtains the desired X estimate, and thus, in view of the above remarks, both integrated local energy decay and non-degenerate boundedness. Note, however, that, in view again of the failure of (109), the spacetime integral in this estimate still degenerates at the horizon, though the boundary term does not.

This weaker version of integrated local energy decay, together with the uniform boundedness, can in turn be used to show decay for the *degenerate* J^T-energy flux through a suitable foliation, as well as pointwise decay, following the new method outlined in Section 10.5 below. Again, however, the degeneration at the horizon requires a modification of this method through the introduction of yet another vector field. See the comments at the end of Section 10.5.

Perhaps the most surprising result of this work, however, is the fact that the above degeneracies in the estimates are in fact **necessary**. Using a hierarchy of conservation laws on the horizon, Aretakis proves that the non-degenerate J^N-energy generically **does not decay** through a foliation $\tilde{\Sigma}_\tau$, and higher order J^N-based energies **blow up**! Thus, **extreme black holes are (mildly) unstable on the event horizon itself!**

In a more recent paper

S Aretakis *Decay of axisymmetric solutions of the wave equation on extreme Kerr backgrounds*, arXiv:1110.2006

Aretakis has extended his stability results to axisymmetric solutions on extremal Kerr. Obtaining analogues of the instability results in the Kerr case remains an open problem.

The non-axisymmetric case comes with yet another difficulty. The main insight leading to the resolution of the decay problem in the full subextremal range, discussed in Section 10.2 above, namely that trapped frequencies are not superradiant, degenerates precisely at extremality! The repercussions of this phenomenon for quantitative decay estimates are yet to be explored.

10.4. Improved decay and non-linear applications. (cf. Open problem 12)

[73] This situation is reminiscent of the original proof of uniform boundedness in [**65**] (i.e. before the argument of Section 3.3 introduced in our later [**68**] was developed) where uniform boundedness was obtained only after obtaining the X estimate. See the discussion in Section 3.4 of these notes. The extremal case thus brings us full circle.

The pointwise estimates of Theorem 4.1 for Schwarzschild yield in particular a uniform decay bound $|\psi| \le Ct^{-1}$, and this rate is sharp *as a uniform decay bound in* t, in view of the behaviour of ϕ along the light cone. The decay obtained in the above theorem is no better, however, in the region $r \le R$, where one expects more decay; it is indeed essential to have this improvement for nonlinear applications.

In Minkowski space, seemingly strong decay results in such a region can be obtained from the fundamental solution, most famously, the strong Huygens principle for solutions arising from data with compact support, which states that for large enough t, the solution vanishes in $r \le R$. As is well-known, however, this level of decay is not "seen" by non-linear problems. The robust measure of decay key to nonlinear stability properties in the most difficult 3 dimensional case is precisely that first captured by weighted commutator estimates introduced by Klainerman. This allowed proving for instance that $|\partial_t \psi| \le Ct^{-5/2}$ for fixed r, where C depends on a suitable initial weighted higher-order energy norm. The significance of this rate is simply that it is greater than 1, and thus, integrable in time. (For some problems, the relevant decay estimate may involve an even slower rate, e.g. $\le Ct^{-2}$–but still integrable!–but *never faster.*) From the modern point of view, these type of estimates thus represent *the sharp robust improved decay result* on Minkowski space.

This problem of improved decay in the black hole setting was taken up by J. Luk. It turned out to be expedient to use a single commutation with the analogue of the scaling vector field on top of the weighted multiplier Z. Results were first obtained for Schwarzschild in:

J. Luk *Improved decay for solutions to the linear wave equation on a Schwarzschild black hole*, Ann. Henri Poincaré **11** (2010), no. 5, 805–880

Results for slowly-rotating Kerr followed in

J. Luk *A vector field method approach to improved decay for solutions to the wave equation on a slowly rotating Kerr black hole*, arXiv:1009.0671

The ultimate test of whether one has "the right" decay-type results is whether they can be used to prove a non-linear stability result by exploiting dispersion. This is indeed accomplished in

J. Luk *The null condition and global existence for nonlinear wave equations on slowly rotating Kerr spacetimes*, arXiv:1009.4109

With the above paper, a certain chapter is closed: In the scalar case, one now understands the dispersive mechanism on black holes sufficiently well to tackle nonlinear stability problems with quadratic nonlinearities in derivatives. But alas, the black hole stability problem is not a scalar problem! For a discussion of progress on understanding its tensorial aspects see Section 10.11.

10.5. A new physical space method for decay. The method of Section 12 of [**65**], streamlined in Section 4.2 of these notes, was already suggestive of the fact that the integrated local energy decay coupled with the well-known behaviour at null infinity together represented the only essential properties required for obtaining the "full" decay results. An additional difficulty, however, whose conceptual origin was not clear, was caused by the weights of the vector field Z near the horizon and the necessity of the positivity of both the associated bulk and boundary terms.

In Schwarzschild, by what appears to be a miracle, the vector field Z was actually well behaved near the horizon (see Section 4.2). Already in slowly rotating Kerr, however, this breaks down, and this fact was responsible for the loss of δ in the argument given in these notes (see Section 5.3.6).

This phenomenon motivated a rethinking of the traditional use of the vector field Z. It turns out that the difficulty of the behaviour of Z near the horizon is in fact completely artificial, and the whole argument can be done in a much more transparent–and, as we shall see, robust–manner with **no weights in t, only weights in r**.

The crux of the new method is to replace Z with a p-hierarchy of r^p-weighted vector field currents which are used in sequence with $p = 2, 1, 0$, coupled at each stage with the boundedness and integrated local energy decay result. The bulk term of the p-current of the hierarchy is related to the boundary term of the $p-1$-current. One obtains thus (after several iterations) quadratic τ^{-2} decay of the energy flux through foliations $\tilde{\Sigma}_\tau$, and from this, the associated pointwise decay bounds by commutations with T and–as systematised in Section 7–N. The power τ^{-2} is dictated by the maximum p which can be taken in the hierarchy, $p = 2$.

The nature of this argument is such that one need not use any information about the geometry in the region of finite r, other than that already encoded in the boundedness and integrated decay statements. This allows one to formulate a "black box" type theorem, stating that *given* a boundedness result to all orders and an integrated decay statement, possibly with finite derivative loss (as one expects when "good" trapping is present), one could obtain all the results traditionally proven through application of the multiplier Z (in fact, the improved decay results of Klainerman's vector field method [**100**] essential for non-linear problems: see below). This argument was first presented in

M. Dafermos and I. Rodnianski, *A new physical-space approach to decay for the wave equation with applications to black hole spacetimes*, in XVIth International Congress on Mathematical Physics, P. Exner (ed.), World Scientific, London, 2009, pp. 421–433

In particular, in view of the integrated decay result of Section 10.2, the above argument applies to Kerr in the full subextremal range $|a| < M$.

Adapting ideas from the work of Luk to the setting of this new argument, Schlue has extended this method (in fact in all dimensions, see Section 10.9!) so as to retrieve the improved decay of Section 10.4. Essentially, upon commutation with weighted vector fields in r (but again, not in t for fixed r), one can extend the p-hierarchy to $p > 2$, allowing for more decay in τ of higher-order energies, from which improved decay follows.

Let us note that this new method has a host of novel applications to the study of linear and nonlinear wave equations on nonstationary perturbations of Minkowski space, boundary value problems, etc. See for example

S. Yang *Global solutions of nonlinear wave equations in time dependent inhomogenous media*, arXiv:1010.4341

Finally, let us explicitly remark that the extremal Reissner–Nordström $Q = M$ or extremal Kerr case $|a| = M$ do **not** satisfy the "black box" assumptions of the new method described above, *precisely due to the degeneration of the estimates at the horizon*. Nonetheless, the method has been extended so as to apply also to this case by Aretakis in the works referred to in Section 10.3 above, by adding to the hierarchy of estimates described above yet another, associated to a vector field P supported near the horizon. The vector fields N, P and T then stand in a hierarchal relation analogous to the p-hierarchy at null infinity.

10.6. Quasinormal modes and Kerr-de-Sitter. Recall our brief discussion of the cosmological case from Section 6. One approach to proving exponential decay in the Schwarzschild-de Sitter spacetime, in the region between the event and cosmological horizons, was by proving resolvent estimates in a strip below the real axis [**24**]. To be extended to the Kerr case, as a first step, one needed to understand the asymptotic distribution of the poles of the resolvent–the so-called quasinormal modes, in the spirit of results of Sa Barreto–Zworski [**135**]. As the ω-dependence of the resolvent is non-standard, even defining these poles requires a new argument. This was accomplished in two beautiful papers of Dyatlov

S. Dyatlov *Quasi-normal modes and exponential decay for the Kerr–de Sitter black hole* Commun. Math. Phys. **306** (2011), 119–163

S. Dyatlov *Asymptotic distribution of quasi-normal modes for Kerr–de Sitter black holes* to appear in Annales Henri Poincaré

where the Schwarzschild–de Sitter picture of [**135, 24**] was reproduced for slowly rotating Kerr–de Sitter black holes, and this was used to show exponential decay type results.

A drawback of the resolvent approach, already in the Schwarzschild–de Sitter case [**24**], is that it required data supported away from the horizons. (See however [**150**].) By combining the approach with the red-shift estimates as introduced in [**65, 68**], Dyatlov was able to remove this limitation, both allowing for general data, and obtaining non-degenerate estimates at and beyond the horizons:

S. Dyatlov *Exponential energy decay for Kerr–de Sitter black holes beyond event horizons*, to appear in Mathematical Research Letters

Another approach to exponential decay on de-Sitter space based on resolvent estimates is given by Vasy:

A. Vasy *Microlocal analysis of asymptotically hyperbolic Kerr-de Sitter spaces* (with an appendix by S. Dyatlov), arXiv:1012.4391

10.7. Fine tails revisited. As we have noted before, the decay results of Section 10.4 and 10.5 are sharp from the point of view of applications to non-linear problems, as they correspond exactly to the full decay results of Klainerman's vector field method [**100**] on Minkowski space.

On the other hand, one may ask to what extent can these results be improved if one is willing to specialise the data to be rapidly decaying (say compactly supported) and very regular (say C^∞), and if one is not so picky about the underlying regularity assumptions on the metric–for instance, if one is only interested in *exactly* Schwarzschild or Kerr spacetimes.

In Minkowski space, under such assumptions one would have the strong Huygens principle. As discussed in Section 4.6, backscattering of low frequencies from far away curvature already suggests that generically one must have at best a polynomially decaying tail. Note, however, that these tails are still "finer" (i.e. they correspond to faster decay) than the improved polynomial decay rates in the interior governed by the vector field method, described in Section 10.4. There is thus a gap between what is sharp from the point of view of the initial data norms of the vector field method and that which may hold for a more restricted class of data.

As discussed in Section 4.6, the first work to obtain a quantitative estimate closing this gap was our work [**63**] on the spherically symmetric Einstein–(Maxwell)–scalar field system, where we showed that if a non-extremal black hole formed, one could estimate the solution in the region $r \leq R$, by $C_\epsilon v^{-3+\epsilon}$, provided that the data initially decayed very fast at spatial infinity. When specialised to the linear problem of spherically symmetric waves on a fixed subextremal Reissner–Nordström background, the result also applies, and C_ϵ can be estimated by a weighted C^1 norm of data.

The above work, which concerns fully dynamic, radiative solutions of an Einstein-matter system, may at first suggest that, indeed, it is purely low-frequency backscattering that determines asymptotics. The spherically symmetric case, however, is anomalous, in particular, because it does not exhibit the phenomenon of trapping, which, as discussed in Section 4.6, effects the nature of any quantitative decay statement.[74] At the time of writing of these lecture notes, it was not clear whether there was any non-spherically symmetric regime where the tails arising from the low-frequency backscattering off far away curvature are not dwarfed by other phenomena.

It turns out, however, that indeed, in various special cases, one can prove rates of decay exactly up to the obstruction from low-frequency backscattering, by first making rigorous some of the low-frequency estimates from the physics literature, and then combining these with quantitative control of trapping, similar to the integrated local energy decay estimates discussed in these notes, so as to control the "totality of high frequencies".

With respect to the first part of the programme, we have already discussed various results concerning the $\ell = 0$ case above and in Section 4.6. This programme was continued in

R. Donninger, W. Schlag and A. Soffer *A proof of Price's law on Schwarzschild black hole manifold for all angular momenta*, Adv. Math. **266** (2011), no. 1, 484–540

[74]The spherically symmetric Einstein–scalar field case is anomalous in a second way, in that the quadratic non-linearities of the Einstein equations do not effect the radiative properties, as these are determined by the scalar field whose dynamics are linear.

where a quantitative estimate is shown for each fixed spherical harmonic number ℓ, coinciding with Price's prediction for $\ell = 0$ in the compactly supported case, but slightly weaker for higher ℓ, but still, increasing with ℓ. Subsequently, in

R. Donninger, W. Schlag and A. Soffer *On pointwise decay for linear waves on a Schwarzschild black hole background*, to appear in Commun. Math. Phys.

it is then shown that one could sum the spherical harmonics *starting from any* ℓ_0 to obtain that the sum still satisfies the faster decay rate shown above associated to ℓ_0. The proof indeed now requires quantitative control of trapping similar to that provided by the integrated local energy decay estimates of Section 4.1. This latter result gives a quantitative formulation of the principle that the totality of "higher spherical modes" indeed decays faster.

Another noteworthy result in the above direction is the quite general result of Tataru:

D. Tataru *Local decay of waves on asymptotically flat stationary spacetimes*, to appear, Amer. J. Math.

which says that for exactly stationary spacetimes, then, *given* a uniform boundedness, integrated local energy decay result and good asymptotics at infinity, one can retrieve the worst-mode decay prediction of Price for very regular initial data that decays rapidly at spatial infinity.

In view of our results described in Section 10.2, the result of the above paper can now be applied to the Kerr family in the whole subextremal range $|a| < M$, just like our own approach of Section 10.5. It cannot be stressed too much that in order to be applicable in the black hole context, the assumptions of the above paper require non-degenerate estimates, in fact to all order, on the horizon, and thus require both the multiplier and commutator propositions of Section 7. In particular, the above paper cannot be applied in the extremal case $|a| = M$, in view of the results of Section 10.3. Nonetheless, it would be interesting to attempt to adapt the approach of the above paper to the extremal case, following the lines of Aretakis's adaptation of our own method of Section 10.5.

The above work of Tataru relied heavily on resolvent estimates and was restricted to exactly stationary spacetimes. A different approach, using the fundamental solution of the standard wave operator on Minkowski space, was given subsequently in

D. Tataru, J. Metcalfe and M. Tohaneanu *Price's law on nonstationary spacetimes*, arXiv:1104.5437

This requires even more restrictive initial data but allows to treat the wave equation on a certain class of dynamical spacetimes, which do not however radiate energy to infinity. Upon imposing the Einstein equations, however, this class essentially reduces to the stationary case.

10.8. Applications of dynamical systems to trapping. A common theme in all the work quantifying the trapping obstruction has been the latter's close relation to geodesic flow. Often this relation is only implicit in the constructions.

It would be nice to make this more explicit, so as in particular to be able to exploit perturbative results in dynamical systems to draw conclusions on decay for waves on say stationary perturbations of Kerr. A first result in this direction is given by very nice work of Wunsch–Zworski:

J. Wunsch and M. Zworski *Resolvent estimates for normally hyperbolic trapped sets*, Ann. Henri Poincaré, to appear

10.9. Higher dimensions. (cf. Open problem 9)

In his Smith–Knight prize essay,

V. Schlue *Linear waves on higher dimensional Schwarzschild black holes* Rayleigh Smith Knight Essay, January 2010, University of Cambridge

Schlue proved the analogue of integrated local energy decay (i.e. the analogue of (39), and then, used this to prove the analogue of Theorem 4.1, by also generalising the 3-dimensional construction of the vector field Z to all higher dimensions. In particular, the details of the scheme described in Section 4.2 of these notes (which differs slightly from the approach [**65**]) are presented there.

In his subsequent

V. Schlue *Linear waves on higher dimensional Schwarzschild black holes*, arXiv:1012.5963

he took the new approach of Section 10.5, generalising it to all dimensions, and extending it so as to yield the improved decay of Luk in the interior region. This argument has far reaching applications beyond the black hole setting. See the discussion of Section 10.5. It remains an open problem, however, to obtain the correct dimensionally dependent improved decay rates, which should become faster with larger n.

Laul and Metcalfe present an independent, alternative construction for the integrated local energy decay part of the above work in the case of higher dimensional Schwarzschild:

P. Laul and J. Metcalfe *Localized energy estimates for wave equations on high dimensional Schwarzschild space-times*, Proc. Amer. Math. Soc., to appear

The Laul–Metcalfe construction has the attractive feature that, following [**115**], it avoids the angular frequency localisation of [**65**] in an alternative way from our own method, introduced in [**67**], of combining multipliers with commutation by angular momentum operators.

10.10. Asymptotically-AdS spacetimes. (cf. Open problem 8)

The mathematical study of the wave and Klein–Gordon equation on general asymptotically-AdS spacetimes was initiated by Holzegel who proved uniform boundedness for solutions if the mass satisfied the Breitenlohner–Freedman bound:

G. Holzegel *On the massive wave equation on slowly rotating Kerr-AdS spacetimes* Commun. Math. Phys. **294** (2009), 169–197

The above problem is non-standard as the underlying spacetime is not globally hyperbolic. The issue of well-posedness must thus also be addressed and such a theorem was indeed obtained in

G. Holzegel *Well-posedness for the massive wave equation on asymptotically anti-de Sitter spacetimes*, arXiv:1103.0710

with a suitable boundary condition at infinity that ensures the finiteness of energy. The above work in particular shed new light on the Breitenlohner-Freedman bound, which now appears as the best constant in a Hardy inequality. An alternative approach to well-posedness has been given by Vasy:

A. Vasy *The wave equation on asymptotically Anti-de Sitter spaces*, to appear in Analysis and PDE

Finally, we mention that there is a range of mass parameters which admit an alternative boundary condition at infinity, and there is work in progress of C. Warnick which obtains well-posedness in that setting as well.

Most recently, in joint work of Holzegel and Smulevici, logarithmic decay has been shown for general solutions of Klein–Gordon on Kerr–AdS.

G. Holzegel and J. Smulevici *Decay properties of Klein–Gordon fields on Kerr–AdS spacetimes*, arXiv:1110.6794

In the Schwarzschild–AdS case, it is in fact shown that individual spherical harmonics decay exponentially. For general solutions made up of infinitely many such spherical harmonics, again, it is only shown that the solution decays logarithmically. **This slow decay result is expected to be sharp as a quantitative measure of decay**, in view of the conjectured existence of a sequence of quasinormal modes ω_i exponentially approaching the real axis as $\mathrm{Re}(\omega_i) \to \pm\infty$.

Previously, Holzegel–Smulevici had investigated the coupled spherically symmetric Einstein–Klein–Gordon system in a series of papers. After settling the well-posedness issue in

G. Holzegel and J. Smulevici *Self-gravitating Klein–Gordon fields in asymptotically Anti-de Sitter spacetimes*, Annales Henri Poincaré, to appear

they prove asymptotic stability of Schwarzschild–AdS in

G. Holzegel and J. Smulevici *Stability of Schwarzschild-AdS for the spherically symmetric Einstein–Klein–Gordon system*, arXiv:1103.3672,

in fact, small perturbations of Schwarzschild–AdS exponentially converge to Schwarzschild–AdS.

The spherically symmetric work was motivated by an older conjecture:

CONJECTURE 10.1 (Dafermos–Holzegel, 2006). *Schwarzschild–AdS is the end-state of generic initial data for the Einstein–Klein–Gordon system under spherical symmetry, including those data which are arbitrarily small.* **In particular, pure AdS would be dynamically unstable.**[75]

This was motivated on the one hand by the existence of an infinite sequence of stationary solutions of the wave equation on pure AdS (and thus the lack of a dispersive mechanism), and on the other hand, the fact that in spherical symmetry the presence of the horizon provides an effective route for dispersion. Following Holzegel–Smulevici's work, numerical evidence for this behaviour was obtained by Bizoń–Rostworowski in

P. Bizoń and A. Rostworowski *On weakly turbulent instability of anti-de Sitter space*, Phys. Rev. Lett. **107**:031102, 2011

The above paper also gives a more detailed heuristic analysis of this instability from the point of second order perturbation theory.

One should not be fooled, however, by the spherically symmetric picture, where trapped surface formation guarantees then exponential convergence to Schwarzschild–AdS! In view of the slow decay result for general solutions of the wave equation Kerr–AdS, this suggests that when non-spherically symmetric perturbations are allowed, then Kerr–AdS should be subject to *the same instability considerations as pure AdS*. In view of this, Holzegel–Smulevici conjecture

CONJECTURE 10.2 (Holzegel–Smulevici). *All asymptotically AdS vacuum spacetimes are non-linearly unstable.*

10.11. Gravitational perturbations. (cf. Open problem 6)

As discussed in Section 3, the wave equation is a "poor man's" linearisation for the Einstein equations themselves. The actual linearisation carries tensorial structure–and the nature of this structure is still poorly understood.

One of the main difficulties of the linearised Einstein equations is that they do not carry an obvious analogue of the energy-momentum tensor from which to construct conserved currents. The situation is actually somewhat clearer when one considers the full Einstein equations, but allows a priori assumptions on the "spin coefficients", which one does not try to retrieve.[76] This approach has been considered by Holzegel in:

G. Holzegel *Ultimately Schwarzschildean spacetimes and the black hole stability problem*, arXiv:1010.3216

In this setting, the curvature tensor of the spacetime satisfies the Bianchi equations and thus admits an energy defined by the Bel-Robinson tensor. (Note in contrast that when linearising the Einstein equations around Schwarzschild or Kerr,

[75]See M. Dafermos, *The Black Hole Stability Problem.* Newton Institute, Cambridge, 2006 `http://www.newton.ac.uk/webseminars/pg+ws/2006/gmx/1010/dafermos/` and M. Dafermos and G. Holzegel, *Dynamic instability of solitons in* 4+1 *dimensional gravity with negative cosmological constant*, unpublished manuscript, 2006

[76]This can be viewed as the "non-linear PDEer's" linearisation, familiar from bootstrap arguments.

the "linearised" curvature tensor will not satisfy the Bianchi equations.) Of course, as this is a fully dynamic spacetime without a Killing field, this energy does not lead to a conserved current, but generates a divergence which can be understood geometrically in terms of contractions with the deformation tensor of a suitable vector field. Using this setup, Holzegel is able to prove a conditional decay result.

The above work of Holzegel contains many other results of independent interest, including, a generalisation of the red-shift estimates of Section 7 for the Einstein equations themselves as well as a generalisation of the method of Section 10.5, using it to capture peeling properties as well as a version of the null condition. In particular, the latter may suggest yet another approach to the proof of stability of Minkowski space.

Another result which, though far easier than the stability problem, gives insight into its novel nonlinear and tensorial aspects, is the problem of constructing non-trivial examples of spacetimes which asymptote to Schwarzschild or Kerr, parameterised by free "scattering" data on the event horizon and null infinity. We have very recently obtained precisely such a result, in collaboration with Holzegel:

M. Dafermos, G. Holzegel, and I. Rodnianski *Construction of ultimately Schwarzschild and Kerr spacetimes*, in preparation

Most interestingly, the above work in particular identifies how to renormalise both the optical structure equations and the Bianchi equations so as to capture approach to a particular Schwarzschild or Kerr solution. Through this renormalisation, energies can be constructed which involve only those quantities that radiate away.

Let us mention also that, like in the stability problem, the above work requires capturing an appropriate version of peeling and the null condition, and this is accomplished directly at the level of the Bianchi equations, using an adaptation of the method of Section 10.5, following also the previous work of Holzegel referred to above.

Finally, an interesting twist in this "scattering" problem is that the redshift, which throughout these notes has always played the role of a stabilising mechanism, now works against us. For in trying to solve the problem backwards, one confronts the positivity computation of Section 7 as a *blue-shift effect*! To counterbalance this effect, in order to construct our spacetimes in the above work, one must impose *exponential* approach to Schwarzschild or Kerr *along* the event horizon and *along* null infinity.[77]

For solutions evolving from generic initial data near Schwarzschild or Kerr, now imposed on an asymptotically flat *Cauchy surface*, on the basis of the conjectured sharpness of the inverse polynomial decay rates for wave equations along[78] $\mathcal{I}^+$ and $\mathcal{H}^+$ obtained as in Section 10.5 or 10.7, one expects that the radiation fields along

[77]We stress however that we are *not* imposing additional decay *towards* null infinity. The decay in r corresponds precisely to the decay one obtains in the "forward problem", and thus the free scattering data of the problem corresponds precisely to the scattering data induced by general solutions of the "forward" problem from the point of view of their functional freedom. Additional decay in r would effectively force the scattering data at null infinity to vanish.

[78]Decay along null infinity or the event horizon is related to the improved decay rates in the interior. We stress again, as in the previous footnote, that this is not the decay rate in r towards null infinity, which is also of course polynomial.

null infinity and the dynamic fields on the event horizon should decay **polynomially**, not exponentially. The estimates of the above work strongly suggest, however, that if one were to start with "generic" such power-law decaying scattering data, and attempt to solve backwards, the solution would **not** exist up to an asymptotically flat Cauchy surface. This in turn suggests that the characterization of the set of solutions arising from regular asymptotically flat Cauchy data is **not** encoded in the falloff rate of scattering data along $\mathcal{I}^+$ and $\mathcal{H}^+$ alone, but in non-local correlations between these two sets of data. Thus, as a matter of principle, the "generic" case from the forward perspective is not easily captured when starting from scattering data at $\mathcal{H}^+$ and $\mathcal{I}^+$, and the type of result proven in the above paper can be expected to be an optimal result of its kind.

Appendix A. Lorentzian geometry

The reader who wishes a formal introduction to Lorentzian geometry can consult [**91**]. For the reader familiar with the concepts and notations of Riemannian geometry, the following remarks should suffice for a quick introduction.

A.1. The Lorentzian signature. Lorentzian geometry is defined as in Riemannian geometry, except that the metric g is not assumed positive definite, but of signature $(-,+,\ldots,+)$. That is to say, we assume that at each point $p \in \mathcal{M}^{n+1}$,[79] we may find a basis $\mathbf{e}_i$ of the tangent space $T_p\mathcal{M}$, $i = 0,\ldots,n$, such that

$$g = -\mathbf{e}_0 \otimes \mathbf{e}_0 + \mathbf{e}_1 \otimes \mathbf{e}_1 + \cdots + \mathbf{e}_n \otimes \mathbf{e}_n.$$

In Riemannian geometry, the $-$ in the first term on the right hand side would be $+$.

A non-zero vector $v \in T_p\mathcal{M}$ is called *timelike*, *spacelike*, or *null*, according to whether $g(v,v) < 0$, $g(v,v) > 0$, or $g(v,v) = 0$. Null and timelike vectors collectively are known as *causal*. There are various conventions for the 0-vector. Let us not concern ourselves with such issues here.

The appellations timelike, spacelike, null are inherited by vector fields and immersed curves by their tangent vectors, i.e. a vector field V is timelike if $V(p)$ is timelike, etc., and a curve γ is timelike if $\dot\gamma$ is timelike, etc. On the other hand, a submanifold $\Sigma \subset \mathcal{M}$ is said to be spacelike if its induced geometry is Riemannian, timelike if its induced geometry is Lorentzian, and null if its induced geometry is degenerate. (Check that these two definitions coincide for embedded curves.) For a codimension-1 submanifold $\Sigma \subset \mathcal{M}$, at every $p \in M$, there exists a non-zero normal n^μ, i.e. a vector in $T_p\mathcal{M}$ such that $g(n,v) = 0$ for all $v \in T_p\Sigma$. It is easily seen that Σ is spacelike iff n is timelike, Σ is timelike iff n is spacelike, and Σ is null iff n is null. Note that in the latter case $n \in T_p\Sigma$. The normal of Σ is thus tangent to Σ.

A.2. Time-orientation and causality. A *time-orientation* on $(\mathcal{M}, g)$ is defined by an equivalence class $[K]$ where K is a continuous timelike vector field, where $K_1 \sim K_2$ if $g(K_1,K_2) < 0$. A Lorentzian manifold admitting a time-orientation is called *time-orientable*, and a triple $(\mathcal{M}, g, [K])$ is said to be a *time-oriented* Lorentzian manifold. Sometimes one reserves the use of the word "spacetime" for such triples. In any case, we shall always consider time-oriented Lorentzian manifolds and often drop explicit mention of the time orientation.

[79] It is conventional to denote the dimension of the manifold by $n+1$.

Given this, we may further partition causal vectors as follows. A causal vector v is said to be *future-pointing* if $g(v, K) < 0$, otherwise *past-pointing*, where K is a representative for the time orientation. As before, these names are inherited by causal curves, i.e. we may now talk of a *future-directed* timelike curve, etc. Given p, we define the *causal future* $J^+(p)$ by

$$J^+(p) = p \cup \{q \in \mathcal{M} : \exists \gamma : [0,1] \to \mathcal{M} : \dot\gamma \text{ future-pointing, causal}\}$$

Similarly, we define $J^-(p)$ where future is replaced by past in the above. If $S \subset \mathcal{M}$ is a set, then we define

$$J^\pm(S) = \cup_{p \in S} J^\pm(p).$$

A.3. Covariant derivatives, geodesics, curvature. The standard local notions of Riemannian geometry carry over. In particular, one defines the Christoffel symbols

$$\Gamma^\mu_{\nu\lambda} = \frac{1}{2} g^{\mu\alpha}(\partial_\nu g_{\alpha\lambda} + \partial_\lambda g_{\nu\alpha} - \partial_\alpha g_{\nu\lambda}),$$

and geodesics $\gamma(t) = (x^\alpha(t))$ are defined as solutions to

$$\ddot{x}^\mu + \Gamma^\mu_{\nu\lambda}\dot{x}^\nu \dot{x}^\lambda = 0.$$

Here $g_{\mu\nu}$ denote the components of g with respect to a local coordinate system x^μ, $g^{\mu\nu}$ denotes the components of the inverse metric, and we are applying the Einstein summation convention where repeated upper and lower indices are summed. The Christoffel symbols allow us to define the *covariant derivative* on (k, l) tensor fields by

$$\nabla_\lambda A^{\nu_1\ldots\nu_k}_{\mu_1\ldots\mu_\ell} = \partial_\lambda A^{\nu_1\ldots\nu_k}_{\mu_1\ldots\mu_\ell} + \sum_{i=1}^{k} \Gamma^{\nu_i}_{\lambda\rho} A^{\nu_1\ldots\rho\ldots\nu_k}_{\mu_1\ldots\mu_\ell} - \sum_{i=1}^{l} \Gamma^{\rho}_{\lambda\mu_i} A^{\nu_1\ldots\nu_k}_{\mu_1\ldots\rho\ldots\mu_\ell}$$

where it is understood that ρ replaces ν_i, μ_i, respectively in the two terms on the right. This defines $(k, l+1)$ tensor. As usual, if we contract this with a vector v at p, then we will denote this operator as ∇_v and we note that this can be defined in the case that the tensor field is defined only on a curve tangent to v at p. We may thus express the geodesic equation as

$$\nabla_{\dot\gamma}\dot\gamma = 0.$$

The *Riemann curvature tensor* is given by

$$R^\mu_{\nu\lambda\rho} \doteq \partial_\lambda \Gamma^\mu_{\rho\nu} - \partial_\rho \Gamma^\mu_{\lambda\nu} + \Gamma^\alpha_{\rho\nu}\Gamma^\mu_{\lambda\alpha} - \Gamma^\alpha_{\lambda\nu}\Gamma^\mu_{\rho\alpha},$$

and the *Ricci* and *scalar* curvatures by

$$R_{\mu\nu} \doteq R^\alpha_{\mu\alpha\nu}, \qquad R \doteq g^{\mu\nu} R_{\mu\nu}.$$

Using the same letter R to denote all these tensors is conventional in relativity, the number of indices indicating which tensor is being referred to. For this reason we will avoid writing "the tensor R". The expression R without indices will always denote the scalar curvature. As usual, we shall also use the letter R with indices to denote the various manifestations of these tensors with indices raised and lowered by the inverse metric and metric, e.g.

$$R_{\mu\nu\lambda\rho} = g_{\mu\alpha} R^\alpha_{\nu\lambda\rho}$$

Note the important formula

$$\nabla_\alpha \nabla_\beta Z_\mu - \nabla_\alpha \nabla_\beta Z_\mu = R_{\sigma\mu\alpha\beta} Z^\sigma$$

We say that an immersed curve $\gamma : I \to \mathcal{M}$ is *inextendible* if there does not exist an immersed curve $\tilde{\gamma} : J \to \mathcal{M}$ where $J \supset I$ and $\tilde{\gamma}|_I = \gamma$.

We say that $(\mathcal{M}, g)$ is *geodesically complete* if for all inextendible geodesics $\gamma : I \to \mathcal{M}$, then $I = \mathbb{R}$. Otherwise, we say that it is *geodesically incomplete*. We can similarly define the notion of spacelike geodesic (in)completeness, timelike geodesic completeness, causal geodesic completeness, etc, by restricting the definition to such geodesics. In the latter two cases, we may further specialise, e.g. to the notion of *future causal geodesic completeness*, by replacing the condition $I = \mathbb{R}$ with $I \supset (a, \infty)$ for some a.

We say that a spacelike hypersurface $\Sigma \subset \mathcal{M}$ is *Cauchy* if every inextendible causal curve in $\mathcal{M}$ intersects it precisely once. A spacetime $(\mathcal{M}, g)$ admitting such a hypersurface is called *globally hyperbolic*. This notion was first introduced by Leray [**112**].

Appendix B. The Cauchy problem for the Einstein equations

We outline here for reference the basic framework of the Cauchy problem for the Einstein equations

$$R_{\mu\nu} - \frac{1}{2} g_{\mu\nu} R + \Lambda g_{\mu\nu} = 8\pi T_{\mu\nu}. \tag{110}$$

Here Λ is a constant known as the *cosmological constant* and $T_{\mu\nu}$ is the so-called energy momentum tensor of matter. We will consider mainly the vacuum case

$$R_{\mu\nu} = \Lambda g_{\mu\nu}, \tag{111}$$

where the system closes in itself. If the reader wants to set $\Lambda = 0$, he should feel free to do so. To illustrate the case of matter, we will consider the example of a scalar field.

B.1. The constraint equations. Let Σ be a spacelike hypersurface in $(\mathcal{M}, g)$, with future directed unit timelike normal N. By definition, Σ inherits a Riemannian metric from g. On the other hand, we can define the so-called second fundamental form of Σ to be the symmetric covariant 2-tensor in $T\Sigma$ defined by

$$K(u, v) = -g(\nabla_u V, N)$$

where V denotes an arbitrary extension of v to a vector field along Σ, and ∇ here denotes the connection of g. As in Riemannian geometry, one easily shows that the above indeed defines a tensor on $T\Sigma$, and that it is symmetric.

Suppose now $(\mathcal{M}, g)$ satisfies (110) with some tensor $T_{\mu\nu}$. With Σ as above, let $\bar{g}_{ab}$, $\bar{\nabla}$, K_{ab} denote the induced metric, connection, and second fundamental form, respectively, of Σ. Let barred quantities and Latin indices refer to tensors, curvature, etc., on Σ, and let $\Pi^\nu_a(p)$ denote the components of the pullback map $T^*\mathcal{M} \to T^*\Sigma$. It follows that

$$\bar{R} + (K^a_a)^2 - K^a_b K^b_a = 16\pi\, T_{\mu\nu} n^\mu n^\nu + 2\Lambda, \tag{112}$$

$$\nabla_b K^b_a - \nabla_a K^b_b = 16\pi\, \Pi^\nu_a T_{\mu\nu} n^\mu. \tag{113}$$

To see this, one derives as in Riemannian geometry the Gauss and Codazzi equations, take traces, and apply (110).

B.2. Initial data. It is clear that (112), (113) are *necessary conditions* on the induced geometry of a spacelike hypersurface Σ so as to arise as a hypersurface in a spacetime satisfying (110). As we shall see, immediately, they will also be sufficient conditions for solving the initial value problem.

B.2.1. *The vacuum case.* Let Σ be a 3-manifold, $\bar{g}$ a Riemannian metric on Σ, and K a symmetric covariant 2-tensor. We shall call $(\Sigma, \bar{g}, K)$ a *vacuum initial data set with cosmological constant* Λ if (112)–(113) are satisfied with $T_{\mu\nu} = 0$. Note that in this case, equations (112)–(113) refer only to Σ, $\bar{g}$, K.

B.2.2. *The case of matter.* Let us here provide only the case for the Einstein-scalar field case. Here, the system is (110) coupled with

$$\Box_g \psi = 0, \tag{114}$$

$$T_{\mu\nu} = \partial_\mu \psi \partial_\nu \psi - \frac{1}{2} g_{\mu\nu} \nabla^\alpha \psi \nabla_\alpha \psi. \tag{115}$$

First note that were Σ a spacelike hypersurface in a spacetime $(\mathcal{M}, g)$ satisfying the Einstein-scalar field system with massless scalar field ψ, and n^μ were the future-directed normal, then setting $\psi' = n^\mu \partial_\mu \phi$, $\psi = \phi|_\Sigma$ we have

$$T_{\mu\nu} n^\mu n^\nu = \frac{1}{2}((\psi')^2 + \bar{\nabla}^a \psi \bar{\nabla}_a \psi),$$

$$\Pi_a^\nu T_{\mu\nu} n^\mu = \psi' \bar{\nabla}_a \psi,$$

where latin indices and barred quantities refer to Σ and its induced metric and connection.

This motivates the following: Let Σ be a 3-manifold, $\bar{g}$ a Riemannian metric on Σ, K a symmetric covariant 2-tensor, and $\psi : \Sigma \to \mathbb{R}$, $\psi' : \Sigma \to \mathbb{R}$ functions. We shall call $(\Sigma, \bar{g}, K)$ an *Einstein-scalar field initial data set with cosmological constant* Λ if (112)–(113) are satisfied replacing $T_{\mu\nu} n^\mu n^\nu$ with $\frac{1}{2}((\psi')^2 + \bar{\nabla}^a \psi \bar{\nabla}_a \psi)$, and replacing $\Pi_a^\nu T_{\mu\nu} n^\mu$ with $\psi' \bar{\nabla}_a \psi$.

Note again that with the above replacements the equations (112)–(113) do not refer to an ambient spacetime $\mathcal{M}$. See [**36**] for the construction of solutions to this system.

B.2.3. *Asymptotic flatness and the positive mass theorem.* The study of the Einstein constraint equations is non-trivial!

Let us refer in this section to a triple $(\Sigma, \bar{g}, K)$ where Σ is a 3-manifold, $\bar{g}$ a Riemannian metric, and K a symmetric two-tensor on Σ as an *initial data set*, even though we have not specified a particular closed system of equations. An initial data set $(\Sigma, \bar{g}, K)$ is *strongly asymptotically flat with one end* if there exists a compact set $\mathcal{K} \subset \Sigma$ and a coordinate chart on $\Sigma \setminus \mathcal{K}$ which is a diffeomorphism to the complement of a ball in $\mathbb{R}^3$, and for which

$$g_{ab} = \left(1 + \frac{2M}{r}\right) \delta_{ab} + o_2(r^{-1}), \qquad k_{ab} = o_1(r^{-2}),$$

where δ_{ab} denotes the Euclidean metric and r denotes the Euclidean polar coordinate.

In appropriate units, M is the "mass" measured by asymptotic observers, when comparing to Newtonian motion in the frame δ_{ab}. On the other hand, under the assumption of a global coordinate system well-behaved at infinity, M can be computed by integration of the t_0^0 component of a certain pseudotensor[80] added to T_0^0. In this manifestation, the quantity $E = M$ is known as the *total energy.*[81] This relation was first studied by Einstein and is discussed in Weyl's book Raum-Zeit-Materie [**154**]. If one looks at E for a family of hypersurfaces with the above asymptotics, then E is conserved.

A celebrated theorem of Schoen-Yau [**137, 138**] (see also [**156**]) states

THEOREM B.1. *Let $(\Sigma, \bar{g}, K)$ be strongly asymptotically flat with one end and satisfy* (112), (113) *with $\Lambda = 0$, and where $T_{\mu\nu}n^\mu n^\nu$, $\Pi_a^\nu T_{\mu\nu}n^\mu$ are replaced by the scalar μ and the tensor J_a, respectively, defined on Σ, such that moreover $\mu \geq \sqrt{J^a J_a}$. Suppose moreover the asymptotics are strengthened by replacing $o_2(r^{-1})$ by $O_4(r^{-2})$ and $o_1(r^{-2})$ by $O_3(r^{-3})$. Then $M \geq 0$ and $M = 0$ iff Σ embeds isometrically into $\mathbb{R}^{3+1}$ with induced metric $\bar{g}$ and second fundamental form K.*

The assumption $\mu \geq \sqrt{J^a J_a}$ holds if the matter satisfies the *dominant energy condition* [**91**]. In particular, it holds for the Einstein scalar field system of Section B.2.2, and (of course) for the vacuum case. The statement we have given above is weaker than the full strength of the Schoen-Yau result. For the most general assumptions under which mass can be defined, see [**9**].

One can define the notion of *strongly asymptotically flat with k ends* by assuming that there exists a compact $\mathcal{K}$ such that $\Sigma \setminus \mathcal{K}$ is a disjoint union of k regions possessing a chart as in the above definition. The Cauchy surface Σ of Schwarzschild of Kerr with $0 \leq |a| < M$, can be chosen to be strongly asymptotically flat with 2-ends. The mass of both ends coincides with the parameter M of the solution.

The above theorem applies to this case as well for the parameter M associated to any end. If $M = 0$ for one end, then it follows by the rigidity statement that there is only one end. Note why Schwarzschild with $M < 0$ does not provide a counterexample.

The association of "naked singularities" with negative mass Schwarzschild gave the impression that the positive energy theorem protects against naked singularities. This is not true! See the examples discussed in Section 2.6.2.

In the presence of black holes, one expects a strengthening of the lower bound on mass in Theorem B.1 to include a term related to the square root of the area of a cross section of the horizon. Such inequalities were first discussed by Penrose [**127**] with the Bondi mass in place of the mass defined above. All inequalities of this type are often called *Penrose inequalities*. It is not clear what this term should be, as the horizon is only identifiable after global properties of the maximal development have been understood. Thus, one often replaces this area in the conjectured inequality with the area of a suitably defined apparent horizon. Such a statement has indeed been obtained in the so-called Riemannian case (corresponding to $K = 0$) where the relevant notion of apparent horizon coincides with that of minimal surface. See the important papers of Huisken–Ilmanen [**95**] and Bray [**25**].

---

[80]This is subtle: The Einstein vacuum equations arise from the Hilbert Lagrangian $\mathcal{L}(g) = \int R$ which is 2nd order in the metric. In local coordinates, the highest order term is a divergence, and the Lagrangian can thus be replaced by a new Lagrangian which is 1st order in the metric. The resultant Lagrangian density, however, is no longer coordinate invariant. The quantity t_0^0 now arises from "Noether's theorem" [**124**]. See [**49**] for a nice discussion.

[81]With the above asymptotics, the so-called linear momentum vanishes. Thus, in this case "mass" and energy are equivalent.

B.3. The maximal development. Let $(\Sigma, \bar{g}, K)$ denote a smooth vacuum initial data set with cosmological constant Λ. We say that a smooth spacetime $(\mathcal{M}, g)$ is a smooth *development of initial data* if

(1) $(\mathcal{M}, g)$ satisfies the Einstein vacuum equations (4) with cosmological constant Λ.
(2) There exists a smooth embedding $i : \Sigma \to \mathcal{M}$ such that $(\mathcal{M}, g)$ is globally hyperbolic with Cauchy surface $i(\Sigma)$, and $\bar{g}$, K are the induced metric and second fundamental form, respectively.

The original local existence and uniqueness theorems were proven in 1952 by Choquet-Bruhat [**33**].[82] In modern language, they can be formulated as follows

THEOREM B.2. *Let $(\Sigma, \bar{g}, K)$ be as in the statement of the above theorem. Then there exists a smooth development $(\mathcal{M}, g)$ of initial data.*

THEOREM B.3. *Let $\mathcal{M}$, $\widetilde{\mathcal{M}}$ be two smooth developments of initial data. Then there exists a third development $\mathcal{M}'$ and isometric embeddings $j : \mathcal{M}' \to \mathcal{M}$, $\tilde{j} : \mathcal{M}' \to \widetilde{\mathcal{M}}$ commuting with i, $\tilde{i}$.*

Application of Zorn's lemma, the above two theorems and simple facts about Lorentzian causality yields:

THEOREM B.4. *(Choquet-Bruhat–Geroch [**35**]) Let $(\Sigma, \bar{g}, K)$ denote a smooth vacuum initial data set with cosmological constant Λ. Then there exists a unique development of initial data $(\mathcal{M}, g)$ satisfying the following maximality statement: If $(\widetilde{\mathcal{M}}, \tilde{g})$ satisfies (1), (2) with embedding $\tilde{i}$, then there exists an isometric embedding $j : \widetilde{\mathcal{M}} \to \mathcal{M}$ such that j commutes with $\tilde{i}$.*

The spacetime $(\mathcal{M}, g)$ is known as the *maximal development* of $(\Sigma, \bar{g}, K)$. The spacetime $\mathcal{M} \cap J^+(\Sigma)$ is known as the *maximal future development* and $\mathcal{M} \cap J^-(\Sigma)$ the *maximal past development.*

We have formulated the above theorems in the class of smooth initial data. They are of course proven in classes of finite regularity. There has been much recent work in proving a version of Theorem B.2 under minimal regularity assumptions. The current state of the art requires only $\bar{g} \in H^{2+\epsilon}$, $K \in H^{1+\epsilon}$. See [**102**].

We leave as an **exercise** formulating the analogue of Theorem B.4 for the Einstein-scalar field system (110), (114), (115), where the notion of initial data set is that given in Section B.2.2.

B.4. Harmonic coordinates and the proof of local existence. The statements of Theorems B.2 and B.3 are coordinate independent. Their proofs, however, require fixing a gauge which determines the form of the metric functions in coordinates from initial data. The classic gauge is the so-called *harmonic* gauge[83]. Here the coordinates x^μ are required to satisfy

$$\Box_g x^\mu = 0. \tag{116}$$

Equivalently, this gauge is characterized by the condition

$$g^{\mu\nu}\Gamma^\alpha_{\mu\nu} = 0. \tag{117}$$

[82]Then called Fourès-Bruhat.
[83]also known as wave coordinates

A linearised version of these coordinates was used by Einstein [**74**] to predict gravitational waves. It appears that de Donder [**70**] was the first to consider harmonic coordinates in general. These coordinates are discussed extensively in the book of Fock [**82**].

The result of Theorem B.3 actually predates Theorem B.2, and in some form was first proven by Stellmacher [**143**]. Given two developments $(\mathcal{M}, g)$, $(\widetilde{\mathcal{M}}, \tilde{g})$ one constructs for each harmonic coordinates x^μ, $\tilde{x}^\mu$ adapted to Σ, such that $g_{\mu\nu} = \tilde{g}_{\mu\nu}$, $\partial_\lambda g_{\mu\nu} = \partial_\lambda \tilde{g}_{\mu\nu}$ along Σ. In these coordinates, the Einstein vacuum equations can be expressed as

$$\Box_g g^{\mu\nu} = Q^{\mu\nu,\alpha\beta}_{\iota\kappa\lambda\rho\sigma\tau}\, g^{\iota\kappa}\, \partial_\alpha g^{\lambda\rho}\, \partial_\beta etag^{\sigma\tau} \tag{118}$$

for which uniqueness follows from general results of Schauder [**136**]. This theorem gives in addition a domain of dependence property.[84]

Existence for solutions of the system (118) with smooth initial data would also follow from the results of Schauder [**136**]. This does not immediately yield a proof of Theorem B.2, because one does not have a priori the spacetime metric g so as to impose (116) or (117)! The crucial observation is that if (117) is true "to first order" on Σ, and g is defined to be the unique solution to (118), then (117) will hold, and thus, g will solve (110). Thus, to prove Theorem B.2, it suffices to show that one can arrange for (117) to be true "to first order" initially. Choquet-Bruhat [**33**] showed that this can be done precisely when the constraint equations (112)–(113) are satisfied with vanishing right hand side. Interestingly, to obtain existence for (118), Choquet-Bruhat's proof [**33**] does not in fact appeal to the techniques of Schauder [**136**], but, following Sobolev, rests on a Kirchhoff formula representation of the solution. Recently, new representations of this type have found applications to refined extension criteria [**103**].

An interesting feature of the classical existence and uniqueness proofs is that Theorem B.3 requires more regularity than Theorem B.2. This is because solutions of (116) are a priori only as regular as the metric. This difficulty has recently been overcome in [**129**].

B.5. Stability of Minkowski space. The most celebrated global result on the Einstein equations is the stability of Minkowski space, first proven in monumental work of Christodoulou and Klainerman [**51**]:

THEOREM B.5. *Let $(\Sigma, \bar{g}, K)$ be a strongly asymptotically flat vacuum initial data set, assumed sufficiently close to Minkowski space in a weighted sense. Then the maximal development is geodesically complete, and the spacetime approaches Minkowski space (with quantitative decay rates) in all directions. Moreover, a complete future null infinity $\mathcal{I}^+$ can be attached to the spacetime such that $J^-(\mathcal{I}^+) = \mathcal{M}$.*

The above theorem also allows one to rigorously define the laws of gravitational radiation. These laws are nonlinear even at infinity. Theorem B.5 led to the discovery of Christodoulou's memory effect [**42**].

A new proof of a version of stability of Minkowski space using harmonic coordinates has been given in [**113**]. This has now been extended in various directions

[84]There is even earlier work on uniqueness in the analytic category going back to Hilbert, appealing to Cauchy-Kovalevskaya. Unfortunately, nature is not analytic; in particular, one cannot infer the domain of dependence property from those considerations.

in [**34**]. The original result [**51**] was extended to the Maxwell case in the Ph.D. thesis of Zipser [**157**]. Bieri [**13**] has very recently given a proof of a version of stability of Minkowski space under weak asymptotics and regularity assumptions, following the basic setup of [**51**].

There was an earlier semi-global result of Friedrich [**83**] where initial data were prescribed on a hyperboloidal initial hypersurface meeting $\mathcal{I}^+$.

A common misconception is that it is the positivity of mass which is somehow responsible for the stability of Minkowski space. The results of [**113**] for this are very telling, for they apply not only to the Einstein-vacuum equations, but also to the Einstein-scalar field system of Section B.2.2, including the case where the definition of the energy-momentum tensor (115) is replaced with its negative. Minkowski space is then not even a local minimiser for the mass functional in the class of perturbations allowed! Nonetheless, by the results of [**113**], Minkowski space is still stable in this context.

Another point which cannot be overemphasised: It is essential that the smallness in (B.5) concern a weighted norm. Compare with the results of Section 2.8.

Stability of Minkowski space is the only truly global result on the maximal development which has been obtained for asymptotically flat initial data without symmetry. There are a number of important results applicable in cosmological settings, due to Friedrich [**83**], Andersson-Moncrief [**3**], and most recently Ringstrom [**134**].

Other than this, our current global understanding of solutions to the Einstein equations (in particular all work on the cosmic censorship conjectures) has been confined to solutions under symmetry. We have given many such references in the asymptotically flat setting in the course of Section 2. The cosmological setting is beyond the scope of these notes, but we refer the reader to the recent review article and book of Rendall [**132, 133**] for an overview and many references.

Appendix C. The divergence theorem

Let $(\mathcal{M}, g)$ be a spacetime, and let Σ_0, Σ_1 be homologous spacelike hypersurfaces with common boundary, bounding a spacetime region $\mathcal{B}$, with $\Sigma_1 \subset J^+(\Sigma_0)$. Let n_0^μ, n_1^μ denote the future unit normals of Σ_0, Σ_1 respectively, and let P_μ denote a one-form. Under our convention on the signature, the divergence theorem takes the form

$$\int_{\Sigma_1} P_\mu n_1^\mu + \int_{\mathcal{B}} \nabla^\mu P_\mu = \int_{\Sigma_0} P_\mu n_0^\mu, \tag{119}$$

where all integrals are with respect to the *induced volume form*.

This is defined as follows. The volume form of spacetime is

$$\sqrt{-\det g}\,dx^0 \ldots dx^n$$

where $\det g$ denotes the determinant of the matrix $g_{\alpha\beta}$ in the above coordinates. The induced volume form of a spacelike hypersurface is defined as in Riemannian geometry.

We will also consider the case where (part of) Σ_1 is null. Then, we choose arbitrarily a future-directed null generator n_1^Σ for Σ_1 arbitrarily and define the volume element so that the divergence theorem applies. For instance the divergence theorem in the region $\mathcal{R}(\tau', \tau'')$ (described in the lectures) for an arbitrary current

P_μ then takes the form

$$\int_{\Sigma_{\tau''}} P_\mu n^\mu_{\Sigma_{\tau''}} + \int_{\mathcal{H}(\tau',\tau'')} P_\mu n^\mu_{\mathcal{H}} + \int_{\mathcal{R}(\tau',\tau'')} \nabla^\mu P_\mu = \int_{\Sigma_{\tau'}} P_\mu n^\mu_{\Sigma_{\tau'}},$$

where the volume elements are as described.

Note how the form of this theorem can change depending on sign conventions regarding the directions of the normal, the definition of the divergence and the signature of the metric.

Appendix D. Vector field multipliers and their currents

Let ψ be a solution of

$$\Box_g \psi = 0 \tag{120}$$

on a Lorentzian manifold $(\mathcal{M}, g)$. Define

$$\mathbf{T}_{\mu\nu}(\psi) = \partial_\mu\psi\partial_\nu\psi - \frac{1}{2}g_{\mu\nu}\partial^\alpha\psi\partial_\alpha\psi \tag{121}$$

We call $T_{\mu\nu}$ the *energy-momentum* tensor of ψ.[85] Note the symmetry property

$$\mathbf{T}_{\mu\nu} = \mathbf{T}_{\nu\mu}.$$

The wave equation (120) implies

$$\nabla^\mu \mathbf{T}_{\mu\nu} = 0. \tag{122}$$

Given a vector field V^μ, we may define the associated currents

$$J^V_\mu(\psi) = V^\nu \mathbf{T}_{\mu\nu}(\psi) \tag{123}$$

$$K^V = {}^V\pi_{\mu\nu}\mathbf{T}^{\mu\nu}(\psi) \tag{124}$$

where ${}^V\pi$ is the deformation tensor defined by

$${}^V\pi_{\mu\nu} = \frac{1}{2}\nabla_{(\mu}V_{\nu)} = \frac{1}{2}(\mathcal{L}_V g)_{\mu\nu}.$$

The identity (122) gives

$$\nabla^\mu J^V_\mu(\psi) = K^V(\psi).$$

Note that $J^V_\mu(\psi)$ and $K^V(\psi)$ both depend only on the 1-jet of ψ, yet the latter is the divergence of the former. Applying the divergence theorem (119), this allows one to relate quantities of the same order.

The existence of a tensor $\mathbf{T}_{\mu\nu}(\psi)$ satisfying (122) follows from the fact that equation (120) derives from a Lagrangian of a specific type. These issues were first systematically studied by Noether [**124**]. For more general such Lagrangian theories, two currents J_μ, K with $\nabla^\mu J_\mu = K$, both depending only on the 1-jet, but not necessarily arising from $\mathbf{T}_{\mu\nu}$ as above, are known as *compatible currents*. These have been introduced and classified by Christodoulou [**46**].

[85] Note that this is the same expression that appears on the right hand side of (110) in the Einstein-scalar field system. See Section B.2.2.

Appendix E. Vector field commutators

PROPOSITION E.0.1. *Let ψ be a solution of the equation of the scalar equation*

$$\Box_g \psi = f,$$

and X be a vectorfield. Then

$$\Box_g(X\psi) = X(f) + 2\,{}^X\pi^{\alpha\beta}\nabla_\alpha\nabla_\beta\psi + \left(2(\nabla^\alpha\,{}^X\pi_{\alpha\mu}) - (\nabla_\mu\,{}^X\pi^\alpha_\alpha)\right)\nabla^\mu\psi.$$

PROOF. To show this we write

$$X(\Box_g\psi) = \mathcal{L}_X(g^{\alpha\beta}\nabla_\alpha\nabla_\beta\psi) = -2\,{}^X\pi^{\alpha\beta}\nabla_\alpha\nabla_\beta\psi + g^{\alpha\beta}\mathcal{L}_X(\nabla_\alpha\nabla_\beta\psi).$$

Furthermore,

$$\mathcal{L}_X(\nabla_\alpha\nabla_\beta\psi) - \nabla_\alpha\mathcal{L}_X\nabla_\beta\psi = -\left((\nabla_\beta\,{}^X\pi_{\alpha\mu}) - (\nabla_\mu\,{}^X\pi_{\beta\alpha}) + (\nabla_\alpha\,{}^X\pi_{\mu\beta})\right)\nabla^\mu\psi$$

and

$$\mathcal{L}_X\nabla_\beta\psi = \nabla_X\nabla_\beta\psi + \nabla_\beta X^\mu\nabla_\mu\psi = \nabla_\beta(X\psi).$$

□

Appendix F. Some useful Schwarzschild computations

In this section, $(\mathcal{M}, g)$ refers to maximal Schwarzschild with $M > 0$, $\mathcal{Q} = \mathcal{M}/SO(3)$, $\mathcal{I}^\pm$, $J^\mp(\mathcal{I}^\pm)$ are as defined in Section 2.4.

F.1. Schwarzschild coordinates (r,t). The coordinates are (r,t) and the metric takes the form

$$-(1-2M/r)dt^2 + (1-2M/r)^{-1}dr^2 + r^2 d\sigma_{\mathbb{S}^2}.$$

These coordinates can be used to cover any of the four connected components of $\mathcal{Q}\setminus\mathcal{H}^\pm$. In particular, the region $J^-(\mathcal{I}^+_A)\cap J^+(\mathcal{I}^-_A)$ (where $\mathcal{I}^\pm_A$ correspond to a pair of connected components of $\mathcal{I}^\pm$ sharing a limit point in the embedding) is covered by a Schwarzschild coordinate system where $2M < r < \infty$, $-\infty < t < \infty$. Note that r has an invariant characterization namely $r(x) = \sqrt{\mathrm{Area(S)}/4\pi}$ where S is the unique group orbit of the SO(3) action containing x.[86]

The hypersurface $\{t = c\}$ in the Schwarzschild coordinate region $J^-(\mathcal{I}^+_A)\cap J^+(\mathcal{I}^-_A)$ extends regularly to a hypersurface with boundary in $\mathcal{M}$ where the boundary is precisely $\mathcal{H}^+\cap\mathcal{H}^-$.

The coordinate vector field ∂_t is Killing (and extends to the globally defined Killing field T).

In a slight abuse of notation, we will often extend Schwarzschild coordinate notation to $\mathcal{D}$, the closure of $J^-(\mathcal{I}^+_A)\cap J^+(\mathcal{I}^-_A)$. For instance, we may talk of the vector field ∂_t "on" $\mathcal{H}^\pm$, or of $\{t = c\}$ having boundary $\mathcal{H}^+\cap\mathcal{H}^-$, etc.

[86]Compare with the Minkowski case $M = 0$ where the SO(3) action is of course not unique.

F.2. Regge-Wheeler coordinates (r^*, t)**.** Here t is as before and

$$r^* = r + 2M\log(r - 2M) - 3M - 2M\log M \tag{125}$$

and the metric takes the form

$$-(1 - 2M/r)(-dt^2 + (dr^*)^2) + r^2 d\sigma_{\mathbb{S}^2}$$

where r is defined implicitly by (125). A coordinate chart defined in $-\infty < r^* < \infty$, $-\infty < t < \infty$ covers $J^-(\mathcal{I}_A^+) \cap J^+(\mathcal{I}_A^-)$.

The constant renormalisation of the coordinate is taken so that $r^* = 0$ at the photon sphere, where $r = 3M$.

Note the explicit form of the wave operator

$$\Box_g \psi = -(1 - 2M/r)^{-1}(\partial_t^2 \psi - r^{-2}\partial_{r^*}(r^2 \partial_{r^*}\psi)) + \nabla\!\!\!/^A \nabla\!\!\!/_A \psi$$

where $\nabla\!\!\!/$ denotes the induced covariant derivative on the group orbit spheres.

Similar warnings of abuse of notation apply, for instance, we may write $\partial_t = \partial_{r*}$ on $\mathcal{H}^+$.

F.3. Double null coordinates (u, v)**.** Our convention is to define

$$u = \frac{1}{2}(t - r^*),$$
$$v = \frac{1}{2}(t + r^*).$$

The metric takes the form

$$-4(1 - 2M/r)du dv + r^2 d\sigma_{\mathbb{S}^2}$$

and $J^-(\mathcal{I}_A^+) \cap J^+(\mathcal{I}_A^-)$ is covered by a chart $-\infty < u < \infty$, $-\infty < v < \infty$.

The usual comments about abuse of notation hold, in particular, we may now parametrize $\mathcal{H}^+ \cap \mathcal{D}$ with $\{\infty\} \times [-\infty, \infty)$ and similarly $\mathcal{H}^- \cap \mathcal{D}$ with $(-\infty, \infty] \times \{-\infty\}$, and write $\partial_v(-\infty, v) = \partial_t(-\infty, v)$, $\partial_u(-\infty, v) = 0$.

Note that the vector field $(1 - 2M/r)^{-1}\partial_u$ extends to a regular vector null field across $\mathcal{H}^+ \setminus \mathcal{H}^-$. Thus, with the basis ∂_v, $(1 - 2M/r)^{-1}\partial_u$, one can choose regular vector fields near $\mathcal{H}^+ \setminus \mathcal{H}^-$ without changing to regular coordinates. In practice, this can be convenient.

References

[1] S. Alinhac *Energy multipliers for perturbations of Schwarzschild metric* preprint, 2008

[2] H. Andréasson *On the Buchdahl inequality for spherically symmetric static shells* Comm. Math. Phys. **274** (2007), 399–408

[3] L. Andersson and V. Moncrief *Future complete vacuum spacetimes* In: The Einstein equations and the large slace behavior of gravitational fields, 299–330, Birkhäuser, Basel, 2004

[4] A. Bachelot *Gravitational scattering of electromagnetic field by Schwarzschild black-hole* Annales de l'I. H. P., section A, **54** (1991), p. 261–320

[5] A. Bachelot *Asymptotic completeness for the Klein-Gordon equation on the Schwarzschild metric*, Ann. Inst. H. Poincaré Phys. Théor. **16** (1994), no. 4, 411–441

[6] A. Bachelot *The Dirac system on the anti-de Sitter universe* Commun. Math. Phys. **283** (2008), 127–167

[7] A. Bachelot and A. Motet-Bachelot *Les résonances d'un trou noir de Schwarzschild* Ann. Inst. H. Poincaré Phys. Théor. **59** (1993), no. 1, 3–68.

[8] J. M. Bardeen, B. Carter, and S. W. Hawking *The four laws of black hole mechanics* Commun. Math. Phys. **31** (1973), 161

[9] R. Bartnik *The mass of an asymptotically flat manifold* Comm. Pure Appl. Math. **39** (1986), 661–693

[10] R. Beig and J. M. Heinzle *CMC-Slicings of Kottler-Schwarzschild-de Sitter Cosmologies*, Comm. Math. Phys., **260** (2005), 673–709

[11] J. D. Bekenstein *Generalized second law of thermodynamics in black hole physics*, Phys. Rev. D **9** (1974), 3292

[12] I. Bengtsson and J. M. M. Senovilla *A note on trapped surfaces in the Vaidya solution* arXiv:0809.2213

[13] L. Bieri *An extension of the stability theorem of the Minkowski space in general relativity*, Thesis, ETH, 2007

[14] H. Buchdahl *General relativistic fluid spheres* Phys. Rev. **116** (1959), 1027–1034

[15] P. Bizon, T. Chmaj, A. Rostworowski *Late-time tails of a self-gravitating scalar field revisited*, in preparation

[16] P. Bizon, T. Chmaj, B. Schmidt *Critical behavior in vacuum gravitational collapse in* $4+1$ *dimensions* Phys. Rev. Lett. **95** (2005), no. 7, 071102, 4 pp.

[17] P. Bizon, B. Schmidt *How to bypass Birkhoff through extra dimensions: a simple framework for investigating the gravitational collapse in vacuum* Internat. J. Modern Phys. D **15** (2006), 2217–2222

[18] P. Blue *Decay of the Maxwell field on the Schwarzschild manifold*, arXiv:0710.4102v1

[19] P. Blue and A. Soffer *Semilinear wave equations on the Schwarzschild manifold. I. Local decay estimates*, Adv. Differential Equations **8** (2003), no. 5, 595–614

[20] P. Blue and A. Soffer *Errata for "Global existence and scattering for the nonlinear Schrodinger equation on Schwarzschild manifolds", "Semilinear wave equations on the Schwarzschild manifold I: Local Decay Estimates", and "The wave equation on the Schwarzschild metric II: Local Decay for the spin 2 Regge Wheeler equation"*, gr-qc/0608073, 6 pages

[21] P. Blue and A. Soffer *Phase space analysis on some black hole manifolds*, preprint

[22] P. Blue and J. Sterbenz *Uniform decay of local energy and the semi-linear wave equation on Schwarzschild space* Comm. Math. Phys. **268** (2006), no. 2, 481–504

[23] P. Blue and J. Sterbenz *A note on the decay properties of linear and non-linear scalar fields on the Schwarzschild/Kruskal space-time*, preprint

[24] J.-F. Bony and D. Häfner *Decay and non-decay of the local energy for the wave equation in the de Sitter-Schwarzschild metric*, preprint 2007

[25] H. Bray *Proof of the Riemannian Penrose inequality using the positive mass theorem* J. Differential Geom. **59** (2001), 177–267

[26] P. Breitenlohner and D. Z. Freedman *Stability in gauged extended supergravity* Ann. Phys. **144** (1982), 249–281

[27] V. Cardoso and O. J. C. Dias *Small Kerr-anti-de Sitter black holes are unstable* Phys. Rev. D **70**, 084011 (2004)

[28] B. Carter *Black hole equilibrium states*, in Black Holes (Les Houches Lectures), edited B. S. DeWitt and C. DeWitt (Gordon and Breach, New York, 1972).

[29] B. Carter *Global structure of the Kerr family of gravitational fields* Phys. Rev. **174** (1968), 1559–1571

[30] B. Carter *Hamilton-Jacobi and Schrödinger separable solutions of Einstein's equations* Comm. Math. Phys. **10** (1968), 280–310

[31] E. S. C. Ching, P. T. Leung, W. M. Suen, and K. Young *Wave propagation in gravitational systems* Phys. Rev. D **52** (1995)

[32] Z. W. Chong, G. W. Gibbons, H. Lu, and C. N. Pope *Separability and Killing tensors in Kerr-Taub-NUT-de Sitter metrics in higher dimensions* Phys. Lett. **B609** 124–132 (2005)

[33] Y. Choquet-Bruhat *Théoreme d'existence pour certains systèmes d'équations aux dérivées partielles non linéaires* Acta Math. **88** (1952), 141–225

[34] Y. Choquet-Bruhat, P.T. Chruściel, and J. Loizelet *Global solutions of the Einstein-Maxwell equations in higher dimensions* Classical and Quantum Gravity **23** (2006), 7383–7394

[35] Y. Choquet-Bruhat and R. Geroch *Global aspects of the Cauchy problem in general relativity* Comm. Math. Phys. **14** (1969), 329–335

[36] Y. Choquet-Bruhat, J. Isenberg, D. Pollack *The constraint equations for the Einstein-scalar field system on compact manifolds* Classical Quantum Gravity **24** (2007), 809–828

[37] M. W. Choptuik *Critical phenomena in gravitational collapse* In XIth International Congress of Mathematical Physics (Paris, 1994), 599–605, Int. Press, Cambridge, MA, 1995.

[38] D. Christodoulou *Reversible and irreversible transformations in black-hole physics* Phys. Rev. Lett. **25** (1970), 1596–1597
[39] D. Christodoulou *Violation of cosmic censorship in the gravitational collapse of a dust cloud* Comm. Math. Phys. **93** (1984), no. 2, 171–195
[40] D. Christodoulou *A mathematical theory of gravitational collapse* Comm. Math. Phys. **109** (1987), 613–647
[41] D. Christodoulou *The formation of black holes and singularities in spherically symmetric gravitational collapse* Comm. Pure Appl. Math. **44** (1991), no. 3, 339–373
[42] D. Christodoulou *The nonlinear nature of gravitation and gravitational wave experiments* Phys. Rev. Letters **67** (1991) no 12, 1486–1489
[43] D. Christodoulou *Examples of naked singularity formation in the gravitational collapse of a scalar field* Ann. of Math. **140** (1994), no. 3, 607–653.
[44] D. Christodoulou *Self-gravitating relativistic fluids: a two-phase model* Arch. Rat. Mech. Anal. **130** (1995), 343–400
[45] D. Christodoulou *The instability of naked singularities in the gravitational collapse of a scalar field* Ann. of Math. **149** (1999), no. 1, 183–217
[46] D. Christodoulou *The action principle and partial differential equations*, Ann. Math. Studies No. 146, 1999
[47] D. Christodoulou *On the global initial value problem and the issue of singularities* Class. Quantum Gravity **16** (1999) A23–A35
[48] D. Christodoulou *The global initial value problem in general relativity* 9th Marcel Grosmann Meeting (Rome 2000), 44–54, World Sci. Publishing, 2002
[49] D. Christodoulou *Mathematical Problems of General Relativity Theory*, ETH-Zürich lecture notes, 2003
[50] D. Christodoulou and S. Klainerman *Asymptotic properties of linear field equations in Minkowski space* Comm. Pure Appl. Math. **43** (1990), no. 2, 137–199.
[51] D. Christodoulou and S. Klainerman *The global nonlinear stability of the Minkowski space* Princeton University Press, 1993
[52] D. Christodoulou *The formation of shocks in 3-dimensional fluids* EMS Monographs in Mathematics. European Mathematical Society (EMS), Zürich, 2007.
[53] D. Christodoulou *The formation of black holes in general relativity* arXiv:0805.3880
[54] P. Chruściel *On uniqueness in the large of solutions of Einstein's equations ("strong cosmic censorship")* Proceedings of the Center for Mathematical Analysis **27**, Australian National University (1991)
[55] P. Chruściel *Black holes* Lect. Notes Phys. **604** (2002), 61–102
[56] P. Chruściel and D. Pollack *Singular Yamabe metrics and initial data with exactly Kottler-Schwarzschild-de Sitter ends* Annales Henri Poincaré **9** (2008) 639–654
[57] P. Chruściel and J. L. Costa *On uniqueness of stationary vacuum black holes* arXiv:0806.0016
[58] M. Dafermos *Stability and instability of the Cauchy horizon for the spherically symmetric Einstein-Maxwell-scalar field equations* Ann. of Math. **158** (2003), no. 3, 875–928
[59] M. Dafermos *The interior of charged black holes and the problem of uniqueness in general relativity* Comm. Pure App. Math. **58** (2005), no. 4, 445–504
[60] M. Dafermos *Spherically symmetric spacetimes with a trapped surface* Class. Quantum Grav. **22** (2005), no. 11, 2221–2232
[61] M. Dafermos *Black hole formation from a complete regular past* to appear, Comm. Math. Phys.
[62] M. Dafermos and G. Holzegel *On the nonlinear stability of higher dimensional triaxial Bianchi-IX black holes* Adv. Theor. Math. Phys. **10** (2006), no. 4, 503–523
[63] M. Dafermos and I. Rodnianski *A proof of Price's law for the collapse of a self-gravitating scalar field*, Invent. Math. **162** (2005), 381–457
[64] M. Dafermos and I. Rodnianski *Small-amplitude nonlinear waves on a black hole background* J. Math. Pures Appl. **84** (2005), 1147–1172
[65] M. Dafermos and I. Rodnianski *The redshift effect and radiation decay on black hole spacetimes*, gr-qc/0512119
[66] M. Dafermos and I. Rodnianski *The wave equation on Schwarzschild-de Sitter spacetimes*, arXiv:0709.2766v1 [gr-qc]

[67] M. Dafermos and I. Rodnianski *A note on energy currents and decay for the wave equation on a Schwarzschild background*, arXiv:0710.0171v1 [math.AP]
[68] M. Dafermos and I. Rodnianski *A proof of the uniform boundedness of solutions to the wave equation on slowly rotating Kerr backgrounds*, arXiv:0805.4309v1 [gr-qc]
[69] M. Dafermos and I. Rodnianski *Polynomial decay for solutions of the wave equation on slowly rotating Kerr backgrounds*, in preparation
[70] T. de Donder *La gravifique Einsteinienne* Annales de l'Observatoire Royal de Belgique, Brussels, 1921.
[71] S. Detweiler *Klein-Gordon equation and rotating black holes* Phys. Rev. D **22** (1980), 2323–2326
[72] J. Dimock *Scattering for the Wave equation on the Schwarzschild metric*, Gen. Rel. Grav. **17** (1985), 353–369
[73] J. Dimock and B. Kay *Classical and quantum scattering theory for linear scalar field on the Schwarzschild metric I* Ann. Phys., **175** (1987), 366–426
[74] A. Einstein *Über Gravitationswellen* Königlich Preussiche Akademie der Wissenschaften Berlin Sitzungberichte (1918) 154-167
[75] A. Einstein *On a stationary system with spherical symmetry consisting of many gravitating masses.* Ann. Math. **40** (1939), 922–936
[76] J. Eisenstaedt *The Early Interpretation of the Schwarzschild Solution* In: Einstein and the History of General Relativity (ed. by D. Howard and J. Stachel), Birkhaeuser, 1989.
[77] R. Emparan and H. Reall *Black Holes in Higher Dimensions* Living Rev. Relativity **11**, (2008), 6.
[78] F. Finster, N. Kamran, J. Smoller, S.-T. Yau *The long-time dynamics of Dirac particles in the Kerr-Newman black hole geometry* Adv. Theor. Math. Phys. **7** (2003), 25–52
[79] F. Finster, N. Kamran, J. Smoller, S. T. Yau *Decay of solutions of the wave equation in Kerr geometry* Comm. Math. Phys. **264** (2006), 465–503
[80] F. Finster, N. Kamran, J. Smoller, S.-T. Yau *Erratum: Decay of solutions of the wave equation in Kerr geometry* Comm. Math. Phys., online first
[81] F. Finster and H. Schmid *Spectral estimates and non-selfadjoint perturbations of spheroidal wave operators* J. Reine Angew. Math. **601** (2006), 71–107
[82] V. Fock *The theory of space, time and gravitation* The Macmillan Co., New York, 1964
[83] H. Friedrich *The existence of n-geodesically complete or future complete solutions of Einstein's field equations with smooth asymptotic structure* Comm. Math. Phys. **107** (1986), no. 4, 587–609
[84] G.W. Gibbons *Anti-de-Sitter spacetime and its uses* In "Mathematical and quantum aspects of relativity and cosmology" (Pythagoreon, 1998), Lecture Notes in Phys. **537**, Springer-Verlag, 102–142, 2000.
[85] G. W. Gibbons and S. W. Hawking *Cosmological event horizons, thermodynamics, and particle creation* Phys. Rev. D. **15** (1977), no. 10, 2738–2751
[86] G. W. Gibbons, C. Warnick and M. Werner *Light-bending in Schwarzschild-de Sitter: projective geometry of the optical metric*, arXiv:0808.3074 [gr-qc]
[87] C. Gundlach and J. M. Martín-García *Critical Phenomena in Gravitational Collapse* Living Rev. Relativity **10**, (2007), 5
[88] C. Gundlach, R. Price, and J. Pullin *Late-time behavior of stellar collapse and explosions. I. Linearized perturbations*, Phys. Rev. D **49** (1994) 883–889
[89] D. Häfner *Sur la théorie de la diffusion pour l'équation de Klein-Gordon dans la métrique de Kerr* Dissertationes Mathematicae **421** (2003)
[90] D. Häfner and J.-P. Nicolas *Scattering of massless Dirac fields by a Kerr black hole* Rev. Math. Phys. **16**(1) (2004), 29–123
[91] S. W. Hawking and G. F. R. Ellis *The large scale structure of space-time* Cambridge Monographs on Mathematical Physics, No. 1. Cambridge University Press, London-New York, 1973
[92] M. Heusler *Black hole uniqueness theorems* Cambridge Lecture Notes in Physics, No. 6 Cambridge University Press, Cambridge, 1996
[93] G. Holzegel *The stability problem for black holes* Ph. D. Dissertation, University of Cambridge, 2008
[94] G. Holzegel *Stability and decay-rates for the Schwarzschild metric under biaxial perturbations* arXiv:0808.3246 [gr-qc]

[95] G. Huisken and T. Ilmanen *The inverse mean curvature flow and the Riemannian Penrose inequality* J. Differential Geom. **59** (2001), 353–437

[96] A. Ionescu and S. Klainerman *On the uniqueness of smooth, stationary black holes in vacuum* arXiv:0711.0040

[97] P. Karageorgis and J. Stalker *Sharp bounds on $2m/r$ for static spherical objects*, arXiv:0707.3632v2 [gr-qc]

[98] B. Kay and R. Wald *Linear stability of Schwarzschild under perturbations which are non-vanishing on the bifurcation 2-sphere* Classical Quantum Gravity **4** (1987), no. 4, 893–898

[99] R. Kerr *Gravitational field of a spinning mass as an example of algebraically special metrics* Phys. Rev. Lett. **11** (1963) 237–238

[100] S. Klainerman *Uniform decay estimates and the Lorentz invariance of the classical wave equation* Comm. Pure Appl. Math. **38** (1985), 321–332

[101] S. Klainerman *The null condition and global existence to nonlinear wave equations* Nonlinear systems of partial differential equations in applied mathematics, Part 1 (Santa Fe, N.M., 1984), 293–326, Lectures in Appl. Math. **23** Amer. Math. Soc., Providence, RI, 1986.

[102] S. Klainerman and I. Rodnianski *Rough solutions of the Einstein vacuum equations* Ann. of Math. **161** (2005), 1143–1193

[103] S. Klainerman and I. Rodnianski *A Kirchoff-Sobolev parametrix for the wave equation and applications* J. Hyperbolic Differ. Equ. **4** (2007), 401–433

[104] K. Kokkotas and B. Schmidt *Quasi-normal modes of stars and black holes* Living Rev. Relativity **2** (1999)

[105] F. Kottler *Über die physikalischen Grundlagen der Einsteinschen Gravitationstheorie*, Ann. Phys. **56** 401–462 (1918)

[106] J. Kronthaler *Decay rates for spherical scalar waves in the Schwarzschild geometry* arXiv:0709.3703

[107] M. Kruskal *Maximal extension of Schwarzschild metric* Phys. Rev. **119** (1960), 1743–1745

[108] D. Kubiznak *Hidden symmetries of higher-dimensional rotating black holes* Ph.D. Thesis, University of Alberta, September 2008

[109] H. K. Kunduri, J. Lucietti and H. S. Reall *Gravitational perturbations of higher dimensional rotating black holes: tensor perturbations*, arXiv:hep-th/0606076v3

[110] I. Laba and A. Soffer *Global existence and scattering for the nonlinear Schrödinger equation on Schwarzschild manifolds* Helv. Phys. Acta **72** (1999), no. 4, 272–294

[111] G. Lemaitre *L'Univers en Expansion* Publication du Laboratoire d'Astronomie et de Géodésie de l'Université de Louvain **9** (1932), 171–205

[112] J. Leray *Hyperbolic differential equations* The Institute for Advanced Study, Princeton, N. J., 1953.

[113] H. Lindblad and I. Rodnianski *The global stability of Minkowski space-time in harmonic gauge* to appear, Ann. of Math.

[114] M. Machedon and J. Stalker *Decay of solutions to the wave equation on a spherically symmetric background*, preprint

[115] J. Marzuola, J. Metcalfe, D. Tataru, M. Tohaneanu *Strichartz estimates on Schwarzschild black hole backgrounds* arXiv:0802.3942

[116] J. Metcalfe *Strichartz estimates on Schwarzschild space-times* Oberwolfach Reports **44** (2007), 8–11.

[117] C. W.Misner, K. S. Thorne, and J. A. Wheeler *Gravitation* W. H. Freeman and Co., San Francisco, Calif., 1973

[118] V. Moncrief and D. Eardley *The global existence problem and cosmic censorship in general relativity* Gen. Rel. Grav. **13** (1981), 887–892

[119] C. S. Morawetz *The limiting amplitude principle* Comm. Pure Appl. Math. **15** (1962) 349–361

[120] R. C. Myers and M. J. Perry *Black holes In higher dimensional space-times* Ann. Phys. (N.Y.) **172** (1986), 304

[121] J.-P. Nicolas *Non linear Klein-Gordon equation on Schwarzschild-like metrics* J. Math. Pures Appl. **74** (1995), 35–58.

[122] J.-P. Nicolas *Opérateur de diffusion pou le système de Dirac en métrique de Schwarzschild* C. R. Acad. Sci. Paris Sér. I Math. **318** (1994), 729–734

[123] H. Nussbaumer and L. Bieri *Discovering the Expanding Universe* Cam. Univ. Press, to appear 2009

[124] E. Noether *Invariante Variationsprobleme* Nachr. d. König. Gesellsch. d. Wiss. zu Göttingen, Math-phys. Klasse (1918) 235–257

[125] J. R. Oppenheimer and H. Snyder *On continued gravitational contraction* Phys. Rev. **56** (1939), 455–459

[126] R. Penrose *Gravitational collapse and space-time singularities* Phys. Rev. Lett. **14**, 57–59

[127] R. Penrose *Gravitational collapse: the role of general relativity* Rev. del Nuovo Cimento **1**, (1969) 272–276

[128] R. Penrose *Singularities and time asymmetry* In "General Relativity–an Einstein Survey" S. Hawking, W. Israel ed., Cambridge University Press, Cambridge, 1979

[129] F. Planchon and I. Rodnianski, *On uniqueness for the Cauchy problem in general relativity,* preprint

[130] R. Price *Nonspherical perturbations of relativistic gravitational collapse. I. Scalar and gravitational perturbations* Phys. Rev. D (3) **5** (1972), 2419–2438

[131] T. Regge and J. Wheeler *Stability of a Schwarzschild singularity* Phys. Rev. **108** (1957), 1063–1069

[132] A. Rendall *Theorems on existence and global dynamics for the Einstein equations* Living Rev. Relativity **8** (2005), 6

[133] A. Rendall *Partial differential equations in general relativity* Oxford Graduate Texts in Mathematics **16** Oxford University Press, Oxford, 2008

[134] H. Ringström *Future stability of the Einstein-non linear scalar field system* Invent. Math. **173** (2008), no. 1, 123–208

[135] A. Sá Barreto and M. Zworski *Distribution of resonances for spherical black holes* Math. Res. Lett. **4** (1997), no. 1, 103–121

[136] J. Schauder *Das Anfangswertproblem einer quasilinearen hyperbolischen Differentialgleichung zweiter Ordnung in beliebiger Anzahl von unabhängigen Veränderlichen* Fundam. Math. **24** (1935)

[137] R. Schoen and S.-T. Yau *On the proof of the positive mass conjecture in general relativity* Comm. Math. Phys. **16** (1979) 45–76

[138] R. Schoen and S.-T. Yau *Proof of the positive mass theorem. II.* Comm. Math. Phys. **79** (1981), no. 2, 231–260

[139] K. Schwarzschild *Über das Gravitationsfeld eines Massenpunktes nach der Einsteinschen Theorie* Sitzungsber. d. Preuss. Akad. d. Wissenschaften **1** (1916), 189–196

[140] K. Schwarzschild *Über das Gravitationsfeld einer Kugel aus inkompressibler Flüssigkeit nach der Einsteinschen Theorie* Sitzungsber. d. Preuss. Akad. d. Wissenschaften, Berlin (1916)

[141] J. A. Smoller, A. G. Wasserman, and S.-T. Yau *Einstein-Yang/Mills black hole solutions* Chen Ning Yang, 209–220, Int. Press, Cambridge, MA, 1995.

[142] C. Sogge *Lectures on nonlinear wave equations* International Press, Boston, 1995

[143] K. Stellmacher *Zum Anfangswertproblem der Gravitationsgleichungen* Math. Ann. **115** (1938), 136–152

[144] J. Stewart *Solutions of the wave equation on a Schwarzschild space-time with localised energy* Proc. Roy. Soc. London Ser. A **424** (1989), 239–244

[145] M. Struwe *Wave maps with and without symmetries*, Clay lecture notes

[146] J. L. Synge *The gravitational field of a particle* Proc. Roy. Irish Acad. **53** (1950), 83–114

[147] J. L. Synge *Relativity: the special theory* North-Holland Publishing Co., Amsterdam, 1956.

[148] P. K. Townsend *Black holes*, arXiv:gr-qc/9707012v1, 145 pages

[149] F. Twainy *The Time Decay of Solutions to the Scalar Wave Equation in Schwarzschild Background* Thesis. San Diego: University of California 1989

[150] A. Vasy *The wave equation on asymptotically de Sitter-like spaces* Oberwolfach Reports **41** (2007), 2388–2392

[151] R. M. Wald *Note on the stability of the Schwarzschild metric* J. Math. Phys. **20** (1979), 1056–1058

[152] R. Wald *General relativity* University of Chicago Press, Chicago, 1984

[153] M. Walker and R. Penrose *On quadratic first integrals of the geodesic equations for type* 22 *spacetimes* Comm. Math. Phys. **18** (1970), 265–274

[154] H. Weyl *Raum, Zeit, Materie* Springer, Berlin, 1919

[155] H. Weyl *Über die statischen kugelsymmetrischen Lösungen von Einsteins 'kosmologischen' Gravitationsgleichungen*, Phys. Z. **20** (1919), 31–34

[156] E. Witten *A new proof of the positive energy theorem* Comm. Math. Phys. **80** (1981), 381–402

[157] N. Zipser *The global nonlinear stability of the trivial solution of the Einstein-Maxwell equations* Ph. D. Thesis, Harvard University, 2000

[158] T. J. M. Zouros and D. M. Eardley *Instabilities of massive scalar perturbations of a rotating black hole* Ann. of Phys. **118** (1976), 139

Department of Pure Mathematics and Mathematical Statistics, University of Cambridge, Wilberforce Road, Cambridge CB3 0WB United Kingdom
E-mail address: m.dafermos@dpmms.cam.ac.uk

Princeton University, Department of Mathematics, Fine Hall, Washington Road, Princeton, New Jersey 08544
E-mail address: irod@princeton.edu

Clay Mathematics Proceedings
Volume **17**, 2013

The Theory of Nonlinear Schrödinger Equations

G. Staffilani

Contents

1. Introduction

The title of these lecture notes is certainly too ambitious. In fact here we will mainly consider semilinear Schrödinger initial value problems (IVP)

$$\begin{cases} iu_t + \frac{1}{2}\Delta u = \lambda|u|^{p-1}u, \\ u(x,0) = u_0(x) \end{cases} \tag{1}$$

where $\lambda = \pm 1$, $p > 1$, $u : \mathbb{R} \times M \to \mathbb{C}$, and M is a manifold[1]. Even in this relatively special case we will not be able to mention all the findings and results concerning the initial value problem (1) and for this we apologize in advance.

Schrödinger equations are classified as *dispersive* partial differential equations and the justification for this name comes from the fact that if no boundary conditions are imposed their solutions tend to be waves which spread out spatially. But what does this mean mathematically? A simple and complete mathematical

2010 *Mathematics Subject Classification.* Primary 35Axx, 35Bxx, 35Exx, 35Gxx, 35Lxx, 35Qxx, 42-XX.

G.S. is supported in part by N.S.F. Grant DMS-0602678.

[1]In most cases M is the Euclidean space $\mathbb{R}^n$ and only at the end we will mention some results and references when M is a different kind of manifold.

characterization of the word *dispersion* is given to us for example by R. Palais in [**66**]. Although his definition is given for one dimensional waves, the concept is expressed so clearly that it is probably a good idea to follow almost[2] literally his explanation: *"Let us [next] consider* linear *wave equations of the form*

$$u_t + P\left(\frac{\partial}{\partial x}\right) u = 0,$$

where P is polynomial. Recall that a solution $u(t,x)$, which Fourier transform is of the form $e^{i(kx-\omega t)}$, is called a plane-wave solution; k is called the wave number *(waves per unit of length) and ω the (angular)* frequency. *Rewriting this in the form $e^{ik(x-(\omega/k)t)}$, we recognize that this is a traveling wave of velocity $\frac{\omega}{k}$. If we substitute this $u(t,x)$ into our wave equation, we get a formula determining a unique frequency $\omega(k)$ associated to any wave number k, which we can write in the form*

$$\frac{\omega(k)}{k} = \frac{1}{ik}P(ik). \tag{2}$$

This is called the "dispersive relation" for this wave equation. Note that it expresses the velocity for the plane-wave solution with wave number k. For example, $P(\frac{\partial}{\partial x}) = c\frac{\partial}{\partial x}$ gives the linear advection equation $u_t + cu_x = 0$, which has the dispersion relation $\frac{\omega}{k} = c$, showing of course that all plane-wave solutions travel at the same velocity c, and we say that we have trivial dispersion in this case. On the other hand if we take $P\left(\frac{\partial}{\partial x}\right) = -\frac{i}{2}\left(\frac{\partial}{\partial x}\right)^2$, then our wave equation is $iu_t + \frac{1}{2}u_{xx} = 0$, which is the linear Schrödinger equation, and we have the non-trivial dispersion relation $\frac{\omega}{k} = \frac{k}{2}$. In this case, plane waves of large wave-number (and hence high frequency) are traveling much faster than low-frequency waves. The effect of this is to "broaden a wave packet". That is, suppose our initial condition is $u_0(x)$. We can use the Fourier transform[3] to write u_0 in the form

$$u_0(x) = \int \widehat{u_0}(k)e^{ikx}\, dk,$$

and then, by superposition, the solution to our wave equation will be

$$u(t,x) = \int \widehat{u_0}(k)e^{ik(x-(\omega(k)/k)t)}\, dk.$$

Suppose for example that our initial wave form is a highly peaked Gaussian. Then in the case of the linear advection equation all the Fourier modes travel together at the same speed and the Gaussian lump remains highly peaked over time. On the other hand, for the linearized Schrödinger equation the various Fourier modes all travel at different velocities, so after time they start canceling each other by destructive interference, and the original sharp Gaussian quickly broadens".

As one can imagine dispersive equations are proposed as descriptions of certain phenomena that occur in nature. But it turned out that some of these equations appear also in more abstract mathematical areas like algebraic geometry [**46**], and certainly we are not in the position to discuss this beautiful part of mathematics here.

[2]R. Palais actually uses the Airy equation as an example, while we use the linear Schrödinger equation to be consistent with the topic of the lectures.

[3]In these lectures we will ignore the absolute constants that may appear in other definitions for the Fourier transform.

The questions that we will address here are more phenomenological. Assume that a profile of a wave is given at time $t = 0$, (initial data). Is it possible to prove that there exists a unique wave that "lives" for an interval of time $[0, T]$, that satisfies the equation, and that at time $t = 0$ has the assigned profile? What kind of properties does the wave have at later times? Does it "live" for all times or does it "blowup" in finite time?

Our intuition tells us that, if we start with *nice* and *small* initial data, then all the questions above should be easier to answer. This is indeed often true. In general in this case one can prove that the wave exists for all times, it is unique and its "size", measured taking into account the order of smoothness, can be controlled in a reasonable way. But what happens when we are not in this advantageous setting? These lecture notes are devoted to the understanding of how much of the above is still true when we consider *large* data and *long* interval of times. To be able to give a rigorous setting for the study of the initial value problem in (1) and to avoid any confusion in the future we need a strong mathematical definition for *well-posedness*. We consider the general initial value problem of type

$$\begin{cases} \partial_t u + P_m(\partial_{x_1}, \dots, \partial_{x_n})u + N(u, \partial_x^\alpha u) = 0, \\ u(x, 0) = u_0(x), \qquad x \in \mathbb{R}^n (\text{ or } x \in \mathbb{T}^n),\ t \in \mathbb{R}, \end{cases} \tag{3}$$

where $m \in \mathbb{N}$, $P_m(\partial_{x_1}, \dots, \partial_{x_n})$ is a differential operator with constant coefficients of order m and $N(u, \partial_x^\alpha u)$ is the nonlinear part of the equation, that is a nonlinear function that depends on u and derivatives of u up to order $m - 1$. The function $u_0(x)$ is the initial condition or initial profile, and most of the time is called initial data. Above we pointed out the fact that finding a solution for an IVP strongly depends on the regularity one asks for the solution itself. So we first have to decide how we "measure" the regularity of a function. The most common way of doing so is to decide where the weak derivatives of the function "live". It is indeed time to recall the definition of Sobolev spaces[4].

DEFINITION 1.1. We say that a function $f \in H^k(\mathbb{R}^n), k \in \mathbb{N}$ if f and all its partial derivatives up to order k are in L^2. We recall that $H^k(\mathbb{R}^n)$ is a Banach space with the norm

$$\|f\|_{H^k} = \sum_{|\alpha|=0}^{k} \|\partial_x^\alpha f\|_{L^2},$$

where $\alpha(\alpha_1, \dots, \alpha_n)$ and $|\alpha| = \sum_{i=1}^n \alpha_i$ is its length.

We also recall here the definition of the Fourier transform.

DEFINITION 1.2. Assume $f \in L^2(\mathbb{R}^n)$, then the Fourier transform of f is defined as

$$\hat{f}(\xi) = \frac{1}{(2\pi)^n} \int_{\mathbb{R}^n} e^{i\langle x, \xi\rangle} f(x)\, dx$$

where $\langle\cdot\rangle$ is the inner product in $\mathbb{R}^n$. We also have an inverse Fourier formula

$$f(x) = \int_{\mathbb{R}^n} e^{-i\langle x, \xi\rangle} \hat{f}(x)\, dx.$$

[4]In more sophisticated instances one replaces Sobolev spaces with different ones, like L^p spaces, Hölder spaces, and so on.

If the function is defined on the torus $\mathbb{T}^n$ then the Fourier transform is defined as

$$\hat{f}(k) = \frac{1}{(2\pi)^n} \int_{\mathbb{T}^n} e^{i\langle x,k\rangle} f(x)\, dx$$

and the inverse Fourier formula is

$$f(x) = \sum_{k\in\mathbb{Z}} e^{-i\langle x,k\rangle} \hat{f}(k).$$

REMARK 1.3. Because $\widehat{\partial_x^\alpha f}(\xi) = (i\xi)^\alpha \hat{f}(\xi)$, it is easy to see that $f \in H^k(\mathbb{R}^n)$ if and only if

$$\int_{\mathbb{R}^n} |\hat{f}(\xi)|^2 (1+|\xi|)^{2k}\, d\xi < \infty,$$

and moreover

$$\left(\int_{\mathbb{R}^n} |\hat{f}(\xi)|^2 (1+|\xi|)^{2k}\, d\xi\right)^{1/2} \sim \|f\|_{H^k}.$$

Then we can generalize our notion of Sobolev space and define $H^s(\mathbb{R}^n), s \in \mathbb{R}$ as the set of functions such that

$$\int_{\mathbb{R}^n} |\hat{f}(\xi)|^2 (1+|\xi|)^{2s}\, d\xi < \infty.$$

Also $H^s(\mathbb{R}^n)$ is a Banach space with norm

$$\left(\int_{\mathbb{R}^n} |\hat{f}(\xi)|^2 (1+|\xi|)^{2s}\, d\xi\right)^{1/2} \sim \|f\|_{H^s}.$$

Sometimes it is useful to use the *homogeneous* Sobolev space $\dot{H}^s(\mathbb{R}^n)$. This is the space of functions such that

$$\int_{\mathbb{R}^n} |\hat{f}(\xi)|^2 |\xi|^{2s}\, d\xi < \infty.$$

Clearly all these observations can be made for Sobolev spaces in $\mathbb{T}^n$, except that in this case $\dot{H}^s(\mathbb{T}^n)$ and $H^s(\mathbb{T}^n)$ coincides.

We use $\|f\|_{L^p}$ to denote the $L^p(\mathbb{R}^n)$ norm. We often need mixed norm spaces, so for example, we say that $f \in L^q_t L^p_x$ if $\|(\|f(t,x)\|_{L^p_x})\|_{L^q_t} < \infty$. Here we also use the Sobolev space $W^{1,p}$, that is functions, that together with their gradient, belong to the space L^p. Finally, for a fixed interval of time $[0,T]$ and a Banach space of functions Z, we denote with $C([0,T], Z)$ the space of continuous maps from $[0,T]$ to Z.

We are now ready to give a first definition of well-posedness. We will give a more refined one later in Subsection 3.10.

DEFINITION 1.4. We say that the IVP (3) is locally well-posed (l.w.p) in H^s if, given $u_0 \in H^s$, there exist T, a Banach space of functions $X_T \subset C([-T,T]; H^s)$ and a unique $u \in X_T$ which solves (3). Moreover we ask that there is continuity with respect to the initial data in the appropriate topology. We say that (3) is globally well-posed (g.w.p) in H^s if the definition above is satisfied in any interval of time $[-T,T]$.

REMARK 1.5. The intervals of time are symmetric about the origin because the problems that we study here, that are of type (1), are all time reversible (i.e. if $u(t,x)$ is a solution, then so is $-u(x,-t)$).

We end this introduction with some notations. Throughout the notes we use C to denote various constants. If C depends on other quantities as well, this will be indicated by explicit subscripting, e.g. $C_{\|u_0\|_2}$ will depend on $\|u_0\|_2$. We use $A \lesssim B$ to denote an estimate of the form $A \leq CB$, where C is an absolute constant. We use $a+$ and $a-$ to denote expressions of the form $a+\varepsilon$ and $a-\varepsilon$, for some $0<\varepsilon \ll 1$.

I would like to thank Jim Colliander, Mark Keel, Hideo Takaoka and Terry Tao for the wonderful collaboration we had for several years: almost all the material I am presenting here comes from joint papers with them. I would like to thank my student Vedran Sohinger, who read these notes in a very early stage and pointed out several typos and inconsistencies. A special thanks also to the careful referee who corrected several typos and inconsistencies. Finally a very warm thank you to the Clay Institute and ETH, for their support and hospitality, and to all the people who attended the lectures, without you the summer school would have been impossible[5].

2. The Linear Schrödinger Equation in $\mathbb{R}^n$: Dispersive and Strichartz Estimates

In this lecture we introduce some of the most important estimates relative to the linear Schrödinger IVP

$$\begin{cases} iv_t + \frac{1}{2}\Delta v = 0, \\ v(x,0) = u_0(x). \end{cases} \tag{4}$$

It is important to understand as much as possible the solution v of (4) that we will denote with $v(t,x) = S(t)u_0(x)$, since by the Duhamel principle one can write the solution of the associated forced or nonlinear problem

$$\begin{cases} iu_t + \frac{1}{2}\Delta u = F(u), \\ u(x,0) = u_0(x). \end{cases} \tag{5}$$

as

$$u(t,x) = S(t)u_0 - i\int_0^t S(t-t')F(u(t'))\,dt'. \tag{6}$$

PROBLEM 2.1. *Prove the Duhamel Principle* (6).

The solution of the linear problem (4) is easily computable by taking Fourier transform. In fact by fixing the frequency ξ problem (4) transforms into the ODE

$$\begin{cases} i\hat{v}_t(t,\xi) - \frac{1}{2}|\xi|^2\hat{v}(t,\xi) = 0, \\ \hat{v}(\xi,0) = \hat{u}_0(\xi) \end{cases} \tag{7}$$

and we can write its solution as

$$\hat{v}(t,\xi) = e^{-i\frac{1}{2}|\xi|^2 t}\hat{u}_0(\xi).$$

[5]These lecture notes were written in 2008, since then enormous progress has been made in several of the problems introduced here. In particular I would like to mention the complete solution of the L^2-critical Schrdinger problem, see [**59, 60**] and [**37**].

In general the solution $v(t,x)$ above is denoted by $S(t)u_0$, where $S(t)$ is called the Schrödinger group. If we define, in the distributional sense,

$$K_t(x) = \frac{1}{(\pi i t)^{n/2}} e^{i\frac{|x|^2}{2t}}$$

then we have

$$S(t)u_0(x) = e^{it\Delta}u_0(x) = u_0 \star K_t(x) = \frac{1}{(\pi i t)^{n/2}} \int e^{i\frac{|x-y|^2}{2t}} u_0(y)\, dy \tag{8}$$

PROBLEM 2.2. *Prove, in the sense of distributions, that the inverse Fourier transform of* $e^{-i\frac{1}{2}|\xi|^2 t}$ *is* $K_t(x) = \frac{1}{(\pi i t)^{n/2}} e^{i\frac{|x|^2}{2t}}$.

As mentioned already

$$\widehat{S(t)u_0}(\xi) = e^{-i\frac{1}{2}|\xi|^2 t}\hat{u}_0(\xi), \tag{9}$$

and this last one can be interpreted as saying that the solution $S(t)u_0$ above is the adjoint of the Fourier transform restricted on the paraboloid $P = \{(\xi, |\xi|^2)$ for $\xi \in \mathbb{R}^n\}$. This remark, strictly linked to (8) and (9), can be used to prove a variety of very deep estimates for $S(t)u_0$, see for example [**71**]. For example from (8) we immediately have the so called *Dispersive Estimate*

$$\|S(t)u_0\|_{L^\infty} \lesssim \frac{1}{t^{n/2}}\|u_0\|_{L^1}. \tag{10}$$

From (9) instead we have the conservation of the homogeneous Sobolev norms[6]

$$\|S(t)u_0\|_{\dot{H}^s} = \|u_0\|_{\dot{H}^s}, \tag{11}$$

for all $s \in \mathbb{R}$. Interpolating (10) with (11) when $s = 0$ and using a so called TT^* argument one can prove the non-endpoint Strichartz estimates in Theorem 2.3 below. The endpoint estimate is due to Tao and Keel who use a more sophisticated argument [**49**]. See [**73**] for some concise proofs, and [**19**] for a complete list of authors who contributed to the final version of the following theorem.

THEOREM 2.3 (Strichartz Estimates for the Schrödinger operator). *Fix* $n \geq 1$. *We call a pair* (q,r) *of exponents admissible if* $2 \leq q, r \leq \infty$, $\frac{2}{q} + \frac{n}{r} = \frac{n}{2}$ *and* $(q,r,n) \neq (2,\infty,2)$. *Then for any admissible exponents* (q,r) *and* $(\tilde{q},\tilde{r})$ *we have the homogeneous Strichartz estimate*

$$\|S(t)u_0\|_{L^q_t L^r_x(\mathbb{R}\times\mathbb{R}^n)} \lesssim \|u_0\|_{L^2_x(\mathbb{R}^n)} \tag{12}$$

and the inhomogeneous Strichartz estimate

$$\left\| \int_0^t S(t-t')F(t')\, dt' \right\|_{L^q_t L^r_x(\mathbb{R}\times\mathbb{R}^n)} \lesssim \|F\|_{L^{\tilde{q}'}L^{\tilde{r}'}_x(\mathbb{R}\times\mathbb{R}^n)}, \tag{13}$$

where $\frac{1}{\tilde{q}} + \frac{1}{\tilde{q}'} = 1$ *and* $\frac{1}{\tilde{r}} + \frac{1}{\tilde{r}'} = 1$.

To finish this lecture we would like to present a *refined* bilinear Strichartz estimate due originally to Bourgain in [**9**] (see also [**12**]).

[6]We will see later that the L^2 norm is conserved also for the nonlinear problem (1).

THEOREM 2.4. *Let $n \geq 2$. For any spacetime slab $I_* \times \mathbb{R}^n$, any $t_0 \in I_*$, and for any $\delta > 0$, we have*

$$\|uv\|_{L^2_t L^2_x(I_*\times\mathbb{R}^n)} \leq C(\delta)(\|u(t_0)\|_{\dot H^{-1/2+\delta}} + \|(i\partial_t + \frac{1}{2}\Delta)u\|_{L^1_t \dot H_x^{-1/2+\delta}}) \times (\|v(t_0)\|_{\dot H^{\frac{n-1}{2}-\delta}} + \|(i\partial_t + \frac{1}{2}\Delta)v\|_{L^1_t \dot H_x^{\frac{n-1}{2}-\delta}}). \tag{14}$$

This estimate is very useful when u is high frequency and v is low frequency, as it moves plenty of derivatives onto the low frequency term. This estimate shows in particular that there is little interaction between high and low frequencies. One can also check easily that when $n = 2$ one recovers the $L^4_t L^4_x$ Strichartz estimate contained in Theorem 2.3 above.

PROOF. We fix δ, and allow our implicit constants to depend on δ. We begin by addressing the homogeneous case, with $u(t) := e^{it\frac{1}{2}\Delta}\zeta$ and $v(t) := e^{it\frac{1}{2}\Delta}\psi$ and consider the more general problem of proving

$$\|uv\|_{L^2_{t,x}} \lesssim \|\zeta\|_{\dot H^{\alpha_1}}\|\psi\|_{\dot H^{\alpha_2}}. \tag{15}$$

Scaling invariance for this estimate[7] demands that $\alpha_1 + \alpha_2 = \frac{n}{2} - 1$. Our first goal is to prove this for $\alpha_1 = -\frac{1}{2} + \delta$ and $\alpha_2 = \frac{n-1}{2} - \delta$. The estimate (15) may be recast using duality and renormalization as

$$\int g(\xi_1 + \xi_2, |\xi_1|^2 + |\xi_2|^2)|\xi_1|^{-\alpha_1}\widehat{\zeta}(\xi_1)|\xi_2|^{-\alpha_2}\widehat{\psi}(\xi_2)d\xi_1 d\xi_2 \lesssim \|g\|_{L^2(\mathbb{R}\times\mathbb{R}^n)}\|\zeta\|_{L^2(\mathbb{R}^n)}\|\psi\|_{L^2(\mathbb{R}^n)}. \tag{16}$$

Since $\alpha_2 \geq \alpha_1$, we may restrict our attention to the interactions with $|\xi_1| \geq |\xi_2|$. Indeed, in the remaining case we can multiply by $(\frac{|\xi_2|}{|\xi_1|})^{\alpha_2-\alpha_1} \geq 1$ to return to the case under consideration. In fact, we may further restrict our attention to the case where $|\xi_1| > 4|\xi_2|$ since, in the other case, we can move the frequencies between the two factors and reduce to the case where $\alpha_1 = \alpha_2$, which can be treated by $L^4_{t,x}$ Strichartz estimates[8] when $n \geq 2$. Next, we decompose $|\xi_1|$ dyadically and $|\xi_2|$ in dyadic multiples of the size of $|\xi_1|$ by rewriting the quantity to be controlled as (N, Λ dyadic):

$$\sum_N \sum_\Lambda \int\int g_N(\xi_1 + \xi_2, |\xi_1|^2 + |\xi_2|^2)|\xi_1|^{-\alpha_1}\widehat{\zeta_N}(\xi_1)|\xi_2|^{-\alpha_2}\widehat{\psi_{\Lambda N}}(\xi_2)d\xi_1 d\xi_2.$$

Note that subscripts on g, ζ, ψ have been inserted to evoke the localizations to $|\xi_1 + \xi_2| \sim N, |\xi_1| \sim N, |\xi_2| \sim \Lambda N$, respectively. Note that in the situation we are considering here, namely $|\xi_1| \geq 4|\xi_2|$, we have that $|\xi_1 + \xi_2| \sim |\xi_1|$ and this explains why g may be so localized.

By renaming components, we may assume that $|\xi_1^1| \sim |\xi_1|$ and $|\xi_2^1| \sim |\xi_2|$. Write $\xi_2 = (\xi_2^1, \underline{\xi_2})$. We now change variables by writing $u = \xi_1 + \xi_2$, $v = |\xi_1|^2 + |\xi_2|^2$ and

[7]Here we use the fact that if v is solution to the linear Schrödinger equation, then $v_\lambda(t,x) = v(\frac{x}{\lambda}, \frac{t}{\lambda^2})$ is also solution.

[8]In one dimension $n = 1$, Lemma 2.4 fails when u, v have comparable frequencies, but continues to hold when u, v have separated frequencies; see [**24**] for further discussion.

$dudv = Jd\xi_2^1 d\xi_1$. A calculation then shows that $J = |2(\xi_1^1 \pm \xi_2^1)| \sim |\xi_1|$. Therefore, upon changing variables in the inner two integrals, we encounter

$$\sum_N N^{-\alpha_1} \sum_{\Lambda \leq 1} (\Lambda N)^{-\alpha_2} \int_{\mathbb{R}^{n-1}} \int_{\mathbb{R}} \int_{\mathbb{R}^n} g_N(u,v) H_{N,\Lambda}(u,v,\underline{\xi_2})\, dudvd\underline{\xi_2}$$

where

$$H_{N,\Lambda}(u,v,\underline{\xi_2}) = \frac{\widehat{\zeta_N}(\xi_1)\widehat{\psi_{\Lambda N}}(\xi_2)}{J}.$$

We apply Cauchy-Schwarz on the u, v integration and change back to the original variables to obtain

$$\sum_N N^{-\alpha_1} \|g_N\|_{L^2} \sum_{\Lambda \leq 1} (\Lambda N)^{-\alpha_2} \int_{\mathbb{R}^{n-1}} \left[\int_{\mathbb{R}} \int_{\mathbb{R}^n} \frac{|\widehat{\zeta_N}(\xi_1)|^2 |\widehat{\psi_{\Lambda N}}(\xi_2)|^2}{J} d\xi_1 d\xi_2^1 \right]^{\frac{1}{2}} d\underline{\xi_2}.$$

We recall that $J \sim N$ and use Cauchy-Schwarz in the $\underline{\xi_2}$ integration, keeping in mind the localization $|\xi_2| \sim \Lambda N$, to get

$$\sum_N N^{-\alpha_1 - \frac{1}{2}} \|g_N\|_{L^2} \sum_{\Lambda \leq 1} (\Lambda N)^{-\alpha_2 + \frac{n-1}{2}} \|\widehat{\zeta_N}\|_{L^2} \|\widehat{\psi_{\Lambda N}}\|_{L^2}.$$

Choose $\alpha_1 = -\frac{1}{2} + \delta$ and $\alpha_2 = \frac{n-1}{2} - \delta$ with $\delta > 0$ to obtain

$$\sum_N \|g_N\|_{L^2} \|\widehat{\zeta_N}\|_{L^2} \sum_{\Lambda \leq 1} \Lambda^{\delta} \|\widehat{\psi_{\Lambda N}}\|_{L^2}$$

which may be summed up, after using the Schwarz inequality, and the Plancherel theorem will give the claimed homogeneous estimate.

We turn our attention to the inhomogeneous estimate (14). For simplicity we set $F := (i\partial_t + \Delta)u$ and $G := (i\partial_t + \Delta)v$. Then we use Duhamel's formula (6) to write

$$u = e^{i(t-t_0)\Delta}u(t_0) - i\int_{t_0}^t e^{i(t-t')\Delta}F(t')\,dt', \quad v = e^{i(t-t_0)\Delta}v(t_0) - i\int_{t_0}^t e^{i(t-t')\Delta}G(t').$$

We obtain[9]

$$\begin{aligned}
\|uv\|_{L^2} \lesssim & \left\| e^{i(t-t_0)\Delta}u(t_0) e^{i(t-t_0)\Delta}v(t_0) \right\|_{L^2} \\
& + \left\| e^{i(t-t_0)\Delta}u(t_0) \int_{t_0}^t e^{i(t-t')\Delta}G(t')\,dt' \right\|_{L^2} \\
& + \left\| e^{i(t-t_0)\Delta}v(t_0) \int_{t_0}^t e^{i(t-t')\Delta}F(t')dt' \right\|_{L^2} \\
& + \left\| \int_{t_0}^t e^{i(t-t')\Delta}F(t')dt' \int_{t_0}^t e^{i(t-t'')\Delta}G(x,t'')\,dt'' \right\|_{L^2} \\
:= & I_1 + I_2 + I_3 + I_4.
\end{aligned}$$

The first term was treated in the first part of the proof. The second and the third are similar so we consider only I_2. Using the Minkowski inequality we have

$$I_2 \lesssim \int_{\mathbb{R}} \|e^{i(t-t_0)\Delta}u(t_0)e^{i(t-t')\Delta}G(t')\|_{L^2}\,dt',$$

[9] Alternatively, one can absorb the homogeneous components $e^{i(t-t_0)\Delta}u(t_0)$, $e^{i(t-t_0)\Delta}v(t_0)$ into the inhomogeneous term by adding an artificial forcing term of $\delta(t-t_0)u(t_0)$ and $\delta(t-t_0)v(t_0)$ to F and G respectively, where δ is the Dirac delta.

and in this case the theorem follows from the homogeneous estimate proved above. Finally, again by Minkowski's inequality we have

$$I_4 \lesssim \int_{\mathbb{R}} \int_{\mathbb{R}} \|e^{i(t-t')\Delta} F(t') e^{i(t-t'')\Delta} G(t'')\|_{L^2_x} dt' dt'',$$

and the proof follows by inserting in the integrand the homogeneous estimate above. □

REMARK 2.5. In the situation where the initial data are dyadically localized in frequency space, the estimate (15) is valid [**9**] at the endpoint $\alpha_1 = -\frac{1}{2}, \alpha_2 = \frac{n-1}{2}$. Bourgain's argument also establishes the result with $\alpha_1 = -\frac{1}{2} + \delta, \alpha_2 = \frac{n-1}{2} + \delta$, which is not scale invariant. However, the full estimate fails at the endpoint.

PROBLEM 2.6. *Consider the following two questions:*

(1) *Prove that the full estimate at the endpoint is false by calculating the left and right sides of* (16) *in the situation where* $\widehat{\zeta_1} = \chi_{R_1}$ *with* $R_1 = \{\xi : \xi_1 = Ne^1 + O(N^{\frac{1}{2}})\}$ *(where* e^1 *denotes the first coordinate unit vector),* $\widehat{\psi_2}(\xi_2) = |\xi_2|^{-\frac{n-1}{2}} \chi_{R_2}$ *where* $R_2 = \{\xi_2 : 1 \ll |\xi_2| \ll N^{\frac{1}{2}}, \xi_2 \cdot e^1 = O(1)\}$ *and* $g(u,v) = \chi_{R_0}(u,v)$ *with* $R_0 = \{(u,v) : u = Ne^1 + O(N^{\frac{1}{2}}), v = |u|^2 + O(N)\}$.
(2) *Use the same counterexample to show that the estimate*

$$\|u\overline{v}\|_{L^2_{t,x}} \lesssim \|\zeta\|_{\dot{H}^{\alpha}_1} \|\psi\|_{\dot{H}^{\alpha}_2},$$

where $u(t) = e^{it\Delta}\zeta$, $v(t) = e^{it\Delta}\psi$, *also fails at the endpoint.*

3. The Nonlinear Schrödinger Equation (NLS) in $\mathbb{R}^n$: Conservation Laws, Classical Morawetz and Virial Identity, Invariances for the Equation

In this section we consider the (NLS) IVP (1) and we formally talk about the solution $u(t,x)$ as an object that exists, is smooth etc. Of course to be able to use whatever we say here later we will need to work on making this formal assumption true!

Given an equation it is always a good idea to read as much as possible out of it. So one should always ask what are the rigid constraints that an equation imposes on its solutions a-priori. Here we will look at conservation laws (in this case integrals involving the solution that are independent of time), some inequalities (or monotonicity formulas) that a solution has to satisfy, symmetries and invariances that a solution to (1) can be subject to. All three of these elements are somehow related (see for example Noether's theorem [**73**]) and here we will not even attempt to discuss ALL the possible connections. It is true though that in describing these important features of the equation one often has to recall some basic principles/quantities coming from physics like conservation of mass, energy and momentum, the notion of density, interaction of particles, resonance etc.

3.1. Conservation laws. A simple way to interpret physically the function $u(t,x)$ solving a Schrödinger equation is to think about $|u(t,x)|^2$ as the particle density at place x and at time t. Then it shouldn't come as a surprise that the density, momentum and energy are conserved in time. More precisely if we introduce the *pseudo-stress-energy tensor* $T_{\alpha,\beta}$ for $\alpha, \beta = 0, 1, ..., n$ then

$$
\begin{aligned}
(17)\, T_{00} &= |u|^2 \quad \text{(mass density)} \\
(18)\, T_{0j} &= T_{j0} = Im(\bar{u}\partial_{x_j} u) \quad \text{(momentum density)} \\
(19)\, T_{jk} &= Re(\partial_{x_j} u \overline{\partial_{x_k} u}) - \frac{1}{4}\delta_{j,k}\Delta(|u|^2) + \lambda\frac{p-1}{p+1}\delta_{jk}|u|^{p+1} \quad \text{(stress tensor)}
\end{aligned}
$$

then by using the equation one can show that

$$\partial_t T_{00} + \partial_{x_j} T_{0j} = 0 \quad \text{and} \quad \partial_t T_{j0} + \partial_{x_k} T_{jk} = 0 \tag{20}$$

for all $j, k = 1, ..., n$.

PROBLEM 3.2. *Prove* (20) *using the equation.*

The conservation laws summarized in (20) are said to be *local* in the sense that they hold pointwise in the physical space. Clearly by integrating in space and assuming that u vanishes at infinity one also has the conserved integrals

$$m(t) = \int T_{00}(t, x)\, dx = \int |u|^2(t, x)\, dx \quad \text{(mass)} \tag{21}$$

$$p_j(t) = -\int T_{0j}(t, x) = -\int Im(\bar{u}\partial_{x_j} u)\, dx \quad \text{(momentum).} \tag{22}$$

We observe here that the stress tensor in (19) is not conserved, but it plays an important role in some "sophisticated" monotonicity formulas involving the solution u. To obtain the conservation of energy $E(t)$ we need to remember that the total energy of a system at time t is

$$E(t) = K(t) + P(t)$$

the sum of kinetic and potential energy. In our case

$$K(t) = \frac{1}{2}\int |\nabla u|^2(t, x)\, dx \quad \text{and} \quad P(t) = \frac{2\lambda}{p+1}\int |u(t, x)|^{p+1}\, dx$$

and hence

$$E(t) = \frac{1}{2}\int |\nabla u|^2(t, x)\, dx + \frac{2\lambda}{p+1}\int |u(t, x)|^{p+1}\, dx = E(0). \tag{23}$$

We immediately observe that now the sign of λ plays a very important role since by picking $\lambda = -1$ one can produce a negative energy. We will discus this later in greater details.

PROBLEM 3.3. *Prove the conservation of energy* (23) *by using the equation.*

As we will see, to have an a-priori control in time of an energy like in (23) when $\lambda = 1$ is an essential tool in order to prove that a solution exists for all times. But it is also true that often this is not sufficient. This is indeed the case when the problem is *critical* [10]. We need then other a-priori controls on norms for the solution u. This is the content of the next subsection.

[10]The notion of criticality will be introduced below.

3.4. Viriel and Classical Morawetz Identities. The Viriel identity was first introduced by Glassey [**40**] to show blowup for certain focusing ($\lambda = -1$) NLS problems. The classical[11] Morawetz identity was introduced instead by Morawetz in the context of the wave equations [**64**]. In the NLS case it was introduced by Lin and Strauss [**61**]. Morawetz type identities are particularly useful in the defocusing setting ($\lambda = 1$).

In general these identities are used in order to show that a positive quantity (often a norm) involving the solution u has a monotonic behavior in time. Monotonic quantities are used systematically in the context of elliptic equations and although both the Viriel and Morawetz estimates go back to the 70's only recently they have been used, together with their variations, in a surprisingly powerful way in the context of dispersive equations.

Suppose that a function $a(x)$ is measuring a particular quantity for our system[12] and we want to look at its overage value and in particular at its change in time. To do so we integrate $a(x)$ against the mass density tensor in (17) and we compute using (20) and integration by parts

$$\partial_t \int a(x)|u|^2(t,x)\,dx = \int \partial_{x_j} a(x) Im(\bar{u}\partial_{x_j} u)(t,x)\,dx. \tag{24}$$

At this stage there is no obvious sign for the right hand side of the equality. The integrals appearing above have special names. In fact we can introduce the following definition:

DEFINITION 3.5. Given the IVP (1), we define the associated Virial potential

$$V_a(t) = \int a(x)|u(t,x)|^2 dx \tag{25}$$

and the associated Morawetz action

$$M_a(t) = \int \partial_{x_j} a(x)\mathrm{Im}(\overline{u}\partial_{x_j} u)dx. \tag{26}$$

By taking the second derivative in time and by using again (20), we obtain

$$\begin{aligned}\partial_t^2 V_a(t) &= \partial_t^2 \int a(x)|u|^2(t,x)\,dx = \partial_t M_a(t) = \int (\partial_{x_j}\partial_{x_k} a(x)) Re(\partial_{x_j} u \partial_{x_k}\overline{u})\,dx \\ &+ \frac{\lambda(p-1)}{p+1}\int |u(t,x)|^{p+1}\Delta a(x)\,dx - \frac{1}{4}|u|^2(t,x)\Delta^2 a(x)\,dx.\end{aligned}$$

Now let's make a particular choice for $a(x)$.

- If $a(x) = |x|^2$, then $\Delta^2 a(x) = 0$ and $\Delta a(x) = 2n$ so

$$\partial_t^2 \int |x|^2|u|^2(t,x)\,dx = 4E + \frac{2\lambda}{p+1}[n(p-1)-4]\int |u|^{p+1}\,dx. \tag{27}$$

REMARK 3.6. For example in the focusing case $\lambda = -1$, when $n = 3$ and $p > \frac{7}{3}$, if one starts with $E < 0$, then the function $f(t) = \int |x|^2|u|^2(t,x)\,dx$ is concave down and positive ($f'(t)$ is monotone decreasing). Hence there exists $T^* < \infty$ such that there the function cannot

[11]Here we talk about *classical* Morawetz type identities in order to distinguish them from the Interaction Morawetz ones.

[12]For example $a(x)$ could represent the distance to a particular point, or the characteristic function of a particular domain.

longer exists. This was in fact the original argument of Glassey to show the existence of blowup time for certain focusing NLS equations.

- If $a(x) = |x|$, then (24) becomes

$$\partial_t \int |x||u|^2(t,x)\,dx = \int Im(\bar{u}\frac{x}{|x|}\cdot\nabla u)(t,x)\,dx, \tag{28}$$

and from here

$$\begin{aligned}(29)\partial_t M_{|x|} &= \partial_t \int Im(\bar{u}\frac{x}{|x|}\cdot\nabla u)(t,x)\,dx = \int \frac{|\nabla\!\!\!/ u(t,x)|^2}{|x|}\,dx \\ &+ \frac{2(n-1)(p-1)\lambda}{p+1}\int \frac{|u(t,x)|^{p+1}}{|x|}\,dx - \frac{1}{4}\int |u(t,x)|^2(\Delta^2|x|)\,dx,\end{aligned}$$

where $\nabla\!\!\!/ u := \nabla - \frac{x}{|x|}(\frac{x}{|x|}\cdot\nabla)$ denotes the angular gradient of u.

PROBLEM 3.7. *Above we used $a(x) = |x|$ which is clearly non smooth at zero. Check that if we take $n \geq 3$ and we replace $|x|$ with $\sqrt{x^2+\epsilon^2}$ and let $\epsilon \to 0$, then the identity (29) is correct.*

One can then compute that for $n \geq 3$, $(\Delta^2|x|) \leq 0$ in the sense of distributions. As a consequence, in the defocusing case $\lambda = 1$, after integrating in time over an interval $[t_0, t_1]$ one has

(30)
$$\int_{t_0}^{t_1} \frac{|\nabla\!\!\!/ u(t,x)|^2}{|x|}\,dx, \int_{t_0}^{t_1}\int \frac{|u(t,x)|^{p+1}}{|x|}\,dx \lesssim \sup_{[t_0,t_1]}\left|\int Im(\bar{u}\frac{x}{|x|}\cdot\nabla u)(t,x)\,dx\right|.$$

One can easily estimate the right hand side as

$$\sup_{[t_0,t_1]}\left|\int Im(\bar{u}\frac{x}{|x|}\cdot\nabla u)(t,x)\,dx\right| \lesssim \|u_0\|_{L^2}E^{1/2}$$

by using both conservation of mass and energy. But if less regularity is preferable then one can use the Hardy inequality (see Lemma A.10 in [**73**]) as in Lemma 6.9 that will be introduced later in Section 6, to obtain

$$\int_{t_0}^{t_1} \frac{|\nabla\!\!\!/ u(t,x)|^2}{|x|}\,dx, \int_{t_0}^{t_1}\int \frac{|u(t,x)|^{p+1}}{|x|}\,dx \lesssim \sup_{[t_0,t_1]}\|u(t)\|^2_{H^{1/2}}, \tag{31}$$

where now the disadvantage is the fact that the $H^{1/2}$ norm of u is not uniformly bounded in time.

3.8. Invariances and symmetries. In this section we only list invariances and symmetries but we do not attempt to describe their usefulness and applications except for one of them that we will start using in today's lecture.

(1) **Scaling Symmetry:** If u solve the IVP (1) then

$$u_\mu(t,x) = \mu^{-\frac{2}{p-1}}u\left(\frac{t}{\mu^2},\frac{x}{\mu}\right) \quad \text{and} \quad u_{\mu,0}(x) = \mu^{-\frac{2}{p-1}}u_0\left(\frac{x}{\mu}\right) \tag{32}$$

solves the IVP for any $\mu \in \mathbb{R}$.

(2) **Galilean Invariance:** If u is again a solution to (1) then

$$e^{ix\cdot v}e^{it|v|^2/2}u(t,x-vt) \quad \text{with initial data} \quad e^{ix\cdot v}u_0(x)$$

for every $v \in \mathbb{R}^n$ also solves the same IVP.

(3) **Obvious Symmetries:** Time and space translation invariance, spatial rotation, phase rotation symmetry $e^{i\theta}u$, time reversal.

(4) **Pseudo-conformal Symmetry:** In the case $p = 1 + \frac{4}{n}$, if u is solution for (1) then also

$$\frac{1}{|t|^{n/2}} u\left(\frac{1}{t}, \frac{x}{t}\right) e^{i|x|^2/2t} \tag{33}$$

for $t \neq 0$ is solution to the same equation.

We now concentrate on the **scaling symmetry** and we show how this can be used to understand for which nonlinerity (or for which $p > 1$) the problem of well-posedness is most difficult to address.

If we compute $\|u_{\mu,0}\|_{\dot{H}^s}$ we see that

$$\|u_{\mu,0}\|_{\dot{H}^s} = \mu^{-s+s_c}\|u_0\|_{\dot{H}^s}, \tag{34}$$

where

$$s_c = \frac{n}{2} - \frac{2}{p-1}.$$

Let us consider the rescaled initial data $u_{\mu,0}$ and the associated solution $u_\mu(t, x)$ that is now defined in the time interval $[0, \mu^2 T]$. From (34) it is clear that if we take $\mu \to +\infty$ then

(1) if $s > s_c$ (**sub-critical case**) the norm of the initial data can be made small while at the same time the interval of time is made longer: our intuition says that this is the best possible setting for well-posedness,

(2) if $s = s_c$ (**critical case**) the norm is invariant while the interval of time is made longer. This looks like a problematic situation.

(3) if $s < s_c$ (**super-critical case**) the norm grows as the time interval gets longer. Scaling is obviously against us.

In order to have a better intuition for scaling that also relates the dispersive part of the solution Δu with the nonlinear part of it $|u|^{p-1}u$, we use an informal argument as in [**73**]. Let's consider a special type of initial wave u_0. We want u_0 such that its support in Fourier space is localized at a large frequency $N \gg 1$, its support in space is inside a Ball of radius $1/N$ and its amplitude is A. Here we are making the assumption that scaling is the only symmetry that could interfere with a behavior that goes from linear to nonlinear, but in general this is not the only one. We have

$$\|u_0\|_{L^2} \sim AN^{-n/2}, \quad \|u_0\|_{\dot{H}^s} \sim AN^{s-n/2}.$$

If we want $\|u_0\|_{\dot{H}^s}$ small then we need to ask that $A \ll N^{n/2-s}$. Now under this restriction we want to compare the linear term Δu with the nonlinear part $|u|^{p-1}u$:

$$|\Delta u| \sim AN^2 \quad \text{while} \quad |u|^p \sim A^p.$$

From here if $AN^2 \gg A^p$ we believe that the linear behavior would win, alternatively the nonlinear one would. Putting everything together we have that

$$A^{p-1} \ll N^2 \quad \text{and} \quad A \ll N^{n/2-s} \Longrightarrow s > s_c \quad \text{(more linear)} \tag{35}$$

$$A^{p-1} \gg N^2 \quad \text{and} \quad A \gg N^{n/2-s} \Longrightarrow s < s_c \quad \text{(more nonlinear)}. \tag{36}$$

As announced at the beginning the so called "scaling argument" presented here should only be used as a guideline since in delivering it we make a purely formal calculation. On the other hand in some cases ill-posedness results below critical exponent have been obtained (see for example [**22, 23**]).

PROBLEM 3.9. *Prove the conservation of mass using Fourier transform for the IVP* (1) *when* $n = 1$ *and* $p = 3$.

3.10. Definition of well-posedness. We conclude this lecture by giving the precise definition of local and global well-posedness for an initial value problem, which in this case we will specify to be of type (1).

DEFINITION 3.11 (Well-posedness). We say that the IVP (1) is *locally well-posed (l.w.p)* in $H^s(\mathbb{R}^n)$ if for any ball B in the space $H^s(\mathbb{R}^n)$ there exist a time T and a Banach space of functions $X \subset L^\infty([-T,T], H^s(\mathbb{R}^n))$ such that for each initial data $u_0 \in B$ there exists a unique solution $u \in X \cap C([-T,T], H^s(\mathbb{R}^n))$ for the integral equation

$$u(t,x) = S(t)u_0 - i\lambda \int_0^t S(t-t')|u|^{p-1}u(t'))\,dt'. \tag{37}$$

Furthermore the map $u_0 \to u$ is continuous as a map from H^s into $C([-T,T], H^s(\mathbb{R}^n))$. If uniqueness is obtained in $C([-T,T], H^s(\mathbb{R}^n))$, then we say that local well -posedness is *unconditional.*

If this hold for all $T \in \mathbb{R}$ then we say that the IVP is *globally well-posed (g.w.p).*

REMARK 3.12. Our notion of global well-posedness does not require that $\|u(t)\|_{H^s(\mathbb{R}^n)}$ remains uniformly bounded in time. In fact, unless $s = 0, 1$ and one can use the conservation of mass or energy, it is not a triviality to show such an uniform bound. This can be obtained as a consequence of scattering, when scattering is available. In general this is a question related to *weak turbulence theory.*

4. Local and global well-posedness for the $H^1(R^n)$ subcritical NLS

Our intuition suggests that if one assumes enough regularity then l.w.p. should be true basically for any $p > 1$. We do not prove this here but one can check this in [**19, 73**], or use the argument that we will present below and the fact that for $s > n/2$ the space H^s is an algebra to obtain this result directly. Here we consider instead the IVP (1) with a nonlinearity that is H^1 subcritical, that is $1 < p < 1 + \frac{4}{n-2}$ for $n \geq 3$ and $1 < p < \infty$ for $n = 1, 2$. To prove l.w.p for $H^s(\mathbb{R}^n)$, the general strategy that we will follow is based on the contraction method. This method is based on these four steps:

(1) Definition of the operator
$$L(v) = \chi(t/T)S(t)u_0 + c\chi(t/T)\int_0^t S(t-t')|v|^{p-1}v(t'))\,dt'$$
where $\chi(r)$ denotes a smooth nonnegative bump even function, supported on $-2 \leq r \leq 2$ and satisfying $\chi(r) = 1$ for $-1 \leq r \leq 1$.
(2) Definition of a Banach space X such that $X \subset L^\infty([-T,T], H^s(\mathbb{R}^n))$.
(3) Proof of the fact that for any ball $B \subset H^s(\mathbb{R}^n)$, there exist T and a ball $B_X \subset X$ such that the operator L sends B_X into itself and it is a contraction there.
(4) Extension of the uniqueness result in B_X to a unique result in the whole space X.

We observe that the continuity with respect to the initial data will be a consequence of the fact that the solution is found through a contraction argument. In fact in this case we obtain way more than just continuity.

PROBLEM 4.1. *Discuss the regularity of the map* $u_0 \to u$ *from* H^s *into* $L^\infty([-T,T], H^s(\mathbb{R}^n))$ *when l.w.p. is proved by contraction method.*

We state the main theorem (for a complete list of authors who contributed to the final version of this theorem see [**19**]):

THEOREM 4.2. *Assume that* $1 < p < 1 + \frac{4}{n-2}$ *for* $n \geq 3$ *and* $1 < p < \infty$ *for* $n = 1, 2$. *Then the IVP* (1) *is l.w.p in* $H^s(\mathbb{R}^n)$ *for all* $s_c < s \leq 1$, *where* $s_c = \frac{n}{2} - \frac{2}{p-1}$. *Moreover if the nonlinearity is algebraic, that is* $n = 2, 3$ *and* $p = 3$, *then there is persistence of regularity, that is if* $u_0 \in H^m$, $m \geq 1$ *then the solution* $u(t) \in H^m(\mathbb{R}^n))$, *for all* t *in its time of existence. If in* (1) *we assume that* $\lambda = 1$ *(defocusing) then the IVP is globally well-posed for* $s = 1$.

Here we prove a less general version of this theorem, namely that under the conditions given above on p there is g.w.p in H^1. We do not prove l.w.p. for $s_c < s \leq 1$ since we would need to introduce a product rule for fractional derivatives and it would become too technical.

Our starting point is the definition of a Banach space X based on the norms we introduced with the Strichartz estimates.

DEFINITION 4.3. Assume $I = [-T, T]$ is fixed. The space $S^0(I \times \mathbb{R}^n)$ is the closure of the Schwartz functions under the norm

$$\|f\|_{S^0(I\times\mathbb{R}^n)} = \sup_{(q,r) \text{ admissible}} \|f\|_{L^q_t L^r_x}.$$

We then define the space $S^1(I \times \mathbb{R}^n)$ where the closure is taken with respect to the norm

$$\|f\|_{S^1(I\times\mathbb{R}^n)} = \|f\|_{S^0(I\times\mathbb{R}^n)} + \|\nabla f\|_{S^0(I\times\mathbb{R}^n)}.$$

PROOF. We consider the operator Lv and using (12) and (13) we obtain

$$\|Lv\|_{S^1(I\times\mathbb{R}^n)} \leq C_1\|u_0\|_{H^1} + C_2\||v|^{p-1}(|v| + |\nabla v|)\|_{L^{q'}_t L^{r'}_x}, \tag{38}$$

where (q, r) is a Strichartz admissible pair. Below we will only estimate the term in the right hand side of 38 that contains the gradient. To treat the other term one can use interpolation and Sobolev embedding theorem. The best couple to use in this context is the one that solves the system

$$\frac{2}{q} + \frac{n}{r} = \frac{n}{2} \quad \text{Strichartz Condition} \tag{39}$$

$$(p-1)\left(\frac{1}{r} - \frac{s}{n}\right) = \frac{1}{r'} - \frac{1}{r}, \tag{40}$$

and the meaning of the second equation will become clear below. The solutions to the system is

$$\frac{1}{r} = \frac{1}{(p+1)} + \frac{(p-1)}{(p+1)}\frac{s}{n} \quad \text{and} \quad \frac{1}{q} = \frac{(p-1)(n-2s)}{4(p+1)}.$$

From here it follows that[13]

$$\frac{1}{q'} > \frac{p}{q} \quad \Longrightarrow \quad s > s_c = \frac{n}{2} - \frac{2}{p-1}.$$

[13] As mentioned above here we only address l.w.p. in H^1, but it is clear that if one uses fractional derivatives and (41) l.w.p in H^s, $s > s_c$ can also be obtained based on the fact that r and q are given in terms of s and $s > s_c$.

Then by Hölder inequality repeated

$$\||v|^{p-1}|\nabla v|\|_{L_t^{q'}L_x^{r'}} \leq T^\alpha \||v|^{p-1}|\nabla v|\|_{L_t^{q/p}L_x^{r'}} \leq \|\nabla v\|_{L_t^q L_x^r}\|v\|^{p-1}_{L_t^q L_x^{\tilde r}}$$

where $\frac{1}{\tilde r} = \frac{1}{r} - \frac{s}{n}$. By Sobolev embedding

$$\|v\|_{L_t^q L_x^{\tilde r}} \lesssim \|(1+\Delta^{\frac{s}{2}})v\|_{L_t^q L_x^r}, \tag{41}$$

and since we are assuming that we are in the H^1 subcritical regime $1 < p < 1 + \frac{4}{n-2}$ it also follows that $s \leq 1$ and as a consequence

$$\||v|^{p-1}|\nabla v|\|_{L_t^{q'}L_x^{r'}} \leq T^\alpha \|v\|^p_{S^1}.$$

We can now conclude that

$$\|Lv\|_{S^1(I\times\mathbb{R}^n)} \leq C_1\|u_0\|_{H^1} + C_2 T^\alpha \|v\|^p_{S^1}. \tag{42}$$

With similar arguments one also obtains

$$\|Lv - Lw\|_{S^1(I\times\mathbb{R}^n)} \leq C_2 T^\alpha (\|v\|^{p-1}_{S^1} + \|w\|^{p-1}_{S^1})\|v-w\|_{S^1}. \tag{43}$$

We are now ready to set up the contraction: pick $R = 2C_1\|u_0\|_{H^1}$ and T such that

$$C_2 T^\alpha R^{p-1} < \frac{1}{2} \iff T \lesssim \|u_0\|_{H^1}^{\frac{1-p}{\alpha}}, \tag{44}$$

then clearly from (42), (43) and (44) it follows that $L : B_R \to B_R$, where B_R is the ball centered at zero and radius R in S^1, and L is a contraction. There is a unique fixed point $u \in B_R$ that is in fact a solution to our integral equation. The next two properties for u that we need to show are continuity with respect to time, that is $u \in C([-T,T], H^1)$ and uniqueness in the whole space S^1. The first is left to the reader since it is a simple consequence of the representation of u through the Duhamel formula (6). For the second we assume that there exists another solution $\tilde u \in S^1$ for the IVP (1). Using again the Duhamel formula for both u and $\tilde u$ and the estimates presented above for Lv we obtain that on an interval of time δ

$$\|u-\tilde u\|_{S^1_\delta} \leq C_2\delta^\alpha(\|\tilde u\|^{p-1}_{S^1_T} + \|u|^{p-1}_{S^1_T})\|u-\tilde u\|_{S^1_\delta}$$

where here we use the lover index δ or T to stress that in the first case the space S^1 is relative to the interval $[-\delta,\delta]$ and in the second to $[-T,T]$. Since u and $\tilde u$ are fixed we can introduce

$$M = \max(\|\tilde u\|^{p-1}_{S^1_T} + \|u|^{p-1}_{S^1_T})$$

and if δ is small enough in terms of C_2, α and M we obtain

$$\|u-\tilde u\|_{S^1_\delta} \leq \frac{1}{2}\|u-\tilde u\|_{S^1_\delta}$$

which forces $u = \tilde u$ in $[-\delta,\delta]$. To cover the whole interval $[-T,T]$ then one iterates this argument $\frac{T}{\delta}$ times and the conclusion follows.

Before going to the proof of g.w.p we would like to consider the question of **propagation of regularity**. As mentioned above with this we mean the answer to the following question: assume that in (1), with the restrictions on p above, we start with $u_0 \in H^m$, $m \geq 1$. Is it true that the unique solution $u \in S^1$ also belongs to H^m at any later time $t \in [0,T]$? The answer to this depends on the regularity of the non-linear term, more precisely the regularity of the function $f(z) = |z|^{p-1}z$. This function is not C^∞ for all p, hence one cannot expect propagation of regularity for all p in the considered range. On the other hand if f is algebraic, namely when $p-1 = 2k$ for some $k \in \mathbb{N}$, then propagation of regularity follows from the estimates we presented above. Briefly we can go back to (42) and if we repeat the same argument we obtain that for the solution u that we already found using only H^1 regularity we also have

$$\|D^m u\|_{S^0} \leq C_1\|u_0\|_{H^m} + C_2T^\alpha\|u\|_{S^1}^{2k}\|D^m u\|_{S^0}$$

because when we apply the operator D^m the term with $D^m u$ appears linearly[14]. Since we already know that $C_2T^\alpha\|u\|_{S^1}^{2k} \leq \frac{1}{2}$ we then obtain[15] that

$$\|D^m u\|_{S^0} \lesssim \|u_0\|_{H^m}.$$

We are now ready for the iteration of the local in time solution u to a uniformly global one[16]. The first step is to go back to (44) and notice that T depends on the H^1 norm of the initial data. From the previous lecture we learned that for a *smooth*[17] solution u to (1) the conservation of the energy and mass gives an a priori uniform bound

$$\|u(t)\|_{H^1} \leq C^*(\|u_0\|_{H^1}),$$

so if we take now $T^* \sim (C^*)^{\frac{1-p}{\alpha}}$ we can repeat the argument above with no changes. In particular when we get to time T^* we can apply the argument again with the new initial data $u(T^*)$ and the same T^* will work. In this way we can cover the whole time real lime and well-posedness becomes global. But in the argument we just outlined there is a *caviat* in the sense that if $u_0 \in H^1$ we do not have a *smooth* solution u. This obstacle can be overcome by introducing various smoothing tools. The precise argument can be found in [**19**]. □

REMARK 4.4. We are not addressing in this first part of the course g.w.p. for the focusing NLS (1) even in the subcritical case. In order to address this issue we need to introduce stationary solutions (or solitons) and this will be done later.

REMARK 4.5. By carefully keeping track of the various exponents that have been introduced in order to get to (42) one can see that for the **critical** H^1 problem, that is $p = 1 + \frac{4}{n-2}$, the estimates are border line. In fact one gets

$$\|Lv\|_{S^1(I\times\mathbb{R}^n)} \leq C_1\|u_0\|_{H^1} + C_2\|v\|_{S^1}^p. \tag{45}$$

The main difference between this and (42) is that there is no time factor appearing in the right hand side. This of course makes the contraction more difficult to attain by shrinking the time. On the other hand if one starts with small data $\|u_0\|_{H^1} \leq \epsilon$ and calls now $R = 2C_1\epsilon$, then a sufficient condition on ϵ to have a contraction would be

$$C_2R^{p-1} = C_2(2C_1\epsilon)^{p-1} \leq \frac{1}{2}.$$

[14]Here we are cheating a little since we are ignoring the mixed lower order derivatives. For this reason the constant C_2 is the same as the one in (42). If one does this calculation correctly then that constant C_2 will need to be replaced by a larger one, which will shrink the time T. To cover the whole interval $[-T, T]$ then one uses the iteration we introduced while proving uniqueness in S^1.

[15]Here we are cheating again in the sense that in principle we cannot even talk about $D^m u$ since we don't know yet that this expression makes sense. The rigorous procedure tells us to start with a smooth and decaying approximation of the initial data, the associated solution exists and is unique. Only at this point one can use the argument proposed here to get the uniform bound independent of the approximation.

[16]This argument only works when a uniform H^1 bound in time for the solution is available, for example in the defocusing case or when the L^2 norm of the initial data is small enough.

[17]Here with smooth we also mean zero at infinity.

This would also guarantee a uniform global solution in H^1.

One could ask if at least l.w.p could be still achieved for large data. The following theorem gives a positive answer.

THEOREM 4.6 (L.w.p. for H^1 critical NLS). *Assume that $p = 1 + \frac{4}{n-2}$ and $u_0 \in H^1$. Assume also that there exists T such that*

$$\|S(t)u_0\|_{L^{\frac{2(n+2)}{n-2}}_{[-T,T]} W_x^{1,\frac{2n(n+2)}{n^2+4}}} \leq \epsilon \tag{46}$$

for ϵ small enough. Then (1) is H^1 well posed in $[-T,T]$.

PROOF. We first notice that the pair $(\frac{2(n+2)}{n-2}, \frac{2n(n+2)}{n^2+4})$ is Strichartz admissible. We define the new space $\tilde{S}^1$ using the following norm

$$\|f\|_{\tilde{S}^1} := T\|f\|_{S^1} + \|f\|_{L^{\frac{2(n+2)}{n-2}}_{[-T,T]} W_x^{1,\frac{2n(n+2)}{n^2+4}}},$$

The idea is to use a contraction method in this space based on the smallness assumption (46). As we did in the proof of Theorem 4.2 we estimate Lv in the space $\tilde{S}^1$:

$$\begin{aligned} \|Lv\|_{\tilde{S}^1} &\lesssim T\|u_0\|_{H^1} + \|S(t)u_0\|_{L^{\frac{2(n+2)}{n-2}}_{[-T,T]} W_x^{1,\frac{2n(n+2)}{n^2+4}}} \\ &+ \||v|^{\frac{4}{n-2}}(|v| + |\nabla v|)\|_{L^{\tilde{q}'}_{[-T,T]} L^{\tilde{r}'}_x} \end{aligned}$$

Now we pick the Strichartz pair $(\tilde{q}, \tilde{r}) = (2, \frac{2n}{n-2})$ and we obtain by Hölder

$$\||v|^{\frac{4}{n-2}}(|v| + |\nabla v|)\|_{L^{\tilde{q}'}_{[-T,T]} L^{\tilde{r}'}_x} \lesssim \|v\|^{\frac{4}{n-2}}_{L^{\frac{2(n+2)}{n-2}}_{[-T,T]} L_x^{\frac{2(n+2)}{n-2}}} \|v\|_{L^{\frac{2(n+2)}{n-2}}_{[-T,T]} W_x^{1,\frac{2n(n+2)}{n^2+4}}}.$$

By the Sobolev embedding theorem we then have

$$\|v\|_{L^{\frac{2(n+2)}{n-2}}_{[-T,T]} L_x^{\frac{2(n+2)}{n-2}}} \lesssim \|v\|_{L^{\frac{2(n+2)}{n-2}}_{[-T,T]} W_x^{1,\frac{2n(n+2)}{n^2+4}}},$$

hence the final bound

$$\|Lv\|_{\tilde{S}^1} \lesssim T\|u_0\|_{H^1} + \|S(t)u_0\|_{L^{\frac{2(n+2)}{n-2}}_{[-T,T]} W_x^{1,\frac{2n(n+2)}{n^2+4}}} + \|v\|^{1+\frac{4}{n-2}}_{L^{\frac{2(n+2)}{n-2}}_{[-T,T]} W_x^{1,\frac{2n(n+2)}{n^2+4}}}. \tag{47}$$

Now if T is small enough, in particular $T \sim \epsilon\|u_0\|^{-1}_{H^1}$, using (46), we deduce from (47) that

$$\|Lv\|_{\tilde{S}^1} \leq 2C_0\epsilon + C_1\|v\|^{1+\frac{4}{n-2}}_{L^{\frac{2(n+2)}{n-2}}_{[-T,T]} W_x^{1,\frac{2n(n+2)}{n^2+4}}}.$$

We then take a ball B of radius $R = 4C_0\epsilon$ and if ϵ is small enough then L sends B into itself and it is a contraction. The rest is now routine. This argument proved the theorem in the interval of time of length approximately $\epsilon\|u_0\|^{-1}_{H^1}$. In order to cover an arbitrary interval $[-T,T]$, then one has to use again the conservation of energy and mass that gives a uniform bound on $\|u\|_{H^1}$. □

REMARK 4.7. We have the following two facts:

(1) By the homogeneous Strichartz estimate (12) it follows that

$$\|S(t)u_0\|_{L^{\frac{2(n+2)}{n-2}}_{[-T,T]} W_x^{1,\frac{2n(n+2)}{n^2+4}}} \lesssim \|u_0\|_{H^1}$$

hence we recover above the small data g.w.p we discussed in Remark 4.5.

(2) Given any data $u_0 \in H^1$, again by (12) we have

$$\|S(t)u_0\|_{L^{\frac{2(n+2)}{n-2}}_{[-T,T]} W_x^{1,\frac{2n(n+2)}{n^2+4}}} \leq C,$$

so we can use the time integral to claim that for T small enough (46) is satisfied. This gives l.w.p. but it is important to notice that in this case $T = T(u_0)$ depends also on the profile of the initial data, not only on its H^1 norm.

The next theorem gives a sort of criteria for the g.w.p. of the H^1 critical NLS. It says that if a certain Strichartz norm of the solution (actually any of them would do!) stays a-priori bounded, then g.w.p. follows.

THEOREM 4.8 (G.w.p. for H^1 critical NLS with $L_t^{\frac{2(n+2)}{n-2}} L_x^{\frac{2(n+2)}{n-2}}$ bound). *Assume that $p = 1 + \frac{4}{n-2}$ and $u_0 \in H^1$. Assume also the a priori estimate*

$$\|u\|_{L^{\frac{2(n+2)}{n-2}}_{[-T,T]} L_x^{\frac{2(n+2)}{n-2}}} \leq C \tag{48}$$

for any solution u to (1) with $p = 1+\frac{4}{n-2}$. Then this IVP is H^1 globally well posed.

PROOF. Fix ϵ to be determined later. Also assume that our until data belongs to $H^k, k \geq 1$. Using (48) we can find finitely many intervals of time $I_1, ..., I_M$ such that

$$\|u\|_{L^{\frac{2(n+2)}{n-2}}_{I_j} L_x^{\frac{2(n+2)}{n-2}}} \leq \epsilon \tag{49}$$

for all $j = 1, .., M$. The goal here is to prove that as a consequence of (49) one actually has the stronger bound

$$\|u\|_{S^k_{I_j}} \leq C, \quad \text{for all} \quad k \geq 1, \tag{50}$$

for all $j = 1, .., M$ and putting all the intervals together

$$\|u\|_{S^k} \leq C, \quad \text{for all} \quad k \geq 1. \tag{51}$$

How do we use now this bound? We consider a method that is know as the *Energy Method.* This argument is based on a priori global bounds of high Sobolev norms, see for example [**38**] for details. In our case, if we start with data in $H^k, k \gg 1$, the bound (51) in particular gives a uniform bound of the solution in H^k, not just in H^1, which we knew as a consequence of the conservation of Hamiltonian and mass. This is enough to show that there is a unique, classical global solution for our initial value problem. If the initial data is only in H^1 then an approximation by data in $H^k, k \gg 1$ can be used and a continuity argument concludes the proof.

It is now time to prove (50). Using estimates like the ones in the proof of Theorem 4.6 this time applied to the Duhamel representation of a solution u we have

$$\|u\|_{S^1_{I_j}} \leq C_1\|u_0\|_{H^1} + C_2\|u\|^{\frac{4}{n-2}}_{L^{\frac{2(n+2)}{n-2}}_{I_j} L_x^{\frac{2(n+2)}{n-2}}} \|u\|_{S^1_{I_j}} \leq C_1\|u_0\|_{H^1} + C_2\epsilon^{\frac{4}{n-2}}\|u\|_{S^1_{I_j}}$$

and if $C_2\epsilon^{\frac{4}{n-2}} < 1/2$ then (50) follows. □

We end this section by announcing that similar theorems, replacing H^1 with L^2 are available for the L^2 subcritical NLS, that is when $1 < p < 1 + \frac{4}{n}$. We do not list them here, but they can be found in [**73**].

5. Global well-posedness for the $H^1(R^n)$ subcritical NLS and the "I-method"

We learned during last lecture that for the H^1 subcritical NLS, i.e. $1 < p < 1 + \frac{4}{n-2}$ and hence $s_c < 1$, l.w.p for (1), either focusing or defocusing, is available in $H^s(\mathbb{R}^n)$ for any $s, s_c \leq s \leq 1$. We also learned that if $s = 1$, in the defocusing case, uniform g.w.p is a consequence of the conservation of mass and energy. We then ask: if $0 \leq s_c < s < 1$ is the defocusing NLS problem globally well posed in H^s? This problem is particularly interesting when we consider the L^2 critical NLS, i.e. $s_c = 0$ and $p = 1 + \frac{4}{n}$. In this case the L^2 norm cannot be used to iterate the l.w.p. since the time interval of existence also depends on the profile of the initial data. It is clear then that this is a difficult question since we are in a regime when the conservation of the L^2 norm is too little of an information and the conservation of the Hamiltonian cannot be used since the data has not enough regularity. It was exactly to answer these kinds of questions that the "I-method" [**24, 25, 26, 27, 50, 51**] was invented. Unfortunately the method is quite technical to be applied in higher dimensions in its full strength. The results that we will report below are not optimal and in general they concern the L^2 critical case $p = 1 + \frac{4}{n}$ since that one is the most interesting, but similar results are available for the general H^1 subcritical case when $s_c < 1$ (see [**20, 78**]). We will list below the state of the art at this point for this problem for the L^2 critical case. We will give references but we will not prove these theorems in full generality. At the end of this lecture we will prove a weaker result than the one stated here when $n = 2$, see Theorem 5.2. We should also say here that if one assumes radial symmetry, then the L^2 critical NLS for $n \geq 2$ has been proved to be globally well-posed both in the defocusing and focusing case with the assumption that the mass of the initial data is strictly less than the mass of the stationary solution. These results are contained in a series of very recent and deep papers [**59, 60, 75, 76**], see also [**58**]. The point here is instead to address the question of global well-posedness without assuming radial symmetry and to present the "I-method".

THEOREM 5.1 (G.w.p for (1) with $\lambda = 1$, $p = 1 + \frac{4}{n}$ and $n \geq 3$). *The initial value problem* (1) *with* $\lambda = 1$, $p = 1 + \frac{4}{n}$ *is globally well-posed in* $H^s(\mathbb{R}^n)$, *for any* $1 \geq s > \frac{\sqrt{7}-1}{3}$ *when* $n = 3$, *and for any* $1 \geq s > \frac{-(n-2)+\sqrt{(n-2)^2+8(n-2)}}{4}$ *for* $n \geq 4$.

Here we have to assume that $s \leq 1$ since in general the non smoothness of the nonlinearity doesn't allow us to prove persistence of regularity. The proof of this theorem can be found in [**34**].

THEOREM 5.2 (G.w.p for (1) with $\lambda = 1$, $p = 1 + \frac{4}{n}$ and $n = 2$). *The initial value problem* (1) *with* $\lambda = 1, n = 2$ *and* $p = 3$ *is globally well-posed in* $H^s(\mathbb{R}^2)$, *for any* $1 > s > \frac{2}{5}$. *Moreover the solution satisfies*

$$\sup_{[0,T]} \|u(t)\|_{H^s} \leq C(1+T)^{\frac{3s(1-s)}{2(5s-2)}}, \tag{52}$$

where the constant C depends only on the index s and $\|u_0\|_{L^2}$.

Here the theorem is stated only for $s < 1$ since we already know that global well-posedness for $s \geq 1$ follows from conservation of mass and energy as explained in the previous lecture[18].

For the proof of Theorem 5.2 see [**32**]. The argument is based on a combination of the "I-method" as in [**25, 26, 28**] and a refined two dimensional Morawetz interaction inequality. This combination first appeared in [**39**].

Finally we recall the result for the L^2 critical problem for $n = 1$:

THEOREM 5.3 (G.w.p for (1) with $\lambda = 1$, $p = 1 + \frac{4}{n}$ and $n = 1$). *The initial value problem* (1) *with $\lambda = 1, n = 1$ and $p = 5$ is globally well-posed in $H^s(\mathbb{R})$, for any $1 > s > \frac{1}{3}$. Moreover the solution satisfies*

$$\sup_{[0,T]} \|u(t)\|_{H^s} \leq C(1+T)^{\frac{s(1-s)}{2(3s-1)}}, \tag{53}$$

where the constant C depends only on the index s and $\|u_0\|_{L^2}$.

For the proof of this theorem see [**36**].

As promised we sketch now the proof of a weaker result than the one reported in Theorem 5.2, namely g.w.p. for $s > 4/7$. This proof is a summary of the work that appeared in [**26**]. Since below we will often refer to a particular IVP we write it here once for all

$$\begin{cases} iu_t + \frac{1}{2}\Delta u = |u|^2 u, \\ u(x,0) = u_0(x). \end{cases} \tag{54}$$

To start the argument we need to introduce some notations and state some lemmas.

We will use the weighted Sobolev norms,

$$||\psi||_{X_{s,b}} \equiv ||\langle\xi\rangle^s \langle\tau - |\xi|^2\rangle^b \tilde{\psi}(\xi,\tau)||_{L^2(\mathbb{R}^n \times \mathbb{R})}. \tag{55}$$

Here $\tilde{\psi}$ is the space-time Fourier transform of ψ. We will need local-in-time estimates in terms of truncated versions of the norms (55),

$$||f||_{X^\delta_{s,b}} \equiv \inf_{\psi = f \text{on } [0,\delta]} ||\psi||_{X^\delta_{s,b}}. \tag{56}$$

We will often use the notation $\frac{1}{2}+ \equiv \frac{1}{2} + \epsilon$ for some universal $0 < \epsilon \ll 1$. Similarly, we shall write $\frac{1}{2}- \equiv \frac{1}{2} - \epsilon$, and $\frac{1}{2} -- \equiv \frac{1}{2} - 2\epsilon$.

For a Schrödinger admissible pair (q,r) we have what we will call the $L^q_t L^r_x$ Strichartz estimate:

$$||\phi||_{L^q_t L^r_x(\mathbb{R}^{n+1})} \lesssim ||\phi||_{X_{0,\frac{1}{2}+}}, \tag{57}$$

which can be proved to be a consequence of (55).

Finally, we will need a refined version of these estimates due to Bourgain [**9**].

[18]It is an open problem to obtain a polynomial bound like in (52) for this problem when $s > 1$ and the data are not radial. In fact if if $p > 3$ a uniform bound follows from scattering. But scattering is still an open problem for general data for the L^2 critical NLS. We should also stress that these kinds of polynomial bounds for higher Sobolev norms are particularly interesting since they are related to the *weak turbulence theory*, a topic that we will not address here.

LEMMA 5.4. *Let $\psi_1, \psi_2 \in X^{\delta}_{0,\frac{1}{2}+}$ be supported on spatial frequencies $|\xi| \sim N_1, N_2$, respectively. Then for $N_1 \leq N_2$, one has*

$$||\psi_1 \cdot \psi_2||_{L^2([0,\delta]\times\mathbb{R}^2)} \lesssim \left(\frac{N_1}{N_2}\right)^{\frac{1}{2}} ||\psi_1||_{X^{\delta}_{0,\frac{1}{2}+}} ||\psi_2||_{X^{\delta}_{0,\frac{1}{2}+}}. \tag{58}$$

In addition, (58) *holds (with the same proof) if we replace the product $\psi_1 \cdot \psi_2$ on the left with either $\overline{\psi}_1 \cdot \psi_2$ or $\psi_1 \cdot \overline{\psi}_2$.*

This lemma is a consequence of Theorem 2.4.

PROBLEM 5.5. *Show how to deduce* (57) *and* (58).
Hint: *Consider the space of frequencies both in time and space. Partition it into parabolic strips of approximate unit size. On each of these strips a function ψ can be viewed as a solution of the linear problem. Use the appropriate Strichartz or improved Strichartz on each of them and then sum with the appropriate weight.*

For rough initial data, with $s < 1$, the energy is infinite, and so the conservation law (23) is meaningless. Instead, here we use the fact that a smoothed version of the solution of the IVP (54) has a finite energy which is *almost* conserved in time. We express this 'smoothed version' as follows.

Given $s < 1$ and a parameter $N \gg 1$, define the multiplier operator

$$\widehat{I_N f}(\xi) \equiv m_N(\xi)\hat{f}(\xi), \tag{59}$$

where the multiplier $m_N(\xi)$ is smooth, radially symmetric, nonincreasing in $|\xi|$ and

$$m_N(\xi) = \begin{cases} 1 & |\xi| \leq N \\ \left(\frac{N}{|\xi|}\right)^{1-s} & |\xi| \geq 2N. \end{cases} \tag{60}$$

For simplicity, we will eventually drop the N from the notation, writing I and m for (59) and (60). Note that for solution and initial data u, u_0 of (54), the quantities $||u||_{H^s(\mathbb{R}^n)}(t)$ and $E(I_N u)(t)$ (see (23)) can be compared,

$$E(I_N u)(t) \leq \left(N^{1-s}||u(\cdot,t)||_{\dot{H}^s(\mathbb{R}^2)}\right)^2 + ||u(t,\cdot)||^4_{L^4(\mathbb{R}^2)}, \tag{61}$$

$$||u(\cdot,t)||^2_{H^s(\mathbb{R}^2)} \lesssim E(I_N u)(t) + ||u_0||^2_{L^2(\mathbb{R}^2)}. \tag{62}$$

Indeed, the $\dot{H}^1(\mathbb{R}^2)$ component of the left hand side of (61) is bounded by the right side by using the definition of I_N and by considering separately those frequencies $|\xi| \leq N$ and $|\xi| \geq N$. The L^4 component of the energy in (61) is bounded by the right hand side of (61) by using (for example) the Hörmander-Mikhlin multiplier theorem. The bound (62) follows quickly from (60) and L^2 conservation (21) by considering separately the $\dot{H}^s(\mathbb{R}^2)$ and $L^2(\mathbb{R}^2)$ components of the left hand side of (62).

To prove our result, we may assume that $u_0 \in C_0^\infty(\mathbb{R}^2)$, and show that the resulting global-in-time solution grows at most polynomially in the H^s norm,

$$||u(t,\cdot)||_{H^s(\mathbb{R}^2)} \leq C_1 t^M + C_2, \tag{63}$$

where the constants C_1, C_2, M depend only on $||u_0||_{H^s(\mathbb{R}^2)}$ and not on higher regularity norms of the smooth data. The result then follows immediately from (63), the local-in-time theory discussed in the previous lecture, and a standard density argument.

By (62), it suffices to show

$$E(I_N u)(t) \lesssim (1+t)^{2M}. \tag{64}$$

for some $N = N(t)$. (See (71), (72) below for the definition of N and the growth rate M we eventually establish). The following proposition, represents an "almost conservation law" and will yield (64).

PROPOSITION 5.6. *Given $s > \frac{4}{7}, N \gg 1$, and initial data $u_0 \in C_0^\infty(\mathbb{R}^2)$ (see preceding remark) with $E(I_N\phi_0) \leq 1$, then there exists a $\delta = \delta(||u_0||_{L^2(\mathbb{R}^2)}) > 0$ so that the solution*

$$u(x,t) \in C([0,\delta], H^s(\mathbb{R}^2))$$

of (54) *satisfies*

$$E(I_N u)(t) = E(I_N u)(0) + O(N^{-\frac{3}{2}+}), \tag{65}$$

for all $t \in [0,\delta]$.

We first show that Proposition 5.6 implies (64). Recall that the initial value problem here has a scaling symmetry, and is H^s-subcritical when $1 > s > 0$, and $n = 2$. That is, if u is a solution, so too

$$u_\lambda(x,t) := \frac{1}{\lambda}u(\frac{x}{\lambda}, \frac{t}{\lambda^2}). \tag{66}$$

Using (61), the following energy can be made arbitrarily small by taking λ large,

$$E(I_N u_{\lambda,0}) \leq \left((N^{2-2s})\lambda^{-2s} + \lambda^{-2}\right) \cdot (1 + ||u_0||_{H^s(\mathbb{R}^2)})^4 \tag{67}$$

$$\leq C_0(N^{2-2s}\lambda^{-2s}) \cdot (1 + ||u_0||_{H^s(\mathbb{R}^2)})^4. \tag{68}$$

It is important to remark that since the problem is L^2 critical, $\|u_0\|_{L^2} \sim \|u_{\lambda,0}\|_{L^2}$. Assuming $N \gg 1$ is given[19], we choose our scaling parameter $\lambda = \lambda(N, ||u_0||_{H^s(\mathbb{R}^2)})$

$$\lambda = N^{\frac{1-s}{s}} \left(\frac{1}{2C_0}\right)^{-\frac{1}{2s}} \cdot \left(1 + ||u_0||_{H^s(\mathbb{R}^2)}\right)^{\frac{2}{s}} \tag{69}$$

so that $E(I_N u_{\lambda,0}) \leq \frac{1}{2}$. We may now apply Proposition 5.6 to the scaled initial data $u_{\lambda,0}$, and in fact we may reapply this proposition until the size of $E(I_N u_\lambda)(t)$ reaches 1, that is at least $C_1 \cdot N^{\frac{3}{2}-}$ times. Hence

$$E(I_N u_\lambda)(C_1 N^{\frac{3}{2}-}\delta) \sim 1. \tag{70}$$

We now have to undo the scaling: given any $T_0 \gg 1$, we establish the polynomial growth (64) from (70) by first choosing our parameter $N \gg 1$ so that

$$T_0 \sim \frac{N^{\frac{3}{2}-}}{\lambda^2} C_1 \cdot \delta \sim N^{\frac{7s-4}{2s}-}, \tag{71}$$

where we've kept in mind (69). Note the exponent of N on the right of (71) is positive provided $s > \frac{4}{7}$, hence the definition of N makes sense for arbitrary T_0. In two space dimensions,

$$E(I_N u)(t) = \lambda^2 E(I_N u_\lambda)(\lambda^2 t).$$

[19]The parameter N will be chosen shortly.

We use (69), (70), and (71) to conclude that for $T_0 \gg 1$,

$$E(I_N u)(T_0) \leq C_2 T_0^{\frac{1-s}{\frac{7}{4}s-1}+}, \tag{72}$$

where N is chosen as in (71) and $C_2 = C_2(||u_0||_{H^s(\mathbb{R}^2)}, \delta)$. Together with (62), the bound (72) establishes the desired polynomial bound (63).

It remains then to prove Proposition 5.6. We will need the following modified version of the usual local existence theorem, wherein we control for small times the smoothed solution in the $X^{\delta}_{1,\frac{1}{2}+}$ norm.

PROPOSITION 5.7. *Assume $\frac{4}{7} < s < 1$ and we are given data for the IVP (54) with $E(Iu_0) \leq 1$. Then there is a constant $\delta = \delta(||u_0||_{L^2(\mathbb{R}^2)})$ so that the solution u obeys the following bound on the time interval $[0, \delta]$,*

$$||Iu||_{X^{\delta}_{1,\frac{1}{2}+}} \lesssim 1. \tag{73}$$

PROOF. We mimic the typical iteration argument showing local existence. We will need the following three estimates involving the $X_{s,\delta}$ spaces (55) and functions $F(x,t), f(x)$. (Throughout this section, the implicit constants in the notation $\lesssim$ are independent of δ.)

$$||S(t)f||_{X^{\delta}_{1,\frac{1}{2}+}} \lesssim ||f||_{H^1(\mathbb{R}^2)}, \tag{74}$$

$$\left\| \int_0^t S(t-\tau)F(x,\tau)d\tau \right\|_{X_{1,\frac{1}{2}+}} \lesssim ||F||_{X^{\delta}_{1,-\frac{1}{2}+}}, \tag{75}$$

$$||F||_{X^{\delta}_{1,-b}} \lesssim \delta^P ||F||_{X^{\delta}_{1,-\beta}}, \tag{76}$$

where in (76) we have $0 < \beta < b < \frac{1}{2}$, and $P = \frac{1}{2}(1 - \frac{\beta}{b}) > 0$. The bounds (74), (75) are analogous to estimates (3.13), (3.15) in [**55**]. As for (76), by duality it suffices to show

$$||F||_{X^{\delta}_{-1,\beta}} \lesssim \delta^P ||F||_{X^{\delta}_{-1,b}}.$$

Interpolation gives

$$||F||_{X^{\delta}_{-1,\beta}} \lesssim ||F||^{(1-\frac{\beta}{b})-}_{X^{\delta}_{-1,0}} \cdot ||F||^{\frac{\beta}{b}}_{X^{\delta}_{-1,b}}.$$

As $b \in (0, \frac{1}{2})$, arguing exactly as on page 771 of [**33**],

$$||F||_{X^{\delta}_{-1,0}} \lesssim \delta^{\frac{1}{2}} ||F||_{X^{\delta}_{-1,b}},$$

and (76) follows.

Duhamel's principle gives us

$$\begin{aligned}
||Iu||_{X^{\delta}_{1,\frac{1}{2}+}} &= \left\| S(t)(Iu_0) + \int_0^t S(t-\tau)I(u\bar{u}u)(\tau)d\tau \right\|_{X^{\delta}_{1,\frac{1}{2}+}} \\
&\lesssim ||Iu_0||_{H^1(\mathbb{R}^2)} + ||I(u\bar{u}u)||_{X^{\delta}_{1,-\frac{1}{2}+}} \\
&\lesssim ||Iu_0||_{H^1(\mathbb{R}^2)} + \delta^{\epsilon} ||I(u\bar{u}u)||_{X^{\delta}_{1,-\frac{1}{2}++}}, \qquad (77)
\end{aligned}$$

where $-\frac{1}{2}{+}{+}$ is a real number slightly larger than $-\frac{1}{2}+$ and $\epsilon > 0$. By the definition of the restricted norm (56),

$$||Iu||_{X^{\delta}_{1,\frac{1}{2}+}} \lesssim ||Iu_0||_{H^1(\mathbb{R}^2)} + \delta^{\epsilon}||I(\psi\overline{\psi}\psi)||_{X_{1,-\frac{1}{2}++}}, \tag{78}$$

where the function ψ agrees with u for $t \in [0, \delta]$, and

$$||Iu||_{X^{\delta}_{1,\frac{1}{2}+}} \sim ||I\psi||_{X_{1,\frac{1}{2}+}}. \tag{79}$$

We will show shortly that

$$||I(\psi\overline{\psi}\psi)||_{X_{1,-\frac{1}{2}++}} \lesssim ||I\psi||^3_{X_{1,\frac{1}{2}+}}. \tag{80}$$

Setting then $Q(\delta) \equiv ||Iu(t)||_{X^{\delta}_{1,\frac{1}{2}+}}$, the bounds (77), (79) and (80) yield

$$Q(\delta) \lesssim ||Iu_0||_{H^1(\mathbb{R}^2)} + \delta^{\epsilon}(Q(\delta))^3. \tag{81}$$

Note

$$||Iu_0||_{H^1(\mathbb{R}^2)} \lesssim (E(Iu_0))^{\frac{1}{2}} + ||u_0||_{L^2(\mathbb{R}^2)} \lesssim 1 + ||u_0||_{L^2(\mathbb{R}^2)}. \tag{82}$$

As Q is continuous in the variable δ, a bootstrap argument yields (73) from (81), (82).

It remains to show (80). Using the interpolation lemma of [**31**], it suffices to show

$$||\psi\overline{\psi}\psi||_{X_{s,-\frac{1}{2}++}} \lesssim ||\psi||^3_{X_{s,\frac{1}{2}+}}, \tag{83}$$

for all $\frac{4}{7} < s < 1$. By duality and a "Leibniz" rule[20], (83) follows from

(84)

$$\left| \int_{\mathbb{R}} \int_{\mathbb{R}^2} (\langle \nabla \rangle^s u_1) \overline{u_2} u_3 u_4 dx dt \right| \lesssim ||u_1||_{X_{s,\frac{1}{2}+}} \cdot ||u_2||_{X_{s,\frac{1}{2}+}} \cdot ||u_3||_{X_{s,\frac{1}{2}+}} ||u_4||_{X_{0,\frac{1}{2}--}}.$$

Note that since the factors in the integrand on the left here will be taken in absolute value, the relative placement of complex conjugates is irrelevant. Use Hölder's inequality on the left side of (84), taking the factors in, respectively, $L^4_{x,t}$, $L^4_{x,t}$, $L^6_{x,t}$ and $L^3_{x,t}$. Using a Strichartz inequality,

$$\begin{aligned} ||\langle \nabla \rangle^s u_1||_{L^4_{x,t}(\mathbb{R}^{2+1})} &\lesssim ||\langle \nabla \rangle^s u_1||_{X_{0,\frac{1}{2}+}} \\ &= ||u_1||_{X_{s,\frac{1}{2}+}}, \end{aligned}$$

and

$$\begin{aligned} ||u_2||_{L^4_{x,t}(\mathbb{R}^{2+1})} &\lesssim ||u_2||_{X_{0,\frac{1}{2}+}} \\ &\lesssim ||u_2||_{X_{s,\frac{1}{2}+}}. \end{aligned}$$

The bound for the third factor uses Sobolev embedding and the $L^6_t L^3_x$ Strichartz estimate,

$$\begin{aligned} ||u_3||_{L^6_t L^6_x(\mathbb{R}^{2+1})} &\lesssim ||\langle \nabla \rangle^{\frac{1}{3}} u_3||_{L^6_t L^3_x(\mathbb{R}^{2+1})} \\ &\lesssim ||\langle \nabla \rangle^{\frac{1}{3}} u_3||_{X_{0,\frac{1}{2}+}} \\ &\leq ||u_3||_{X_{s,\frac{1}{2}+}}. \end{aligned}$$

---

[20] By this, we mean the operator $\langle D \rangle^s$ can be distributed over the product by taking Fourier transform and using $\langle \xi_1 + \ldots \xi_4 \rangle^s \lesssim \langle \xi_1 \rangle^s + \ldots \langle \xi_4 \rangle^s$.

It remains to bound $||u_4||_{L^3(\mathbb{R}^{2+1})}$. Interpolating between $||u_4||_{L^2_tL^2_x} \leq ||u_4||_{X_{0,0}}$ and the Strichartz estimate $||u_4||_{L^4_tL^4_x} \lesssim ||u_4||_{X_{0,\frac{1}{2}+}}$ yields

$$||u_4||_{L^3_tL^3_x} \lesssim ||u_4||_{X_{0,\frac{1}{2}--}}.$$

This completes the proof of (84), and hence Proposition 5.7. □

Before we proceed to the proof of Proposition 5.6 we would like to present the proof of conservation of mass[21] for (54) using Fourier transform. Understanding this proof is fundamental to understand the types of cancelations that will make $E(Iu)$ almost conserved.

PROPOSITION 5.8. *Assume that u is a solution to* (54) *smooth and decaying at infinity. Then* $\|u(t)\|^2_{L^2} = \|u_0\|^2_{L^2}$.

PROOF. We write this L^2 norm using Plancherel formula

$$\|u(t)\|^2_{L^2} = \int \hat{u}(\xi,t)\bar{\hat{u}}(\xi,t)\,d\xi$$

Using the equation we then have

$$\begin{aligned}
\frac{d}{dt}\|u(t)\|^2_{L^2} &= 2Re\int (\hat{u}(\xi,t))_t\bar{\hat{u}}(\xi,t)\,d\xi \\
&= -Im\int |\xi|^2\hat{u}(\xi,t)\bar{\hat{u}}(\xi,t)\,d\xi - 2Im\int \widehat{u^2\bar{u}}(\xi)\bar{\hat{u}}(\xi,t)\,d\xi \\
&= -2Im\int_{\xi_1+\xi_2+\xi_3-\xi=0} \hat{u}(\xi_1)\bar{\hat{u}}(-\xi_2)\hat{u}(\xi_3)\bar{\hat{u}}(\xi)\,d\xi d\xi_1 d\xi_2 d\xi_3 \\
&= -2Im\int_{\xi_1+\xi_2+\xi_3+\xi_4=0} \hat{u}(\xi_1)\bar{\hat{u}}(-\xi_2)\hat{u}(\xi_3)\bar{\hat{u}}(-\xi_4)\,d\xi_1 d\xi_2 d\xi_3 d\xi_4
\end{aligned}$$

and by symmetry

$$\begin{aligned}
&2Im\int_{\xi_1+\xi_2+\xi_3+\xi_4=0} \hat{u}(\xi_1)\bar{\hat{u}}(-\xi_2)\hat{u}(\xi_3)\bar{\hat{u}}(-\xi_4)\,d\xi_1 d\xi_2 d\xi_3 d\xi_4 = \\
&Im\int_{\xi_1+\xi_2+\xi_3+\xi_4=0} \hat{u}(\xi_1)\bar{\hat{u}}(-\xi_2)\hat{u}(\xi_3)\bar{\hat{u}}(-\xi_4)\,d\xi_1 d\xi_2 d\xi_3 d\xi_4 \\
+\ &Im\int_{\xi_1+\xi_2+\xi_3+\xi_4=0} \hat{u}(-\xi_2)\bar{\hat{u}}(\xi_1)\hat{u}(-\xi_4)\bar{\hat{u}}(\xi_3)\,d\xi_1 d\xi_2 d\xi_3 d\xi_4 = 0
\end{aligned}$$

□

PROBLEM 5.9. *Prove the conservation of energy* (23) *by using Fourier transform.*

[21] Actually showing the proof of conservation of energy would be even more appropriate here since in Proposition 5.6 we will be dealing with an energy instead of a mass, but clearly for the mass the calculation is less involved and the ideas are still present in full power!

PROOF OF PROPOSITION 5.6. The usual energy (23) is shown to be conserved by differentiating in time, integrating by parts, and using the equation (54),

$$\begin{aligned}\partial_t E(u) &= \mathrm{Re}\int_{\mathbb{R}^2} \overline{u_t}(|u|^2u - \Delta u)dx\\ &= \mathrm{Re}\int_{\mathbb{R}^2} \overline{u_t}(|u|^2u - \Delta u - iu_t)dx\\ &= 0.\end{aligned}$$

We follow the same strategy to estimate the growth of $E(Iu)(t)$,

$$\begin{aligned}\partial_t E(Iu)(t) &= \mathrm{Re}\int_{\mathbb{R}^2} \overline{I(u)_t}(|Iu|^2Iu - \Delta Iu - iIu_t)dx\\ &= \mathrm{Re}\int_{\mathbb{R}^2} \overline{I(u)_t}(|Iu|^2Iu - I(|u|^2u))dx,\end{aligned}$$

where in the last step we've applied I to (54). When we integrate in time and apply the Parseval formula[22] it remains for us to bound

$$E(Iu(\delta)) - E(Iu(0)) = \int_0^\delta \int_{\sum_{j=1}^4 \xi_j=0} \left(1 - \frac{m(\xi_2+\xi_3+\xi_4)}{m(\xi_2)\cdot m(\xi_3)\cdot m(\xi_4)}\right) \widehat{\overline{I\partial_t u}}(\xi_1)\widehat{Iu}(\xi_2)\widehat{\overline{Iu}}(\xi_3)\widehat{Iu}(\xi_4). \tag{85}$$

The reader may ignore the appearance of complex conjugates here and in the sequel, as they have no impact on the availability of estimates, (see e.g. Lemma 5.4 above). We include the complex conjugates for completeness.

We use the equation to substitute for $\partial_t I(u)$ in (85). Our aim is to show that

$$\mathrm{Term}_1 + \mathrm{Term}_2 \lesssim N^{-\frac{3}{2}+}, \tag{86}$$

where the two terms on the left are

$$\mathrm{Term}_1 \equiv \left|\int_0^\delta \int_{\sum_{i=1}^4 \xi_i=0} \left(1 - \frac{m(\xi_2+\xi_3+\xi_4)}{m(\xi_2)m(\xi_3)m(\xi_4)}\right) \widehat{(\overline{\Delta Iu})}(\xi_1)\cdot \widehat{Iu}(\xi_2)\cdot \widehat{\overline{Iu}}(\xi_3)\cdot \widehat{Iu}(\xi_4)\right| \tag{87}$$

$$\mathrm{Term}_2 \equiv \left|\int_0^\delta \int_{\sum_{i=1}^4 \xi_i=0} \left(1 - \frac{m(\xi_2+\xi_3+\xi_4)}{m(\xi_2)m(\xi_3)m(\xi_4)}\right) \widehat{(\overline{I(|u|^2u)})}(\xi_1)\cdot \widehat{Iu}(\xi_2)\cdot \widehat{\overline{Iu}}(\xi_3)\cdot \widehat{Iu}(\xi_4)\right|. \tag{88}$$

From this point on the proof proceeds with a case by case analysis based on the relative magnitude of various frequencies. The basic cancellation of the type we presented in the proof of Proposition 5.8 are fundamental as is the fact that the multiplier is smooth. We send the reader to the original paper for a complete proof. □

REMARK 5.10. Here we only gave an idea of the "I-method". One can implement it in more effective ways by defining formally families of energies that, if controlled analytically, are proved to be more and more *almost conserved.* This was in fact the case for the one dimensional derivative NLS [**24, 25**] and the KdV [**27**] for example. Unfortunately controlling these families of energies becomes more

---

[22] That is, $\int_{\mathbb{R}^n} f_1(x)f_2(x)f_3(x)f_4(x)dx = \int_{\xi_1+\xi_2+\xi_3+\xi_4=0} \hat{f}_1(\xi_1)\hat{f}_2(\xi_2)\hat{f}_3(\xi_3)\hat{f}_4(\xi_4)$ where $\int_{\sum_i \xi_i=0}$ here denotes integration with respect to the hyperplane's measure $\delta_0(\xi_1+\xi_2+\xi_3+\xi_4)d\xi_1 d\xi_2 d\xi_3 d\xi_4$, with δ_0 the one dimensional Dirac mass.

difficult in higher dimensions since orthogonality issues start appearing, see for example [**30**].

6. Interaction Morawetz estimates and scattering

In the last lecture we discussed the question of global well-posedness. Once one can prove that given an initial data a unique solution evolving from that data exists for all times it becomes natural to ask how this solution looks like as $t \to \pm\infty$. The theory that addresses these questions is called *scattering theory.* In order to put scattering in a more general context we need a few definitions. We will give them by assuming that the solution for (1) is defined globally in time with respect to the energy space H^1, but it will be easy to generalize them when more general Sobolev spaces are considered.

DEFINITION 6.1 (Scattering). Given a global solution $u \in H^1$ to (1) we say that u *scatters* to $u_+ \in H^1$ if

$$\|u(t) - S(t)u_+\|_{H^1} \longrightarrow 0 \quad \text{as} \quad t \to +\infty. \tag{89}$$

Clearly a similar definition is given if $t \to -\infty$.

REMARK 6.2. Using the properties of the group $S(t)$ it is easy to see that (89) is equivalent to

$$\|S(-t)u(t) - u_+\|_{H^1} \longrightarrow 0 \quad \text{as} \quad t \to +\infty. \tag{90}$$

Since by the Duhamel formula (6)

$$S(-t)u(t) - u_+ = u_0 - u_+ - i\int_0^t S(-t')|u(t')|^{p-1}u(t')\,dt'$$

it is clear that scattering is **equivalent** to showing that the improper time integral

$$\int_0^\infty S(-t')|u(t')|^{p-1}u(t')\,dt'$$

converges in H^1 and in particular this will give the formula for u_+, i.e.

$$u_+ = u_0 - i\int_0^\infty S(-t')|u(t')|^{p-1}u(t')\,dt'. \tag{91}$$

One can also consider an inverse problem: assume $u_+ \in H^1$, can we find an initial data $u_0 \in H^1$ such that the global solution u for (1) scatters to u_+?

DEFINITION 6.3 (Wave Operator). Assume that for any $u_+ \in H^1$ there exists a unique $u_0 \in H^1$ such that the solution u to (1) scatters to u_+ in the sense of (91). Then we define the wave operator

$$\Omega^+ : H^1 \longrightarrow H^1 \quad \text{such that} \quad \Omega^+(u_+) = u_0.$$

In order to prove the existence of Ω^+ it is useful to write the solution u in terms of u_+. In fact using the Duhamel representation (6) and (91) above we can write

$$u(t) = S(t)u_+ - i\int_t^\infty S(t-t')(|u(t')|^{p-1}u(t')\,dt', \tag{92}$$

and being able to define Ω^+ is equivalent to being able to define (92) for $t = 0$.

REMARK 6.4. From the two definitions given above it is clear that proving scattering is equivalent to proving that the wave operator Ω^+ is invertible. In this case we also say that we have *Asymptotic Completeness.*

At first, from the definitions, it is not clear what is harder to prove, if existence of the wave operator or asymptotic completeness. But in practice the former is easier. One of the reasons is that the existence of the wave operator usually follows from the strong[23] dispersive estimates (10) and from iteration of local well-posedness. On the other hand to prove scattering one needs global space time bounds that are very difficult to get. Here we only address the question of existence of the wave operator (see [**19**]) briefly in Theorem 6.14, but we will concentrate on the scattering issue much more. The bibliography on scattering is quite large (see for example [**19**] for a good list of results), but certainly the work of Ginibre and Velo (see for example [**42**]) takes a special stand in it. But in this lecture we will take a different and more recent approach that is based on the so called *Interaction Morawetz Estimates* [**28, 73, 78**].

6.5. Interaction Morawetz Estimates. At this point there are several ways one can present these estimates: as weighted overages of the classical Morawetz estimates presented in Section 3 [**28, 78**], as classical Morawetz estimates applied to tensors of solutions to (1) [**20, 44, 45**], or as more general and refined calculations dealing with vector fields [**32, 68**]. Here we describe the first one, which was also the original one given in 3 dimensions[24].

In the following we introduce an interaction potential generalization of the classical Morawetz action and associated inequalities. We first recall the standard Morawetz action centered at a point and the proof that this action is monotonically increasing with time when the nonlinearity is defocusing. The interaction generalization is introduced in the second subsection. The key consequence of the analysis in this section is the $L^4_{x,t}$ estimate (116).

The discussion in this section will be carried out in the context of the following generalization of (1):

$$i\partial_t u + \alpha\Delta u = \mu f(|u|^2)u, \qquad u : \mathbb{R}\times\mathbb{R}^3 \longmapsto \mathbb{C}, \tag{93}$$

$$u(0) = u_0. \tag{94}$$

Here f is a smooth function $f : \mathbb{R}^+ \longmapsto \mathbb{R}^+$ and α and μ are real constants that permit us to easily distinguish in the analysis below those terms arising from the Laplacian or the nonlinearity. We also define $F(z) = \int_0^z f(s)ds$.

We will use polar coordinates $x = r\omega$, $r > 0, \omega \in S^2$, and write Δ_ω for the Laplace-Beltrami operator on S^2. For ease of reference below, we record some alternate forms of the equation in (93):

$$u_t = i\alpha\Delta u - i\mu f(|u|^2)u, \tag{95}$$

$$\overline{u}_t = -i\alpha\Delta\overline{u} + i\mu f(|u|^2)\overline{u}, \tag{96}$$

$$u_t = i\alpha u_{rr} + i\frac{2\alpha}{r}u_r + i\frac{\alpha}{r^2}\Delta_\omega u - i\mu f(|u|^2)u, \tag{97}$$

---

[23]Especially in higher dimensions.

[24]The reader will see below that for $n = 1, 2$ the argument breaks down. In fact for $n = 1$ one needs to use tensors of solutions [**20**] and for $n = 2$ one either is happy with a local in time estimate [**39**] or needs to introduce a much more refined argument [**32**]. For $n > 3$ the argument below can be used but the estimates are less "clean" than the $L^4_t L^4_x$ norm we find below. But some use of standard harmonic analysis leads to a better space time estimates which is as good as the one we prove here [**70, 77, 78**].

$$(ru_t) = i\alpha(ru)_{rr} + i\frac{\alpha}{r}\Delta_\omega u - i\mu r f(|u|^2)u, \tag{98}$$

$$(r\overline{u}_t) = -i\alpha(r\overline{u})_{rr} - i\frac{\alpha}{r}\Delta_\omega \overline{u} + i\mu f(|u|^2)\overline{u}. \tag{99}$$

6.6. Standard Morawetz action and inequalities. We will call the following quantity the *Morawetz action centered at 0* for the solution u of (93) and this should be compared with (29),

$$M_0[u](t) = \int_{\mathbb{R}^3} \mathrm{Im}[\bar{u}(t,x)\nabla u(t,x)] \cdot \frac{x}{|x|} dx. \tag{100}$$

We check using the equation that,

$$\partial_t(|u|^2) = -2\alpha\nabla \cdot \mathrm{Im}[\overline{u}(t,x)\nabla u(t,x)], \tag{101}$$

hence we may interpret M_0 as the spatial average of the radial component of the L^2-mass current. We might expect that M_0 will increase with time if the wave u scatters since such behavior involves a broadening redistribution of the L^2-mass. The following proposition of Lin and Strauss [**61**] that is equivalent to (29), indeed gives $\frac{d}{dt}M_0[u](t) \geq 0$ for defocusing equations.

PROPOSITION 6.7. [**61**] *If u solves* (93)-(94) *then the Morawetz action at 0 satisfies the identity*

$$\begin{aligned} \partial_t M_0[u](t) = 4\pi\alpha|u(t,0)|^2 &+ \int_{\mathbb{R}^3} \frac{2\alpha}{|x|}|\not{\nabla}_0 u(t,x)|^2 dx \\ &+ \mu \int_{\mathbb{R}^3} \frac{2}{|x|}\left\{|u|^2 f(|u|^2)(t) - F(|u|^2)\right\} dx, \end{aligned} \tag{102}$$

where $\not{\nabla}_0$ is the angular component of the derivative,

$$\not{\nabla}_0 u = \nabla u - \frac{x}{|x|}(\frac{x}{|x|}\cdot \nabla u). \tag{103}$$

In particular, M_0 is an increasing function of time if the equation (93) *satisfies the repulsivity condition,*

$$\mu\left\{|u|^2 f(|u|^2)(t) - F(|u|^2)\right\} \geq 0. \tag{104}$$

Note that for pure power potentials $F(x) = \frac{2}{p+1}x^{\frac{p+1}{2}}$, where the nonlinear term in (93) is $|u|^{p-1}u$, the function $|u|^2 f(|u|^2) - F(|u|^2) = \frac{p-1}{2}F(|u|^2)$. Hence condition (104) holds.

We may center the above argument at any other point $y \in \mathbb{R}^3$ with corresponding results. Toward this end, define the *Morawetz action centered at y* to be,

$$M_y[u](t) = \int_{\mathbb{R}^3} \mathrm{Im}[\overline{u}(x)\nabla u(x)] \cdot \frac{x-y}{|x-y|} dx. \tag{105}$$

We shall often drop the u from this notation, as we did previously in writing $M_0(t)$.

COROLLARY 6.8. *If u solves* (93) *the Morawetz action at y satisfies the identity*

$$\begin{aligned} \frac{d}{dt}M_y &= 4\pi\alpha|u(t,y)|^2 + \int_{\mathbb{R}^3} \frac{2\alpha}{|x-y|}|\not{\nabla}_y u(t,x)|^2 dx \\ &+ \int_{\mathbb{R}^3} \frac{2\mu}{|x-y|}\left\{|u|^2 f(|u|^2) - F(|u|^2)\right\} dx, \end{aligned} \tag{106}$$

where $\not\nabla_y u \equiv \nabla u - \frac{x-y}{|x-y|}\left(\frac{x-y}{|x-y|}\cdot\nabla u\right)$. *In particular,* M_y *is an increasing function of time if the nonlinearity satisfies the repulsivity condition* (104).

Corollary 6.8 shows that a solution is, on average, repulsed from any fixed point y in the sense that $M_y[u](t)$ is increasing with time.

For our scattering results, we'll need the following pointwise bound for $M_y[u](t)$.

LEMMA 6.9. *Assume* u *is a solution of* (93) *and* $M_y[u](t)$ *as in* (105). *Then,*

$$|M_y[u](t)| \lesssim \|u(t)\|^2_{\dot{H}^{\frac{1}{2}}_x}. \tag{107}$$

PROOF. Without loss of generality we take $y = 0$. This is a refinement of the easy bound using Cauchy-Schwarz $|M_y[u](t)| \lesssim \|u(t)\|_{L^2_x}\|\nabla u(t)\|_{L^2_x}$. By duality

$$|\,\mathrm{Im}\int_{\mathbb{R}^3}\overline{u(x,t)}\partial_r u(x,t)dx\,| \leq \|u\|_{\dot{H}^{\frac{1}{2}}(\mathbb{R}^3)}\cdot\|\partial_r u\|_{\dot{H}^{-\frac{1}{2}}(\mathbb{R}^3)}.$$

It suffices to show $\|\partial_r u\|_{\dot{H}^{-\frac{1}{2}}(\mathbb{R}^3)} \leq \|u\|_{\dot{H}^{-\frac{1}{2}}(\mathbb{R}^3)}$. By duality and the definition $\partial_r \equiv \frac{x}{|x|}\cdot\nabla$, it remains to prove,

$$\|\frac{x}{|x|}f\|_{\dot{H}^{\frac{1}{2}}(\mathbb{R}^3)} \leq \|f\|_{\dot{H}^{\frac{1}{2}}(\mathbb{R}^3)}, \tag{108}$$

for any f for which the right hand side is finite. Inequality (108) follows from interpolating between the following two bounds,

$$\|\frac{x}{|x|}f\|_{L^2(\mathbb{R}^3)} \leq \|f\|_{L^2(\mathbb{R}^3)}$$

$$\|\frac{x}{|x|}f\|_{\dot{H}^1(\mathbb{R}^3)} \lesssim \|f\|_{H^1(\mathbb{R}^3)}$$

the first of which is trivial, the second of which follows from Hardy's inequality,

$$\begin{aligned}\|\nabla\left(\frac{x}{|x|}f\right)\|_{L^2} &\leq \|\frac{x}{|x|}\cdot\nabla f\|_{L^2} + \|\frac{1}{|x|}f\|_{L^2}\\ &\lesssim \|\nabla f\|_{L^2}.\end{aligned}$$

□

The well-known Morawetz-type inequalities, so useful in proving local decay or scattering for (93), arise by integrating the identity (102) or (106) in time. For nonlinear Schrödinger equations, this argument appears in the work of Lin and Strauss [**61**], who cite as motivation earlier work on Klein-Gordon equations by Morawetz [**64**].

COROLLARY 6.10 (Morawetz estimate centered at y.). *Suppose* u *solves* (93)-(94). *Then for any* $y \in \mathbb{R}^3$,

$$\begin{aligned}2\sup_{t\in[0,T]}\|u(t)\|^2_{\dot{H}^{\frac{1}{2}}_x} &\gtrsim 4\pi\alpha\int_0^T|u(t,y)|^2dt + \int_0^T\int_{\mathbb{R}^3}\frac{2\alpha}{|x-y|}|\not\nabla_y u(t,x)|^2dxdt\\ &+\int_0^T\int_{\mathbb{R}^3}\frac{2\mu}{|x-y|}\left\{|u|^2f(|u|^2) - F(|u|^2)\right\}dxdt.\end{aligned} \tag{109}$$

Assuming (93) has a repulsive nonlinearity as in (104), all terms on the right side of the inequality (109) are positive. The inequality therefore gives in particular a bound uniform in T for the quantity $\int_0^T \int_{\mathbb{R}^3} \frac{|u(t,x)|^4}{|x-y|} dxdt$, for solutions u of the defocusing (1), when $p = 3$.

In their proof of scattering in the energy space for the cubic defocusing problem (1), Ginibre and Velo [**42**] combine this relatively localized[25] decay estimate with a bound surrogate for finite propagation speed in order to show the solution is in certain global-in-time Lebesgue spaces $L^q([0,\infty), L^r(\mathbb{R}^3))$. Scattering follows rather quickly, as will be shown later.

In the following section, we show how to establish an unweighted, global in time Lebesgue space bound directly. The argument below involves the identity (106), but our estimate arises eventually from the linear part of the equation, more specifically from the first term on the right of (106), rather than the third (nonlinearity) term.

6.11. Morawetz interaction potential. Given a solution u of (93), we define the *Morawetz interaction potential* to be

$$M(t) = \int_{\mathbb{R}^3} |u(t,y)|^2 M_y(t) dy. \tag{110}$$

The bound (107) immediately implies

$$|M(t)| \lesssim \|u(t)\|_{L^2}^2 \|u(t)\|_{\dot{H}_x^{\frac{1}{2}}}^2 . \tag{111}$$

If u solves (93) then the identity (106) gives us the following identity for $\frac{d}{dt}M(t)$,

$$\begin{aligned} \frac{d}{dt}M(t) &= 4\pi\alpha \int_y |u(y)|^4 dy + \int_{\mathbb{R}^3}\int_{\mathbb{R}^3} \frac{2\alpha}{|x-y|} |u(y)|^2 |\nabla\!\!\!/_y u(x)|^2 dxdy \\ &+ \int_{\mathbb{R}^3}\int_{\mathbb{R}^3} \frac{2\mu}{|x-y|} |u(y)|^2 \left\{ |u(x)|^2 f(|u(x)|^2) - F(|u(x)|^2) \right\} dxdy \\ &+ \int_{\mathbb{R}^3} \partial_t(|u(t,y)|^2)\, M_y(t) dy. \end{aligned} \tag{112}$$

We write the right side of (112) as $I + II + III + IV$, and work now to rewrite this as a sum involving nonnegative terms.

PROPOSITION 6.12. *Referring to the terms comprising* (112), *we have*

$$IV \geq -II. \tag{113}$$

Consequently, solutions of (93) *satisfy*

$$\begin{aligned} \frac{d}{dt}M(t) &\geq 4\pi\alpha \int_{\mathbb{R}^3} |u(t,y)|^4 dy \\ &+ \int_{\mathbb{R}^3}\int_{\mathbb{R}^3} \frac{2\mu}{|x-y|} |u(t,y)|^2 \left\{ |u|^2 f(|u|^2) - F(|u|^2) \right\} dxdy. \end{aligned} \tag{114}$$

In particular, $M(t)$ is monotone increasing for equations with repulsive nonlinearities.

[25]The bound mentioned here may be considered localized since it implies decay of the solution near the fixed point y, but doesn't preclude the solution staying large at a point which moves rapidly away from y, for example.

Assuming Proposition 6.12 for the moment, we combine (111) and (114) to obtain the following estimate which plays the major new role in our scattering analysis below.

COROLLARY 6.13. *Take u to be a smooth solution to the initial value problem (93)-(94) above, under the repulsivity assumption (104). Then we have the following interaction Morawetz inequalities,*

$$(115)\qquad 2\|u(0)\|_{L^2}^2 \sup_{t\in[0,T]} \|u(t)\|_{\dot{H}_x^{\frac{1}{2}}}^2 \gtrsim 4\pi\alpha \int_0^T \int_{\mathbb{R}^3} |u(t,y)|^4 dydt$$
$$+ \int_0^T \int_y \int_x \frac{2\mu}{|x-y|} |u(t,y)|^2 \left\{|u|^2 f(|u|^2) - F(|u|^2)\right\}(t,x) dxdydt.$$

In particular, we obtain the following spacetime $L^4([0,T]\times\mathbb{R}^3)$ estimate,

$$(116)\qquad \int_0^T \int_{\mathbb{R}^3} |u(t,y)|^4 dydt \leq C\|u_0\|_{L^2(\mathbb{R}^3)}^2 \sup_{t\in[0,T]} \|u(t)\|_{\dot{H}_x^{\frac{1}{2}}}^2,$$

where C is independent of T.

Of course, for solutions of the defocusing IVP (1) starting from finite energy initial data, the right side of (116) is uniformly bounded by energy considerations - leading to a rather direct proof of the result in [**42**] of scattering in the energy space that we will present below.

PROOF. We now turn to the proof of Proposition 6.12. Use (101) to write

$$\begin{aligned} IV &= -\int_{\mathbb{R}_y^3} \nabla \cdot \mathrm{Im}[2\alpha\overline{u}(y)\nabla u(y)] M_y(t) dy \\ &= -\int_y \int_x \partial_{y_l} \mathrm{Im}[2\alpha\overline{u}(y)\partial_{y_l} u(y)]\ \mathrm{Im}[\overline{u}(x)\frac{x_m - y_m}{|x-y|}\partial_{x_m} u(x)] dxdy, \end{aligned}$$

where repeated indices are implicitly summed. We integrate by parts in y, moving the leading ∂_{y_l} to the unit vector $\frac{x-y}{|x-y|}$. Note that,

$$(117)\qquad \partial_{y_l}\left(\frac{x_m - y_m}{|x-y|}\right) = \frac{-\delta_{lm}}{|x-y|} + \frac{(x_l - y_l)(x_m - y_m)}{|x-y|^3}.$$

Write $\mathbf{p}(x) = \mathrm{Im}[\overline{u}(x)\nabla u(x)]$ for the mass current at x and use (117) to obtain

$$(118)\qquad IV = -2\alpha \int_y \int_x \left[\mathbf{p}(y)\cdot\mathbf{p}(x) - (\mathbf{p}(y)\cdot\frac{x-y}{|x-y|})(\mathbf{p}(x)\cdot\frac{x-y}{|x-y|})\right]\frac{dxdy}{|x-y|}.$$

The preceding integrand has a natural geometric interpretation. We are removing the inner product of the components of $\mathbf{p}(y)$ and $\mathbf{p}(x)$ parallel to the vector $\frac{x-y}{|x-y|}$ from the full inner product of $\mathbf{p}(y)$ and $\mathbf{p}(x)$. This amounts to taking the inner product of $\pi_{(x-y)^\perp}\mathbf{p}(y)\cdot\pi_{(x-y)^\perp}\mathbf{p}(x)$ where we have introduced the projections onto the subspace of $\mathbb{R}^3$ perpendicular to the vector $\frac{x-y}{|x-y|}$. But

$$(119)\qquad \begin{aligned} |\pi_{(x-y)^\perp}\mathbf{p}(y)| &= |\mathbf{p}(y) - \frac{x-y}{|x-y|}(\frac{x-y}{|x-y|}\cdot\mathbf{p}(y))| \\ &= |\mathrm{Im}[\overline{u}(y)\nabla\!\!\!/_x u(y)| \leq |u(y)|\cdot|\nabla\!\!\!/_x u(y)|. \end{aligned}$$

A similar identity and inequality holds upon switching the roles of x and y in (119). We have thus shown that

$$IV \geq -2\alpha \int_y \int_x |u(x)| \cdot |\nabla\!\!\!/_y u(x)| \cdot |u(y)| \cdot |\nabla\!\!\!/_x u(y)| \frac{dxdy}{|x-y|}. \tag{120}$$

The conclusion (113) follows by applying the elementary bound $|ab| \leq \frac{1}{2}(a^2+b^2)$ with $a = |u(y)| \cdot |\nabla\!\!\!/_y u(x)|$ and $b = |u(x)| \cdot |\nabla\!\!\!/_x u(y)|$. □

We now state the following theorem as an example of how to use Morawetz interaction estimates in order to prove scattering

THEOREM 6.14. *Consider the cubic, defocusing, NLS* (1) *in* $\mathbb{R}^3$ *with* H^1 *initial data. Then the wave operator exists and there is asymptotic completeness.*

REMARK 6.15. Theorem 6.14 is not the best known result for this cubic NLS. In fact in [**28**] this same IVP was considered and the $L^4_t L^4_x$ Morawetz estimate was used to prove scattering below H^1. For other H^1 subcritical scattering results one should also consult [**78**] when $n \geq 3$, [**32**] when $n = 2$ and [**20**] when $n = 1$. In these cases if one wants to show scattering with regularity $s < 1$, for example when $n = 3$ in [**28**], the argument is more complicated than the one described for H^1 since one has to prove that the H^s norm of the solution is bounded by using the "I-method" as in Section 5. The basic idea though is the same.

PROOF. **Existence of Ω^+:** we go back to the formula (92). The idea is to go first from $t = +\infty$ to $t = T$ for some $T > 0$ using some smallness and then solve the problem in the finite interval of time backward from T to 0.

We know already in what kind of spaces we can argue by contraction method: the space S^1 containing all the admissible Strichartz norms of the function and its derivatives and possibly also those that are embedded into these norms by the Sobolev theorem. But in this case there is one more request that we want to make. We want a smallness assumption, possibly obtained by shrinking the time interval or better by taking the time interval at infinity where the "tail" of the function lives. For this reason we should avoid any norm that contains a L^∞_t. So we proceed in two steps first we consider the smaller space $\tilde{S^1}$ given by the norm

$$\|f\|_{\tilde{S^1}} = \|f\|_{L^5_t L^5_x} + \|f\|_{L^{10/3}_t W^{1,10/3}_x}.$$

Notice that by Sobolev

$$\|f\|_{L^5_t L^5_x} \lesssim \|f\|_{L^5_t W^{1,30/11}_x}$$

and $(5, 30/11)$ is a Strichartz admissible pair. It follows that if $u_+ \in H^1$ then by (12)

$$\|S(t)u_+\|_{\tilde{S^1}_{[T,\infty)}} \leq \epsilon \tag{121}$$

for T large enough. From (92) if we define

$$Lv(t) = S(t)u_+ + i\int_t^\infty S(t-t')(|v(t')|^2 v(t')\, dt', \tag{122}$$

and we use (13), where we pick the couple $(\tilde{q}, \tilde{r}) = (10/3, 10/3)$, we have

$$\begin{aligned} \|Lv\|_{\tilde{S^1}_{[T,\infty)}} &\leq \epsilon + C\||v|^2(|v| + |\nabla v|)\|_{L^{10/7}_{[T,\infty)} L^{10/7}_x} \\ &\leq \epsilon + C\|v\|^2_{L^5_{[T,\infty)} L^5_x} \|v\|_{L^{10/3}_{[T,\infty)} W^{1,10/3}_x} = \epsilon + C\|v\|^3_{\tilde{S^1}_{[T,\infty)}}. \end{aligned} \tag{123}$$

With a similar estimate

$$\|Lv - Lw\|_{\tilde{S^1}_{[T,\infty)}} \leq C(\|v\|_{\tilde{S^1}_{[T,\infty)}} + \|w\|_{\tilde{S^1}_{[T,\infty)}})\|v - w\|_{\tilde{S^1}_{[T,\infty)}}. \tag{124}$$

Thanks to the presence of ϵ one can proceed with the contraction argument. This would give a solution in $[T, \infty)$, which in particular has the property that

$$\|u\|_{\tilde{S^1}_{[T,\infty)}} \lesssim \epsilon. \tag{125}$$

But we didn't prove that this solution is in $C([T,\infty), H^1)$ for example. To do this we need to go back and estimate the solution u in the Strichartz space $S^1_{[T,\infty)}$. We in fact have by (12) and (13)

$$\|u\|_{S^1} \leq C\|u_+\|_{H^1} + C\||u|^2(|v| + |\nabla u|)\|_{L^{10/7}_{[T,\infty)}L^{10/7}_x}$$

and from (125)

$$\|u\|_{S^1} \leq C\|u_+\|_{H^1} + C\|u\|^3_{\tilde{S^1}_{[T,\infty)}} \lesssim \|u_+\|_{H^1},$$

and we are done in the interval $[T, \infty)$.

We now need to proceed from $t = T$ back to $t = 0$. Since the problem is subcritical, an iteration of local well-posedness like we presented in Section 5, using the conservation of the energy and mass, will suffice to cover the finite interval $[0, T]$.

Invertibility of Ω^+: This is the proof of scattering and we need to go back to (91). From here we see that we only need to show that the integral involving the global solution u

$$\int_0^\infty S(t)(|u|^2u)(t)\,dt$$

converges in H^1. By the dual of the homogeneous Strichartz estimate (12) we have that

$$\begin{aligned}\left\|\int_0^\infty S(t)|u|^2u(t)\,dt\right\|_{H^1} &\lesssim \||u|^2(|u| + |\nabla u|)\|_{L^{10/7}_t L^{10/7}_x} \\ &\lesssim C\|u\|^2_{L^5_t L^5_x}\|u\|_{L^{10/3}_t W^{1,10/3}_x} \lesssim \|u\|^3_{S^1}.\end{aligned}$$

Clearly to conclude it would be enough to show that $\|u\|_{S^1} \leq C$. This is in fact proved in the following proposition. □

PROPOSITION 6.16. *Assume that u is the H^1 global solution to the cubic, defocusing NLS in $\mathbb{R}^3$. Then*

$$\|u\|_{S^1} \leq C.$$

PROOF. We first observe that (116) provides a bound in $L^4_t L^4_x$. It is to be noted that in $\mathbb{R}^3$ this norm is not an admissible Strichartz norm so we need to do a bit more work. We start by picking $\epsilon \ll 1$ to be defined later and intervals of time I_k, $k = 1, ..., M < \infty$ such that

$$\|u\|_{L^4_{I_k} L^4_x} \leq \epsilon, \tag{126}$$

for all $k = 1, ..., M$. We now work on each separate interval and at the end we put everything back together. Since for now I_k is fixed we drop the index k and we set $I = [a, b]$. By the Duhamel principle and (12) and (13) we have as above

$$\|u\|_{S^1_I} \lesssim \|u(a)\|_{H^1} + \|u\|^2_{L^5_I L^5_x}\|u\|_{L^{10/3}_I W^{1,10/3}_x}. \tag{127}$$

It is important to notice that $10/3 < 4 < 5 < 10$, where $(10/3, 10/3)$ is an admissible pair in the L^2 sense and $(10, 10)$ is admissible in the H^1 sense since by Sobolev

$$\|u\|_{L_t^{10}L_x^{10}} \leq \|u\|_{L_t^{10}W_x^{1,30/13}},$$

and $(10, 30/13)$ is an admissible pair. It follows by interpolation and (126) that

$$\|u\|_{L_I^5 L_x^5} \lesssim \epsilon^{\alpha}\|u\|_{S_I^1}^{1-\alpha},$$

for some $\alpha > 0$. As a consequence (127) gives

$$\|u\|_{S_I^1} \lesssim \|u(a)\|_{H^1} + \epsilon^{2\alpha}\|u\|_{S_I^1}^{3-2\alpha},$$

and since the H^1 norm is uniformly bounded by energy and mass we have

$$\|u\|_{S_I^1} \lesssim 1 + \epsilon^{2\alpha}\|u\|_{S_I^1}^{3-2\alpha}. \tag{128}$$

We now use a continuity argument. Set $X(t) = \|u\|_{S^1_{[a,a+t]}}$. One can easily prove that $X(t)$ is continuous. From (128) we have

$$X(t) \lesssim 1 + \epsilon^{2\alpha}X(t)^{3-2\alpha}.$$

Then if ϵ is small enough there exist $X_0 < X_1, X_1 \gg 1$ such that either $X(t) \leq X_0$ or $X(t) \geq X_1$. But since $X(0) \lesssim 1$ and $X(t)$ is continuous it follows that $X(t) \leq X_0$ for all $t \in I$. This conclusion can be made for all I_k, $k = 1, ..., M$ and this concludes the proof. □

7. Global well-posedness for the $H^1(\mathbb{R}^n)$ critical NLS -Part I

We recall that the H^1 critical exponent for (1) is $p = 1 + \frac{4}{n-2}$. We also recall the following theorem that can be basically completely proved using either directly or indirectly theorems and arguments already presented in Section 5 and Section 6:

THEOREM 7.1 (Local or global small data well-posedness for the H^1 critical NLS). *We have the following two results:*

(1) *For any $u_0 \in H^1$ there exist $T = T(u_0)$ and a unique solution $u \in S^1_{[T,T]}$ to (1) with $p = 1 + \frac{4}{n-2}$ and $\mu = \pm 1$. Moreover there is continuity with respect to the initial data.*
(2) *There exists ϵ small enough such that for any u_0, $\|u_0\|_{H^1} \leq \epsilon$ there exists a unique global solution $u \in S^1$ to (1) with $p = 1 + \frac{4}{n-2}$ and $\mu = \pm 1$. Moreover there is continuity with respect to the initial data and scattering in the sense that there exists $u_\pm \in H^1$ such that*

$$\|u(t) - S(t)u_\pm\|_{H^1} \longrightarrow 0 \quad as \quad t \to \pm\infty.$$

PROOF. It is clear that the part about well-posedness is a summary of what has been proved in Section 5. The part about scattering instead can be proved as in Section 6 and by simply observing that Proposition 6.16 follows directly from the well-posedness proof thanks to the small data assumption. □

REMARK 7.2. We first remark that this theorem doesn't *see* the focusing or defocusing nature of the equation. This clearly means that in Theorem 8.1 the NLS is treated as a "small" perturbation of the linear problem. Due to the criticality of the problem and hence the fact that T depends also on the profile of the initial data an iteration argument based on the conservation of mass and energy is not possible. It is also clear that even increasing the regularity of the data the large data problem doesn't become any easier.

The first breakthrough on this problem is due to Bourgain [**13**]. He considers the defocusing case with $n = 3, 4$ and assumes radial symmetry for the problem. He proves the second part of Theorem 8.1 for arbitrarily large radially symmetric data. Here we summarize the main steps of Bourgain's proof for $n = 3$, which doesn't really do justice to the novelty and depth of the proof itself. The background argument is done by induction on the size of the energy E, the only quantity, besides the mass that here doesn't play much of a role, that remains controlled over time. From Theorem 8.1 the first step of the induction (small E) is in place. Let's now assume the second induction assumption that if $E < E_0$, for E_0 arbitrarily large, then the theorem is true. We take $E = E_0$ and we want to prove that also in this case the theorem is true. One first shows that the theorem follows if and only if the norm $L_t^{10}L_x^{10}$ of the solution remains bounded (see Theorem 4.8). Then the proof proceeds by contradiction. One supposes that there is a solution u such that $\|u\|_{L_t^{10}L_x^{10}}$ is arbitrarily large and $E = E_0$. The heart of the proof is on showing that at some time t_0 there is concentration of the H^1 norm: there exists a small ball B_0 centered at the origin such that $\|u(t_0)\|_{H^1(B_0)} > \delta$, and this ball is "sufficiently isolated" from the rest of the solution . It is here that the radial assumption is used. At this point one restarts the evolution at time t_0 by splitting the data as

$$\psi_0 = u(t_0)\chi_{B_0} \quad \text{and} \quad \psi_1 = u(t_0)(1 - \chi_{B_0}),$$

where χ_{B_0} is the indicator function for the ball B_0, and evolving ψ_0 with NLS and ψ_1 with a difference equation so that the sum of the two evolutions give the solution to NLS. Since now $\psi_0 \in H^1$ and $x\psi \in L^2$ it follows[26] that the evolution v of ψ_0 is global in time. Moreover since $E(\psi_0) \sim \delta^2$ it follows that $E(\psi_1) < E_0 - \delta^2$. Hence for the difference equation we are in the induction assumption. This is not quite like to have the equation under the induction assumption, but with some relatively straightforward perturbation theory[27] one also gets that the evolution w of ψ_1 is global. Hence we have a global evolution for the solution $u = v + w$ to NLS and as a consequence a uniform bound for $\|u\|_{L_t^{10}L_x^{10}}$ which is a contradiction.

Almost at the same time, with the same radial symmetry assumption above, Grillakis [**43**] proved a slighter weaker result than Bourgain's, namely existence and uniqueness for smooth global solution. It took few more years to remove the radial assumption and obtain the following theorem and its corollary [**29**]:

THEOREM 7.3. *For any u_0 with finite energy, $E(u_0) < \infty$, there exists a unique[28] global solution $u \in C_t^0(\dot{H}_x^1) \cap L_{t,x}^{10}$ to (1) with $p = 5, n = 3, \mu = 1$ such that*

$$\int_{-\infty}^{\infty}\int_{\mathbb{R}^3} |u(t,x)|^{10}\, dxdt \leq C(E(u_0)). \tag{129}$$

for some constant $C(E(u_0))$ that depends only on the energy.

---

[26]This result is for example proved in [**19**] as a consequence of the pseudo-conformal transformation and a monotonicity formula linked to it.

[27]That works tanks to the fact that the ball is "sufficiently" isolated from the rest of the solution.

[28]In fact, uniqueness actually holds in the larger space $C_t^0(\dot{H}_x^1)$ (thus eliminating the constraint that $u \in L_{t,x}^{10}$) [**29**].

As one can see from Theorem 4.8 and from the arguments in Section 6, the $L^{10}_{t,x}$ bound above also gives scattering and and persistence of regularity:

COROLLARY 7.4. *Let u_0 have finite energy. Then there exist finite energy solutions $u_\pm(t,x)$ to the free Schrödinger equation $(i\partial_t + \Delta)u_\pm = 0$ such that*

$$\|u_\pm(t) - u(t)\|_{\dot{H}^1} \to 0 \text{ as } t \to \pm\infty.$$

Furthermore, the maps $u_0 \mapsto u_\pm(0)$ are homeomorphisms from $\dot{H}^1(\mathbb{R}^3)$ to $\dot{H}^1(\mathbb{R}^3)$. Finally, if $u_0 \in H^s$ for some $s > 1$, then $u(t) \in H^s$ for all time t, and one has the uniform bound

$$\sup_{t\in\mathbb{R}} \|u(t)\|_{H^s} \leq C(E(u_0), s)\|u_0\|_{H^s}.$$

Most of the rest of this lecture and Section 8 will be devoted to give an idea of the proof for Theorem 8.3. Still for the defocusing case and for $n > 3$ we recall first the result of Tao [**74**], where an equivalent of Theorem 8.3 is proved still under the radial assumption, the result of Ryckman and Visan [**70**] for $n = 4$, where the radial assumption is removed, and finally the full generalization for any $n \geq 5$ by Visan [**77**].

The situation in the focusing case was first considered successfully by Kenig and Merle. They prove the following theorem [**52**]:

THEOREM 7.5. *Assume that $E(u_0) < E(W)$, $\|u_0\|_{\dot{H}^1} < \|W\|_{\dot{H}^1}$, where $n = 3, 4, 5$ and u_0 is radial and W is the stationary solution. Then the solution u to the critical H^1 focusing IVP (1) with data u_0 at $t = 0$ is defined for all time and there exists $u_\pm \in \dot{H}^1$ such that*

$$\|S(t)u_\pm - u(t)\|_{\dot{H}^1} \to 0 \text{ as } t \to \pm\infty.$$

Moreover for u_0 radial, $E(u_0) < E(W)$, but $\|u_0\|_{\dot{H}^1} > \|W\|_{\dot{H}^1}$, the solution must break down in finite time.

This result has been extended in every dimension $n \geq 3$ and for general data in [**57**]. Moreover a similar result has been proved by Kenig and Merle for the critical wave equation without the radial assumption [**53**], see also [**54**]. The proof of Theorem 8.5 introduces a new point of view for these problems. Using a concentration-compactness argument the authors reduce matters to a rigidity theorem, which is proved with the aid of a localized Virial identity (in the spirit of Merle [**62, 63**]). The radiality enters only in the proof of the rigidity theorem. In the case of the critical wave equation other consideration of *elliptic nature* are used to remove the radial assumption. The authors also use in their approach a profile decomposition proved in the context of the Schrödinger equation by Keraani [**56**]. For a more elaborate discussion one should consult [**58**].

7.6. Idea of the proof of Theorem 8.3. To give a complete proof of this theorem in less than two lectures is impossible, so we will first outline the idea of the proof and then we only show rigorously few parts of it.

First the naive approach: we follow the strategy of induction/contraddiction introduced by Bourgain. We define E_{crit} the critical energy below which the $L^{10}_t L^{10}_x$ norm of a solutions stays bounded by some constant depending on the energy. We then identify a smooth *minimal energy blow up solution* u of energy E_{crit} such that

$$\|u\|_{L^{10}_t L^{10}_x} > M, \tag{130}$$

where M is as large as we please. For this solution we then show a series of properties that at the end will actually give

$$\|u\|_{L_t^{10}L_x^{10}} \leq C(E_{crit}), \tag{131}$$

contradicting (149).

This is in order the summary of the properties we prove for the *minimal energy blow up solution* on a fixed (compact) interval of time I:

(1) **Frequency and space localization:** For each $t \in I$ there exists $N(t) > 0$ and $x(t) \in \mathbb{R}^3$ such that $\hat{u}(t)$ is *mostly* supported at frequency of size proportional to $N(t)$ and $u(t)$ is mostly supported on a ball centered at $x(t)$ and radius proportional to $\frac{1}{N(t)}$. To prove the frequency localization part one uses the intuition that the *minimal energy blow up solution* u, at a given time t_0, cannot have two components u_- and u_+ which Fourier transforms are supported respectively in $|\xi| \leq N$ and $|\xi| \geq KN$, $K \ll 1$, and such that both pieces carrie a large amount of energy. The reason for this is that the energy relative to u_- will make the energy relative to u_+ smaller than E_{crit} and viceversa. Hence both u_- and u_+ can flow globally. On the other hand if K is large enough their nonlinear interaction is basically negligible, hence perturbation theory says that $u \sim u_- + u_+$, hence u exists globally and its $L_t^{10}L_x^{10}$ norm is uniformly bounded, a contradiction. A similar, but just a bit more complicated, argument gives also space localization.

(2) **Frequency localized interaction Morawetz inequality:** As we mentioned several times whenever a problem is not a perturbation of the linear one, like the critical ones for example, in order to obtain a global statement we need to have a global space-time bound. We learned that the Morawetz estimates for the defocusing problem and the Viriel identity for the focusing one are the types of estimates that we want to have. Bourgain in fact used the classical Morawetz estimate that appears in (30) with $p = 5$. Here the presence of the denominator forced the radial symmetry. In our argument instead we would like to use the Interaction Morawetz estimate (116). This is weaker in the sense that we only have the fourth power, but it is also stronger since we do not have a denominator. We keep in mind that our final goal is to show boundedness of the $L_I^{10}L_x^{10}$ norm of the *minimal energy blow up solution* u so we need to upgrade the $L_I^4L_x^4$ norm. We believe that for the low frequencies, where the energy is very small thanks to localization, Strichartz estimates will be enough to give us the bound in the $L_I^{10}L_x^{10}$ norm. For the high frequencies we also have small energy, but we expect that the Strichartz estimates are too weak here. So the idea is to first prove (116) for the high frequency part of the solution. We have for all $N_* < N_{min}$

$$\int_I \int |P_{\geq N_*} u(t,x)|^4 \, dxdt \lesssim \eta_1 N_*^{-3}, \tag{132}$$

where $N_{min} = \inf_{t \in I} N(t)$ for which one can prove $N_{min} > 0$ and η_1 is a small quantity. Note that the quantities appearing in the right hand side of (151) are independent of I.

(3) **Uniform boundedness of time interval I:** Assuming that $N(t)$ doesn't run to infinity, use the $L^4_I L^4_x$ bound, which is uniform in I, to get a uniform bound on the length of time interval I itself. With this information now, since most of the solution remains on a uniformly bounded frequency window, perturbation will provide the final uniform bound for the $L^{10}_I L^{10}_x$ norm.

(4) **Uniform Boundedness of $N(t)$:** We mentioned above that there exists N_{min} such that $0 < N_{min} \leq N(t)$, and this in not hard to prove. In fact by rescaling[29] one can assume that

$$N_{min} = 1.$$

The difficult part is to show that there exists $N_{max} < \infty$ such that

$$N(t) \leq N_{max}.$$

Again by contradiction one assumes that given $R \gg 1$ there exists t_R such that $N(t_R) > R$ and by definition most of the energy is located on frequencies $R < N(t_R) \lesssim |\xi|$. But then one can prove by a simple application of the "I-method" that although the energy has migrated on very large frequencies, some *littering* of mass has been left on medium frequencies. But mass on medium frequencies is equivalent to energy, hence there is some significant energy left over on medium frequencies. If then R is large enough these two pieces of the solution u, the one at very high frequencies and the one at medium frequencies, are very separated and each has a significant amount of energy. But this cannot happen for an *energy critical blow up solution*, as discussed above. Hence N_{max} must be bounded.

In order to proceed with the outline given above we use heavily Strichartz estimates (12) and (13), the improved bilinear estimate (14) and multilinear estimates of different kinds. A very important tool that was mentioned often above is the theory of perturbation that in practice is made of a serious of perturbation lemmas. These lemmas are particularly useful when we have to claim that if u is a solution to NLS and v is a solution to an equation which is a small perturbation of NLS, then u and v are close to each other and if one exists the other does too. Here we report two examples of such lemmas.

LEMMA 7.7 (Short-time perturbations). *Let I be a compact interval, and let $\tilde{u}$ be a function on $I \times \mathbb{R}^3$ which is a near-solution to (1) with $p = 5$ and $\mu = 1$ in the sense that*

$$(i\partial_t + \frac{1}{2}\Delta)\tilde{u} = |\tilde{u}|^4\tilde{u} + e \tag{133}$$

for some function e. Suppose that we also have the energy bound

$$\|\tilde{u}\|_{L^\infty_t \dot{H}^1_x(I \times \mathbb{R}^3)} \leq E$$

for some $E > 0$. Let $t_0 \in I$, and let $u(t_0)$ be close to $\tilde{u}(t_0)$ in the sense that

$$\|u(t_0) - \tilde{u}(t_0)\|_{\dot{H}^1_x} \leq E' \tag{134}$$

[29] Since the problem is H^1 critical and we only use the energy, nothing will change by rescaling!

for some $E' > 0$. Assume also that we have the smallness conditions

$$\|\nabla \tilde{u}\|_{L_t^{10} L_x^{30/13}(I \times \mathbb{R}^3)} \leq \epsilon_0 \tag{135}$$

$$\|\nabla e^{i(t-t_0)\Delta}(u(t_0) - \tilde{u}(t_0))\|_{L_t^{10} L_x^{30/13}(I \times \mathbb{R}^3)} \leq \epsilon \tag{136}$$

$$\|\nabla e\|_{L_t^2 L_x^{6/5}} \leq \epsilon \tag{137}$$

for some $0 < \epsilon < \epsilon_0$, where ϵ_0 is some constant $\epsilon_0 = \epsilon_0(E, E') > 0$.

We conclude that there exists a solution u to (1) with $p = 5$ and $\mu = 1$ on $I \times \mathbb{R}^3$ with the specified initial data $u(t_0)$ at t_0, and furthermore

$$\|u - \tilde{u}\|_{\dot{S}^1(I \times \mathbb{R}^3)} \lesssim E' \tag{138}$$

$$\|u\|_{\dot{S}^1(I \times \mathbb{R}^3)} \lesssim E' + E \tag{139}$$

$$\|u - \tilde{u}\|_{L_{t,x}^{10}(I \times \mathbb{R}^3)} \lesssim \|\nabla(u - \tilde{u})\|_{L_t^{10} L_x^{30/13}(I \times \mathbb{R}^3)} \lesssim \epsilon \tag{140}$$

$$\|\nabla(i\partial_t + \frac{1}{2}\Delta)(u - \tilde{u})\|_{L_t^2 L_x^{6/5}(I \times \mathbb{R}^3)} \lesssim \epsilon. \tag{141}$$

Note that $u(t_0) - \tilde{u}(t_0)$ is allowed to have large energy, albeit at the cost of forcing ϵ to be smaller, and worsening the bounds in (157). From the Strichartz estimate (12), we see that the hypothesis (155) is redundant if one is willing to take $E' = O(\varepsilon)$.

PROOF. By the well-posedness theory presented in Section 5, it suffice to prove (157) - (160) as *a priori* estimates[30]. We establish these bounds for $t \geq t_0$, since the corresponding bounds for the $t \leq t_0$ portion of I are proved similarly.

First note that the Strichartz estimate (12) and (13) give,

$$\|\tilde{u}\|_{\dot{S}^1(I \times \mathbb{R}^3)} \lesssim E + \|\tilde{u}\|_{L_{t,x}^{10}(I \times \mathbb{R}^3)} \cdot \|\tilde{u}\|^4_{\dot{S}^1(I \times \mathbb{R}^3)} + \varepsilon.$$

By (154) and Sobolev embedding we have $\|\tilde{u}\|_{L_{t,x}^{10}(I \times \mathbb{R}^3)} \lesssim \varepsilon_0$. A standard continuity argument in I then gives (if ε_0 is sufficiently small depending on E)

$$\|\tilde{u}\|_{\dot{S}^1(I \times \mathbb{R}^3)} \lesssim E. \tag{142}$$

Define $v := u - \tilde{u}$. For each $t \in I$ define the quantity

$$S(t) := \|\nabla(i\partial_t + \frac{1}{2}\Delta)v\|_{L_t^2 L_x^{6/5}([t_0,t] \times \mathbb{R}^3)}.$$

From using again Strichartz estimates and the definition of S^1, (155), we have

$$\begin{aligned} \|\nabla v\|_{L_t^{10} L_x^{30/13}([t_0,t] \times \mathbb{R}^3)} &\lesssim \|\nabla(v - e^{i(t-t_0)\frac{1}{2}\Delta}v(t_0))\|_{L_t^{10} L_x^{30/13}([t_0,t] \times \mathbb{R}^3)} \\ &\quad + \|\nabla e^{i(t-t_0)\frac{1}{2}\Delta}v(t_0)\|_{L_t^{10} L_x^{30/13}([t_0,t] \times \mathbb{R}^3)} \\ &\lesssim \|v - e^{i(t-t_0)\frac{1}{2}\Delta}v(t_0)\|_{\dot{S}^1([t_0,t] \times \mathbb{R}^3)} + \varepsilon \end{aligned} \tag{143}$$

$$\lesssim S(t) + \varepsilon. \tag{144}$$

On the other hand, since v obeys the equation

$$(i\partial_t + \frac{1}{2}\Delta)v = |\tilde{u} + v|^4(\tilde{u} + v) - |\tilde{u}|^4\tilde{u} - e = \sum_{j=1}^{5} \emptyset(v^j \tilde{u}^{5-j}) - e$$

[30]That is, we may assume the solution u already exists and is smooth on the entire interval I.

where $\emptyset(v_1, v_2, v_3, v_4, v_5)$ denotes any combination of v_i and $\bar{v_j}$. By some standard multilinear estimates, (154), (156), (163) then

$$S(t) \lesssim \varepsilon + \sum_{j=1}^{5} (S(t) + \varepsilon)^j \varepsilon_0^{5-j}.$$

If ε_0 is sufficiently small, a standard continuity argument then yields the bound $S(t) \lesssim \varepsilon$ for all $t \in I$. This gives (160), and (159) follows from (163). Applying Strichartz inequalities again, (153) we then conclude (157) (if ε is sufficiently small), and then from (161) and the triangle inequality we conclude (158). □

We will actually be more interested in iterating the above lemma to deal with the more general situation of near-solutions with finite but arbitrarily large $L^{10}_{t,x}$ norms.

LEMMA 7.8 (Long-time perturbations). *Let I be a compact interval, and let $\tilde{u}$ be a function on $I \times \mathbb{R}^3$ which obeys the bounds*

$$\|\tilde{u}\|_{L^{10}_{t,x}(I \times \mathbb{R}^3)} \leq M \tag{145}$$

and

$$\|\tilde{u}\|_{L^\infty_t \dot{H}^1_x(I \times \mathbb{R}^3)} \leq E \tag{146}$$

for some $M, E > 0$. Suppose also that $\tilde{u}$ is a near-solution to (1) with $p = 5$ and $\mu = 1$ in the sense that it solves (152) for some e. Let $t_0 \in I$, and let $u(t_0)$ be close to $\tilde{u}(t_0)$ in the sense that

$$\|u(t_0) - \tilde{u}(t_0)\|_{\dot{H}^1_x} \leq E'$$

for some $E' > 0$. Assume also that we have the smallness conditions,

$$\|\nabla e^{i(t-t_0)\frac{1}{2}\Delta}(u(t_0) - \tilde{u}(t_0))\|_{L^{10}_t L^{30/13}_x(I \times \mathbb{R}^3)} \leq \varepsilon \tag{147}$$

$$\|\nabla e\|_{L^2_t L^{6/5}_x(I \times \mathbb{R}^3)} \leq \varepsilon$$

for $0 < \varepsilon < \varepsilon_1$, where ε_1 is some constant $\varepsilon_1 = \varepsilon_1(E, E', M) > 0$. We conclude there exists a solution u to (1) with $p = 5$ and $\mu = 1$ on $I \times \mathbb{R}^3$ with the specified initial data $u(t_0)$ at t_0, and furthermore

$$\|u - \tilde{u}\|_{\dot{S}^1(I \times \mathbb{R}^3)} \leq C(M, E, E')$$

$$\|u\|_{\dot{S}^1(I \times \mathbb{R}^3)} \leq C(M, E, E')$$

$$\|u - \tilde{u}\|_{L^{10}_{t,x}(I \times \mathbb{R}^3)} \leq \|\nabla(u - \tilde{u})\|_{L^{10}_t L^{30/13}_x(I \times \mathbb{R}^3)} \leq C(M, E, E')\varepsilon.$$

Once again, the hypothesis (166) is redundant by the Strichartz estimate if one is willing to take $E' = O(\varepsilon)$; however it will be useful in our applications to know that this Lemma can tolerate a perturbation which is large in the energy norm but whose free evolution is small in the $L^{10}_t \dot{W}^{1,30/13}_x$ norm.

This lemma is already useful in the $e = 0$ case, as it says that one has local well-posedness in the energy space whenever the $L^{10}_{t,x}$ norm is bounded; in fact one has locally Lipschitz dependence on the initial data. For similar perturbative results see [**13**], [**12**].

PROOF. As in the previous proof, we may assume that t_0 is the lower bound of the interval I. Let $\varepsilon_0 = \varepsilon_0(E, 2E')$ be as in Lemma 8.7. (We need to replace E' by the slightly larger $2E'$ as the $\dot{H}^1$ norm of $u - \tilde{u}$ is going to grow slightly in time.)

The first step is to establish a $\dot{S}^1$ bound on $\tilde{u}$. Using (164) we may subdivide I into $C(M, \varepsilon_0)$ time intervals such that the $L^{10}_{t,x}$ norm of $\tilde{u}$ is at most ε_0 on each such interval. By using (165) and Strichartz estimates, as in the proof of (161), we see that the $\dot{S}^1$ norm of $\tilde{u}$ is $O(E)$ on each of these intervals. Summing up over all the intervals we conclude

$$\|\tilde{u}\|_{\dot{S}^1(I\times\mathbb{R}^3)} \leq C(M, E, \varepsilon_0)$$

and in particular

$$\|\nabla\tilde{u}\|_{L^{10}_t L^{30/13}_x(I\times\mathbb{R}^3)} \leq C(M, E, \varepsilon_0).$$

We can then subdivide the interval I into $N \leq C(M, E, \varepsilon_0)$ subintervals $I_j \equiv [T_j, T_{j+1}]$ so that on each I_j we have,

$$\|\nabla\tilde{u}\|_{L^{10}_t L^{30/13}_x(I_j\times\mathbb{R}^3)} \leq \varepsilon_0.$$

We can then verify inductively using Lemma 8.7 for each j that if ε_1 is sufficiently small depending on ε_0, N, E, E', then we have

$$\begin{aligned}
\|u - \tilde{u}\|_{\dot{S}^1(I_j\times\mathbb{R}^3)} &\leq C(j)E' \\
\|u\|_{\dot{S}^1(I_j\times\mathbb{R}^3)} &\leq C(j)(E' + E) \\
\|\nabla(u - \tilde{u})\|_{L^{10}_t L^{30/13}_x(I_j\times\mathbb{R}^3)} &\leq C(j)\varepsilon \\
\|\nabla(i\partial_t + \frac{1}{2}\Delta)(u - \tilde{u})\|_{L^2_t L^{6/5}_x(I_j\times\mathbb{R}^3)} &\leq C(j)\varepsilon
\end{aligned}$$

and hence by Strichartz we have

$$\begin{aligned}
&\|\nabla e^{i(t-T_{j+1})\frac{1}{2}\Delta}(u(T_{j+1}) - \tilde{u}(T_{j+1}))\|_{L^{10}_t L^{30/13}_x(I\times\mathbb{R}^3)} \\
&\quad \leq \|\nabla e^{i(t-T_j)\frac{1}{2}\Delta}(u(T_j) - \tilde{u}(T_j))\|_{L^{10}_t L^{30/13}_x(I\times\mathbb{R}^3)} + C(j)\varepsilon
\end{aligned}$$

and

$$\|u(T_{j+1}) - \tilde{u}(T_{j+1})\|_{\dot{H}^1} \leq \|u(T_j) - \tilde{u}(T_j)\|_{\dot{H}^1} + C(j)\varepsilon$$

allowing one to continue the induction (if ε_1 is sufficiently small depending on E, N, E', ε_0, then the quantity in (153) will not exceed $2E'$). The claim follows. □

8. Global well-posedness for the $H^1(\mathbb{R}^n)$ critical NLS -Part I

We recall that the H^1 critical exponent for (1) is $p = 1 + \frac{4}{n-2}$. We also recall the following theorem that can be basically completely proved using either directly or indirectly theorems and arguments already presented in Section 5 and Section 6:

THEOREM 8.1 (Local or global small data well-posedness for the H^1 critical NLS). *We have the following two results:*

(1) *For any $u_0 \in H^1$ there exist $T = T(u_0)$ and a unique solution $u \in S^1_{[T,T]}$ to (1) with $p = 1 + \frac{4}{n-2}$ and $\mu = \pm 1$. Moreover there is continuity with respect to the initial data.*

(2) *There exists* ϵ *small enough such that for any* u_0, $\|u_0\|_{H^1} \leq \epsilon$ *there exists a unique global solution* $u \in S^1$ *to* (1) *with* $p = 1 + \frac{4}{n-2}$ *and* $\mu = \pm 1$. *Moreover there is continuity with respect to the initial data and scattering in the sense that there exists* $u_\pm \in H^1$ *such that*

$$\|u(t) - S(t)u_\pm\|_{H^1} \longrightarrow 0 \quad as \quad t \to \pm\infty.$$

PROOF. It is clear that the part about well-posedness is a summary of what has been proved in Section 5. The part about scattering instead can be proved as in Section 6 and by simply observing that Proposition 6.16 follows directly from the well-posedness proof thanks to the small data assumption. □

REMARK 8.2. We first remark that this theorem doesn't *see* the focusing or defocusing nature of the equation. This clearly means that in Theorem 8.1 the NLS is treated as a "small" perturbation of the linear problem. Due to the criticality of the problem and hence the fact that T depends also on the profile of the initial data an iteration argument based on the conservation of mass and energy is not possible. It is also clear that even increasing the regularity of the data the large data problem doesn't become any easier.

The first breakthrough on this problem is due to Bourgain [**13**]. He considers the defocusing case with $n = 3, 4$ and assumes radial symmetry for the problem. He proves the second part of Theorem 8.1 for arbitrarily large radially symmetric data. Here we summarize the main steps of Bourgain's proof for $n = 3$, which doesn't really do justice to the novelty and depth of the proof itself. The background argument is done by induction on the size of the energy E, the only quantity, besides the mass that here doesn't play much of a role, that remains controlled over time. From Theorem 8.1 the first step of the induction (small E) is in place. Let's now assume the second induction assumption that if $E < E_0$, for E_0 arbitrarily large, then the theorem is true. We take $E = E_0$ and we want to prove that also in this case the theorem is true. One first shows that the theorem follows if and only if the norm $L^{10}_t L^{10}_x$ of the solution remains bounded (see Theorem 4.8). Then the proof proceeds by contradiction. One supposes that there is a solution u such that $\|u\|_{L^{10}_t L^{10}_x}$ is arbitrarily large and $E = E_0$. The heart of the proof is on showing that at some time t_0 there is concentration of the H^1 norm: there exists a small ball B_0 centered at the origin such that $\|u(t_0)\|_{H^1(B_0)} > \delta$, and this ball is "sufficiently isolated" from the rest of the solution . It is here that the radial assumption is used. At this point one restarts the evolution at time t_0 by splitting the data as

$$\psi_0 = u(t_0)\chi_{B_0} \quad \text{and} \quad \psi_1 = u(t_0)(1 - \chi_{B_0}),$$

where χ_{B_0} is the indicator function for the ball B_0, and evolving ψ_0 with NLS and ψ_1 with a difference equation so that the sum of the two evolutions give the solution to NLS. Since now $\psi_0 \in H^1$ and $x\psi \in L^2$ it follows[31] that the evolution v of ψ_0 is global in time. Moreover since $E(\psi_0) \sim \delta^2$ it follows that $E(\psi_1) < E_0 - \delta^2$. Hence for the difference equation we are in the induction assumption. This is not quite like to have the equation under the induction assumption, but with some relatively straightforward perturbation theory[32] one also gets that the evolution w of ψ_1 is

[31]This result is for example proved in [**19**] as a consequence of the pseudo-conformal transformation and a monotonicity formula linked to it.

[32]That works tanks to the fact that the ball is "sufficiently" isolated from the rest of the solution.

global. Hence we have a global evolution for the solution $u = v + w$ to NLS and as a consequence a uniform bound for $\|u\|_{L^{10}_t L^{10}_x}$ which is a contradiction.

Almost at the same time, with the same radial symmetry assumption above, Grillakis [**43**] proved a slighter weaker result than Bourgain's, namely existence and uniqueness for smooth global solution. It took few more years to remove the radial assumption and obtain the following theorem and its corollary [**29**]:

THEOREM 8.3. *For any* u_0 *with finite energy,* $E(u_0) < \infty$, *there exists a unique*[33] *global solution* $u \in C^0_t(\dot{H}^1_x) \cap L^{10}_{t,x}$ *to* (1) *with* $p = 5, n = 3, \mu = 1$ *such that*

$$\int_{-\infty}^{\infty} \int_{\mathbb{R}^3} |u(t,x)|^{10}\, dxdt \leq C(E(u_0)). \tag{148}$$

for some constant $C(E(u_0))$ *that depends only on the energy.*

As one can see from Theorem 4.8 and from the arguments in Section 6, the $L^{10}_{t,x}$ bound above also gives scattering and and persistence of regularity:

COROLLARY 8.4. *Let* u_0 *have finite energy. Then there exist finite energy solutions* $u_\pm(t,x)$ *to the free Schrödinger equation* $(i\partial_t + \Delta)u_\pm = 0$ *such that*

$$\|u_\pm(t) - u(t)\|_{\dot{H}^1} \to 0 \text{ as } t \to \pm\infty.$$

Furthermore, the maps $u_0 \mapsto u_\pm(0)$ *are homeomorphisms from* $\dot{H}^1(\mathbb{R}^3)$ *to* $\dot{H}^1(\mathbb{R}^3)$. *Finally, if* $u_0 \in H^s$ *for some* $s > 1$, *then* $u(t) \in H^s$ *for all time* t, *and one has the uniform bound*

$$\sup_{t\in\mathbb{R}} \|u(t)\|_{H^s} \leq C(E(u_0), s)\|u_0\|_{H^s}.$$

Most of the rest of this lecture and Section 8 will be devoted to give an idea of the proof for Theorem 8.3. Still for the defocusing case and for $n > 3$ we recall first the result of Tao [**74**], where an equivalent of Theorem 8.3 is proved still under the radial assumption, the result of Ryckman and Visan [**70**] for $n = 4$, where the radial assumption is removed, and finally the full generalization for any $n \geq 5$ by Visan [**77**].

The situation in the focusing case was first considered successfully by Kenig and Merle. They prove the following theorem [**52**]:

THEOREM 8.5. *Assume that* $E(u_0) < E(W), \|u_0\|_{\dot{H}^1} < \|W\|_{\dot{H}^1}$, *where* $n = 3, 4, 5$ *and* u_0 *is radial and* W *is the stationary solution. Then the solution* u *to the critical* H^1 *focusing IVP* (1) *with data* u_0 *at* $t = 0$ *is defined for all time and there exists* $u_\pm \in \dot{H}^1$ *such that*

$$\|S(t)u_\pm - u(t)\|_{\dot{H}^1} \to 0 \text{ as } t \to \pm\infty.$$

Moreover for u_0 *radial,* $E(u_0) < E(W)$, *but* $\|u_0\|_{\dot{H}^1} > \|W\|_{\dot{H}^1}$, *the solution must break down in finite time.*

This result has been extended in every dimension $n \geq 3$ and for general data in [**57**]. Moreover a similar result has been proved by Kenig and Merle for the critical wave equation without the radial assumption [**53**], see also [**54**]. The proof of Theorem 8.5 introduces a new point of view for these problems. Using a concentration-compactness argument the authors reduce matters to a rigidity theorem, which is

[33] In fact, uniqueness actually holds in the larger space $C^0_t(\dot{H}^1_x)$ (thus eliminating the constraint that $u \in L^{10}_{t,x}$) [**29**].

proved with the aid of a localized Virial identity (in the spirit of Merle [**62, 63**]). The radiality enters only in the proof of the rigidity theorem. In the case of the critical wave equation other consideration of *elliptic nature* are used to remove the radial assumption. The authors also use in their approach a profile decomposition proved in the context of the Schrödinger equation by Keraani [**56**]. For a more elaborate discussion one should consult [**58**].

8.6. Idea of the proof of Theorem 8.3. To give a complete proof of this theorem in less than two lectures is impossible, so we will first outline the idea of the proof and then we only show rigorously few parts of it.

First the naive approach: we follow the strategy of induction/contraddiction introduced by Bourgain. We define E_{crit} the critical energy below which the $L^{10}_t L^{10}_x$ norm of a solutions stays bounded by some constant depending on the energy. We then identify a smooth *minimal energy blow up solution* u of energy E_{crit} such that

$$\|u\|_{L^{10}_t L^{10}_x} > M, \tag{149}$$

where M is as large as we please. For this solution we then show a series of properties that at the end will actually give

$$\|u\|_{L^{10}_t L^{10}_x} \leq C(E_{crit}), \tag{150}$$

contradicting (149).

This is in order the summary of the properties we prove for the *minimal energy blow up solution* on a fixed (compact) interval of time I:

(1) **Frequency and space localization:** For each $t \in I$ there exists $N(t) > 0$ and $x(t) \in \mathbb{R}^3$ such that $\hat{u}(t)$ is *mostly* supported at frequency of size proportional to $N(t)$ and $u(t)$ is mostly supported on a ball centered at $x(t)$ and radius proportional to $\frac{1}{N(t)}$. To prove the frequency localization part one uses the intuition that the *minimal energy blow up solution* u, at a given time t_0, cannot have two components u_- and u_+ which Fourier transforms are supported respectively in $|\xi| \leq N$ and $|\xi| \geq KN$, $K \ll 1$, and such that both pieces carrie a large amount of energy. The reason for this is that the energy relative to u_- will make the energy relative to u_+ smaller than E_{crit} and viceversa. Hence both u_- and u_+ can flow globally. On the other hand if K is large enough their nonlinear interaction is basically negligible, hence perturbation theory says that $u \sim u_- + u_+$, hence u exists globally and its $L^{10}_t L^{10}_x$ norm is uniformly bounded, a contradiction. A similar, but just a bit more complicated, argument gives also space localization.

(2) **Frequency localized interaction Morawetz inequality:** As we mentioned several times whenever a problem is not a perturbation of the linear one, like the critical ones for example, in order to obtain a global statement we need to have a global space-time bound. We learned that the Morawetz estimates for the defocusing problem and the Viriel identity for the focusing one are the types of estimates that we want to have. Bourgain in fact used the classical Morawetz estimate that appears in (30) with $p = 5$. Here the presence of the denominator forced the radial symmetry. In our argument instead we would like to use the Interaction Morawetz estimate (116). This is weaker in the sense that we only have the fourth power, but it is also stronger since we do not have a denominator. We

keep in mind that our final goal is to show boundedness of the $L_I^{10}L_x^{10}$ norm of the *minimal energy blow up solution* u so we need to upgrade the $L_I^4L_x^4$ norm. We believe that for the low frequencies, where the energy is very small thanks to localization, Strichartz estimates will be enough to give us the bound in the $L_I^{10}L_x^{10}$ norm. For the high frequencies we also have small energy, but we expect that the Strichartz estimates are too weak here. So the idea is to first prove (116) for the high frequency part of the solution. We have for all $N_* < N_{min}$

$$\int_I \int |P_{\geq N_*} u(t,x)|^4 \, dxdt \lesssim \eta_1 N_*^{-3}, \tag{151}$$

where $N_{min} = \inf_{t\in I} N(t)$ for which one can prove $N_{min} > 0$ and η_1 is a small quantity. Note that the quantities appearing in the right hand side of (151) are independent of I.

(3) **Uniform boundedness of time interval I:** Assuming that $N(t)$ doesn't run to infinity, use the $L_I^4L_x^4$ bound, which is uniform in I, to get a uniform bound on the length of time interval I itself. With this information now, since most of the solution remains on a uniformly bounded frequency window, perturbation will provide the final uniform bound for the $L_I^{10}L_x^{10}$ norm.

(4) **Uniform Boundedness of $N(t)$:** We mentioned above that there exists N_{min} such that $0 < N_{min} \leq N(t)$, and this in not hard to prove. In fact by rescaling[34] one can assume that

$$N_{min} = 1.$$

The difficult part is to show that there exists $N_{max} < \infty$ such that

$$N(t) \leq N_{max}.$$

Again by contradiction one assumes that given $R \gg 1$ there exists t_R such that $N(t_R) > R$ and by definition most of the energy is located on frequencies $R < N(t_R) \lesssim |\xi|$. But then one can prove by a simple application of the "I-method" that although the energy has migrated on very large frequencies, some *littering* of mass has been left on medium frequencies. But mass on medium frequencies is equivalent to energy, hence there is some significant energy left over on medium frequencies. If then R is large enough these two pieces of the solution u, the one at very high frequencies and the one at medium frequencies, are very separated and each has a significant amount of energy. But this cannot happen for an *energy critical blow up solution*, as discussed above. Hence N_{max} must be bounded.

In order to proceed with the outline given above we use heavily Strichartz estimates (12) and (13), the improved bilinear estimate (14) and multilinear estimates of different kinds. A very important tool that was mentioned often above is the theory of perturbation that in practice is made of a serious of perturbation lemmas. These lemmas are particularly useful when we have to claim that if u is a solution to NLS and v is a solution to an equation which is a small perturbation of NLS, then u and v are close to each other and if one exists the other does too. Here we report two examples of such lemmas.

[34]Since the problem is H^1 critical and we only use the energy, nothing will change by rescaling!

LEMMA 8.7 (Short-time perturbations). *Let I be a compact interval, and let $\tilde{u}$ be a function on $I \times \mathbb{R}^3$ which is a near-solution to (1) with $p = 5$ and $\mu = 1$ in the sense that*

$$(i\partial_t + \frac{1}{2}\Delta)\tilde{u} = |\tilde{u}|^4\tilde{u} + e \tag{152}$$

for some function e. Suppose that we also have the energy bound

$$\|\tilde{u}\|_{L_t^\infty \dot{H}_x^1(I\times\mathbb{R}^3)} \leq E$$

for some $E > 0$. Let $t_0 \in I$, and let $u(t_0)$ be close to $\tilde{u}(t_0)$ in the sense that

$$\|u(t_0) - \tilde{u}(t_0)\|_{\dot{H}_x^1} \leq E' \tag{153}$$

for some $E' > 0$. Assume also that we have the smallness conditions

$$\|\nabla\tilde{u}\|_{L_t^{10}L_x^{30/13}(I\times\mathbb{R}^3)} \leq \epsilon_0 \tag{154}$$

$$\|\nabla e^{i(t-t_0)\Delta}(u(t_0) - \tilde{u}(t_0))\|_{L_t^{10}L_x^{30/13}(I\times\mathbb{R}^3)} \leq \epsilon \tag{155}$$

$$\|\nabla e\|_{L_t^2 L_x^{6/5}} \leq \epsilon \tag{156}$$

for some $0 < \epsilon < \epsilon_0$, where ϵ_0 is some constant $\epsilon_0 = \epsilon_0(E, E') > 0$.

We conclude that there exists a solution u to (1) with $p = 5$ and $\mu = 1$ on $I \times \mathbb{R}^3$ with the specified initial data $u(t_0)$ at t_0, and furthermore

$$\|u - \tilde{u}\|_{\dot{S}^1(I\times\mathbb{R}^3)} \lesssim E' \tag{157}$$

$$\|u\|_{\dot{S}^1(I\times\mathbb{R}^3)} \lesssim E' + E \tag{158}$$

$$\|u - \tilde{u}\|_{L_{t,x}^{10}(I\times\mathbb{R}^3)} \lesssim \|\nabla(u - \tilde{u})\|_{L_t^{10}L_x^{30/13}(I\times\mathbb{R}^3)} \lesssim \epsilon \tag{159}$$

$$\|\nabla(i\partial_t + \frac{1}{2}\Delta)(u - \tilde{u})\|_{L_t^2 L_x^{6/5}(I\times\mathbb{R}^3)} \lesssim \epsilon. \tag{160}$$

Note that $u(t_0) - \tilde{u}(t_0)$ is allowed to have large energy, albeit at the cost of forcing ϵ to be smaller, and worsening the bounds in (157). From the Strichartz estimate (12), we see that the hypothesis (155) is redundant if one is willing to take $E' = O(\varepsilon)$.

PROOF. By the well-posedness theory presented in Section 5, it suffice to prove (157) - (160) as *a priori* estimates[35]. We establish these bounds for $t \geq t_0$, since the corresponding bounds for the $t \leq t_0$ portion of I are proved similarly.

First note that the Strichartz estimate (12) and (13) give,

$$\|\tilde{u}\|_{\dot{S}^1(I\times\mathbb{R}^3)} \lesssim E + \|\tilde{u}\|_{L_{t,x}^{10}(I\times\mathbb{R}^3)} \cdot \|\tilde{u}\|^4_{\dot{S}^1(I\times\mathbb{R}^3)} + \varepsilon.$$

By (154) and Sobolev embedding we have $\|\tilde{u}\|_{L_{t,x}^{10}(I\times\mathbb{R}^3)} \lesssim \varepsilon_0$. A standard continuity argument in I then gives (if ε_0 is sufficiently small depending on E)

$$\|\tilde{u}\|_{\dot{S}^1(I\times\mathbb{R}^3)} \lesssim E. \tag{161}$$

Define $v := u - \tilde{u}$. For each $t \in I$ define the quantity

$$S(t) := \|\nabla(i\partial_t + \frac{1}{2}\Delta)v\|_{L_t^2 L_x^{6/5}([t_0,t]\times\mathbb{R}^3)}.$$

[35]That is, we may assume the solution u already exists and is smooth on the entire interval I.

From using again Strichartz estimates and the definition of S^1, (155), we have

$$\begin{aligned}
\|\nabla v\|_{L_t^{10}L_x^{30/13}([t_0,t]\times\mathbb{R}^3)} &\lesssim \|\nabla(v - e^{i(t-t_0)\frac{1}{2}\Delta}v(t_0))\|_{L_t^{10}L_x^{30/13}([t_0,t]\times\mathbb{R}^3)} \\
&\quad + \|\nabla e^{i(t-t_0)\frac{1}{2}\Delta}v(t_0)\|_{L_t^{10}L_x^{30/13}([t_0,t]\times\mathbb{R}^3)} \\
&\lesssim \|v - e^{i(t-t_0)\frac{1}{2}\Delta}v(t_0)\|_{\dot{S}^1([t_0,t]\times\mathbb{R}^3)} + \varepsilon \qquad (162)\\
&\lesssim S(t) + \varepsilon. \qquad (163)
\end{aligned}$$

On the other hand, since v obeys the equation

$$(i\partial_t + \frac{1}{2}\Delta)v = |\tilde{u}+v|^4(\tilde{u}+v) - |\tilde{u}|^4\tilde{u} - e = \sum_{j=1}^{5} \emptyset(v^j\tilde{u}^{5-j}) - e$$

where $\emptyset(v_1, v_2, v_3, v_4, v_5)$ denotes any combination of v_i and $\bar{v_j}$. By some standard multilinear estimates, (154), (156), (163) then

$$S(t) \lesssim \varepsilon + \sum_{j=1}^{5}(S(t)+\varepsilon)^j\varepsilon_0^{5-j}.$$

If ε_0 is sufficiently small, a standard continuity argument then yields the bound $S(t) \lesssim \varepsilon$ for all $t \in I$. This gives (160), and (159) follows from (163). Applying Strichartz inequalities again, (153) we then conclude (157) (if ε is sufficiently small), and then from (161) and the triangle inequality we conclude (158). □

We will actually be more interested in iterating the above lemma to deal with the more general situation of near-solutions with finite but arbitrarily large $L^{10}_{t,x}$ norms.

LEMMA 8.8 (Long-time perturbations). *Let I be a compact interval, and let $\tilde{u}$ be a function on $I \times \mathbb{R}^3$ which obeys the bounds*

$$\|\tilde{u}\|_{L^{10}_{t,x}(I\times\mathbb{R}^3)} \le M \qquad (164)$$

and

$$\|\tilde{u}\|_{L_t^\infty \dot{H}_x^1(I\times\mathbb{R}^3)} \le E \qquad (165)$$

for some $M, E > 0$. Suppose also that $\tilde{u}$ is a near-solution to (1) with $p = 5$ and $\mu = 1$ in the sense that it solves (152) for some e. Let $t_0 \in I$, and let $u(t_0)$ be close to $\tilde{u}(t_0)$ in the sense that

$$\|u(t_0) - \tilde{u}(t_0)\|_{\dot{H}_x^1} \le E'$$

for some $E' > 0$. Assume also that we have the smallness conditions,

$$\begin{aligned}
\|\nabla e^{i(t-t_0)\frac{1}{2}\Delta}(u(t_0) - \tilde{u}(t_0))\|_{L_t^{10}L_x^{30/13}(I\times\mathbb{R}^3)} &\le \varepsilon \qquad (166)\\
\|\nabla e\|_{L_t^2L_x^{6/5}(I\times\mathbb{R}^3)} &\le \varepsilon
\end{aligned}$$

for $0 < \varepsilon < \varepsilon_1$, where ε_1 is some constant $\varepsilon_1 = \varepsilon_1(E, E', M) > 0$. We conclude there exists a solution u to (1) with $p = 5$ and $\mu = 1$ on $I \times \mathbb{R}^3$ with the specified initial data $u(t_0)$ at t_0, and furthermore

$$\begin{aligned}
\|u - \tilde{u}\|_{\dot{S}^1(I\times\mathbb{R}^3)} &\le C(M, E, E') \\
\|u\|_{\dot{S}^1(I\times\mathbb{R}^3)} &\le C(M, E, E') \\
\|u - \tilde{u}\|_{L^{10}_{t,x}(I\times\mathbb{R}^3)} \le \|\nabla(u - \tilde{u})\|_{L_t^{10}L_x^{30/13}(I\times\mathbb{R}^3)} &\le C(M, E, E')\varepsilon.
\end{aligned}$$

Once again, the hypothesis (166) is redundant by the Strichartz estimate if one is willing to take $E' = O(\varepsilon)$; however it will be useful in our applications to know that this Lemma can tolerate a perturbation which is large in the energy norm but whose free evolution is small in the $L_t^{10}\dot{W}_x^{1,30/13}$ norm.

This lemma is already useful in the $e = 0$ case, as it says that one has local well-posedness in the energy space whenever the $L_{t,x}^{10}$ norm is bounded; in fact one has locally Lipschitz dependence on the initial data. For similar perturbative results see [**13**], [**12**].

PROOF. As in the previous proof, we may assume that t_0 is the lower bound of the interval I. Let $\varepsilon_0 = \varepsilon_0(E, 2E')$ be as in Lemma 8.7. (We need to replace E' by the slightly larger $2E'$ as the $\dot{H}^1$ norm of $u - \tilde{u}$ is going to grow slightly in time.)

The first step is to establish a $\dot{S}^1$ bound on $\tilde{u}$. Using (164) we may subdivide I into $C(M, \varepsilon_0)$ time intervals such that the $L_{t,x}^{10}$ norm of $\tilde{u}$ is at most ε_0 on each such interval. By using (165) and Strichartz estimates, as in the proof of (161), we see that the $\dot{S}^1$ norm of $\tilde{u}$ is $O(E)$ on each of these intervals. Summing up over all the intervals we conclude

$$\|\tilde{u}\|_{\dot{S}^1(I\times\mathbb{R}^3)} \leq C(M, E, \varepsilon_0)$$

and in particular

$$\|\nabla\tilde{u}\|_{L_t^{10}L_x^{30/13}(I\times\mathbb{R}^3)} \leq C(M, E, \varepsilon_0).$$

We can then subdivide the interval I into $N \leq C(M, E, \varepsilon_0)$ subintervals $I_j \equiv [T_j, T_{j+1}]$ so that on each I_j we have,

$$\|\nabla\tilde{u}\|_{L_t^{10}L_x^{30/13}(I_j\times\mathbb{R}^3)} \leq \varepsilon_0.$$

We can then verify inductively using Lemma 8.7 for each j that if ε_1 is sufficiently small depending on ε_0, N, E, E', then we have

$$\begin{aligned}
\|u - \tilde{u}\|_{\dot{S}^1(I_j\times\mathbb{R}^3)} &\leq C(j)E' \\
\|u\|_{\dot{S}^1(I_j\times\mathbb{R}^3)} &\leq C(j)(E' + E) \\
\|\nabla(u - \tilde{u})\|_{L_t^{10}L_x^{30/13}(I_j\times\mathbb{R}^3)} &\leq C(j)\varepsilon \\
\|\nabla(i\partial_t + \frac{1}{2}\Delta)(u - \tilde{u})\|_{L_t^2L_x^{6/5}(I_j\times\mathbb{R}^3)} &\leq C(j)\varepsilon
\end{aligned}$$

and hence by Strichartz we have

$$\begin{aligned}
&\|\nabla e^{i(t-T_{j+1})\frac{1}{2}\Delta}(u(T_{j+1}) - \tilde{u}(T_{j+1}))\|_{L_t^{10}L_x^{30/13}(I\times\mathbb{R}^3)} \\
&\quad\leq \|\nabla e^{i(t-T_j)\frac{1}{2}\Delta}(u(T_j) - \tilde{u}(T_j))\|_{L_t^{10}L_x^{30/13}(I\times\mathbb{R}^3)} + C(j)\varepsilon
\end{aligned}$$

and

$$\|u(T_{j+1}) - \tilde{u}(T_{j+1})\|_{\dot{H}^1} \leq \|u(T_j) - \tilde{u}(T_j)\|_{\dot{H}^1} + C(j)\varepsilon$$

allowing one to continue the induction (if ε_1 is sufficiently small depending on E, N, E', ε_0, then the quantity in (153) will not exceed $2E'$). The claim follows. $\square$

9. The periodic NLS

So far we only talked about the Schrödinger equation on $\mathbb{R}^n$, and one can certainly define this equation in more general manifolds M by replacing the usual Laplacian Δ with the Laplace-Beltrami operator Δ_M. In recent years there has been a flurry of activity concerning well-posedness and blow up of the IVP (1) on different manifolds, see for example in the setting of compact Riemannian manifolds $(M, \mathbf{g})$ [**8, 7, 17, 18**]. In this case the conclusions are generally weaker than those in Euclidean spaces: there is no scattering to linear solutions, or some other type of asymptotic control of the nonlinear evolution as $t \to \infty$. Moreover, in certain cases such as the spheres $\mathbb{S}^n$, the well-posedness theory requires sufficiently subcritical nonlinearities, due to concentration of certain spherical harmonics, see [**16**]. The situation is different when we are in the setting of symmetric spaces of noncompact type[36]. The simplest such spaces are the hyperbolic spaces $\mathbb{H}^n$, $n \geq 2$. On hyperbolic spaces one can in fact prove *stronger* theorems than on Euclidean spaces. For the linear flow one can exhibit a larger class of global in time Strichartz estimates [**1, 47**], (for radial functions these were already proved in [**3, 4, 5, 67**]). For the nonlinear flow with $N(u) = u|u|^{p-1}$ one can prove noneuclidean Morawetz inequalities, and scattering in H^1 in the full subcritical range $p \in (1, 1 + 4/(n-2))$, [**47**]. These stronger theorems are possible because of the more robust geometry at infinity of noncompact symmetric spaces compared to Euclidean spaces; for example, the scattering result for the nonlinear Schrödinger equation can be interpreted as the absence of long range effects of the nonlinearity.

Here we cannot clearly address all the work mentioned above, but instead we will consider the spacial case of the periodic NLS (1), or in other words the problem on the torus $\mathbb{T}^n$. The first work on the periodic NLS with non smooth data goes back to Bourgain [**8**]. Since we already learned from Section 2 that the first step to take is to analyze in the best possible way the linear problem, we will do this now. We cannot hope to prove Strichartz estimates starting from a dispersive estimate since there is no dispersion here in the sense introduced in Section 2. This is because the periodic condition at the boundary does not allow the solution to decay in time. So one needs to use a different analysis. We start by saying that the torus that will be considered here is the one on which

$$\widehat{\Delta_{\mathbb{T}^n} f}(k) = \sum_{i=1}^{n} k_i^2 \hat{f}(k).$$

The situation is very different if instead one consider general[37] tori $\tilde{\mathbb{T}}^n$ where

$$\widehat{\Delta_{\tilde{\mathbb{T}}^n} f}(k) = (\sum_{i=1}^{n} a_i^2 k_i^2) \hat{f}(k), \tag{167}$$

where $a_i^2 > 0$ for $i = 1, .., n$. in this case the theorems below are either not proved or the results are much weaker, [**15**].

Let's go back to $\mathbb{T}^n$. We will show here only one bilinear estimate that is particularly instructive:

[36]The symmetric spaces of noncompact type are simply connected Riemannian manifolds of nonpositive sectional curvature, without Euclidean factors, and for which every geodesic symmetry defines an isometry.

[37]These are also called *irrational tori.*

THEOREM 9.1. *Assume ϕ_i has Fourier transform supported at frequency N_i for $i = 1, 2$ and that $S(t)\phi_i$ is the linear solution for the linear IVP (4) on $\mathbb{T}^2$ with data ϕ_i . Then if $N_1 \geq N_2$, for any $\epsilon > 0$ we have*

$$\|\chi(t)S(t)\phi_1\chi(t)\overline{S(t)\phi_2}\|_{L^2_tL^2_{\mathbb{T}^2}} \lesssim N_2^\epsilon \|\phi_1\|_{L^2}\|\phi_2\|_{L^2}, \tag{168}$$

where $\chi(t)$ is a smooth cut off function in time near $t = 0$.

This theorem is only part of a more general conjecture of Bourgian [**8**] (see also [**41**]) that we now recall. Assume that ϕ is supported at frequency N and assume that

$$\|\chi(t)S(t)\phi\|_{L^r_tL^r_{\mathbb{T}^n}} \leq K(n, r, N)\|\phi\|_{L^2_{\mathbb{T}^n}},$$

then we have the following estimates for $K(n, r, N)$

CONJECTURE 9.2. *With the above assumptions*

$$K(n, r, N) \quad < \quad C_r \quad \textit{for} \quad r < \frac{2(n+2)}{n} \tag{169}$$

$$K(n, r, N) \quad \ll \quad N^\epsilon \quad \textit{for} \quad r = \frac{2(n+2)}{n} \tag{170}$$

$$K(n, r, N) \quad < \quad C_rN^{\frac{n}{2}-\frac{n+2}{r}} \quad \textit{for} \quad r > \frac{2(n+2)}{n} \tag{171}$$

For a partial proof of this conjecture see [**8**].

REMARK 9.3. It is important to note that (168) can also be read as the $L^4_{[-1,1]}L^4_x$ Strichartz estimate, since in fact $(4, 4)$ is an admissible pair in this case. Based on this and on the techniques to prove well-posedness in Section 4, we can immediately deduce for example that for the H^1 subcritical IVP (1) in $\mathbb{T}^2$ l.w.p. is available for $0 < s \leq 1$ when the nonlinearity is not algebraic and $0 < s$ when it is. Also it should be stressed that l.w.p. for $s = 0$ cannot be proved using (170) because the loss of regularity represented by N^ϵ. It should be said that this loss can be proved to be even smaller, of the order of $\log(N)$, see footnote at the end of the lecture.

PROBLEM 9.4. *Prove that there exists ϕ such that*

$$\|\chi(t)S(t)\phi\|_{L^4_tL^4_{\mathbb{T}^2}} \sim \log(N)\|\phi\|_{L^2_{\mathbb{T}^2}},$$

*(see [**41**]).*

The proof of (168) is based on some number theoretic facts that we recall in the following three lemmas; see also related estimates in the work of Bourgain [**7, 15**] and [**35**].

The following lemma is known as **Pick's Lemma** [**69**]:

LEMMA 9.5. *Let Ar be the area of a simply connected lattice polygon. Let E denote the number of lattice points on the polygon edges and I the number of lattice points in the interior of the polygon. Then*

$$Ar = I + \frac{1}{2}E - 1.$$

LEMMA 9.6. *Let $\mathcal{C}$ be a circle of radius R. If γ is an arc on $\mathcal{C}$ of length $|\gamma| < \left(\frac{3}{4}R\right)^{1/3}$, then γ contains at most 2 lattice points.*

PROOF. We prove the lemma by contradiction. Assume that there are 3 lattice points P_1, P_2 and P_3 on an arc $\gamma = AB$ of $\mathcal{C}$, and denote by $T(P_1, P_2, P_3)$ the triangle with vertices P_1, P_2 and P_3. Then, by Lemma 9.5 we have

$$\text{Area of } T(P_1, P_2, P_3) = I + \frac{1}{2}E - 1 \geq I + \frac{3}{2} - 1 = I + \frac{1}{2} \geq \frac{1}{2}.$$

We shall prove that under the assumption that $|\gamma| < \left(\frac{3}{4}R\right)^{1/3}$, then

$$\text{Area of } T(P_1, P_2, P_3) < \frac{1}{2}, \tag{172}$$

hence γ must contain at most two lattice points.

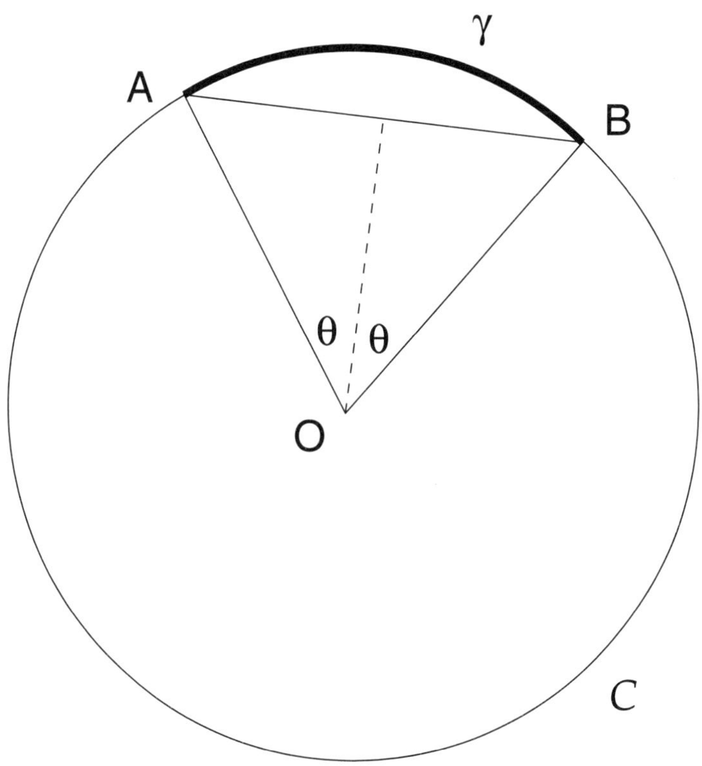

FIGURE 1. Triangle area.

We observe that (see Figure 1)

$$\text{Area of the sector } ABO = R^2\theta,$$

$$\text{Area of the triangle } ABO = R^2 \sin\theta\cos\theta.$$

Hence, for any P_1, P_2, P_3 on γ we have

$$\text{Area of } T(P_1, P_2, P_3) \leq R^2\theta - R^2\sin\theta\cos\theta = R^2(\theta - \frac{1}{2}\sin(2\theta)). \tag{173}$$

One can easily check that

$$\theta - \frac{1}{2}\sin(2\theta) \leq \frac{2}{3}\theta^3. \tag{174}$$

Thus (173), (174) and the fact that $|\gamma| = 2R\theta$ imply that

$$\text{Area of } T(P_1, P_2, P_3) \leq \frac{2}{3}R^2\theta^3 = \frac{1}{12}R^2(|\gamma|R^{-1})^3 < \frac{1}{2},$$

where to obtain the last inequality we used the assumption that $|\gamma| < \left(\frac{3}{4}R\right)^{1/3}$. Therefore (172) is proved. □

Also we recall the following result of Gauss, see, for example [**48**]

LEMMA 9.7. *Let K be a convex domain in $\mathbb{R}^2$. If*

$$N(\lambda) = \#\{\mathbb{Z}^2 \cap \lambda K\},$$

then, for $\lambda \ll 1$

$$N(\lambda) = \lambda^2|K| + O(\lambda),$$

where $|K|$ denotes the area of K and $\#A$ denotes the number of points of a set A.

We are now ready for the proof of Theorem 9.1

PROOF. Let ψ be a positive even Schwartz function such that $\psi = \hat{\chi}$. Then we have (here we use for simplicity $\int dk = \sum_k$)

$$\begin{aligned}
B &= \|\chi(t)(S(t)\phi_1)\,\chi(t)\overline{(S(t)\phi_2)}\|_{L^2_tL^2_x} \\
&= \left\| \int_{k=k_1+k_2,\ \tau=\tau_1+\tau_2} \widehat{\phi_1}(k_1)\widehat{\phi_2}(k_2)\psi(\tau_1 - k_1^2)\ \psi(\tau_2 - k_2^2)\ dk_1\ dk_2\ d\tau_1\ d\tau_2 \right\|_{L^2_\tau L^2_k} \\
&\lesssim \left\| \left(\int_{k=k_1+k_2} \widetilde{\psi}(\tau - k_1^2 - k_2^2)\ dk_1\ dk_2 \right)^{1/2} \times \right. \\
&\qquad \left. \times \left(\int_{k=k_1+k_2} \widetilde{\psi}(\tau - k_1^2 - k_2^2)\ |\widehat{\phi}_1(k_1)|^2\ |\widehat{\phi}_2(k_2)|^2\ dk_1\ dk_2 \right)^{1/2} \right\|_{L^2_\tau L^2_k},
\end{aligned} \tag{175}$$

where to obtain (175) we used Cauchy-Schwartz and the following definition of $\widetilde{\psi} \in \mathcal{S}$

$$\int_{\tau=\tau_1+\tau_2} \psi(\tau_1 - k_1^2)\ \psi(\tau_2 - k_2^2)\ d\tau_1 d\tau_2 = \widetilde{\psi}(\tau - k_1^2 - k_2^2).$$

An application of Hölder gives us the following upper bound on (175)

$$M \left\| \left(\int_{k=k_1+k_2} \widetilde{\psi}(\tau - k_1^2 - k_2^2)\ |\widehat{\phi}_1(k_1)|^2\ |\widehat{\phi}_2(k_2)|^2\ dk_1\ dk_2 \right)^{1/2} \right\|_{L^2_\tau L^2_k}, \tag{176}$$

where

$$M = \left\| \int_{k=k_1+k_2} \widetilde{\psi}(\tau - k_1^2 - k_2^2)\ dk_1\ dk_2 \right\|^{1/2}_{L^\infty_\tau L^\infty_k}.$$

Now by integration in τ followed by Fubini in k_1, k_2 and two applications of Plancharel we have

$$\begin{aligned}
&\left\| \left(\int_{k=k_1+k_2} \widetilde{\psi}(\tau - k_1^2 - k_2^2)\ |\widehat{\phi}_1(k_1)|^2\ |\widehat{\phi}_2(k_2)|^2\ dk_1\ dk_2 \right)^{1/2} \right\|_{L^2_{k,\tau}} \\
&\lesssim \|\phi_1\|_{L^2_x} \|\phi_2\|_{L^2_x},
\end{aligned}$$

which combined with (175), (176) gives

$$B \lesssim M\|\phi_1\|_{L^2_x}\|\phi_2\|_{L^2_x}. \tag{177}$$

We find an upper bound on M as follows:

$$M \lesssim \left(\sup_{\tau,k} \#S\right)^{\frac{1}{2}}, \tag{178}$$

where

$$S = \{k_1 \in \mathbb{Z}^2 \mid |k_1| \sim N_1,\ |k - k_1| \ \sim N_2,\ |k|^2 - 2k_1 \cdot (k - k_1) = \tau + O(1)\}.$$

For notational purposes, let us rename $k_1 = z$, that is

$$S = \{z \in \mathbb{Z}^2 \mid |z| \sim N_1,\ |k - z| \ \sim N_2,\ |k|^2 + 2|z|^2 - 2k \cdot z = \tau + O(1)\}.$$

Let z_0 be an element of S i.e.

$$|z_0| \sim N_1,\ \ |k - z_0| \sim N_2, \tag{179}$$

and

$$|k|^2 + 2|z_0|^2 - 2k \cdot z_0 = \tau + O(1). \tag{180}$$

In order to obtain an upper bound on $\#S$, we shall count the number of l's $\in \mathbb{Z}^2$ such that $z_0 + l \in S$ where z_0 satisfies (179) - (180). Thus such l's must satisfy

$$|z_0 + l| \sim N_1,\ \ |z_0 + l - k| \sim N_2, \tag{181}$$

and

$$|k|^2 + 2\,|z_0 + l|^2 - 2k \cdot (z_0 + l) = \tau + O(1). \tag{182}$$

However by (180) we can rewrite the left hand side of (182) as follows

$$\begin{aligned} &|k|^2 + 2\,|z_0 + l|^2 - 2k \cdot (z_0 + l) \\ &= |k|^2 + 2|z_0|^2 + 2\,|l|^2 + 4z_0 \cdot l - 2k \cdot z_0 - 2k \cdot l \\ &= \tau + O(1) + 2\,|l|^2 + 4z_0 \cdot l - 2k \cdot l. \end{aligned}$$

Therefore (182) holds if

$$|l|^2 + 2l \cdot (z_0 - \frac{k}{2}) = O(1). \tag{183}$$

Moreover, (179) and (181) yield

$$|l| = |l + z_0 - k - z_0 + k| \lesssim N_2 + N_2,$$

that is

$$|l| \lesssim N_2. \tag{184}$$

Finally we observe that (179) together with the assumption that $N_1 >> N_2$ implies that

$$N_1 \sim N_1 - N_2 \sim \left|\left|\frac{z_0}{2} - \frac{k}{2}\right| - \left|\frac{z_0}{2}\right|\right| \leq \left|z_0 - \frac{k}{2}\right| \leq \left|\frac{z_0}{2} - \frac{k}{2}\right| + \left|\frac{z_0}{2}\right| \sim N_2 + N_1 \lesssim N_1,$$

i.e.

$$\left|z_0 - \frac{k}{2}\right| \sim N_1. \tag{185}$$

Hence, it suffices to count the $l's \in \mathbb{Z}^2$ satisfying (183) and (184) where z_0 is such that (185) holds.

Let $w = (a, b)$ denote the vector $z_0 - \frac{k}{2}$. Thus we need to count the number of points in the set A

(186) $$A = \{l \in \mathbb{Z}^2 : \bigl|\, |l|^2 + 2l \cdot w \,\bigr| = O(1),\ |l| \lesssim N_2,\ |w| \sim N_1\},$$

or equivalently,

(187) $$A = \{(x, y) \in \mathbb{Z}^2 : \bigl|x^2 + y^2 + 2(ax + by)\bigr| \leq c,\ x^2 + y^2 \leq (\sigma_2 N_2)^2,\ a^2 + b^2 \sim N_1^2\},$$

for some $c, \sigma_2 > 0$. Let $\mathcal{C}_-$, $\mathcal{C}_+$ be the following circles,

$$\mathcal{C}_- : (x + a)^2 + (y + b)^2 = -c + (a^2 + b^2)$$
$$\mathcal{C}_+ : (x + a)^2 + (y + b)^2 = c + (a^2 + b^2)$$

and for any integer n, let $\mathcal{C}_n$ be the circle

$$\mathcal{C}_n : (x + a)^2 + (y + b)^2 = n + (a^2 + b^2).$$

Finally, let $\mathcal{D}$ denote the disk

$$\mathcal{D} : x^2 + y^2 \leq (\sigma_2 N_2)^2.$$

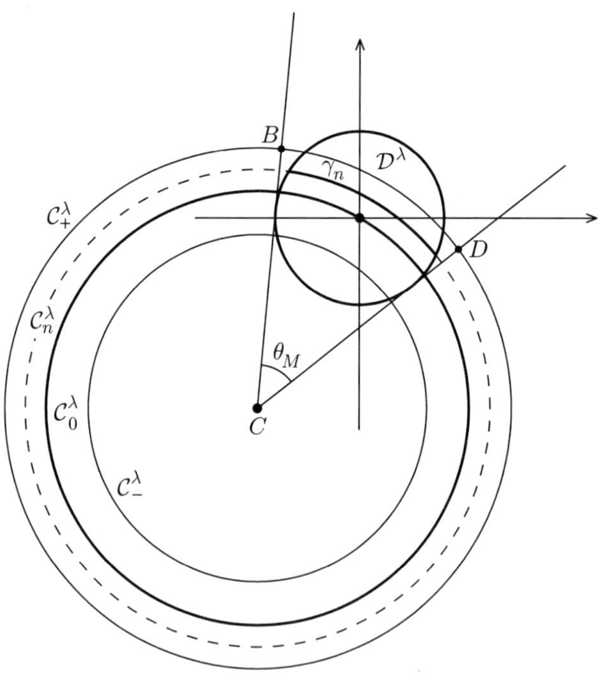

FIGURE 2. Circular sector (here ignore λ).

We need to count the number of lattice points inside $\mathcal{D}$ that are on arcs of circles $\mathcal{C}_n$, with

$$-c \leq n \leq c.$$

Precisely, the total number of lattice point in A can be bounded from above by

$$2c \times \#(\mathcal{C}_n \cap \mathcal{D}). \tag{188}$$

Denote by γ_n the arc of circle $\mathcal{C}_n$ which is contained in $\mathcal{D}$. Notice that (see Figure 2)

$$|\gamma_n| \leq R_M \theta_M \tag{189}$$

where $R_M = \sqrt{c + \sigma_1 N_1^2}$ for some constant $\sigma_1 > 0$, and θ_M is the angle between the line segment CB and CD, which lie along the tangent lines from $C = (-a, -b)$ to the circle $x^2 + y^2 = (\sigma_2 N_2)^2$. Hence,

$$\sin \theta_M \leq \sigma \frac{N_2}{N_1},$$

for some constant $\sigma > 0$. Since $N_1 \gg N_2$, we can assume that $\sin \theta_M > \frac{1}{2}\theta_M$. Hence,

$$\theta_M < 2\sigma \frac{N_2}{N_1}. \tag{190}$$

In order to count efficiently the number of lattice points on each γ_n, we distinguish two cases based on the application of Lemma 9.6.

Case 1: $2\sigma \frac{N_2}{N_1} < \left(\frac{3}{4}\right)^{\frac{1}{3}} R_M^{-\frac{2}{3}}$.

In this case (189)-(190) guarantee that the hypothesis of Lemma 9.6 is satisfied by each arc of circle γ_n. Hence, on each γ_n there are at most two lattice points.

Case 2: $2\sigma \frac{N_2}{N_1} \geq \left(\frac{3}{4}\right)^{\frac{1}{3}} R_M^{-\frac{2}{3}}$.

In this case we approximate the number of lattice points on γ_n by the number[38] of lattice points on $\mathcal{C}_n$ (see for example [**6, 8**]):

$$\#\mathcal{C}_n \lesssim R_M^{\epsilon} \sim (N_1)^{\epsilon} \lesssim (N_2)^{3\epsilon} \tag{191}$$

for any $\epsilon > 0$.

Combining the estimate in (188), **Case 1** and **Case 2** we conclude that

$$\#S \lesssim 1 + N_2^{\epsilon},$$

for any $\epsilon > 0$. Since $N_2 \geq 1$, together with (178), this implies that

$$M \lesssim N_2^{\epsilon},$$

for all positive ϵ's. Hence (168) follows. □

[38] Actually by Gauss Theorem one can get a even better logarithmic estimate in terms of the radius.

References

[1] J.-P. Anker and V. Pierfelice. *Nonlinear Schrödinger equation on real hyperbolic spaces*, Ann. Inst. H. Poincaré Anal. Non Linéaire 26 (2009), no. 5, 1853–1869.

[2] R. Anton. *Strichartz Inequalities for Lipschitz Metrics on Manifolds and Nonlinear Schrodinger Equation on Domains*, Bull. Soc. Math. France 136 (2008), no. 1, 27–65.

[3] V. Banica. *The nonlinear Schrödinger equation on the hyperbolic space*, Comm. P.D.E. **32** (2007), no. 10, 1643-1677.

[4] V. Banica, R. Carles, and G. Staffilani. *Scattering theory for radial nonlinear Schrödinger equations on hyperbolic space*, Geom. Funct. Anal. 18 (2008), no. 2, 367–399.

[5] V. Banica, R. Carles, and T. Duyckaerts *On scattering for NLS: from Euclidean to hyperbolic space*, Discrete Contin. Dyn. Syst. 24 (2009), no. 4, 1113–1127.

[6] E. Bombieri and J. Pila, *The number of integral points on arcs and ovals,* Duke Math. J. **59** (1989), no. 2, 337–357.

[7] J. Bourgain. *Exponential sums and nonlinear Schrödinger equations*, Geom. Funct. Anal. **3** (1993), 157–178.

[8] J. Bourgain. *Fourier transform restriction phenomena for certain lattice subsets and applications to nonlinear evolution equations I,II*, Geom. Funct. Anal., **3** (1993), 107–156, 209–262.

[9] J. Bourgain. *Refinements of Strichartz' inequality and applications to 2D-NLS with critical nonlinearity.* Internat. Math. Res. Notices, no. **5**, (1998) 253–283.

[10] J. Bourgain, *Growth of Sobolev norms in linear Schrödinger equations with quasi-periodic potential*, Comm. Math. Phys. **204** (1999), no. 1, 207–247.

[11] J. Bourgain, *On growth of Sobolev norms in linear Schrödinger equations with smooth time dependent potential*, J. Anal. Math. **77** (1999), 315–348.

[12] J. Bourgain. *New global well-posedness results for nonlinear Schrödinger equations*, AMS Publications, (1999).

[13] J. Bourgain. *Global well-posedness of defocusing 3D critical NLS in the radial case*, JAMS **12** (1999), 145-171.

[14] J. Bourgain, W.-M. Wang, *Diffusion bound for a nonlinear Schrödinger equation*, Mathematical aspects of nonlinear dispersive equations, 21–42, Ann. of Math. Stud., **163**, Princeton Univ. Press, Princeton, NJ, 2007.

[15] J. Bourgain. *On Strichartz' inequalities and the nonlinear Schrödinger equation on irrational tori*, Ann. of Math. Stud., 163, Princeton Univ. Press, Princeton, NJ, 2007.

[16] N. Burq, P. Gérard, and N. Tzvetkov. *An instability property of the nonlinear Schrödinger equation on S^d*, Math. Res. Lett. **9** (2002), 323–335.

[17] N. Burq, P. Gérard, and N. Tzvetkov. *Strichartz inequalities and the nonlinear Schrödinger equation on compact manifolds*, Amer. J. Math. **126** (2004), 569–605.

[18] N. Burq, P. Gérard, and N. Tzvetkov. *Bilinear eigenfunction estimates and the nonlinear Schrödinger equation on surfaces*, Invent. Math. **159** (2005), 187–223.

[19] T. Cazenave. *Semilinear Schrödinger equations*, Courant Lecture Notes in Mathematics, **10**. New York University, Courant Institute of Mathematical Sciences, AMS, (2003).

[20] J. Colliander, J. Holmer, M. Visan and X. Zhang. *Global existence and scattering for rough solutions to generalized nonlinear Schrödinger equations on* $\mathbb{R}$, Comm. Pure Appl. Anal. **7**, (2008), 467–489.

[21] T. Cazenave and F. Weissler, *The Cauchy problem for the critical nonlinear Schrödinger equation in H^s*, Nonlinear Anal. **14** (1990), 807–836.

[22] M.Christ, J. Colliander and T.Tao. *Asymptotics, frequency modulation, and low regularity ill-posedness for canonical defocusing equations*, Amer. J. Math. **124** (2003), 1235–1293.

[23] M. Christ, J. Colliander and T.Tao *Instability of the periodic nonlinear Schrodinger equation*, Preprint.

[24] J. Colliander, M. Keel, G. Staffilani, H. Takaoka, T. Tao. *Global well-posedness for Schrödinger equations with derivative*, Siam J. Math. **33** (2001), 649–669.

[25] J. Colliander, M. Keel, G. Staffilani, H. Takaoka, T. Tao. *A refined global well-posedness result for Schrödinger equations with derivative,* SIAM J. Math. Anal. 34 (2002), no.3, 649-669

[26] J. Colliander, M. Keel, G. Staffilani, H. Takaoka, T. Tao. *Almost conservation laws and global rough solutions to a nonlinear Schrödinger equation*, Math. Research Letters, 9 (2002), 659-682.

[27] J.Colliander, M.Keel, G.Staffilani, H.Takaoka, T.Tao. *Sharp global well-posedness for KdV and modified KdV on* **R** *and* **T**, J. Amer.Math.Soc. 16(2003), 705-749

[28] J. Colliander, M. Keel, G. Staffilani, H. Takaoka and T. Tao. *Global existence and scattering for rough solutions to a nonlinear Schödinger equations on* $\mathbb{R}^3$, C. P. A. M. **57** (2004), no. 8, 987–1014.

[29] J. Colliander, M. Keel, G. Staffilani, H. Takaoka, and T. Tao. *Global well-posedness and scattering in the energy space for the critical nonlinear Schrödinger equation in* $\mathbb{R}^3$, Ann. of Math. (2) 167 (2008), no. 3, 767–865.

[30] J. Colliander, M. Keel, G. Staffilani, H. Takaoka, and T. Tao. *Resonant decompositions and the I-method for cubic nonlinear Schrodinger on* $\mathbb{R}^2$, Discrete Contin. Dyn. Syst. 21 (2008), no. 3, 665–686.

[31] J. Colliander, M. Keel, G. Staffilani, H. Takaoka, T. Tao *Sharp multi-linear periodic KdV estimates and applications*, J. Amer. Math. Soc. 16 (2003), no. 3, 705–749 (electronic).

[32] J. Colliander, M. Grillakis and N. Tzirakis. *Improved interaction Morawetz inequalities for the cubic nonlinear Schrödinger equation on* $\mathbb{R}^2$, Int. Math. Res. Not. (2007).

[33] J. Colliander, G. Staffilani, H. Takaoka, *Global wellposedness for KdV below* L^2, Math. Res. Lett. **6** (1999), no. 5-6, 755–778.

[34] D. De Silva, N. Pavlovic and N. Tzirakis. *Global well-posedness for the* L^2 *critical nonlinear Schrödinger equation in higher dimensions,* Commun. Pure Appl. Anal. 6 (2007), no. 4, 1023–1041.

[35] D. De Silva, N. Pavlovic, G. Staffilani and N. Tzirakis. *Global well-posedness for the periodic Schrödinger equation in 1D and 2D*, Discr. Cont. Dyn. Syst. 19, (2007), 37–65.

[36] D. De Silva, N. Pavlovic, G. Staffilani and N. Tzirakis. *Global well-posedness and polynomial bounds for the defocusing* L^2 *critical nonlinear Schrödinger equation in* $\mathbb{R}$, Comm. Partial Differential Equations 33 (2008), no. 7-9, 1395–1429.

[37] B. Dodson. *Global well-posedness and scattering for the mass critical nonlinear Schrödinger equation with mass below the mass of the ground state,* preprint, see Arxiv.org.

[38] L. C. Evans. *Partial Differential Equations: Second Edition* (Graduate Studies in Mathematics).

[39] Y. Fang and M. Grillakis. *On the global existence of rough solutions of the cubic defocusing Schrödinger equation in* $\mathbb{R}^{2+1}$, J. Hyper. Diff. Eqs., 4, (2007), 233–257.

[40] R. Glassey. *On the blow up of solutions to the Cauchy problem for nonlinera Schrödinger operators*, J. Math. Phys. **8**, (1977), 1794–1797.

[41] J. Ginibre. *Le probléme de Cauchy pour des EDP semi-linéaires préiodiques en variables d'espace (d'aprés Bourgain).* (French. French summary) [The Cauchy problem for periodic semilinear PDE in space variables (after Bourgain)] Seminaire Bourbaki, Vol. 1994/95. Asterisque No. 237 (1996), Exp. No. 796, 4, 163–187.

[42] J. Ginibre and G. Velo. *Scattering theory in the energy space for a class of nonlinear Schrödinger equations*, J. Math. Pure Appl. **64**, (1985), 363–401.

[43] M. Grillakis. *On nonlinear Schrödinger equations.* , Comm. Partial Differential Equations **25** (2000), no. 9-10, 1827–1844.

[44] A. Hassell, T. Tao and J. Wunsch. *A Strichartz inequality for the Schroedinger equation on non-trapping asymptotically conic manifolds*, Comm. part. Diff. Equ. , 30, (2005), 157–205.

[45] A. Hassell, T. Tao and J. Wunsch. *Sharp Strichartz estimates on non-trapping asymptotically conic manifolds*, Amer. J. Math., 128, (2006), 963–1024.

[46] N.J. Hitchin, G.B. Segal, R.S. Ward *Integrable Systems*, Oxford Graduate Texts in Mathematics, **4** (1999).

[47] A. D. Ionescu and G.Staffilani. *Semilinear Schrödinger Flows on Hyperbolic Spaces: Scattering in* H^1, Math. Ann. 345 (2009), no. 1, 133–158.

[48] A. Iosevich, E. Sawyer and A. Seeger, *Mean square discrepancy bounds for the number of lattice points in large convex bodies* J. Anal. Math. **87** (2002), 209–230.

[49] M. Keel and T. Tao. *Endpoint Strichartz estimates*, Amer. J. Math. **120** (1998), 955–980.

[50] M. Keel and T. Tao, *Global well-posedness for large data for the Maxwell-Klein-Gordon equation below the energy norm,* preprint.

[51] M. Keel and T. Tao, *Local and global well-posedness of wave maps on* $\mathbb{R}^{1+1}$ *for rough data,* Intl. Math. Res. Notices **21** (1998), 1117-1156.

[52] C. E. Kenig and F. Merle. *Global well-posedness, scattering and blow-up for the energy-critical, focusing, non-linear Schrödinger equation in the radial case*, Inventiones Math. 166, (2006), 645–675.
[53] C. E.Kenig and F. Merle. *Global well-posedness, scattering and blow-up for the energy critical focusing non-linear wave equation*, Sem. Equ. Der. Partielles, Ecole Polytech., Palaiseau (2007)
[54] C. E. Kenig and F. Merle. *Scattering for $H^1/2$ bounded solutions to the cubic, defocusing NLS in 3 dimensions.* Trans. Amer. Math. Soc. 362 (2010), no. 4, 1937–1962.
[55] C. Kenig, G. Ponce, L. Vega, *A Bilinear Estimate With Applications to the KdV Equation*, Journal of the A.M.S., **9** (1996), 573–603.
[56] S. Keraani. *On the defect of compactness for the Strichartz estimates of the Schrödinger equations*, J. Differential Equations, **175**, 2001, 2, 353–392.
[57] R. Killip, M. Visan. *The focusing energy-critical nonlinear Schrödinger equation in dimensions five and higher,* Amer. J. Math. 132 (2010), no. 2, 361–424.
[58] R. Killip and M. Visan *Nonlinear Schrödinger equations at critical regularity* Clay Lecture Notes, 2008.
[59] R. Killip, M. Visan and X.Zhang. *The mass-critical nonlinear Schrödinger equation with radial data in dimensions three and higher*, Anal. PDE 1 (2008), no. 2, 229–266.
[60] R. Killip, T. Tao and M. Visan *The cubic nonlinear Schrödinger equation in two dimensions with radial data*, J. Eur. Math. Soc. (JEMS) 11 (2009), no. 6, 1203–1258.
[61] J. Lin and W. Strauss. *Decay and scattering of solutions of a nonlinear Schrödinger equation*, J. Funct. Anal. **30**, 1978, 245–263.
[62] F. Merle. *Determination of blow-up solutions with minimal mass for nonlinear Schrödinger equations with critical power*, Duke Math. J., **69**, 1993, 2, 427–454.
[63] F. Merle. *On uniqueness and continuation properties after blow-up time of self-similar solutions of nonlinear Schrdinger equation with critical exponent and critical mass*, Comm. Pure Appl. Math. **45**,(1992), no. 2, 203–254.
[64] C. Morawetz. *Time decay for the nonlinear Klein-Gordon equation*, Proc. Roy. Soc. A **306**, 1968, 291–296.
[65] T. Ozawa. *On the nonlinear Schrödinger equations of derivative type*, Indiana Univ. Math. J. **45** (1996), 137–163.
[66] R. Palais. *The symmetries of solitons*, Bull. of AMS, **34** (1997), 339–403.
[67] V. Pierfelice. *Weighted Strichartz estimates for the Schrödinger and wave equations on Damek-Ricci spaces*, Math. Z. 260 (2008), no. 2, 377–392.
[68] F. Planchon and L. Vega. *Bilinear virial identities and applications*, Ann. Sci. Éc. Norm. Supér. (4) 42 (2009), no. 2, 261–290.
[69] G Pick, *Geometrisches zur Zahlenlehre,* Sitzungber, Lotos (Prague) **19** (1899), 311–319.
[70] E. Ryckman, M. Visan. ıGlobal well-posedness and scattering for the defocusing energy-critical nonlinear Schrödinger equation in $\mathbb{R}^{1+4}$, Amer. J. Math. 129, (2007), 34–56.
[71] E. Stein. *Harmonic Analysis*, Princeton University Press, (1993).
[72] G. Staffilani, *On the growth of high Sobolev norms of solutions for KdV and Schrödinger equations*, Duke Math. J. **86** (1997), no. 1, 109–142.
[73] T. Tao. *Nonlinear Dispersive Equations: local and global analysis*, CBMS, **106** (2006).
[74] T. Tao. *Global well-posedness and scattering for the higher-dimensional energy-critical nonlinear Schrodinger equation for radial data.* New York J. Math. 11 (2005), 57–80.
[75] T. Tao, M. Visan and X. Zhang. *Global well-posedness and scattering for the defocusing mass-critical nonlinear Schrd̈inger equation for radial data in high dimensions*, Duke Math. J. 140 (2007), 165–202.
[76] T. Tao, M. Visan and X. Zhang. *Minimal-mass blowup solutions of the mass-critical NLS*, preprint, see Arxiv.org.
[77] M. Visan. *The defocusing energy-critical nonlinear Schrödinger equation in higher dimensions*, Duke Math. J. 138 (2007), 281–374.

[78] M. Visan and X. Zhang *Global well-posedness and scattering for a class of nonlinera Schrödinger equations below the energy space*, Differential Integral Equations 22 (2009), no. 1-2, 99–124.

[79] W.-M. Wang, *Logarithmic bounds on Sobolev norms for the time dependent linear Schrödinger equation*, Comm. Partial Differential Equations, 33 (2008), no. 10-12, 2164–2179.

Department of Mathematics, Massachusetts Institute of Technology, Cambridge, Massachusetts

E-mail address: gigliola@math.mit.edu

Clay Mathematics Proceedings
Volume **17**, 2013

On the Singularity Formation for the Nonlinear Schrödinger Equation

Pierre Raphaël

These notes are an introduction to the qualitative description of singularity formation for the nonlinear Schrödinger equation. Part of the material was presented during the 2008 Clay summer school on Nonlinear Evolution Equations at the ETH Zurich. The manuscript has been enriched with additions in 2012 in order to give a more accurate view on this very active research field and present a number of open problems.

We consider the semi linear Schrödinger equation

$$(0.1)\qquad (NLS)\quad \begin{cases} iu_t = -\Delta u - |u|^{p-1}u, & (t,x)\in[0,T)\times\mathbb{R}^N \\ u(0,x) = u_0(x), & u_0:\mathbb{R}^N\to\mathbb{C} \end{cases}$$

with $u_0 \in H^1 = \{u, \nabla u \in L^2(\mathbb{R}^N)\}$ in dimension $N \geq 1$ and for energy subcritical nonlinearities:

$$(0.2)\qquad 1 < p < 2^* - 1 \text{ with } 2^* = \begin{cases} +\infty & \text{for } N = 1,2 \\ \frac{2N}{N-2} & \text{for } N \geq 3 \end{cases}.$$

where 2^* is the Sobolev exponent of the injection $\dot{H}^1 \hookrightarrow L^{2^*}$. The case $p = 3$ appears in various areas of physics: for the propagation of waves in non linear media and optical fibers for $N = 1$, the focusing of laser beams for $N = 2$, the Bose-Einstein condensation phenomenon for $N = 3$, see the monograph [**106**] for a more systematic introduction to this physical aspect of the problem.

Our aim is to develop tools for the qualitative description of the flow for data in the energy space H^1, and this includes long time existence, scattering or formation of singularities. The possibility of finite time blow up corresponding to a self focusing of the nonlinear wave and the concentration of energy will be of particular interest to us. Note that (NLS) is an infinite dimensional Hamiltonian system without any space localization property and infinite speed of propagation. It is in this context together with the critical generalized (gKdV) equation[1] one of the few examples where blow up is known to occur. For (NLS), an elementary proof of existence of blow up solutions is known since the 60's but is based on energy constraints and is not constructive. In particular, *no qualitative information of any*

2010 *Mathematics Subject Classification.* Primary 35Q51, 35Q55.
[1]see (4.22).

type on the blow up dynamics is obtained this way. In fact, the theory of global existence or blow up for (NLS) as known up to now is intimately connected to the theory of ground states or solitons which are special periodic in time solutions to the Hamiltonian system. A central question is the stability of these solutions and the description of the flow around them which has attracted a considerable amount of work for the past thirty years.

These notes are organized as follows.

In the first section, we recall the main standard results about subcritical non linear Schrödinger equations and in particular the existence and orbital stability of soliton like solutions which relies on nowadays standard variational tools. In section 2, we introduce the blow up problem and present some of the very few general results known on the singularity formation in this case, and this includes old results from the 50's and very recent ones. Section 3 focuses onto the mass critical problem $p = 1 + \frac{4}{N}$ and we extend in the critical blow up regime the subcritical variational theory of ground states. In section 4, we present the state of the art on the question of description of the flow near the ground state for mass critical problems, including recent complete answers for the generalized (gKdV) problem. In section 5, we present a detailed proof of the pioneering result obtained in collaboration with F.Merle in [**71**], [**72**] on the derivation of the sharp log-log upper bound on blow up rate for a suitable class of initial data near the ground state solitary wave.

We expect the presentation to be essentially self contained provided the prior knowledge of standard tools in the study of non linear PDE's, in particular Sobolev embeddings.

1. The subcritical problem

We recall in this section the main classical facts regarding the global well posedness in the energy space of (NLS), and the main variational tools at the heart of the proof of the existence and stability of special periodic solutions: the ground state solitary waves.

1.1. Global well posedness in the subcritical case. Let us consider the general non linear Schrödinger equation:

$$\left\{ \begin{array}{l} iu_t = -\Delta u - |u|^{p-1}u \\ u(0,x) = u_0(x) \in H^1 \end{array} \right. \tag{1.1}$$

with p satisfying the energy subcriticality assumption (0.2). The local well posedness of (1.1) in H^1 is a result of Ginibre, Velo, [**23**], see also [**31**]. Thus, for $u_0 \in H^1$, there exists $0 < T \leq +\infty$ such that $u(t) \in \mathcal{C}([0,T), H^1)$. Moreover, the life time of the solution can be proved to be lower bounded by a function depending on the H^1 size of the solution only, $T(u_0) \geq f(\|u_0\|_{H^1})$, and hence there holds the blow up alternative:

$$T < +\infty \quad \text{implies} \quad \lim_{t \to T} \|u(t)\|_{H^1} = +\infty. \tag{1.2}$$

We refer to [**11**] for a complete introduction to the Cauchy theory. To prove the global existence of the solution, it thus suffices to control the size of the solution in H^1. This is achieved in some cases using the invariants of the flow. Indeed, the following H^1 quantities are conserved:

- L^2-norm:

$$\int |u(t,x)|^2 = \int |u_0(x)|^2; \tag{1.3}$$

- Energy -or Hamiltonian-:

$$E(u(t,x)) = \frac{1}{2}\int |\nabla u(t,x)|^2 - \frac{1}{p+1}\int |u(t,x)|^{p+1} = E(u_0); \tag{1.4}$$

- Momentum:

$$Im\left(\int \nabla u \overline{u}(t,x)\right) = Im\left(\int \nabla u_0 \overline{u_0}(x)\right). \tag{1.5}$$

Note that the growth condition on the non linearity (0.2) ensures from Sobolev embedding that the energy is well defined, and this is why H^1 is referred to as the energy space. These invariants are related to the group of symmetry of (1.1) in H^1:

- Space-time translation invariance: if $u(t,x)$ solves (1.1), then so does $u(t+t_0, x+x_0)$, $t_0 \in \mathbb{R}$, $x_0 \in \mathbb{R}^N$.
- Phase invariance: if $u(t,x)$ solves (1.1), then so does $u(t,x)e^{i\gamma}$, $\gamma \in \mathbb{R}$.
- Scaling invariance: if $u(t,x)$ solves (1.1), then so does $u_\lambda(t,x) = \lambda^{\frac{2}{p-1}} u(\lambda^2 t, \lambda x)$, $\lambda > 0$.
- Galilean invariance: if $u(t,x)$ solves (1.1), then so does $u(t, x-\beta t)e^{i\frac{\beta}{2}\cdot(x-\frac{\beta}{2}t)}$, $\beta \in \mathbb{R}^N$.

Let us point out that this group of H^1 symmetries is the same like for the *linear* Schrödinger equation -up to the conformal invariance to which we will come back later-.

The *critical space* is a fundamental phenomenological number for the analysis and is defined as the number of derivatives in L^2 which are left invariant by the scaling symmetry of the flow:

$$\|u_\lambda(t)\|_{\dot{H}^{s_c}} = \|u(\lambda^2 t)\|_{\dot{H}^{s_c}} \quad \text{for} \quad s_c = \frac{N}{2} - \frac{2}{p-1}. \tag{1.6}$$

Observe that $s_c < 1$ from (0.2).

A direct consequence of the Cauchy theory, the conservation laws and Sobolev embeddings is the celebrated global existence result:

THEOREM 1.1 (Global wellposedness in the subcritical case). *Let $N \geq 1$ and $1 < p < 1 + \frac{4}{N}$ -equivalently $s_c < 0$-, then all solutions to (1.1) are global and bounded in H^1.*

PROOF OF THEOREM 1.1. By L^2 conservation: $\|u(t)\|_{L^2} = \|u_0\|_{L^2}$. Moreover, the Gagliardo-Nirenberg interpolation estimate:

$$\forall v \in H^1, \quad \int |v|^{p+1} \leq C(N,p)\left(\int |\nabla v|^2\right)^{\frac{N(p-1)}{4}} \left(\int |v|^2\right)^{\frac{p+1}{2} - \frac{N(p-1)}{4}} \tag{1.7}$$

applied to $v = u(t)$ implies using the conservation of the energy and the L^2 norm:

$$\forall t \in [0,T), \quad E_0 \geq \frac{1}{2}\left[\int |\nabla v|^2 - C(u_0)\left(\int |\nabla v|^2\right)^{\frac{N(p-1)}{4}}\right].$$

The subcriticality assumption $p < 1 + \frac{4}{N}$ now implies an a priori bound on the H^1 norm which concludes the proof of Theorem 1.1. □

The critical exponent

$$p = 1 + \frac{4}{N} \quad \text{ie} \quad s_c = 0$$

arises from this analysis and corresponds to the so-called L^2 *or mass critical case*. It is the smallest power nonlinearity for which blow up can occur and corresponds to an exact balance between the kinetic and potential energies under the constraint of conserved L^2 mass. The L^2 supercritical -and energy subcritical cases- correspond to

$$1 + \frac{4}{N} < p < 2^* - 1 \quad \text{ie} \quad 0 < s_c < 1.$$

1.2. The solitary wave. A fundamental feature of the focusing (NLS) problem is the existence of time periodic solutions. Indeed,

$$u(t,x) = \phi(x)e^{it}$$

is an H^1 solution to (1.1) iff ϕ solves the nonlinear elliptic equation:

$$\Delta\phi - \phi + \phi|\phi|^{p-1} = 0, \quad \phi \in H^1(\mathbb{R}^N). \tag{1.8}$$

There are various ways to construct solutions to (1.8), the simplest one being to look for radial solutions via a shooting method, [**4**].

PROPOSITION 1.2 (Existence of solitary waves). *(i) For $N = 1$, all solutions to (1.8) are translates of*

$$Q(x) = \left(\frac{p+1}{2\cosh^2\left(\frac{(p-1)x}{2}\right)}\right)^{p-1}. \tag{1.9}$$

(ii) For $N \geq 2$, there exist a sequence of radial solutions $(Q_n)_{n\geq 0}$ with increasing L^2 norm such that Q_n vanishes n times on $\mathbb{R}^N$.

The exact structure of the set of solutions to (1.8) is not known in dimension $N \geq 2$. An important rigidity property however which combines nonlinear elliptic techniques and ODE techniques is *the uniqueness of the nonnegative* solution to (1.8).

PROPOSITION 1.3 (Uniqueness of the ground state). *All solutions to*

$$\Delta\phi - \phi + \phi|\phi|^{p-1} = 0, \quad \phi \in H^1(\mathbb{R}^N), \quad \phi(x) > 0 \tag{1.10}$$

*are a translate of an exponentially decreasing $\mathcal{C}^2$ radial profile $Q(r)$ ([**22**]) which is the unique nonnegative radially symmetric solution to (1.8) ([**42**]). Q is the so called ground state solution.*

The uniqueness is thus the consequence of two facts. A positive decaying at infinity solution to (1.10) is necessarily radially symmetric with respect to a point, this is a very deep and non trivial result due to Gidas, Ni, Nirenberg [**22**] which relies on the maximum principle. And then there is uniqueness of the radial decaying positive solution in the ODE sense. The original -and delicate- proof of this last fact by Kwong [**42**] has been revisited by MacLeod [**52**] and is very nicely presented in the Appendix of Tao [**107**]. We also refer to [**48**] for a beautiful extension of uniqueness methods to nonlocal problems where the ODE approach fails.

Let us now observe that we may let the full group of symmetries of (1.1) act on the solitary wave $u(t,x) = Q(x)e^{it}$ to get a $2N+2$ parameters family of solitary waves: for $(\lambda_0, x_0, \gamma_0, \beta) \in \mathbb{R}_+^* \times \mathbb{R}^N \times \mathbb{R} \times \mathbb{R}^N$,

$$u(t,x) = \lambda_0^{\frac{2}{p-1}} Q(\lambda_0(x+x_0) - \lambda_0^2 \beta t) e^{i\lambda_0^2 t} e^{i\gamma_0} e^{i\frac{\beta}{2}\cdot(\lambda_0(x+x_0)-\lambda_0^2\beta t)}.$$

These waves are moving according to the free Galilean motion and oscillating at a phase related to their size: the larger the λ_0, the wilder the oscillations in time. An explicit computation reveals that the solitary wave can be made arbitrarily small in H^1 in the subcritical regime $s_c < 0$ only.

1.3. Orbital stability of the ground states in the subcritical case. We address in this section the question of the stability of the ground state solitary wave $u(t,x) = Q(x)e^{it}$, $Q > 0$, as a solution to (1.1) in the mass subcritical case

$$1 < p < 1 + \frac{4}{N}, \quad s_c < 0. \tag{1.11}$$

Let us first observe that two trivial instabilities are given by the symmetries of the equation:

- Scaling instability: $\forall \lambda > 0$, the solution to (1.1) with initial data $u_0(x) = \lambda^{\frac{2}{p-1}} Q(\lambda x)$ is $u(t,x) = \lambda^{\frac{2}{p-1}} Q(\lambda x) e^{i\lambda^2 t}$.
- Galilean instability: $\forall \beta > 0$, the solution to (1.1) with initial data $u_0(x) = Q(x)e^{i\beta}$ is $u(t,x) = Q(x-\beta t)e^{it+\frac{\beta}{2}\cdot(x-\frac{\beta}{2}t)}$.

In both cases,

$$\sup_{t\in\mathbb{R}} |u(t,x) - Q(x)e^{it}| > |Q(x)|$$

and thus the solution does not stay uniformly close to Q. Cazenave and Lions [**12**] proved that these trivial instabilities are the only ones in the mass ubcritical setting: this is the celebrated *orbital stability* of the ground state solitary wave.

THEOREM 1.4 (Orbital stability of the ground state, [**12**]). *Let $N \geq 1$ and p satisfy (1.11). For all $\varepsilon > 0$, there exists $\delta(\varepsilon)$ such that the following holds true. Let $u_0 \in H^1$ with*

$$\|u_0 - Q\|_{H^1} < \delta(\varepsilon),$$

then there exist a translation shift $x(t) \in \mathcal{C}^0(\mathbb{R}, \mathbb{R}^N)$ and a phase shift $\gamma(t) \in \mathcal{C}^0(\mathbb{R}, \mathbb{R})$ such that:

$$\forall t \in \mathbb{R}, \quad \|u(t,x) - Q(x - x(t))e^{i\gamma(t)}\|_{H^1} < \varepsilon.$$

The strength -and the weakness- of the proof is that it relies only on the conservation laws and the *variational characterization of the ground state solitary wave.*

This study falls into the classical sets of *concentration compactness techniques* as introduced by Lions in [**50**],[**51**]. Given $\lambda > 0$, we let

$$Q_\lambda(x) = \lambda^{\frac{2}{p-1}} Q(\lambda x).$$

The following variational result immediately implies Theorem 1.4:

PROPOSITION 1.5 (Description of the minimizing sequences). *Let $N \geq 1$ and p satisfy (1.11). Let $M > 0$ be fixed.*
(i) Variational characterization of Q: The minimization problem

$$I(M) = \inf_{\|u\|_{L^2}=M} E(u) \tag{1.12}$$

is attained on the family

$$Q_{\lambda(M)}(\cdot - x_0)e^{i\gamma_0}, \quad x_0 \in \mathbb{R}^N, \gamma_0 \in \mathbb{R},$$

where $\lambda(M)$ is the unique scaling such that $\|Q_{\lambda(M)}\|_{L^2} = M$.
(ii) Description of the minimizing sequences: Any minimizing sequence v_n to (1.12) is relatively compact in H^1 up to translation and phase shifts, that is up to a subsequence:

$$v_n(\cdot + x_n)e^{i\gamma_n} \to Q_{\lambda(M)} \quad in \quad H^1.$$

The fact that Proposition 1.5 implies Theorem 1.4 is now a simple consequence of the conservation laws and is left to the reader. The next section is devoted to the proof of Proposition 1.5.

1.4. The concentration compactness argument. The first key to the proof of Proposition 1.5 is the description of the lack of compactness in $\mathbb{R}^N$ of the Sobolev injection $H^1 \hookrightarrow L^{p+1}$, $2 \leq p+1 < 2^*$. This description is a consequence of Lions' concentration compactness Lemma. Let us recall that the injection is compact on a smooth bounded domain. Note also that the injection is still compact when restricted to radial functions in dimension $N \geq 2$. Here one uses the estimate:

$$u^2(r) = -\int_r^{+\infty} u(s)u'(s)ds \ \text{ and thus } \ \|u\|_{L^\infty(r\geq R)} \leq \frac{C}{R^{\frac{N-1}{2}}} \|\nabla u\|_{L^2}^{\frac{1}{2}} \|u\|_{L^2}^{\frac{1}{2}}$$

so that any H^1 bounded sequence of radially symmetric functions is L^{p+1} compact. This would considerably simplify the proof of Proposition 1.5 when restricting the problem to radially symmetric functions. In general, there holds the following:

PROPOSITION 1.6 (Description of the lack of compactness of $H^1 \hookrightarrow L^q$). *Let a sequence $u_n \in H^1$ with*

$$\|u_n\|_{L^2} = M, \quad \|\nabla u_n\|_{L^2} \leq C, \tag{1.13}$$

Then there exists a subsequence u_{n_k} such that one of the following three scenario occurs:
(i) Compactness: $\exists y_k \in \mathbb{R}^N$ such that

$$\forall 2 \leq q < 2^*, \quad u_{n_k}(\cdot + y_k) \to u \quad in \quad L^q. \tag{1.14}$$

(ii) Vanishing:

$$\forall 2 < q < 2^*, \quad u_{n_k} \to 0 \quad in \quad L^q. \tag{1.15}$$

(iii) Dichotomy: $\exists v_k, w_k$, $\exists 0 < \alpha < 1$ *such that* $\forall 2 \leq q < 2^*$:

$$\left\{\begin{array}{l} Supp(v_k) \cap Supp(w_k) = \emptyset, \quad dist(Supp(v_k), Supp(w_k)) \to +\infty, \\ \|v_k\|_{H^1} + \|w_k\|_{H^1} \leq C, \\ \|v_k\|_{L^2} \to \alpha M, \quad \|w_k\|_{L^2} \to (1-\alpha)M, \\ \lim_{k\to+\infty} \left|\int |u_{n_k}|^q - \int |v_k|^q - \int |w_k|^q\right| = 0, \\ \liminf_{k\to+\infty} \int |\nabla u_{n_k}|^2 - \int |\nabla v_k|^2 - \int |\nabla w_k|^2 \geq 0. \end{array}\right. \tag{1.16}$$

REMARK 1.7. *The key in the dichotomy case is that there is no loss of potential energy during the splitting in space of* u_{n_k} *into two bumps* v_k, w_k *which support go away from each other, while on the other hand only a lower semi continuity bound can be derived for the kinetic energy.*

REMARK 1.8. *The case dichotomy corresponds to the localization of the first bubble of concentration. One can then continue the extraction iteratively and obtain the profile decomposition of the sequence* u_n, *see P. Gerard* [**21**], *Hmidi, Keraani* [**28**] *for a very elegant proof.*

The proof of Proposition 1.6 is given in Appendix A. We now show how the description of the lack of compactness of the Sobolev injection is a powerful tool for the study of variational problems.

PROOF OF PROPOSITION 1.5. **step1** Computation of $I(M)$. Let $I(M)$ be given by (1.12). We claim that

$$-\infty < I(M) = M^{\frac{2(1-s_c)}{|s_c|}} I(1) < 0. \tag{1.17}$$

Indeed, $I(M) > -\infty$ follows directly form the Gagliardo-Nirenberg inequality (1.7) and the subcriticality condition (1.11). The computation of the nonpositive value of the infimum follows from the scaling properties of the problem. First, given $u \in H^1$ with $\|u\|_{L^2} = 1$, we use the L^2 scaling

$$v_\lambda(x) = \lambda^{\frac{N}{2}} u(\lambda x)$$

to get:

$$E(v_\lambda) = \lambda^2 \left[\frac{1}{2}\int |\nabla u|^2 - \frac{1}{(p+1)\lambda^{(p-1)|s_c|}} \int |u|^{p+1}\right].$$

Letting $\lambda \to 0$ yields $I(1) < 0$. The homogeneity in M of $I(M)$ is derived using the scaling of the equation

$$v_\lambda(x) = \lambda^{\frac{2}{p-1}} u(\lambda x), \quad \|v_\lambda\|_{L^2} = \lambda^{|s_c|}\|u\|_{L^2}, \quad E(v_\lambda) = \lambda^{2(1-s_c)} E(u),$$

which yields the claim.
Let now u_n be a minimizing sequence for $I(M)$. Then u_n is bounded in H^1 from (1.7) and satisfies the assumptions of Proposition 1.5, and we now examine the various scenario:

step 2 Vanishing cannot occur. Otherwise, from (1.15):

$$I(M) = \lim_{k\to+\infty} E(u_{n_k}) \geq \liminf_{k\to+\infty} \frac{1}{2}\int |\nabla u_{n_k}|^2 \geq 0$$

which contradicts (1.17).

step 3 Dichotomy cannot occur. Otherwise, from (1.16), we have sequences v_k, w_k and $0 < \alpha < 1$ such that

$$\|v_k\|_{L^2} = \alpha M, \quad \|w_k\|_{L^2} = (1-\alpha)M$$

and

$$I(M) \geq \liminf_{k\to+\infty} E(v_k) + \liminf_{k\to+\infty} E(w_k).$$

In particular, this implies:

$$I(M) \geq I(\alpha M) + I((1-\alpha)M) \tag{1.18}$$

and thus from (1.17):

$$1 \leq \alpha^{\frac{2(1-s_c)}{|s_c|}} + (1-\alpha)^{\frac{2(1-s_c)}{|s_c|}} \quad \text{for some } \ 0 < \alpha < 1.$$

Now a straightforward convexity argument implies from $\frac{2(1-s_c)}{|s_c|} > 1$ that $\alpha = 0$ or $\alpha = 1$, a contradiction.

step 4 Conclusion. We conclude that only compactness occurs ie

$$u_{n_k}(\cdot + x_k) \to u \ \text{ in } \ L^{p+1}.$$

Observe then from the strong L^{p+1} convergence and the lower semicontinuity of the $\dot{H}^1$ norm that u attains the infimum:

$$\|u\|_{L^2} = M, \quad E(u) = I(M).$$

It thus remains to characterize the infimum. We claim that:

$$u(x) = Q_{\lambda(M)}(\cdot + x_0)e^{i\gamma_0} \tag{1.19}$$

which concludes the proof of Proposition 1.6.
Proof of (1.19): First observe from $\int |\nabla|u||^2 \leq \int |\nabla u|^2$ that $v = |u|$ is a minimizer with $v \geq 0$. From standard Euler Lagrange theory, v solves

$$\Delta v + v|v|^{p-1} = \mu v, \quad v \in H^1.$$

The Lagrange multiplier, which a priori depends on v, can be computed by multiplying the equation by v and then $y \cdot \nabla v$ (Pohozaev integration) leading to:

$$\mu = \mu(M) = \frac{N+2-p(N-2)}{2M\left(\frac{N(p-1)}{4} - 1\right)} I(M) > 0.$$

We now observe by rescaling that $w(x) = \lambda^{\frac{2}{p-1}} v(\lambda x)$ with $\lambda = \sqrt{\mu}$ satisfies

$$\Delta w - w + w|w|^{p-1} = 0, \quad w \in H^1(\mathbb{R}^N), \quad w \geq 0,$$

and w non zero. From the uniqueness statement of Proposition 1.3, this yields:

$$w(x) = Q(x - x_0),$$

and hence $v(x) = Q_{\lambda(M)}(x - x_0)$. This implies in particular that v does not vanish which together with $\int |\nabla u|^2 = \int |\nabla |u||^2$ -because they both are minimizers- implies[2]

$$u(x) = |u(x)|e^{i\gamma_0} = Q_{\lambda(M)}(x - x_0)e^{i\gamma_0},$$

and (1.19) is proved. □

[2]see for example[**49**].

Furthers comments

1. More general nonlinearities: The proof we have presented reproduces the original argument by Cazenave, Lions [**12**] and heavily relies onto the specific scaling properties of the nonlinearity. The advantage of this argument is to completely avoid the linearization near the ground state, but the prize to pay is the proof of global estimates like (1.18) which may be non trivial in the absence of symmetries. Another approach to stability proceeds by brute force linearization and the derivation of suitable coercivity properties of the linearized operator close to the ground state as for example done in Grillakis, Shatah, Strauss [**26**] to treat more general nonlinearities. We also refer to [**44**], [**46**], [**47**] for analogue results for gravitational kinetic equations which display a similar structure.

2. Asymptotic stability: An important question is to know whether, when stability holds, asymptotic stability also holds, that is do solutions asymptotically converge to the ground state in some local norm in space as $t \to +\infty$? This kind of property corresponds to a form of asymptotic irreversibility of the flow. This is an extremely delicate problem which has attracted a considerable amount of work for the past ten years. For some specific type of nonlinearities, asymptotic stability holds due to a fine tuning mechanism known as the "Fermi Golden Rule", see Soffer, Weinstein [**105**], Rodnianski, Soffer, Schlag [**102**], Sulem, Buslaev [**10**], Sigal, Zhou [**20**]. However, the case of pure power is still open because essentially small solitons are delicate to deal with. Indeed, in the pure power case, a soliton Q_λ can be made arbitrarily small in H^1 and not disperse. Moreover, one should keep in mind that the asymptotic stability is *false* in the completely integrable case $N = 1$, $p = 3$, see [**112**].

3. Generic long time dynamics: In general, one expects the long time behavior of the solution to correspond to a splitting of the solution into a non dispersive part corresponding to a sum of decoupled solitary waves moving at different speeds and a radiative part which disperses -ie goes to 0 in L^∞ say-. Such a general behavior has been proved in the integrable case for the KdV system

$$(KdV) \quad \begin{cases} u_t + (u_{xx} + u^2)_x = 0, & (t,x) \in [0,T) \times \mathbb{R}, \\ u(0,x) = u_0(x), & u_0 : \mathbb{R}^N \to \mathbb{R}, \end{cases}$$

but complete integrability plays a very specific role here. See Rodnianski, Soffer, Schlag [**102**], Martel, Merle, Tsai [**63**], for the case of non integrable (NLS) systems but with specific nonlinearities. One should think here that in general, even the simpler question of the orbital stability of the multisolitary wave in the pure power case for (NLS) is open.

2. The blow up problem

We focus in this section on the NLS problem (1.1) with mass critical/super critical and energy subcritical nonlinearities, or equivalently according to (1.6):

$$0 \le s_c < 1, \quad 1 + \frac{4}{N} \le p < 2^* - 1.$$

Our aim is to collect old and new results regarding the qualitative description of blow up solutions which involves so far many open problems.

2.1. Existence of blow up solutions: the virial law. The Cauchy theory ensures global existence for small data in H^1 but for large data, the Gagliardo Nirenberg inequality (1.7) does not suffice anymore to ensure global existence. A well known global obstructive argument known as the virial law allows one to very easily prove the existence of finite time blow up solutions.

THEOREM 2.1 (Virial blow up for $E_0 < 0$). *Let $u_0 \in \Sigma = H^1 \cap \{xu \in L^2\}$ with*

$$E_0 < 0,$$

then the corresponding solution to (1.1) blows up in finite time $0 < T < +\infty$.

PROOF OF THEOREM 2.1. Integrating by parts in (1.1), we find:

$$\frac{d^2}{dt^2}\int |x|^2|u(t,x)|^2dx = 4N(p-1)E_0 - \frac{16s_c}{N-2s_c}\int |\nabla u|^2 \leq 4N(p-1)E_0 \tag{2.1}$$

from $s_c \geq 0$. Hence from $E_0 < 0$, the positive quantity $\int |x|^2|u(t,x)|^2dx$ lies below an inverted parabola and hence the solution cannot exist for all times. □

This blow up argument is extraordinary because it provides a blow up criterion based essentially on a pure Hamiltonian information $E_0 < 0$, which applies to arbitrarily large initial data in H^1. In particular, it exhibits an *open* region of the energy space -up to extra integrability condition- where blow up is proven to be a stable phenomenon. While it may seem at first hand to be very specific to the (NLS) problem, this kind of convexity argument is very common for parabolic or wave type problems, see for example [**30**], kinetic problems [**25**], or even compressible Euler equations, [**104**]. However, it has has two major weaknesses:
(i) It heavily relies on a very specific algebra and hence is very unstable by perturbation of the equation. It thus is completely unable to predict blow up even in situations where it is strongly expected. A typical case is for example (NLS) on a domain with Dirichlet boundary conditions, [**96**].
(ii) More fundamentally, this argument is *purely obstructive* in nature and says very little a priori on the singularity formation. In fact the blow up time formally predicted which is the time of vanishing of the variance $\int |x|^2|u|^2$ is almost never correct, solutions generically blow up before.

2.2. Scaling lower bound on blow up rate. In the setting of arbitrarily large initial data, little is known regarding the description of the singularity formation. This is mainly a consequence of the fact that the virial blow up argument does not provide any insight into the blow up dynamics. More generally, the a priori control of the blow up speed $\|\nabla u(t)\|_{L^2}$ which plays a fundamental role for the classification of blow up dynamics for example for the heat or the wave equation, is poorly understood. However a general lower bound on the blow up rate holds as a very simple consequence of the scaling invariance of the problem:

PROPOSITION 2.2 (Scaling lower bound on blow up rate). *Let $N \geq 1$, $0 \leq s_c < 1$. Let $u_0 \in H^1$ such that the corresponding solution $u(t)$ to (1.1) blows up in finite time $0 < T < +\infty$, then there holds:*

$$\forall t \in [0,T), \quad \|\nabla u(t)\|_{L^2} \geq \frac{C(u_0)}{(T-t)^{\frac{1-s_c}{2}}}. \tag{2.2}$$

PROOF OF PROPOSITION 2.2. We give the proof for $s_c = 0$ which is elementary and based on the scaling invariance of the equation and the local well posedness theory in H^1. The proof for $s_c > 0$ is similar and requires the Cauchy theory in $\dot{H}^{s_c} \cap \dot{H}^1$, see [**76**]. Consider for fixed $t \in [0, T)$

$$v^t(\tau, z) = \|\nabla u(t)\|_{L^2}^{-\frac{N}{2}} u\left(t + \|\nabla u(t)\|_{L^2}^{-2}\tau, \|\nabla u(t)\|_{L^2}^{-1} z\right).$$

v^t is a solution to (1.1) by scaling invariance. We have $\|\nabla v^t(0)\|_{L^2} = 1$, $\|v^t\|_{L^2} = \|u_0\|_{L^2}$, and thus by the resolution of the Cauchy problem locally in time by fixed point argument, there exists $\tau_0 > 0$ independent of t such that v^t is defined on $[0, \tau_0]$. Therefore, $t + \|\nabla u(t)\|_{L^2}^{-2}\tau_0 \leq T$ which is the desired result. □

One can ask for the sharpness of the bound (2.2), or equivalently for the existence of self similar solutions in the energy space, i.e. solutions which blow according to the scaling law

$$\|\nabla u(t)\|_{L^2} \sim \frac{1}{(T-t)^{\frac{1-s_c}{2}}}. \tag{2.3}$$

For $s_c = 0$, it is an important open problem, [**7**]. It is however proved in [**97**], [**74**] that the lower bound (2.2) is *not sharp* for data near the ground state in connection with the log log law, see Theorem 4.3. On the contrary, for $s_c > 0$, a stable self similar blow up regime in the sense of (2.3) is observed numerically, [**106**], and a rigorous derivation of these solutions is obtained in collaboration with Merle and Szeftel in [**78**] for slightly super critical problems:

THEOREM 2.3 (Existence and stability of self similar solutions, [**78**]). *Let $1 \leq N \leq 5$ and $0 < s_c \ll 1$. Then there exists an open set of initial data $u_0 \in H^1$ such that the corresponding solution to* (1.1) *blows up with in finite time $T = T(u_0) < +\infty$ with the self similar speed:*

$$\|\nabla u(t)\|_{L^2} \sim \frac{1}{(T-t)^{\frac{1-s_c}{2}}}.$$

The extension of this result to the full critical range $s_c < 1$ is an important open problem, in particular to address the physical case $N = p = 3$, $s_c = \frac{1}{2}$, but is confronted to the construction and the understanding of the stationary self similar profiles which is poorly understood, see [**78**] for a further discussion.

2.3. On concentration of the critical norm. A second general phenomenon of finite blow up solutions is the *concentration of the critical norm.* The first result of this type goes back to Merle, Tsutsumi, [**81**] in the radial case, and generalized by Nawa, [**92**], for the mass critical NLS.

THEOREM 2.4 (L^2 concentration phenomenon for $s_c = 0$, [**81**], [**92**]). *Let $s_c = 0$. Let $u_0 \in H^1$ such that the corresponding solution $u(t)$ to* (1.1) *blows up in finite time $0 < T < +\infty$. Then there exists $x(t) \in \mathcal{C}^0([0,T)\mathbb{R}^N)$ such that:*

$$\forall R > 0, \quad \liminf_{t \to T} \int_{|x - x(t)| \leq R} |u(t,x)|^2 dx \geq \int Q^2. \tag{2.4}$$

Theorem 2.4 relies on the sharp variational characterization of the ground state solitary wave Q and we therefore postpone the proof to section 3.1. We refer to [**108**] for an extension to critical regularity $u_0 \in L^2$. Two natural questions following Theorem 2.4 are still open in the general case:

(i) Does the function $x(t)$ have a limit as $t \to T$ defining then at least one exact blow up point in space where L^2 concentration takes place?
(ii) Which is the exact amount of mass focused by the blow up dynamic?

An explicit construction of blow up solutions due to Merle, [**64**], is the following: let k points $(x_i)_{1\leq i\leq k} \in \mathbb{R}^N$, then there exists a blow up solution $u(t)$ which blows up in finite time $0 < T < +\infty$ exactly at these k points and accumulates exactly the mass:

$$|u(t)|^2 \rightharpoonup \Sigma_{1\leq i\leq k}\|Q\|_{L^2}^2\delta_{x=x_i} \quad \text{as} \quad t \to T,$$

in the sense of measures. A general conjecture concerning L^2 concentration is formulated in [**75**] and states that a blow up solution focuses a quantized and universal amount of mass at a finite number of points in $\mathbb{R}^N$, the rest of the L^2 mass being purely dispersed. The exact statement which is directly related to the soliton resolution conjecture is the following:

Conjecture (*): *Let $u(t) \in H^1$ be a solution to (1.1) which blows up in finite time $0 < T < +\infty$. Then there exist $(x_i)_{1\leq i\leq L} \in \mathbb{R}^N$ with $L \leq \frac{\int |u_0|^2}{\int Q^2}$, and $u^* \in L^2$ such that: $\forall R > 0$,*

$$u(t) \to u^* \quad in \quad L^2(\mathbb{R}^N - \bigcup_{1\leq i\leq L} B(x_i, R))$$

$$and \quad |u(t)|^2 \rightharpoonup \Sigma_{1\leq i\leq L} m_i\delta_{x=x_i} + |u^*|^2 \quad with \quad m_i \in [\int Q^2, +\infty).$$

Let us now address the same question of the behavior of the critical norm for the super critical NLS $0 < s_c < 1$. There is no simple a priori lower bound like for (2.2) for the critical norm $\|u(t)\|_{\dot{H}^{s_c}}$ which is invariant by the scaling symmetry of the flow. Moreover, a major difference between the mass critical problem and the super critical problem is that the critical norm is conserved by the flow for $s_c = 0$ only, and this leads to dramatic differences in the blow up dynamics. We for example proved in [**76**] that for radial data the critical norm not only concentrates at blow up, it explodes:

THEOREM 2.5 (Blow up of the critical norm, [**76**]). *Let $0 < s_c < 1$, $p < 5$ and $N \geq 2$. There exists a universal constant $\gamma = \gamma(N,p) > 0$ such that the following holds true. Let $u_0 \in H^1$ with radial symmetry and assume that the corresponding solution to (1.1) blows up in finite time $T < +\infty$. Then there holds the lower bound for t close enough to T:*

$$\|u(t)\|_{\dot{H}^{s_c}} \geq |\log(T-t)|^{\gamma(N,p)}.$$

Related results were proved for the Navier Stokes equation [**16**], and are a first step towards the understanding of the formation of the blow up bubble. Note that the logarithmic lower bound can be proved to be sharp in some regimes, [**78**], but there also exist regimes where the critical norm blows up polynomially, [**80**]. The regimes $N = 1, 2$ with $p \geq 5$ are still open, as well as the general non radial case. The proof relies on the quantification of a Liouville type theorem, see [**38**] for recent extensions to the wave equation.

2.4. A sharp upper bound on blow up rate. We now address the question of upper bounds on blow up rate for general solutions. A simple observation by Merle is that for $0 < s_c < 1$, the brute force time integration of the virial law (2.1) not only implies finite time blow up for $E_0 < 0$, it also immediately yields an upper bound on the blow up rate for any finite time blow up solution:

THEOREM 2.6 (General upper bound on blow up rate). *Let $0 < s_c < 1$ and $u_0 \in \Sigma$ such that the corresponding solution to* (1.1) *blows up in finite time $0 < T < +\infty$, then:*

$$\int_0^T (T-t)\|\nabla u(t)\|_{L^2}^2 dt < +\infty. \tag{2.5}$$

Note that in particular on a subsequence

$$\|\nabla u(t_n)\|_{L^2}(T-t_n) \to 0 \quad \text{as} \quad t_n \to T.$$

Interestingly enough, this bound fails for $s_c = 0$, see (3.10), and in fact *there exists no known upper bound on blow up rate in the mass critical case* which is one of the reason why the mass critical problem is in some sense more degenerate[3]. For $0 < s_c < 1$, we observed in collaboration with Merle and Szeftel [**80**] that a relatively elementary argument based on a localization of the virial identity as initiated in [**76**] implies an improved upper bound for u_0 radial.

THEOREM 2.7 (Sharp upper bound on blow up rate for radial data, [**80**]). *Let*

$$N \geq 2, \quad 0 < s_c < 1, \quad p < 5.$$

Let the interpolation number[4]

$$\alpha = \frac{5-p}{(p-1)(N-1)}. \tag{2.6}$$

Let $u_0 \in H^1$ with radial symmetry and assume that the corresponding solution $u \in \mathcal{C}([0,T), H^1)$ of (1.1) *blows up in finite time $T < +\infty$. Then there holds the space time upper bound:*

$$\int_t^T (T-\tau)\|\nabla u(\tau)\|_{L^2}^2 d\tau \leq C(u_0)(T-t)^{\frac{2\alpha}{1+\alpha}}. \tag{2.7}$$

This implies in particular

$$\|\nabla u(t_n)\|_{L^2} \lesssim \frac{1}{(T-t_n)^{\frac{1}{1+\alpha}}}$$

on a subsequence $t_n \to T$. Note that it would be very interesting to obtain the pointwise bound for all times.

Before proving Theorem 2.7 which relies on a sharp localization of the virial law, let us say that we do not know if the bound (2.5) is sharp. However, we claim that the general bound for radial data (2.7) *is indeed sharp* and saturated on a new class of blow up solutions: the collapsing ring profiles.

---

[3]The example of the (gKdV) problem and Theorem 4.9 indicate that there may be no bound...
[4]Observe that $0 < \alpha < 1$.

THEOREM 2.8 (Collapsing ring solutions, [**80**]). *Let*

$$N \geq 2, \quad 0 < s_c < 1, \quad p < 5,$$

let $0 < \alpha < 1$ *be given by* (2.6) *and the Galilean shift:*

$$\beta_\infty = \sqrt{\frac{5-p}{p+3}}.$$

Let Q *be the one dimensional mass subcritical ground state* (1.9). *Then there exists a time* $\underline{t} < 0$ *and a solution* $u \in \mathcal{C}([\underline{t}, 0), H^1)$ *of* (1.1) *with radial symmetry which blows up at time* $T = 0$ *according to the following dynamics. There exist geometrical parameters* $(r(t), \lambda(t), \gamma(t)) \in \mathbb{R}_+^* \times \mathbb{R}_+^* \times \mathbb{R}$ *such that:*

$$u(t,r) - \frac{1}{\lambda^{\frac{2}{p-1}}(t)} \left[Qe^{-i\beta_\infty y}\right] \left(\frac{r - r(t)}{\lambda(t)}\right) e^{i\gamma(t)} \to 0 \quad in \quad L^2(\mathbb{R}^N). \tag{2.8}$$

The blow up speed, the radius of concentration and the phase drift are given by the asymptotic laws:

$$r(t) \sim |t|^{\frac{\alpha}{1+\alpha}}, \quad \lambda(t) \sim |t|^{\frac{1}{1+\alpha}}, \quad \gamma(t) \sim |t|^{-\frac{1-\alpha}{1+\alpha}} \quad as \quad t \uparrow 0. \tag{2.9}$$

Moreover, the blow up speed admits the equivalent:

$$\|\nabla u(t)\|_{L^2} \sim \frac{1}{(T-t)^{\frac{1}{1+\alpha}}} \quad as \quad t \uparrow 0. \tag{2.10}$$

Comments on the result:

1. Standing and collapsing ring: The construction of ring solutions started in [**98**], [**100**] for $p = 5$ in dimension $N \geq 2$ where we constructed *standing ring* blow up solutions which concentrate on a standing sphere $r = 1$ at the speed given by the log-log law (4.14). The idea is that the geometry of the blow up set given by a standing sphere allows one to reduce the leading order blow up dynamics to the one dimension quintic NLS which is the mass critical one for $p = 5$. This has been further extended to other geometries in higher dimensions [**29**], [**114**]. Then in the breakthrough paper [**17**], Fibich, Gavish and Wang extended formally the construction to $3 < p < 5$ in dimension $N = 2$ and observed numerically the collapsing ring solutions which existence is made rigorous in [**80**]. Note that the collapsing ring is expected to be stable by radial perturbation of the data, but this is still an open problem.

2. Mass concentration: The ring solutions have a quite unexpected blow up behavior. Indeed, despite the fact that the problem is mass super critical, the structure (2.8) coupled with the speeds (2.9) imply the *concentration of the* L^2 *mass*

$$|u(t)|^2 \rightharpoonup \|Q\|_{L^2}^2 \delta_{x=0} \quad \text{as} \quad t \uparrow 0. \tag{2.11}$$

A contrario the self similar blow up solutions of Theorem 2.3 constructed in [**78**] have a strong limit in L^2 at blow up time. In fact, by rescaling, we can let the amount of concentrated mass in (2.11) be arbitrary, and hence the expected quantization of Conjecture (*) for the mass critical problem does not hold here. In some sense, the proof of Theorem 2.8 amounts showing that *in the ring regime*, the super critical problem can be treated as a mass critical problem. Moreoever, this is the first construction of blow up solutions for a large set of super critical regimes including the physical one $N = p = 3$.

We now turn to the proof of the sharp upper bound (2.7) which relies on a suitably localized virial identity in the continuation of [**76**].

PROOF OF THEOREM 2.7. **step 1** Localized virial identity. Let $N \geq 2$, $0 < s_c < 1$ and $u \in \mathcal{C}([0,T), H^1)$ be a radially symmetric finite time blow up solution $0 < T < +\infty$. Pick a time $t_0 < T$ and a radius $0 < R = R(t_0) \ll 1$ to be chosen. Let $\chi \in \mathcal{C}_c^\infty(\mathbb{R}^N)$ and recall the localized virial identity[5] for radial solutions:

$$\frac{1}{2}\frac{d}{d\tau}\int \chi|u|^2 = Im\left(\int \nabla\chi \cdot \nabla u\overline{u}\right), \tag{2.12}$$

$$\frac{1}{2}\frac{d}{d\tau}Im\left(\int \nabla\chi\cdot\nabla u\overline{u}\right) = \int \chi''|\nabla u|^2 - \frac{1}{4}\int \Delta^2\chi|u|^2 - \left(\frac{1}{2} - \frac{1}{p+1}\right)\int \Delta\chi|u|^{p+1}.$$

Applying with $\chi = \psi_R = R^2\psi(\frac{x}{R})$ where $\psi(x) = \frac{|x|^2}{2}$ for $|x| \leq 2$ and $\psi(x) = 0$ for $|x| \geq 3$, we get:

$$\begin{aligned}
&\frac{1}{2}\frac{d}{d\tau}Im\left(\int \nabla\psi_R\cdot\nabla u\overline{u}\right)\\
= &\int \psi''(\frac{x}{R})|\nabla u|^2 - \frac{1}{4R^2}\int \Delta^2\psi(\frac{x}{R})|u|^2 - \left(\frac{1}{2} - \frac{1}{p+1}\right)\int \Delta\psi(\frac{x}{R})|u|^{p+1}\\
\leq &\int |\nabla u|^2 - N\left(\frac{1}{2} - \frac{1}{p+1}\right)\int |u|^{p+1} + C\left[\frac{1}{R^2}\int_{2R\leq|x|\leq 3R}|v|^2 + \int_{|x|\geq R}|u|^{p+1}\right].
\end{aligned}$$

Now from the conservation of the energy:

$$\int |u|^{p+1} = \frac{p+1}{2}\int |\nabla u|^2 - (p+1)E(u_0)$$

from which

$$\int |\nabla u|^2 - N\left(\frac{1}{2} - \frac{1}{p+1}\right)\int |u|^{p+1} = \frac{N(p-1)}{2}E(u_0) - \frac{2s_c}{N-2s_c}\int |\nabla u|^2,$$

and thus:

$$\begin{aligned}
&\frac{2s_c}{N-2s_c}\int |\nabla u|^2 + \frac{1}{2}\frac{d}{d\tau}Im\left(\int \nabla\psi_R\cdot\nabla u\overline{u}\right)\\
\lesssim &\left[|E_0| + \int_{|x|\geq R}|u|^{p+1} + \frac{1}{R^2}\int_{2R\leq|x|\leq 3R}|u|^2\right] \qquad (2.13)\\
\leq &\ C(u_0)\left[1 + \frac{1}{R^2} + \int_{|x|\geq R}|u|^{p+1}\right]
\end{aligned}$$

from the energy and L^2 norm conservations.

step 2 Radial Gagliardo-Nirenberg interpolation estimate. In order to control the outer nonlinear term in (2.13), we recall the radial interpolation bound:

$$\|u\|_{L^\infty(r\geq R)} \leq \frac{\|\nabla u\|_{L^2}^{\frac{1}{2}}\|u\|_{L^2}^{\frac{1}{2}}}{R^{\frac{N-1}{2}}},$$

[5]see [**76**] for further details.

which together with the L^2 conservation law ensures:

$$\begin{aligned}\int_{|x|\geq R}|u|^{p+1} &\leq \|u\|^{p-1}_{L^\infty(r\geq R)}\int |u|^2 \leq \frac{C(u_0)}{R^{\frac{(N-1)(p-1)}{2}}}\|\nabla u\|^{\frac{p-1}{2}}_{L^2}\\ &\leq \delta\frac{2s_c}{N-2s_c}\int|\nabla u|^2+\frac{C}{\delta R^{\frac{2(N-1)(p-1)}{(5-p)}}}\\ &= \delta\frac{2s_c}{N-2s_c}\int|\nabla u|^2+\frac{C}{\delta R^{\frac{2}{\alpha}}}\end{aligned}$$

where we used Hölder for $p<5$ and the definition of α (2.6). Injecting this into (2.13) yields for $\delta>0$ small enough using $R\ll 1$ and $0<\alpha<1$:

$$\frac{s_c}{N-2s_c}\int|\nabla u|^2+\frac{d}{d\tau}Im\left(\int\nabla\psi_R\cdot\nabla u\overline{u}\right)\leq\frac{C(u_0,p)}{R^{\frac{2}{\alpha}}} \tag{2.14}$$

step 3 Time integration. We now integrate (2.14) twice in time on $[t_0,t_2]$ using (2.12). This yields up to constants using Fubini in time:

$$\begin{aligned}&\int\psi_R|u(t_2)|^2+\int_{t_0}^{t_2}(t_2-t)\|\nabla u(t)\|^2_{L^2}dt\\ \lesssim\ &\frac{(t_2-t_0)^2}{R^{\frac{2}{\alpha}}}+(t_2-t_0)\left|Im\left(\int\nabla\psi_R\cdot\nabla u\overline{u}\right)(t_0)\right|+\int\psi_R|u(t_0)|^2\\ \leq\ &C(u_0)\left[\frac{(t_2-t_0)^2}{R^{\frac{2}{\alpha}}}+R(t_2-t_0)\|\nabla u(t_0)\|_{L^2}+R^2\|u_0\|^2_{L^2}\right].\end{aligned}$$

We now let $t\to T$. We conclude that the integral in the left hand side converges and

$$\int_{t_0}^{T}(T-t)\|\nabla u(t)\|^2_{L^2}dt\leq C(u_0)\left[\frac{(T-t_0)^2}{R^{\frac{2}{\alpha}}}+R(T-t_0)\|\nabla u(t_0)\|_{L^2}+R^2\right]. \tag{2.15}$$

We now optimize in R by choosing:

$$\frac{(T-t_0)^2}{R^{\frac{2}{\alpha}}}=R^2\ \text{ ie }\ R(t_0)=(T-t_0)^{\frac{\alpha}{1+\alpha}}.$$

(2.15) now becomes:

$$\begin{aligned}\int_{t_0}^{T}(T-t)\|\nabla u(t)\|^2_{L^2}dt &\leq C(u_0)\left[(T-t_0)^{\frac{2\alpha}{1+\alpha}}+(T-t_0)^{\frac{\alpha}{1+\alpha}}(T-t_0)\|\nabla u(t_0)\|_{L^2}\right]\\ &\leq C(u_0)(T-t_0)^{\frac{2\alpha}{1+\alpha}}+(T-t_0)^2\|\nabla u(t_0)\|^2_{L^2}. \qquad (2.16)\end{aligned}$$

In order to integrate this differential inequality, let

$$g(t_0)=\int_{t_0}^{T}(T-t)\|\nabla u(t)\|^2_{L^2}dt, \tag{2.17}$$

then (2.16) means:

$$g(t)\leq C(T-t)^{\frac{2\alpha}{1+\alpha}}-(T-t)g'(t)$$

ie

$$\left(\frac{g}{T-t}\right)'=\frac{1}{(T-t)^2}((T-t)g'+g)\leq\frac{C(u_0)}{(T-t)^{2-\frac{2\alpha}{1+\alpha}}}.$$

Integrating this in time yields

$$\frac{g(t)}{T-t} \lesssim 1 + \frac{1}{(T-t)^{1-\frac{2\alpha}{1+\alpha}}} \quad \text{ie} \quad g(t) \lesssim (T-t)^{\frac{2\alpha}{1+\alpha}}$$

for t close enough to T, which together with (2.17) yields (2.7). □

2.5. More blow up problems. The study of singularity formation for nonlinear dispersive equations has experienced a substantial acceleration since the end of the 1990's in particular in the continuation of the pioneering breakthrough works by Merle and Zaag on the nonlinear heat equation [**83**], [**84**], [**85**], and Martel and Merle on the mass critical (gKdV) problem [**55**], [**56**], [**57**], [**58**], [**59**]. The analysis has spread to various other problems and led to the development of new tools. It is not the aim of these notes to give a complete account of the existing literature, but we would like to point out the deep unity between some of these recent works. One particularly active direction or research is on *energy critical models* $s_c = 1$ which surprisingly enough display a similar structure like the mass critical problem, even though essential new phenomenons occur. This includes energy critical wave or heat problems, or more geometric problems like wave and Schrödinger maps for which the sole existence of blow up solutions in the critical regimes has been a long standing open problem. Among the key results obtained in the past ten years, let us mention some dynamical constructions: the first construction of blow up solutions for the energy critical wave map problem by Krieger, Schlag, Tataru [**41**], the derivation of the stable regime for the wave map jointly with Rodnianski [**99**], the first construction of blow up bubble for the Schrödinger map problem and the discovery of the rotational instability jointly with Merle and Rodnianski [**77**]. Moreover, a new generation of *classification theorems* have occurred in the direction of the multi solitary wave resolution conjecture, see in particular Duyckaerts, Kenig, Merle [**15**] for the energy critical nonlinear wave equation and the spectacular series of works by Merle and Zaag [**86**], [**87**], [**88**], [**89**], [**90**] which give the first complete classification of all blow up regimes for a nonlinear wave equation.

3. The mass critical problem

We focus in this section and for the rest of these notes onto the L^2 critical case

$$p = 1 + \frac{4}{N}, \quad s_c = 0.$$

which is the smallest power nonlinearity for which blow up occurs. We will show that a large part of the orbital stability theory developed for subcritical problems still applies in some generalized sense and provides some essential information on the structure of the blow up bubble. We will in particular show that there exists a *sharp* criterion for global existence, Theorem 3.5, and obtain the first dynamical informations on the structure of the singularity formation which are mostly a consequence of the variational characterization of the ground state solitary wave.

3.1. Variational characterization of the ground state. The minimization problem (1.12) is no longer adapted to the critical problem due to the L^2 scaling invariance

$$u_\lambda(t,x) = \lambda^{\frac{N}{2}} u(\lambda^2 t, \lambda x). \tag{3.1}$$

Indeed, one easily proves that $I(M) = 0$ for $M << 1$ and $I(M) = -\infty$ for $M \gg 1$. In fact, as observed by Weinstein [**111**], the L^2 criticality of (1.1) corresponds to an exact balance between the kinetic and potential energies which can be quantified through the knowledge of the sharp constant in the Gagliardo-Nirenberg inequality (1.7).

PROPOSITION 3.1 (Sharp Gagliardo-Nirenberg estimate, [**111**]). *Let the H^1 functional:*

$$J(v) = \frac{(\int |\nabla v|^2)(\int |v|^2)^{\frac{2}{N}}}{\int |v|^{2+\frac{4}{N}}}. \tag{3.2}$$

The minimization problem

$$\min_{v \in H^1, \ v \neq 0} J(v)$$

is attained on the three parameters family:

$$\lambda_0^{\frac{N}{2}} Q(\lambda_0 x + x_0) e^{i\gamma_0}, \quad (\lambda_0, x_0, \gamma_0) \in \mathbb{R}_*^+ \times \mathbb{R}^N \times \mathbb{R},$$

where Q is the unique ground state solution to:

$$\begin{cases} \Delta Q - Q + Q^{1+\frac{4}{N}} = 0, \quad Q > 0, \quad Q \text{ radial} \\ Q(r) \to 0 \quad \text{as} \quad r \to +\infty. \end{cases} \tag{3.3}$$

In particular, there holds the following Gagliardo-Nirenberg inequality with best constant:

$$\forall v \in H^1, \quad E(v) \geq \frac{1}{2} \int |\nabla v|^2 \left(1 - \left(\frac{\|v\|_{L^2}}{\|Q\|_{L^2}}\right)^{\frac{4}{N}}\right). \tag{3.4}$$

While $E(Q) = I(M) < 0$ in the subcritical case, we have in the critical case [6]

$$E(Q) = 0.$$

A reformulation of (3.4) which is very useful is the following variational characterization of Q:

PROPOSITION 3.2 (Variational characterization of the ground state). *Let $v \in H^1$ such that*

$$\int |v|^2 = \int Q^2 \quad \text{and} \quad E(v) = 0,$$

then

$$v(x) = \lambda_0^{\frac{N}{2}} Q(\lambda_0 x + x_0) e^{i\gamma_0},$$

for some parameters $\lambda_0 \in \mathbb{R}_+^*$, $x_0 \in \mathbb{R}^N$, $\gamma_0 \in \mathbb{R}$.

To sum up, the situation is as follows: let $v \in H^1$, then if $\|v\|_{L^2} < \|Q\|_{L^2}$, the kinetic energy dominates the potential energy and (3.4) yields $E(v) > C(v) \int |\nabla v|^2$ and the energy is in particular non negative; at the critical mass level $\|v\|_{L^2} = \|Q\|_{L^2}$, the only zero energy function is Q up to the symmetries of scaling, phase and translation which generate the three dimensional manifold of minimizers of (3.2). For $\|v\|_{L^2} > \|Q\|_{L^2}$, the sign of the energy is no longer prescribed.

[6] This can be seen for example by multiplying the Q equation by $\frac{N}{2}Q + y \cdot \nabla Q$ and integrating by parts.

REMARK 3.3. *Remark that on the contrary to the subcritical case, the scaling (3.1) leaves the L^2 norm invariant and hence there are no small solitary waves in the critical case.*

A simple consequence of the sharp lower bound (3.4) is the concentration of the mass at blow up given by Theorem 2.4.

PROOF OF THEOREM 2.4. The proof is purely variational. We prove the result in the radial case for $N \geq 2$. The general case follows from concentration compactness techniques, see **[91]**, **[28]**. Let $u_0 \in H^1$ radial and assume that the corresponding solution $u(t)$ to (1.1) blows up at time $0 < T < +\infty$, or equivalently:

$$\lim_{t \to T} \|\nabla u(t)\|_{L^2} = +\infty. \tag{3.5}$$

We need to prove (2.4) and argue by contradiction: assume that for some $R > 0$ and $\varepsilon > 0$, there holds on some sequence $t_n \to T$,

$$\lim_{n \to +\infty} \int_{|y| \leq R} |u(t_n, y)|^2 dy \leq \int Q^2 - \varepsilon. \tag{3.6}$$

Let us rescale the solution by its size and set:

$$\lambda(t_n) = \frac{1}{\|\nabla u(t_n)\|_{L^2}}, \quad v_n(y) = \lambda^{\frac{N}{2}}(t_n) u(t_n, \lambda(t_n) y),$$

then from explicit computation:

$$\|\nabla v_n\|_{L^2} = 1 \text{ and } E(v_n) = \lambda^2(t_n) E(u). \tag{3.7}$$

First observe that v_n is H^1 bounded and we may assume on a sequence $n \to +\infty$:

$$v_n \rightharpoonup V \text{ in } H^1.$$

We first claim that V is non zero. Indeed, from (3.5), (3.7) and the conservation of the energy for $u(t)$, $E(v_n) \to 0$ as $n \to +\infty$, and thus:

$$\frac{1}{2 + \frac{4}{N}} \int |v_n|^{2+\frac{4}{N}} = \frac{1}{2} \int |\nabla v_n|^2 - E(v_n) = \frac{1}{2} - E(v_n) \to \frac{1}{2} \text{ as } n \to +\infty.$$

Now from the compact embedding of $H^1_{radial} \hookrightarrow L^{2+\frac{4}{N}}$, $v_n \to V$ in $L^{2+\frac{4}{N}}$ up to a subsequence, and thus $\frac{1}{2+\frac{4}{N}} \int |V|^{2+\frac{4}{N}} \geq \frac{1}{2}$ and V is non zero. Moreover, from the weak H^1 convergence and the strong $L^{2+\frac{4}{N}}$ convergence,

$$E(V) \leq \liminf_{n \to +\infty} E(v_n) = 0.$$

Last, we have from (3.5), (3.6) and the weak H^1 convergence: $\forall A > 0$

$$\begin{aligned}
\int_{|y| \leq A} |V(y)|^2 dy &\leq \liminf_{n \to +\infty} \int_{|y| \leq A} |v_n(y)|^2 dy \leq \lim_{n \to +\infty} \int_{|y| \leq \frac{R}{\lambda(t_n)}} |v(t_n, y)|^2 dy \\
&= \lim_{n \to +\infty} \int_{|x| \leq R} |u(t_n, x)|^2 dx \leq \int Q^2 - \varepsilon.
\end{aligned}$$

Thus $\int |V|^2 \leq \int Q^2 - \varepsilon$ which together with V non zero and $E(V) \leq 0$ contradicts the sharp Gagliardo-Nirenberg inequality (3.4). □

The proof in the non radial case has been simplified by Hmidi, Keraani **[28]**, which derived the following optimal result from concentration compactness - more precisely profile decomposition- techniques:

LEMMA 3.4. *Let a sequence $u_n \in H^1$ with*

$$\limsup_{n\to+\infty} \|\nabla u_n\|_{L^2} \leq \|\nabla Q\|_{L^2}, \quad \limsup_{n\to+\infty} \|u_n\|_{L^{2+\frac{4}{N}}} \geq \|Q\|_{L^{2+\frac{4}{N}}},$$

then there exists $x_n \in \mathbb{R}^N$ and $V \in H^1$ such that up to a subsequence:

$$v_n(\cdot + x_n) \rightharpoonup V \quad in \ H^1 \quad with \quad \|V\|_{L^2} \geq \|Q\|_{L^2}.$$

3.2. The sharp global wellposedness criterion. A generalization of Theorem 1.1 has been obtained by Weinstein [**111**]:

THEOREM 3.5 (Global well posedness for subcritical mass, [**111**]). *Let $u_0 \in H^1$ with $\|u_0\|_{L^2} < \|Q\|_{L^2}$, the corresponding solution $u(t)$ to (1.1) is global and bounded in H^1. More precisely, the solution scatters as $t \pm \infty$.*

PROOF OF THEOREM 3.5. From the conservation of the L^2 norm, $\|u(t)\|_{L^2} < \|Q\|_{L^2}$ for all $t \in [0,T)$, and thus an a priori bound on $\|u(t)\|_{H^1}$ follows from the conservation of the energy and the sharp Gagliardo-Nirenberg inequality (3.4) applied to $v = u(t)$. The scattering claim is easily proved for $u_0 \in \Sigma = H^1 \cap \{xu \in L^2\}$ using the explicit pseudo conformal symmetry: if $u(t,x)$ is a solution to (1.1), then so is

$$v(t,x) = \frac{1}{|t|^{\frac{N}{2}}} u(\frac{-1}{t}, \frac{x}{t}) e^{i\frac{|x|^2}{4t}}. \tag{3.8}$$

The pseudo conformal symmetry is a well known symmetry of the linear Schrödinger flow and a symmetry of the nonlinear problem in the mass critical case only. It is moreover an L^2 isometry and thus applying Weinstein's criterion to v ensures that v has a limit in Σ as $t \uparrow 0$, and hence u scatters as $t \to +\infty$ as readily seen on (3.8). The case when $u_0 \in L^2$ only is considerably more delicate and relies on the rigidity theorem approach developed by Kenig, Merle [**33**], see Killip, Tao, Visan, Li, Zhang [**35**], [**36**], [**37**] and references therein, Dodson [**14**]. □

A spectacular feature is that Weinstein's criterion for global existence is sharp. On the one hand, from (3.3),

$$W(t,x) = Q(x)e^{it}$$

is a gobal solution to (1.1) with critical mass $\|W\|_{L^2} = \|Q\|_{L^2}$ which does not disperse. One should thus think of $\|Q\|_{L^2}$ as the minimal amount of mass required to avoid complete dispersion of the wave, and the solitary wave is the *smallest non linear object* for which dispersion and concentration exactly balance each other.

Observe now that the pseudo conformal symmetry (3.8) applied to the solitary wave solution $u(t,x) = Q(x)e^{it}$ yields the *explicit minimal mass blow up element*:

$$S(t,x) = \frac{1}{|t|^{\frac{N}{2}}} Q(\frac{x}{t}) e^{-i\frac{|x|^2}{4t} + \frac{i}{t}} \tag{3.9}$$

which scatters as $t \to -\infty$, and blows up at the origin at the speed

$$\|\nabla S(t)\|_{L^2} \sim \frac{1}{|t|} \tag{3.10}$$

by concentrating its mass:

$$|S(t)|^2 \rightharpoonup \|Q\|^2_{L^2} \delta_{x=0} \quad \text{as} \quad t \uparrow 0. \tag{3.11}$$

REMARK 3.6. *For the mass critical NLS, the sharp threshold for global existence and for scattering are therefore the same. This in fact an exceptional case induced by the Laplace operator and the Galilean symmetry -which is again an L^2 isometry-. For a more general dispersion of the type $(-\Delta)^\alpha$, these threshold are not the same,* [**39**].

3.3. Orbital stability of the ground state. More can be said on the structure of the singularity formation, and in particular on the blow up profile for initial data with L^2 mass just above the critical mass required for blow up:

$$u_0 \in \mathcal{B}_{\alpha^*} = \{u_0 \in H^1 \text{ with } \int Q^2 \leq \int |u_0|^2 \leq \int Q^2 + \alpha^*\} \tag{3.12}$$

for some parameter $\alpha^* > 0$ small enough. This situation is moreover conjectured to locally describe the generic blow up dynamic around one blow up point.

Let us recall that $E(Q) = 0$ together with the virial blow up result of Theorem 2.1 imply the instability of the solitary wave $Q(x)e^{it}$. We claim however that the orbital stability of Q may be retrieved in some sense according to the following generalization of Theorem 1.4:

THEOREM 3.7 (Orbital stability in the critical case). *Let $N \geq 1$. For all $\alpha^* > 0$ small enough, there exists $\delta(\alpha^*)$ with $\delta(\alpha^*) \to 0$ as $\alpha^* \to 0$ such that the following holds true. Let $u_0 \in H^1$ with*

$$\int |u_0|^2 \leq \int Q^2 + \alpha^*, \quad E(u) \leq \alpha^* \int |\nabla u|^2, \tag{3.13}$$

and let $u(t)$ be the corresponding solution to (1.1) with life time $0 < T \leq +\infty$, then there exist $(x(t), \gamma(t)) \in \mathcal{C}^0([0,T), \mathbb{R}^N \times \mathbb{R})$ such that:

$$\forall t \in [0,T), \quad \|\lambda^{\frac{N}{2}}(t) u(t, \lambda(t)x + x(t)) e^{-i\gamma(t)} - Q\|_{H^1} < \delta(\alpha^*). \tag{3.14}$$

Note that a finite time blow up solution with small super critical mass automatically satisfies (3.13) near blow up time, and hence it is close to the ground state in H^1 up to the set of H^1 symmetries. This property is again purely based on the conservation laws and the variational characterization of Q, and not on refined properties of the flow.

PROOF OF THEOREM 3.7. Equivalently, we need to prove the following: let a sequence $u_n \in H^1$ with

$$\|u_n\|_{L^2} \to \|Q\|_{L^2}, \quad \limsup_{n\to+\infty} \frac{E(u_n)}{\|\nabla u_n\|_{L^2}^2} \leq 0, \tag{3.15}$$

let

$$v_n = \lambda_n^{\frac{N}{2}} u(\lambda_n x) \text{ with } \lambda_n = \frac{\|\nabla Q\|_{L^2}}{\|\nabla u_n\|_{L^2}}, \tag{3.16}$$

then there exist $x_n \in \mathbb{R}^N$, $\gamma_n \in \mathbb{R}$ such that:

$$v_n(\cdot + x_n)e^{i\gamma_n} \to Q \text{ in } H^1 \text{ as } n \to +\infty. \tag{3.17}$$

Indeed, observe from (3.15) and (3.16) that

$$\|v_n\|_{L^2} \to \|Q\|_{L^2}, \quad \|\nabla v_n\|_{L^2} = \|\nabla Q\|_{L^2}, \quad \limsup_{n\to+\infty} E(v_n) \leq 0.$$

We now apply Proposition 1.6 to v_n. If vanishing occurs, then up to a subsequence, we have for n large enough:

$$E(v_n) \geq \frac{\|\nabla Q\|_{L^2}^2}{4}$$

which contradicts $\limsup_{n\to+\infty} E(v_n) \leq 0$. If dichotomy occurs, then there exist w_k, z_k and $0 < \alpha < 1$ such that

$$\|w_k\|_{L^2} \to \alpha\|Q\|_{L^2}, \quad \|z_k\|_{L^2} \to (1-\alpha)\|Q\|_{L^2} \quad \text{and} \quad 0 \geq \limsup_{k\to+\infty}(E(w_k)+E(z_k)).$$

But from the sharp Gagliardo-Nirenberg inequality (3.4) applied to w_k and z_k, this implies

$$\|\nabla w_k\|_{L^2} + \|\nabla z_k\|_{L^2} \to 0 \quad \text{as} \quad k \to +\infty$$

and thus

$$\|v_{n_k}\|_{L^{2+\frac{4}{N}}} \to 0 \quad \text{as} \quad k \to +\infty,$$

and we are back to the vanishing case. Hence compactness occurs and

$$v_n(\cdot + x_n) \to V \quad \text{strongly in} \quad L^{2+\frac{4}{N}}, L^2$$

up to a subsequence. But then $E(v) \leq 0$ and $\|V\|_{L^2} = \|Q\|_{L^2}$ imply from (3.4) and Proposition 3.2 that $V(x) = Q(x+x_0)e^{i\gamma_0}$. This in turns implies $E(V) = 0$ and thus $|\nabla v_n(\cdot + x_n)|_{L^2}^2 \to |\nabla Q|_{L^2}^2$ which impies (3.17). □

4. Dynamical construction of blow up solutions

We give in this section an overview on the known results on singularity formation in the mass critical case which go beyond the pure variational analysis of the previous section and rely on an explicit construction of blow up solutions for data near the ground state. This kind of question still attracts a considerable amount of interest, and we shall not be able to give a complete overview of the existing literature in these notes. We shall only give some key results in connection in particular with the question of the description of the flow near the ground state solitary wave which is the first nonlinear object.

4.1. Minimal mass blow up. Initial data $u_0 \in H^1$ with subcritical mass $\|u_0\|_{L^2} < \|Q\|_{L^2}$ generate global bounded solutions from Theorem 3.5. Moreoever, there exists an explicit minimal mass blow up element $S(t)$ induced by the pseudo conformal symmetry (3.8) and explicitly given by (3.9). The existence of the minimal element plays a distinguished role in the Kenig Merle approach to global existence [**33**]. An essential feature of (3.9) is that $S(t)$ is *compact* up to the symmetries of the flow, meaning that all the mass is put into the singularity formation. The basic intuition is that such a behavior is very special, and minimal elements should be classified[7]. This was proved using the pseudo conformal symmetry in a seminal work by Merle:

THEOREM 4.1 (Classification of the minimal mass blow up solution, [**66**]). *Let $u_0 \in H^1$ with*

$$\|u_0\|_{L^2} = \|Q\|_{L^2}.$$

[7]This is a dispersive intuition which for example is completely false in the parabolic setting, [**5**].

Assume that the corresponding solution to (1.1) blows up in finite time $0 < T < +\infty$. Then

$$u(t) = S(t)$$

up to the symmetries.

Before giving the proof of Merle's classification Theorem, let us say that the question of the existence of minimal elements in various settings has been a long standing open problem, mostly due to the fact that the existence of the minimal element for NLS relies entirely on the exceptional pseudo conformal symmetry. Merle in [**67**] considered the inhomogeneous problem

$$i\partial_t u + \Delta u + k(x)u|u|^2 = 0, \quad x \in \mathbb{R}^2$$

which breaks the full symmetry group, and obtains for non smooth k *non existence* results of minimal elements. A contrario and more recently, a sharp criterion for the existence and uniqueness of minimal solutions is derived in collaboration with Szeftel in [**101**] which relies on a dynamical construction and new Lypapounov rigidity functionals at the minimal mass level. A further extension to non local dispersion can be found in [**39**] which shows that minimal mass blow up is in fact the generic situation, and has little to do with the pseudo conformal symmetry, see also [**2**] for an extension to curved backgrounds, and Theorem 4.6 for the case of the critical (gKdV).

PROOF OF THEOREM 4.1. This is the first proof of classification of minimal elements in the Schrödinger setting. We advise the reader to compare it with the proof of the Liouville theorem in [**33**] and observe the deep unity of both arguments. The original proof by Merle [**66**] has been further simplified by Banica [**1**] and Hmidi, Keraani [**27**], and it is the proof we present now.

step 1 Compactness of the flow in H^1 up to scaling. Let u as in the hypothesis of the Theorem with blow up time $0 < T < \infty$. Let

$$\lambda(t) = \frac{|\nabla Q|_{L^2}}{\|\nabla u(t)\|_{L^2}} \to 0 \text{ as } t \to T.$$

Then

$$v(t,x) = \lambda^{\frac{N}{2}}(t)u(t, \lambda(t)x + x(t))$$

satisfies:

$$\|\nabla v(t)\|_{L^2} = \|\nabla Q\|_{L^2}, \quad \lim_{t\to T} E(v) = 0, \quad \|v(t)\|_{L^2} = \|Q\|_{L^2}.$$

Arguing as for the proof of Theorem 3.7, we conclude from standard concentration compactness techniques and the variational characterization of the ground state that:

$$v(t, x + x(t))e^{i\gamma(t)} \to Q \text{ in } H^1 \text{ as } t \to T. \tag{4.1}$$

step 2 A refined Cauchy-Schwarz for critical mass functions. For $\|w\|_{L^2} < \|Q\|_{L^2}$, the energy controls the kinetic energy from (3.4). This controls fails for $\|w\|_{L^2} = \|Q\|_{L^2}$ but can be retrieved in some weak sense. Indeed, Banica observed the following: let a smooth real valued ψ and $w \in H^1$ with $\|w\|_{L^2} = \|Q\|_{L^2}$, then:

$$\left|\int Im(\nabla\psi \cdot \nabla w\overline{w})\right|^2 \lesssim \sqrt{E(w)}\left(\int |\nabla\psi|^2|w|^2\right)^{\frac{1}{2}}. \tag{4.2}$$

Indeed, for any $a > 0$,

$$\|we^{ia\psi}\|_{L^2} = \|Q\|_{L^2} \text{ and thus } E(we^{ia\psi}) \geq 0$$

and the result follows by expanding in a.

step 3 L^2 compactness of u and control of the concentration point. We now claim that u is L^2 compact: $\forall \varepsilon > 0$, $\exists R > 0$ such that

$$\forall t \in [0,T), \quad \int_{|x|\geq R} |u(t,x)|^2 dx < \varepsilon. \tag{4.3}$$

Indeed, pick ε small enough, For $R > 0$, let $\chi_R(x) = \chi(\frac{x}{R})$ where χ is a smooth radial cut off function with $\chi(r) = 0$ for $r \leq \frac{1}{2}$, $\chi(r) = 1$ for $r \geq 1$. Then integrating by parts in (1.1) and using (4.2), we get:

$$\left|\frac{1}{2}\frac{d}{dt}\int \chi_R|u|^2\right| = \left|Im\int \nabla\chi_R \cdot \nabla u\overline{u}\right| \leq C\sqrt{E(u)}\left(\int |\nabla\chi_R|^2|u|^2\right)^{\frac{1}{2}} \leq \frac{C}{R}\sqrt{E_0}\|u_0\|_{L^2}$$

where we used the conservation of energy and L^2 norm in the last step. Integrating in time on $[0,T]$ and using $T < +\infty$ yields (4.3).
Now observe that (4.1) and (4.3) automatically imply a localization of the concentration point:

$$\forall t \in [0,T), \quad |x(t)| \leq C(u_0). \tag{4.4}$$

step 4 $u \in \Sigma$. From (4.4) and up to a translation in space, we may consider a sequence of times $t_n \to T$ such that

$$x(t_n) \to 0 \in \mathbb{R}^N.$$

From (4.1), (4.3):

$$|u(t_n,x)|^2 \rightharpoonup \left(\int |Q|^2\right)\delta_0 \text{ as } t_n \to T. \tag{4.5}$$

This means that at time T, all the mass is at the origin. Even though there is no finite speed of propagation for (NLS), the idea is to integrate backwards from the singularity to conclude that this implies that there was not much mass initially at infinity, that is

$$u_0 \in \Sigma = H^1 \cap \{xu\} \in L^2. \tag{4.6}$$

This step is very important and corresponds to a non trivial *gain of regularity* for the asymptotic object which is a direct consequence of its non dispersive behavior. Let a smooth radial cut off function $\psi(r) = r^2$ for $r \leq 1$, $\psi(r) = 8$ for $r \geq 2$ and such that $|\nabla\psi|^2 \leq C\psi$. Let $A > 0$ and $\psi_A(r) = A^2\psi(\frac{r}{A})$, then:

$$|\nabla\psi_A|^2 \lesssim \psi_A. \tag{4.7}$$

Then integrating by parts in (1.1), we have using (4.2) and (4.7):

$$\begin{aligned}\left|\frac{1}{2}\frac{d}{dt}\int \psi_A|u|^2\right| &= \left|Im\int (\nabla\psi_A \cdot \nabla u\overline{u})\right| \lesssim \sqrt{E_0}\left(\int |\nabla\psi_A|^2|u|^2\right)^{\frac{1}{2}} \\ &\lesssim \sqrt{E_0}\left(\int \psi_A|u|^2\right)^{\frac{1}{2}}\end{aligned}$$

or equivalently:

$$\left|\frac{d}{dt}\sqrt{\int \psi_A |u|^2}\right| \lesssim \sqrt{E_0}. \tag{4.8}$$

Now observe from (4.5) that

$$\int \psi_A |u(t_n)|^2 \to 0 \quad \text{as} \quad t_n \to T.$$

Integrating (4.8) on $[t, t_n]$ and letting $t_n \to T$, we thus get:

$$\forall t \in [0,T), \quad \sqrt{\int \psi_A |u(t)|^2} \leq C(E_0)(T-t).$$

Note that the right hand side of the above expression is independent of A. We may thus let $A \to \infty$ and conclude to an even more precise version of (4.6):

$$\forall t \in [0,T), \quad u(t) \in \Sigma \quad \text{with} \quad \int |x|^2 |u(t,x)|^2 dx \to 0 \quad \text{as} \quad t \to T. \tag{4.9}$$

step 5 Pseudo-conformal transformation. The conclusion of the proof is pure magic. It relies on the following completely general fact. Let $u(t)$ be a solution to (1.1) leaving on $[0,T)$, then

$$v(t,x) = \left(\frac{T}{T+t}\right)^{\frac{N}{2}} u\left(\frac{tT}{T+t}, \frac{Tx}{T+t}\right) e^{i\frac{|x|^2}{4(T+t)}}$$

is a solution to (1.1) with

$$\|v\|_{L^2} = \|u\|_{L^2} \quad \text{and} \quad E(v) = \frac{1}{8}\lim_{t \to T} \int |x|^2 |u(t,x)|^2 dx.$$

Applying this to u and using (4.9), this implies that

$$\|v\|_{L^2} = \|u\|_{L^2} = \|Q\|_{L^2} \quad \text{and} \quad E(v) = 0.$$

From Proposition 3.2, $v = Q$ up to the symmetries of the flow, and this concludes the proof of Theorem 4.1. □

4.2. Log log blow up. The only explicit blow up solution we have encountered so far is the minimal mass blow up bubble (3.9). This bubble is intrinsically unstable because a mass subcritical perturbation leads to a globally defined solution. The question of the description of *stable* blow up bubbles has attracted a considerable attention which started in the 80's with the development of sharp numerical methods and the prediction of the "log-log law" for NLS by Landman, Papanicoalou, Sulem, Sulem [**43**].

To simplify the presentation, let us restrict our attention with mass just above the minimal required for singularity formation

$$u_0 \in \mathcal{B}_{\alpha^*} = \left\{u_0 \in H^1 \quad \text{with} \quad \|Q\|_{L^2} < \|u_0\|_{L^2} < \|Q\|_{L^2} + \alpha^*\right\}, \quad 0 < \alpha^* \ll 1. \tag{4.10}$$

A general and fundamental open problem is to completely describe the flow for such initial data which in some sense corresponds according to the scattering statement of Theorem 3.5 to the first non linear zone. The generalized orbital stability statement of Theorem 3.7 ensures that under (4.10), if u blows up at $T < +\infty$. then for t close enough to T, the solution must admit a nonlinear decomposition

$$u(t,x) = \frac{1}{\lambda(t)^{\frac{N}{2}}}(Q+\varepsilon)(t, \frac{x-x(t)}{\lambda(t)})e^{i\gamma(t)}, \tag{4.11}$$

where

$$\|\varepsilon(t)\|_{H^1} \leq \delta(\alpha^*), \quad \lambda(t) \sim \frac{1}{\|\nabla u(t)\|_{L^2}}. \tag{4.12}$$

This decomposition implies that in any blow up regime, the ground state solitary wave Q is a good approximation of the blow up profile, and this is the starting point for a perturbative analysis. The sharp description of the blow up bubble now relies on the extraction of the finite dimensional and possibly universal dynamic for the evolution of the geometrical parameters $(\lambda(t), x(t), \gamma(t))$ which is coupled to the infinite dimensional dispersive dynamic driving the small excess of mass $\varepsilon(t)$.

REMARK 4.2. *An illuminating computation is to reformulate* (3.9) *for the minimal blow up element in terms of* (4.11)*:*

$$\lambda(t) = |t|, \quad \varepsilon(t,y) = Q(y)\left(e^{-i\frac{b(t)|y|^2}{4}} - 1\right), \quad b(t) = |t|.$$

All possible regimes of $\lambda(t)$ are not known, but some progress has been done on the understanding of stable and threshold dynamics. The following Theorem summarizes the series of results obtained in [**71**], [**72**], [**73**], [**74**], [**75**], [**97**]:

THEOREM 4.3 ([**71**], [**72**], [**73**], [**74**], [**75**], [**97**]). *Let $N \leq 5$. There exists a universal constant $\alpha^* > 0$ such that the following holds true. Let $u_0 \in \mathcal{B}_{\alpha^*}$ and $u \in \mathcal{C}([0,T), H^1)$, $0 < T \leq +\infty$ be the corresponding solution to* (1.1).
(i) Sharp L^2 concentration: Assume $T < +\infty$, then there exist parameters $(\lambda(t), x(t), \gamma(t)) \in \mathcal{C}^1([0,T), \mathbb{R}^*_+ \times \mathbb{R}^N \times \mathbb{R})$ and an asymptotic profile $u^* \in L^2$ such that

$$u(t) - \frac{1}{\lambda(t)^{\frac{N}{2}}} Q\left(\frac{x-x(t)}{\lambda(t)}\right) e^{i\gamma(t)} \to u^* \ \ in \ \ L^2 \ \ as \ \ t \to T, \tag{4.13}$$

and the blow up point is finite:

$$x(t) \to x(T) \in \mathbb{R}^N \ \ as \ \ t \to T.$$

(ii) Classification of the speed: Under (i), the solution is either in the log-log regime

$$\lambda(t)\sqrt{\frac{\log|\log(T-t)|}{T-t}} \to \sqrt{2\pi} \ \ as \ \ t \to T \tag{4.14}$$

and then the asymptotic profile is not smooth:

$$u^* \notin H^1 \ \ and \ \ u^* \notin L^p \ \ for \ \ p > 2, \tag{4.15}$$

or there holds the sharp lower bound

$$\lambda(t) \lesssim C(u_0)(T-t) \tag{4.16}$$

and the improved regularity:

$$u^* \in H^1. \tag{4.17}$$

(iii) Sufficient condition for log-log blow up: Assume $E_0 < 0$, then the solution blows un finite time $T < +\infty$ in the log log regime (4.14).

(iv) H^1 stability of the log log blow up: More generally, the set of initial data in $\mathcal{B}_{\alpha^*}$ such that the corresponding solution to (1.1) blows up in finite time with the log-log law (4.14) is open in H^1.

Comments on the result:

1. The log log law. The log log law (4.14) of stable blow up was first proposed in the pioneering formal and numerical work [**43**]. The first rigorous construction of such a solution is due to Galina Perelman [**95**] in dimension $N = 1$. The proof of Theorem 4.3 involves a mild coercivity property of the linearized operator close to Q, see the Spectral Property 5.6, which is proved in dimension $N = 1$ in [**71**] and checked numerically in an elementary way in [**18**] for $N \leq 5$. Here we face the difficulty that there is no explicit formula for the ground state in dimensions $N \geq 2$.

2. Upper bound on the blow up speed: There exists *no upper bound of no type* on the blow up speed $\|\nabla u(t)\|_{L^2}$ in the mass critical case, even for data $u_0 \in \mathcal{B}_{\alpha^*}$ only. The lower bound (4.16) is sharp and saturated by the minimal blow up element $S(t)$. The derivation of slower blow up, which through the pseudo conformal symmetry is equivalent to the construction of infinite time grow up solutions, is linked to the description of the flow near the ground state which is still incomplete for (NLS). The intuition is led here by the recent classification results obtained for the mass critical KdV problem which we present in section 4.5.

3. Quantization of the blow up mass: The strong convergence (4.13) gives a complete description of the blow up bubble in the scaling invariance space and implies in particular that the mass which is put into the singularity formation is quantized

$$|u(t)|^2 \rightharpoonup \|Q\|_{L^2}^2 \delta_{x=x(T)} + |u^*|^2 \ \text{ as } \ t \to T, \ \ |u^*|^2 \in L^1$$

which shows the validity of the conjecture (*) for near minimal mass blow up solutions. This kind of general asymptotic simplification theorem started in the dispersive setting in the pioneering works by Martel and Merle [**55**] , and was recently propagated to impressive classification result -without assumption of size on the data- for energy critical wave equations [**15**]. Underlying the convergence (4.13) is the asymptotic stability statement of the solitary wave as the universal attractor of all blow up solutions which in the language (4.11) means

$$\varepsilon(t, x) \to 0 \ \text{ as } \ t \to T \ \text{ in } \ L^2_{\text{loc}}.$$

In fact, there are steps in the proof of Theorem 4.3 and the derivation of either upper bounds or lower bounds on the blow up rate is intimately connected to the question of dispersion for the excess of mass $\varepsilon(t, x)$.

4. Asymptotic profile: The regularity of the asymptotic profile u^* sees the change of regime because in the stable log log regime, the singular and regular parts of the solution are very much coupled, while they are more separated in any other regimes.

4.3. Threshold dynamics. We still consider small super critical mass initial data $u_0 \in \mathcal{B}_{\alpha^*}$. Theorem 4.3 describes the stable log log blow up. The explicit minimal mass blow up given by (4.3) does not belong to this class and is unstable.

Bourgain and Wang [8] observed however that $S(t)$ can be stabilized on a finite codimensional manifold, and they do so by integrating the flow backwards from the singularity. The excess of mass in this regime corresponds to a *flat and smooth* asymptotic profile. More precisely, let $N = 1, 2$, fix the origin as the blow up point and let a limiting profile $u^* \in H^1$ such that

$$\frac{d^i}{dx^i}u^*(0) = 0, \quad 1 \le i \le A, \quad A \gg 1, \tag{4.18}$$

then one can build a solution to (1.1) which blows up at $t = 0$ at $x = 0$ and satisfies:

$$u(t) - S(t) \to u^* \ \text{ in } H^1 \ \text{ as } t \uparrow 0. \tag{4.19}$$

We refer to [40] for a further discussion on the manifold construction. Note that this produces blow up solutions with super critical mass $\|u_0\|_{L^2} > \|Q\|_{L^2}$ which saturate the lower bound (4.16):

$$\|\nabla u(t)\|_{L^2} \sim \frac{1}{T-t}.$$

Also for small L^2 perturbation of $S(-1)$, the Bourgain Wang solution blows up at $t = 0$ but is global and scatters as $t \to -\infty$, simply because $S(t)$ scatters as $t \to -\infty$, and scattering is an L^2 stable behavior[8].

We proved in collaboration with Merle and Szeftel in [79] that these solutions sit on the border between the two open sets of solutions which scatter to the left as $t \to -\infty$ and respectively are global to the right and scatter as $t \to +\infty$, and blow up in finite time *in the log log regime.*

THEOREM 4.4 (Strong instability of Bourgain Wang solutions, [79]). *Let $N = 1, 2$. Let u^* be a smooth radially symmetric satisfying the degeneracy at blow up point* (4.18). *Let $u^0_{BW} \in \mathcal{C}((-\infty, 0), H^1)$ be the corresponding Bourgain-Wang. solution. Then there exists a continuous map*

$$\Gamma : [-1, 1] \to \Sigma$$

such that the following holds true. Given $\eta \in [-1, 1]$, let $u_\eta(t)$ be the solution to (1.1) *with data $u_\eta(-1) = \Gamma(\eta)$, then:*

- *$\Gamma(0) = u^0_{BW}(-1)$ ie $\forall t < 0$, $u_{\eta=0}(t) = u^0_{BW}(t)$ is the Bourgain Wang solution on $(-\infty, 0)$ with blow up profile $S(t)$ and regular part u^*;*
- *$\forall \eta \in (0, 1]$, $u_\eta \in \mathcal{C}(\mathbb{R}, \Sigma)$ is global in time and scatters forward and backwards;*
- *$\forall \eta \in [-1, 0)$, $u_\eta \in \mathcal{C}((-\infty, T^*_\eta), \Sigma)$ scatters to the left and blows up in finite time $T^*_\eta < 0$ on the right in the log-log regime* (4.14) *with*

$$T^*_\eta \to 0 \quad as \quad \eta \to 0. \tag{4.20}$$

Note that this theorem describes the flow near the Bourgain Wang solution along one instability solution. A major open problem in the field is to describe the flow near the ground state Q. Theorem 4.4 is a first step towards the description of the flow near the Bourgain Wang solutions which itself is a very interesting open problem.

---

[8]This is a simple consequence of Strichartz estimates and the L^2 critical Cauchy theory of Cazenave-Weissler [13].

4.4. Structural instability of the log-log law. Another model with fundamental physical relevance, [**106**], is the Zakharov system in dimensions $N = 2, 3$:

$$\left\{ \begin{array}{l} iu_t = -\Delta u + nu \\ \frac{1}{c_0^2} n_{tt} = \Delta n + \Delta |u|^2 \end{array} \right. \tag{4.21}$$

for some fixed constant $0 < c_0 < +\infty$. In the limit $c_0 \to +\infty$, we formally recover (1.1). In dimension $N = 2$, this system displays a variational structure like (1.1), even though the scaling symmetry is destroyed by the wave coupling. In particular, a virial law in the spirit of (2.1) holds and yields finite time blow up for radial non positive energy initial data, see Merle [**69**]. Moreover, a one parameter family of blow up solutions has been constructed as a continuation of the exact $S(t)$ solution for (1.1), see Glangetas, Merle, [**24**]. These explicit solutions have blow up speed:

$$\|\nabla u(t)\|_{L^2} \sim \frac{C(u_0)}{T-t}$$

and appear to be *stable* from numerics, see Papanicolaou, Sulem, Sulem, Wang, [**94**]. Now from Merle, [**68**], *all finite time blow up solutions to (*4.21*) satisfy*

$$\|\nabla u(t)\|_{L^2} \geq \frac{C(u_0)}{T-t}.$$

In particular, there will be no log-log blow up solutions for (4.21). This fact suggests that in some sense, the Zakharov system provides a much more stable and robust blow up dynamics than its asymptotic limit (NLS). This fact enlightens the belief that the log-log law heavily relies on the specific algebraic structure of (1.1), and some non linear degeneracy properties will indeed be at the heart of our understanding of the blow up dynamics. Let us insist that the fine study of the singularity formation for the Zakharov system is mostly open, and in some sense it is the first towards the understanding of more physical and complicated systems related to Maxwell's equations.

4.5. Classification of the flow near Q: the case of the generalized KdV. We present in this section the recent series of results [**62**], [**61**], [**60**] which give a complete description of the flow near the ground for an L^2 critical problem: the generalized KdV equation

$$(gKdV) \left\{ \begin{array}{l} \partial_t u + (u_{xx} + u^5)_x = 0 \\ u_{|t=0} = u_0 \end{array} \right. , \quad (t, x) \in \mathbb{R} \times \mathbb{R}. \tag{4.22}$$

This problem admits the same L^2 norm and energy conservation laws like (NLS), and the same mass critical scaling. The solitary wave is here a traveling wave solution

$$u(t, x) = Q(x - t)$$

where Q is the one dimensional ground state

$$Q(x) = \left(\frac{3}{\text{ch}^2(2x)} \right)^{\frac{1}{4}}.$$

This model problem has been thoroughly studied by Martel and Merle in the pioneering breakthrough works [**55**], [**56**],[**57**],[**58**], [**59**]

as a toy model for which the pseudo conformal symmetry and the associated virial algebra are lost. The long standing open problem of the existence of blow up solutions was solved in [**70**], but the structure of the singularity formation was still

only poorly understood. We give in the series of works [**62**], [**61**], [**60**] a complete description of the flow near the ground state and expect that the obtained picture is canonical.
More precisely, let the set of initial data

$$\mathcal{A} = \left\{ u_0 = Q + \varepsilon_0 \text{ with } \|\varepsilon_0\|_{H^1} < \alpha_0 \text{ and } \int_{y>0} y^{10}\varepsilon_0^2 < 1 \right\},$$

and consider the L^2 tube around the family of solitary waves

$$\mathcal{T}_{\alpha^*} = \left\{ u \in H^1 \text{ with } \inf_{\lambda_0>0,\, x_0\in\mathbb{R}} \|u - \frac{1}{\lambda_0^{\frac{1}{2}}} Q\left(\frac{\cdot - x_0}{\lambda_0}\right)\|_{L^2} < \alpha^* \right\}.$$

We first claim the rigidity of the dynamics for data in $\mathcal{A}$:

THEOREM 4.5 (Rigidity of the flow in $\mathcal{A}$, [**62**]). *Let $0 < \alpha_0 \ll \alpha^* \ll 1$ and $u_0 \in \mathcal{A}$. Let $u \in \mathcal{C}([0,T), H^1)$ be the corresponding solution to* (4.22). *Then one of the following three scenarios occurs:*
(Blow up)*: the solution blows up in finite time $0 < T < +\infty$ in the universal regime*

$$\|u(t)\|_{H^1} = \frac{\ell(u_0) + o(1)}{T - t} \quad \text{as } t \to T, \quad \ell(u_0) > 0. \tag{4.23}$$

(Soliton)*: the solution is global $T = +\infty$ and converges asymptotically to a solitary wave.*
(Exit)*: the solution leaves the tube $\mathcal{T}_{\alpha^*}$ at some time $0 < t_u^* < +\infty$.*
Moreover, the scenarios (Blow up) *and* (Exit) *are stable by small perturbation of the data in $\mathcal{A}$.*

In other words, we obtain a complete classification of solutions with data in $\mathcal{A}$ which remain close in the L^2 critical sense to the manifold of solitary waves. It remains to understand the long time dynamics in the (Exit) regime. The first step is the *existence and uniqueness* of a minimal blow up element which is the generalization of the $S(t)$ dynamics for (NLS):

THEOREM 4.6 (Existence and uniqueness of the minimal mass blow up element, [**61**]).
(i) Existence. *There exists a solution $\tilde{S}(t) \in \mathcal{C}((0,+\infty), H^1)$ to* (4.22) *with minimal mass $\|\tilde{S}(t)\|_{L^2} = \|Q\|_{L^2}$ which blows up backward at the origin at the speed*

$$\|\nabla \tilde{S}(t)\|_{L^2} \sim \frac{1}{t} \quad \text{as } t \downarrow 0,$$

and is globally defined on the right in time.
(ii) Uniqueness. *Let $u_0 \in H^1$ with $\|u_0\|_{L^2} = \|Q\|_{L^2}$ and assume that the corresponding solution $u(t)$ to* (4.22) *blows up in finite time. Then*

$$u \equiv S$$

up to the symmetries of the flow.

In other words, we recover Merle's result in the absence of pseudo conformal symmetry, and the proof is here completely dynamical and deeply related to the analysis of the inhomogeneous NLS model in [**101**]. We now claim that $\tilde{S}$ is the *universal attractor* of all solutions in the (Exit) regime.

THEOREM 4.7 (Description of the (Exit) scenario, [**61**]). *Let $u(t)$ be a solution of* (4.22) *corresponding to the* (Exit) *scenario in Theorem* 4.6 *and let $t_u^* \gg 1$ be the corresponding exit time. Then there exist $\tau^* = \tau^*(\alpha^*)$ (independent of u) and (λ_u^*, x_u^*) such that*

$$\left\|(\lambda_u^*)^{\frac{1}{2}} u\left(t_u^*, \lambda_u^* x + x_u^*\right) - \tilde{S}(\tau^*, x)\right\|_{L^2} \leq \delta_I(\alpha_0),$$

where $\delta_I(\alpha_0) \to 0$ as $\alpha_0 \to 0$.

In fact a solution at the (Exit) time acquires a *specific* profile with a large defocusing spreading $\lambda_u^* \gg 1$ -coherent with dispersion-. Understanding the flow for u after the (Exit) is now equivalent to controlling the flow of $\tilde{S}(t)$ for large times. For (NLS), we can see on the formula (3.9) that $S(t)$ blows up at $t = 0$ and *scatters* as $t \to +\infty$. For (gKdV), we know from Theorem 4.6 that $\tilde{S}(t)$ is global as $t \to +\infty$, but scattering is not known. We however expect that this holds true, in which case because scattering is open in L^2 thanks to the Kenig, Ponce, Vega L^2 critical theory [**34**], we obtain the following:

COROLLARY 4.8. *Assume that $S(t)$ scatters as $t \to +\infty$. Then any solution in the* (Exit) *scenario is global for positive time and scatters as $t \to +\infty$.*

It is important to notice that the above results rely on the *explicit* computation of the solution in the various regimes, and not on algebraic virial type identities. Indeed we introduce the nonlinear decomposition of the flow

$$u(t,x) = \frac{1}{\lambda(t)^{\frac{1}{2}}}(Q + \varepsilon)\left(t, \frac{x - x(t)}{\lambda(t)}\right)$$

and show that to leading order, $\lambda(t)$ obeys the dynamical system

$$\lambda_{tt} = 0, \quad \lambda(0) = 1. \tag{4.24}$$

The three regimes (Exit), (Blow up), (Soliton) now correspond respectively to $\lambda_t(0) > 0$, $\lambda_t(0) < 0$ and the threshold dynamic $\lambda_t(0) = 0$.

Our last result shows that the universality of the leading order ODE (4.24) is valid *under the decay assumption $u_0 \in \mathcal{A}$ only*, and indeed the tail of slowly decaying data can interact with the solitary wave which for (KdV) is moving to the right, and this may lead to new exotic singular regimes:

THEOREM 4.9 (Exotic blow up regimes, [**60**]).
(i) Blow up in finite time: *for any $\nu > \frac{11}{13}$, there exists $u \in C((0,1], H^1)$ solution to* (4.22) *which blows up at $t = 0$ with speed*

$$\|u_x(t)\|_{L^2} \sim t^{-\nu} \quad as \quad t \to 0^+. \tag{4.25}$$

(ii) Blow up in infinite time: *there exists $u \in C([1,+\infty), H^1)$ solution of* (4.22) *growing up at $+\infty$ with speed*

$$\|u_x(t)\|_{L^2} \sim e^t \quad as \quad t \to +\infty. \tag{4.26}$$

For any $\nu > 0$, there exists $u \in C([1,+\infty), H^1)$ solution of (4.22) *blowing up at $+\infty$ with*

$$\|u_x(t)\|_{L^2} \sim t^{\nu} \quad as \quad t \to +\infty. \tag{4.27}$$

Such solutions can be constructed arbitrarily close in H^1 to the ground state solitary wave.

Note that this implies in particular that blow up can be arbitrarily slow.

We expect that the (KdV) picture is fairly general, and Theorem 4.4 is a first step towards a similar description for the mass critical NLS. Let also mention that in super critical regimes and large dimensions, Nakanishi and Schlag have obtained a related classification of the flow near the solitary wave which in particular involves a complete description of the scattering zone and its boundary.

5. The log log upper bound on blow up rate

Our aim in this section is to present a self contained proof of the first result contained in Theorem 4.3 for the mass critical problem and for small super critical mass initial data.

THEOREM 5.1 ([**71**],[**72**]). *Let $N \leq 4$. There exist universal constants $\alpha^*, C^* > 0$ such that the following holds true. Given $u_0 \in \mathcal{B}_{\alpha^*}$ with*

$$E_G(u) = E(u) - \frac{1}{2}\left(\frac{Im(\int \nabla u \overline{u})}{|u|_{L^2}}\right)^2 < 0, \tag{5.1}$$

then the corresponding solution $u(t)$ to (1.1) blows up in finite time $0 < T < +\infty$ and there holds for t close to T:

$$\|\nabla u(t)\|_{L^2} \leq C^* \left(\frac{\log|\log(T-t)|}{T-t}\right)^{\frac{1}{2}}. \tag{5.2}$$

This theorem is the first fundamental improvement on the virial law: it not only shows blow up in finite time of non positive energy solutions, it also gives an upper bound on the blow up rate which in particular rules out the $S(t)$ type of dynamic. Moreover the steps of the proof are in some sense canonical for our study.

The heart of our analysis will be to exhibit as a consequence of dispersive properties of (1.1) close to Q strong rigidity constraints for the dynamics of non positive energy solutions. These will in turn imply monotonicity properties, that is the existence of a Lyapounov function. The corresponding estimates will then allow us to prove blow up in a dynamical way and the sharp upper bound on the blow up speed will follow.

5.1. Existence of the geometrical decomposition. Let an initial data $u_0 \in \mathcal{B}_{\alpha^*}$ with $E_G(u_0) < 0$. First observe that up to a fixed Galilean transform, we may equivalently assume

$$E(u_0) < 0 \text{ and } Im \int \nabla u \overline{u_0} = 0. \tag{5.3}$$

Proposition 3.7 thus applies and implies for $t \in [0, T)$ the existence of a geometrical decomposition

$$u(t,x) = \frac{1}{\lambda_0^{\frac{N}{2}}(t)}(Q + \varepsilon_0)(t, \frac{x - x_0(t)}{\lambda_0(t)})e^{i\gamma_0(t)}, \quad \|\varepsilon_0\|_{H^1} \leq \delta(\alpha^*).$$

Let us observe that this geometrical decomposition is by no mean unique. Nevertheless, one can freeze and regularize this decomposition by choosing a set of orthogonality conditions on the excess of mass: this is the modulation argument which will be examined later on. Let us so far assume that we have a smooth decomposition of the solution: $\forall t \in [0, T)$,

$$u(t,x) = \frac{1}{\lambda(t)^{\frac{N}{2}}}(Q+\varepsilon)(t, \frac{x-x(t)}{\lambda(t)})e^{i\gamma(t)} \tag{5.4}$$

with

$$\lambda(t) \sim \frac{C}{\|\nabla u(t)\|_{L^2}} \quad \text{and} \quad \|\varepsilon(t)\|_{H^1} \leq \delta(\alpha^*) \to 0 \quad \text{as} \quad \alpha^* \to 0.$$

To study the blow up dynamic is now equivalent to understanding the coupling between the finite dimensional dynamic which governs the evolution of the geometrical parameters $(\lambda(t), \gamma(t), x(t))$ and the infinite dimensional dispersive dynamic which drives the excess of mass $\varepsilon(t)$.

To enlighten the main issues, let us rewrite (1.1) in the so-called rescaled variables. Let us introduce the rescaled time:

$$s(t) = \int_0^t \frac{d\tau}{\lambda^2(\tau)}.$$

It is elementary to check that whatever is the blow up behavior of $u(t)$, one always has:

$$s([0,T)) = \mathbb{R}^+.$$

Let us set:

$$v(s,y) = e^{i\gamma(t)}\lambda(t)^{\frac{N}{2}}u(t, \lambda(t)x + x(t)).$$

For a given function f, we introduce the generator of L^2 scaling

$$\Lambda f = \frac{N}{2}f + y \cdot \nabla f$$

then from direct computation, $u(t,x)$ solves (1.1) on $[0,T)$ iff $v(s,y)$ solves: $\forall s \geq 0$,

$$iv_s + \Delta v - v + v|v|^{\frac{4}{N}} = i\frac{\lambda_s}{\lambda}\Lambda v + i\frac{x_s}{\lambda} \cdot \nabla v + \tilde{\gamma}_s v, \tag{5.5}$$

where $\tilde{\gamma} = -\gamma - s$. Now $v(s,y) = Q(y) + \varepsilon(s,y)$ and we linearize (5.5) close to Q. The obtained system has the form:

$$i\varepsilon_s + L\varepsilon = i\frac{\lambda_s}{\lambda}\Lambda Q + \gamma_s Q + i\frac{x_s}{\lambda} \cdot \nabla Q + R(\varepsilon), \tag{5.6}$$

$R(\varepsilon)$ formally quadratic in ε, and $L = (L_+, L_-)$ is the matrix linearized operator closed to Q which has components:

$$L_+ = -\Delta + 1 - \left(1 + \frac{4}{N}\right)Q^{\frac{4}{N}}, \quad L_- = -\Delta + 1 - Q^{\frac{4}{N}}.$$

A standard approach is to think of equation (5.6) in the following way: it is essentially a linear equation forced by terms depending on the law for the geometrical parameters. The classical study of this kind of system relies on the understanding of the dispersive properties of the propagator e^{isL} of the linearized operator close to Q. In particular, one needs to exhibit its spectral structure. This has been partially done by Weinstein, [**110**], using the variational characterization of Q. The result is the following: L is a non self adjoint operator with a generalized eigenspace at zero. The eigenmodes are explicit and generated by the symmetries of the problem:

$$L_+(\Lambda Q) = -2Q \ \ \text{(scaling invariance)}, \ \ L_+(\nabla Q) = 0 \ \ \text{(translation invariance)},$$

$$L_-(Q) = 0 \text{ (phase invariance)}, \quad L_-(yQ) = -2\nabla Q \text{ (Galilean invariance)}.$$

An additional relation is induced by the pseudo-conformal symmetry:

$$L_-(|y|^2 Q) = -4\Lambda Q,$$

and this in turns implies the existence of an additional mode ρ solution to

$$L_+\rho = -|y|^2 Q.$$

These explicit directions induce "growing" solutions to the homogeneous linear equation $i\partial_s\varepsilon + L\varepsilon = 0$. More precisely, there exists a (2N+3) dimensional space S spanned by the above directions such that $H^1 = M \oplus S$ with $|e^{isL}\varepsilon|_{H^1} \leq C$ for $\varepsilon \in M$ and $|e^{isL}\varepsilon|_{H^1} \sim s^3$ for $\varepsilon \in S$. As each symmetry is at the heart of a growing direction, a first idea is to use the symmetries from modulation theory to a priori ensure that ε is orthogonal to S. Roughly speaking, the strategy to construct blow up solutions is then: chose the parameters λ, γ, x so as to get good a priori dispersive estimates on ε in order to build it from a fixed point scheme. Now the fundamental problem is that one has (2N+2) symmetries, but (2N+3) bad modes in the set S. Both constructions in [**8**] and [**95**] develop non trivial strategies to overcome this intrinsic difficulty of the problem.

Our strategy will be more non linear. On the basis of the decomposition (5.4)*, we will prove bounds on ε induced by the virial structure* (2.1)*. The proof will rely on non linear degeneracies of the structure of* (1.1) *around Q. Using then the Hamiltonian information $E_0 < 0$, we will inject these estimates into the finite dimensional dynamic which governs $\lambda(t)$ -which measures the size of the solution- and prove rigidity properties of Lyapounov type. This will then allow us to prove finite time blow up together with the control of the blow up speed.*

5.2. Choice of the blow up profile. Before exhibiting the modulation theory type of arguments, we present in this subsection a formal discussion regarding explicit solutions of equation (5.5) which is inspired from a discussion in [**106**]. This corresponds to a finite dimensional reduction of the problem which actually computes the leading order terms of the solution.

First, let us observe that the key geometrical parameter is λ which measures the size of the solution. Let us then set

$$-\frac{\lambda_s}{\lambda} = b$$

and look for solutions to a simpler version of (5.5):

$$iv_s + \Delta v - v + ib\left(\frac{N}{2}v + y\cdot\nabla v\right) + v|v|^{\frac{4}{N}} = 0.$$

From the orbital stability property, we want solutions which remain close to Q in H^1. Let us look for solutions of the form $v(s,y) = Q_{b(s)}(y)$ where the mappings $b \to Q_b$ and the law for $b(s)$ are the unknown. We think of b as remaining uniformly small and $Q_{b=0} = Q$. Injecting this ansatz into the equation, we get:

$$i\frac{db}{ds}\left(\frac{\partial \overline{Q}_b}{\partial b}\right) + \Delta\overline{Q}_{b(s)} - \overline{Q}_{b(s)} + ib(s)\left(\frac{N}{2}\overline{Q}_{b(s)} + y\cdot\nabla\overline{Q}_{b(s)}\right) + \overline{Q}_{b(s)}|\overline{Q}_{b(s)}|^{\frac{4}{N}} = 0.$$

To handle the linear group, we let $\overline{P}_{b(s)} = e^{i\frac{b(s)}{4}|y|^2}\overline{Q}_{b(s)}$ and solve:

$$(5.7)\quad i\frac{db}{ds}\left(\frac{\partial \overline{P}_b}{\partial b}\right) + \Delta \overline{P}_{b(s)} - \overline{P}_{b(s)} + \left(\frac{db}{ds} + b^2(s)\right)\frac{|y|^2}{4}\overline{P}_{b(s)} + \overline{P}_{b(s)}|\overline{P}_{b(s)}|^{\frac{4}{N}} = 0.$$

A remarkable fact related to the specific algebraic structure of (1.1) around Q is that (5.7) admits three solutions:

- The first one is $(b(s), \overline{P}_{b(s)}) = (0, Q)$, that is the ground state itself. This is just a consequence of the scaling invariance.
- The second one is $(b(s), \overline{P}_{b(s)}) = (\frac{1}{s}, Q)$. This non trivial solution is a rewriting of the explicit critical mass blow up solution $S(t)$ and is induced by the pseudo-conformal symmetry.
- The third one is given by $(b(s), \overline{P}_{b(s)}) = (b, \overline{P}_b)$ for some fixed non zero constant b and $\overline{P}_b$ satisfies:

$$(5.8)\quad \Delta \overline{P}_b - \overline{P}_b + \frac{b^2}{4}|y|^2\overline{P}_b + \overline{P}_b|\overline{P}_b|^{\frac{4}{N}} = 0.$$

This corresponds to self similar profiles. Indeed, recall that $b = -\frac{\lambda_s}{\lambda}$, so if b is frozen, we have from $\frac{ds}{dt} = \frac{1}{\lambda^2}$:

$$b = -\frac{\lambda_s}{\lambda} = -\lambda\lambda_t \ \text{ ie } \ \lambda(t) = \sqrt{2b(T-t)},$$

this is the scaling law for the blow up speed.

Now a crucial point again is -[**103**]- that the solutions to (5.8) never belong to L^2 from a logarithmic divergence at infinity:

$$|P_b(y)| \sim \frac{C(P_b)}{|y|^{\frac{N}{2}}} \quad \text{as} \quad |y| \to +\infty.$$

This behavior is a consequence of the oscillations induced by the linear group after the turning point $|y| \geq \frac{2}{|b|}$. Nevertheless, in the ball $|y| < \frac{2}{|b|}$, the operator $-\Delta + 1 - \frac{b^2|y|^2}{4}$ is coercive, and no oscillations will take place in this zone.

Because we track a log-log correction to the self similar law as an upper bound on the blow up speed, the profiles $\overline{Q}_b = e^{-i\frac{b}{4}|y|^2}\overline{P}_b$ with $\overline{P}_b$ solving (5.8) are natural candidates as refinements of the Q profile in the geometrical decomposition (4.11). Nevertheless, as they are not in L^2, we need to build a smooth localized version avoiding the non L^2 tail, what according to the above discussion is doable in the coercive zone $|y| < \frac{2}{|b|}$.

PROPOSITION 5.2 (Localized self similar profiles). *There exist universal constants $C > 0$, $\eta^* > 0$ such that the following holds true. For all $0 < \eta < \eta^*$, there exist constants $\nu^*(\eta) > 0$, $b^*(\eta) > 0$ going to zero as $\eta \to 0$ such that for all $|b| < b^*(\eta)$, let*

$$R_b = \frac{2}{|b|}\sqrt{1-\eta}, \quad R_b^- = \sqrt{1-\eta}R_b,$$

$B_{R_b} = \{y \in \mathbb{R}^N, \ \ |y| \leq R_b\}$. *Then there exists a unique radial solution Q_b to*

$$\begin{cases} \Delta Q_b - Q_b + ib\left(\frac{N}{2}Q_b + y\cdot\nabla Q_b\right) + Q_b|Q_b|^{\frac{4}{N}} = 0, \\ P_b = Q_b e^{i\frac{b|y|^2}{4}} > 0 \ \ in \ \ B_{R_b}, \\ Q_b(0) \in (Q(0) - \nu^*(\eta), Q(0) + \nu^*(\eta)), \quad Q_b(R_b) = 0. \end{cases}$$

Moreover, let a smooth radially symmetric cut-off function $\phi_b(x) = 0$ *for* $|x| \geq R_b$ *and* $\phi_b(x) = 1$ *for* $|x| \leq R_b^-$, $0 \leq \phi_b(x) \leq 1$ *and set*

$$\tilde{Q}_b(r) = Q_b(r)\phi_b(r),$$

then

$$\tilde{Q}_b \to Q \quad as \quad b \to 0$$

in some very strong sense, and $\tilde{Q}_b$ *satisfies*

$$\Delta \tilde{Q}_b - \tilde{Q}_b + ib(\tilde{Q}_b)_1 + \tilde{Q}_b|\tilde{Q}_b|^{\frac{4}{N}} = -\Psi_b \tag{5.9}$$

with

$$Supp(\Psi) \subset \{R_b^- \leq |y| \leq R_b\} \quad and \quad |\Psi_b|_{C^1} \leq e^{-\frac{C}{|b|}}.$$

Eventually, $\tilde{Q}_b$ *has supercritical mass:*

$$\int |\tilde{Q}_b|^2 = \int Q^2 + c_0 b^2 + o(b^2) \quad as \quad b \to 0 \tag{5.10}$$

for some universal constant $c_0 > 0$.

The meaning of this proposition is that one can build localized profiles $\tilde{Q}_b$ on the ball B_{R_b} which are a smooth function of b and approximate Q in a very strong sense as $b \to 0$, and these profiles satisfy the self similar equation up to an exponentially small term Ψ_b supported around the turning point $\frac{2}{b}$. The proof of this Proposition uses standard variational tools in the setting of non linear elliptic problems. In fact, the implicit function theorem would do the job as well, see [**95**].

Now one can think of making a formal expansion of $\tilde{Q}_b$ in terms of b, and the first term is non zero:

$$\frac{\partial \tilde{Q}_b}{\partial b}_{|b=0} = -\frac{i}{4}|y|^2 Q.$$

However, the energy of $\tilde{Q}_b$ is degenerated in b at all orders:

$$|E(\tilde{Q}_b)| \leq e^{-\frac{C}{|b|}}, \tag{5.11}$$

for some universal constant $C > 0$.

The existence of a one parameter family of profiles satisfying the self similar equation up to an exponentially small term and having an exponentially small energy is an algebraic property of the structure of (1.1) *around* Q *which is at the heart of the existence of the log-log regime.*

5.3. Modulation theory. We are now in position to exhibit the sharp decomposition needed for the proof of the log-log upper bound. From Theorem 3.7 and the proximity of $\tilde{Q}_b$ to Q in H^1, the solution $u(t)$ to (1.1) is for all time close to the four dimensional manifold

$$\mathcal{M} = \{e^{i\gamma}\lambda^{\frac{N}{2}}\tilde{Q}_b(\lambda y + x), \quad (\lambda, \gamma, x, b) \in \mathbb{R}_+^* \times \mathbb{R} \times \mathbb{R}^N \times \mathbb{R}\}.$$

We now sharpen the decomposition according to the following Lemma. In the sequel, we let

$$\varepsilon = \varepsilon_1 + i\varepsilon_2$$

be the real and imaginary parts decomposition.

LEMMA 5.3 (Non linear modulation of the solution close to $\mathcal{M}$). *There exist $\mathcal{C}^1$ functions of time $(\lambda, \gamma, x, b) : [0, T) \to (0, +\infty) \times \mathbb{R} \times \mathbb{R}^N \times \mathbb{R}$ such that:*

$$\forall t \in [0, T), \quad \varepsilon(t, y) = e^{i\gamma(t)} \lambda^{\frac{N}{2}}(t) u(t, \lambda(t)y + x(t)) - \tilde{Q}_{b(t)}(y) \tag{5.12}$$

satisfies:

(i)

$$\left(\varepsilon_1(t), \Lambda\Sigma_{b(t)}\right) + \left(\varepsilon_2(t), \Lambda\Theta_{b(t)}\right) = 0, \tag{5.13}$$

$$\left(\varepsilon_1(t), y\Sigma_{b(t)}\right) + \left(\varepsilon_2(t), y\Theta_{b(t)}\right) = 0, \tag{5.14}$$

$$-\left(\varepsilon_1(t), \Lambda^2\Theta_{b(t)}\right) + \left(\varepsilon_2(t), \Lambda^2\Sigma_{b(t)}\right) = 0, \tag{5.15}$$

$$-\left(\varepsilon_1(t), \Lambda\Theta_{b(t)}\right) + \left(\varepsilon_2(t), \Lambda\Sigma_{b(t)}\right) = 0, \tag{5.16}$$

where $\varepsilon = \varepsilon_1 + i\varepsilon_2$, $\tilde{Q}_b = \Sigma_b + i\Theta_b$ in terms of real and imaginary parts;

(ii) $|1 - \lambda(t)\frac{\|\nabla u(t)\|_{L^2}}{|\nabla Q|_{L^2}}| + \|\varepsilon(t)\|_{H^1} + |b(t)| \leq \delta(\alpha^*)$ *with* $\delta(\alpha^*) \to 0$ *as* $\alpha^* \to 0$.

Let us insist onto the fact that the reason for this precise choice of orthogonality conditions is a fundamental issue which will be addressed in the next section.

PROOF OF LEMMA 5.3. This Lemma follows the standard frame of modulation theory and is obtained from Theorem 3.7 using the implicit function theorem. From Theorem 3.7, there exist parameters $\gamma_0(t) \in \mathbb{R}$ and $x_0(t) \in \mathbb{R}^N$ such that with $\lambda_0(t) = \frac{|\nabla Q|_{L^2}}{\|\nabla u(t)\|_{L^2}}$,

$$\forall t \in [0, T), \quad \left| Q - e^{i\gamma_0(t)} \lambda_0(t)^{\frac{N}{2}} u(\lambda_0(t)x + x_0(t)) \right|_{H^1} < \delta(\alpha^*)$$

with $\delta(\alpha^*) \to 0$ as $\alpha^* \to 0$. Now we sharpen this decomposition using the fact that $\tilde{Q}_b \to Q$ in H^1 as $b \to 0$, i.e. we chose $(\lambda(t), \gamma(t), x(t), b(t))$ close to $(\lambda_0(t), \gamma_0(t), x_0(t), 0)$ such that

$$\varepsilon(t, y) = e^{i\gamma(t)} \lambda^{1/2}(t) u(t, \lambda(t)y + x(t)) - \tilde{Q}_{b(t)}(y)$$

is small in H^1 and satisfies suitable orthogonality conditions (5.13), (5.14), (5.15) and (5.16).The existence of such a decomposition is a consequence of the implicit function Theorem. For $\delta > 0$, let $V_\delta = \{v \in H^1(\mathbb{C}); \ |v - Q|_{H^1} \leq \delta\}$, and for $v \in H^1(\mathbb{C})$, $\lambda_1 > 0$, $\gamma_1 \in \mathbb{R}$, $x_1 \in \mathbb{R}^N$, $b \in \mathbb{R}$ small, define

$$\varepsilon_{\lambda_1, \gamma_1, x_1, b}(y) = e^{i\gamma_1} \lambda_1^{\frac{N}{2}} v(\lambda_1 y + x_1) - \tilde{Q}_b. \tag{5.17}$$

We claim that there exists $\overline{\delta} > 0$ and a unique C^1 map : $V_{\overline{\delta}} \to (1 - \overline{\lambda}, 1 + \overline{\lambda}) \times (-\overline{\gamma}, \overline{\gamma}) \times B(0, \overline{x}) \times (-\overline{b}, \overline{b})$ such that if $v \in V_{\overline{\delta}}$, there is a unique $(\lambda_1, \gamma_1, x_1, b)$ such that $\varepsilon_{\lambda_1, \gamma_1, x_1, b} = (\varepsilon_{\lambda_1, \gamma_1, x_1, b})_1 + i(\varepsilon_{\lambda_1, \gamma_1, x_1, b})_2$ defined as in (5.17) satisfies

$$\rho^1(v) = ((\varepsilon_{\lambda_1, \gamma_1, x_1, b})_1, \Lambda\Sigma_b) + ((\varepsilon_{\lambda_1, \gamma_1, x_1, b})_2, \Lambda\Theta_b) = 0,$$
$$\rho^2(v) = ((\varepsilon_{\lambda_1, \gamma_1, x_1, b})_1, y\Sigma_b) + ((\varepsilon_{\lambda_1, \gamma_1, x_1, b})_2, y\Theta_b) = 0,$$
$$\rho^3(v) = -\left((\varepsilon_{\lambda_1, \gamma_1, x_1, b})_1, \Lambda^2\Theta_b\right) + \left((\varepsilon_{\lambda_1, \gamma_1, x_1, b})_2, \Lambda^2\Sigma_b\right) = 0,$$
$$\rho^4(v) = ((\varepsilon_{\lambda_1, \gamma_1, x_1, b})_1, \Lambda\Theta_b) - ((\varepsilon_{\lambda_1, \gamma_1, x_1, b})_2, \Lambda\Sigma_b) = 0.$$

Moreover, there exists a constant $C_1 > 0$ such that if $v \in V_{\overline{\delta}}$, then $|\varepsilon_{\lambda_1, \gamma_1, x_1}|_{H^1} + |\lambda_1 - 1| + |\gamma_1| + |x_1| + |b| \leq C_1\overline{\delta}$. Indeed, we view the above functionals $\rho^1, \rho^2, \rho^3, \rho^4$ as functions of $(\lambda_1, \gamma_1, x_1, b, v)$. We first compute at $(\lambda_1, \gamma_1, x_1, b, v) = (1, 0, 0, 0, v)$:

$$\frac{\partial \varepsilon_{\lambda_1,\gamma_1,x_1,b}}{\partial x_1} = \nabla v, \quad \frac{\partial \varepsilon_{\lambda_1,\gamma_1,x_1,b}}{\partial \lambda_1} = \frac{N}{2}v + x\cdot\nabla v,$$
$$\frac{\partial \varepsilon_{\lambda_1,\gamma_1,x_1,b}}{\partial \gamma_1} = iv, \quad \frac{\partial \varepsilon_{\lambda_1,\gamma_1,x_1,b}}{\partial b} = -\left(\frac{\partial \tilde{Q}_b}{\partial b}\right)_{|b=0}.$$

Now recall that $(\tilde{Q}_b)_{|b=0} = Q$ and $\left(\frac{\partial \tilde{Q}_b}{\partial b}\right)_{|b=0} = -i\frac{|y|^2}{4}Q$. Therefore, we obtain at the point $(\lambda_1,\gamma_1,x_1,b,v) = (1,0,0,0,Q)$,

$$\frac{\partial \rho^1}{\partial \lambda_1} = |\Lambda Q|_2^2, \quad \frac{\partial \rho^1}{\partial \gamma_1} = 0, \quad \frac{\partial \rho^1}{\partial x_1} = 0, \frac{\partial \rho^1}{\partial b} = 0,$$
$$\frac{\partial \rho^2}{\partial \lambda_1} = 0, \quad \frac{\partial \rho^2}{\partial \gamma_1} = 0, \quad \frac{\partial \rho^2}{\partial x_1} = -\frac{1}{2}|Q|_2^2, \frac{\partial \rho^2}{\partial b} = 0,$$
$$\frac{\partial \rho^3}{\partial \lambda_1} = 0, \quad \frac{\partial \rho^3}{\partial \gamma_1} = -|\Lambda Q|_2^2, \quad \frac{\partial \rho^3}{\partial x_1} = 0, \frac{\partial \rho^3}{\partial b} = 0,$$
$$\frac{\partial \rho^4}{\partial \lambda_1} = 0, \quad \frac{\partial \rho^4}{\partial \gamma_1} = 0, \quad \frac{\partial \rho^4}{\partial x_1} = 0, \frac{\partial \rho^4}{\partial b} = \frac{1}{4}|yQ|_2^2.$$

The Jacobian of the above functional is non zero, thus the implicit function Theorem applies and conclusion follows. □

Let us now write down the equation satisfied by ε in rescaled variables. To simplify notations, we note

$$\tilde{Q}_b = \Sigma + \Theta$$

in terms of real and imaginary parts. We have: $\forall s \in \mathbb{R}_+$, $\forall y \in \mathbb{R}^N$,

$$\text{(5.18)}\quad b_s\frac{\partial \Sigma}{\partial b} + \partial_s\varepsilon_1 - M_-(\varepsilon) + b\Lambda\varepsilon_1 = \left(\frac{\lambda_s}{\lambda} + b\right)\Lambda\Sigma + \tilde{\gamma}_s\Theta + \frac{x_s}{\lambda}\cdot\nabla\Sigma + \left(\frac{\lambda_s}{\lambda} + b\right)\Lambda\varepsilon_1 + \tilde{\gamma}_s\varepsilon_2 + \frac{x_s}{\lambda}\cdot\nabla\varepsilon_1 + Im(\Psi) - R_2(\varepsilon)$$

$$\text{(5.19)}\quad b_s\frac{\partial \Theta}{\partial b} + \partial_s\varepsilon_2 + M_+(\varepsilon) + b\Lambda\varepsilon_2 = \left(\frac{\lambda_s}{\lambda} + b\right)\Lambda\Lambda\Theta - \tilde{\gamma}_s\Sigma + \frac{x_s}{\lambda}\cdot\nabla\Theta + \left(\frac{\lambda_s}{\lambda} + b\right)\Lambda\varepsilon_2 - \tilde{\gamma}_s\varepsilon_1 + \frac{x_s}{\lambda}\cdot\nabla\varepsilon_2 - Re(\Psi) + R_1(\varepsilon),$$

with $\tilde{\gamma}(s) = -s - \gamma(s)$. The linear operator close to $\tilde{Q}_b$ is now a deformation of the linear operator L close to Q and is $M = (M_+, M_-)$ with

$$M_+(\varepsilon) = -\Delta\varepsilon_1 + \varepsilon_1 - \left(\frac{4\Sigma^2}{N|\tilde{Q}_b|^2} + 1\right)|\tilde{Q}_b|^{\frac{4}{N}}\varepsilon_1 - \left(\frac{4\Sigma\Theta}{N|\tilde{Q}_b|^2}|\tilde{Q}_b|^{\frac{4}{N}}\right)\varepsilon_2,$$
$$M_-(\varepsilon) = -\Delta\varepsilon_2 + \varepsilon_2 - \left(\frac{4\Theta^2}{N|\tilde{Q}_b|^2} + 1\right)|\tilde{Q}_b|^{\frac{4}{N}}\varepsilon_2 - \left(\frac{4\Sigma\Theta}{N|\tilde{Q}_b|^2}|\tilde{Q}_b|^{\frac{4}{N}}\right)\varepsilon_1.$$

The formally quadratic in ε interaction terms are:

$$R_1(\varepsilon) = (\varepsilon_1+\Sigma)|\varepsilon+\tilde{Q}_b|^{\frac{4}{N}} - \Sigma|\tilde{Q}_b|^{\frac{4}{N}} - \left(\frac{4\Sigma^2}{N|\tilde{Q}_b|^2} + 1\right)|\tilde{Q}_b|^{\frac{4}{N}}\varepsilon_1 - \left(\frac{4\Sigma\Theta}{N|\tilde{Q}_b|^2}|\tilde{Q}_b|^{\frac{4}{N}}\right)\varepsilon_2,$$

$$R_2(\varepsilon) = (\varepsilon_2+\Theta)|\varepsilon+\tilde{Q}_b|^{\frac{4}{N}} - \Theta|\tilde{Q}_b|^{\frac{4}{N}} - \left(\frac{4\Theta^2}{N|\tilde{Q}_b|^2}+1\right)|\tilde{Q}_b|^{\frac{4}{N}}\varepsilon_2 - \left(\frac{4\Sigma\Theta}{N|\tilde{Q}_b|^2}|\tilde{Q}_b|^{\frac{4}{N}}\right)\varepsilon_1.$$

Two natural estimates may now be performed:

- First, we may rewrite the conservation laws in the rescaled variables and linearize the obtained identities close to Q. This will give crucial degeneracy estimates on some specific order one in ε scalar products.
- Next, we may inject the orthogonality conditions of Lemma 5.3 into the equations (5.18), (5.19). This will compute the geometrical parameters in their differential form $\frac{\lambda_s}{\lambda}, \tilde{\gamma}_s, \frac{x_s}{\lambda}, b_s$ in terms of ε: these are the so called modulation equations. This step requires estimating the non linear interaction terms. A crucial point here is to use the fact that the ground state Q is exponentially decreasing in space.

The outcome is the following:

LEMMA 5.4 (First estimates on the decomposition). *We have for all $s \geq 0$:*
(i) Estimates induced by the conservation of the energy and the momentum:

$$|(\varepsilon_1, Q)| \leq \delta(\alpha^*)\left(\int |\nabla\varepsilon|^2 + \int |\varepsilon|^2 e^{-|y|}\right)^{\frac{1}{2}} + e^{-\frac{C}{|b|}} + C\lambda^2|E_0|, \tag{5.20}$$

$$|(\varepsilon_2, \nabla Q)| \lesssim \delta(\alpha^*)\left(\int |\nabla\varepsilon|^2 + \int |\varepsilon|^2 e^{-|y|}\right)^{\frac{1}{2}}. \tag{5.21}$$

(ii) Estimate on the geometrical parameters in differential form:

$$\left|\frac{\lambda_s}{\lambda} + b\right| + |b_s| + |\tilde{\gamma}_s| \lesssim \left(\int |\nabla\varepsilon|^2 + \int |\varepsilon|^2 e^{-|y|}\right)^{\frac{1}{2}} + e^{-\frac{C}{|b|}}, \tag{5.22}$$

$$\left|\frac{x_s}{\lambda}\right| \lesssim \delta(\alpha^*)\left(\int |\nabla\varepsilon|^2 + \int |\varepsilon|^2 e^{-|y|}\right)^{\frac{1}{2}} + e^{-\frac{C}{|b|}}, \tag{5.23}$$

where $\delta(\alpha^*) \to 0$ as $\alpha^* \to 0$.

REMARK 5.5. *The exponentially small term in the degeneracy estimate (*5.20*) is in fact related to the value of $E(\tilde{Q}_b)$, so we use here in a fundamental way the non linear degeneracy estimate (*5.11*).*

Comments on Lemma 5.4:

1. $\dot{H}^1$ norm: The norm which appears in the estimates of Lemma 5.4 is essentially a local norm in space. The conservation of the energy indeed relates the $\int |\nabla\varepsilon|^2$ norm with the local norm. These two norms will turn out to play an equivalent role in the analysis. A key is that no global L^2 norm is needed so far.

2. Degeneracy of the translation shift: Comparing estimates (5.22) and (5.23), we see that the term induced by translation invariance is smaller than the ones induced by scaling and phase invariances. This non trivial fact is an outcome of our use of the Galilean transform to ensure the zero momentum condition (5.3).

5.4. The virial type dispersive estimate. We now turn to the proof of the dispersive virial type inequality at the heart of the proof of the log-log upper bound. This information will be obtained as a consequence of the virial structure of (1.1) in Σ.

Let us first recall that the virial identity (2.1) corresponds to two identities:

$$\frac{d^2}{dt^2}\int |x|^2|u|^2 = 4\frac{d}{dt}Im(\int x\cdot\nabla u\overline{u}) = 16E_0. \tag{5.24}$$

We want to understand what information can be extracted from this dispersive information in the variables of the geometrical decomposition.

To clarify the claim, let us consider an ε solution to the linear homogeneous equation

$$i\partial_s\varepsilon + L\varepsilon = 0 \tag{5.25}$$

where $L = (L_+, L_-)$ is the linearized operator close to Q. A dispersive information on ε may be extracted using a similar virial law like (2.1):

$$\frac{1}{2}\frac{d}{ds}Im(\int y\cdot\nabla\varepsilon\overline{\varepsilon}) = H(\varepsilon,\varepsilon), \tag{5.26}$$

where $H(\varepsilon,\varepsilon) = (\mathcal{L}_1\varepsilon_1,\varepsilon_1) + (\mathcal{L}_2\varepsilon_2,\varepsilon_2)$ is a Schrödinger type quadratic form decoupled in the real and imaginary parts with explicit Schrödinger operators:

$$\mathcal{L}_1 = -\Delta + \frac{2}{N}\left(\frac{4}{N}+1\right)Q^{\frac{4}{N}-1}y\cdot\nabla Q \ , \ \mathcal{L}_2 = -\Delta + \frac{2}{N}Q^{\frac{4}{N}-1}y\cdot\nabla Q.$$

Note that both these operators are of the form $-\Delta + V$ for some smooth well localized time independent potential $V(y)$, and thus from standard spectral theory, they both have a finite number of negative eigenvalues, and then continuous spectrum on $[0,+\infty)$. A simple outcome is then that given an $\varepsilon\in H^1$ which is orthogonal to all the bound states of $\mathcal{L}_1,\mathcal{L}_2$, then $H(\varepsilon,\varepsilon)$ is coercive, that is

$$H(\varepsilon,\varepsilon) \geq \delta_0\left(\int|\nabla\varepsilon|^2 + \int|\varepsilon|^2e^{-|y|}\right)$$

for some universal constant $\delta_0 > 0$. Now assume that for some reason -it will be in our case a consequence of modulation theory and the conservation laws-, ε is indeed for all times orthogonal to the bound states -and resonances...-, then injecting the coercive control of $H(\varepsilon,\varepsilon)$ into (5.26) yields:

$$\frac{1}{2}\frac{d}{ds}Im(\int y\cdot\nabla\varepsilon\overline{\varepsilon}) \geq \delta_0\left(\int|\nabla\varepsilon|^2 + \int|\varepsilon|^2e^{-|y|}\right). \tag{5.27}$$

Integrating this in time yields a standard dispersive information: a space time norm is controlled by a norm in space.

We want to apply this strategy to the full ε equation. There are two main obstructions.

First, it is not reasonable to assume that ε is orthogonal to the exact bound states of H. In particular, due to the right hand side in the ε equation, other second order terms will appear which will need be controlled. We thus have to exhibit a set of orthogonality conditions which ensures both the coercivity of the quadratic form

H and the control of these other second order interactions. Note that the number of orthogonality conditions we can ensure on ε is the number of symmetries plus the one from b. A first key is the following Spectral Property which has been proved in dimension $N = 1$ in [**71**] using the explicit value of Q and checked numerically for $N = 2, 3, 4$.

PROPOSITION 5.6 (Spectral Property). *Let $N = 1, 2, 3, 4$. There exists a universal constant $\delta_0 > 0$ such that $\forall \varepsilon = \varepsilon_1 + i\varepsilon_2 \in H^1$,*

$$\begin{aligned} H(\varepsilon,\varepsilon) &\geq \delta_0\left(\int |\nabla\varepsilon|^2 + \int |\varepsilon|^2 e^{-|y|}\right) - \frac{1}{\delta_0}\{(\varepsilon_1, Q)^2 + (\varepsilon_1, \Lambda Q)^2 + (\varepsilon_1, yQ)^2 \\ (5.28)\qquad &+ (\varepsilon_2, \Lambda Q)^2 + (\varepsilon_2, \Lambda^2 Q)^2 + (\varepsilon_2, \nabla Q)^2\}. \end{aligned}$$

To prove this property amounts first counting exactly the number of negative eigenvalues of each Schrödinger operator, and then prove that the specific chosen set of orthogonality conditions, which is not exactly the set of the bound states, is enough to ensure the coercivity of the quadratic form. Both these issues appear to be non trivial when Q is not explicit, but obvious to check numerically through the drawing of a small number (less than 10) explicit curves.

Then, the second major obstruction is the fact that the right hand side $Im(\int y \cdot \nabla\varepsilon\overline{\varepsilon})$ in (5.27) is an unbounded function of ε in H^1. This is a priori a major obstruction to the strategy, *but an additional non linear algebra inherited from the virial law (2.1) rules out this difficulty.*

The formal computation is as follows. Given a function $f \in \Sigma$, we let $\Phi(f) = Im(\int y \cdot \nabla f \overline{f})$. According to (5.26), we want to compute $\frac{d}{ds}\Phi(\varepsilon)$. Now from (5.24) and the conservation of the energy:

$$\forall t \in [0, T), \quad \Phi(u(t)) = 4E_0 t + c_0$$

for some constant c_0. The key observation is that the quantity $\Phi(u)$ is scaling, phase and also translation invariant from zero momentum assumption (5.3). Using (5.12), we get:

$$\forall t \in [0, T), \quad \Phi(\varepsilon + \tilde{Q}_b) = 4E_0 t + c_0.$$

We now expand this according to:

$$\Phi(\varepsilon + \tilde{Q}_b) = \Phi(\tilde{Q}_b) - 2(\varepsilon_2, \Lambda\Sigma) + 2(\varepsilon_1, \Lambda\Theta) + \Phi(\varepsilon).$$

A simple algebra yields:

$$\Phi(\tilde{Q}_b) = -\frac{b}{2}|y\tilde{Q}_b|_2^2 \sim -Cb$$

for some universal constant $C > 0$. Next, from the choice of orthogonality condition (5.16),

$$(\varepsilon_2, \Lambda\Sigma) - (\varepsilon_1, \Lambda\Theta) = 0.$$

We thus get using $\frac{dt}{ds} = \lambda^2$:

$$(\Phi(\varepsilon))_s \sim 4\lambda^2 E_0 + Cb_s.$$

In other words, to compute the a priori unbounded quantity $(\Phi(\varepsilon))_s$ *for the full non linear equation* is from the virial law equivalent to computing the time derivative of b_s, what of course makes now perfectly sense in H^1.

The virial dispersive structure on $u(t)$ in Σ thus induces a dispersive structure in $L^2_{loc} \cap \dot{H}^1$ on $\varepsilon(s)$ for the full non linear equation.

The key dispersive virial estimate is now the following.

PROPOSITION 5.7 (Local viriel estimate in ε). *There exist universal constants $\delta_0 > 0$, $C > 0$ such that for all $s \geq 0$, there holds:*

$$b_s \geq \delta_0 \left(\int |\nabla \varepsilon|^2 + \int |\varepsilon|^2 e^{-|y|} \right) - \lambda^2 E_0 - e^{-\frac{C}{|b|}}. \tag{5.29}$$

PROOF OF PROPOSITION 5.7. Using the heuristics, we can compute in a suitable way b_s using the orthogonality condition (5.16). The computation -see Lemma 5 in [**72**]- yields:

$$\begin{aligned}\frac{1}{4}|yQ|_2^2 b_s &= H(\varepsilon,\varepsilon) + 2\lambda^2|E_0| - \frac{x_s}{\lambda}\cdot\{(\varepsilon_2, \nabla\Lambda\Sigma) - (\varepsilon_1, \nabla\Lambda\Theta)\} \\ &- \left(\frac{\lambda_s}{\lambda} + b\right)\{(\varepsilon_2, \Lambda^2\Sigma) - (\varepsilon_1, \Lambda^2\Theta)\} - \tilde{\gamma}_s\{(\varepsilon_1, \Lambda\Sigma) + (\varepsilon_2, \Lambda\Theta)\} \\ &- (\varepsilon_1, Re\Lambda\Psi)) - (\varepsilon_2, Im(\Lambda\Psi)) + (l.o.t),\end{aligned} \tag{5.30}$$

where the lower order terms may be estimated from the smallness of ε in H^1:

$$|l.o.t| \leq \delta(\alpha^*) \left(\int |\nabla \varepsilon|^2 + \int |\varepsilon|^2 e^{-|y|} \right).$$

We now explain how the choice of orthogonality conditions and the conservation laws allow us to deduce (5.29).

step 1 Modulation theory for phase and scaling. The choice of orthogonality conditions (5.15), (5.13) has been made to cancel the two second order in ε scalar products in (5.30):

$$\left(\frac{\lambda_s}{\lambda} + b\right)\{(\varepsilon_2, \Lambda^2\Sigma) - (\varepsilon_1, \Lambda^2\Theta)\} + \tilde{\gamma}_s\{(\varepsilon_1, \Lambda\Sigma) + (\varepsilon_2, \Lambda\Theta)\} = 0.$$

step 2 Elliptic estimate on the quadratic form H. We now need to control the negative directions in the quadratic form as given by Proposition 5.6. The directions $(\varepsilon_1, \Lambda Q)$, (ε_1, yQ), $(\varepsilon_2, \Lambda^2 Q)$ and $(\varepsilon_2, \Lambda Q)$ are treated thanks to the choice of orthogonality conditions and the closeness of $\tilde{Q}_b$ to Q for $|b|$ small. For example,

$$\begin{aligned}(\varepsilon_2, \Lambda Q)^2 &= |\{(\varepsilon_2, \Lambda Q - \Lambda\Sigma) + (\varepsilon_1, \Lambda\Theta)\} + (\varepsilon_2, \Lambda\Sigma) - (\varepsilon_1, \Lambda\Theta)|^2 \\ &= |(\varepsilon_2, \Lambda Q - \Lambda\Sigma) + (\varepsilon_1, \Lambda\Theta)|^2\end{aligned}$$

so that

$$(\varepsilon_2, \Lambda Q)^2 \leq \delta(\alpha^*)(\int |\nabla\varepsilon|^2 + \int |\varepsilon|^2 e^{-|y|}).$$

Similarly, we have:

$$(\varepsilon_1, yQ)^2 + (\varepsilon_2, \Lambda^2 Q)^2 + (\varepsilon_1, \Lambda Q)^2 \leq \delta(\alpha^*)(\int |\nabla\varepsilon|^2 + \int |\varepsilon|^2 e^{-|y|}). \tag{5.31}$$

The negative direction $(\varepsilon_1, Q)^2$ is treated from the conservation of the energy which implied (5.20). The direction $(\varepsilon_2, \nabla Q)$ is treated from the zero momentum condition which ensured (5.21). Putting this together yields:

$$(\varepsilon_1, Q)^2 + (\varepsilon_2, \nabla Q)^2 \leq \delta(\alpha^*)\left(\int |\nabla\varepsilon|^2 + \int |\varepsilon|^2 e^{-|y|} + \lambda^2|E_0|\right) + e^{-\frac{C}{|b|}}.$$

step 3 Modulation theory for translation and use of Galilean invariance. The Galilean invariance has been used to ensure the zero momentum condition (5.3) which in turn led together with the choice of orthogonality condition (5.14) to the degeneracy estimate (5.23):

$$|\frac{x_s}{\lambda}| \lesssim \delta(\alpha^*)(\int |\nabla\varepsilon|^2 + \int |\varepsilon|^2 e^{-|y|})^{\frac{1}{2}} + e^{-\frac{C}{|b|}}.$$

Therefore, we estimate the term induced by translation invariance in (5.30) as

$$\left|\frac{x_s}{\lambda} \cdot \{(\varepsilon_2, \nabla\Lambda\Sigma) - (\varepsilon_1, \nabla\Lambda\Theta)\}\right| \lesssim \delta(\alpha^*)\left(\int |\nabla\varepsilon|^2 + \int |\varepsilon|^2 e^{-|y|}\right) + e^{-\frac{C}{|b|}}.$$

step 4 Conclusion. Injecting these estimates into the elliptic estimate (5.28) yields so far:

$$b_s \geq \delta_0\left(\int |\nabla\varepsilon|^2 + \int |\varepsilon|^2 e^{-|y|}\right) - 2\lambda^2 E_0 - e^{-\frac{C}{|b|}} - \frac{1}{\delta_0}(\lambda^2 E_0)^2.$$

We now use in a crucial way *the sign of the energy* $E_0 < 0$ and the smallness $\lambda^2|E_0| \leq \delta(\alpha^*)$ which is a consequence of the conservation of the energy to conclude. □

5.5. Monotonicity and control of the blow up speed. The virial dispersive estimate (5.29) means a control of the excess of mass ε by an exponentially small correction in b in time averaging sense. More specifically, this means that in rescaled variables, the solution writes $\tilde{Q}_b + \varepsilon$ where $\tilde{Q}_b$ is the regular deformation of Q and the rest is in a suitable norm exponentially small in b. This is thus an expansion of the solution with respect to an internal parameter in the problem: b.

This virial control is the first dispersive estimate for the infinite dimensional dynamic driving ε. Observe that it means little by itself if nothing is known about $b(t)$. We shall now inject this information into the finite dimensional dynamic driving the geometrical parameters. The outcome will be *a rigidity property for the parameter* $b(t)$ *which will in turn imply the existence of a Lyapounov functional in the problem.* This step will again heavily rely on the conservation of the energy.

We start with exhibiting the rigidity property which proof is a maximum principle type of argument.

PROPOSITION 5.8 (Rigidity property for b). *$b(s)$ vanishes at most once on* $\mathbb{R}_+$.

Note that the existence of a quantity with prescribed sign in the description of the dynamic is unexpected. Indeed, b is no more then the projection of some a priori highly oscillatory function onto a prescribed direction. It is a very specific feature of the blow up dynamic that this projection has a fixed sign.

Proof of Proposition 5.8. Assume that there exists some time $s_1 \geq 0$ such that $b(s_1) = 0$ and $b_s(s_1) \leq 0$, then from (5.29), $\varepsilon(s_1) = 0$. Thus from the conservation of the L^2 norm and $\tilde{Q}_{b(s_1)} = Q$, we conclude $\int |u_0|^2 = \int Q^2$ what contradicts the strictly negative energy assumption. □

The next step is to get the exact sign of b. This is done by injecting the virial dispersive information (5.29) into the modulation equation for the scaling parameter what will yield

$$-\frac{\lambda_s}{\lambda} \sim b. \tag{5.32}$$

The key rigidity property is the following:

Proposition 5.9 (Rigidity of the flow). *There exists a time $s_0 \geq 0$ such that*

$$\forall s > s_0, \quad b(s) > 0.$$

Moreover, the size of the solution is in this regime an almost Lyapounov functional in the sense that:

$$\forall s_2 \geq s_1 \geq s_0, \quad \lambda(s_2) \leq 2\lambda(s_1). \tag{5.33}$$

Proof of Proposition 5.9. **step 1** Equation for the scaling parameter. The modulation equation for the scaling parameter λ inherited from choice of orthogonality condition (5.13) implied control (5.22):

$$\left|\frac{\lambda_s}{\lambda} + b\right| \lesssim \left(\int |\nabla\varepsilon|^2 + \int |\varepsilon|^2 e^{-|y|}\right)^{\frac{1}{2}} + e^{-\frac{C}{|b|}},$$

which implies (5.32) in a weak sense. Nevertheless, this estimate is not good enough to possibly use the virial estimate (5.29). We claim using extra degeneracies of the equation that (5.22) can be improved for:

$$\left|\frac{\lambda_s}{\lambda} + b\right| \lesssim \left(\int |\nabla\varepsilon|^2 + \int |\varepsilon|^2 e^{-|y|}\right) + e^{-\frac{C}{|b|}} \tag{5.34}$$

step 2 Use of the virial dispersive relation and the rigidity property. We now inject the virial dispersive relation (5.29) into (5.34) to get:

$$\left|\frac{\lambda_s}{\lambda} + b\right| \lesssim b_s + e^{-\frac{C}{|b|}}.$$

We integrate this inequality in time to get: $\forall 0 \leq s_1 \leq s_2$,

$$\left|\log\left(\frac{\lambda(s_2)}{\lambda(s_1)}\right) + \int_{s_1}^{s_2} b(s)ds\right| \leq \frac{1}{4} + \int_{s_1}^{s_2} e^{-\frac{C}{|b(s)|}} ds. \tag{5.35}$$

The key is now to use the rigidity property of Proposition 5.8 to ensure that $b(s)$ has a fixed sign for $s \geq \tilde{s}_0$, and thus: $\forall s \geq \tilde{s}_0$,

$$\left|\int_{s_1}^{s_2} e^{-\frac{C}{|b(s)|}} ds\right| \leq \frac{1}{2}\left|\int_{s_1}^{s_2} b(s)ds\right|. \tag{5.36}$$

step 3 b is positive for s large enough. Assume that $\left|\int_0^{+\infty} b(s)ds\right| < +\infty$, then b has a fixed sign for $s \geq \tilde{s}_0$ and $|b_s| \leq C$, and thus: $b(s) \to 0$ as $s \to +\infty$. Now from (5.35) and (5.36), this implies that $|\log(\lambda(s))| \leq C$ as $s \to +\infty$, and in particular $\lambda(s) \geq \lambda_0 > 0$ for s large enough. Injecting this into virial control (5.29) for s large enough yields:

$$b_s \geq \frac{1}{2}|E_0|\lambda_0^2.$$

Integrating this on large time intervals contradicts the uniform boundedness of b. Here we have used again the assumption $E_0 < 0$. We thus have proved: $\left|\int_0^{+\infty} b(s)ds\right| = +\infty$.
Now assume that $b(s) < 0$ for all $s \geq \tilde{s}_1$, then from (5.35) and (5.36) again, we conclude that $\log(\lambda(s)) \to +\infty$ as $s \to +\infty$. Now from $\lambda(t) \sim \frac{1}{\|\nabla u(t)\|_{L^2}}$, this yields $\|\nabla u(t)\|_{L^2} \to 0$ as $t \to T$. But from Gagliardo-Nirenberg inequality and the conservation of the energy and the L^2 mass, this implies $E_0 = 0$, contradicting again the assumption $E_0 < 0$.

step 4 Almost monotonicity of the norm. We now are in position to prove (5.33). Indeed, injecting the sign of b into (5.35) and (5.36) yields in particular: $\forall s_0 \leq s_1 \leq s_2$,

$$\frac{1}{4} + \frac{1}{2}\int_{s_1}^{s_2} b(s)ds \leq -\log\left(\frac{\lambda(s_2)}{\lambda(s_1)}\right) \leq \frac{1}{4} + 2\int_{s_1}^{s_2} b(s)ds, \tag{5.37}$$

and thus:

$$\forall s_0 \leq s_1 \leq s_2, \quad -\log\left(\frac{\lambda(s_2)}{\lambda(s_1)}\right) \geq \frac{1}{4},$$

what yields (5.33). This concludes the proof of Proposition 5.9.

Note that from the above proof, we have obtained $\int_0^{+\infty} b(s)ds = +\infty$, and thus from (5.37):

$$\lambda(s) \to 0 \quad \text{as} \quad s \to \infty, \tag{5.38}$$

that is finite or infinite time blow up. *On the contrary to the virial argument, the blow up proof is no longer obstructive but completely dynamical, and relies mostly on the rigidity property of Proposition* 5.8. □

Let us now conclude the proof of Theorem 5.1. We need to prove finite time blow up together with the log-log upper bound (5.2) on blow up rate.

PROOF OF THEOREM 5.1. **step 1** Lower bound on $b(s)$. We claim: there exist some universal constant $C > 0$ and some time $s_1 > 0$ such that $\forall s \geq s_1$,

$$Cb(s) \geq \frac{1}{\log|\log(\lambda(s))|}. \tag{5.39}$$

Indeed, first recall (5.29). Now that we know the sign of $b(s)$ for $s \geq s_0$ from Proposition 5.9, and we may thus view (5.29) as a differential inequality for b for $s > s_0$:

$$b_s \geq -e^{-\frac{C}{b}} \geq -b^2 e^{-\frac{C}{2b}} \text{ ie } -\frac{b_s}{b^2}e^{\frac{C}{2b}} \leq 1.$$

We integrate this inequality from the non vanishing property of b and get for $s \geq \tilde{s}_1$ large enough:

$$e^{\frac{C}{b(s)}} \leq s + e^{\frac{C}{b(1)}} \lesssim s \text{ ie } b(s) \gtrsim \frac{1}{\log(s)}. \tag{5.40}$$

We now recall (5.37) on the time interval $[\tilde{s}_1, s]$:

$$\frac{1}{2}\int_{\tilde{s}_1}^{s} b \leq -\log(\frac{\lambda(s)}{\lambda(\tilde{s}_1)}) + \frac{1}{4} \leq -2\log(\lambda(s))$$

for $s \geq \tilde{s}_2$ large enough from $\lambda(s) \to 0$ as $s \to +\infty$. Inject (5.40) into the above inequality, we get for $s \geq \tilde{s}_3$

$$\frac{s}{\log(s)} \lesssim \int_{\tilde{s}_2}^{s} \frac{d\tau}{\log(\tau)} \leq \frac{1}{4}\int_{\tilde{s}_2}^{s} b \leq -\log(\lambda(s)) \text{ ie } |\log(\lambda(s))| \gtrsim \frac{s}{\log(s)}$$

and thus for s large

$$\log|\log(\lambda(s))| \geq \log(s) - \log(\log(s)) \geq \frac{1}{2}\log(s)$$

and conclusion follows from (5.40). This concludes the proof of (5.39).

step 2 Finite time blow up and control of the blow up speed. We first use the finite or infinite time blow up result (5.38) to consider a sequence of times $t_n \to T \in [0, +\infty]$ defined for n large such that

$$\lambda(t_n) = 2^{-n}.$$

Let $s_n = s(t_n)$ the corresponding sequence and $\bar{t}$ such that $s(\bar{t}) = s_0$ given by Proposition 5.9. Note that we may assume $n \geq \bar{n}$ such that $t_n \geq \bar{t}$. Remark that $0 < t_n < t_{n+1}$ from (5.33), and so $0 < s_n < s_{n+1}$. Moreover, there holds from (5.33)

$$(5.41) \qquad \forall s \in [s_n, s_{n+1}], \;\; 2^{-n-1} \leq \lambda(s) \leq 2^{-(n-1)}.$$

We now claim that (5.2) follows from a control from above of the size of the intervals $[t_n, t_{n+1}]$ for $n \geq \bar{n}$.
Let $n \geq \bar{n}$. (5.39) implies

$$\int_{s_n}^{s_{n+1}} \frac{ds}{\log|\log(\lambda(s))|} \lesssim \int_{s_n}^{s_{n+1}} b(s)ds.$$

(5.37) with $s_1 = s_n$ and $s_2 = s_{n+1}$ yields:

$$\frac{1}{2}\int_{s_n}^{s_{n+1}} b(s) \leq \frac{1}{4} - |yQ|_{L^2}^2 \log(\frac{\lambda(s_{n+1})}{\lambda(s_n)}) \lesssim 1.$$

Therefore,

$$\forall n \geq \bar{n}, \quad \int_{s_n}^{s_{n+1}} \frac{ds}{\log|\log(\lambda(s))|} \lesssim 1.$$

Now we change variables in the integral at the left of the above inequality according to $\frac{ds}{dt} = \frac{1}{\lambda^2(s)}$ and estimate with (5.41):

$$1 \gtrsim \int_{s_n}^{s_{n+1}} \frac{ds}{\log|\log(\lambda(s))|} = \int_{t_n}^{t_{n+1}} \frac{dt}{\lambda^2(t)\log|\log(\lambda(t))|}$$
$$\geq \frac{1}{10\lambda^2(t_n)\log|\log(\lambda(t_n))|}\int_{t_n}^{t_{n+1}} dt$$

so that

$$t_{n+1} - t_n \lesssim \lambda^2(t_n)\log|\log(\lambda(t_n))|.$$

From $\lambda(t_n) = 2^{-n}$ and summing the above inequality in n, we first get

$$T < +\infty$$

and

$$\begin{aligned} T - t_n &\lesssim \sum_{k \geq n} 2^{-2k} \log(k) = \sum_{n \leq k \leq 2n} 2^{-2k} \log(k) + \sum_{k \geq 2n} 2^{-2k} \log(k) \\ &\lesssim 2^{-2n} \log(n) + 2^{-4n} \log(2n) \sum_{k \geq 0} 2^{-2k} \frac{\log(2n+k)}{\log(2n)} \\ &\lesssim 2^{-2n} \log(n) + 2^{-4n} \log(n) \lesssim 2^{-2n} \log(n) \lesssim \lambda^2(t_n) \log|\log(\lambda(t_n))|. \end{aligned}$$

From the monotonicity of λ (5.33), we extend the above control to the whole sequence $t \geq \bar{t}$. Let $t \geq \bar{t}$, then $t \in [t_n, t_{n+1}]$ for some $n \geq \bar{n}$, and from $\frac{1}{2}\lambda(t_n) \leq \lambda(t) \leq 2\lambda(t_n)$, we conclude

$$\lambda^2(t) \log|\log(\lambda(t))| \gtrsim \lambda^2(t_n) \log|\log(\lambda(t_n))| \gtrsim T - t_n \gtrsim T - t.$$

Now remark that the function $f(x) = x^2 \log|\log(x)|$ is non decreasing in a neighborhood at the right of $x = 0$, and moreover

$$\begin{aligned} & f\left(\frac{C}{2}\sqrt{\frac{T-t}{\log|\log(T-t)|}}\right) \\ & \qquad = \frac{C^2}{4} \frac{(T-t)}{\log|\log(T-t)|} \log\left|\log\left(C\sqrt{\frac{T-t}{\log|\log(T-t)|}}\right)\right| \leq C(T-t) \end{aligned}$$

for t close enough to T, so that we get for some universal constant C^*:

$$f(\lambda(t)) \geq f\left(C^*\sqrt{\frac{T-t}{\log|\log(T-t)|}}\right) \quad \text{ie} \quad \lambda(t) \geq C^*\sqrt{\frac{T-t}{\log|\log(T-t)|}}$$

and (5.2) is proved.

□

Appendix

This Appendix is devoted to the proof of the concentration compactness Lemma, i.e. Proposition 1.6. We follow Cazenave [**11**].

Proof of Proposition 1.6. . Let $u_n \in H^1$ be as in the hypothesis of Proposition 1.6.

step 1 Concentration function. Let the sequence of concentration functions:

$$\rho_n(R) = \sup_{y \in \mathbb{R}^N} \int_{B(y,R)} |u_n(x)|^2 dx.$$

The following facts are elementary and left to the reader:

- Monotonicity: $\forall n \geq 0$, $\rho_n(R)$ is a nondecreasing function of R.
- The concentration point is attained:

$$\forall R > 0, \quad \forall n \geq 0, \quad \exists y_n(R) \in \mathbb{R}^N \quad \text{such that} \quad \rho_n(R) = \int_{B(y_n(R),R)} |u_n(x)|^2 dx.$$

- Uniform Hölder continuity: $\exists C, \alpha > 0$ independent of n such that

$$\forall R_1, R_2 > 0, \quad \forall n \geq 0, \quad |\rho_n(R_1) - \rho_n(R_2)| \leq C|R_1^N - R_2^N|^{\alpha}. \tag{5.42}$$

This last fact is a simple consequence of the H^1 bound (1.13).

step 2 Limit of concentration functions. From (5.42) and Ascoli's theorem, there exists a subsequence $n_k \to +\infty$ and a nondecreasing limit ρ such that

$$\forall R > 0, \quad \lim_{k \to +\infty} \rho_{n_k}(R) = \rho(R). \tag{5.43}$$

Let now

$$\mu = \lim_{R \to +\infty} \liminf_{n \to +\infty} \rho_n(R).$$

By definition, there exists $R_k \to +\infty$ such that

$$\lim_{k \to +\infty} \rho_{n_k}(R_k) = \mu.$$

We now claim some stability of the sequence R_k which is a very general and simple fact but crucial for the rest of the argument:

$$\mu = \lim_{k \to +\infty} \rho_{n_k}(R_k) = \lim_{k \to +\infty} \rho_{n_k}(\frac{R_k}{2}) = \lim_{R \to +\infty} \rho(R). \tag{5.44}$$

Proof of (5.44): First oberve from the monotonicity of ρ_{n_k} that

$$\limsup_{k \to +\infty} \rho_{n_k}(\frac{R_k}{2}) \leq \limsup_{k \to +\infty} \rho_{n_k}(R_k) = \mu. \tag{5.45}$$

For the other sense, we argue as follows. For every $R > 0$, there holds:

$$\rho(R) = \liminf_{k \to +\infty} \rho_{n_k}(R) \geq \liminf_{n \to +\infty} \rho_n(R)$$

and thus:

$$\lim_{R \to +\infty} \rho(R) \geq \mu. \tag{5.46}$$

Eventually, for any $R > 0$, we have $\frac{R_k}{2} \geq R$ for k large enough and thus:

$$\rho_{n_k}(\frac{R_k}{2}) \geq \rho_{n_k}(R).$$

Letting $k \to +\infty$ implies:

$$\forall R > 0, \quad \lim_{k \to +\infty} \rho_{n_k}(\frac{R_k}{2}) \geq \rho(R).$$

Letting now $R > 0$ yields:

$$\lim_{k \to +\infty} \rho_{n_k}(\frac{R_k}{2}) \geq \lim_{R \to +\infty} \rho(R) \geq \mu$$

where we used (5.46) in the last step. This together with (5.45) concludes the proof of (5.44).
The proof now proceed by making an hypothesis on μ.

Step 3: $\mu = 0$ is vanishing. Assume $\mu = 0$. Then from (5.44), $\lim_{R \to +\infty} \rho(R) = 0$. But ρ is nondecreasing positive so: $\forall R > 0$, $\rho(R) = 0$. In particular, $\rho(1) = 0$ and thus

$$\lim_{k \to +\infty} \rho_{n_k}(1) = \lim_{k \to +\infty} \sup_{y \in \mathbb{R}^N} \int_{B(y,1)} |u_{n_k}|^2 = 0. \tag{5.47}$$

We claim that this together with the H^1 bound on u_{n_k} implies (1.15). There is a slight difficulty here which is that we need to pass from a local information -vanishing on every ball- to a global information -vanishing of the global L^q norm-. This relies on a refinement of the Gagliardo Nirenberg interpolation inequality. Indeed, we claim that

$$\forall u \in H^1, \quad \int |u|^{2+\frac{4}{N}} \lesssim \|u\|_{H^1}^2 \|u\|_{L^2}^{\frac{4}{N}} \tag{5.48}$$

can be refined for:

$$\forall u \in H^1, \quad \int |u|^{2+\frac{4}{N}} \lesssim \|u\|_{H^1}^2 \left[\sup_{y\in\mathbb{R}^N} \int_{B(y,1)} |u|^2\right]^{\frac{2}{N}}. \tag{5.49}$$

This together with (5.47) implies

$$u_{n_k} \to 0 \text{ in } L^{2+\frac{4}{N}} \text{ as } k \to +\infty$$

and (1.15) follows by interpolation using the global H^1 bound.
Proof of (5.49): Let a partition of $\mathbb{R}^d$ with disjoint rectangles Q_j of side $\frac{1}{2}$. Assume $N \geq 3$ and write Hölder noticing:

$$\frac{1}{2+\frac{4}{N}} = \frac{\alpha}{2} + \frac{1-\alpha}{\frac{2N}{N-2}} \text{ with } \alpha = \frac{2}{N+2}$$

so that

$$\|u\|_{L^{2+\frac{4}{N}}(Q_i)} \lesssim \|u\|_{L^2(Q_j)}^{\alpha} \|u\|_{L^{2^*}(Q_j)}^{1-\alpha}$$

and hence using Sobolev in Q_j:

$$\|u\|_{L^{2+\frac{4}{d}}}^{2+\frac{4}{d}} \lesssim \|u\|_{L^2(Q_j)}^{\frac{4}{d}} \|u\|_{H^1(Q_j)}^2$$

where the Sobolev constant does not depend on j thanks to the translation invariance of Lebesgue's mesure. We may now sum on the disjoint cubes:

$$\begin{aligned}\int |u|^{2+\frac{4}{N}} dx &= \Sigma_{j\geq 1} \int_{Q_j} |u|^{2+\frac{4}{N}} dx \lesssim \left[\sup_{j\geq 1} \|u\|_{L^2(Q_j)}^2\right]^{\frac{2}{d}} \Sigma_{j\geq 1} \|u\|_{H^1(Q_j)}^2 \\ &= \left[\sup_{j\geq 1} \|u\|_{L^2(Q_j)}^2\right]^{\frac{2}{N}} \|u\|_{H^1}^2\end{aligned}$$

and (5.49) is proved. The cases $N = 1, 2$ is similar and left to the reader.

Step 4: $\mu = M$ is compactness. Let n_k be the sequence satisfying (5.43). For $R > 0$, let $y_k(R)$ such that

$$\rho_{n_k}(R) = \int_{B(y_k(R),R)} |u_{n_k}(x)|^2 dx. \tag{5.50}$$

Pick $\varepsilon > 0$. Then from (5.44), there exist $R_0, R(\varepsilon)$ such that

$$\rho(R_0) > \frac{M}{2}, \quad \rho(R(\varepsilon)) > M - \varepsilon.$$

Hence there exists $k_0(\varepsilon)$ such that $\forall k \geq k_0(\varepsilon)$,

$$\rho_{n_k}(R_0) = \int_{B(y_k(R_0),R_0)} |u_{n_k}|^2 > \frac{M}{2}, \quad \rho_{n_k}(R(\varepsilon)) = \int_{B(y_k(R(\varepsilon)),R(\varepsilon))} |u_{n_k}|^2 > M - \varepsilon.$$

But the total L^2 mass being M, this implies that the balls $B(y_k(R_0), R_0)$ and $B(y_k(R(\varepsilon)), R(\varepsilon))$ cannot be disjoint. Hence -draw a picture- we can find $R_1(\varepsilon)$ such that:

$$\forall \varepsilon > 0, \ \ \forall k \geq k_0(\varepsilon), \ \ \int_{B(y_k(R_0), R_1(\varepsilon))} |u_{n_k}|^2 \geq M - \varepsilon.$$

By possibly raising the value of $R_1(\varepsilon)$ for the values $k \in [1, k_0(\varepsilon)]$, this implies that the sequence $v_k = u_{n_k}(\cdot + y_k(R_0))$ is L^2 compact:

$$\forall \varepsilon > 0, \ \ \exists R_2(\varepsilon) > 0 \ \text{ such that } \ \forall k \geq 1, \ \ \int_{|y| \geq R_2(\varepsilon)} |v_k(y)|^2 dy < \varepsilon.$$

The compactness of the embedding $H^1 \hookrightarrow L^2(B(0, R(\varepsilon)))$ then implies that v_k a Cauchy sequence in L^2, and the H^1 boundedness now implies (1.14) by interpolation.

Step 5: $0 < \mu < M$ is dichotomy. Let again (n_k, R_k) satisfying (5.43), (5.44). Then we can write:

$$u_{n_k} = v_k + w_k + z_k$$

with

$$v_k = u_{n_k} \mathbf{1}_{|y - y_k(\frac{R_k}{2})| \leq \frac{R_k}{2}}, \quad w_k = u_{n_k} \mathbf{1}_{|y - y_k(\frac{R_k}{2})| \geq R_k}, \quad z_k = u_{n_k} \mathbf{1}_{\frac{R_k}{2} < |y - y_k(\frac{R_k}{2})| < R_k}.$$

The key is to observe from (5.50) and (5.44) that:

$$\begin{aligned} \int |z_k|^2 &= \int_{B(y_k(\frac{R_k}{2}), R_k)} |u_{n_k}|^2 - \int_{B(y_k(\frac{R_k}{2}), \frac{R_k}{2})} |u_{n_k}|^2 \\ &\leq \rho_{n_k}(R_k) - \int_{B(y_k(\frac{R_k}{2}), \frac{R_k}{2})} |u_{n_k}|^2 = \rho_{n_k}(R_k) - \rho_{n_k}(\frac{R_k}{2}) \\ &\to 0 \ \text{ as } \ k \to +\infty. \end{aligned}$$

The claim dichotomy now follows by taking smooth cut off in the localization. The L^p norm of z_k will go to zero using the vanishing of the L^2 norm and the global H^1 bound, and the error introduced by localization will be treated using $R_k \to +\infty$. This is left to the reader.
This concludes the proof of Proposition 1.6. □

References

[1] Banica, V., Remarks on the blow-up for the Schrödinger equation with critical mass on a plane domain, Ann. Sc. Norm. Super. Pisa Cl. Sci. (5) 3 (2004), no. 1, 139–170.
[2] Boulenger, T., PhD Thesis, Paris XI (2012).
[3] Berestycki, H.; Lions, P.-L., Nonlinear scalar field equations. I. Existence of a ground state. Arch. Rational Mech. Anal. 82 (1983), no. 4, 313–345.
[4] Berestycki, H.; Lions, P.-L.; Peletier, L. A., An ODE approach to the existence of positive solutions for semilinear problems in R^N. Indiana Univ. Math. J. 30 (1981), no. 1, 141–157.
[5] Blanchet, A.; Carrillo, J.A.; Masmoudi, N., Infinite time aggregation for the critical Patlak-Keller-Segel model in $\mathbb{R}^2$, Comm. Pure Appl. Math. 61 (2008), no. 10, 1449–1481.
[6] Bourgain, J., Refinements of Strichartz' inequality and applications to 2D-NLS with critical nonlinearity, Internat. Math. Res. Notices 1998, no. 5, 253–283
[7] Bourgain, J., Problems in Hamiltonian PDE's, Visions in mathematics towards 2000, Proceedings of the meeting held at Tel Aviv University, Tel Aviv, August 25–September 3, 1999. Edited by N. Alon, J. Bourgain, A. Connes, M. Gromov and V. Milman, Geom. Funct. Anal Special Volume (2000), 32–56.

[8] Bourgain, J.; Wang, W., Construction of blowup solutions for the nonlinear Schrödinger equation with critical nonlinearity. Ann. Scuola Norm. Sup. Pisa Cl. Sci. (4) 25 (1997), no. 1-2, 197–215 (1998).

[9] Burq, N.; Gérard, P.; Tzvetkov, N., Two singular dynamics of the nonlinear Schrödinger equation on a plane domain, Geom. Funct. Anal. 13 (2003), no. 1, 1–19.

[10] Buslaev, V.S.; Sulem, C., On asymptotic stability of solitary waves for nonlinear Schrödinger equations, Ann. Inst. H. Poincaré Anal. Non Linéaire 20 (2003), no. 3, 419–475.

[11] Cazenave, Th.; Semilinear Schrödinger equations, Courant Lecture Notes in Mathematics, 10, NYU, CIMS, AMS 2003.

[12] Cazenave, T; Lions, P.-L, Orbital stability of standing waves for some nonlinear Schrödinger equations. Comm. Math. Phys. 85 (1982), no. 4, 549–561.

[13] Cazenave, Th.; Weissler, F., Some remarks on the nonlinear Schrödinger equation in the critical case. Nonlinear semigroups, partial differential equations and attractors (Washington, DC, 1987), 18–29, Lecture Notes in Math., 1394, Springer, Berlin, 1989.

[14] Dodson, B., Global well-posedness and scattering for the mass critical nonlinear Schrödinger equation with mass below the mass of the ground state, arXiv:1104.1114 (2011)

[15] Duyckaerts, T; Kenig, C.E.; Merle, F., Profiles of bounded radial solutions of the focusing, energy-critical wave equation, arXiv:1201.4986 (2012)

[16] Escauriaza, L.; Seregin, G. A.; Svverak, V., $L_{3,\infty}$-solutions of Navier-Stokes equations and backward uniqueness, Uspekhi Mat. Nauk 58 (2003), no. 2(350), 3–44; translation in Russian Math. Surveys 58 (2003), no. 2, 211–250.

[17] Fibich, G.; Gavish, N,; Wang, X-P., Singular ring solutions of critical and supercritical nonlinear Schrödinger equations. Phys. D 231 (2007), no. 1, 55–86.

[18] Fibich, G.; Merle, F.; Raphaël, P., Proof of a spectral property related to the singularity formation for the L2 critical nonlinear Schrödinger equation, Phys. D 220 (2006), no. 1, 1–13.

[19] Frank, R.; Lenzman, E., Uniqueness and Nondegeneracy of Ground States for $(-?)^s Q + Q - Q^{?+1} = 0$ in $\mathbb{R}$ arXiv:1009.4042

[20] Gang, Z.; Sigal, I. M., On soliton dynamics in nonlinear Schrödinger equations, Geom. Funct. Anal. 16 (2006), no. 6, 1377–1390.

[21] Gerard, Description du défaut de compacité de l'injection de Sobolev, ESAIM Control Optom. Calc. Var., 3 (1998), pp 213-233.

[22] Gidas, B.; Ni, W.M.; Nirenberg, L., Symmetry and related properties via the maximum principle, Comm. Math. Phys. **68** (1979), 209—243.

[23] Ginibre, J.; Velo, G., On a class of nonlinear Schrödinger equations. I. The Cauchy problem, general case. J. Funct. Anal. 32 (1979), no. 1, 1–32.

[24] Glangetas, L.; Merle, F., Existence of self-similar blow-up solutions for Zakharov equation in dimension two. I. Comm. Math. Phys. 160 (1994), no. 1, 173–215.

[25] Glassey, R.T.; Schaeffer, J., On symmetric solutions of the relativistic Vlasov-Poisson system, Comm. Math. Phys. 101 (1985), no. 4, 459–473.

[26] Grillakis, M.; Shatah, J.; Strauss, W., Stability theory of solitary waves in the presence of symmetry. I. J. Funct. Anal. 74 (1987), no. 1.

[27] Hmidi, T.; Keraani, S., Blowup theory for the critical nonlinear Schrd̈inger equations revisited, Int. Math. Res. Not. 2005, no. 46, 2815–2828.

[28] Hmidi, T.; Keraani, S., Remarks on the blow up for the L^2 critical nonlinear Schrödinger equation, SIAM J. Math. Anal, vol. 38, no 4, pp 1035-1047.

[29] Holmer, J.; Roudenko, S., A class of solutions to the 3D cubic nonlinear Schrödinger equation that blows up on a circle, Appl. Math. Res. Express. AMRX 2011, no. 1, 23–94.

[30] Karageorgis, P.; Strauss, W.A., Instability of steady states for nonlinear wave and heat equations, J. Differential Equations 241 (2007), no. 1, 184–205.

[31] Kato, T., On nonlinear Schrödinger equations. Ann. Inst. H. Poincaré Phys. Theor. 46 (1987), no. 1, 113–129.

[32] Kavian, O., Introduction à la théorie des points critiques et applications aux problèmes elliptiques, Mathématiques and Applications (Berlin) [Mathematics and Applications], 13. Springer-Verlag, Paris, 1993.

[33] Kenig, C.E.; Merle, F., Global well-posedness, scattering and blow-up for the energy-critical, focusing, non-linear Schrödinger equation in the radial case, Invent. Math. 166 (2006), no. 3, 645–675.

[34] C.E. Kenig, G. Ponce and L. Vega, On the concentration of blow up solutions for the generalized KdV equation critical in L^2, Nonlinear wave equations (Providence, RI, 1998), 131–156, Contemp. Math., **263**, Amer. Math. Soc., Providence, RI, 2000.
[35] Killip, R.; ; Tao, T.; Visan, M., The cubic nonlinear Schrödinger equation in two dimensions with radial data, J. Eur. Math. Soc. (JEMS) 11 (2009), no. 6, 1203–1258.
[36] Killip, R.; Li, D.; Visan, M.; Zhang, X., Characterization of minimal-mass blowup solutions to the focusing mass-critical NLS, SIAM J. Math. Anal. 41 (2009), no. 1, 219–236.
[37] Killip, R.; Visan, M.; Zhang, X.,The mass-critical nonlinear Schrödinger equation with radial data in dimensions three and higher, Anal. PDE 1 (2008), no. 2, 229–266.
[38] Killip, R.; Stovall, B.; Visan, M., Blowup behaviour for the nonlinear Klein–Gordon equation, arXiv:1203.4886 (2012).
[39] Krieger, J.; Lenzmann, E.; Raphaël, P., Nondispersive solutions to the L2-critical half-wave equation, arXiv:1203.2476 (2012)
[40] Krieger, J.; Schlag, J., Non generic blow up solutions for the critical focusing NLS in 1d, J. Eur. Math. Soc. (JEMS) 11 (2009), no. 1, 1–125.
[41] Krieger, J.; Schlag, W.; Tataru, D. Renormalization and blow up for charge one equivariant critical wave maps. Invent. Math. 171 (2008), no. 3, 543–615.
[42] Kwong, M. K., Uniqueness of positive solutions of $\Delta u - u + u^p = 0$ in R^n, Arch. Rational Mech. Anal. 105 (1989), no. 3, 243–266.
[43] Landman, M. J.; Papanicolaou, G. C.; Sulem, C.; Sulem, P.-L., Rate of blowup for solutions of the nonlinear Schrödinger equation at critical dimension. Phys. Rev. A (3) 38 (1988), no. 8, 3837–3843.
[44] Lemou, M.; Méhats, F.; Raphaël, P., On the orbital stability of the ground states and the singularity formation for the gravitational Vlasov-Poisson system, to appear in Arch. Rat. mech. Anal.
[45] Lemou, M.; Méhats, F.; Raphaël, P., Stable self similar blow up dynamics for the relativistic gravitational Vlasov-Poisson system, to appear in Jour. Amer. Math. Soc.
[46] Lemou, M.; Méhats, F.; Raphaël, P., A new variational approach to the stability of gravitational systems. Comm. Math. Phys. 302 (2011), no. 1, 161–224.
[47] Lemou, M.; Méhats, F.; Raphaël, P., Orbital stability of spherical galactic models. Invent. Math. 187 (2012), no. 1, 145–194.
[48] Lenzmann, E.; Rupert, L.F., Uniqueness and Nondegeneracy of Ground States for $(-?)^s Q + Q - Q^{?+1} = 0$ in $\mathbb{R}$, arXiv:1009.4042 (2011)
[49] Lieb, E. H.; Loss, M., Analysis. Second edition. Graduate Studies in Mathematics, 14. American Mathematical Society, Providence, RI, 2001.
[50] Lions, P.-L.; The concentration-compactness principle in the calculus of variations. The locally compact case. I. Ann. Inst. H. Poincaré Anal. Non Linéaire 1 (1984), no. 2, 109–145.
[51] Lions, P.-L.; The concentration-compactness principle in the calculus of variations. The locally compact case. II. Ann. Inst. H. Poincaré Anal. Non Linéaire 1 (1984), no. 4, 223–283.
[52] MacLeod, K., Uniqueness of positive radial solutions of $\Delta + f(u) = 0$ in $\mathbb{R}^N$, II, Trnas.Amer. Math. Soc. 339 (2) (1993), 495-505.
[53] Makino, T., Blowing up solutions of the Euler-Poisson equation for the evolution of gaseous stars, Proceedings of the Fourth International Workshop on Mathematical Aspects of Fluid and Plasma Dynamics (Kyoto, 1991). Transport Theory Statist. Phys. 21 (1992), no. 4-6, 615–624.
[54] Maris, M., Existence of nonstationary bubbles in higher dimensions, J. Math. Pures Appl. (9) 81 (2002), no. 12, 1207–1239.
[55] Martel, Y.; Merle, F., Stability of blow up profile and lower bounds for blow up rate for the critical generalized KdV equation, Ann. of Math. (2) 155 (2002), no.1, 235-280.
[56] Martel, Y.; Merle, F., Nonexistence of blow-up solution with minimal L^2-mass for the critical gKdV equation, Duke Math. J. 115 (2002), no. 2, 385–408.
[57] Y. Martel and F. Merle, A Liouville theorem for the critical generalized Korteweg–de Vries equation, J. Math. Pures Appl. **79** (2000), 339–425.
[58] Y. Martel and F. Merle, Instability of solitons for the critical generalized Korteweg-de Vries equation, Geom. Funct. Anal. **11** (2001), 74–123.
[59] Y. Martel and F. Merle, Blow up in finite time and dynamics of blow up solutions for the L^2-critical generalized KdV equation, J. Amer. Math. Soc. **15** (2002), 617–664.

[60] Martel, Y.; Merle, F.; Raphaël, P.; Blow up for the critical gKdV equation III: exotic regimes, submitted.

[61] Martel, Y.; Merle, F.; Raphaël, P.; Blow up for the critical gKdV equation II: minimal mass blow up, submitted.

[62] Martel, Y.; Merle, F.; Raphaël, P.; Blow up for the critical gKdV equation I: dynamics near the solitary wave, submitted.

[63] Martel, Y.; Merle, F.; Tsai, T.P,, Stability in H^1 of the sum of K solitary waves for some nonlinear Schrödinger equations, Duke Math. J. 133 (2006), no. 3, 405–466.

[64] Merle, F., Construction of solutions with exact k blow up points for the Schrödinger equation with critical power, Comm. Math.Phys. 129 (1990), 223-240.

[65] Merle, F., On uniqueness and continuation properties after blow up time of self similar solutions of nonlinear Schrödinger equation with critical exponent and critical mass, Comm. Pure. Appl. Math (1992), no. 45, 203-254.

[66] Merle, F., Determination of blow-up solutions with minimal mass for nonlinear Schrödinger equations with critical power, Duke Math. J. 69 (1993), no. 2, 427–454.

[67] Merle, F., Nonexistence of minimal blow-up solutions of equations $iu_t = -\Delta u - k(x)u^{4/N}u$ in R^N, Ann. Inst. H. Poincaré Phys. Théor. 64 (1996), no. 1, 33–85.

[68] Merle, F., Lower bounds for the blow up rate of solutions of the Zakharov equations in dimension two, Comm. Pure. Appl. Math. 49 (1996), n0. 8, 765-794.

[69] Merle, F., Blow up results of virial type for Zakharov equations, Comm.Math. Phys. (1996), n0. 2, 205-214.

[70] F. Merle, Existence of blow-up solutions in the energy space for the critical generalized KdV equation. J. Amer. Math. Soc. **14** (2001), 555–578.

[71] Merle, F.; Raphaël, P., Blow up dynamic and upper bound on the blow up rate for critical nonlinear Schrödinger equation, Ann. of Math. (2) 161 (2005), no. 1, 157–222.

[72] Merle, F.; Raphaël, P., Sharp upper bound on the blow up rate for critical nonlinear Schrödinger equation, Geom. Funct. Anal. 13 (2003), 591–642.

[73] Merle, F.; Raphaël, P., On universality of blow-up profile for L^2 critical nonlinear Schrödinger equation. Invent. Math. 156 (2004), no. 3, 565–672.

[74] Merle, F.; Raphaël, P., On a sharp lower bound on the blow up rate for critical nonlinear Schrödinger equation, J. Amer. Math. Soc. 19 (2006), no. 1, 37–90.

[75] Merle, F; Raphaël, P., Profiles and Quantization of the Blow Up Mass for critical nonlinear Schrödinger Equation, Comm. Math. Phys. 253 (2005), no. 3, 675–704.

[76] Merle, F.; Raphaël, P., Blow up of the critical norm for some radial L2 super critical nonlinear Schrödinger equations, Amer. J. Math. 130 (2008), no. 4, 945–978.

[77] Merle, F.; Raphaël, P.; Rodnianski, I., Blow up dynamics for smooth equivariant solutions to the energy critical Schrödinger map, to appear in Invent. Math.

[78] Merle, F.; Raphaël, P., Szeftel, J., Stable self-similar blow-up dynamics for slightly L2 super-critical NLS equations. Geom. Funct. Anal. 20 (2010), no. 4, 1028–1071.

[79] Merle, F.; Raphaël, P., Szeftel, J., The instability of Bourgain-Wang solutions for the L^2 critical NLS, arXiv:1010.5168 (2011)

[80] Merle, F.; Raphaël, P., Szeftel, J., On collapsing ring blow up solutions to the mass super-critical NLS, arXiv:1202.5218 (2012)

[81] Merle, F.; Tsutsumi, Y., L^2 concentration of blow up solutions for the nonlinear Schrödinger equation with critical power nonlinearity, J.Diff.Eq. 84 (1990), 205-214.

[82] Merle, F.: Vega, L., Compactness at blow up time for L^2 solutions ofnthe critical non linear Schrödinger equation in 2D, IMRN 1998, N8, 399-425.

[83] Merle, F.; Zaag, H., Stability of the blow-up profile for equations of the type $u\bar{s}t = \Delta u + |u|^{p-1}u$, Duke Math. J. 86 (1997), no. 1, 143–195.

[84] Merle, F.; Zaag, H., Refined uniform estimates at blow-up and applications for nonlinear heat equations, Geom. Funct. Anal. 8 (1998), no. 6, 1043–1085.

[85] Merle, F.; Zaag, H., Optimal estimates for blowup rate and behavior for nonlinear heat equations, Comm. Pure Appl. Math. 51 (1998), no. 2, 139–196.

[86] Merle, F.; Zaag, H., Openness of the set of non-characteristic points and regularity of the blow-up curve for the 1 D semilinear wave equation, Comm. Math. Phys. 282 (2008), no. 1, 55–86.

[87] Merle, F.; Zaag, H., Existence and universality of the blow-up profile for the semilinear wave equation in one space dimension, J. Funct. Anal. 253 (2007), no. 1, 43–121.

[88] Merle, F.; Zaag, H., On growth rate near the blowup surface for semilinear wave equations, Int. Math. Res. Not. 2005, no. 19, 1127–1155.
[89] Merle, F.; Zaag, H., Determination of the blow-up rate for a critical semilinear wave equation, Math. Ann. 331 (2005), no. 2, 395–416.
[90] Merle, F.; Zaag, H., Determination of the blow-up rate for the semilinear wave equation, Amer. J. Math. 125 (2003), no. 5, 1147–1164.
[91] Nawa, H., "Mass concentration" phenomenon for the nonlinear Schrödinger equation with the critical power nonlinearity, Funkcial. Ekvac. 35 (1992), no. 1, 1–18.
[92] Nawa, H., Asymptotic and limiting profiles of blowup solutions of the nonlinear Schrödinger equation with critical power. Comm. Pure Appl. Math. 52 (1999), no. 2, 193–270.
[93] Ogawa, T.; Tsutsumi, Y., Blow-up of H^1 solution for the nonlinear Schrödinger equation. J. Differential Equations 92 (1991), no. 2, 317–330.
[94] Papanicolaou, G. C.; Sulem, C.; Sulem, P.-L.; Wang, X. P., Singular solutions of the Zakharov equations for Langmuir turbulence. Phys. Fluids B 3 (1991), no. 4, 969–980.
[95] Perelman, G., On the blow up phenomenon for the critical nonlinear Schrödinger equation in 1D, Ann. Henri. Poincaré, 2 (2001), 605-673.
[96] Planchon, F.; Raphaël, P., Existence and stability of the log-log blow-up dynamics for the L^2-critical nonlinear Schrödinger equation in a domain, Ann. Henri Poincaré 8 (2007), no. 6, 1177–1219.
[97] Raphaël, P., Stability of the log-log bound for blow up solutions to the critical nonlinear Schrödinger equation, Math. Ann. 331 (2005), no. 3, 577–609.
[98] Raphaël, P., Existence and stability of a solution blowing up on a sphere for an L2-supercritical nonlinear Schrödinger equation, Duke Math. J. 134 (2006), no. 2, 199–258.
[99] Raphaël, P.; Rodnianski, I., Stable blow up dynamics for critical corotational wave maps and the equivariant Yang Mills problems, Publ. Math. Inst. Hautes Etudes Sci. 115 (2012), 1–122.
[100] Raphaël, P.; Szeftel, J., Standing ring blow up solutions to the N-dimensional quintic nonlinear Schrödinger equation, Comm. Math. Phys. 290 (2009), no. 3, 973–996.
[101] Raphaël, P.; Szeftel, J., Existence and uniqueness of minimal blow-up solutions to an inhomogeneous mass critical NLS., J. Amer. Math. Soc. 24 (2011), no. 2, 471–546.
[102] Rodnianski, I.; Schlag, W.; Soffer, A., Dispersive analysis of charge transfer models, Comm. Pure Appl. Math. 58 (2005), no. 2, 149–216.
[103] Russell, J.; Xingbin, P., On an elliptic equation related to the blow up phenomenon in the nonlinear Schrödinger equation. Proc. Roy. Soc. Ed., **123 A**, 763-782, 1993.
[104] Sideris, T.C., Formation of singularities in three-dimensional compressible fluids, Comm. Math. Phys. 101 (1985), no. 4, 475–485.
[105] Soffer, A.; Weinstein, M. I., Resonances, radiation damping and instability in Hamiltonian nonlinear wave equations, Invent. Math. 136 (1999), no. 1, 9–74.
[106] Sulem, C.; Sulem, P.L., The nonlinear Schrödinger equation. Self-focusing and wave collapse. Applied Mathematical Sciences, 139. Springer-Verlag, New York, 1999.
[107] Tao, T., Nonlinear dispersive equations. Local and global analysis, CBMS Regional Conference Series in Mathematics, 106, 2006.
[108] Tao, T., Global existence and uniqueness results for weak solutions of the focusing mass-critical nonlinear Schrödinger equation, Anal. PDE 2 (2009), no. 1, 61–81.
[109] Weinstein, M.I., Lyapunov stability of ground states of nonlinear dispersive evolution equations, Comm. Pure. Appl. Math. **39** (1986), 51—68.
[110] Weinstein, M.I., Modulational stability of ground states of nonlinear Schrödinger equations, SIAM J. Math. Anal. **16** (1985), 472—491.
[111] Weinstein, M.I., Nonlinear Schrödinger equations and sharp interpolation estimates, Comm. Math. Phys. **87** (1983), 567—576.
[112] Zakharov, V. E.; Shabat, A. B., Exact theory of two-dimensional self-focusing and one-dimensional self-modulation of waves in nonlinear media, Soviet Physics JETP 34 (1972), no. 1, 62–69.

[113] Zakharov, V.E.; Shabat, A.B., Exact theory of two-dimensional self-focusing and one-dimensional self-modulation of waves in non-linear media, Sov. Phys. JETP **34** (1972), 62—69.

[114] Zwiers, I., Standing ring blowup solutions for cubic nonlinear Schrödinger equations, Anal. PDE 4 (2011), no. 5, 677–727.

Université de Nice Sophia Antipolis & Institut Universitaire de France
E-mail address: praphael@math.unice.fr

Clay Mathematics Proceedings
Volume **17**, 2013

Nonlinear Schrödinger Equations at Critical Regularity

Rowan Killip and Monica Vişan

CONTENTS

2010 *Mathematics Subject Classification.* Primary 35Q55.

1. Introduction

We will be discussing the Cauchy problem for the nonlinear Schrödinger equation:

$$\begin{cases} iu_t = -\Delta u + \mu |u|^p u \\ u(t=0,x) = u_0(x). \end{cases} \tag{1.1}$$

Here $u : \mathbb{R} \times \mathbb{R}^d \to \mathbb{C}$ is a complex-valued function of time and space, the Laplacian is in the space variables only, $\mu \in \mathbb{R} \setminus \{0\}$, and $p \geq 0$. By rescaling the values of u, it is possible to restrict attention to the cases $\mu = -1$ or $\mu = +1$; these are known as the *focusing* and *defocusing* equations, respectively.

The class of solutions to (1.1) is left invariant by the scaling

$$u(t,x) \mapsto \lambda^{\frac{2}{p}} u(\lambda^2 t, \lambda x). \tag{1.2}$$

This scaling defines a notion of criticality, specifically, for a given Banach space of initial data u_0, the problem is called *critical* if the norm is invariant under (1.2). The problem is called *subcritical* if the norm of the rescaled solution diverges as $\lambda \to \infty$; if the norm shrinks to zero, then the problem is *supercritical.* Notice that sub-/super-criticality is determined by the response of the norm to the behaviour of u_0 at small length scales, or equivalently, at high-frequencies. This is natural as the low frequencies are comparatively harmless; they are both smooth and slow-moving.

To date, most authors have focused on initial data belonging to L^2_x-based Sobolev spaces

$$\|u_0\|^2_{H^s_x} := \int_{\mathbb{R}^d} |\hat{u}_0(\xi)|^2 \,(1+|\xi|^2)^s \, d\xi \quad \text{or} \quad \|u_0\|^2_{\dot{H}^s_x} := \int_{\mathbb{R}^d} |\hat{u}_0(\xi)|^2 \, |\xi|^{2s} \, d\xi. \tag{1.3}$$

These are known as the inhomogeneous and homogeneous Sobolev spaces, respectively. The latter is better behaved under scaling, which makes it the more natural choice for studying critical problems. Let us pause to reiterate criticality in these terms.

DEFINITION 1.1. Consider the initial value problem (1.1) for $u_0 \in \dot{H}^s_x(\mathbb{R}^d)$. This problem is *critical* when $s = s_c := \frac{d}{2} - \frac{2}{p}$, *subcritical* when $s > s_c$, and *supercritical* when $s < s_c$.

In these notes, we will be focusing on two specific critical problems, which are singled out by the fact that the critical regularity coincides with a conserved quantity. These are the *mass-critical* equation,

$$iu_t = -\Delta u + \mu|u|^{\frac{4}{d}}u, \tag{1.4}$$

which is associated with the conservation of *mass*,

$$M(u(t)) := \int_{\mathbb{R}^d} |u(t,x)|^2\,dx, \tag{1.5}$$

and the *energy-critical* equation (in dimensions $d \geq 3$),

$$iu_t = -\Delta u + \mu|u|^{\frac{4}{d-2}}u, \tag{1.6}$$

which is associated with the conservation of *energy*,

$$E(u(t)) := \int_{\mathbb{R}^d} \tfrac{1}{2}|\nabla u(t,x)|^2 + \mu\tfrac{d-2}{2d}|u(t,x)|^{\frac{2d}{d-2}}\,dx. \tag{1.7}$$

For subcritical equations, the local problem is well understood, because it is amenable to treatment as a perturbation of the linear equation. This has lead to a satisfactory global theory at conserved regularity. A major theme of current research is to understand the global behaviour of subcritical solutions at non-conserved regularity. By comparison, supercritical equations, even at conserved regularity, are terra incognita at present.

To describe the current state of affairs regarding the mass- and energy-critical nonlinear Schrödinger equations we need to introduce a certain amount of vocabulary. We begin with what it means to be a solution of (1.4) or (1.6).

DEFINITION 1.2 (Solution). Let I be an interval containing the origin. A function $u : I \times \mathbb{R}^d \to \mathbb{C}$ is a (strong) *solution* to (1.6) if it lies in the class $C_t^0\dot{H}_x^1$ and obeys the Duhamel formula

$$u(t) = e^{it\Delta}u_0 - i\mu\int_0^t e^{i(t-s)\Delta}|u(s)|^{\frac{4}{d-2}}u(s)\,ds. \tag{1.8}$$

for all $t \in I$. We say that u is a solution to (1.4) if it belongs to both $C_t^0L_x^2$ and $L_{t,\text{loc}}^{2(d+2)/d}L_x^{2(d+2)/d}$ and also obeys

$$u(t) = e^{it\Delta}u_0 - i\mu\int_0^t e^{i(t-s)\Delta}|u(s)|^{\frac{4}{d}}u(s)\,ds. \tag{1.9}$$

For the definition of $L_t^qL_x^r$ see (1.10).

When we say that (1.8) or (1.9) are obeyed, we mean as a weak integral of distributions. Note that in the mass-critical case, the nonlinearity $|u|^{\frac{4}{d}}u$ is not even a distribution for arbitrary $u \in C_t^0L_x^2$ and $d \leq 3$. This is one reason we require u to have some additional spacetime integrability. A second reason (the primary one for $d \geq 4$) is that uniqueness of solutions is not currently known without this hypothesis. The particular spacetime integrability we require holds for solutions of the linear equation (this is Strichartz inequality, Theorem 3.2); moreover, in Section 3 we will show that (1.4) does admit local solutions in this space.

The existence of local solutions, that is, solutions on some small neighbourhood of $t = 0$, was proved by Cazenave and Weissler, [**13, 14**]. Note that in this result, the time of existence depends on the profile of u_0 rather than simply its norm. Indeed, the latter would be inconsistent with scaling invariance.

Primarily, these notes are devoted to global questions, specifically, whether the solution exists forever ($I = \mathbb{R}$) and if it does, what is its asymptotic behaviour as $t \to \pm\infty$. Here are the main notions:

DEFINITION 1.3. A Cauchy problem is called *globally wellposed* if solutions exist for all time, are unique, and depend continuously on the initial data. A stronger notion is that the problem admits *global spacetime bounds.* In the mass-critical case, (1.4), this means that the solution u also obeys

$$\int_{\mathbb{R}} \int_{\mathbb{R}^d} |u(t,x)|^{\frac{2(d+2)}{d}}\, dx\, dt \le C(M(u_0))$$

for some function C. For the analogous notion in the energy-critical case, (1.6), replace u by ∇u and u_0 by ∇u_0. We say that *asymptotic completeness* holds if for each (global) solution u there exist u_+ and u_- so that

$$u(t) - e^{it\Delta}u_+ \to 0 \quad \text{as } t \to \infty \text{ and } \quad u(t) - e^{it\Delta}u_- \to 0 \quad \text{as } t \to -\infty.$$

Note that u_+ and u_- are supposed to lie in the same space as the initial data; convergence is with respect to its norm. A converse notion is the *existence of wave operators.* This means that for each u_+ there is a global solution u of the nonlinear problem so that $u(t) - e^{it\Delta}u_+ \to 0$ and similarly for each u_-. We say *scattering* holds if wave operators exist and are asymptotically complete.

Simple arguments show that scattering follows from global spacetime bounds. In the defocusing case ($\mu = +1$), we believe that critical equations admit global spacetime bounds even when the critical Sobolev norm does not correspond to a conserved quantity. No such bold claim can hold in the focusing case; indeed, there are explicit counterexamples.

As we will discuss in Subsection 4.1, the elliptic problem

$$-\Delta f - |f|^{\frac{4}{d}} f = -f$$

on $\mathbb{R}^d$ admits Schwartz-space solutions. Indeed, there is a unique non-negative spherically symmetric Schwartz solution, which we denote by Q; see [**49, 105**]. This function is known as the *ground state*; it is, at least, the lowest eigenstate of the operator $f \mapsto -\Delta f - Q^{4/d} f$.

Now, $u(t,x) = e^{it}Q(x)$ is a global solution to the mass-critical focusing NLS that manifestly does not obey spacetime bounds, nor does it scatter (cf. (4.28)). Furthermore, by applying the pseudo-conformal identity, (2.12), we may transform this to a solution that blows up in finite time:

$$u(t,x) = (1-t)^{-\frac{d}{2}} e^{-i\frac{|x|^2}{4(1-t)} + i\frac{t}{1-t}}\, Q\big(\tfrac{x}{1-t}\big).$$

By comparison, the work of Cazenave and Weissler mentioned before shows that initial data of sufficiently small mass (that is, L^2_x norm) does lead to global solutions obeying spacetime bounds. Thus one may hope to identify the minimal mass at which such good behaviour first fails; $M(Q)$ is one candidate. Indeed, it is widely believed to be the correct answer:

CONJECTURE 1.4. *For arbitrary initial data $u_0 \in L^2_x(\mathbb{R}^d)$, the defocusing mass-critical nonlinear Schrödinger equation is globally wellposed and solutions obey global spacetime bounds; in particular, scattering holds.*

For the focusing equation, the same conclusions hold for initial data obeying $M(u_0) < M(Q)$.

Perhaps the earliest (and one of the strongest) indications that $M(Q)$ is the correct bound in the focusing case comes from work of Weinstein, [**105**], which proves global well-posedness for H^1_x initial data obeying $M(u_0) < M(Q)$. Recent progress toward settling the conjecture (at critical regularity) is discussed in the next subsection.

Before formulating the analogous conjecture for the energy-critical problem, let us discuss the natural candidate for the role of Q. By a result of Pohožaev, [**68**], the equation $-\Delta f - |f|^{\frac{4}{d-2}} f = -\beta f$ does not have $\dot H^1_x(\mathbb{R}^d)$ solutions for $\beta \neq 0$. When $\beta = 0$, this equation has a very explicit solution, namely,

$$W(x) := \big(1 + \tfrac{1}{d(d-2)}|x|^2\big)^{-\frac{d-2}{2}}.$$

From the elliptic equation, we see that $u(t,x) = W(x)$ is a stationary solution of (1.6). The general belief is that W is the minimal counterexample to global spacetime bounds in the energy-critical setting; however, the way in which it is minimal is more subtle than in the mass-critical setting. Firstly, we should not measure minimality in terms of the energy, (1.7), since the energy can be made arbitrarily negative. An alternative is to consider the kinetic energy,

$$E_0(u(t)) := \int_{\mathbb{R}^d} \tfrac{1}{2}|\nabla u(t,x)|^2\,dx.$$

However, this creates problems of its own since it is not a conserved quantity. The solution we choose (cf. [**38, 44**]) is to assert that the only way a solution can fail to be global and obey spacetime bounds is if its kinetic energy matches (or exceeds) that of W, at least asymptotically:

CONJECTURE 1.5. *For arbitrary initial data* $u_0 \in \dot H^1_x(\mathbb{R}^d)$, *the defocusing energy-critical nonlinear Schrödinger equation is globally wellposed and solutions obey global spacetime bounds; in particular, scattering holds.*

For the focusing equation, we have the following statement: Let $u : I \times \mathbb{R}^d \to \mathbb{C}$ *be a solution to* (1.6) *such that*

$$E_* := \sup_{t\in I} E_0(u(t)) < E_0(W).$$

Then

$$\int_I \int_{\mathbb{R}^d} |u(t,x)|^{\frac{2(d+2)}{d-2}}\,dx\,dt \le C(E_*) < \infty.$$

The defocusing case of this conjecture has been completely resolved, while for the focusing equation only the three- and four-dimensional cases remain open. These results, as well as some of their precursors, are the topic of the next subsection.

1.1. Where are we? And how did we get there? We will not discuss the nonlinear wave equation in these notes; however, it seems appropriate to point out that global well-posedness for the defocusing energy-critical wave equation was proved (after considerable effort) some years before the analogous result for the nonlinear Schrödinger equation; see [**78**] where references to the original papers may be found. Treatment of the focusing energy-critical wave equation is much more recent, [**39**]. There is no analogue of mass conservation for NLW and hence no true analogue of the mass-critical NLS.

Turning now to NLS, we would like to point out two important differences between it and NLW. First, it does not enjoy finite speed of propagation. Second, in the wave case, the natural monotonicity formula (i.e., the Morawetz identity) has critical scaling; this is not the case for NLS. Both differences have had an important effect on how the theory has developed.

In **[6]**, Bourgain considers the two-dimensional mass-critical NLS for inital data in L_x^2. It is shown that in order for a solution to blow up, it must concentrate some finite amount of mass in ever smaller sets (as one approaches the blowup time). Perhaps more important than the result itself were two aspects of the proof: the use of recent progress toward the restriction conjecture (see Conjecture 4.17) and a rather precise form of inverse Strichartz inequality.

Using these ingredients, Merle and Vega **[58]** obtained a concentration compactness principle for the mass-critical NLS in two dimensions. (For the analogous result in other dimensions, see **[4, 12]**.) The formulation mimics results for the wave equation **[3]**, although the proof is very different. The techniques used for the wave equation are better suited to the energy-critical NLS and were used by Keraani **[41]** to obtain a concentration compactness principle for this equation. These concentration compactness principles are discussed in Section 4 and play an important role in the arguments presented in these notes. History, however, took a slightly different route.

The first major step toward verifying either conjecture was Bourgain's proof, **[7]**, of global spacetime bounds for the defocusing energy-critical NLS in three and four dimensions with spherically symmetric data. A major new tool introduced therein was 'induction on energy'. We will now try to convey the outline. The role of the base step is played by the fact that global spacetime bounds are known for small data, say for data with energy less than e_0. Next we choose a small η depending on e_0. If all solutions with energy less than $e_1 := e_0 + \eta$ obey satisfactory spacetime bounds then we are ready to move to the next step. Suppose not, that is, suppose that there is a (local) solution u with enormous spacetime norm, but energy less than e_1. Then, using Morawetz and inverse Strichartz-type inequalities, one may show that the there is a bubble of concentration carrying energy $\gg \eta$ that is protected by a comparatively long time interval over which u has little spacetime norm. If we remove the bubble, we obtain initial data with energy less than e_0 which then leads to a global solution with good bounds (thanks to the inductive hypothesis). Taking advantage of the buffer zone, it is possible to glue the bubble back in without completely destroying this bound. By defining what was meant earlier by 'satisfactory spacetime bound' in an appropriate manner, we reach a contradiction. This proves the result for solutions with energy less than e_1. Next, we turn our attention to solutions with energy less that $e_2 := e_1 + \eta(e_1)$, and so on, and so on.

Concentration results such as those mentioned in the previous paragraph provide important leverage in critical problems; the size of the bubbles they exhibit provide a characteristic length scale. The fact that we are dealing with scale-invariant problems means that any length scale must be dictated by the solution; it cannot be imposed from without. It is only through breaking the scaling symmetry, in a manner such as this, that non-critical tools such as the Morawetz identity can be properly brought to bear.

In [**32**], Grillakis showed global regularity for the three-dimensional energy-critical defocusing NLS with spherically symmetric initial data, that is, he proved that smooth spherically symmetric initial data leads to a global smooth solution. This can be deduced *a posteriori* from [**7**]; however, the argument in [**32**] is rather different. Subsequent progress in the spherically symmetric case, including the treatment of higher dimensions, can be found in [**89**].

The big breakthrough for non-spherically symmetric initial data was made in [**20**]. This paper brought a wealth of new ideas and tools to the problem, of which we will describe just a few. First, the authors use an interaction Morawetz inequality (introduced in [**19**]), which is much better suited to the non-symmetric case than the (Lin–Strauss) Morawetz used in previous works. See Section 7 for a discussion of both.

Unfortunately, the interaction Morawetz identity is further from critical scaling than its predecessor, which necessitates a much stronger form of concentration result. By reaping the ultimate potential of the induction on energy technique, the authors of [**20**] showed that it suffices to consider solutions that are well localized in both space and frequency. Indeed, modulo the action of scaling and space translations, these solutions remain in a very small neighbourhood of a compact set in $\dot H^1_x(\mathbb{R}^3)$.

The argument from [**20**] was generalized to four space dimensions in [**75**] and then to dimensions five and higher in [**103, 104**]. Taken together, these papers resolve the defocusing case of Conjecture 1.5.

In [**42**], Keraani used the concentration compactness statements discussed earlier to show that if the mass-critical NLS did not obey global spacetime bounds, then there is a solution u with minimal mass and infinite spacetime norm. Simple contrapositive would show that there is a sequence of global solutions with mass growing to the minimal value whose spacetime norms diverge to infinity. The point here is that the limit object exists, albeit after passing to a subsequence and performing symmetry operations. An additional immediate consequence of this compactness principle is that the minimal mass blowup solution u is almost periodic modulo symmetries (cf. Definition 5.1). This is a stronger form of concentration result than is provided by the induction on energy technique. We will turn to a more formal comparison shortly. The existence of minimal blowup solutions was adapted to the energy-critical case in [**38**], which is also the first application of this important innovation to the well-posedness problem. The main result of that paper was to prove the focusing case of Conjecture 1.5 for spherically symmetric data in dimensions $d = 3, 4, 5$. This was extended to all dimensions in [**47**]. For general (non-symmetric) data in dimensions five and higher, Conjecture 1.5 was proved in [**44**]. The complete details of this argument will be presented here. The conjecture remains open for $d = 3, 4$.

The difference between the 'minimal blowup solution' strategy and the 'induction on energy' approach is akin to that between the well ordering principle (any non-empty subset of $\{0, 1, 2, \dots\}$ contains a least element) and the principle of induction. By its intrinsically recursive nature, induction is well suited to obtaining concrete bounds and this is, indeed, what the induction on energy approach provides. By contrast, proof by contradiction, which is the basis of the minimal counterexample approach, often leads to cleaner simpler arguments, but can seldom be made effective. These general principles hold true in the NLS setting. The minimal counterexample approach leads to simpler proofs, particularly because it allows for a much more modular approach — induction on energy requires delicately interconnected arguments that cannot be disentangled until the very end — however, it does not seem possible to obtain effective bounds without reverting to the older technology. On pedagogical grounds, we will confine our attention to the minimal counterexample method in these notes.

Perhaps we have done too good a job of distinguishing the two approaches; they are two sides of the same coin: they may look very different, but are built upon the same substrate, namely, improved Strichartz inequalities. These are discussed in Subsection 4.4.

Let us now describe the current state of affairs for the mass-critical equation. Building on developments in the energy-critical case, Conjecture 1.4 has been settled for spherically symmetric data in dimensions two and higher. For the defocusing case, $d \geq 3$, see [**96, 97**]. For $d = 2$, both focusing and defocusing, see [**43**]. The latter argument was adapted to treat the $d \geq 3$ focusing case in [**46**].

With so much of the road left to travel, it would be premature to try to discern what parts of the these works may prove valuable in settling the full conjecture. We present here a number of building blocks taken from those papers that we believe will be useful in the non-symmetric case.

Acknowledgements We are grateful to Jason Murphy, Shuanglin Shao, Betsy Stovall, and Michael Struwe for comments and corrections.

The authors were supported by NSF Grant DMS-0635607 and by the state of New Jersey under the auspices of the Institute for Advanced Study. R. K. was additionally supported by NSF grant DMS-0701085.

Any opinions, findings and conclusions or recommendations expressed are those of the authors and do not reflect the views of the National Science Foundation.

1.2. Notation. We will be regularly referring to the spacetime norms

$$\|u\|_{L_t^q L_x^r(\mathbb{R}\times\mathbb{R}^d)} := \left(\int_{\mathbb{R}} \left[\int_{\mathbb{R}^d} |u(t,x)|^r \, dx\right]^{\frac{q}{r}} dt\right)^{\frac{1}{q}}, \tag{1.10}$$

with obvious changes if q or r is infinity. To save space in in-line formulas, we will abbreviate

$$\|f\|_r := \|f\|_{L_x^r} \quad \text{and} \quad \|u\|_{q,r} := \|u\|_{L_t^q L_x^r}.$$

We write $X \lesssim Y$ to indicate that $X \leq CY$ for some constant C, which is permitted to depend on the ambient spatial dimension, d, without further comment. Other dependencies of C will be indicated with subscripts, for example, $X \lesssim_u Y$. We will write $X \sim Y$ to indicate that $X \lesssim Y \lesssim X$.

We use the 'Japanese bracket' convention: $\langle x \rangle := (1+|x|^2)^{1/2}$ as well as $\langle \nabla \rangle := (1-\Delta)^{1/2}$. Similarly, $|\nabla|^s$ denotes the Fourier multiplier with symbol $|\xi|^s$. These are used to define the Sobolev norms

$$\|f\|_{W^{s,r}} := \|\langle \nabla \rangle^s f\|_{L_x^r}.$$

Our convention for the Fourier transform is

$$\hat{f}(\xi) = (2\pi)^{-\frac{d}{2}} \int_{\mathbb{R}^d} e^{-ix\cdot\xi} f(x)\, dx$$

so that

$$f(x) = (2\pi)^{-\frac{d}{2}} \int_{\mathbb{R}^d} e^{ix\cdot\xi} \hat{f}(\xi)\, d\xi \quad \text{and} \quad \int_{\mathbb{R}^d} |\hat{f}(\xi)|^2\, d\xi = \int_{\mathbb{R}^d} |f(x)|^2\, dx.$$

Notations associated to Littlewood-Paley projections are discussed in Appendix A.

2. Symmetries

2.1. Hamiltonian formulation. As we will see, the nonlinear Schrödinger equation may be viewed as an infinite dimensional Hamiltonian system. In the finite dimensional case, Hamiltonian mechanics has many general theorems of wide applicability. In the PDE setting, however, these tend to become guiding principles with each system requiring its own special treatment; indeed, compare the local theory for ODE with that for PDE. In what follows, we will take a rather formal approach, since it is not difficult to check the conclusions *a posteriori*. In particular, we will allow ourselves a rather fluid notion of phase space. In all cases, it will be a vector space of functions from $\mathbb{R}^d$ into $\mathbb{C}$. If we were working with polynomial nonlinearities, it would be reasonable to use Schwartz space. However, in the case of fractional power nonlinearities, this space is not conserved by the flow; besides, the main goal of these notes is to work in low regularity spaces.

A symplectic form is a closed non-degenerate (anti-symmetric) 2-form on phase space. In particular, it takes two tangent vectors f, g at a point u in phase space and returns a real number. The symplectic form relevant to us is

$$\omega(f,g) := \operatorname{Im} \int_{\mathbb{R}^d} f(x)\overline{g(x)}\,dx.$$

Notice that this implies $q(x) : u \mapsto \operatorname{Re} u(x)$ and $p(x) : u \mapsto \operatorname{Im} u(x)$ are canonically conjugate coordinates (indexed by x). In light of this, we see that (with the sign conventions in [**1**]) the Poisson bracket associated to ω is given by

$$\{G,F\}(u) = \int_{\mathbb{R}^d} \frac{\delta F}{\delta p}\Big|_u(x)\frac{\delta G}{\delta q}\Big|_u(x) - \frac{\delta F}{\delta q}\Big|_u(x)\frac{\delta G}{\delta p}\Big|_u(x)\,dx, \tag{2.1}$$

where the functional derivatives are defined by

$$\lim_{\varepsilon\to 0}\frac{G(u+\varepsilon v)-G(u)}{\varepsilon} = dG\big|_u(v) = \int_{\mathbb{R}^d}\frac{\delta G}{\delta q}\Big|_u(x)\operatorname{Re} v(x) + \frac{\delta G}{\delta p}\Big|_u(x)\operatorname{Im} v(x)\,dx$$

for all $v : \mathbb{R}^d \to \mathbb{C}$. In particular, $\{q(y), p(x)\}(u) = \delta(x-y)$, independent of u, which expresses the fact that these are canonically conjugate coordinates.

For a general real-valued function H defined on phase space, the associated (Hamiltonian) flow is defined by

$$u_t = \nabla_\omega H(u) \quad \text{where the vector field } \nabla_\omega H \text{ is defined by} \quad dH(\cdot) = \omega(\cdot, \nabla_\omega H).$$

A consequence (or alternate definition) is that for any function F on phase space,

$$\tfrac{d}{dt}F(u(t)) = \{F,H\}(u(t)).$$

In particular, $q_t = \frac{\delta H}{\delta p}$ and $p_t = -\frac{\delta H}{\delta q}$, which are the usual form of Hamilton's equations. When needed, we will write $\exp(t\nabla_\omega H)$ for the time-t flow map.

With all these notions in place, we leave the final (indeed central) point to the reader:

EXERCISE. Show that formally, the Hamiltonian

$$H(u) := \int_{\mathbb{R}^d} \tfrac{1}{2}|\nabla u|^2 + \tfrac{\mu}{p+2}|u|^{p+2}\,dx \tag{2.2}$$

leads to the flow

$$iu_t = -\Delta u + \mu |u|^p u. \tag{2.3}$$

2.2. The symmetries. In this subsection, we will list the main symmetries of (2.3), together with a brief discussion of each.

Recall that Noether's Theorem guarantees that there is a bijection between conserved quantities and one-parameter groups of symplectomorphisms preserving the Hamiltonian. Specifically, using the conserved quantity as a Hamiltonian leads to a (symplectic form preserving) flow that conserves the original Hamiltonian. In each case that this theorem is applicable, we will note the corresponding conservation law. Some important symmetries do not preserve the symplectic form and/or the Hamiltonian; nevertheless, we will still be able to find an appropriate substitute for a corresponding conserved quantity.

Time translations. If $u(t)$ is a solution of (2.3), then clearly so is $u(t+\tau)$ for τ fixed. This symmetry is associated with conservation of the Hamiltonian (2.2).

Space translations. It is not difficult to see that both the Hamiltonian (2.2) and the symplectic/Poisson structure are invariant under spatial translations: $u(t,x) \mapsto u(t, x-x_0)$. This symmetry is generated by the total momentum

$$P(u) := \int_{\mathbb{R}^d} 2\operatorname{Im}(\bar{u}\nabla u)\,dx. \tag{2.4}$$

Indeed, given $x_0 \in \mathbb{R}^d$,

$$u(x-x_0) = \big[e^{\frac{1}{4}\nabla_\omega(x_0\cdot P)}u\big](x).$$

The factor 2 has been included in (2.4) to match conventions elsewhere.

Space rotations. Invariance under rotations of the coordinate axes corresponds to the conservation of angular momentum. The later is a tensor with $\binom{d}{2}$ components, indexed by pairs $1 \leq j < k \leq d$:

$$L_{jk}(u) = i\int_{\mathbb{R}^d} \bar{u}[x_j\partial_k u - x_k\partial_j u]\,dx.$$

Concomitant with the non-commutativity of the rotation group $SO(d)$, the components of angular momentum do not all Poisson commute with one another, forming instead, a representation of the Lie algebra $so(d)$.

Phase rotations. The map $u(x) \mapsto e^{i\theta}u(x)$ is a simple form of gauge symmetry. It is connected to the conservation of mass:

$$M(u) := \int_{\mathbb{R}^d} |u|^2\,dx \quad \text{obeys} \quad e^{\tau\nabla_\omega M}u = e^{-2i\tau}u. \tag{2.5}$$

Time reversal. As intuition dictates, one may invert the time evolution by simply reversing all momenta. Given our choice of canonical coordinates, this corresponds to the map $u \mapsto \bar{u}$. We leave the reader to check that

$$e^{t\nabla_\omega H}\bar{u} = \overline{e^{-t\nabla_\omega H}u}.$$

Galilei boosts. A central tenet of mechanics is that the same laws of motion apply in all inertial (non-accelerating) reference frames. Combined with an absolute notion of time, this leads directly to Galilean relativity.

The class of solutions to the nonlinear Schrödinger equation (2.3) is left invariant by Galilei boosts:

$$u(t,x) \mapsto e^{ix\cdot\xi_0 - it|\xi_0|^2}u(t, x-2\xi_0 t), \tag{2.6}$$

where $\xi_0 \in \mathbb{R}^d$ denotes (half the) relative velocity of the two reference frames.

There are two (connected) problems with applying Noether's Theorem in this case: the symmetry explicitly involves time, it is not simply a transformation of phase space, and it does not leave the Hamiltonian invariant (cf. Proposition 2.3 below). As we will explain, the appropriate substitute for a conserved quantity is

$$X(u) := \int_{\mathbb{R}^d} x|u|^2\,dx. \tag{2.7}$$

This represents the location of the centre of mass, at least when $M(u) = 1$.

The time derivative of X is

$$\{X, H\} = P, \quad \text{which implies} \quad \{\{X, H\}, H\} = 0. \tag{2.8}$$

Thus, although it is not conserved, X has a very simple time evolution:

$$X(u(t)) = X(u(0)) + t \cdot P(u(0)).$$

It remains for us to connect X with Galilei boosts. The first indication of this is

$$\big[e^{-\frac{1}{2}\nabla_\omega(\xi_0\cdot X)}u\big](x) = e^{ix\cdot\xi_0}u(x),$$

which reproduces the action of a Galilei boost on the initial data $u(t=0)$. Perhaps this is enough to convince the reader of a connection; however, we wish to use this example to elucidate a little abstract theory. The central tenet is quite simple: One may extend the privileged status of conserved quantities, that is, those obeying $\{F, H\} = 0$, to those functions F that together with H generate a finite-dimensional Lie algebra under the action of the Poisson bracket. The concomitant group multiplication law gives a form of time-dependent symmetry.

Together with the Hamiltonian, X generates a $(2d+2)$-dimensional Lie algebra under the action of the Poisson bracket. The basis vectors are H, M, and X_j, P_j, $1 \leq j \leq d$ and the only non-zero brackets among them are

$$\{X, H\} = P \quad \text{and} \quad \{X_j, P_k\} = 4\delta_{jk}M. \tag{2.9}$$

Note that (X, P, M) form the Heisenberg Lie algebra; indeed, the corresponding flows (on u) exactly reproduce the standard Schrödinger representation of the Heisenberg group. Using the (Lie group) commutation laws induced by (2.9), we obtain

$$e^{t\nabla_\omega H}e^{-\frac{1}{2}\nabla_\omega(\xi_0\cdot X)} = e^{\frac{t}{2}\nabla_\omega(\xi_0\cdot P-|\xi_0|^2M)}e^{-\frac{1}{2}\nabla_\omega(\xi_0\cdot X)}e^{t\nabla_\omega H},$$

which is exactly the statement that (2.6) preserves solutions to (2.3).

Scaling. The scaling symmetry for (2.3) is

$$u(t,x) \mapsto \lambda^{\frac{2}{p}}u(\lambda^2 t, \lambda x). \tag{2.10}$$

This does not preserve the symplectic/Poisson structure, except in the mass-critical ($p = \frac{4}{d}$) case. It does not preserve the Hamiltonian unless $p = \frac{4}{d-2}$, which corresponds to the energy-critical equation.

As noted, the mass-critical scaling does preserve the symplectic/Poisson structure, which guarantees that it is generated by some Hamiltonian flow. A few computations reveal that

$$A(u) := \tfrac{1}{4i}\int_{\mathbb{R}^d} \bar{u}(x\cdot\nabla + \nabla\cdot x)u\,dx = \tfrac{1}{2}\int_{\mathbb{R}^d} x\cdot \operatorname{Im}(\bar{u}\nabla u)\,dx$$

obeys

$$\big[e^{-\tau\nabla_\omega A}u\big](x) = e^{\frac{d}{2}\tau}u(e^\tau x).$$

and further, that

$$\{A, H\} = 2H + \tfrac{\mu(pd-4)}{2(p+2)} \int_{\mathbb{R}^d} |u|^{2+p}\,dx.$$

This is the best substitute we have for a conservation law associated to (2.10). The peculiar combination of kinetic and potential energies on the right-hand side actually turns out to play an important role; see Section 7.

Specializing to the mass-critical or the linear Schrödinger equation, we obtain the simple relation $\{A, H\} = 2H$, which is much more amenable to a Lie-theoretic perspective. In particular,

$$e^{t\nabla_\omega H} e^{-\tau\nabla_\omega A} = e^{-\tau\nabla_\omega A} e^{e^{2\tau} t\nabla_\omega H},$$

which reproduces (2.10).

Lens transformations. An idealized lens advances (or retards) the phase of the incident wave in proportion to the square of the distance to the optical axis. This leads us to consider

$$V(u) := \int_{\mathbb{R}^d} |x|^2 |u|^2\,dx, \tag{2.11}$$

which is the generator of lens transformations:

$$[e^{\tau\nabla_\omega V} u](x) = e^{-2i\tau|x|^2} u(x).$$

The time evolution of V is given by $\{V, H\} = 8A$.

Under the linear or mass-critical nonlinear Schrödinger evolutions, A behaves in a simple manner, as we discussed above. This leads directly to a time-dependent symmetry, known as the *pseudo-conformal symmetry*; see (2.12) below. We leave the computations to the reader's private pleasure:

EXERCISE. In the mass-critical (or linear) case, H, A, V form a three dimensional Lie algebra with relations $\{A, H\} = 2H$, $\{V, H\} = 8A$, and $\{V, A\} = 2V$. By comparing this with matrices of the form

$$\begin{bmatrix} -a & -8v \\ h & a \end{bmatrix},$$

show that this is the Lie algebra of $SL_2(\mathbb{R})$. Use this (or not) to verify that

$$\left[\begin{smallmatrix} \alpha & \beta \\ \gamma & \delta \end{smallmatrix}\right] : \psi(t, x) \mapsto (\beta t + \delta)^{-\frac{d}{2}} e^{\frac{i\beta|x|^2}{4(\beta t+\delta)}} \psi\big(\tfrac{\alpha t+\gamma}{\beta t+\delta}, \tfrac{x}{\beta t+\delta}\big) \tag{2.12}$$

gives an explicit representation of $SL_2(\mathbb{R})$ on the class of mass-critical solutions.

2.3. Group therapy. The main purpose of this subsection is to introduce some notation we will be using for (a subgroup of) the symmetries just introduced. After that, we will record the effect of symmetries on the major conserved quantities.

DEFINITION 2.1 (Mass-critical symmetry group). For any phase $\theta \in \mathbb{R}/2\pi\mathbb{Z}$, position $x_0 \in \mathbb{R}^d$, frequency $\xi_0 \in \mathbb{R}^d$, and scaling parameter $\lambda > 0$, we define the unitary transformation $g_{\theta,x_0,\xi_0,\lambda} : L_x^2(\mathbb{R}^d) \to L_x^2(\mathbb{R}^d)$ by the formula

$$[g_{\theta,\xi_0,x_0,\lambda} f](x) := \frac{1}{\lambda^{d/2}} e^{i\theta} e^{ix\cdot\xi_0} f\big(\frac{x - x_0}{\lambda}\big).$$

We let G be the collection of such transformations. If $u : I \times \mathbb{R}^d \to \mathbb{C}$, we define $T_{g_{\theta,\xi_0,x_0,\lambda}} u : \lambda^2 I \times \mathbb{R}^d \to \mathbb{C}$, where $\lambda^2 I := \{\lambda^2 t : t \in I\}$, by the formula

$$[T_{g_{\theta,\xi_0,x_0,\lambda}} u](t, x) := \frac{1}{\lambda^{d/2}} e^{i\theta} e^{ix\cdot\xi_0} e^{-it|\xi_0|^2} u\left(\frac{t}{\lambda^2}, \frac{x - x_0 - 2\xi_0 t}{\lambda}\right),$$

or equivalently,

$$[T_{g_{\theta,\xi_0,x_0,\lambda}}u](t) = g_{\theta - t|\xi_0|^2,\xi_0,x_0+2\xi_0 t,\lambda}\Big(u(\lambda^{-2}t)\Big).$$

Note that if u is a solution to the mass-critical NLS, then $T_g u$ is also solution and has initial data $g[u(t=0)]$.

DEFINITION 2.2 (Energy-critical symmetry group). For any phase $\theta \in \mathbb{R}/2\pi\mathbb{Z}$, position $x_0 \in \mathbb{R}^d$, and scaling parameter $\lambda > 0$, we define a unitary transformation $g_{\theta,x_0,\lambda} : \dot H^1_x(\mathbb{R}^d) \to \dot H^1_x(\mathbb{R}^d)$ by

$$[g_{\theta,x_0,\lambda}f](x) := \lambda^{-\frac{d-2}{2}} e^{i\theta} f\big(\lambda^{-1}(x-x_0)\big).$$

Let G denote the collection of such transformations. For a function $u : I \times \mathbb{R}^d \to \mathbb{C}$, we define $T_{g_{\theta,x_0,\lambda}}u : \lambda^2 I \times \mathbb{R}^d \to \mathbb{C}$, where $\lambda^2 I := \{\lambda^2 t : t \in I\}$, by the formula

$$[T_{g_{\theta,x_0,\lambda}}u](t,x) := \lambda^{-\frac{d-2}{2}} e^{i\theta} u\big(\lambda^{-2}t, \lambda^{-1}(x-x_0)\big).$$

Note that if u is a solution to the energy-critical NLS, then so is $T_g u$; the latter has initial data $g[u(t=0)]$.

The next proposition shows how the total mass, momentum, and energy are affected by elements of the mass- or energy-critical symmetry groups. In the latter case, we also record the effect of Galilei boosts. Although they have been omitted from the definition of the symmetry group (they will not be required in the concentration compactness step), they are valuable in further simplifying the structure of minimal blowup solutions.

PROPOSITION 2.3 (Mass, Momentum, and Energy under symmetries). *Let g be an element of the mass-critical symmetry group with parameters θ, x, ξ, and λ. Then*

$$\begin{aligned} M(gu_0) &= M(u_0), \quad P(gu_0) = 2\xi M(u) + \lambda^{-1}P(u_0), \\ E(gu_0) &= \lambda^{-2}E(u_0) + \tfrac12 \lambda^{-1}\xi \cdot P(u_0) + \tfrac12 |\xi|^2 M(u_0). \end{aligned} \tag{2.13}$$

The analogous statement for the energy-critical case reads

$$\begin{aligned} M(v_0) &= \lambda^2 M(u_0), \quad P(v_0) = 2\lambda^2 \xi M(u_0) + \lambda P(u_0), \\ E(v_0) &= E(u_0) + \tfrac12 \lambda\xi \cdot P(u_0) + \tfrac12 \lambda^2 |\xi|^2 M(u_0), \end{aligned} \tag{2.14}$$

where $v_0(x) = [e^{-\frac12 \nabla_\omega(\xi\cdot X)} g u_0](x) = e^{ix\cdot\xi}[gu_0](x)$.

COROLLARY 2.4 (Minimal energy in the rest frame). *Let $\tilde u \in L^\infty_t H^1_x$ be a blowup solution to the mass- or energy-critical NLS. Then there is a blowup solution $u \in L^\infty_t H^1_x$, obeying $M(u) = M(\tilde u)$, $E(u) \le E(\tilde u)$, and*

$$P(u(t)) = 2\,\mathrm{Im} \int_{\mathbb{R}^d} \overline{u(t,x)} \nabla u(t,x)\,dx \equiv 0.$$

Note also that $\|\nabla u\|_{\infty,2} \le \|\nabla \tilde u\|_{\infty,2}$.

PROOF. Choose u to be the unique Galilei boost of $\tilde u$ that has zero momentum. All the conclusions now follow quickly from the formulae above. Note that u has minimal energy among all Galilei boosts of $\tilde u$; indeed, this is an expression of the well-know physical fact that the total energy can be decomposed as the energy viewed in the centre of mass frame plus the energy arising from the motion of the center of mass (cf. [**50**, §8]). □

2.4. Complete integrability. The purpose of this subsection is to share an observation of Jürgen Moser: scattering implies complete integrability. This was passed on to us by Percy Deift.

In the finite dimensional setting, a Hamiltonian flow on a $2n$-dimensional phase space is called *completely integrable* if it admits n functionally independent Poisson commuting conserved quantities. An essentially equivalent formulation is the existence of action-angle coordinates (cf. [**1**]). These are a system of canonically conjugate coordinates $I_1, \dots, I_n, \phi_1, \dots, \phi_n$, which is to say

$$\{I_j, I_k\} = \{\phi_j, \phi_k\} = 0 \qquad \{I_j, \phi_k\} = \delta_{jk},$$

so that under the flow,

$$\tfrac{d}{dt} I_j = 0 \quad \text{and} \quad \tfrac{d}{dt}\phi_j = \omega_j(I_1, \dots, I_n).$$

Here $\omega_1, \dots, \omega_n$ are smooth functions.

In what follows, we will exemplify Moser's assertion in the context of the mass-critical defocusing equation. For clarity of exposition, we presuppose the truth of the associated global well-posedness and scattering conjecture. The principal ideas can be applied to any NLS setting.

As we will see in Section 3, we are guaranteed that the wave operator

$$\Omega : u_0 \mapsto u_+ = \lim_{t\to\infty} e^{-it\Delta} u(t)$$

defines a bijection on $L^2_x(\mathbb{R}^d)$; here $u(t)$ denotes the solution of NLS with initial data u_0. In fact, since both the free Schrödinger and the NLS evolutions are Hamiltonian, the wave operator preserves the symplectic form. As the Fourier transform is also bijective and symplectic (both follow from unitarity), so is the combined map

$$\hat\Omega : u_0 \mapsto \widehat{u_+}, \quad \text{which obeys} \quad \big[\hat\Omega(u(t))\big](\xi) = e^{-it|\xi|^2}\widehat{u_+}(\xi).$$

Thus we have found a symplectic map that trivializes the flow; moreover, we have an infinite family of Poisson commuting conserved quantities, namely,

$$u \mapsto \int_{\mathbb{R}^d} g(\xi) |\widehat{u_+}(\xi)|\, d\xi$$

as g varies over real-valued functions in $L^2_\xi(\mathbb{R}^d)$. Lastly, to see that these do indeed Poisson commute and also to exhibit action-angle variables, we note that if we define $I(\xi) = \frac12 |\widehat{u_+}(\xi)|^2$ and $\phi(\xi)$ by $\widehat{u_+}(\xi) = |\widehat{u_+}(\xi)| e^{-i\phi(\xi)}$, then

$$\{I(\xi), I(\eta)\} = \{\phi(\xi), \phi(\eta)\} = 0, \quad \{I(\xi), \phi(\eta)\} = \delta(\xi - \eta),$$
$$\tfrac{d}{dt} I(\xi) = 0, \quad \text{and} \quad \tfrac{d}{dt}\phi(\xi) = |\xi|^2.$$

REMARK. By integrating $|\widehat{u_+}(\xi)|^2$ against appropriate powers of ξ, one obtains conserved quantities that agree with the asymptotic $\dot H^s_x$ norm. For $s = 0$ or $s = 1$, these are exactly the mass and energy. For general values of s, the conserved quantities need not take such a simple (polynomial in u, $\bar u$, and their derivatives) form.

3. The local theory

3.1. Dispersive and Strichartz inequalities. It is not difficult to check (or derive) that the fundamental solution of the heat equation is given by

$$e^{s\Delta}(x,y) = (2\pi)^{-d} \int_{\mathbb{R}^d} e^{i\xi\cdot(x-y) - s|\xi|^2}\, d\xi = (4\pi s)^{-d/2} e^{-|x-y|^2/4s}$$

for all $s > 0$. By analytic continuation, we find the fundamental solution of the free Schrödinger equation:

$$e^{it\Delta}(x,y) = (4\pi it)^{-d/2} e^{i|x-y|^2/4t} \tag{3.1}$$

for all $t \neq 0$. Note that here

$$(4\pi it)^{-d/2} = (4\pi|t|)^{-d/2} e^{-i\pi d\,\mathrm{sign}(t)/4}.$$

From (3.1) one easily derives the standard dispersive inequality

$$\|e^{it\Delta} f\|_{L_x^p(\mathbb{R}^d)} \lesssim |t|^{d(\frac{1}{p}-\frac{1}{2})} \|f\|_{L_x^{p'}(\mathbb{R}^d)} \tag{3.2}$$

for all $t \neq 0$ and $2 \le p \le \infty$, where $\frac{1}{p} + \frac{1}{p'} = 1$.

A different way to express the dispersive effect of the operator $e^{it\Delta}$ is in terms of spacetime integrability. To state the estimates, we first need the following definition.

DEFINITION 3.1 (Admissible pairs). For $d \ge 1$, we say that a pair of exponents (q, r) is *Schrödinger-admissible* if

$$\frac{2}{q} + \frac{d}{r} = \frac{d}{2}, \quad 2 \le q, r \le \infty, \quad \text{and} \quad (d,q,r) \neq (2,2,\infty). \tag{3.3}$$

For a fixed spacetime slab $I \times \mathbb{R}^d$, we define the *Strichartz norm*

$$\|u\|_{S^0(I)} := \sup_{(q,r) \text{ admissible}} \|u\|_{L_t^q L_x^r(I \times \mathbb{R}^d)} \tag{3.4}$$

We write $S^0(I)$ for the closure of all test functions under this norm and denote by $N^0(I)$ the dual of $S^0(I)$.

REMARK. In the case of two space dimensions, the absence of the endpoint requires us to restrict the supremum in (3.4) to a closed subset of admissible pairs. As any reasonable argument only involves finitely many admissible pairs, this is of little consequence.

We are now ready to state the standard Strichartz estimates:

THEOREM 3.2 (Strichartz). *Let $0 \le s \le 1$, let I be a compact time interval, and let $u : I \times \mathbb{R}^d \to \mathbb{C}$ be a solution to the forced Schrödinger equation*

$$iu_t + \Delta u = F.$$

Then,

$$\||\nabla|^s u\|_{S^0(I)} \lesssim \|u(t_0)\|_{\dot H_x^s} + \||\nabla|^s F\|_{N^0(I)}$$

for any $t_0 \in I$.

PROOF. We will treat the non-endpoint cases in Subsection 4.4 following [**28, 83**]. For the endpoint $(q,r) = \big(2, \frac{2d}{d-2}\big)$ in dimensions $d \ge 3$, see [**37**]. For failure of the $d = 2$ endpoint, see [**59**]. This endpoint can be partially recovered in the case of spherically symmetric functions; see [**82, 87**]. □

3.2. The $\dot H^s_x$ critical case. In this subsection we revisit the local theory at critical regularity. Consider the initial-value problem

$$\begin{cases} iu_t + \Delta u = F(u) \\ u(0) = u_0 \end{cases} \tag{3.5}$$

where $u(t,x)$ is a complex-valued function of spacetime $\mathbb{R}\times\mathbb{R}^d$ with $d \geq 1$. Assume that the nonlinearity $F : \mathbb{C} \to \mathbb{C}$ is continuously differentiable and obeys the power-type estimates

$$F(z) = O\big(|z|^{1+p}\big) \tag{3.6}$$

$$F_z(z),\ F_{\bar z}(z) = O\big(|z|^p\big) \tag{3.7}$$

$$F_z(z) - F_z(w),\ F_{\bar z}(z) - F_{\bar z}(w) = O\big(|z-w|^{\min\{p,1\}}(|z|+|w|)^{\max\{0,p-1\}}\big) \tag{3.8}$$

for some $p > 0$, where F_z and $F_{\bar z}$ are the usual complex derivatives

$$F_z := \frac{1}{2}\Big(\frac{\partial F}{\partial x} - i\frac{\partial F}{\partial y}\Big), \quad F_{\bar z} := \frac{1}{2}\Big(\frac{\partial F}{\partial x} + i\frac{\partial F}{\partial y}\Big).$$

For future reference, we record the chain rule

$$\nabla F(u(x)) = F_z(u(x))\nabla u(x) + F_{\bar z}(u(x))\overline{\nabla u(x)}, \tag{3.9}$$

as well as the closely related integral identity

$$F(z)-F(w) = (z-w)\int_0^1 F_z\big(w+\theta(z-w)\big)\,d\theta + \overline{(z-w)}\int_0^1 F_{\bar z}\big(w+\theta(z-w)\big)\,d\theta \tag{3.10}$$

for any $z, w \in \mathbb{C}$; in particular, from (3.7), (3.10), and the triangle inequality, we have the estimate

$$\big|F(z) - F(w)\big| \lesssim |z-w|\big(|z|^p + |w|^p\big). \tag{3.11}$$

The model example of a nonlinearity obeying the conditions above is $F(u) = |u|^p u$, for which the critical homogeneous Sobolev space is $\dot H^{s_c}_x$ with $s_c := \frac{d}{2} - \frac{2}{p}$. The local theory for (3.5) at this critical regularity was developed by Cazenave and Weissler [**13, 14, 15**]. Like them, we are interested in strong solutions to (3.5).

Definition 3.3 (Solution). A function $u : I \times \mathbb{R}^d \to \mathbb{C}$ on a non-empty time interval $0 \in I \subset \mathbb{R}$ is a *solution* (more precisely, a strong $\dot H^{s_c}_x(\mathbb{R}^d)$ solution) to (3.5) if it lies in the class $C^0_t \dot H^{s_c}_x(K\times\mathbb{R}^d) \cap L^{p+2}_t L_x^{\frac{dp(p+2)}{4}}(K\times\mathbb{R}^d)$ for all compact $K \subset I$, and obeys the Duhamel formula

$$u(t) = e^{it\Delta}u(0) - i\int_0^t e^{i(t-s)\Delta}F(u(s))\,ds \tag{3.12}$$

for all $t \in I$. We refer to the interval I as the *lifespan* of u. We say that u is a *maximal-lifespan solution* if the solution cannot be extended to any strictly larger interval. We say that u is a *global solution* if $I = \mathbb{R}$.

Note that for $s_c \in \{0,1\}$, this is slightly different from the definition of solution given in the introduction. However, one of the consequences of the theory developed in this section is that the two notions are equivalent.

THEOREM 3.4 (Standard local well-posedness, [**13, 14, 15**]). *Let $d \geq 1$ and $u_0 \in H_x^{s_c}(\mathbb{R}^d)$. Assume further that $0 \leq s_c \leq 1$. There exists $\eta_0 = \eta_0(d) > 0$ such that if $0 < \eta \leq \eta_0$ and I is a compact interval containing zero such that*

$$\big\| |\nabla|^{s_c} e^{it\Delta} u_0 \big\|_{L_t^{p+2} L_x^{\frac{2d(p+2)}{2(d-2)+dp}} (I\times\mathbb{R}^d)} \leq \eta, \tag{3.13}$$

then there exists a unique solution u to (3.5) on $I \times \mathbb{R}^d$. Moreover, we have the bounds

$$\big\| |\nabla|^{s_c} u \big\|_{L_t^{p+2} L_x^{\frac{2d(p+2)}{2(d-2)+dp}} (I\times\mathbb{R}^d)} \leq 2\eta \tag{3.14}$$

$$\big\| |\nabla|^{s_c} u \big\|_{S^0(I\times\mathbb{R}^d)} \lesssim \big\| |\nabla|^{s_c} u_0 \big\|_{L_x^2} + \eta^{1+p} \tag{3.15}$$

$$\| u \|_{S^0(I\times\mathbb{R}^d)} \lesssim \| u_0 \|_{L_x^2}. \tag{3.16}$$

REMARKS. 1. By Strichartz inequality, we know that

$$\big\| |\nabla|^{s_c} e^{it\Delta} u_0 \big\|_{L_t^{p+2} L_x^{\frac{2d(p+2)}{2(d-2)+dp}} (\mathbb{R}\times\mathbb{R}^d)} \lesssim \big\| |\nabla|^{s_c} u_0 \big\|_{L_x^2}.$$

Thus, (3.13) holds for initial data with sufficiently small norm. Alternatively, by the monotone convergence theorem, (3.13) holds provided I is chosen sufficiently small. Note that by scaling, the length of the interval I depends on the fine properties of u_0, not only on its norm.

2. Note that the initial data in the theorem above is assumed to belong to the *inhomogeneous* Sobolev space $H_x^{s_c}(\mathbb{R}^d)$, as in the work of Cazenave and Weissler. This makes the proof significantly simpler. In the next two subsections, we will present a technique which allows one to show uniform continuous dependence of the solution u upon the initial data u_0 in *critical* spaces. This technique (or indeed, the result) can be used to treat initial data in the *homogeneous* Sobolev space $\dot{H}_x^{s_c}(\mathbb{R}^d)$.

3. The sole purpose of the restriction to $s_c \leq 1$ is to simplify the statement and proof. In any event, it covers the two cases of greatest interest to us, $s_c = 0, 1$.

PROOF. We will essentially repeat the original argument from [**14**]; the fractional chain rule Lemma A.11 leads to some simplifications.

The theorem follows from a contraction mapping argument. More precisely, using the Strichartz estimates from Theorem 3.2, we will show that the solution map $u \mapsto \Phi(u)$ defined by

$$\Phi(u)(t) := e^{it\Delta} u_0 - i \int_0^t e^{i(t-s)\Delta} F(u(s))\, ds,$$

is a contraction on the set $B_1 \cap B_2$ where

$$B_1 := \Big\{ u \in L_t^\infty H_x^{s_c}(I \times \mathbb{R}^d) : \|u\|_{L_t^\infty H_x^{s_c}(I\times\mathbb{R}^d)} \leq 2\|u_0\|_{H_x^{s_c}} + C(d)(2\eta)^{1+p} \Big\}$$

$$B_2 := \Big\{ u \in L_t^{p+2} W_x^{s_c, \frac{2d(p+2)}{2(d-2)+dp}} (I \times \mathbb{R}^d) : \big\| |\nabla|^{s_c} u \big\|_{L_t^{p+2} L_x^{\frac{2d(p+2)}{2(d-2)+dp}} (I\times\mathbb{R}^d)} \leq 2\eta$$

$$\text{and } \|u\|_{L_t^{p+2} L_x^{\frac{2d(p+2)}{2(d-2)+dp}} (I\times\mathbb{R}^d)} \leq 2C(d)\|u_0\|_{L_x^2} \Big\}$$

under the metric given by

$$d(u,v) := \|u - v\|_{L_t^{p+2} L_x^{\frac{2d(p+2)}{2(d-2)+dp}} (I\times\mathbb{R}^d)}.$$

Here $C(d)$ denotes the constant from the Strichartz inequality. Note that the norm appearing in the metric scales like L_x^2; see the second remark above. Note that both B_1 and B_2 are closed (and hence complete) in this metric.

Using Strichartz inequality followed by the fractional chain rule Lemma A.11 and Sobolev embedding, we find that for $u \in B_1 \cap B_2$,

$$\begin{aligned}
&\|\Phi(u)\|_{L_t^\infty H_x^{s_c}(I\times\mathbb{R}^d)} \\
&\quad\leq \|u_0\|_{H_x^{s_c}} + C(d)\big\|\langle\nabla\rangle^{s_c}F(u)\big\|_{L_t^{\frac{p+2}{p+1}}L_x^{\frac{2d(p+2)}{2(d+2)+dp}}(I\times\mathbb{R}^d)} \\
&\quad\leq \|u_0\|_{H_x^{s_c}} + C(d)\big\|\langle\nabla\rangle^{s_c}u\big\|_{L_t^{p+2}L_x^{\frac{2d(p+2)}{2(d-2)+dp}}(I\times\mathbb{R}^d)}\|u\|^p_{L_t^{p+2}L_x^{\frac{dp(p+2)}{4}}(I\times\mathbb{R}^d)} \\
&\quad\leq \|u_0\|_{H_x^{s_c}} + C(d)\big(2\eta + 2C(d)\|u_0\|_{L_x^2}\big)\big\||\nabla|^{s_c}u\big\|^p_{L_t^{p+2}L_x^{\frac{2d(p+2)}{2(d-2)+dp}}(I\times\mathbb{R}^d)} \\
&\quad\leq \|u_0\|_{H_x^{s_c}} + C(d)\big(2\eta + 2C(d)\|u_0\|_{L_x^2}\big)(2\eta)^p
\end{aligned}$$

and similarly,

$$\begin{aligned}
\big\|\Phi(u)\big\|_{L_t^{p+2}L_x^{\frac{2d(p+2)}{2(d-2)+dp}}(I\times\mathbb{R}^d)} &\leq C(d)\|u_0\|_{L_x^2} + C(d)\big\|F(u)\big\|_{L_t^{\frac{p+2}{p+1}}L_x^{\frac{2d(p+2)}{2(d+2)+dp}}(I\times\mathbb{R}^d)} \\
&\leq C(d)\|u_0\|_{L_x^2} + 2C(d)^2\|u_0\|_{L_x^2}(2\eta)^p.
\end{aligned}$$

Arguing as above and invoking (3.13), we obtain

$$\begin{aligned}
\big\||\nabla|^{s_c}\Phi(u)\big\|_{L_t^{p+2}L_x^{\frac{2d(p+2)}{2(d-2)+dp}}(I\times\mathbb{R}^d)} &\leq \eta + C(d)\big\||\nabla|^{s_c}F(u)\big\|_{L_t^{\frac{p+2}{p+1}}L_x^{\frac{2d(p+2)}{2(d+2)+dp}}(I\times\mathbb{R}^d)} \\
&\leq \eta + C(d)(2\eta)^{1+p}.
\end{aligned}$$

Thus, choosing $\eta_0 = \eta_0(d)$ sufficiently small, we see that for $0 < \eta \leq \eta_0$, the functional Φ maps the set $B_1 \cap B_2$ back to itself. To see that Φ is a contraction, we repeat the computations above and use (3.11) to obtain

$$\begin{aligned}
\big\|\Phi(u) - \Phi(v)\big\|_{L_t^{p+2}L_x^{\frac{2d(p+2)}{2(d-2)+dp}}(I\times\mathbb{R}^d)} &\leq C(d)\big\|F(u) - F(v)\big\|_{L_t^{\frac{p+2}{p+1}}L_x^{\frac{2d(p+2)}{2(d+2)+dp}}(I\times\mathbb{R}^d)} \\
&\leq C(d)(2\eta)^p\|u - v\|_{L_t^{p+2}L_x^{\frac{2d(p+2)}{2(d-2)+dp}}(I\times\mathbb{R}^d)}.
\end{aligned}$$

Thus, choosing $\eta_0 = \eta_0(d)$ even smaller (if necessary), we can guarantee that Φ is a contraction on the set $B_1 \cap B_2$. By the contraction mapping theorem, it follows that Φ has a fixed point in $B_1 \cap B_2$. Moreover, noting that Φ maps into $C_t^0 H_x^{s_c}$ (not just $L_t^\infty H_x^{s_c}$), we derive that the fixed point of Φ is indeed a solution to (3.5).

We now turn our attention to the uniqueness statement. Since uniqueness is a local property, it suffices to study a neighbourhood of $t = 0$. By Definition 3.3, any solution to (3.5) belongs to $B_1 \cap B_2$ on some such neighbourhood. Uniqueness thus follows from uniqueness in the contraction mapping theorem.

The claims (3.15) and (3.16) follow from another application of Strichartz inequality, as above. □

We end this section with a collection of statements which encapsulate the local theory for (3.5).

Corollary 3.5 (Local theory, [**13, 14, 15**]). *Let $d \geq 1$ and $u_0 \in H_x^{s_c}(\mathbb{R}^d)$. Assume also that $0 \leq s_c \leq 1$. Then there exists a unique maximal-lifespan solution $u : I \times \mathbb{R}^d \to \mathbb{C}$ to (3.5) with initial data $u(0) = u_0$. This solution also has the following properties:*

• *(Local existence) I is an open neighbourhood of zero.*
• *(Energy and mass conservation) The mass of u is conserved, that is, $M(u(t)) = M(u_0)$ for all $t \in I$. Moreover, if $s_c = 1$ then the energy of u is also conserved, that is, $E(u(t)) = E(u_0)$ for all $t \in I$.*
• *(Blowup criterion) If $\sup I$ is finite, then u blows up forward in time, that is, there exists a time $t \in I$ such that*

$$\|u\|_{L_t^{p+2} L_x^{\frac{pd(p+2)}{4}}([t,\sup I)\times\mathbb{R}^d)} = \infty.$$

A similar statement holds in the negative time direction.
• *(Scattering) If $\sup I = +\infty$ and u does not blow up forward in time, then u scatters forward in time, that is, there exists a unique $u_+ \in H_x^{s_c}(\mathbb{R}^d)$ such that*

$$\lim_{t\to+\infty} \|u(t) - e^{it\Delta}u_+\|_{H_x^{s_c}(\mathbb{R}^d)} = 0. \tag{3.17}$$

Conversely, given $u_+ \in H_x^{s_c}(\mathbb{R}^d)$ there exists a unique solution to (3.5) in a neighbourhood of infinity so that (3.17) holds.
• *(Small data global existence) If $\||\nabla|^{s_c}u_0\|_2$ is sufficiently small (depending on d), then u is a global solution which does not blow up either forward or backward in time. Indeed,*

$$\||\nabla|^{s_c}u\|_{S^0(\mathbb{R})} \lesssim \||\nabla|^{s_c}u_0\|_2. \tag{3.18}$$

• *(Unconditional uniqueness in the energy-critical case) Suppose $s_c = 1$ and $\tilde{u} \in C_t^0 \dot{H}_x^1(J \times \mathbb{R}^d)$ obeys (3.12) and $\tilde{u}(t_0) = u_0$, then $J \subseteq I$ and $\tilde{u} \equiv u$ throughout J.*

PROOF. The corollary is a consequence of Theorem 3.4 and its proof. We leave it as an exercise. □

3.3. Stability: the mass-critical case. An important part of the local well-posedness theory is the study of how the strong solutions built in the previous subsection depend upon the initial data. More precisely, we would like to know whether small perturbations of the initial data lead to small changes in the solution. More generally, we are interested in developing a *stability* theory for (3.5). By stability, we mean the following property: Given an *approximate* solution to (3.5), say $\tilde{u}$ obeying

$$\begin{cases} i\tilde{u}_t + \Delta\tilde{u} = F(\tilde{u}) + e \\ \tilde{u}(0,x) = \tilde{u}_0(x) \end{cases}$$

with e small in a suitable space and $\tilde{u}_0 - u_0$ small in $\dot{H}_x^{s_c}$, then there exists a *genuine* solution u to (3.5) which stays very close to $\tilde{u}$ in critical norms. The question of continuous dependence of the solution upon the initial data corresponds to taking $e = 0$; the case where $e \neq 0$ can be used to consider situations where NLS is only an approximate model for the physical system under consideration.

Although stability is a local question, it plays an important role in all existing treatments of the global well-posedness problem for NLS at critical regularity. It has also proved useful in the treatment of local and global questions for more exotic nonlinearities [**95, 108**].

In these notes, we will only address the stability question for the mass- and energy-critical NLS. The techniques we will employ (particularly, those from the next subsection) can be used to develop a stability theory for the more general equation (3.5). We start with the mass-critical equation, which is the more elementary of the two. That is to say, for the remainder of this subsection we adopt the following

CONVENTION. The nonlinearity F obeys (3.6) through (3.8) and (3.11) with $p = 4/d$.

LEMMA 3.6 (Short-time perturbations, [**95**]). *Let I be a compact interval and let $\tilde{u}$ be an approximate solution to (3.5) in the sense that*

$$(i\partial_t + \Delta)\tilde{u} = F(\tilde{u}) + e,$$

for some function e. Assume that

$$\|\tilde{u}\|_{L_t^\infty L_x^2(I\times\mathbb{R}^d)} \leq M \tag{3.19}$$

for some positive constant M. Let $t_0 \in I$ and let $u(t_0)$ be such that

$$\|u(t_0) - \tilde{u}(t_0)\|_{L_x^2} \leq M' \tag{3.20}$$

for some $M' > 0$. Assume also the smallness conditions

$$\|\tilde{u}\|_{L_{t,x}^{\frac{2(d+2)}{d}}(I\times\mathbb{R}^d)} \leq \varepsilon_0 \tag{3.21}$$

$$\big\|e^{i(t-t_0)\Delta}\big(u(t_0) - \tilde{u}(t_0)\big)\big\|_{L_{t,x}^{\frac{2(d+2)}{d}}(I\times\mathbb{R}^d)} \leq \varepsilon \tag{3.22}$$

$$\|e\|_{N^0(I)} \leq \varepsilon, \tag{3.23}$$

for some $0 < \varepsilon \leq \varepsilon_0$ where $\varepsilon_0 = \varepsilon_0(M, M') > 0$ is a small constant. Then, there exists a solution u to (3.5) on $I \times \mathbb{R}^d$ with initial data $u(t_0)$ at time $t = t_0$ satisfying

$$\|u - \tilde{u}\|_{L_{t,x}^{\frac{2(d+2)}{d}}(I\times\mathbb{R}^d)} \lesssim \varepsilon \tag{3.24}$$

$$\|u - \tilde{u}\|_{S^0(I)} \lesssim M' \tag{3.25}$$

$$\|u\|_{S^0(I)} \lesssim M + M' \tag{3.26}$$

$$\|F(u) - F(\tilde{u})\|_{N^0(I)} \lesssim \varepsilon. \tag{3.27}$$

REMARK. Note that by Strichartz,

$$\big\|e^{i(t-t_0)\Delta}\big(u(t_0) - \tilde{u}(t_0)\big)\big\|_{L_{t,x}^{\frac{2(d+2)}{d}}(I\times\mathbb{R}^d)} \lesssim \|u(t_0) - \tilde{u}(t_0)\|_{L_x^2},$$

so hypothesis (3.22) is redundant if $M' = O(\varepsilon)$.

PROOF. By symmetry, we may assume $t_0 = \inf I$. Let $w := u - \tilde{u}$. Then w satisfies the following initial value problem

$$\begin{cases} iw_t + \Delta w = F(\tilde{u} + w) - F(\tilde{u}) - e \\ w(t_0) = u(t_0) - \tilde{u}(t_0). \end{cases}$$

For $t \in I$ we define

$$A(t) := \big\|F(\tilde{u} + w) - F(\tilde{u})\big\|_{N^0([t_0,t])}.$$

By (3.21),

$$\begin{aligned} A(t) &\lesssim \big\|F(\tilde{u} + w) - F(\tilde{u})\big\|_{L_{t,x}^{\frac{2(d+2)}{d+4}}([t_0,t]\times\mathbb{R}^d)} \\ &\lesssim \|w\|^{1+\frac{4}{d}}_{L_{t,x}^{\frac{2(d+2)}{d}}([t_0,t]\times\mathbb{R}^d)} + \|\tilde{u}\|^{\frac{4}{d}}_{L_{t,x}^{\frac{2(d+2)}{d}}([t_0,t]\times\mathbb{R}^d)} \|w\|_{L_{t,x}^{\frac{2(d+2)}{d}}([t_0,t]\times\mathbb{R}^d)} \end{aligned}$$

$$\lesssim \|w\|^{1+\frac{4}{d}}_{L^{\frac{2(d+2)}{d}}_{t,x}([t_0,t]\times\mathbb{R}^d)} + \varepsilon_0^{\frac{4}{d}}\|w\|_{L^{\frac{2(d+2)}{d}}_{t,x}([t_0,t]\times\mathbb{R}^d)}. \tag{3.28}$$

On the other hand, by Strichartz, (3.22), and (3.23), we get

$$\begin{aligned}\|w\|_{L^{\frac{2(d+2)}{d}}_{t,x}([t_0,t]\times\mathbb{R}^d)} &\lesssim \|e^{i(t-t_0)\Delta}w(t_0)\|_{L^{\frac{2(d+2)}{d}}_{t,x}([t_0,t]\times\mathbb{R}^d)} + A(t) + \|e\|_{N^0([t_0,t])}\\ &\lesssim A(t)+\varepsilon. \end{aligned}\tag{3.29}$$

Combining (3.28) and (3.29), we obtain

$$A(t) \lesssim (A(t)+\varepsilon)^{1+\frac{4}{d}} + \varepsilon_0^{\frac{4}{d}}(A(t)+\varepsilon).$$

A standard continuity argument then shows that if ε_0 is taken sufficiently small,

$$A(t)\lesssim \varepsilon \quad \text{for any} \quad t\in I,$$

which implies (3.27). Using (3.27) and (3.29), one easily derives (3.24). Moreover, by Strichartz, (3.20), (3.23), and (3.27),

$$\|w\|_{S^0(I)} \lesssim \|w(t_0)\|_{L^2_x} + \|F(\tilde u + w) - F(\tilde u)\|_{N^0(I)} + \|e\|_{N^0(I)} \lesssim M' + \varepsilon,$$

which establishes (3.25) for $\varepsilon_0 = \varepsilon_0(M')$ sufficiently small.

To prove (3.26), we use Strichartz, (3.19), (3.20), (3.27), and (3.21):

$$\begin{aligned}\|u\|_{S^0(I)} &\lesssim \|u(t_0)\|_{L^2_x} + \|F(u)\|_{N^0(I)}\\ &\lesssim \|\tilde u(t_0)\|_{L^2_x} + \|u(t_0)-\tilde u(t_0)\|_{L^2_x} + \|F(u)-F(\tilde u)\|_{N^0(I)} + \|F(\tilde u)\|_{N^0(I)}\\ &\lesssim M + M' + \varepsilon + \|\tilde u\|^{1+\frac{4}{d}}_{L^{\frac{2(d+2)}{d}}_{t,x}(I\times\mathbb{R}^d)}\\ &\lesssim M + M' + \varepsilon + \varepsilon_0^{1+\frac{4}{d}}.\end{aligned}$$

Choosing $\varepsilon_0 = \varepsilon_0(M, M')$ sufficiently small, this finishes the proof of the lemma. □

Building upon the previous result, we are now able to prove stability for the mass-critical NLS.

THEOREM 3.7 (Mass-critical stability result, [**95**]). *Let I be a compact interval and let $\tilde u$ be an approximate solution to (3.5) in the sense that*

$$(i\partial_t + \Delta)\tilde u = F(\tilde u) + e,$$

for some function e. Assume that

$$\|\tilde u\|_{L^\infty_t L^2_x(I\times\mathbb{R}^d)} \le M \tag{3.30}$$

$$\|\tilde u\|_{L^{\frac{2(d+2)}{d}}_{t,x}(I\times\mathbb{R}^d)} \le L, \tag{3.31}$$

for some positive constants M and L. Let $t_0 \in I$ and let $u(t_0)$ obey

$$\|u(t_0) - \tilde u(t_0)\|_{L^2_x} \le M' \tag{3.32}$$

for some $M' > 0$. Moreover, assume the smallness conditions

$$\left\|e^{i(t-t_0)\Delta}\big(u(t_0)-\tilde u(t_0)\big)\right\|_{L^{\frac{2(d+2)}{d}}_{t,x}(I\times\mathbb{R}^d)} \le \varepsilon \tag{3.33}$$

$$\|e\|_{N^0(I)} \le \varepsilon, \tag{3.34}$$

for some $0 < \varepsilon \leq \varepsilon_1$ *where* $\varepsilon_1 = \varepsilon_1(M, M', L) > 0$ *is a small constant. Then, there exists a solution* u *to* (3.5) *on* $I \times \mathbb{R}^d$ *with initial data* $u(t_0)$ *at time* $t = t_0$ *satisfying*

$$\|u - \tilde{u}\|_{L_{t,x}^{\frac{2(d+2)}{d}}(I\times\mathbb{R}^d)} \leq \varepsilon C(M, M', L) \tag{3.35}$$

$$\|u - \tilde{u}\|_{S^0(I)} \leq C(M, M', L)M' \tag{3.36}$$

$$\|u\|_{S^0(I)} \leq C(M, M', L). \tag{3.37}$$

PROOF. Subdivide I into $J \sim (1 + \frac{L}{\varepsilon_0})^{\frac{2(d+2)}{d}}$ subintervals $I_j = [t_j, t_{j+1}]$, $0 \leq j < J$, such that

$$\|\tilde{u}\|_{L_{t,x}^{\frac{2(d+2)}{d}}(I_j\times\mathbb{R}^d)} \leq \varepsilon_0,$$

where $\varepsilon_0 = \varepsilon_0(M, 2M')$ is as in Lemma 3.6. We need to replace M' by $2M'$ as the mass of the difference $u - \tilde{u}$ might grow slightly in time.

By choosing ε_1 sufficiently small depending on J, M, and M', we can apply Lemma 3.6 to obtain for each j and all $0 < \varepsilon < \varepsilon_1$

$$\begin{aligned}
\|u - \tilde{u}\|_{L_{t,x}^{\frac{2(d+2)}{d}}(I_j\times\mathbb{R}^d)} &\leq C(j)\varepsilon \\
\|u - \tilde{u}\|_{S^0(I_j)} &\leq C(j)M' \\
\|u\|_{S^0(I_j)} &\leq C(j)(M + M') \\
\|F(u) - F(\tilde{u})\|_{N^0(I_j)} &\leq C(j)\varepsilon,
\end{aligned}$$

provided we can prove that analogues of (3.32) and (3.33) hold with t_0 replaced by t_j. In order to verify this, we use an inductive argument. By Strichartz, (3.32), (3.34), and the inductive hypothesis,

$$\begin{aligned}
\|u(t_j) - \tilde{u}(t_j)\|_{L_x^2} &\lesssim \|u(t_0) - \tilde{u}(t_0)\|_{L_x^2} + \|F(u) - F(\tilde{u})\|_{N^0([t_0,t_j])} + \|e\|_{N^0([t_0,t_j])} \\
&\lesssim M' + \sum_{k=0}^{j-1} C(k)\varepsilon + \varepsilon.
\end{aligned}$$

Similarly, by Strichartz, (3.33), (3.34), and the inductive hypothesis,

$$\begin{aligned}
&\big\|e^{i(t-t_j)\Delta}\big(u(t_j) - \tilde{u}(t_j)\big)\big\|_{L_{t,x}^{\frac{2(d+2)}{d}}(I_j\times\mathbb{R}^d)} \\
&\lesssim \big\|e^{i(t-t_0)\Delta}\big(u(t_0) - \tilde{u}(t_0)\big)\big\|_{L_{t,x}^{\frac{2(d+2)}{d}}(I_j\times\mathbb{R}^d)} + \|e\|_{N^0([t_0,t_j])} \\
&\qquad + \|F(u) - F(\tilde{u})\|_{N^0([t_0,t_j])} \\
&\lesssim \varepsilon + \sum_{k=0}^{j-1} C(k)\varepsilon.
\end{aligned}$$

Choosing ε_1 sufficiently small depending on J, M, and M', we can guarantee that the hypotheses of Lemma 3.6 continue to hold as j varies. □

3.4. Stability: the energy-critical case. In this subsection we address the stability theory for the energy-critical NLS, that is, we adopt the following

CONVENTION. The nonlinearity F obeys (3.6) through (3.8) and (3.11) with $p = 4/(d-2)$ and $d \geq 3$.

To motivate the approach we will take, let us consider the question of continuous dependence of the solution upon the initial data. To make things as simple as possible, let us choose initial data $u_0, \tilde{u}_0 \in H_x^1$ which are small:

$$\|u_0\|_{\dot{H}_x^1} + \|\tilde{u}_0\|_{\dot{H}_x^1} \leq \eta_0.$$

By Corollary 3.5, if η_0 is sufficiently small, there exist unique global solutions u and $\tilde{u}$ to (3.5) with initial data u_0 and $\tilde{u}_0$, respectively; moreover, they satisfy

$$\|\nabla u\|_{S^0(\mathbb{R})} + \|\nabla \tilde{u}\|_{S^0(\mathbb{R})} \lesssim \eta_0.$$

We would like to see that if u_0 and $\tilde{u}_0$ are *close* in $\dot{H}_x^1$, say $\|\nabla(u_0 - \tilde{u}_0)\|_2 \leq \varepsilon \ll \eta_0$, then u and $\tilde{u}$ remain close in *energy-critical* norms, measured in terms of ε, not η_0. An application of Strichartz inequality combined with the bounds above yields

$$\|\nabla(u - \tilde{u})\|_{S^0(\mathbb{R})} \lesssim \|\nabla(u_0 - \tilde{u}_0)\|_{L_x^2} + \eta_0^{\frac{4}{d-2}} \|\nabla(u - \tilde{u})\|_{S^0(\mathbb{R})} + \eta_0 \|\nabla(u - \tilde{u})\|_{S^0(\mathbb{R})}^{\frac{4}{d-2}}.$$

If $4/(d-2) \geq 1$, a simple bootstrap argument will imply continuous dependence of the solution upon the initial data. However, this will not work if $4/(d-2) < 1$, that is, if $d > 6$. The obstacle comes from the last term above; tiny numbers become much larger when raised to a fractional power. Ultimately, the problem stems from the fact that in high dimensions the derivative maps F_z and $F_{\bar{z}}$ are merely Hölder continuous rather than Lipschitz. The remedy is to work in spaces with fractional derivatives (rather than a full derivative), while still maintaining criticality with respect to the scaling. This is the approach taken by Tao and Visan [**94**], who proved stability for the energy-critical NLS in all dimensions $d \geq 3$ (see also [**20, 75**] for earlier treatments in dimensions $d = 3, 4$). A similar technique was employed by Nakanishi [**64**] for the energy-critical Klein-Gordon equation in high dimensions.

Here we present a small improvement upon the results obtained in [**94**] made possible by the fractional chain rule for fractional powers; see Lemma A.12. The proof is rather involved and will occupy the remainder of this subsection. It is joint work with Xiaoyi Zhang (unpublished).

THEOREM 3.8 (Energy-critical stability result). *Let I be a compact time interval and let $\tilde{u}$ be an approximate solution to (3.5) on $I \times \mathbb{R}^d$ in the sense that*

$$i\tilde{u}_t + \Delta\tilde{u} = F(\tilde{u}) + e$$

for some function e. Assume that

$$\|\tilde{u}\|_{L_t^\infty \dot{H}_x^1(I \times \mathbb{R}^d)} \leq E \tag{3.38}$$

$$\|\tilde{u}\|_{L_{t,x}^{\frac{2(d+2)}{d-2}}(I \times \mathbb{R}^d)} \leq L \tag{3.39}$$

for some positive constants E and L. Let $t_0 \in I$ and let $u(t_0)$ obey

$$\|u(t_0) - \tilde{u}(t_0)\|_{\dot{H}_x^1} \leq E' \tag{3.40}$$

for some positive constant E'. Assume also the smallness conditions

$$\left\|e^{i(t-t_0)\Delta}\big(u(t_0) - \tilde{u}(t_0)\big)\right\|_{L_{t,x}^{\frac{2(d+2)}{d-2}}(I \times \mathbb{R}^d)} \leq \varepsilon \tag{3.41}$$

$$\|\nabla e\|_{N^0(I)} \leq \varepsilon \tag{3.42}$$

for some $0 < \varepsilon < \varepsilon_1 = \varepsilon_1(E, E', L)$. *Then, there exists a unique strong solution* $u : I \times \mathbb{R}^d \mapsto \mathbb{C}$ *to* (3.5) *with initial data* $u(t_0)$ *at time* $t = t_0$ *satisfying*

$$\|u - \tilde{u}\|_{L_{t,x}^{\frac{2(d+2)}{d-2}}(I\times\mathbb{R}^d)} \lesssim C(E, E', L)\varepsilon^c \tag{3.43}$$

$$\|\nabla(u - \tilde{u})\|_{\dot{S}^0(I)} \lesssim C(E, E', L)\, E' \tag{3.44}$$

$$\|\nabla u\|_{\dot{S}^0(I)} \lesssim C(E, E', L), \tag{3.45}$$

where $0 < c = c(d) < 1$.

REMARK. The result in [**94**] assumes

$$\Big(\sum_{N\in 2^{\mathbb{Z}}} \big\|\nabla P_N e^{i(t-t_0)\Delta}(u(t_0) - \tilde{u}(t_0))\big\|^2_{L_t^{\frac{2(d+2)}{d-2}} L_x^{\frac{2d(d+2)}{d^2+4}}(I\times\mathbb{R}^d)} \Big)^{1/2} \leq \varepsilon$$

in place of (3.41). Note that by Sobolev embedding, this is a strictly stronger requirement.

One of the consequences of the theorem above is a local well-posedness statement in energy-critical norms. More precisely, in Theorem 3.4 and Corollary 3.5 one can remove the assumption that the initial data belongs to L_x^2, since every $\dot{H}_x^1$ function is well approximated by H_x^1 functions. Alternatively, one may use the techniques we present to prove the following corollary directly. The approach we have chosen is motivated by the desire to introduce the difficulties one at a time.

COROLLARY 3.9 (Local well-posedness). *Let* I *be a compact time interval,* $t_0 \in I$, *and let* $u_0 \in \dot{H}_x^1(\mathbb{R}^d)$. *Assume that*

$$\|u_0\|_{\dot{H}_x^1} \leq E.$$

Then for any $\varepsilon > 0$ *there exists* $\delta = \delta(E, \varepsilon) > 0$ *such that if*

$$\big\|e^{i(t-t_0)\Delta}u_0\big\|_{L_{t,x}^{\frac{2(d+2)}{d-2}}(I\times\mathbb{R}^d)} < \delta,$$

then there exists a unique solution u *to* (3.5) *with initial data* u_0 *at time* $t = t_0$. *Moreover,*

$$\|u\|_{L_{t,x}^{\frac{2(d+2)}{d-2}}(I\times\mathbb{R}^d)} \leq \varepsilon \quad \textit{and} \quad \|\nabla u\|_{S^0(I)} \leq 2E.$$

We now turn our attention to the proof of Theorem 3.8. Let us first introduce the spaces we will use; as mentioned above, these are critical with respect to scaling and have a small fractional number of derivatives. Throughout the remainder of this subsection, for any time interval I we will use the abbreviations

$$\begin{aligned} \|u\|_{X^0(I)} &:= \|u\|_{L_t^{\frac{d(d+2)}{2(d-2)}} L_x^{\frac{2d^2(d+2)}{(d+4)(d-2)^2}}(I\times\mathbb{R}^d)} \\ \|u\|_{X(I)} &:= \big\||\nabla|^{\frac{4}{d+2}} u\big\|_{L_t^{\frac{d(d+2)}{2(d-2)}} L_x^{\frac{2d^2(d+2)}{d^3-4d+16}}(I\times\mathbb{R}^d)} \\ \|F\|_{Y(I)} &:= \big\||\nabla|^{\frac{4}{d+2}} F\big\|_{L_t^{\frac{d}{2}} L_x^{\frac{2d^2(d+2)}{d^3+4d^2+4d-16}}(I\times\mathbb{R}^d)}. \end{aligned} \tag{3.46}$$

First, we connect the spaces in which the solution to (3.5) is measured to the spaces in which the nonlinearity is measured. As usual, this is done via a Strichartz inequality; we reproduce the standard proof.

LEMMA 3.10 (Strichartz estimate). *Let I be a compact time interval containing t_0. Then*

$$\Big\| \int_{t_0}^{t} e^{i(t-s)\Delta} F(s)\, ds \Big\|_{X(I)} \lesssim \|F\|_{Y(I)}.$$

PROOF. By the dispersive estimate (3.2),

$$\big\|e^{i(t-s)\Delta}F(s)\big\|_{L_x^{\frac{2d^2(d+2)}{d^3-4d+16}}} \lesssim |t-s|^{-\frac{d^2+2d-8}{d(d+2)}} \|F(s)\|_{L_x^{\frac{2d^2(d+2)}{d^3+4d^2+4d-16}}}.$$

An application of the Hardy-Littlewood-Sobolev inequality yields

$$\Big\|\int_{t_0}^{t} e^{i(t-s)\Delta}F(s)ds\Big\|_{L_t^{\frac{d(d+2)}{2(d-2)}} L_x^{\frac{2d^2(d+2)}{d^3-4d+16}}(I\times\mathbb{R}^d)} \lesssim \|F\|_{L_t^{\frac{d}{2}} L_x^{\frac{2d^2(d+2)}{d^3+4d^2+4d-16}}(I\times\mathbb{R}^d)}.$$

As the differentiation operator $|\nabla|^{\frac{4}{d+2}}$ commutes with the free evolution, we recover the claim. □

We next establish some connections between the spaces defined in (3.46) and the usual Strichartz spaces.

LEMMA 3.11 (Interpolations). *For any compact time interval I,*

$$\|u\|_{X^0(I)} \lesssim \|u\|_{X(I)} \lesssim \|\nabla u\|_{S^0(I)} \tag{3.47}$$

$$\|u\|_{X(I)} \lesssim \|u\|^{\frac{1}{d+2}}_{L_{t,x}^{\frac{2(d+2)}{d-2}}(I\times\mathbb{R}^d)} \|\nabla u\|^{\frac{d+1}{d+2}}_{S^0(I)} \tag{3.48}$$

$$\|u\|_{L_{t,x}^{\frac{2(d+2)}{d-2}}(I\times\mathbb{R}^d)} \lesssim \|u\|^{c}_{X(I)} \|\nabla u\|^{1-c}_{S^0(I)}, \tag{3.49}$$

where $0 < c = c(d) \leq 1$.

PROOF. A simple application of Sobolev embedding yields (3.47).
Using interpolation followed by Sobolev embedding,

$$\begin{aligned}\|u\|_{X(I)} &\lesssim \|u\|^{\frac{1}{d+2}}_{L_{t,x}^{\frac{2(d+2)}{d-2}}(I\times\mathbb{R}^d)} \big\||\nabla|^{\frac{4}{d+1}}u\big\|^{\frac{d+1}{d+2}}_{L_t^{\frac{2d(d+1)(d+2)}{(d-2)(3d+8)}} L_x^{\frac{2d^2(d+1)(d+2)}{d^4+d^3-2d^2+8d+32}}(I\times\mathbb{R}^d)} \\ &\lesssim \|u\|^{\frac{1}{d+2}}_{L_{t,x}^{\frac{2(d+2)}{d-2}}(I\times\mathbb{R}^d)} \|\nabla u\|^{\frac{d+1}{d+2}}_{S^0(I)}.\end{aligned}$$

This settles (3.48).

To establish (3.49), we analyze two cases. When $d = 3$, interpolation yields

$$\|u\|_{L_{t,x}^{\frac{2(d+2)}{d-2}}(I\times\mathbb{R}^d)} \lesssim \|u\|^{\frac{3}{4}}_{X^0(I)} \|u\|^{\frac{1}{4}}_{L_t^\infty L_x^{\frac{2d}{d-2}}(I\times\mathbb{R}^d)}$$

and the claim follows (with $c = \frac{3}{4}$) from (3.47) and Sobolev embedding. For $d \geq 4$, another application of interpolation gives

$$\|u\|_{L_{t,x}^{\frac{2(d+2)}{d-2}}(I\times\mathbb{R}^d)} \lesssim \|u\|^{\frac{2}{d-2}}_{X^0(I)} \|u\|^{\frac{d-4}{d-2}}_{L_t^{\frac{2d}{d-2}} L_x^{\frac{2d^2}{(d-2)^2}}(I\times\mathbb{R}^d)}$$

and the claim follows again (with $c = \frac{2}{d-2}$) from (3.47) and Sobolev embedding. □

Finally, we derive estimates that will help us control the nonlinearity. The main tools we use in deriving these estimates are the fractional chain rules; see Lemmas A.11 and A.12.

LEMMA 3.12 (Nonlinear estimates). *Let I a compact time interval. Then,*

$$\|F(u)\|_{Y(I)} \lesssim \|u\|_{X(I)}^{\frac{d+2}{d-2}} \tag{3.50}$$

and

$$\|F_z(u+v)w\|_{Y(I)} + \|F_{\bar z}(u+v)\bar w\|_{Y(I)} \lesssim \Big(\|u\|_{X(I)}^{\frac{8}{d^2-4}}\|\nabla u\|_{S^0(I)}^{\frac{4d}{d^2-4}} + \|v\|_{X(I)}^{\frac{8}{d^2-4}}\|\nabla v\|_{S^0(I)}^{\frac{4d}{d^2-4}}\Big)\|w\|_{X(I)}. \tag{3.51}$$

PROOF. Throughout the proof, all spacetime norms are on $I \times \mathbb{R}^d$.

Applying Lemma A.11 combined with (3.7) and (3.47) we find

$$\|F(u)\|_{Y(I)} \lesssim \|u\|_{L_t^{\frac{d(d+2)}{2(d-2)}} L_x^{\frac{2d^2(d+2)}{(d-2)^2(d+4)}}}^{\frac{4}{d-2}} \big\||\nabla|^{\frac{4}{d+2}} u\big\|_{L_t^{\frac{d(d+2)}{2(d-2)}} L_x^{\frac{2d^2(d+2)}{d^3-4d+16}}} \lesssim \|u\|_{X(I)}^{\frac{d+2}{d-2}}.$$

This establishes (3.50).

We now turn to (3.51); we only treat the first term on the left-hand side, as the second can be handled similarly. By Lemma A.10 followed by (3.7) and (3.47),

$$\begin{aligned}
&\|F_z(u+v)w\|_{Y(I)}\\
&\quad\lesssim \|F_z(u+v)\|_{L_t^{\frac{d(d+2)}{8}} L_x^{\frac{d^2(d+2)}{2(d-2)(d+4)}}} \big\||\nabla|^{\frac{4}{d+2}} w\big\|_{L_t^{\frac{d(d+2)}{2(d-2)}} L_x^{\frac{2d^2(d+2)}{d^3-4d+16}}}\\
&\qquad + \big\||\nabla|^{\frac{4}{d+2}} F_z(u+v)\big\|_{L_t^{\frac{d(d+2)}{8}} L_x^{\frac{d^2(d+2)}{2d^2+8d-16}}} \|w\|_{X^0(I)}\\
&\quad\lesssim \|u+v\|_{X^0(I)}^{\frac{4}{d-2}} \|w\|_{X(I)} + \big\||\nabla|^{\frac{4}{d+2}} F_z(u+v)\big\|_{L_t^{\frac{d(d+2)}{8}} L_x^{\frac{d^2(d+2)}{2d^2+8d-16}}} \|w\|_{X(I)}.
\end{aligned}$$

Thus, the claim will follow from (3.47), once we establish

$$\begin{aligned}
\big\||\nabla|^{\frac{4}{d+2}} F_z(u+v)\big\|_{L_t^{\frac{d(d+2)}{8}} L_x^{\frac{d^2(d+2)}{2d^2+8d-16}}}&\\
\lesssim \|u\|_{X(I)}^{\frac{8}{d^2-4}}\|\nabla u\|_{S^0(I)}^{\frac{4d}{d^2-4}} &+ \|v\|_{X(I)}^{\frac{8}{d^2-4}}\|\nabla v\|_{S^0(I)}^{\frac{4d}{d^2-4}}.
\end{aligned} \tag{3.52}$$

In dimensions $3 \leq d \leq 5$, this follows from Lemma A.11 and (3.47):

$$\big\||\nabla|^{\frac{4}{d+2}} F_z(u+v)\big\|_{L_t^{\frac{d(d+2)}{8}} L_x^{\frac{d^2(d+2)}{2d^2+8d-16}}} \lesssim \|u+v\|_{X^0(I)}^{\frac{6-d}{d-2}} \|u+v\|_{X(I)} \lesssim \|u+v\|_{X(I)}^{\frac{4}{d-2}}.$$

To derive (3.52) in dimensions $d \geq 6$, we apply Lemma A.12 (with $\alpha := \frac{4}{d-2}$, $s := \frac{4}{d+2}$, and $\sigma := \frac{d}{d+2}$) followed by Hölder's inequality in the time variable, Sobolev embedding, and (3.47):

$$\begin{aligned}
&\big\||\nabla|^{\frac{4}{d+2}} F_z(u+v)\big\|_{L_t^{\frac{d(d+2)}{8}} L_x^{\frac{d^2(d+2)}{2d^2+8d-16}}}\\
&\quad\lesssim \|u+v\|_{L_t^{\frac{d(d+2)}{2(d-2)}} L_x^{\frac{2d^2(d+2)}{(d+4)(d-2)^2}}}^{\frac{8}{d(d-2)}} \big\||\nabla|^{\frac{d}{d+2}}(u+v)\big\|_{L_t^{\frac{d(d+2)}{2(d-2)}} L_x^{\frac{2d^2(d+2)}{d^3+2d^2-12d+16}}}^{\frac{4}{d}}\\
&\quad\lesssim \big\||\nabla|^{\frac{d}{d+2}}(u+v)\big\|_{L_t^{\frac{d(d+2)}{2(d-2)}} L_x^{\frac{2d^2(d+2)}{d^3+2d^2-12d+16}}}^{\frac{4}{d-2}}\\
&\quad\lesssim \|u\|_{X(I)}^{\frac{8}{d^2-4}}\|\nabla u\|_{S^0(I)}^{\frac{4d}{d^2-4}} + \|v\|_{X(I)}^{\frac{8}{d^2-4}}\|\nabla v\|_{S^0(I)}^{\frac{4d}{d^2-4}}.
\end{aligned}$$

This settles (3.52) and hence (3.51). □

We have now all the tools we need to attack Theorem 3.8. As in the mass-critical setting, the stability result for the energy-critical NLS will be obtained iteratively from a short-time perturbation result.

LEMMA 3.13 (Short-time perturbations). *Let I be a compact time interval and let $\tilde{u}$ be an approximate solution to* (3.5) *on $I \times \mathbb{R}^d$ in the sense that*

$$i\tilde{u}_t + \Delta\tilde{u} = F(\tilde{u}) + e$$

for some function e. Assume that

$$\|\tilde{u}\|_{L_t^\infty \dot{H}_x^1(I\times\mathbb{R}^d)} \leq E$$

for some positive constant E. Moreover, let $t_0 \in I$ and let $u(t_0)$ obey

$$\|u(t_0) - \tilde{u}(t_0)\|_{\dot{H}_x^1} \leq E'$$

for some positive constant E'. Assume also the smallness conditions

$$\|\tilde{u}\|_{X(I)} \leq \delta \tag{3.53}$$

$$\big\|e^{i(t-t_0)\Delta}\big(u(t_0) - \tilde{u}(t_0)\big)\big\|_{X(I)} \leq \varepsilon \tag{3.54}$$

$$\|\nabla e\|_{N^0(I)} \leq \varepsilon \tag{3.55}$$

for some small $0 < \delta = \delta(E)$ and $0 < \varepsilon < \varepsilon_0(E, E')$. Then there exists a unique solution $u : I \times \mathbb{R}^d \to \mathbb{C}$ to (3.5) *with initial data $u(t_0)$ at time $t = t_0$ satisfying*

$$\|u - \tilde{u}\|_{X(I)} \lesssim \varepsilon \tag{3.56}$$

$$\|\nabla(u - \tilde{u})\|_{S^0(I)} \lesssim E' \tag{3.57}$$

$$\|\nabla u\|_{S^0(I)} \lesssim E + E' \tag{3.58}$$

$$\|F(u) - F(\tilde{u})\|_{Y(I)} \lesssim \varepsilon \tag{3.59}$$

$$\big\|\nabla\big(F(u) - F(\tilde{u})\big)\big\|_{N^0(I)} \lesssim E'. \tag{3.60}$$

PROOF. We prove the lemma under the additional assumption that $M(u) < \infty$, so that we can rely on Theorem 3.4 to guarantee that u exists. This additional assumption can be removed *a posteriori* by the usual limiting argument: approximate $u(t_0)$ in $\dot{H}_x^1$ by $\{u_n(t_0)\}_n \subseteq H_x^1$ and apply the lemma with $\tilde{u} = u_m$, $u = u_n$, and $e = 0$ to deduce that the sequence of solutions $\{u_n\}_n$ with initial data $\{u_n(t_0)\}_n$ is Cauchy in energy-critical norms and thus convergent to a solution u with initial data $u(t_0)$ which obeys $\nabla u \in S^0(I)$. Thus, it suffices to prove (3.56) through (3.60) as *a priori* estimates, that is we assume that the solution u exists and obeys $\nabla u \in S^0(I)$.

We start by deriving some bounds on $\tilde{u}$ and u. By Strichartz, Lemma 3.11, (3.53), and (3.55),

$$\begin{aligned}
\|\nabla\tilde{u}\|_{S^0(I)} &\lesssim \|\tilde{u}\|_{L_t^\infty \dot{H}_x^1(I\times\mathbb{R}^d)} + \|\nabla F(\tilde{u})\|_{N^0(I)} + \|\nabla e\|_{N^0(I)} \\
&\lesssim E + \|\tilde{u}\|^{\frac{4}{d-2}}_{L_{t,x}^{\frac{2(d+2)}{d-2}}(I\times\mathbb{R}^d)} \|\nabla\tilde{u}\|_{S^0(I)} + \varepsilon \\
&\lesssim E + \delta^{\frac{4c}{d-2}} \|\nabla\tilde{u}\|_{S^0(I)}^{1+\frac{4(1-c)}{d-2}} + \varepsilon,
\end{aligned}$$

where $c = c(d)$ is as in Lemma 3.11. Choosing δ small depending on d, E and ε_0 sufficiently small depending on E, we obtain

$$\|\nabla\tilde{u}\|_{S^0(I)} \lesssim E. \tag{3.61}$$

Moreover, by Lemma 3.10, Lemma 3.12, (3.53), and (3.55),

$$\big\|e^{i(t-t_0)\Delta}\tilde u(t_0)\big\|_{X(I)} \lesssim \|\tilde u\|_{X(I)} + \|F(\tilde u)\|_{Y(I)} + \|\nabla e\|_{N^0(I)} \lesssim \delta + \delta^{\frac{d+2}{d-2}} + \varepsilon \lesssim \delta,$$

provided δ and ε_0 are chosen sufficiently small. Combining this with (3.54) and the triangle inequality, we obtain

$$\big\|e^{i(t-t_0)\Delta}u(t_0)\big\|_{X(I)} \lesssim \delta.$$

Thus, another application of Lemma 3.10 combined with Lemma 3.12 gives

$$\|u\|_{X(I)} \lesssim \big\|e^{i(t-t_0)\Delta}u(t_0)\big\|_{X(I)} + \|F(u)\|_{Y(I)} \lesssim \delta + \|u\|_{X(I)}^{\frac{d+2}{d-2}}.$$

Choosing δ sufficiently small, the usual bootstrap argument yields

$$\|u\|_{X(I)} \lesssim \delta. \tag{3.62}$$

Next we derive the claimed bounds on $w := u - \tilde u$. Note that w is a solution to

$$\begin{cases} iw_t + \Delta w = F(\tilde u + w) - F(\tilde u) - e \\ w(t_0) = u(t_0) - \tilde u(t_0). \end{cases}$$

Using Lemma 3.10 together with Lemma 3.11 and (3.55), we see that

$$\begin{aligned}\|w\|_{X(I)} &\lesssim \big\|e^{i(t-t_0)\Delta}\big(u(t_0) - \tilde u(t_0)\big)\big\|_{X(I)} + \|\nabla e\|_{N^0(I)} + \|F(u) - F(\tilde u)\|_{Y(I)} \\ &\lesssim \varepsilon + \|F(u) - F(\tilde u)\|_{Y(I)}.\end{aligned}$$

To estimate the difference of the nonlinearities, we use Lemma 3.12, (3.53), and (3.61):

$$\begin{aligned}\|F(u) - F(\tilde u)\|_{Y(I)} &\lesssim \Big[\|\tilde u\|_{X(I)}^{\frac{8}{d^2-4}}\|\nabla \tilde u\|_{S^0(I)}^{\frac{4d}{d^2-4}} + \|w\|_{X(I)}^{\frac{8}{d^2-4}}\|\nabla w\|_{S^0(I)}^{\frac{4d}{d^2-4}}\Big]\|w\|_{X(I)} \\ &\lesssim \delta^{\frac{8}{d^2-4}} E^{\frac{4d}{d^2-4}}\|w\|_{X(I)} + \|\nabla w\|_{S^0(I)}^{\frac{4d}{d^2-4}}\|w\|_{X(I)}^{1+\frac{8}{d^2-4}}.\end{aligned} \tag{3.63}$$

Thus, choosing δ sufficiently small depending only on E, we obtain

$$\|w\|_{X(I)} \lesssim \varepsilon + \|\nabla w\|_{S^0(I)}^{\frac{4d}{d^2-4}}\|w\|_{X(I)}^{1+\frac{8}{d^2-4}}. \tag{3.64}$$

On the other hand, by the Strichartz inequality and the hypotheses,

$$\begin{aligned}\|\nabla w\|_{S^0(I)} &\lesssim \|u_0 - \tilde u_0\|_{\dot H^1_x} + \|\nabla e\|_{N^0(I)} + \big\|\nabla\big(F(u) - F(\tilde u)\big)\big\|_{N^0(I)} \\ &\lesssim E' + \varepsilon + \big\|\nabla\big(F(u) - F(\tilde u)\big)\big\|_{N^0(I)}.\end{aligned} \tag{3.65}$$

To estimate the difference of the nonlinearities, we consider low and high dimensions separately. Consider first $3 \le d \le 5$. Using Hölder's inequality followed by Lemma 3.11, (3.53), (3.61), and (3.62),

$$\begin{aligned}\big\|\nabla\big(F(u) - &F(\tilde u)\big)\big\|_{N^0(I)} \\ &\lesssim \big\|\nabla\big(F(u) - F(\tilde u)\big)\big\|_{L_t^{\frac{2d(d+2)}{d^2+2d+4}} L_x^{\frac{2d^2(d+2)}{d^3+4d^2+4d-8}}(I\times\mathbb{R}^d)} \\ &\lesssim \|\nabla\tilde u\|_{S^0(I)}\big(\|u\|_{X^0(I)} + \|\tilde u\|_{X^0(I)}\big)^{\frac{6-d}{d-2}}\|w\|_{X^0(I)} + \|u\|_{X^0(I)}^{\frac{4}{d-2}}\|\nabla w\|_{S^0(I)} \\ &\lesssim \big(E\delta^{\frac{6-d}{d-2}} + \delta^{\frac{4}{d-2}}\big)\|\nabla w\|_{S^0(I)}.\end{aligned} \tag{3.66}$$

Thus, choosing δ small depending only on E, (3.65) implies

$$\|\nabla w\|_{S^0(I)} \lesssim E' + \varepsilon$$

for $3 \leq d \leq 5$. Consider now higher dimensions, that is, $d \geq 6$. Using Hölder's inequality followed by Lemma 3.11, (3.61), and (3.62),

$$\begin{aligned}\big\|\nabla\big(F(u)-F(\tilde u)\big)\big\|_{N^0(I)} &\lesssim \big\|\nabla\big(F(u)-F(\tilde u)\big)\big\|_{L_t^{\frac{2d(d+2)}{d^2+2d+4}} L_x^{\frac{2d^2(d+2)}{d^3+4d^2+4d-8}}(I\times\mathbb{R}^d)}\\ &\lesssim \|\nabla\tilde u\|_{S^0(I)}\|w\|_{X^0(I)}^{\frac{4}{d-2}} + \|u\|_{X^0(I)}^{\frac{4}{d-2}}\|\nabla w\|_{S^0(I)}\\ &\lesssim E\|w\|_{X(I)}^{\frac{4}{d-2}} + \delta^{\frac{4}{d-2}}\|\nabla w\|_{S^0(I)}. \qquad (3.67)\end{aligned}$$

Therefore, taking δ sufficiently small, (3.65) implies

$$\|\nabla w\|_{S^0(I)} \lesssim E' + \varepsilon + E\|w\|_{X(I)}^{\frac{4}{d-2}}$$

for $d \geq 6$. Collecting the estimates for low and high dimensions (and choosing $\varepsilon_0 = \varepsilon_0(E')$ sufficiently small), we obtain

$$\|\nabla w\|_{S^0(I)} \lesssim E' + E\|w\|_{X(I)}^{\frac{4}{d-2}} \qquad (3.68)$$

for all $d \geq 3$.

Combining (3.64) with (3.68), the usual bootstrap argument yields (3.56) and (3.57), provided ε_0 is chosen sufficiently small depending on E and E'. By the triangle inequality, (3.57) and (3.61) imply (3.58).

Claims (3.59) and (3.60) follow from (3.63), (3.66), and (3.67) combined with (3.56) and (3.57), provided we take δ sufficiently small depending on E and ε_0 sufficiently small depending on E, E'. □

We are finally in a position to prove the energy-critical stability result.

Proof of Theorem 3.8. Our first goal is to show

$$\|\nabla\tilde u\|_{S^0(I)} \leq C(E, L). \qquad (3.69)$$

Indeed, by (3.39) we may divide I into $J_0 = J_0(L,\eta)$ subintervals $I_j = [t_j, t_{j+1}]$ such that on each spacetime slab $I_j \times \mathbb{R}^d$

$$\|\tilde u\|_{L_{t,x}^{\frac{2(d+2)}{d-2}}(I_j\times\mathbb{R}^d)} \leq \eta$$

for a small constant $\eta > 0$ to be chosen in a moment. By the Strichartz inequality combined with (3.38) and (3.42),

$$\begin{aligned}\|\nabla\tilde u\|_{S^0(I_j)} &\lesssim \|\tilde u(t_j)\|_{\dot H_x^1} + \|\nabla e\|_{N^0(I_j)} + \|\nabla F(\tilde u)\|_{N^0(I_j)}\\ &\lesssim E + \varepsilon + \|\tilde u\|_{L_{t,x}^{\frac{2(d+2)}{d-2}}(I_j\times\mathbb{R}^d)}^{\frac{4}{d-2}} \|\nabla\tilde u\|_{S^0(I_j)}\\ &\lesssim E + \varepsilon + \eta^{\frac{4}{d-2}}\|\nabla\tilde u\|_{S^0(I_j)}.\end{aligned}$$

Thus, choosing $\eta > 0$ small depending on the dimension d and ε_1 sufficiently small depending on E, we obtain

$$\|\nabla\tilde u\|_{S^0(I_j)} \lesssim E.$$

Summing this over all subintervals I_j, we derive (3.69).

Using Lemma 3.11 together with (3.69) and then with (3.40) and (3.41), we obtain

$$\|\tilde u\|_{X(I)} \leq C(E, L) \qquad (3.70)$$

$$\big\|e^{i(t-t_0)\Delta}\big(u(t_0)-\tilde u(t_0)\big)\big\|_{X(I)} \lesssim \varepsilon^{\frac{1}{d+2}}(E')^{\frac{d+1}{d+2}}. \qquad (3.71)$$

By (3.70), we may divide I into $J_1 = J_1(E, L)$ subintervals $I_j = [t_j, t_{j+1}]$ such that on each spacetime slab $I_j \times \mathbb{R}^d$

$$\|\tilde{u}\|_{X(I_j)} \leq \delta$$

for some small $\delta = \delta(E) > 0$ as in Lemma 3.13. Moreover, taking $\varepsilon_1(E, E', L)$ sufficiently small compared to $\varepsilon_0(E, C(J_1)E')$, (3.71) guarantees (3.54) with ε replaced by $\varepsilon^c \ll \varepsilon_0$, where c may be taken equal to $\frac{1}{2(d+2)}$. Note that E' is being replaced by $C(J_1)E'$, as the energy of the difference of the two initial data may increase with each iteration.

Thus, choosing ε_1 sufficiently small (depending on J_1, E, and E'), we may apply Lemma 3.13 to obtain for each $0 \leq j < J_1$ and all $0 < \varepsilon < \varepsilon_1$,

$$\begin{aligned} \|u - \tilde{u}\|_{X(I_j)} &\leq C(j)\varepsilon^c \\ \|u - \tilde{u}\|_{\dot{S}^1(I_j)} &\leq C(j)E' \\ \|u\|_{\dot{S}^1(I_j)} &\leq C(j)(E + E') \\ \|F(u) - F(\tilde{u})\|_{Y(I_j)} &\leq C(j)\varepsilon^c \\ \big\|\nabla\big(F(u) - F(\tilde{u})\big)\big\|_{N^0(I_j)} &\leq C(j)E', \end{aligned} \tag{3.72}$$

provided we can show

$$\big\|e^{i(t-t_j)\Delta}\big(u(t_j) - \tilde{u}(t_j)\big)\big\|_{X(I_j)} \lesssim \varepsilon^c \quad \text{and} \quad \|u(t_j) - \tilde{u}(t_j)\|_{\dot{H}^1_x(\mathbb{R}^d)} \lesssim E' \tag{3.73}$$

for each $0 \leq j < J_1$. By Lemma 3.10 and the inductive hypothesis,

$$\begin{aligned} &\big\|e^{i(t-t_j)\Delta}\big(u(t_j) - \tilde{u}(t_j)\big)\big\|_{X(I_j)} \\ &\qquad \lesssim \big\|e^{i(t-t_0)\Delta}\big(u(t_0) - \tilde{u}(t_0)\big)\big\|_{X(I_j)} + \|\nabla e\|_{N^0(I)} + \|F(u) - F(\tilde{u})\|_{Y([t_0,t_j])} \\ &\qquad \lesssim \varepsilon^c + \varepsilon + \sum_{k=0}^{j-1} C(k)\varepsilon^c. \end{aligned}$$

Similarly, by the Strichartz inequality and the inductive hypothesis,

$$\begin{aligned} &\|u(t_j) - \tilde{u}(t_j)\|_{\dot{H}^1_x} \\ &\qquad \lesssim \|u(t_0) - \tilde{u}(t_0)\|_{\dot{H}^1_x} + \|\nabla e\|_{N^0([t_0,t_j])} + \big\|\nabla\big(F(u) - F(\tilde{u})\big)\big\|_{N^0([t_0,t_j])} \\ &\qquad \lesssim E' + \varepsilon + \sum_{k=0}^{j-1} C(k)E'. \end{aligned}$$

Taking ε_1 sufficiently small depending on J_1, E, and E', we see that (3.73) is satisfied.

Summing the bounds in (3.72) over all subintervals I_j and using Lemma 3.11, we derive (3.43) through (3.45). This completes the proof of the theorem. □

4. A word from our sponsor: Harmonic Analysis

Without doubt, recent progress on nonlinear Schrödinger equations at critical regularity has been made possible by the introduction of important ideas from harmonic analysis, particularly some related to the restriction conjecture.

4.1. The Gagliardo–Nirenberg inequality. The sharp constant for the Gagliardo–Nirenberg inequality was derived by Nagy [**63**], in the one-dimensional setting, and by Weinstein [**105**] for higher dimensions. We begin by recounting this theorem. After that, we will present two applications to nonlinear Schrödinger equations.

THEOREM 4.1 (Sharp Gagliardo–Nirenberg, [**63, 105**]). *Fix $d \geq 1$ and $0 < p < \infty$ for $d = 1, 2$ or $0 < p < \frac{4}{d-2}$ for $d \geq 3$. Then for all $f \in H^1_x(\mathbb{R}^d)$,*

$$\|f\|_{L_x^{p+2}}^{p+2} \leq \tfrac{2(p+2)}{4-p(d-2)}\big(\tfrac{pd}{4-p(d-2)}\big)^{-\frac{pd}{4}}\|Q\|_{L_x^2}^{-p}\|f\|_{L_x^2}^{p+2-\frac{pd}{2}}\|\nabla f\|_{L_x^2}^{\frac{pd}{2}}. \tag{4.1}$$

Here Q denotes the unique positive radial Schwartz solution to $\Delta Q + Q^{p+1} = Q$. Moreover, equality holds in (4.1) if and only if $f(x) = \alpha Q(\lambda(x - x_0))$ for some $\alpha \in \mathbb{C}$, $\lambda \in (0, \infty)$, and $x_0 \in \mathbb{R}^d$.

PROOF. The traditional (non-sharp) Gagliardo–Nirenberg inequality says

$$J(f) := \frac{\|f\|_{L_x^{p+2}}^{p+2}}{\|f\|_{L_x^2}^{p+2-\frac{pd}{2}}\|\nabla f\|_{L_x^2}^{\frac{pd}{2}}} \leq C. \tag{4.2}$$

What we seek here is the optimal constant $C = C_d$ in this inequality. We will present only the proof for $d \geq 2$, following [**105**].

It suffices to consider merely non-negative spherically symmetric functions, since we may replace f by its spherically symmetric decreasing rearrangement f^* (cf. [**54**, §7.17]). The $\dot H^1_x$ norm of f^* is no larger than that of f, while the L^2_x and L_x^{2+p} norms are invariant under $f \mapsto f^*$. Thus $J(f) \leq J(f^*)$.

Let f_n be an optimizing sequence (of non-negative spherically symmetric functions). By rescaling space and the values of the function, we may assume that $\|\nabla f_n\|_2 = \|f_n\|_2 = 1$. We are now ready for the key step in the argument: The embedding $H^1_{\text{rad}} \hookrightarrow L_x^{2+p}$ is compact; see Lemma A.4. Thus we may deduce that, up to a subsequence, f_n converge strongly in L_x^{2+p}. Additionally, since f_n is an optimizing sequence, we can upgrade the weak convergence of f_n in H^1_x (courtesy of Alaoglu's theorem) to strong convergence.

In the previous paragraph, we deduced that optimizers exist, that is, there are functions f maximizing $J(f)$. Moreover, f has been normalized to obey $\|\nabla f\|_2 = \|f\|_2 = 1$, which implies $C_d = \|f\|_{p+2}^{p+2}$. By studying small Schwartz-space perturbations of f, we quickly see that any optimizer f must be a distributional solution to

$$(p+2)f^{1+p} - C_d\{(p+2-\tfrac{pd}{2})f - \tfrac{pd}{2}\Delta f\} = 0. \tag{4.3}$$

This equation can be reduced to $\Delta Q + Q^{p+1} = Q$ by setting

$$f(x) = \alpha^{\frac{1}{p}}Q(\beta^{\frac{1}{2}}x) \quad \text{with } \beta = \tfrac{4-p(d-2)}{pd} \text{ and } \alpha = \tfrac{pd\beta}{2(p+2)}C_d.$$

Taking advantage of $\|f\|_2 = 1$, we may deduce $C_d = \frac{2(p+2)}{4-p(d-2)}\beta^{pd/4}\|Q\|_2^{-p}$.

We now turn to the uniqueness question. It is very tempting to believe that $J(f) \leq J(f^*)$ with equality if and only if $f(x) = e^{i\theta}f^*(x + x_0)$ for some $\theta \in [0, 2\pi)$ and $x_0 \in \mathbb{R}^d$. (This would immediately imply that any optimizer is radially symmetric up to translations.) Alas, it is not true without an additional constraint, for instance, that ∇f^* does not vanish on a set of positive measure; see [**11**]. Fortunately for us, as f^* is a non-zero spherically symmetric solution to (4.3),

∇f^* cannot vanish on a set of positive measure; indeed this is a basic uniqueness property of ODEs.

This leaves us to show uniqueness of positive spherically symmetric solutions of $\Delta Q + Q^{p+1} = Q$, for which we refer the reader to [**49**]. □

REMARK. That rearrangement of a non-spherically-symmetric function may fail to reduce the $\dot H^1_x$ norm can be demonstrated with a simple example, which we will now describe. Let $\phi \in C^\infty(\mathbb{R}^d)$ be supported on $\{|x| \leq 2\}$ and obey $\phi(x) = 1$ when $|x| \leq 1$. The skewed 'wedding cake' $f(x) = \phi(x) + \phi(4(x - x_0))$ with $|x_0| \leq \frac{1}{2}$ has $\dot H^1_x$ norm equal to that of its spherically-symmetric decreasing rearrangement.

The main application of Theorem 4.1 in these notes is embodied by the following

COROLLARY 4.2 (Kinetic energy trapping). *Let $f \in H^1_x(\mathbb{R}^d)$ obey $\|f\|_2 < \|Q\|_2$. Then $\|\nabla f\|_2^2 \lesssim E(f)$, where E denotes the energy associated to the mass-critical focusing NLS. The implicit constant depends only on $\|f\|_2/\|Q\|_2$.*

PROOF. Exercise. □

Combining this with the standard local well-posedness result for subcritical equations and the conservation of mass and energy, we obtain:

COROLLARY 4.3 (Focusing mass-critical NLS in H^1_x, [**105**]). *For initial data $u(0) \in H^1_x$ obeying $\|u(0)\|_2 < \|Q\|_2$, the focusing mass-critical NLS is globally wellposed.*

PROOF. Exercise. □

Note that this result does not claim that these global solutions scatter. Indeed, scaling shows that scattering for H^1_x initial data is essentially equivalent to scattering for general L^2_x initial data.

4.2. Refined Sobolev embedding. In this subsection, we will describe several refinements of the classical Sobolev embedding inequality. The first is the determination of the optimal constant in that inequality. The following theorem is a special case of results of Aubin [**2**] and Talenti [**86**] (see also [**5, 73**]):

THEOREM 4.4 (Sharp Sobolev embedding). *For $d \geq 3$ and $f \in \dot H^1_x(\mathbb{R}^d)$,*

$$\|f\|_{L_x^{\frac{2d}{d-2}}} \leq C_d \|\nabla f\|_{L^2_x} \tag{4.4}$$

with equality if and only if $f = \alpha W(\lambda(x - x_0))$ for some $\alpha \in \mathbb{C}$, $\lambda \in (0, \infty)$, and $x_0 \in \mathbb{R}^d$. Here W denotes

$$W(x) := \big(1 + \tfrac{1}{d(d-2)}|x|^2\big)^{-\frac{d-2}{2}}, \tag{4.5}$$

which is the unique non-negative radial $\dot H^1_x$ solution to $\Delta W + W^{\frac{d+2}{d-2}} = 0$, up to scaling.

In this context, the analogue of Corollary 4.2 is

COROLLARY 4.5 (Energy trapping, [**38**]). *Assume $E(u_0) \leq (1 - \delta_0)E(W)$ for some $\delta_0 > 0$. Then there exists a positive constant δ_1 so that if $\|\nabla u_0\|_2 \leq \|\nabla W\|_2$, then*

$$\|\nabla u_0\|_2^2 \leq (1 - \delta_1)\|\nabla W\|_2^2.$$

Here E denotes the energy functional associated to the focusing energy-critical NLS.

PROOF. Exercise. □

We will discuss the proof of Theorem 4.4 in some detail as it is our first brush with our sworn enemy: scaling invariance. First let us note that the argument used to prove Theorem 4.1 will not work here. For instance, $f_n(x) = n^{(d-2)/2}W(nx)$ is a radial optimizing sequence that does not converge. To put it another way, Lemma A.4 fails for $p = \frac{2d}{d-2}$ because of scaling.

There are several proofs of Theorem 4.4. The textbook [**54**] gives an elegant treatment relying on the connection to the Hardy–Littlewood–Sobolev inequality and a (hidden) conformal symmetry. We will be giving a proof that does not rely heavily on rearrangement ideas, since we wish to introduce some techniques that will be important when we discuss improvements to Strichartz inequality.

Lions gave a rearrangement-free proof of the existence of optimizers as one of the first applications of the concentration compactness principle; see [**56**]. The proof we present is a descendant of the one given there. The philosophy underlying concentration compactness has also led to a second kind of refinement to the classical Sobolev embedding, which has proved valuable in the treatment of the energy-critical NLS. The goal is not to understand the maximal possible value of the ratio $J(f) := \|f\|_{2d/(d-2)} \div \|\nabla f\|_2$, but rather for what kind of functions this is big (or equivalently, for which f it is small). Before giving a precise statement, we quickly introduce some of the ideas that will motivate the formulation. We will then revisit the Gagliardo–Nirenberg inequality from this perspective.

Let $A : X \to Y$ be a linear transformation between two Banach spaces. Recall that A is called compact if for every bounded sequence $f_n \in X$, the sequence Af_n has a convergent subsequence. A slightly more convoluted way of saying this is the following.

EXERCISE. Suppose X is reflexive. Then $A : X \to Y$ is compact if and only if for any bounded sequence $\{f_n\} \subseteq X$ there exists $\phi \in X$ so that along some subsequence $f_n = \phi + r_n$ with $Ar_n \to 0$ in Y. (This may fail if X is not reflexive.)

Even for $2 < q < \frac{2d}{d-2}$, the embedding $H_x^1 \hookrightarrow L_x^q$ is not compact since given any non-zero $f \in H_x^1(\mathbb{R}^d)$, the sequence of translates $f_n(x) = f(x - x_n)$, associated to a sequence $x_n \to \infty$ in $\mathbb{R}^d$, is uniformly bounded in $H_x^1(\mathbb{R}^d)$, but has no L_x^q-convergent subsequence. A first attempt to address this failure of compactness, might be to seek a convergent subsequence from among the translates of the original sequence. This does not quite work as can be seen by considering $f_n(x) = \phi_1(x) + \phi_2(x - x_n)$ for some fixed $\phi_1, \phi_2 \in H_x^1(\mathbb{R}^d)$.

Having just seen the example of a sequence that breaks into two 'bubbles' we may begin to despair that a sequence f_n may break into infinitely many small bubbles dancing around $\mathbb{R}^d$ more or less at random. It is time for some good news: $q > 2$, which is to say that in the inequality

$$\|f\|_{L_x^q} \lesssim \|f\|_{L_x^2}^{1-\theta} \|\nabla f\|_{L_x^2}^{\theta}, \quad \theta = \frac{(q-2)d}{2q},$$

the power of f integrated on the left-hand side is larger than the power of f and ∇f that is integrated on the right-hand side. The significance of this is that the ℓ^q norm of many small numbers is much much smaller than the ℓ^2 norm of the same collection of numbers. Therefore, a large collection of tiny bubbles whose total H_x^1 norm is of order one will have a negligible L_x^q norm.

THEOREM 4.6 (The Gagliardo–Nirenberg inequality: bubble decomposition, [**33**]). *Fix $d \geq 2$, $2 < q < \frac{2d}{d-2}$, and let f_n be a bounded sequence in $H_x^1(\mathbb{R}^d)$. Then there exist $J^* \in \{0, 1, 2, \ldots\} \cup \{\infty\}$, $\{\phi^j\}_{j=1}^{J^*} \subseteq H_x^1$, and $\{x_n^j\}_{j=1}^{J^*} \subseteq \mathbb{R}^d$ so that along some subsequence in n we may write*

$$f_n(x) = \sum_{j=1}^{J} \phi^j(x - x_n^j) + r_n^J(x) \quad \textit{for all } 0 \leq J \leq J^*, \tag{4.6}$$

where

$$\limsup_{J\to J^*} \limsup_{n\to\infty} \|r_n^J\|_{L_x^q} = 0 \tag{4.7}$$

$$\sup_{J} \limsup_{n\to\infty} \left| \|f_n\|_{H_x^1}^2 - \Big(\sum_{j=1}^{J} \|\phi^j\|_{H_x^1}^2 + \|r_n^J\|_{H_x^1}^2 \Big) \right| = 0 \tag{4.8}$$

$$\limsup_{J\to J^*} \left| \limsup_{n\to\infty} \|f_n\|_{L_x^q}^q - \sum_{j=1}^{J} \|\phi^j\|_{L_x^q}^q \right| = 0. \tag{4.9}$$

Moreover, for each $j \neq j'$, we have $|x_n^j - x_n^{j'}| \to \infty$. When J^* is finite, we define $\limsup_{J\to J^*} a(J) := a(J^*)$ for any $a : \{0, 1, \ldots, J^*\} \to \mathbb{R}$.

We will not make use of this result and we leave its proof to the avid reader who wishes to cement their understanding of the methods described in this subsection. Note that ϕ^j represent the bubbles into which the subsequence is decomposing and J^* is their number. They may be regarded as ordered by decreasing H_x^1 norm. The functions r_n^J represent a remainder term, which is guaranteed to be asymptotically irrelevant in L_x^q, but need not converge to zero in H_x^1. This is why r_n^J needs to appear in (4.8), even as $J \to \infty$. Indeed, this is the essence of compactness. Regarding (4.8), we also wish to point out that the divergence of the x_n^j from one another implies that the H_x^1 norms of the individual bubbles decouple. That they also decouple from r_n^J is a more subtle statement. It is an expression of the fact that for each pair $j \leq J$,

$$r_n^J(x + x_n^j) \rightharpoonup 0 \quad \text{weakly in } H_x^1,$$

which is built into the way ϕ^j are chosen. (It can also be derived *a posteriori* from the conclusions of this theorem, cf. [**44**, Lemma 2.10].)

The analogue of Theorem 4.6 for Sobolev embedding reads very similarly; it is merely necessary to incorporate the scaling symmetry.

THEOREM 4.7 (Sobolev embedding: bubble decomposition, [**26**]). *Fix $d \geq 3$ and let f_n be a bounded sequence in $\dot H_x^1(\mathbb{R}^d)$. Then there exist $J^* \in \{0, 1, 2, \ldots\} \cup \{\infty\}$, $\{\phi^j\}_{j=1}^{J^*} \subseteq \dot H_x^1$, $\{x_n^j\}_{j=1}^{J^*} \subseteq \mathbb{R}^d$, and $\{\lambda_n^j\}_{j=1}^{J^*} \subseteq (0, \infty)$ so that along some subsequence in n we may write*

$$f_n(x) = \sum_{j=1}^{J} (\lambda_n^j)^{\frac{2-d}{2}} \phi^j\big((x - x_n^j)/\lambda_n^j\big) + r_n^J(x) \quad \textit{for all } 0 \leq J \leq J^* \tag{4.10}$$

with the following five properties:

$$\limsup_{J\to J^*} \limsup_{n\to\infty} \|r_n^J\|_{L_x^{\frac{2d}{d-2}}} = 0 \tag{4.11}$$

$$\sup_J \limsup_{n\to\infty} \left| \|f_n\|_{\dot H^1_x}^2 - \Big(\|r_n^J\|_{\dot H^1_x}^2 + \sum_{j=1}^J \|\phi^j\|_{\dot H^1_x}^2 \Big) \right| = 0 \tag{4.12}$$

$$\limsup_{J\to J^*} \limsup_{n\to\infty} \left| \|f_n\|_{L_x^{\frac{2d}{d-2}}}^{\frac{2d}{d-2}} - \sum_{j=1}^J \|\phi^j\|_{L_x^{\frac{2d}{d-2}}}^{\frac{2d}{d-2}} \right| = 0 \tag{4.13}$$

$$\liminf_{n\to\infty} \left[\frac{|x_n^j - x_n^{j'}|^2}{\lambda_n^j \lambda_n^{j'}} + \frac{\lambda_n^j}{\lambda_n^{j'}} + \frac{\lambda_n^{j'}}{\lambda_n^j} \right] = \infty \quad \textit{for all } j \neq j' \tag{4.14}$$

$$(\lambda_n^j)^{\frac{d-2}{2}} r_n^J(\lambda_n^j x + x_n^j) \rightharpoonup 0 \quad \textit{weakly in } \dot H_x^1 \textit{ for each } j \leq J. \tag{4.15}$$

Notice that (4.14) says that each pair of bubbles are either widely separated in space or live at very different length scales (or possibly both). This time, we have incorporated the strong form of r_n^J decoupling, (4.15), into the statement of the theorem.

Before embarking on the proofs of Theorems 4.4 and 4.7, let us briefly depart on a small historical excursion. We will, at least, explain why we use the word 'bubble'. In [**76**], Sacks and Uhlenbeck proved the existence of minimal-area spheres in Riemannian manifolds in certain (higher) homotopy classes. They also gave a vivid explanation of why the result is merely for *some* homotopy classes: sometimes the minimal sphere is not really a sphere, but two (or more) spheres joined by one-dimensional geodesic 'umbilical cords'. This obstruction necessitated an ingenious snipping procedure, which can be viewed as an early precursor to the bubble decomposition above. (In this setting, the group of translations is replaced by the group of conformal maps of S^2, that is, of Möbius transformations.)

Minimal surfaces correspond to zero mean curvature. In general, soap films produce surfaces with constant mean curvature. In fact, the mean curvature is proportional to the pressure difference between the two sides; this can be non-zero, as in the case of a spherical bubble. Around the same time as the work of Sacks and Uhlenbeck described above, Wente, [**106**], considered the problem of a large bubble blown on a (comparatively) small wire. He shows that the resulting bubble is asymptotically spherical. The result relies on the extremal property by which the bubble is constructed and, thanks to a subadditivity-type argument deep within the proof, avoids the possibility of multiple bubbles. Consideration of more general (non-extremal) surfaces of constant mean curvature necessitates a full bubble decomposition. This was worked out independently by Brézis and Coron, [**9**], and Struwe, [**85**].

Shortly prior to its appearance in the highly nonlinear setting of constant mean curvature surfaces, Struwe proved a bubble decomposition for the energy-critical elliptic problem $\Delta u + |u|^{\frac{4}{d-2}} u = 0$. This is clearly closely related to Sobolev embedding. Nonetheless, Theorem 4.7 is from [**26**] (building upon some earlier work) as noted above.

As we will see, there is a simple trick for finding the translation parameters x_n^j appearing in (4.10); it uses little more than Hölder's inequality. To deal with the scaling symmetry we need something a little more sophisticated. Littlewood–Paley theory is the natural choice; separating scales is exactly what it does!

Proposition 4.8 (An embedding). *For $d \geq 3$ and $f \in \mathcal{S}(\mathbb{R}^d)$,*

$$\|f\|_{L_x^{\frac{2d}{d-2}}} \lesssim \|\nabla f\|_{L_x^2}^{\frac{d-2}{d}} \cdot \sup_{N \in 2^{\mathbb{Z}}} \|f_N\|_{L_x^{\frac{2d}{d-2}}}^{\frac{2}{d}}. \tag{4.16}$$

PROOF. First we give the proof for $d \geq 4$. The key ingredient is the well-known estimate for the Littlewood–Paley square function, Lemma A.7, which we use in the first step. We also use Bernstein's inequality, Lemma A.6.

$$
\begin{aligned}
\|f\|_{L_x^{\frac{2d}{d-2}}}^{\frac{2d}{d-2}} &\lesssim \int_{\mathbb{R}^d} \Big(\sum_M |f_M|^2\Big)^{\frac{d}{2(d-2)}} \Big(\sum_N |f_N|^2\Big)^{\frac{d}{2(d-2)}}\, dx \\
&\lesssim \sum_{M\leq N} \int_{\mathbb{R}^d} |f_M|^{\frac{d}{d-2}} |f_N|^{\frac{d}{d-2}}\, dx \\
&\lesssim \Big(\sup_{K\in 2^{\mathbb{Z}}} \|f_K\|_{L_x^{\frac{2d}{d-2}}}\Big)^{\frac{4}{d-2}} \sum_{M\leq N} \|f_M\|_{L_x^{\frac{2d}{d-4}}} \|f_N\|_{L_x^2} \\
&\lesssim \Big(\sup_{K\in 2^{\mathbb{Z}}} \|f_K\|_{L_x^{\frac{2d}{d-2}}}\Big)^{\frac{4}{d-2}} \sum_{M\leq N} M^{-1}N^{-1} \|\nabla f_M\|_{L_x^{\frac{2d}{d-4}}} \|\nabla f_N\|_{L_x^2} \\
&\lesssim \Big(\sup_{K\in 2^{\mathbb{Z}}} \|f_K\|_{L_x^{\frac{2d}{d-2}}}\Big)^{\frac{4}{d-2}} \sum_{M\leq N} MN^{-1} \|\nabla f_M\|_{L_x^2} \|\nabla f_N\|_{L_x^2} \\
&\lesssim \Big(\sup_{K\in 2^{\mathbb{Z}}} \|f_K\|_{L_x^{\frac{2d}{d-2}}}\Big)^{\frac{4}{d-2}} \Big(\sum_{K\in 2^{\mathbb{Z}}} \|\nabla f_K\|_{L_x^2}^2\Big).
\end{aligned}
$$

In passing from the first line to the second, we used that $\frac{d}{2(d-2)} \leq 1$, which is the origin of the restriction $d \geq 4$. To treat three dimensions, one modifies the argument as follows:

$$
\begin{aligned}
\|f\|_{L_x^6}^6 &\lesssim \int_{\mathbb{R}^d} \Big(\sum_K |f_K|^2\Big)\Big(\sum_M |f_M|^2\Big)\Big(\sum_N |f_N|^2\Big)\, dx \\
&\lesssim \sum_{K\leq M\leq N} \|f_K\|_{L_x^6} \|f_K\|_{L_x^\infty} \|f_M\|_{L_x^6}^2 \|f_N\|_{L_x^3} \|f_N\|_{L_x^6} \\
&\lesssim \Big(\sup_{L\in 2^{\mathbb{Z}}} \|f_L\|_{L_x^6}^4\Big) \sum_{K\leq M\leq N} K^{\frac{3}{2}} N^{\frac{1}{2}} \|f_K\|_{L_x^2} \|f_N\|_{L_x^2} \\
&\lesssim \Big(\sup_{L\in 2^{\mathbb{Z}}} \|f_L\|_{L_x^6}^4\Big) \sum_{K\leq M\leq N} K^{\frac{1}{2}} N^{-\frac{1}{2}} \|\nabla f_K\|_{L_x^2} \|\nabla f_N\|_{L_x^2},
\end{aligned}
$$

which leads to (4.16) via Schur's test and other elementary considerations. □

Our next result introduces the important idea of inverse inequalities. The content of such inequalities is as follows: if a bounded sequence in some strong norm (e.g. $\dot H_x^1$) does not converge weakly to zero in a weaker norm (e.g., $L_x^{2d/(d-2)}$), then this can be attributed to the sequence containing a bubble of concentration. While we have not seen the following precise statement in print, it is a natural off-shoot of existing ideas.

PROPOSITION 4.9 (Inverse Sobolev Embedding). *Fix $d \geq 3$ and let $\{f_n\} \subseteq \dot H_x^1(\mathbb{R}^d)$. If*

$$
\lim_{n\to\infty} \|f_n\|_{\dot H_x^1(\mathbb{R}^d)} = A \quad \text{and} \quad \liminf_{n\to\infty} \|f_n\|_{L_x^{\frac{2d}{d-2}}(\mathbb{R}^d)} = \varepsilon, \tag{4.17}
$$

then there exist a subsequence in n, $\phi \in \dot H^1_x(\mathbb{R}^d)$, $\{\lambda_n\} \subseteq (0,\infty)$, *and* $\{x_n\} \subseteq \mathbb{R}^d$ *so that along the subsequence, we have the following three properties:*

$$\lambda_n^{\frac{d-2}{2}} f_n(\lambda_n x + x_n) \rightharpoonup \phi(x) \quad \textit{weakly in } \dot H^1_x(\mathbb{R}^d) \tag{4.18}$$

$$\lim_{n\to\infty}\Big[\big\|f_n(x)\big\|^2_{\dot H^1_x} - \big\|f_n(x) - \lambda_n^{\frac{2-d}{2}}\phi\big(\lambda_n^{-1}(x-x_n)\big)\big\|^2_{\dot H^1_x}\Big] = \|\phi\|^2_{\dot H^1_x} \gtrsim A^2\big(\tfrac{\varepsilon}{A}\big)^{\frac{d^2}{2}} \tag{4.19}$$

$$\limsup_{n\to\infty}\Big\|f_n(x) - \lambda_n^{\frac{2-d}{2}}\phi\big(\lambda_n^{-1}(x-x_n)\big)\Big\|^{\frac{2d}{d-2}}_{L_x^{\frac{2d}{d-2}}(\mathbb{R}^d)} \le \varepsilon^{\frac{2d}{d-2}}\Big[1 - c\big(\tfrac{\varepsilon}{A}\big)^{\frac{d(d+2)}{2}}\Big]. \tag{4.20}$$

Here c *is a (dimension-dependent) constant.*

Proof. By passing to a subsequence, we may assume that $\|f_n\|_{\frac{2d}{d-2}} \to \varepsilon$ from the very beginning. This will not be important until we turn our attention to (4.20).

By Proposition 4.8, there exists $\{N_n\} \subseteq 2^{\mathbb{Z}}$ so that

$$\liminf_{n\to\infty} \|P_{N_n} f_n\|_{L_x^{\frac{2d}{d-2}}(\mathbb{R}^d)} \gtrsim \varepsilon^{\frac{d}{2}} A^{-\frac{d-2}{2}}.$$

We set $\lambda_n = N_n^{-1}$. To find x_n, we use Hölder's inequality:

$$\begin{aligned}
\varepsilon^{\frac{d}{2}} A^{-\frac{d-2}{2}} &\lesssim \liminf_{n\to\infty}\big\|P_{N_n} f_n\big\|_{L_x^{\frac{2d}{d-2}}(\mathbb{R}^d)}\\
&\lesssim \liminf_{n\to\infty}\big\|P_{N_n} f_n\big\|^{\frac{d-2}{d}}_{L^2_x(\mathbb{R}^d)}\big\|P_{N_n} f_n\big\|^{\frac{2}{d}}_{L^\infty_x(\mathbb{R}^d)}\\
&\lesssim \liminf_{n\to\infty}\big(AN_n^{-1}\big)^{\frac{d-2}{d}}\big\|P_{N_n} f_n\big\|^{\frac{2}{d}}_{L^\infty_x(\mathbb{R}^d)}.
\end{aligned}$$

That is, there exists $x_n \in \mathbb{R}^d$ so that

$$\liminf_{n\to\infty} N_n^{\frac{2-d}{2}}\big|[P_{N_n} f_n](x_n)\big| \gtrsim \varepsilon^{\frac{d^2}{4}} A^{1-\frac{d^2}{4}}. \tag{4.21}$$

Having chosen the parameters λ_n and x_n, Alaoglu's theorem guarantees that (4.18) holds for some subsequence in n and some $\phi \in \dot H^1_x$. To see that ϕ is non-zero, let us write k for the convolution kernel of the Littlewood–Paley projection onto frequencies of size one. That is, let $k := P_1\delta_0$. Using (4.21) we obtain

$$\begin{aligned}
|\langle k, \phi\rangle| &= \lim_{n\to\infty}\bigg|\int_{\mathbb{R}^d} \bar k(x) N_n^{-\frac{d-2}{2}} f_n(x_n + N_n^{-1}x)\,dx\bigg|\\
&= \lim_{n\to\infty} N_n^{\frac{2-d}{2}}\bigg|\int_{\mathbb{R}^d} N_n^d \bar k\big(N_n(y-x_n)\big) f_n(y)\,dy\bigg|\\
&= \lim_{n\to\infty} N_n^{\frac{2-d}{2}}\big|[P_{N_n} f_n](x_n)\big|\\
&\gtrsim \varepsilon^{\frac{d^2}{4}} A^{1-\frac{d^2}{4}}.
\end{aligned}$$

This implies that $\|\nabla\phi\|_2 \gtrsim \|\phi\|_{\frac{2d}{d-2}} \gtrsim \varepsilon^{\frac{d^2}{4}} A^{1-\frac{d^2}{4}}$. To deduce (4.19) we apply the following basic Hilbert-space fact:

$$g_n \rightharpoonup g \quad\Longrightarrow\quad \|g_n\|^2 - \|g - g_n\|^2 \to \|g\|^2 \tag{4.22}$$

with $g_n := \lambda_n^{\frac{d-2}{2}} f_n(\lambda_n x + x_n)$.

To obtain (4.20), we are going to need to work a little harder (cf. the warning below). First we note that since g_n is bounded in $\dot H^1_x(\mathbb{R}^d)$, we may pass to a further subsequence so that $g_n \to \phi$ in L_x^2-sense on any compact set (via the Rellich–Kondrashov Theorem). By passing to yet another subsequence, we can then guarantee that $g_n \to \phi$ almost everywhere in $\mathbb{R}^d$. Thus we may apply Lemma A.5 to obtain

$$\limsup_{n\to\infty}\Big\|\lambda_n^{\frac{d-2}{2}} f_n(\lambda_n x + x_n) - \phi(x)\Big\|_{L_x^{\frac{2d}{d-2}}(\mathbb{R}^d)}^{\frac{2d}{d-2}} = \varepsilon^{\frac{2d}{d-2}} - \big\|\phi\big\|_{L_x^{\frac{2d}{d-2}}(\mathbb{R}^d)}^{\frac{2d}{d-2}}.$$

This gives (4.20) after taking into account the invariance of the norm under symmetries. □

WARNING. It is very tempting to believe that extracting a bubble automatically reduces the $L_x^q(\mathbb{R}^d)$ norm, which is to say that some adequate analogue of (4.22) holds outside of Hilbert spaces. This is not the case; indeed, for $1 \leq q < \infty$,

$$\Big(g_n \rightharpoonup g \text{ in } L_x^q \;\Rightarrow\; \limsup\big[\|g_n\|_{L_x^q} - \|g_n - g\|_{L_x^q}\big] \geq 0\Big) \Rightarrow q = 2. \tag{4.23}$$

To see this, it suffices to consider the case where g_n and g are supported on the same unit cube and where g is equal to a constant there. Under these restrictions, (4.23) reduces to the following probabilistic statement:

$$\Big(\mathbb{E}\{|X|^q\} \geq \mathbb{E}\{|X - \mathbb{E}(X)|^q\} \text{ for all random variables } X\Big) \Rightarrow q = 2.$$

This in turn can be verified by a random variable taking only two values. Indeed, let X be the random variable defined by $X = 2$ with probability p and $X = -1$ with probability $1 - p$ and consider p close to $\frac{1}{3}$.

With Proposition 4.9 in hand, we will be able to quickly complete the

PROOF OF THEOREM 4.7. As $\|\nabla f_n\|_2$ is a bounded sequence, we may pass to a subsequence so that it converges. Applying Proposition 4.9 recursively leads to

$$\begin{aligned} f_n^1 &:= f_n(x) - (\lambda_n^1)^{\frac{2-d}{2}}\phi^1\big((x - x_n^1)/\lambda_n^1\big)\\ f_n^2 &:= f_n^1(x) - (\lambda_n^2)^{\frac{2-d}{2}}\phi^2\big((x - x_n^2)/\lambda_n^2\big)\\ &\;\;\vdots\\ f_n^{j+1} &:= f_n^j(x) - (\lambda_n^j)^{\frac{2-d}{2}}\phi^j\big((x - x_n^j)/\lambda_n^j\big), \end{aligned}$$

where in passing from each iteration to the next we successively require n to lie in an ever smaller (infinite!) subset of the integers. This process terminates (and J^* is finite) as soon as we have $\liminf_{n\to\infty}\|f_n^{j_0}\|_{\frac{2d}{d-2}} = 0$; indeed, $J^* = j_0$. In this case we restrict n to lie in the final subsequence. If instead $J^* = \infty$, we simply restrict n to lie in the diagonal subsequence.

Setting $r_n^0 := f_n$ and $r_n^J := f_n^J$ for $1 \leq J \leq J^*$, it remains to check the various conclusions of the theorem. Equation (4.11) is inherited directly from (4.20). We turn now to (4.14); this is a consequence of (4.18) and the fact that (by our choice of J^*) all ϕ^j are non-zero. Claim (4.15) follows from (4.14) and (4.20). Next, by approximating ϕ^j by C_c^∞ functions, it is not difficult to deduce (4.13) from (4.11) and (4.14). Lastly, (4.12) follows from (4.14) and (4.15) together with (4.22). □

PROOF OF THEOREM 4.4. The key point is to show the existence of optimizers; once this is known, one may repeat the arguments from Theorem 4.1.

Let f_n be a maximizing sequence for the ratio

$$J(f) := \|f\|_{L_x^{\frac{2d}{d-2}}}^{\frac{2d}{d-2}} \div \|\nabla f\|_{L_x^2}^{\frac{2d}{d-2}}$$

with $\|\nabla f_n\|_2 \equiv 1$. Applying Theorem 4.7 and passing to the requisite subsequence, we find

$$\sup_f J(f) = \lim_{n\to\infty} J(f_n) = \sum_{j=1}^{\infty} \|\phi^j\|_{L_x^{\frac{2d}{d-2}}}^{\frac{2d}{d-2}} \leq \sup_f J(f) \sum_{j=1}^{\infty} \|\nabla\phi^j\|_{L_x^2}^{\frac{2d}{d-2}}. \tag{4.24}$$

We also find $\sum_{j=1}^{\infty} \|\nabla \phi^j\|_2^2 \leq 1$, where the inequality stems from the omission of r_n^J. Combining these two observations with $\frac{2d}{d-2} > 2$, we see that only one of the ϕ^j may have non-zero norm; indeed, we must also have $\|\nabla\phi^j\|_2 = 1$. Thus f_n can be made to converge strongly by applying symmetries to each function. This confirms the existence of an optimizer. □

While Proposition 4.8 seems a little odd, it is well suited to proving Theorem 4.7, as we saw. To finish this subsection, we will describe some more natural improved Sobolev embeddings. These are expressed in terms of Besov norms,

$$\|f\|_{\dot{B}^s_{p,q}} := \Big(\sum_{N\in 2^{\mathbb{Z}}} \|N^s f_N\|_{L_x^p}^q \Big)^{\frac{1}{q}},$$

though we will not presuppose any familiarity with Besov spaces. The following result is a strengthening of Sobolev embedding in terms of Besov spaces (cf. [**48**, p. 56] or [**99**, p. 170]):

PROPOSITION 4.10 (Besov embedding). *For $d \geq 3$ and $f \in \mathcal{S}(\mathbb{R}^d)$,*

$$\|f\|_{L_x^{\frac{2d}{d-2}}}^{\frac{2d}{d-2}} \lesssim \sum_{N\in 2^{\mathbb{Z}}} \|N f_N\|_{L_x^2}^{\frac{2d}{d-2}} \sim \sum_{N\in 2^{\mathbb{Z}}} \|\nabla f_N\|_{L_x^2}^{\frac{2d}{d-2}} \tag{4.25}$$

That is, $\dot{B}^1_{2,2d/(d-2)} \hookrightarrow L_x^{2d/(d-2)}$.

PROOF. Exercise: prove this result by mimicking the proof of Proposition 4.8. □

By applying Hölder's inequality to the sum over $2^{\mathbb{Z}}$, we see that this proposition directly implies $\dot{B}^1_{2,q} \hookrightarrow L_x^{2d/(d-2)}$ for any $q \leq \frac{2d}{d-2}$ (e.g., $q = 2$ corresponds to the usual Sobolev embedding). Larger values of q are forbidden, as can be seen by considering a linear combination of many many bumps that are well separated both in space and in characteristic length scale. In this sense, the embedding given above is sharp.

The following variant of Proposition 4.10 forms the basis for the proof of Theorem 4.7 in [**26**]; see [**26**, Proposition 3.1] or [**27**, Théorème 1].

COROLLARY 4.11 (Interpolated Besov embedding, [**27**]). *For $d \geq 3$ and $f \in \mathcal{S}(\mathbb{R}^d)$,*

$$\|f\|_{L_x^{\frac{2d}{d-2}}} \lesssim \|f\|_{\dot{H}_x^1}^{1-\frac{2}{d}} \cdot \sup_{N\in 2^{\mathbb{Z}}} \|\nabla f_N\|_{L_x^2}^{\frac{2}{d}} \sim \|f\|_{\dot{B}^1_{2,2}}^{1-\frac{2}{d}} \|f\|_{\dot{B}^1_{2,\infty}}^{\frac{2}{d}}. \tag{4.26}$$

PROOF. Exercise ×2: deduce this from Proposition 4.10 and then independently from Proposition 4.8. □

Note that relative to Proposition 4.8, the only difference is that the supremum factor contains the $\dot H^1_x$ norm rather than the $L_x^{2d/(d-2)}$ norm. It is this change that allowed us to include (4.20) in Proposition 4.9, which in turn simplified the proof of Theorem 4.7.

4.3. In praise of stationary phase. Although we are blessed with a simple exact formula for the kernel of the free propagator $e^{it\Delta}$,

$$e^{it\Delta}(x,y) = (2\pi)^{-d}\int_{\mathbb{R}^d} e^{i\xi\cdot(x-y)-it|\xi|^2}\,d\xi = (4\pi i t)^{-d/2}e^{i|x-y|^2/4t}, \tag{4.27}$$

many of its properties are more clearly visible from the method of stationary phase.

Our first result is perhaps the best known of this genre. The name we use originates in optics, where it describes diffraction patterns in the (monochromatic) paraxial approximation. In particular, it shows how a laser pointer can be used to draw Fourier transforms.

LEMMA 4.12 (Fraunhofer formula). *For $\psi \in L^2_x(\mathbb{R}^d)$ and $t \to \pm\infty$,*

$$\big\|[e^{it\Delta}\psi](x) - (2it)^{-\frac{d}{2}}e^{i|x|^2/4t}\hat\psi\big(\tfrac{x}{2t}\big)\big\|_{L^2_x} \to 0. \tag{4.28}$$

PROOF. While this asymptotic is most easily understood in terms of stationary phase, the simplest proof dodges around this point. By (4.27), we have the identity

$$\begin{aligned} \text{LHS}(4.28) &= \Big\|(4\pi i t)^{-\frac{d}{2}}\int_{\mathbb{R}^d} e^{i|x-y|^2/4t}[1-e^{-i|y|^2/4t}]\psi(y)\,dy\Big\|_{L^2_x} \\ &= \Big\|\int_{\mathbb{R}^d} e^{it\Delta}(x,y)\,[1-e^{-i|y|^2/4t}]\psi(y)\,dy\Big\|_{L^2_x} \\ &= \big\|[1-e^{-i|y|^2/4t}]\psi(y)\big\|_{L^2_y}. \end{aligned} \tag{4.29}$$

The result now follows from the dominated convergence theorem. □

The Fraunhofer formula clearly shows that wave packets centered at frequency ξ travel with velocity 2ξ. That is, the *group* velocity is 2ξ, in the usual jargon. By comparison, plane wave solutions, $e^{i\xi\cdot(x-\xi t)}$, travel at the *phase* velocity ξ. As one last piece of jargon, we define the *dispersion relation*: it is the relation $\omega = \omega(\xi)$, so that plane wave solutions take the form $e^{i\xi\cdot x - i\omega t}$. In particular, for the Schrödinger equation, $\omega = |\xi|^2$.

The remaining two results in this subsection are both expressions of the dispersive nature of the free propagator, that is, of the fact that different frequencies travel at different speeds. In the first instance, this is quite clear. The second result shows that high-frequency waves spend little time near the spatial origin.

LEMMA 4.13 (Kernel estimates). *For any $m \geq 0$, the kernel of the linear propagator obeys the following estimates:*

$$|(P_N e^{it\Delta})(x,y)| \lesssim_m \begin{cases} |t|^{-d/2} & : |x-y| \sim N|t| \geq N^{-1} \\ \dfrac{N^d}{\langle N^2 t\rangle^m \langle N|x-y|\rangle^m} & : \textit{otherwise}. \end{cases} \tag{4.30}$$

PROOF. Exercise in stationary phase. □

PROPOSITION 4.14 (Local Smoothing, [**21, 79, 100**]). *Fix $\varphi \in C_c^\infty(\mathbb{R}^d)$. Then for all $f \in L_x^2(\mathbb{R}^d)$ and $R > 0$,*

$$\int_{\mathbb{R}}\int_{\mathbb{R}^d} \big|\big[|\nabla|^{\frac{1}{2}} e^{it\Delta} f\big](x)\big|^2 \varphi(x/R)\,dx\,dt \lesssim_\varphi R\|f\|^2_{L_x^2(\mathbb{R}^d)} \tag{4.31}$$

and so,

$$\int_{\mathbb{R}}\int_{\mathbb{R}^d} \big|\big[|\nabla|^{\frac{1}{2}} e^{it\Delta} f\big](x)\big|^2 \langle x\rangle^{-1-\varepsilon}\,dx\,dt \lesssim_\varepsilon \|f\|^2_{L_x^2(\mathbb{R}^d)} \tag{4.32}$$

for any $\varepsilon > 0$.

PROOF. Both (4.31) and (4.32) follow from the same argument (though the second can also be deduced from the first by summing over dyadic R): Given $a : \mathbb{R}^d \to [0, \infty)$,

$$\iint \big|\big[|\nabla|^{\frac{1}{2}} e^{it\Delta} f\big](x)\big|^2 a(x)\,dx\,dt \sim \iint \frac{|\xi|^{\frac{1}{2}}|\eta|^{\frac{1}{2}}}{|\xi| + |\eta|} \hat{a}(\eta - \xi)\delta(|\xi| - |\eta|)\hat{f}(\xi)\overline{\hat{f}(\eta)}\,d\xi\,d\eta.$$

The result now follows from Schur's test. □

EXERCISE. Show that for $d \geq 2$, one may make the replacement $|\nabla| \mapsto \langle\nabla\rangle$ in (4.32) provided one also requires $\varepsilon \geq 1$.

The next result is Lemma 3.7 from [**41**] extended to all dimensions. This will be used in the proof of Lemma 5.7. We give a quantitative proof.

COROLLARY 4.15. *Given $\phi \in \dot{H}_x^1(\mathbb{R}^d)$,*

$$\|\nabla e^{it\Delta}\phi\|^3_{L^2_{t,x}([-T,T]\times\{|x|\leq R\})} \lesssim T^{\frac{2}{d+2}} R^{\frac{3d+2}{2(d+2)}} \|e^{it\Delta}\phi\|_{L_{t,x}^{2(d+2)/(d-2)}} \|\nabla\phi\|^2_{L_x^2}.$$

PROOF. Given $N > 0$, Hölder's and Bernstein's inequalities imply

$$\begin{aligned}\|\nabla e^{it\Delta}\phi_{<N}\|_{L^2_{t,x}([-T,T]\times\{|x|\leq R\})} &\lesssim T^{2/(d+2)} R^{2d/(d+2)} \|e^{it\Delta}\nabla\phi_{<N}\|_{L_{t,x}^{2(d+2)/(d-2)}} \\ &\lesssim T^{2/(d+2)} R^{2d/(d+2)}\, N\, \|e^{it\Delta}\phi\|_{L_{t,x}^{2(d+2)/(d-2)}}.\end{aligned}$$

On the other hand, the high frequencies can be estimated using local smoothing:

$$\begin{aligned}\|\nabla e^{it\Delta}\phi_{\geq N}\|_{L^2_{t,x}([-T,T]\times\{|x|\leq R\})} &\lesssim R^{1/2}\||\nabla|^{1/2}\phi_{\geq N}\|_{L_x^2} \\ &\lesssim N^{-1/2}R^{1/2}\|\nabla\phi\|_{L_x^2}.\end{aligned}$$

The result now follows by optimizing the choice of N. □

4.4. Improved Strichartz inequalities. Let us begin by recalling the original Strichartz inequality in a slightly different formulation (cf. Theorem 3.2).

THEOREM 4.16 (Strichartz). *Fix $2 \leq q, r, \tilde{q}, \tilde{r} \leq \infty$ with $\frac{2}{q} + \frac{d}{r} = \frac{2}{\tilde{q}} + \frac{d}{\tilde{r}} = \frac{d}{2}$. If $d = 2$, we also require that $q, \tilde{q} > 2$. Then*

$$\big\|e^{it\Delta}u_0\big\|_{L_t^q L_x^r(\mathbb{R}\times\mathbb{R}^d)} \lesssim \|u_0\|_{L_x^2(\mathbb{R}^d)} \tag{4.33}$$

$$\Big\|\int_{\mathbb{R}} e^{-it\Delta}F(t)\,dt\Big\|_{L_x^2(\mathbb{R}^d)} \lesssim \|F\|_{L_t^{q'}L_x^{r'}(\mathbb{R}\times\mathbb{R}^d)} \tag{4.34}$$

$$\Big\|\int_{s<t} e^{i(t-s)\Delta}F(s)\,ds\Big\|_{L_t^q L_x^r(\mathbb{R}\times\mathbb{R}^d)} \lesssim \|F\|_{L_t^{\tilde{q}'}L_x^{\tilde{r}'}(\mathbb{R}\times\mathbb{R}^d)} \tag{4.35}$$

for all $u_0 \in \mathcal{S}(\mathbb{R}^d)$ and $F \in \mathcal{S}(\mathbb{R}\times\mathbb{R}^d)$.

PROOF. We treat the case $q, \tilde{q} > 2$. The endpoint case is more involved; see [**37**].

The linear operators in (4.33) and (4.34) are adjoints of one another; thus, by the method of TT^* both will follow once we prove

$$\Big\| \int_{\mathbb{R}} e^{i(t-s)\Delta} F(s)\, ds \Big\|_{L_t^q L_x^r(\mathbb{R}\times\mathbb{R}^d)} \lesssim \|F\|_{L_t^{q'} L_x^{r'}(\mathbb{R}\times\mathbb{R}^d)}. \tag{4.36}$$

By the dispersive estimate (3.2) and then the Hardy-Littlewood-Sobolev inequality, we have

$$\text{LHS}(4.36) \lesssim \Big\| \int_{\mathbb{R}} |t-s|^{\frac{d}{r}-\frac{d}{2}} \|F(s)\|_{L_x^{r'}}\, ds \Big\|_{L_t^q(\mathbb{R})} \lesssim \text{RHS}(4.36).$$

The argument just presented also covers (4.35) in the case $q = \tilde{q}$, $r = \tilde{r}$. To go beyond this case, it helps to consider the estimate in dualized form:

$$\Big| \iint_{s<t} \langle e^{i(t-s)\Delta} F(s), G(t)\rangle\, ds\, dt \Big| \lesssim \|F\|_{L_t^{q'} L_x^{r'}(\mathbb{R}\times\mathbb{R}^d)} \|G\|_{L_t^{\tilde{q}'} L_x^{\tilde{r}'}(\mathbb{R}\times\mathbb{R}^d)}. \tag{4.37}$$

The case $\tilde{q} = \infty$, $\tilde{r} = 2$ follows from (4.34):

$$\text{LHS}(4.37) \leq \Big\| \int_{s<t} e^{i(t-s)\Delta} F(s)\, ds \Big\|_{L_t^\infty L_x^2} \|G\|_{L_t^1 L_x^2} \lesssim \|F\|_{L_t^{q'} L_x^{r'}} \|G\|_{L_t^1 L_x^2}$$

Interpolating between this and the case $q = \tilde{q}$ mentioned above proves (4.35) for all exponents where $q \leq \tilde{q}$. The other case may be deduced symmetrically. □

The main purpose of this subsection is to discuss some variants and extensions of Theorem 4.16. While (4.33) and (4.34) do not hold for any larger class of exponents, (4.35) does. Indeed, this fact plays an important role in the proof of the endpoint case, [**37**]. We have seen one instance of this already, namely, Lemma 3.10. For the largest set of exponents currently known (and a discussion of counterexamples), see [**25, 101**].

One may also consider changing the norm on the right-hand side of (4.33). Placing u_0 in an L_x^p space, brings us back to the dispersive estimate, (3.2). Asking for bounds in terms of $\hat{u}_0$ leads us directly to a profound question:

CONJECTURE 4.17 (Stein's Restriction Conjecture, [**80**]).

$$\|e^{it\Delta} f\|_{L_{t,x}^q(\mathbb{R}\times\mathbb{R}^d)} \lesssim \|\hat{f}\|_{L_\xi^p(\mathbb{R}^d)} \tag{4.38}$$

provided $\frac{d+2}{d} p' = q > \frac{2(d+1)}{d}$.

Despite intensive effort, this conjecture remains unresolved except when $d = 1$, [**24, 109**]. To date, the best result we know is that the conjecture holds for $q > \frac{2(d+3)}{d+1}$, [**88**]. The proof of this takes advantage of a certain bilinear estimate, which we reproduce below as Theorem 4.20.

A variety of bilinear estimates have played an important role in the treatment of mass- and energy-critical NLS. The first such estimate we give appears as [**66**, Theorem 2] in the one dimensional setting, as [**6**, Lemma 111] for $d = 2$, and as [**20**, Lemma 3.4] for general dimensions. We postpone further discussion until after Corollary 4.19.

THEOREM 4.18 (Bilinear Strichartz I, [**6, 20, 66**]). *Fix* $d \geq 1$ *and* $M \leq N$, *then*

$$\big\| [e^{it\Delta} P_M f][e^{it\Delta} P_N g] \big\|_{L_{t,x}^2(\mathbb{R}\times\mathbb{R}^d)} \lesssim M^{\frac{d-1}{2}} N^{-\frac{1}{2}} \|f\|_{L_x^2(\mathbb{R}^d)} \|g\|_{L_x^2(\mathbb{R}^d)} \tag{4.39}$$

When $d = 1$ we require $M \leq \frac{1}{4}N$, so that $P_N P_M = 0$.

PROOF. For $M \sim N$ and $d \neq 1$, the result follows from the $L_x^2 \to L_t^4 L_x^{\frac{2d}{d-1}}$ Strichartz inequality and Bernstein.

Turning to the case $M \leq \frac{1}{4}N$, we note that by duality and the Parseval identity, it suffices to show

$$\Big|\iint_{\mathbb{R}^d\times\mathbb{R}^d} F(|\xi|^2+|\eta|^2, \xi+\eta)\widehat{f_M}(\xi)\widehat{g_N}(\eta)\,d\xi\,d\eta\Big| \lesssim M^{\frac{d-1}{2}}N^{-\frac{1}{2}}\|F\|_{L^2_{\omega,\xi}(\mathbb{R}^{1+d})}\|\hat f\|_{L^2_\xi(\mathbb{R}^d)}\|\hat g\|_{L^2_\xi(\mathbb{R}^d)}. \tag{4.40}$$

Indeed, by breaking the region of integration into several pieces (and rotating the coordinate system appropriately), we may restrict the region of integration to a set where $\eta_1 - \xi_1 \gtrsim N$. Next, we make the change of variables $\zeta = \xi + \eta$, $\omega = |\xi|^2 + |\eta|^2$, and $\beta = (\xi_2, \dots, \xi_d)$. Note that $|\beta| \lesssim M$ while the Jacobian is $J \sim N^{-1}$. Using this information together with Cauchy–Schwarz:

$$\begin{aligned}
\mathrm{LHS}(4.40) &= \Big|\iiint F(\omega,\zeta)\widehat{f_M}(\xi)\widehat{g_N}(\eta)J\,d\omega\,d\zeta\,d\beta\Big| \\
&\leq \|F\|_{L^2_{\omega,\xi}(\mathbb{R}^{1+d})}\int\Big[\iint |\widehat{f_M}(\xi)|^2|\widehat{g_N}(\eta)|^2J^2\,d\omega\,d\zeta\Big]^{\frac{1}{2}}\,d\beta \\
&\lesssim \|F\|_{L^2_{\omega,\xi}(\mathbb{R}^{1+d})}M^{\frac{d-1}{2}}\Big(\iiint |\widehat{f_M}(\xi)|^2|\widehat{g_N}(\eta)|^2J^2\,d\omega\,d\zeta\,d\beta\Big)^{\frac{1}{2}} \\
&\lesssim \|F\|_{L^2_{\omega,\xi}(\mathbb{R}^{1+d})}M^{\frac{d-1}{2}}\Big(\iint |\widehat{f_M}(\xi)|^2|\widehat{g_N}(\eta)|^2N^{-1}\,d\xi\,d\eta\Big)^{\frac{1}{2}},
\end{aligned}$$

which implies (4.39). □

COROLLARY 4.19 (Bilinear Strichartz, II). *Let M, N, and d be as above. Given any spacetime slab $I \times \mathbb{R}^d$, any $t_0 \in I$, and any functions u, v defined on $I \times \mathbb{R}^d$,*

$$\begin{aligned}
\|(P_{\geq N}u)(P_{\leq M}v)\|_{L^2_{t,x}} \lesssim M^{\frac{d-1}{2}}N^{-\frac{1}{2}}&\Big(\|P_{\geq N}u(t_0)\|_{L^2_x} + \|(i\partial_t+\Delta)P_{\geq N}u\|_{L_{t,x}^{\frac{2(d+2)}{d+4}}}\Big) \\
&\times\Big(\|P_{\leq M}v(t_0)\|_{L^2_x} + \|(i\partial_t+\Delta)P_{\leq M}v\|_{L_{t,x}^{\frac{2(d+2)}{d+4}}}\Big),
\end{aligned}$$

where all spacetime norms are taken over $I \times \mathbb{R}^d$.

PROOF. See [**104**, Lemma 2.5], which builds on earlier versions in [**8, 20**]. □

We now embark on a brief discussion of Theorem 4.18. The total power of M and N in (4.39) is dictated by scaling; the point here is that we can skew it heavily in favour of M, thereby obtaining smallness when $M \ll N$. Results of this type have played a vital role in the treatment of mass- and energy-critical NLS, because they have made it possible to 'break' the scaling symmetry. More precisely, Theorem 4.18 shows that interactions between widely separated scales are suppressed, thus, ultimately, permitting one to focus on a single scale at a time. We have already seen a related example of such spontaneous symmetry breaking in the previous subsection (and will see another shortly), namely, that individual optimizers in the Sobolev embedding inequality fail to be dilation/translation invariant; indeed, they have a very definite location and intrinsic length scale.

The particular bilinear estimate given in Theorem 4.18 has proved more useful for the energy-critical NLS than for the mass-critical problem. For the mass-critical NLS, we need a different kind of bilinear estimate:

THEOREM 4.20 (Bilinear Restriction, [**88**]). *Let $f, g \in L_x^2(\mathbb{R}^d)$. Suppose that for some $c > 0$,*

$$N := \operatorname{dist}(\operatorname{supp} \hat{f}, \operatorname{supp} \hat{g}) \geq c \max\{\operatorname{diam}(\operatorname{supp} \hat{f}), \operatorname{diam}(\operatorname{supp} \hat{g})\}.$$

Then for $q > \frac{d+3}{d+1}$,

$$\big\| [e^{it\Delta} f][e^{it\Delta} g] \big\|_{L^q_{t,x}} \lesssim_c N^{d-\frac{d+2}{q}} \|f\|_{L_x^2} \|g\|_{L_x^2}$$

REMARKS. 1. For a fuller discussion of this result and its context, see [**88, 93**]. In particular, we note that Theorem 4.20 was conjectured by Klainerman and Machedon and that Tao indicates that his work was inspired by the analogous result for the wave equation, [**107**].

2. For $q = \frac{d+2}{d}$ (or greater) this follows from Theorem 4.16 (and Bernstein). The point here is that some $q < \frac{d+2}{d}$ are allowed.

3. Whether the theorem remains true for $q = \frac{d+3}{d+1}$ is currently open (except when $d = 1$); however it does fail for q smaller (cf. [**93**, §2.7]). The picture to have in mind is of one train overtaking another: two wave packets that are long in the common direction of propagation (though not so large in the transverse direction) travelling at different speeds. More precisely, consider

$$f = \delta^{\frac{d+1}{2}} \phi(\delta^2 x_1) \phi(\delta x_2) \cdots \phi(\delta x_d) \quad \text{and} \quad g = \delta^{\frac{d+1}{2}} e^{ix_1} \phi(\delta^2 x_1) \phi(\delta x_2) \cdots \phi(\delta x_d)$$

with $\hat{\phi} \in C^\infty(\mathbb{R})$ of compact support and $\delta \downarrow 0$. Note that if the wave packets are made more slender in the transverse direction, they will disperse too quickly.

We will not even attempt to outline the proof of Theorem 4.20; however, we will endeavour to provide a reasonable description of how it is used in the treatment of NLS. To do this, we need to introduce the standard family of dyadic cubes, which we do next. After that, we give an immediate corollary of Theorem 4.20, using this new vocabulary.

DEFINITION 4.21. Given $j \in \mathbb{Z}$, we write $\mathcal{D}_j = \mathcal{D}_j(\mathbb{R}^d)$ for the set of all dyadic cubes of side-length 2^j in $\mathbb{R}^d$:

$$\mathcal{D}_j = \Big\{ \prod_{l=1}^{d} \big[2^j k_l, 2^j(k_l + 1)\big) \subseteq \mathbb{R}^d : k \in \mathbb{Z}^d \Big\}.$$

We also write $\mathcal{D} = \cup_j \mathcal{D}_j$. Given $Q \in \mathcal{D}$, we define f_Q by $\hat{f}_Q = \chi_Q \hat{f}$.

COROLLARY 4.22. *Suppose $Q, Q' \in \mathcal{D}$ with*

$$\operatorname{dist}(Q, Q') \gtrsim \operatorname{diam}(Q) = \operatorname{diam}(Q'),$$

then for some $p < 2$ (indeed, an interval of such p)

$$\big\| [e^{it\Delta} f_Q][e^{it\Delta} f_{Q'}] \big\|_{L_{t,x}^{\frac{d^2+3d+1}{d(d+1)}}} \lesssim |Q|^{1-\frac{2}{p}-\frac{1}{d^2+3d+1}} \|\hat{f}\|_{L_\xi^p(Q)} \|\hat{f}\|_{L_\xi^p(Q')}.$$

PROOF. The result follows from interpolating between Theorem 4.20 and

$$\big\| [e^{it\Delta} f][e^{it\Delta} g] \big\|_{L^\infty_{t,x}} \lesssim \|\hat{f}\|_{L^1_\xi} \|\hat{g}\|_{L^1_\xi},$$

which is a consequence of the fact that the Fourier transform maps $L^1_\xi \to L^\infty_x$. □

Our next theorem is clearly a strengthening of Theorem 4.16 (apply Hölder's inequality inside the second factor in (4.42)). The name is taken from the standard notation for the norm appearing on the right-hand side in (4.41). It was first proved in the case $d = 2$; see [**62**, Theorem 4.2]. For higher dimensions, see [**4**, Theorem 1.2] and for $d = 1$, see [**12**, Proposition 2.1].

THEOREM 4.23 (X^q_p Strichartz, [**4, 12, 62**]). *Given $f \in \mathcal{S}$, $\frac{1}{2} < \frac{1}{p} < \frac{1}{2} + \frac{1}{(d+1)(d+2)}$, and $\frac{p}{2} < \beta < 1$,*

$$\big\|e^{it\Delta} f\big\|_{L_{t,x}^{\frac{2(d+2)}{d}}(\mathbb{R}^{1+d})} \lesssim \Big[\sum_{Q\in\mathcal{D}} \Big(|Q|^{\frac{1}{2}-\frac{1}{p}} \big\|\hat{f}\big\|_{L^p_\xi(Q)}\Big)^{\frac{2(d+2)}{d}}\Big]^{\frac{d}{2(d+2)}} \tag{4.41}$$

$$\lesssim \|f\|^{\beta}_{L^2_x(\mathbb{R}^d)} \Big[\sup_{Q\in\mathcal{D}} |Q|^{\frac{1}{2}-\frac{1}{p}} \big\|\hat{f}\big\|_{L^p_\xi(Q)}\Big]^{1-\beta}. \tag{4.42}$$

Recall that this sum is over all dyadic cubes Q of all sizes.

We will not prove this result; however, the proof of Proposition 4.24 below is closely modelled on the argument given in [**4**]. This proposition is a small tweaking of (the proof of) (4.42) so as to exhibit the supremum of a spacetime norm.

PROPOSITION 4.24. *Let $q = \frac{2(d^2+3d+1)}{d^2}$. Then*

$$\big\|e^{it\Delta} f\big\|_{L_{t,x}^{\frac{2(d+2)}{d}}(\mathbb{R}^{1+d})} \lesssim \|f\|^{\frac{d+1}{d+2}}_{L^2_x(\mathbb{R}^d)} \Big(\sup_{Q\in\mathcal{D}} |Q|^{\frac{d+2}{dq}-\frac{1}{2}} \big\|e^{it\Delta} f_Q\big\|_{L^q_{t,x}(\mathbb{R}^{1+d})}\Big)^{\frac{1}{d+2}}. \tag{4.43}$$

PROOF. As noted above, we will be mimicking [**4**], albeit with a small twist. The first part of the argument is based on the proof of their Theorem 1.2.

Given distinct $\xi, \xi' \in \mathbb{R}^d$, there is a unique maximal pair of dyadic cubes $Q \ni \xi$ and $Q' \ni \xi'$ obeying

$$|Q| = |Q'| \quad \text{and} \quad \operatorname{dist}(Q, Q') \geq 4\operatorname{diam}(Q). \tag{4.44}$$

Let $\mathcal{F}$ denote the family of all such pairs as $\xi \neq \xi'$ vary over $\mathbb{R}^d$. According to this definition,

$$\sum_{(Q,Q')\in\mathcal{F}} \chi_Q(\xi)\chi_{Q'}(\xi') = 1 \quad \text{for a.e. } (\xi, \xi') \in \mathbb{R}^d \times \mathbb{R}^d. \tag{4.45}$$

Note that since Q and Q' are maximal, $\operatorname{dist}(Q, Q') \leq 10\operatorname{diam}(Q)$. In addition, this shows that given Q there are a bounded number of Q' so that $(Q, Q') \in \mathcal{F}$, that is,

$$\forall Q \in \mathcal{D}, \quad \#\{Q' : (Q, Q') \in \mathcal{F}\} \lesssim 1. \tag{4.46}$$

In view of (4.45), we can write

$$[e^{it\Delta} f]^2 = \sum_{(Q,Q')\in\mathcal{F}} [e^{it\Delta} f_Q][e^{it\Delta} f_{Q'}],$$

which clearly brings Corollary 4.22 into the game. Treating the sum via the triangle inequality is not a winning play; we need to do a bit better. The key point is to look at the spacetime Fourier supports of the products on the right-hand side. As we will see, their dilates have bounded overlap.

Given $F : \mathbb{R} \times \mathbb{R}^d \to \mathbb{C}$ we write

$$\hat{F}(\omega, \xi) = (2\pi)^{-\frac{d+1}{2}} \int_{\mathbb{R}^d} \int_{\mathbb{R}} e^{i\omega t - i\xi\cdot x} F(t, x)\, dt\, dx.$$

With this convention,

$$\operatorname{supp}\big([e^{it\Delta}f_Q][e^{it\Delta}f_{Q'}]\widehat{\ }\big) \subseteq R(Q+Q') \tag{4.47}$$

where $Q+Q'$ denotes the Minkowski (or 'all pairs') sum and R denotes an associated parallelepiped that we will now define. Given a cube Q'' in $\mathbb{R}^d$ (and $Q+Q'$ is a cube), we define

$$R(Q'') = \Big\{(\omega,\eta) : \eta \in Q'' \text{ and } 2 \leq \frac{\omega - \frac{1}{2}|c(Q'')|^2 - c(Q'')\cdot[\eta - c(Q'')]}{\operatorname{diam}(Q'')^2} \leq 19\Big\}$$

where $c(Q'')$ denotes the center of the cube Q''. To verify (4.47) we merely need to note that for $\xi \in Q$ and $\xi' \in Q'$,

$$\begin{aligned}|\xi|^2 + |\xi'|^2 &= \tfrac{1}{2}|\xi+\xi'|^2 + \tfrac{1}{2}|\xi-\xi'|^2\\ &= \tfrac{1}{2}|c(Q+Q')|^2 + c(Q+Q')\cdot[\xi+\xi'-c(Q+Q')]\\ &\qquad + \tfrac{1}{2}|\xi+\xi'-c(Q+Q')|^2 + \tfrac{1}{2}|\xi-\xi'|^2,\end{aligned}$$

$|\xi+\xi'-c(Q+Q')| \leq \operatorname{diam}(Q)$, and $4\operatorname{diam}(Q) \leq |\xi-\xi'| \leq 12\operatorname{diam}(Q)$. We also remind the reader that $\operatorname{diam}(Q+Q') = \operatorname{diam}(Q) + \operatorname{diam}(Q') = 2\operatorname{diam}(Q)$.

Before we can turn to the analytical portion of the argument, we still need to control the overlap of the Fourier supports, or rather, of the enclosing parallelepipeds. We claim that for any $\alpha \leq 1.01$,

$$\sup_{\omega,\eta} \sum_{(Q,Q')\in\mathcal{F}} \chi_{\alpha R(Q+Q')}(\omega,\eta) \lesssim 1, \tag{4.48}$$

where αR denotes the α-dilate of R with the same center. To see this, we argue as follows: Given $(\omega,\eta) \in \alpha R(Q+Q')$, a few computations show that $\operatorname{diam}(Q)^2 \sim \omega - \frac{1}{2}|\eta|^2$, which allows us to identify the size of Q to within a bounded number of dyadic generations. This then gives an upper bound on the distance between Q and Q'. Lastly, since $\eta \in \alpha(Q+Q')$ we may deduce that both Q and Q' must lie within $O(\operatorname{diam} Q)$ of $\frac{1}{2}\eta$. To recap, each (ω,η) belongs to a bounded number of $\alpha R(Q+Q')$, which is exactly (4.48).

With the information we have gathered together, we are now ready to begin estimating the right-hand side of (4.43). For $d \geq 2$, may apply Lemma A.9, Hölder's inequality, Corollary 4.22, and (4.46) as follows:

$$\begin{aligned}\big\|e^{it\Delta}f\big\|_{L_{t,x}^{\frac{2(d+2)}{d}}}^{\frac{2(d+2)}{d}} &= \Big\|\sum_{(Q,Q')\in\mathcal{F}} [e^{it\Delta}f_Q][e^{it\Delta}f_{Q'}]\Big\|_{L_{t,x}^{\frac{d+2}{d}}}^{\frac{d+2}{d}}\\ &\lesssim \sum_{(Q,Q')\in\mathcal{F}} \big\|[e^{it\Delta}f_Q][e^{it\Delta}f_{Q'}]\big\|_{L_{t,x}^{\frac{d+2}{d}}}^{\frac{d+2}{d}}\\ &\lesssim \sum_{(Q,Q')\in\mathcal{F}} \big\|e^{it\Delta}f_Q\big\|_{L_{t,x}^q}^{\frac{1}{d}} \big\|e^{it\Delta}f_{Q'}\big\|_{L_{t,x}^q}^{\frac{1}{d}} \big\|[e^{it\Delta}f_Q][e^{it\Delta}f_{Q'}]\big\|_{L_{t,x}^{\frac{d^2+3d+1}{d(d+1)}}}^{\frac{d+1}{d}}\\ &\lesssim \Big(\sup_{Q\in\mathcal{D}} |Q|^{\frac{d+2}{dq}-\frac{1}{2}}\big\|e^{it\Delta}f_Q\big\|_{L_{t,x}^q}\Big)^{\frac{2}{d}} \cdot \sum_{Q\in\mathcal{D}} \Big(|Q|^{-\frac{2-p}{p}}\|\hat f\|_{L_\xi^p(Q)}^2\Big)^{\frac{d+1}{d}}\end{aligned}$$

for some $p<2$. While the final inequality obtained above holds when $d=1$, the argument needs minor modifications (cf. the first inequality). In this case, one should use (A.2) in place of Lemma A.9; we leave the details to the reader.

In order to complete the proof of the proposition, we need to show that the sum given above can be bounded in terms of the L^2_ξ norm of $\hat f$. Once again we turn to [**4**] for advice, this time, to the proof of their Theorem 1.3 (see also [**8**, p. 37] for the case $d = 2$).

The key idea is to break $\hat f$ into two pieces, depending on the size of Q:

$$\hat f(\xi) = \chi_{\{|\hat f|\geq 2^{-jd/2}\}}(\xi)\hat f(\xi) + \chi_{\{|\hat f|\leq 2^{-jd/2}\}}(\xi)\hat f(\xi) =: \hat f^j(\xi) + \hat f_j(\xi).$$

Here and below we assume (without loss of generality) that f is L^2_x-normalized; otherwise the size of f has to be incorporated into the height of this splitting, with concomitant detriment to readability.

For the first piece, we need only use the fact that $p < 2$:

$$\begin{aligned}
\sum_{j\in\mathbb{Z}}\sum_{Q\in\mathcal{D}_j}\Big(|Q|^{-\frac{2-p}{p}}\|\hat f^j\|^2_{L^p_\xi(Q)}\Big)^{\frac{d+1}{d}} &\lesssim \Big(\sum_{j\in\mathbb{Z}}\sum_{Q\in\mathcal{D}_j}|Q|^{-\frac{2-p}{2}}\|\hat f^j\|^p_{L^p_\xi(Q)}\Big)^{\frac{2(d+1)}{pd}}\\
&\lesssim \Big(\int_{\mathbb{R}^d}\sum_{j:|\hat f|\geq 2^{-jd/2}} 2^{-jd\frac{2-p}{2}}\big|\hat f(\xi)\big|^p\,d\xi\Big)^{\frac{2(d+1)}{pd}}\\
&\lesssim \Big(\int_{\mathbb{R}^d}\big|\hat f(\xi)\big|^2\,d\xi\Big)^{\frac{2(d+1)}{pd}} \lesssim 1.
\end{aligned}$$

For the second piece, we lead off with Hölder's inequality:

$$\begin{aligned}
\sum_{j\in\mathbb{Z}}\sum_{Q\in\mathcal{D}_j}\Big(|Q|^{-\frac{2-p}{p}}\|\hat f_j\|^2_{L^p_\xi(Q)}\Big)^{\frac{d+1}{d}} &\lesssim \sum_{j\in\mathbb{Z}}\sum_{Q\in\mathcal{D}_j}|Q|^{\frac{1}{d}}\|\hat f_j\|^{\frac{2(d+1)}{d}}_{L_\xi^{\frac{2(d+1)}{d}}(Q)}\\
&\lesssim \int_{\mathbb{R}^d}\sum_{j:|\hat f|\leq 2^{-jd/2}}\big(2^{-\frac{jd}{2}}\big)^{-\frac{2}{d}}\big|\hat f(\xi)\big|^{\frac{2(d+1)}{d}}\,d\xi\\
&\lesssim \int_{\mathbb{R}^d}\big|\hat f(\xi)\big|^2\,d\xi \lesssim 1.
\end{aligned}$$

This completes the proof of (4.43). □

We are now ready to state our preferred form of inverse Strichartz inequality. For other variants, see for example, [**6**, §§2–3], [**58**, Theorem 1], [**92**, Appendix A].

PROPOSITION 4.25 (Inverse Strichartz Inequality). *Fix $d \geq 1$ and $\{f_n\} \subseteq L^2_x(\mathbb{R}^d)$. Suppose that*

$$\lim_{n\to\infty}\|f_n\|_{L^2_x(\mathbb{R}^d)} = A \quad \textit{and} \quad \lim_{n\to\infty}\|e^{it\Delta}f_n\|_{L^{\frac{2(d+2)}{d}}_{t,x}(\mathbb{R}^{1+d})} = \varepsilon.$$

Then there exist a subsequence in n, $\phi \in L^2_x(\mathbb{R}^d)$, $\{\lambda_n\} \subseteq (0,\infty)$, $\{\xi_n\} \subseteq \mathbb{R}^d$, and $\{(t_n, x_n)\} \subseteq \mathbb{R}^{1+d}$ so that along the subsequence, we have the following:

$$\lambda_n^{\frac{d}{2}}e^{-i\xi_n\cdot(\lambda_n x+x_n)}[e^{it_n\Delta}f_n](\lambda_n x + x_n) \rightharpoonup \phi(x) \quad \textit{weakly in } L^2_x(\mathbb{R}^d) \tag{4.49}$$

$$\lim_{n\to\infty}\|f_n\|^2_{L^2_x} - \|f_n - \phi_n\|^2_{L^2_x} = \|\phi\|^2_{L^2_x} \gtrsim A^2\big(\tfrac{\varepsilon}{A}\big)^{2(d+1)(d+2)} \tag{4.50}$$

$$\limsup_{n\to\infty}\big\|e^{it\Delta}(f_n-\phi_n)\big\|^{\frac{2(d+2)}{d}}_{L^{\frac{2(d+2)}{d}}_{t,x}(\mathbb{R}^{1+d})} \leq \varepsilon^{\frac{2(d+2)}{d}}\Big[1 - c\big(\tfrac{\varepsilon}{A}\big)^{\beta}\Big], \tag{4.51}$$

where c and β are (dimension-dependent) constants and

$$\phi_n(x) := e^{-it_n\Delta}[g_{0,\xi_n,x_n,\lambda_n}\phi](x) = \lambda_n^{-\frac{d}{2}}\, e^{-it_n\Delta}\big[e^{i\xi_n\cdot}\phi\big(\lambda_n^{-1}(\cdot - x_n)\big)\big](x). \tag{4.52}$$

PROOF. By Proposition 4.24, there exists $\{Q_n\} \subseteq \mathcal{D}$ so that

$$\varepsilon^{(d+2)}A^{-(d+1)} \lesssim \liminf_{n\to\infty} |Q_n|^{\frac{d+2}{dq}-\frac12}\|e^{it\Delta}(f_n)_{Q_n}\|_{L^q_{t,x}(\mathbb{R}^{1+d})} \tag{4.53}$$

where $q = 2(d^2+3d+1)/d^2$. We choose λ_n^{-1} to be the side-length of Q_n, which implies $|Q_n| = \lambda_n^{-d}$. We also set $\xi_n := c(Q_n)$, that is, the centre of this cube.

Next we determine x_n and t_n. By Hölder's inequality,

$$\begin{aligned}
&\liminf_{n\to\infty} |Q_n|^{\frac{d+2}{dq}-\frac12}\|e^{it\Delta}(f_n)_{Q_n}\|_{L^q_{t,x}(\mathbb{R}^{1+d})}\\
&\qquad\lesssim \liminf_{n\to\infty} |Q_n|^{\frac{d+2}{dq}-\frac12}\|e^{it\Delta}(f_n)_{Q_n}\|_{L^{\frac{2(d+2)}{d}}_{t,x}(\mathbb{R}^{1+d})}^{\frac{d(d+2)}{d^2+3d+1}}\|e^{it\Delta}(f_n)_{Q_n}\|_{L^{\infty}_{t,x}(\mathbb{R}^{1+d})}^{\frac{d+1}{d^2+3d+1}}\\
&\qquad\lesssim \liminf_{n\to\infty} \lambda_n^{\frac{d}{2}-\frac{d+2}{q}}\varepsilon^{\frac{d(d+2)}{d^2+3d+1}}\|e^{it\Delta}(f_n)_{Q_n}\|_{L^{\infty}_{t,x}(\mathbb{R}^{1+d})}^{\frac{d+1}{d^2+3d+1}}.
\end{aligned}$$

Thus by (4.53), there exists $\{(t_n,x_n)\} \subseteq \mathbb{R}^{1+d}$ so that

$$\liminf_{n\to\infty} \lambda_n^{\frac{d}{2}}\big|\big[e^{it_n\Delta}(f_n)_{Q_n}\big](x_n)\big| \gtrsim \varepsilon^{(d+1)(d+2)}A^{-(d^2+3d+1)}. \tag{4.54}$$

Having selected our symmetry parameters, weak compactness of $L^2_x(\mathbb{R}^d)$ (i.e. Alaoglu's theorem) guarantees that (4.49) holds for some $\phi \in L^2_x(\mathbb{R}^d)$ and some subsequence in n. Our next job is to show that ϕ carries non-trivial norm.

Define h so that $\hat{h}$ is the characteristic function of the cube $[-\frac12,\frac12)^d$. From (4.54) we obtain

$$\begin{aligned}
|\langle h,\phi\rangle| &= \lim_{n\to\infty}\left|\int \bar{h}(x)\lambda_n^{\frac{d}{2}}e^{-i\xi_n\cdot(\lambda_n x+x_n)}[e^{it_n\Delta}f_n](\lambda_n x+x_n)\,dx\right|\\
&= \lim_{n\to\infty}\lambda_n^{\frac{d}{2}}\Big|\big[e^{it_n\Delta}(f_n)_{Q_n}\big](x_n)\Big|\\
&\gtrsim \varepsilon^{(d+1)(d+2)}A^{-(d^2+3d+1)}, \qquad (4.55)
\end{aligned}$$

which quickly implies (4.50) as seen in the proof of Proposition 4.9. This leaves us to consider (4.51). First we claim that after passing to a subsequence,

$$e^{it\Delta}\Big[\lambda_n^{\frac{d}{2}}e^{-i\xi_n\cdot(\lambda_n x+x_n)}[e^{it_n\Delta}f_n](\lambda_n x+x_n)\Big] \to e^{it\Delta}\phi(x) \quad \text{for a.e. } (t,x)\in\mathbb{R}^{1+d}.$$

Indeed, this follows from the local smoothing estimate, Proposition 4.14, and the Rellich–Kondrashov Theorem. Thus by applying Lemma A.5 and transferring the symmetries, we obtain

$$\|e^{it\Delta}f_n\|_{L^{\frac{2(d+2)}{d}}_{t,x}(\mathbb{R}^{1+d})}^{\frac{2(d+2)}{d}} - \|e^{it\Delta}(f_n-\phi_n)\|_{L^{\frac{2(d+2)}{d}}_{t,x}(\mathbb{R}^{1+d})}^{\frac{2(d+2)}{d}} - \|e^{it\Delta}\phi_n\|_{L^{\frac{2(d+2)}{d}}_{t,x}(\mathbb{R}^{1+d})}^{\frac{2(d+2)}{d}} \to 0.$$

The requisite lower bound on the right-hand side follows from (4.55). □

Note that one may replace (4.49) by weak convergence of the free evolutions:

EXERCISE. Let $\{f_n\}$ be a bounded sequence $L^2_x(\mathbb{R}^d)$. Show that $f_n \rightharpoonup f$ weakly in $L^2_x(\mathbb{R}^d)$ if and only if $e^{it\Delta}f_n \rightharpoonup e^{it\Delta}f$ weakly in $L_x^{2(d+2)/d}(\mathbb{R}\times\mathbb{R}^d)$.

The next two theorems are Strichartz analogues of the bubble decomposition discussed in the previous subsection. This kind of result was introduced by Bahouri and Gérard [**3**] in the context of the wave equation; we will follow their nomenclature and refer to it as a 'profile decomposition'. What we will present here are the mass- and energy-critical analogues of the linear profile decomposition given in that paper. Analogues of the nonlinear version appear in the proofs of Propositions 5.3 and 5.6.

The mass-critical linear profile decomposition was first proved in the case of two space dimensions. This is a result of Merle and Vega [**58**]; see also [**6**, §§2–3] for results of a very similar spirit. Carles and Keraani treated the one-dimensional case [**12**, Theorem 1.4]. The result was obtained for general dimension by Begout and Vargas [**4**]. We remind the reader that the definition of the symmetry group G associated to the mass-critical equation can be found in Subsection 2.3.

THEOREM 4.26 (Mass-critical linear profile decomposition, [**4, 12, 58**]). *Let u_n be a bounded sequence in $L_x^2(\mathbb{R}^d)$. Then (after passing to a subsequence if necessary) there exist $J^* \in \{0,1,\dots\} \cup \{\infty\}$, functions $\{\phi^j\}_{j=1}^{J^*} \subseteq L_x^2(\mathbb{R}^d)$, group elements $\{g_n^j\}_{j=1}^{J^*} \subseteq G$, and times $\{t_n^j\}_{j=1}^{J^*} \subseteq \mathbb{R}$ so that defining w_n^J by*

$$u_n = \sum_{j=1}^{J} g_n^j e^{it_n^j\Delta}\phi^j + w_n^J, \tag{4.56}$$

we have the following properties:

$$\lim_{J\to J^*} \limsup_{n\to\infty} \|e^{it\Delta}w_n^J\|_{L_{t,x}^{\frac{2(d+2)}{d}}} = 0 \tag{4.57}$$

$$e^{-it_n^j\Delta}\big[(g_n^j)^{-1}w_n^J\big] \rightharpoonup 0 \quad \text{weakly in } L_x^2(\mathbb{R}^d) \text{ for each } j \le J, \tag{4.58}$$

$$\sup_J \lim_{n\to\infty} \Big[\|u_n\|_{L_x^2(\mathbb{R}^d)}^2 - \sum_{j=1}^{J} \|\phi^j\|_{L_x^2(\mathbb{R}^d)}^2 - \|w_n^J\|_{L_x^2(\mathbb{R}^d)}^2\Big] = 0 \tag{4.59}$$

and lastly, for $j \neq k$ and $n \to \infty$,

$$\begin{aligned} \frac{\lambda_n^j}{\lambda_n^k} + \frac{\lambda_n^k}{\lambda_n^j} + \lambda_n^j\lambda_n^k|\xi_n^j - \xi_n^k|^2 + \frac{|t_n^j(\lambda_n^j)^2 - t_n^k(\lambda_n^k)^2|}{\lambda_n^j\lambda_n^k} \\ + \frac{|x_n^j - x_n^k - 2t_n^j(\lambda_n^j)^2(\xi_n^j - \xi_n^k)|^2}{\lambda_n^j\lambda_n^k} \to \infty. \end{aligned} \tag{4.60}$$

Here λ_n^j, ξ_n^j, x_n^j are the parameters associated to g_n^j (the θ parameter is zero).

PROOF. Exercise: mimic the proof of Theorem 4.7 using Proposition 4.25 in place of Proposition 4.9. Note that the order of the propagator and the symmetries is changed in (4.56) relative to (4.52). As a result, the meaning of x_n^j and t_n^j has also changed relative to the parameters appearing in Proposition 4.25; indeed, the change can be deduced from

$$e^{-it_n\Delta}[g_{0,\xi_n,x_n,\lambda_n}\phi](x) = g_{t_n|\xi_n|^2,\xi_n,x_n-2t_n\xi_n,\lambda_n}\big[e^{-it_n(\lambda_n)^{-2}\Delta}\phi\big](x).$$

In addition, there is also a change in the sign of t_n^j. □

The analogue of (4.13) can be added to the conclusions of Theorem 4.26, which is to say that the profiles also decouple in the symmetric Strichartz norm; indeed, this follows *a posteriori* from (4.57) and (4.60). We will not need this fact.

The linear profile decomposition in the energy-critical case was proved by Keraani [41]. As in the treatment of the wave equation [3], the original argument used refinements of Sobolev embedding rather than of Strichartz inequality.

THEOREM 4.27 (Energy-critical linear profile decomposition, [41]). *Fix $d \geq 3$ and let $\{u_n\}_{n\geq 1}$ be a sequence of functions bounded in $\dot H^1_x(\mathbb{R}^d)$. Then, after passing to a subsequence if necessary, there exist $J^* \in \{0,1,\ldots\}\cup\{\infty\}$, functions $\{\phi^j\}_{j=1}^{J^*} \subset \dot H^1_x(\mathbb{R}^d)$, group elements $\{g_n^j\}_{j=1}^{J^*} \subset G$, and times $\{t_n^j\}_{j=1}^{J^*} \subset \mathbb{R}$ such that for each $1 \leq J \leq J^*$, we have the decomposition*

$$u_n = \sum_{j=1}^{J} g_n^j e^{it_n^j\Delta}\phi^j + w_n^J \tag{4.61}$$

with the following properties:

$$\lim_{J\to J^*} \limsup_{n\to\infty} \bigl\| e^{it\Delta} w_n^J \bigr\|_{L_{t,x}^{\frac{2(d+2)}{d-2}}(\mathbb{R}\times\mathbb{R}^d)} = 0 \tag{4.62}$$

$$e^{-it_n^j\Delta}\bigl[(g_n^j)^{-1} w_n^J\bigr] \rightharpoonup 0 \quad \text{weakly in } \dot H^1_x(\mathbb{R}^d) \text{ for each } j \leq J \tag{4.63}$$

$$\lim_{n\to\infty} \Bigl[\|\nabla u_n\|_2^2 - \sum_{j=1}^{J} \|\nabla \phi^j\|_2^2 - \|\nabla w_n^J\|_2^2 \Bigr] = 0 \tag{4.64}$$

and for each $j \neq k$,

$$\frac{\lambda_n^j}{\lambda_n^k} + \frac{\lambda_n^k}{\lambda_n^j} + \frac{|x_n^j - x_n^k|^2}{\lambda_n^j \lambda_n^k} + \frac{\bigl|t_n^j(\lambda_n^j)^2 - t_n^k(\lambda_n^k)^2\bigr|}{\lambda_n^j\lambda_n^k} \to \infty \quad \text{as } n \to \infty, \tag{4.65}$$

where λ_n^j and x_n^j are the symmetry parameters associated to g_n^j by Definition 2.2; the θ parameter is identically zero.

PROOF. Exercise. Deduce this result from Theorem 4.26. Note that the disappearance of the Galilei boosts can be attributed to the absence of a gradient in (4.62).

The original approach taken by Keraani involves interpolation, Theorem 4.7, and a Strichartz inequality with unequal space and time exponents. See [41] for more information on how this can be done. □

4.5. Radial Improvements. Most problems related to critical NLS have first been solved in the case of spherically symmetric data. This allows one to take advantage of stronger harmonic analysis tools, some of which we record below. In truth, however, the greatest advantage really appears in the nonlinear analysis.

LEMMA 4.28 (Weighted Radial Strichartz, [43]). *Let $F \in L_{t,x}^{2(d+2)/(d+4)}(\mathbb{R}\times\mathbb{R}^d)$ and $u_0 \in L_x^2(\mathbb{R}^d)$ be spherically symmetric. Then,*

$$u(t) := e^{i(t-t_0)\Delta}u_0 - i\int_{t_0}^{t} e^{i(t-t')\Delta}F(t')\,dt'$$

obeys the estimate

$$\Bigl\| |x|^{\frac{2(d-1)}{q}} u \Bigr\|_{L_t^q L_x^{\frac{2q}{q-4}}(\mathbb{R}\times\mathbb{R}^d)} \lesssim \|u_0\|_{L_x^2(\mathbb{R}^d)} + \|F\|_{L_{t,x}^{\frac{2(d+2)}{d+4}}(\mathbb{R}\times\mathbb{R}^d)}$$

for all $4 \leq q \leq \infty$.

PROOF. For $q = \infty$, this corresponds to the trivial endpoint in the Strichartz inequality. We will only prove the result for the $q = 4$ endpoint, since the remaining cases then follow by interpolation.

As in the proof of the Strichartz inequality, the method of TT^* together with Hardy–Littlewood–Sobolev inequality reduce matters to proving that

$$\big\| |x|^{\frac{d-1}{2}} e^{it\Delta} |x|^{\frac{d-1}{2}} g \big\|_{L_x^\infty(\mathbb{R}^d)} \lesssim |t|^{-\frac{1}{2}} \|g\|_{L_x^1(\mathbb{R}^d)} \tag{4.66}$$

for all radial functions g.

Let P_{rad} denote the projection onto radial functions, which commutes with the free propagator. Then

$$[e^{it\Delta} P_{\text{rad}}](x,y) = (4\pi i t)^{-\frac{d}{2}} e^{i\frac{|x|^2+|y|^2}{4t}} \int_{S^{d-1}} e^{-i\frac{|y|\omega\cdot x}{2t}} \, d\sigma(\omega),$$

where $d\sigma$ denotes the uniform probability measure on the unit sphere S^{d-1}. Using stationary phase (or properties of Bessel functions), one sees that

$$\big|[e^{it\Delta} P_{\text{rad}}](x,y)\big| \lesssim |t|^{-\frac{d}{2}} \big(\tfrac{|y||x|}{|t|}\big)^{-\frac{d-1}{2}} \lesssim |t|^{-\frac{1}{2}} |x|^{-\frac{d-1}{2}} |y|^{-\frac{d-1}{2}}.$$

The radial dispersive estimate (4.66) now follows easily. □

The last two results are taken from the thesis work of Shuanglin Shao.

THEOREM 4.29 (Shao's Strichartz Estimate, [**77**, Corollary 6.2]). *If $f \in L_x^2(\mathbb{R}^d)$ is spherically symmetric with $d \geq 2$, then*

$$\|P_N e^{it\Delta} f\|_{L^q_{t,x}(\mathbb{R}\times\mathbb{R}^d)} \lesssim_q N^{\frac{d}{2}-\frac{d+2}{q}} \|f\|_{L_x^2(\mathbb{R}^d)}, \tag{4.67}$$

provided $q > \frac{4d+2}{2d-1}$.

The new point is that q can go below $2(d+2)/d$, which is the exponent given by Theorem 4.16. The Knapp counterexample (a wave packet whose momentum is concentrated in a single direction) shows that such an improvement is not possible without the radial assumption. Spherical symmetry also allows for stronger bilinear estimates, extending both Theorem 4.18 and Theorem 4.20. We record here only a special case of [**77**, Corollary 6.5]:

THEOREM 4.30 (Shao's Bilinear Estimate, [**77**, Corollary 6.5]). *Fix $d \geq 2$ and $f, g \in L_x^2(\mathbb{R}^d)$ spherically symmetric. Then*

$$\big\| [e^{it\Delta} f_{\leq 1}][e^{it\Delta} g_N] \big\|_{L^q_{t,x}} \lesssim N^{d-\frac{d+2}{q}} \|f\|_{L_x^2} \|g\|_{L_x^2}$$

for any $\frac{2(d+2)}{2d+1} < q \leq 2$ and $N \geq 4$.

5. Minimal blowup solutions

The purpose of this section is to prove that if the global well-posedness and scattering conjectures were to fail, then one could construct *minimal* counterexamples. These counterexamples are *minimal blowup solutions* and enjoy a wealth of properties, all of which are consequences of their minimality.

The discovery that such minimal blowup solutions would exist was made by Keraani [**42**, Theorem 1.3] in the context of the mass-critical equation. This was later adapted to the energy-critical setting by Kenig and Merle, [**38**].

We would also like to mention that earlier works on the energy-critical NLS (see [**7, 20, 75, 104**]) proposed *almost*-minimal blowup solutions as counterexamples to the global well-posedness and scattering conjecture. These solutions were shown to have space and frequency localization properties similar to (but slightly weaker than) those of the minimal blowup solutions. In fact, on a technical level, the tools involved in obtaining both types of counterexamples are closely related. However, while the earlier methods have the advantage of being quantitative, they add significantly to the complexity of the argument.

In these notes, we will only prove the existence of minimal blowup solutions for the mass- and energy-critical nonlinear Schrödinger equations. However, using the arguments presented below (especially those for the energy-critical NLS), one can construct minimal blowup solutions for the more general equation (3.5); see [**40**] for one such example.

5.1. The mass-critical NLS. In the defocusing case, $\mu = +1$, Conjecture 1.4 says that all solutions obey spacetime bounds depending only on the mass. With this in mind, let

$$L^+(M) := \sup\{S_I(u) : u : I \times \mathbb{R}^d \to \mathbb{C} \text{ such that } M(u) \leq M\},$$

where the supremum is taken over all solutions $u : I \times \mathbb{R}^d \to \mathbb{C}$ to the defocusing mass-critical NLS and

$$S_I(u) := \int_I \int_{\mathbb{R}^d} |u(t,x)|^{\frac{2(d+2)}{d}}\,dx\,dt.$$

Note that $L^+ : [0,\infty) \to [0,\infty]$ is nondecreasing and, by Theorem 3.7, continuous. Thus, failure of Conjecture 1.4 (in the defocusing case) is equivalent to the existence of a *critical mass*, $M_c \in (0,\infty)$, so that

$$L^+(M) < \infty \quad \text{for } M < M_c \quad \text{and} \quad L^+(M) = \infty \quad \text{for } M \geq M_c.$$

Similarly, in the focusing case, $\mu = -1$, we may define $L^- : \big[0, M(Q)\big] \to [0,\infty]$ by

$$L^-(M) := \sup\{S_I(u) : u : I \times \mathbb{R}^d \to \mathbb{C} \text{ such that } M(u) \leq M\},$$

where the supremum is again taken over all solutions of the focusing equation. Much as before, failure of Conjecture 1.4 corresponds to the existence of a critical mass $M_c \in (0, M(Q))$, where L^- changes from being finite to infinite.

Note that the explicit solution $u(t,x) = e^{it}Q(x)$ shows that $L^-(M(Q)) = \infty$. Note also that from the local well-posedness theory (see Corollary 3.5),

$$L^+(M) + L^-(M) \lesssim M^{\frac{d+2}{d}} \quad \text{for} \quad M \leq \eta_0, \tag{5.1}$$

where $\eta_0 = \eta_0(d)$ is the threshold from the small data theory.

In order to treat the focusing and defocusing equations in as uniform a manner as possible, we adopt the following convention.

Convention. We write L for $L^\pm$ with the understanding that $L = L^+$ in the defocusing case and $L = L^-$ in the focusing case.

By the discussion above, we see that any initial data u_0 with $M(u_0) < M_c$ must give rise to a global solution, which obeys

$$S_{\mathbb{R}}(u) \leq L\big(M(u_0)\big).$$

This fact plays much the same role as the inductive hypothesis in the induction on mass/energy approach.

Our goals for this subsection are firstly, to show that if Conjecture 1.4 fails, then there exists a blowup solution u to (1.4) whose mass is exactly equal to the critical mass M_c and secondly, to derive some of its properties. In order to state the precise result, we need the following important concept:

DEFINITION 5.1 (Almost periodicity modulo symmetries). Fix μ and $d \geq 1$. A solution u to the mass-critical NLS (1.4) with lifespan I is said to be *almost periodic modulo symmetries* if there exist (possibly discontinuous) functions $N : I \to \mathbb{R}^+$, $\xi : I \to \mathbb{R}^d$, $x : I \to \mathbb{R}^d$ and a function $C : \mathbb{R}^+ \to \mathbb{R}^+$ such that

$$\int_{|x-x(t)|\geq C(\eta)/N(t)} |u(t,x)|^2 \, dx + \int_{|\xi-\xi(t)|\geq C(\eta)N(t)} |\hat{u}(t,\xi)|^2 \, d\xi \leq \eta$$

for all $t \in I$ and $\eta > 0$. We refer to the function N as the *frequency scale function* for the solution u, ξ is the *frequency center function*, x is the *spatial center function*, and C is the *compactness modulus function*. Furthermore, if we can select $x(t) = \xi(t) = 0$, then we say that u is *almost periodic modulo scaling*.

REMARKS. 1. The parameter $N(t)$ measures the frequency scale of the solution at time t, and $1/N(t)$ measures the spatial scale; see [**43, 96, 97**] for further discussion. Note that we have the freedom to modify $N(t)$ by any bounded function of t, provided that we also modify the compactness modulus function C accordingly. In particular, one could restrict $N(t)$ to be a power of 2 if one wished, although we will not do so here. Alternatively, the fact that the solution trajectory $t \mapsto u(t)$ is continuous in $L^2_x(\mathbb{R}^d)$ can be used to show that the functions N, ξ, x may be chosen to depend continuously on t (cf. Lemma 5.18).

2. One can view $\xi(t)$ and $x(t)$ as roughly measuring the (normalised) momentum and center-of-mass, respectively, at time t, although as u is only assumed to lie in $L^2_x(\mathbb{R}^d)$, these latter quantities are not quite rigourously defined.

3. By Proposition A.1, a family of functions is precompact in $L^2_x(\mathbb{R}^d)$ if and only if it is norm-bounded and there exists a compactness modulus function C so that

$$\int_{|x|\geq C(\eta)} |f(x)|^2 \, dx + \int_{|\xi|\geq C(\eta)} |\hat{f}(\xi)|^2 \, d\xi \leq \eta$$

for all functions f in the family. Thus, an equivalent formulation of Definition 5.1 is as follows: u is almost periodic modulo symmetries if and only if there exists a compact subset K of $L^2_x(\mathbb{R}^d)$ such that the orbit $\{u(t) : t \in I\}$ is contained inside $GK := \{gf : g \in G, f \in K\}$. This perspective also clarifies why we use the term 'almost periodic'.

We are now ready to state the main result of this subsection.

THEOREM 5.2 (Reduction to almost periodic solutions, [**42, 96**]). *Fix μ and d and suppose that Conjecture* 1.4 *failed for this choice. Then there exists a maximal-lifespan solution u with mass $M(u) = M_c$, which is almost periodic modulo symmetries and which blows up both forward and backward in time.*

REMARK. If we consider Conjecture 1.4 in the case of spherically symmetric data ($d \geq 2$), then the conclusion may be strengthened to almost periodicity modulo scaling, that is, $x(t) \equiv 0 \equiv \xi(t)$. This is the greatest advantage in restricting to such data.

The proof of Theorem 5.2 rests on the following key proposition, asserting a certain compactness (modulo symmetries) in sequences of solutions with mass converging to the critical mass from below.

PROPOSITION 5.3 (Palais–Smale condition modulo symmetries, [**96**]). *Fix μ and d, and suppose that Conjecture 1.4 failed for this choice. Let $u_n : I_n \times \mathbb{R}^d \to \mathbb{C}$ be a sequence of solutions and $t_n \in I_n$ a sequence of times such that $M(u_n) \leq M_c$, $M(u_n) \to M_c$, and*

$$\lim_{n\to\infty} S_{\geq t_n}(u_n) = \lim_{n\to\infty} S_{\leq t_n}(u_n) = +\infty. \tag{5.2}$$

Then the sequence $Gu_n(t_n)$ has a subsequence which converges in the $G\backslash L_x^2(\mathbb{R}^d)$ topology.

REMARK. The hypothesis (5.2) asserts that the sequence u_n asymptotically blows up both forward and backward in time. Both components of this hypothesis are essential, as can be seen by considering the pseudo-conformal transformation of the ground state, which only blows up in one direction (and whose orbit is non-compact in the other direction, even after quotienting out by G).

PROOF. Using the time-translation symmetry of (1.4), we may take $t_n = 0$ for all n; thus, we may assume

$$\lim_{n\to\infty} S_{\geq 0}(u_n) = \lim_{n\to\infty} S_{\leq 0}(u_n) = +\infty. \tag{5.3}$$

Applying Theorem 4.26 to the bounded sequence $u_n(0)$ (passing to a subsequence if necessary), we obtain the linear profile decomposition

$$u_n(0) = \sum_{j=1}^{J} g_n^j e^{it_n^j\Delta}\phi^j + w_n^J \tag{5.4}$$

with the stated properties.

By refining the subsequence once for each j and using a standard diagonalisation argument, we may assume that for each j the sequence t_n^j, $n = 1, 2, \dots$ converges to some time $t^j \in [-\infty, +\infty]$. If $t^j \in (-\infty, +\infty)$, we may shift ϕ^j by the linear propagator $e^{it^j\Delta}$, and so assume that $t^j = 0$. Moreover, we may assume that $t_n^j \equiv 0$, since the error $e^{it_n^j\Delta}\phi^j - \phi^j$ may be absorbed into w_n^J; this will not significantly affect the scattering size of the linear evolution of w_n^J, thanks to the Strichartz inequality and the L_x^2-continuity of the free propagator. Thus, for each j either $t_n^j \equiv 0$ or $t_n^j \to \pm\infty$ as $n \to \infty$.

We now define a nonlinear profile $v^j : I^j \times \mathbb{R}^d \to \mathbb{C}$ associated to ϕ^j and depending on the limiting value of t_n^j, as follows:

- If $t_n^j \equiv 0$, we define v^j to be the maximal-lifespan solution with initial data $v^j(0) = \phi^j$.
- If $t_n^j \to \infty$, we define v^j to be the maximal-lifespan solution which scatters forward in time to $e^{it\Delta}\phi^j$.
- If $t_n^j \to -\infty$, we define v^j to be the maximal-lifespan solution which scatters backward in time to $e^{it\Delta}\phi^j$.

Finally, for each $j, n \geq 1$ we define $v_n^j : I_n^j \times \mathbb{R}^d \to \mathbb{C}$ by

$$v_n^j(t) := T_{g_n^j}\big[v^j(\cdot + t_n^j)\big](t),$$

where $I_n^j := \{t \in \mathbb{R} : (\lambda_n^j)^{-2}t + t_n^j \in I^j\}$. Each v_n^j is a solution to (1.4) with initial data $v_n^j(0) = g_n^j v^j(t_n^j)$. Note that for each J, we have

$$u_n(0) - \sum_{j=1}^{J} v_n^j(0) - w_n^J \longrightarrow 0 \quad \text{in } L_x^2 \text{ as } n \to \infty, \tag{5.5}$$

by virtue of the way v_n^j is constructed.

From Theorem 4.26 we have the mass decoupling

$$\sum_{j=1}^{J^*} M(\phi^j) \le \limsup_{n\to\infty} M(u_n(0)) \le M_c \tag{5.6}$$

and in particular, $\sup_j M(\phi^j) \le M_c$.

Case I: Suppose first that

$$\sup_j M(\phi^j) \le M_c - \varepsilon \tag{5.7}$$

for some $\varepsilon > 0$; we will eventually show that this leads to a contradiction. Indeed, by the discussion at the beginning of this subsection it follows that in this case, all v_n^j are defined globally in time and obey the estimates

$$M(v_n^j) = M(\phi^j) \le M_c - \varepsilon$$

and (in view of (5.1))

$$S(v_n^j) \le L(M(\phi^j)) \lesssim M(\phi^j)^{\frac{d+2}{d}} \lesssim M(\phi^j). \tag{5.8}$$

We will eventually derive a bound on the scattering size of u_n, thus contradicting (5.3). In order to achieve this, we will use the stability result Theorem 3.7. To this end, we define an approximate solution

$$u_n^J(t) := \sum_{j=1}^{J} v_n^j(t) + e^{it\Delta} w_n^J. \tag{5.9}$$

Note that by the asymptotic orthogonality conditions in Theorem 4.26, followed by (5.8) and (5.6),

$$\begin{aligned}\lim_{J\to J^*} \limsup_{n\to\infty} S(u_n^J) &\le \lim_{J\to J^*} \limsup_{n\to\infty} S\Big(\sum_{j=1}^{J} v_n^j\Big)\\ &= \lim_{J\to J^*} \limsup_{n\to\infty} \sum_{j=1}^{J} S(v_n^j) \lesssim \lim_{J\to J^*} \sum_{j=1}^{J} M(\phi^j) \lesssim M_c.\end{aligned} \tag{5.10}$$

We will show that u_n^J is indeed a good approximation to u_n for n, J sufficiently large.

LEMMA 5.4 (Asymptotic agreement with initial data). *For any $J \ge 1$ we have*

$$\lim_{n\to\infty} M\big(u_n^J(0) - u_n(0)\big) = 0.$$

PROOF. This follows from (5.5), (5.4), and (5.9). □

LEMMA 5.5 (Asymptotic solution to the equation). *We have*

$$\lim_{J\to J^*} \limsup_{n\to\infty} \big\|(i\partial_t + \Delta)u_n^J - F(u_n^J)\big\|_{L_{t,x}^{\frac{2(d+2)}{d+4}}(\mathbb{R}\times\mathbb{R}^d)} = 0.$$

PROOF. By the definition of u_n^J, we have

$$(i\partial_t + \Delta)u_n^J = \sum_{j=1}^{J} F(v_n^j)$$

and so, by the triangle inequality, it suffices to show that

$$\lim_{J\to J^*} \limsup_{n\to\infty} \big\| F(u_n^J - e^{it\Delta}w_n^J) - F(u_n^J)\big\|_{L_{t,x}^{\frac{2(d+2)}{d+4}}(\mathbb{R}\times\mathbb{R}^d)} = 0$$

and

$$\lim_{n\to\infty} \Big\| F\Big(\sum_{j=1}^{J} v_n^j\Big) - \sum_{j=1}^{J} F(v_n^j)\Big\|_{L_{t,x}^{\frac{2(d+2)}{d+4}}(\mathbb{R}\times\mathbb{R}^d)} = 0 \quad \text{for all } J \geq 1.$$

That the first limit is zero follows fairly quickly from the asymptotically vanishing scattering size of $e^{it\Delta}w_n^J$ together with (5.10); indeed, one need only invoke (3.11) and Hölder's inequality. To see that the second limit is zero, we use the elementary inequality

$$\big|F(\sum_{j=1}^{J} z_j) - \sum_{j=1}^{J} F(z_j)\big| \leq C_{J,d} \sum_{j\neq j'} |z_j||z_{j'}|^{\frac{4}{d}},$$

for some $C_{J,d} < \infty$, (5.8), and the asymptotic orthogonality of the v_n^j provided by (4.60) from Theorem 4.26. □

We are now in a position to apply the stability result Theorem 3.7. Let $\delta > 0$ be a small number. Then, by the above two lemmas, we have

$$M\big(u_n^J(0) - u_n(0)\big) + \big\|(i\partial_t + \Delta)u_n^J - F(u_n^J)\big\|_{L_{t,x}^{\frac{2(d+2)}{d+4}}(\mathbb{R}\times\mathbb{R}^d)} \leq \delta,$$

provided J is sufficiently large (depending on δ) and n is sufficiently large (depending on J, δ). Invoking (5.10), we may apply Theorem 3.7 (for δ chosen small enough depending on M_c) to deduce that u_n exists globally and

$$S_{\mathbb{R}}(u_n) \lesssim M_c.$$

This contradicts (5.3).

Case II: The only remaining possibility is that (5.7) fails for every $\varepsilon > 0$, and thus

$$\sup_j M(\phi^j) = M_c.$$

Comparing this with (5.6), we see $J^* = 1$, that is, there is only one bubble. Consequently, the profile decomposition simplifies to

$$u_n(0) = g_n e^{it_n\Delta}\phi + w_n \tag{5.11}$$

for some sequence $t_n \in \mathbb{R}$ such that either $t_n \equiv 0$ or $t_n \to \pm\infty$, $g_n \in G$, some ϕ of mass $M(\phi) = M_c$, and some w_n with $M(w_n) \to 0$ (and hence $S(e^{it\Delta}w_n) \to 0$) as $n \to \infty$ (this is from (4.59)). By applying the symmetry operation $T_{g_n^{-1}}$ to u_n, which does not affect the hypotheses of Proposition 5.3, we may take all g_n to be the identity, and thus

$$M\big(u_n(0) - e^{it_n\Delta}\phi\big) \to 0 \text{ as } n \to \infty.$$

If $t_n \equiv 0$, then $u_n(0)$ converge in $L_x^2(\mathbb{R}^d)$ to ϕ, and thus $Gu_n(0)$ converge in $G\backslash L_x^2(\mathbb{R}^d)$, as desired. So the only remaining case is when $t_n \to \pm\infty$; we shall assume that $t_n \to \infty$, as the other case is similar. By the Strichartz inequality we have

$$S_{\mathbb{R}}(e^{it\Delta}\phi) < \infty$$

and hence, by time-translation invariance and monotone convergence,

$$\lim_{n\to\infty} S_{\geq 0}(e^{it\Delta}e^{it_n\Delta}\phi) = 0.$$

As the action of G preserves linear solutions of the Schrödinger equation, we have $e^{it\Delta}g_n = T_{g_n}e^{it\Delta}$; as T_{g_n} preserves the scattering norm S (as well as $S_{\geq 0}$ and $S_{\leq 0}$), we deduce

$$\lim_{n\to\infty} S_{\geq 0}(e^{it\Delta}g_n e^{it_n\Delta}\phi) = 0.$$

Since $S(e^{it\Delta}w_n) \to 0$ as $n \to \infty$, we see from (5.11) that

$$\lim_{n\to\infty} S_{\geq 0}(e^{it\Delta}u_n(0)) = 0.$$

Applying Theorem 3.7 (using 0 as the approximate solution and $u_n(0)$ as the initial data), we conclude that

$$\lim_{n\to\infty} S_{\geq 0}(u_n) = 0.$$

But this contradicts one of the estimates in (5.3). A similar argument, using the other half of (5.3), allows us to exclude the possibility that $t_n \to -\infty$. This concludes the proof of Proposition 5.3. □

We are finally ready to extract the minimal-mass blowup solution to (1.4).

Proof of Theorem 5.2. By the definition of the critical mass M_c (and the continuity of L), we can find a sequence $u_n : I_n \times \mathbb{R}^d \to \mathbb{C}$ of maximal-lifespan solutions with $M(u_n) \leq M_c$ and $\lim_{n\to\infty} S(u_n) = +\infty$. By choosing $t_n \in I_n$ to be the median time of the $L_{t,x}^{2(d+2)/d}$ norm of u_n (cf. the "middle third" trick in [**7**]), we can thus arrange that (5.2) holds. By time-translation invariance we may take $t_n = 0$.

Invoking Proposition 5.3 and passing to a subsequence if necessary, we find group elements $g_n \in G$ such that $g_n u_n(0)$ converges strongly in $L_x^2(\mathbb{R}^d)$ to some function $u_0 \in L_x^2(\mathbb{R}^d)$. By applying the group action T_{g_n} to the solutions u_n we may take g_n to all be the identity; thus, $u_n(0)$ converge strongly in $L_x^2(\mathbb{R}^d)$ to u_0. In particular this implies $M(u_0) \leq M_c$.

Let $u : I \times \mathbb{R}^n \to \mathbb{C}$ be the maximal-lifespan solution to (1.4) with initial data $u(0) = u_0$ as given by Corollary 3.5. We claim that u blows up both forward and backward in time. Indeed, if u does not blow up forward in time (say), then $[0, +\infty) \subseteq I$ and $S_{\geq 0}(u) < \infty$. By Theorem 3.7, this implies that for sufficiently large n, we have $[0, +\infty) \subseteq I_n$ and

$$\limsup_{n\to\infty} S_{\geq 0}(u_n) < \infty,$$

contradicting (5.2). By the definition of M_c, this forces $M(u_0) \geq M_c$ and hence $M(u_0)$ must be exactly M_c.

It remains to show that the solution u is almost periodic modulo G. Consider an arbitrary sequence $u(t'_n)$ in the orbit $\{u(t) : t \in I\}$. Now, since u blows up both forward and backward in time, but is locally in $L_{t,x}^{2(d+2)/d}$, we have

$$S_{\geq t'_n}(u) = S_{\leq t'_n}(u) = \infty.$$

Applying Proposition 5.3 once again, we see that $Gu(t'_n)$ has a convergent subsequence in $G\backslash L_x^2(\mathbb{R}^d)$. Thus, the orbit $\{Gu(t) : t \in I\}$ is precompact in $G\backslash L_x^2(\mathbb{R}^d)$, as desired. □

5.2. The energy-critical NLS. In this subsection, we outline the proof of the existence of a minimal kinetic energy blowup solution to the energy-critical NLS (1.6). The argument we present is from [**44**], which builds upon earlier work by Kenig and Merle [**38**]. The fact that the kinetic energy is not a conserved quantity for (1.6) introduces several difficulties over the material presented in the previous subsection. We will elaborate upon them at the appropriate time.

Let us start by investigating what the failure of Conjecture 1.5 would imply.

If $\mu = +1$, for any $0 \leq E_0 < \infty$, we define

$$L^+(E_0) := \sup\{S_I(u) : u : I \times \mathbb{R}^d \to \mathbb{C} \text{ such that } \sup_{t\in I} \|\nabla u(t)\|_2^2 \leq E_0\},$$

where the supremum is taken over all solutions $u : I \times \mathbb{R}^d \to \mathbb{C}$ to (1.6). Throughout this subsection we will use the notation

$$S_I(u) := \int_I \int_{\mathbb{R}^d} |u(t,x)|^{\frac{2(d+2)}{d-2}} \, dx \, dt$$

for the scattering size of u on an interval I. Note that this is an energy-critical Strichartz norm.

Similarly, if $\mu = -1$, for any $0 \leq E_0 \leq \|\nabla W\|_2^2$, we define

$$L^-(E_0) := \sup\{S_I(u) : u : I \times \mathbb{R}^d \to \mathbb{C} \text{ such that } \sup_{t\in I} \|\nabla u(t)\|_2^2 \leq E_0\},$$

where the supremum is again taken over all solutions $u : I \times \mathbb{R}^d \to \mathbb{C}$ to (1.6).

Thus, $L^+ : [0, \infty) \to [0, \infty]$ and $L^- : [0, \|\nabla W\|_2^2] \to [0, \infty]$ are non-decreasing functions with $L^-(\|\nabla W\|_2^2) = \infty$. Moreover, from the local well-posedness theory (see Corollary 3.5),

$$L^+(E_0) + L^-(E_0) \lesssim E_0^{\frac{d+2}{d-2}} \quad \text{for} \quad E_0 \leq \eta_0,$$

where $\eta_0 = \eta_0(d)$ is the threshold from the small data theory.

From the stability result Theorem 3.8, we see that L^+ and L^- are continuous. Therefore, there must exist a unique *critical kinetic energy* E_c such that $0 < E_c \leq \infty$ if $\mu > 0$ and $0 < E_c \leq \|\nabla W\|_2^2$ if $\mu < 0$ and such that $L^\pm(E_0) < \infty$ for $E_0 < E_c$ and $L^\pm(E_0) = \infty$ for $E_0 \geq E_c$. To ease notation, we adopt the same convention as in the mass-critical case:

CONVENTION. We write L for $L^\pm$ with the understanding that $L = L^+$ in the defocusing case and $L = L^-$ in the focusing case.

By the discussion above, we see that if $u : I \times \mathbb{R}^d \to \mathbb{C}$ is a maximal-lifespan solution to (1.6) such that $\sup_{t\in I} \|\nabla u(t)\|_2^2 < E_c$, then u is global and

$$S_{\mathbb{R}}(u) \leq L\big(\sup_{t\in I} \|\nabla u(t)\|_2^2\big).$$

Failure of Conjecture 1.5 is equivalent to $0 < E_c < \infty$ in the defocusing case and $0 < E_c < \|\nabla W\|_2^2$ in the focusing case.

Just as in the mass-critical case, the extraction of a minimal blowup solution will be a consequence of the following key compactness result.

PROPOSITION 5.6 (Palais–Smale condition modulo symmetries, [**44**]). *Fix μ and $d \geq 3$. Let $u_n : I_n \times \mathbb{R}^d \mapsto \mathbb{C}$ be a sequence of solutions to* (1.6) *such that*

$$\limsup_{n\to\infty} \sup_{t\in I_n} \|\nabla u_n(t)\|_2^2 = E_c \tag{5.12}$$

and

$$\lim_{n\to\infty} S_{\geq t_n}(u_n) = \lim_{n\to\infty} S_{\leq t_n}(u_n) = \infty.$$

for some sequence of times $t_n \in I_n$. Then the sequence $u_n(t_n)$ has a subsequence which converges in $\dot H^1_x(\mathbb{R}^d)$ modulo symmetries.

PROOF. Using the time-translation symmetry of the equation (1.6), we may set $t_n = 0$ for all $n \geq 1$. Thus,

$$\lim_{n\to\infty} S_{\geq 0}(u_n) = \lim_{n\to\infty} S_{\leq 0}(u_n) = \infty. \tag{5.13}$$

Applying the linear profile decomposition Theorem 4.27 to the sequence $u_n(0)$ (which is bounded in $\dot H^1_x(\mathbb{R}^d)$ by (5.12)) and passing to a subsequence if necessary, we obtain the decomposition

$$u_n(0) = \sum_{j=1}^J g_n^j e^{it_n^j\Delta}\phi^j + w_n^J.$$

Arguing as in the proof of Proposition 5.3, we may assume that for each $j \geq 1$ either $t_n^j \equiv 0$ or $t_n^j \to \pm\infty$ as $n \to \infty$. Continuing as there, we define the nonlinear profiles $v^j : I^j \times \mathbb{R}^d \to \mathbb{C}$ and $v_n^j : I_n^j \times \mathbb{R}^d \to \mathbb{C}$.

By the asymptotic decoupling of the kinetic energy, there exists $J_0 \geq 1$ such that

$$\|\nabla\phi^j\|_2^2 \leq \eta_0 \quad \text{for all} \quad j \geq J_0,$$

where $\eta_0 = \eta_0(d)$ is the threshold for the small data theory. Hence, by Corollary 3.9, for all $n \geq 1$ and all $j \geq J_0$ the solutions v_n^j are global and moreover,

$$\sup_{t\in\mathbb{R}} \|\nabla v_n^j(t)\|_2^2 + S_{\mathbb{R}}(v_n^j) \lesssim \|\nabla\phi^j\|_2^2. \tag{5.14}$$

At this point the proof of the Palais–Smale condition for the energy-critical NLS starts to diverge from that in the mass-critical case. Indeed, as the kinetic energy is not a conserved quantity, even if $v_n^j(0) = g_n^j v^j(t_n^j)$ has kinetic energy less than the critical value E_c, this does not guarantee the same will hold throughout the lifespan of v_n^j and in particular, it does not guarantee global existence nor global spacetime bounds. As a consequence, we must actively search for a profile responsible for the asymptotic blowup (5.13). As we will see shortly, the existence of at least one such profile is a consequence of the stability result Theorem 3.8 and the asymptotic orthogonality of the profiles given by Theorem 4.27.

LEMMA 5.7 (At least one bad profile). *There exists $1 \leq j_0 < J_0$ such that*

$$\limsup_{n\to\infty} S_{[0,\sup I_n^{j_0})}(v_n^{j_0}) = \infty.$$

PROOF. We argue by contradiction. Assume that for all $1 \leq j < J_0$,

$$\limsup_{n\to\infty} S_{[0,\sup I_n^{j})}(v_n^{j}) < \infty. \tag{5.15}$$

In particular, this implies $\sup I_n^j = \infty$ for all $1 \le j < J_0$ and all sufficiently large n. Combining (5.15) with (5.14), and then using (5.12),

$$\sum_{j\ge 1} S_{[0,\infty)}(v_n^j) \lesssim 1 + \sum_{j\ge J_0} \|\nabla\phi^j\|_2^2 \lesssim 1 + E_c \tag{5.16}$$

for all n sufficiently large.

Using the estimates above and the stability result Theorem 3.8, we will derive a bound on the scattering size of u_n (for n sufficiently large), thus contradicting (5.13). To this end, we define the approximate solution

$$u_n^J(t) := \sum_{j=1}^J v_n^j(t) + e^{it\Delta} w_n^J.$$

Note that by (5.16) and the asymptotic vanishing of the scattering size of $e^{it\Delta}w_n^J$,

$$\begin{aligned}\lim_{J\to J^*}\limsup_{n\to\infty} S_{[0,\infty)}(u_n^J) &\lesssim \lim_{J\to J^*}\limsup_{n\to\infty}\Big(S_{[0,\infty)}\big(\sum_{j=1}^J v_n^j\big) + S_{[0,\infty)}\big(e^{it\Delta}w_n^J\big)\Big)\\ &\lesssim \lim_{J\to J^*}\limsup_{n\to\infty}\sum_{j=1}^J S_{[0,\infty)}(v_n^j) \lesssim 1 + E_c.\end{aligned}\tag{5.17}$$

The next two lemmas show that u_n^J is indeed a good approximation to u_n for n and J sufficiently large.

LEMMA 5.8 (Asymptotic agreement with initial data). *For any $J \ge 1$ we have*

$$\lim_{n\to\infty}\big\|u_n^J(0) - u_n(0)\big\|_{\dot H_x^1(\mathbb{R}^d)} = 0.$$

PROOF. Exercise: mimic the proof of Lemma 5.4. □

LEMMA 5.9 (Asymptotic solution to the equation). *We have*

$$\lim_{J\to J^*}\limsup_{n\to\infty}\big\|\nabla\big[(i\partial_t + \Delta)u_n^J - F(u_n^J)\big]\big\|_{L_{t,x}^{\frac{2(d+2)}{d+4}}([0,\infty)\times\mathbb{R}^d)} = 0.$$

PROOF. Exercise: mimic the proof of Lemma 5.5. There is one new difficulty, namely, one needs to show that

$$\lim_{J\to J^*}\limsup_{n\to\infty}\|v_n^j\nabla e^{it\Delta}w_n^J\|_{L_{t,x}^{\frac{d+2}{d-1}}([0,\infty)\times\mathbb{R}^d)} = 0$$

for each $j \le J$. After transferring symmetries to w_n^J, this follows from Corollary 4.15. □

We are now in a position to apply the stability result Theorem 3.8. Indeed, invoking the two lemmas above and (5.17), we conclude that for n sufficiently large,

$$S_{[0,\infty)}(u_n) \lesssim 1 + E_c,$$

thus contradicting (5.13). This finishes the proof of Lemma 5.7. □

Returning to the proof of Proposition 5.6 and rearranging the indices, we may assume that there exists $1 \le J_1 < J_0$ such that

$$\limsup_{n\to\infty} S_{[0,\sup I_n^j)}(v_n^j) = \infty \text{ for } 1 \le j \le J_1 \quad\text{and}\quad \limsup_{n\to\infty} S_{[0,\infty)}(v_n^j) < \infty \text{ for } j > J_1.$$

Passing to a subsequence in n, we can guarantee that $S_{[0,\sup I_n^1)}(v_n^1) \to \infty$.

At this point our enemy scenario is that consisting of two or more profiles that take turns at driving the scattering norm of u_n to infinity. In order to finish the proof of the Palais–Smale condition, we have to prove that only one profile is responsible for the asymptotic blowup (5.13). In order to achieve this, we have to prove kinetic energy decoupling for the nonlinear profiles for large periods of time, large enough that the kinetic energy of v_n^1 has achieved the critical kinetic energy.

For each $m, n \geq 1$ let us define an integer $j(m,n) \in \{1, \dots, J_1\}$ and an interval K_n^m of the form $[0, \tau]$ by

$$\sup_{1 \leq j \leq J_1} S_{K_n^m}(v_n^j) = S_{K_n^m}(v_n^{j(m,n)}) = m. \tag{5.18}$$

By the pigeonhole principle, there is a $1 \leq j_1 \leq J_1$ so that for infinitely many m one has $j(m,n) = j_1$ for infinitely many n. Note that the infinite set of n for which this holds may be m-dependent. By reordering the indices, we may assume that $j_1 = 1$. Then, by the definition of the critical kinetic energy, we obtain

$$\limsup_{m \to \infty} \limsup_{n \to \infty} \sup_{t \in K_n^m} \|\nabla v_n^1(t)\|_2^2 \geq E_c. \tag{5.19}$$

On the other hand, by virtue of (5.18), all v_n^j have finite scattering size on K_n^m for each $m \geq 1$. Thus, by the same argument used in Lemma 5.7, we see that for n and J sufficiently large, u_n^J is a good approximation to u_n on each K_n^m. More precisely,

$$\lim_{J \to J^*} \limsup_{n \to \infty} \|u_n^J - u_n\|_{L_t^\infty \dot{H}_x^1(K_n^m \times \mathbb{R}^d)} = 0 \tag{5.20}$$

for each $m \geq 1$.

Our next result proves asymptotic kinetic energy decoupling for u_n^J.

LEMMA 5.10 (Kinetic energy decoupling for u_n^J). *For all $J \geq 1$ and $m \geq 1$,*

$$\limsup_{n \to \infty} \sup_{t \in K_n^m} \left| \|\nabla u_n^J(t)\|_2^2 - \sum_{j=1}^{J} \|\nabla v_n^j(t)\|_2^2 - \|\nabla w_n^J\|_2^2 \right| = 0.$$

PROOF. Fix $J \geq 1$ and $m \geq 1$. Then, for all $t \in K_n^m$,

$$\begin{aligned} \|\nabla u_n^J(t)\|_2^2 &= \langle \nabla u_n^J(t), \nabla u_n^J(t) \rangle \\ &= \sum_{j=1}^{J} \|\nabla v_n^j(t)\|_2^2 + \|\nabla w_n^J\|_2^2 + \sum_{j \neq j'} \langle \nabla v_n^j(t), \nabla v_n^{j'}(t) \rangle \\ &\quad + \sum_{j=1}^{J} \big(\langle \nabla e^{it\Delta} w_n^J, \nabla v_n^j(t) \rangle + \langle \nabla v_n^j(t), \nabla e^{it\Delta} w_n^J \rangle \big). \end{aligned}$$

To prove Lemma 5.10, it thus suffices to show that for all sequences $t_n \in K_n^m$,

$$\langle \nabla v_n^j(t_n), \nabla v_n^{j'}(t_n) \rangle \to 0 \quad \text{as } n \to \infty \tag{5.21}$$

and

$$\langle \nabla e^{it_n\Delta} w_n^J, \nabla v_n^j(t_n) \rangle \to 0 \quad \text{as } n \to \infty \tag{5.22}$$

for all $1 \leq j, j' \leq J$ with $j \neq j'$. We will only demonstrate the latter, which requires (4.63); the former can be deduced in much the same manner using the asymptotic orthogonality of the nonlinear profiles.

By a change of variables,

$$\big\langle \nabla e^{it_n\Delta} w_n^J, \nabla v_n^j(t_n)\big\rangle = \big\langle \nabla e^{it_n(\lambda_n^j)^{-2}\Delta}[(g_n^j)^{-1}w_n^J], \nabla v^j\big(\tfrac{t_n}{(\lambda_n^j)^2}+t_n^j\big)\big\rangle. \tag{5.23}$$

As $t_n \in K_n^m \subseteq [0, \sup I_n^j)$ for all $1 \le j \le J_1$, we have $t_n(\lambda_n^j)^{-2} + t_n^j \in I^j$ for all $j \ge 1$. Recall that I^j is the maximal lifespan of v^j; for $j > J_1$ this is $\mathbb{R}$. By refining the sequence once for every j and using the standard diagonalisation argument, we may assume $t_n(\lambda_n^j)^{-2} + t_n^j$ converges for every j.

Fix $1 \le j \le J$. If $t_n(\lambda_n^j)^{-2} + t_n^j$ converges to some point τ^j in the interior of I^j, then by the continuity of the flow, $v^j\big(t_n(\lambda_n^j)^{-2} + t_n^j\big)$ converges to $v^j(\tau^j)$ in $\dot H_x^1(\mathbb{R}^d)$. On the other hand,

$$\limsup_{n\to\infty}\big\|e^{it_n(\lambda_n^j)^{-2}\Delta}[(g_n^j)^{-1}w_n^J]\big\|_{\dot H_x^1(\mathbb{R}^d)} = \limsup_{n\to\infty}\|w_n^J\|_{\dot H_x^1(\mathbb{R}^d)} \lesssim E_c. \tag{5.24}$$

Combining this with (5.23), we obtain

$$\begin{aligned}\lim_{n\to\infty}\big\langle \nabla e^{it_n\Delta} w_n^J, \nabla v_n^j(t_n)\big\rangle &= \lim_{n\to\infty}\big\langle \nabla e^{it_n(\lambda_n^j)^{-2}\Delta}[(g_n^j)^{-1}w_n^J], \nabla v^j(\tau^j)\big\rangle\\ &= \lim_{n\to\infty}\big\langle \nabla e^{-it_n^j\Delta}[(g_n^j)^{-1}w_n^J], \nabla e^{-i\tau^j\Delta}v^j(\tau^j)\big\rangle.\end{aligned}$$

Invoking (4.63), we deduce (5.22).

Consider now the case when $t_n(\lambda_n^j)^{-2} + t_n^j$ converges to $\sup I^j$. Then we must have $\sup I^j = \infty$ and v^j scatters forward in time. This is clearly true if $t_n^j \to \infty$ as $n \to \infty$; in the other cases, failure would imply

$$\limsup_{n\to\infty} S_{[0,t_n]}(v_n^j) = \limsup_{n\to\infty} S_{\left[t_n^j,\, t_n(\lambda_n^j)^{-2}+t_n^j\right]}(v^j) = \infty,$$

which contradicts $t_n \in K_n^m$. Therefore, there exists $\phi^j \in \dot H_x^1(\mathbb{R}^d)$ such that

$$\lim_{n\to\infty}\Big\|v^j\big(t_n(\lambda_n^j)^{-2}+t_n^j\big) - e^{i\left(t_n(\lambda_n^j)^{-2}+t_n^j\right)\Delta}\phi^j\Big\|_{\dot H_x^1(\mathbb{R}^d)} = 0.$$

Together with (5.23), this yields

$$\lim_{n\to\infty}\big\langle \nabla e^{it_n\Delta} w_n^J, \nabla v_n^j(t_n)\big\rangle = \lim_{n\to\infty}\big\langle \nabla e^{-it_n^j\Delta}[(g_n^j)^{-1}w_n^J], \nabla\phi^j\big\rangle,$$

which by (4.63) implies (5.22).

Finally, we consider the case when $t_n(\lambda_n^j)^{-2} + t_n^j$ converges to $\inf I^j$. Since $t_n(\lambda_n^j)^{-2} \ge 0$ and $\inf I^j < \infty$ for all $j \ge 1$ we see that t_n^j does not converge to $+\infty$. Moreover, if $t_n^j \equiv 0$, then $\inf I^j < 0$; as $t_n(\lambda_n^j)^{-2} \ge 0$, we see that t_n^j cannot be identically zero. This leaves $t_n^j \to -\infty$ as $n \to \infty$. Thus $\inf I^j = -\infty$ and v^j scatters backward in time to $e^{it\Delta}\phi^j$. We obtain

$$\lim_{n\to\infty}\Big\|v^j\big(t_n(\lambda_n^j)^{-2}+t_n^j\big) - e^{i\left(t_n(\lambda_n^j)^{-2}+t_n^j\right)\Delta}\phi^j\Big\|_{\dot H_x^1(\mathbb{R}^d)} = 0,$$

which by (5.23) implies

$$\lim_{n\to\infty}\big\langle \nabla e^{it_n\Delta} w_n^J, \nabla v_n^j(t_n)\big\rangle = \lim_{n\to\infty}\big\langle \nabla e^{-it_n^j\Delta}[(g_n^j)^{-1}w_n^J], \nabla\phi^j\big\rangle.$$

Invoking (4.63) once again, we derive (5.22).

This finishes the proof of Lemma 5.10. □

Returning to the proof of Proposition 5.6 and using (5.12) and (5.20) together with Lemma 5.10, we find

$$E_c \geq \limsup_{n\to\infty} \sup_{t\in K_n^m} \|\nabla u_n(t)\|_2^2 = \lim_{J\to\infty} \limsup_{n\to\infty} \Big\{ \|\nabla w_n^J\|_2^2 + \sup_{t\in K_n^m} \sum_{j=1}^{J} \|\nabla v_n^j(t)\|_2^2 \Big\}$$

for each $m \geq 1$. Invoking (5.19), we thus obtain the simplified decomposition

$$u_n(0) = g_n e^{i\tau_n \Delta}\phi + w_n \tag{5.25}$$

for some $g_n \in G$, $\tau_n \in \mathbb{R}$, and some functions $\phi, w_n \in \dot{H}_x^1(\mathbb{R}^d)$ with $w_n \to 0$ strongly in $\dot{H}_x^1(\mathbb{R}^d)$. Moreover, the sequence τ_n obeys $\tau_n \equiv 0$ or $\tau_n \to \pm\infty$.

If $\tau_n \equiv 0$, (5.25) immediately implies that $u_n(0)$ converge modulo symmetries to ϕ, which proves Proposition 5.6 in this case. Finally, arguing as in the proof of the Palais–Smale condition in the mass-critical case, one shows that this is the only possible case, that is, τ_n cannot converge to either ∞ or $-\infty$.

This completes the proof of Proposition 5.6. □

With the Palais–Smale condition in place, we can now extract a minimal blowup solution, very much as we did in the previous subsection. Let us first revisit the definition of almost periodicity in the energy-critical context.

DEFINITION 5.11 (Almost periodicity modulo symmetries). Fix μ and $d \geq 3$. A solution u to the energy-critical NLS (1.6) with lifespan I and uniformly bounded kinetic energy is said to be *almost periodic modulo symmetries* if there exist (possibly discontinuous) functions $N : I \to \mathbb{R}^+$, $x : I \to \mathbb{R}^d$, and a function $C : \mathbb{R}^+ \to \mathbb{R}^+$ such that

$$\int_{|x-x(t)|\geq C(\eta)/N(t)} |\nabla u(t,x)|^2 \, dx + \int_{|\xi|\geq C(\eta)N(t)} |\xi \hat{u}(t,\xi)|^2 \, d\xi \leq \eta$$

for all $t \in I$ and $\eta > 0$. We refer to the function N as the *frequency scale function* for the solution u, x is the *spatial center function*, and C is the *compactness modulus function*.

REMARK. Comparing Definitions 5.1 and 5.11, we see that there are two differences. The first is that in the energy-critical case, compactness is in $\dot{H}_x^1$ rather than in L_x^2. A deeper difference is the absence of Galilei boosts among the symmetry parameters in the energy-critical case. While Galilei boosts leave the mass and the equation invariant, they modify the energy (cf. Proposition 2.3); boundedness of the kinetic energy implies $|\xi(t)|/N(t) = O(1)$, which allows us to take $\xi(t) \equiv 0$ in the definition above, modifying the compactness modulus function if necessary.

We are now ready to introduce the central result of this subsection.

THEOREM 5.12 (Reduction to almost periodic solutions, [**44**]). *Fix μ and $d \geq 3$ and suppose that Conjecture* 1.5 *failed for this choice of μ and d. Then there exists a maximal-lifespan solution $u : I \times \mathbb{R}^d \to \mathbb{C}$ to* (1.6) *such that* $\sup_{t\in I} \|\nabla u(t)\|_2^2 = E_c$, *$u$ is almost periodic modulo symmetries and blows up both forward and backward in time.*

PROOF. Exercise. □

5.3. Almost periodic solutions. In this subsection, we continue our study of solutions to (1.4) and (1.6) that are almost periodic modulo symmetries. We record basic properties of the frequency scale function $N(t)$, spatial center function $x(t)$, and frequency center function $\xi(t)$. Most of the material we present is taken from [**43**].

LEMMA 5.13 (Quasi-uniqueness of $N(t), x(t), \xi(t)$). *Let u be a non-zero solution to* (1.4) *with lifespan I, which is almost periodic modulo symmetries with parameters $N(t), x(t), \xi(t)$ and compactness modulus function C, and also almost periodic modulo symmetries with parameters $N'(t), x'(t), \xi'(t)$ and compactness modulus function C'. Then we have*

$$N(t) \sim_{u,C,C'} N'(t), \quad |x(t) - x'(t)| \lesssim_{u,C,C'} \frac{1}{N(t)}, \quad |\xi(t) - \xi'(t)| \lesssim_{u,C,C'} N(t)$$

for all $t \in I$. A similar result holds for almost periodic solutions to (1.6).

PROOF. Let u be a solution to (1.4). We turn to the first claim and notice that by symmetry, it suffices to establish the bound $N'(t) \lesssim_{u,C,C'} N(t)$.

Fix t and let $\eta > 0$ to be chosen later. By Definition 5.1 we have

$$\int_{|x-x'(t)|\geq C'(\eta)/N'(t)} |u(t,x)|^2\, dx \leq \eta$$

and

$$\int_{|\xi-\xi(t)|\geq C(\eta)N(t)} |\hat{u}(t,\xi)|^2\, d\xi \leq \eta.$$

We split $u := u_1 + u_2$, where $u_1(t,x) := u(t,x)\chi_{|x-x'(t)|\geq C'(\eta)/N'(t)}$ and $u_2(t,x) := u(t,x)\chi_{|x-x'(t)|<C'(\eta)/N'(t)}$. Then, by Plancherel's theorem we have

$$\int_{\mathbb{R}^d} |\hat{u}_1(t,\xi)|^2\, d\xi \lesssim \eta, \tag{5.26}$$

while from Cauchy-Schwarz we have

$$\sup_{\xi\in\mathbb{R}^d} |\hat{u}_2(t,\xi)|^2 \lesssim_{\eta,C'} M(u)N'(t)^{-d}.$$

Integrating the last inequality over the ball $|\xi - \xi(t)| \leq C(\eta)N(t)$ and invoking (5.26), we conclude that

$$\int_{\mathbb{R}^d} |\hat{u}(t,\xi)|^2\, d\xi \lesssim \eta + O_{\eta,C,C'}(M(u)N(t)^d N'(t)^{-d}).$$

Thus, by Plancherel and mass conservation,

$$M(u) \lesssim \eta + O_{\eta,C,C'}(M(u)N(t)^d N'(t)^{-d}).$$

Choosing η to be a small multiple of $M(u)$ (which is non-zero by hypothesis), we obtain the first claim.

The last two claims now follow from a quick inspection of Definition 5.1. □

To describe how the symmetry parameters depend on u, we use the natural notion of convergence for solutions:

DEFINITION 5.14 (Convergence of solutions). Let $u_n : I_n \times \mathbb{R}^d \to \mathbb{C}$ be a sequence of solutions to the mass-critical NLS, let $u : I \times \mathbb{R}^d \to \mathbb{C}$ be another solution, and let K be a compact time interval. We say that u_n *converge uniformly to u* on K if $K \subset I$, $K \subset I_n$ for all sufficiently large n, and u_n converges strongly to

u in $C_t^0 L_x^2(K\times\mathbb{R}^d)\cap L_{t,x}^{2(d+2)/d}(K\times\mathbb{R}^d)$ as $n\to\infty$. We say that u_n *converge locally uniformly to* u if u_n converges uniformly to u on every compact interval $K\subset I$.

In the energy-critical case, we ask that $u_n\to u$ on $K\times\mathbb{R}^d$ in the $C_t^0\dot H_x^1\cap L_{t,x}^{2(d+2)/(d-2)}$ topology.

LEMMA 5.15 (Quasi-continuous dependence of $N(t), x(t), \xi(t)$ on u). *Let u_n be a sequence of solutions to* (1.4) *with lifespans I_n, which are almost periodic modulo symmetries with parameters $N_n(t), x_n(t), \xi_n(t)$ and compactness modulus function $C:\mathbb{R}^+\to\mathbb{R}^+$, independent of n. Suppose that u_n converge locally uniformly to a non-zero solution u to* (1.4) *with lifespan I. Then u is almost periodic modulo symmetries with some parameters $N(t), x(t), \xi(t)$ and the same compactness modulus function C. Furthermore, we have*

$$\liminf_{n\to\infty} N_n(t) \lesssim_{u,C} N(t) \lesssim_{u,C} \limsup_{n\to\infty} N_n(t) \tag{5.27}$$

$$\limsup_{n\to\infty} |x_n(t)-x(t)| \lesssim_{u,C} \frac{1}{N(t)} \tag{5.28}$$

$$\limsup_{n\to\infty} |\xi_n(t)-\xi(t)| \lesssim_{u,C} N(t) \tag{5.29}$$

for all $t\in I$. A similar result holds for the energy-critical NLS.

PROOF. We first show that

$$0<\liminf_{n\to\infty} N_n(t) \le \limsup_{n\to\infty} N_n(t) < \infty \tag{5.30}$$

$$\limsup_{n\to\infty} |x_n(t)|N_n(t) + \limsup_{n\to\infty} \frac{|\xi_n(t)|}{N_n(t)} < \infty \tag{5.31}$$

for all $t\in I$. Indeed, if one of the inequalities in (5.30) failed for some t, then (by passing to a subsequence if necessary) $N_n(t)$ would converge to zero or to infinity as $n\to\infty$. Thus, by Definition 5.1, $u_n(t)$ would converge weakly to zero, and hence, by the local uniform convergence, would converge strongly to zero. But this contradicts the hypothesis that u is not identically zero. This establishes (5.30). A similar argument settles (5.31).

From (5.30) and (5.31), we see that for each $t\in I$ the sequences $N_n(t)$, $x_n(t)$, and $\xi_n(t)$ each have at least one limit point, which we denote $N(t)$, $x(t)$, and $\xi(t)$, respectively. Using the local uniform convergence, we easily verify that u is almost periodic modulo symmetries with parameters $N(t), x(t), \xi(t)$ and compactness modulus function C.

It remains to establish (5.27) through (5.29), which we prove by contradiction. Suppose for example that (5.27) failed. Then given any A, there exists a $t\in I$ for which $N_n(t)$ has at least two limit points which are separated by a ratio of at least A, and so u has two frequency scale functions with compactness modulus function C, which are separated by this ratio. This contradicts Lemma 5.13 for A large enough depending on u. Hence (5.27) holds. A similar argument establishes (5.28) and (5.29). □

DEFINITION 5.16 (Normalised solution). Let u be a solution to (1.4), which is almost periodic modulo symmetries with parameters $N(t), x(t), \xi(t)$. We say that u is *normalised* if the lifespan I contains zero and

$$N(0)=1,\quad x(0)=\xi(0)=0.$$

More generally, we can define the *normalisation* of a solution u at a time $t_0 \in I$ by

$$u^{[t_0]} := T_{g_{0,-\xi(t_0)/N(t_0),-x(t_0)N(t_0),N(t_0)}}\big(u(\cdot + t_0)\big). \tag{5.32}$$

Observe that $u^{[t_0]}$ is a normalised solution which is almost periodic modulo symmetries and has lifespan

$$I^{[t_0]} := \{s \in \mathbb{R} : t_0 + sN(t_0)^{-2} \in I\}$$

(so, in particular, $0 \in I^{[t_0]}$). The parameters of $u^{[t_0]}$ are given by

$$\begin{aligned} N^{[t_0]}(s) &:= \frac{N\big(t_0 + sN(t_0)^{-2}\big)}{N(t_0)} \\ \xi^{[t_0]}(s) &:= \frac{\xi\big(t_0 + sN(t_0)^{-2}\big) - \xi(t_0)}{N(t_0)} \\ x^{[t_0]}(s) &:= N(t_0)\big[x\big(t_0 + sN(t_0)^{-2}\big) - x(t_0)\big] - 2\frac{\xi(t_0)}{N(t_0)}s \end{aligned} \tag{5.33}$$

and it has the same compactness modulus function as u. Furthermore, if u is a maximal-lifespan solution then so is $u^{[t_0]}$. A similar definition can be made in the energy-critical case.

LEMMA 5.17 (Compactness of normalized almost periodic solutions). *Let u_n be a sequence of normalised maximal-lifespan solutions to* (1.4) *with lifespans $I_n \ni 0$, which are almost periodic modulo symmetries with parameters N_n, x_n, ξ_n and a uniform compactness modulus function C. Assume that we also have a uniform mass bound*

$$0 < \inf_n M(u_n) \le \sup_n M(u_n) < \infty. \tag{5.34}$$

Then, after passing to a subsequence if necessary, there exists a non-zero maximal-lifespan solution u to (1.4) *with lifespan $I \ni 0$ that is almost periodic modulo symmetries, such that u_n converge locally uniformly to u. A similar statement holds in the energy-critical setting.*

PROOF. By hypothesis and Definition 5.1, we see that for every $\varepsilon > 0$ there exists $R > 0$ such that

$$\int_{|x|\ge R} |u_n(0,x)|^2\, dx \le \varepsilon$$

and

$$\int_{|\xi|\ge R} |\widehat{u_n}(0,\xi)|^2\, d\xi \le \varepsilon$$

for all n. From this, (5.34), and Proposition A.1, we see that the sequence $u_n(0)$ is precompact in the strong topology of $L^2_x(\mathbb{R}^d)$. Thus, by passing to a subsequence if necessary, we can find $u_0 \in L^2_x(\mathbb{R}^d)$ such that $u_n(0)$ converge strongly to u_0 in $L^2_x(\mathbb{R}^d)$. From (5.34) we see that u_0 is not identically zero.

Now let u be the maximal Cauchy development of u_0 from time 0, with lifespan I. By Theorem 3.7, u_n converge locally uniformly to u. The remaining claims now follow from Lemma 5.15. □

LEMMA 5.18 (Local constancy of $N(t), x(t), \xi(t)$). *Let u be a non-zero maximal-lifespan solution to* (1.4) *with lifespan I that is almost periodic modulo symmetries* *with parameters* $N(t), x(t), \xi(t)$*. Then there exists a small number* δ*, depending on* u*, such that for every* $t_0 \in I$ *we have*

$$\left[t_0 - \delta N(t_0)^{-2}, t_0 + \delta N(t_0)^{-2}\right] \subset I \tag{5.35}$$

and

$$\begin{aligned} N(t) \sim_u N(t_0), \qquad |\xi(t) - \xi(t_0)| \lesssim_u N(t_0), \\ \left|x(t) - x(t_0) - 2(t-t_0)\xi(t_0)\right| \lesssim_u N(t_0)^{-1} \end{aligned} \tag{5.36}$$

whenever $|t - t_0| \leq \delta N(t_0)^{-2}$*. The same statement holds for the energy-critical NLS if we set* $\xi(t) \equiv 0$*.*

Proof. Let us first establish (5.35). We argue by contradiction. Assume (5.35) fails. Then, there exist sequences $t_n \in I$ and $\delta_n \to 0$ such that $t_n + \delta_n N(t_n)^{-2} \notin I$ for all n. Define the normalisations $u^{[t_n]}$ of u at time t_n as in (5.32). Then, $u^{[t_n]}$ are maximal-lifespan normalised solutions whose lifespans $I^{[t_n]}$ contain 0 but not δ_n; they are also almost periodic modulo symmetries with parameters given by (5.33) and the same compactness modulus function C as u. Applying Lemma 5.17 (and passing to a subsequence if necessary), we conclude that $u^{[t_n]}$ converge locally uniformly to a maximal-lifespan solution v with some lifespan $J \ni 0$. By the local well-posedness theory, J is open and so contains δ_n for all sufficiently large n. This contradicts the local uniform convergence as, by hypothesis, δ_n does not belong to $I^{[t_n]}$. Hence (5.35) holds.

We now show (5.36). Again, we argue by contradiction, shrinking δ if necessary. Suppose one of the three claims in (5.36) failed no matter how small one selected δ. Then, one can find sequences $t_n, t'_n \in I$ such that $s_n := (t'_n - t_n)N(t_n)^2 \to 0$ but $N(t'_n)/N(t_n)$ converge to either zero or infinity (if the first claim failed) or $|\xi(t'_n) - \xi(t_n)|/N(t_n) \to \infty$ (if the second claim failed) or $|x(t'_n) - x(t_n) - 2(t'_n - t_n)\xi(t_n)|N(t_n) \to \infty$ (if the third claim failed). If we define $u^{[t_n]}$ as before and apply Lemma 5.17 (passing to a subsequence if necessary), we see once again that $u^{[t_n]}$ converge locally uniformly to a maximal-lifespan solution v with some open lifespan $J \ni 0$. But then $N^{[t_n]}(s_n)$ converge to either zero or infinity or $\xi^{[t_n]}(s_n) \to \infty$ or $x^{[t_n]}(s_n) \to \infty$ and thus, by Definition 5.1, $u^{[t_n]}(s_n)$ converge weakly to zero. On the other hand, since s_n converge to zero and $u^{[t_n]}$ are locally uniformly convergent to $v \in C^0_{t,\mathrm{loc}} L^2_x(J \times \mathbb{R}^d)$, we may conclude that $u^{[t_n]}(s_n)$ converge strongly to $v(0)$ in $L^2_x(\mathbb{R}^d)$. Thus $v(0) = 0$ and $M(u^{[t_n]})$ converge to $M(v) = 0$. But since $M(u^{(n)}) = M(u)$, we see that u vanishes identically, a contradiction. Thus (5.36) holds. □

Corollary 5.19 ($N(t)$ at blowup). *Let* u *be a non-zero maximal-lifespan solution to* (1.4) *with lifespan* I *that is almost periodic modulo symmetries with frequency scale function* $N : I \to \mathbb{R}^+$*. If* T *is any finite endpoint of* I*, then* $N(t) \gtrsim_u |T - t|^{-1/2}$*; in particular,* $\lim_{t \to T} N(t) = \infty$*. If* I *is infinite or semi-infinite, then for any* $t_0 \in I$ *we have* $N(t) \gtrsim_u \min\{N(t_0), |t - t_0|^{-1/2}\}$*. The identical statement holds for the energy-critical NLS.*

Proof. This is immediate from (5.35). □

Lemma 5.20 (Local quasi-boundedness of N). *Let* u *be a non-zero solution to the mass-critical NLS with lifespan* I *that is almost periodic modulo symmetries with frequency scale function* $N : I \to \mathbb{R}^+$*. If* K *is any compact subset of* I*, then*

$$0 < \inf_{t \in K} N(t) \leq \sup_{t \in K} N(t) < \infty.$$

The same statement holds in the energy-critical setting.

PROOF. We only prove the first inequality; the other follows similarly.

We argue by contradiction. Suppose that the first inequality fails. Then, there exists a sequence $t_n \in K$ such that $\lim_{n\to\infty} N(t_n) = 0$ and hence, by Definition 5.1, $u(t_n)$ converge weakly to zero. Since K is compact, we can assume t_n converge to a limit $t_0 \in K$. As $u \in C_t^0 L_x^2(K \times \mathbb{R}^d)$, we see that $u(t_n)$ converge strongly to $u(t_0)$. Thus $u(t_0)$ must be zero, contradicting the hypothesis. □

LEMMA 5.21 (Strichartz norms via $N(t)$). *Let u be a non-zero solution to the mass-critical NLS with lifespan I that is almost periodic modulo symmetries with parameters $N(t), x(t), \xi(t)$. If J is any subinterval of I, then*

$$\int_J N(t)^2\,dt \lesssim_u \int_J\int_{\mathbb{R}^d} |u(t,x)|^{\frac{2(d+2)}{d}}\,dx\,dt \lesssim_u 1 + \int_J N(t)^2\,dt. \tag{5.37}$$

Similarly, if u is a non-zero solution to the energy-critical NLS on $I \times \mathbb{R}^d$ that is almost periodic modulo symmetries with parameters $N(t), x(t)$, then

$$\int_J N(t)^2\,dt \lesssim_u \int_J\int_{\mathbb{R}^d} |u(t,x)|^{\frac{2(d+2)}{d-2}}\,dx\,dt \lesssim_u 1 + \int_J N(t)^2\,dt$$

for any subinterval $J \subset I$.

PROOF. We consider the mass-critical case; the claim in the energy-critical case can be proved similarly. Let u be a solution to (1.4) as in the statement of the lemma. We first prove

$$\int_J\int_{\mathbb{R}^d} |u(t,x)|^{\frac{2(d+2)}{d}}\,dx\,dt \lesssim_u 1 + \int_J N(t)^2\,dt. \tag{5.38}$$

Let $0 < \eta < 1$ be a small parameter to be chosen momentarily and partition J into subintervals I_j so that

$$\int_{I_j} N(t)^2\,dt \leq \eta; \tag{5.39}$$

this requires at most $\eta^{-1} \times \mathrm{RHS}(5.38)$ many intervals.

For each j, we may choose $t_j \in I_j$ so that

$$N(t_j)^2|I_j| \leq 2\eta. \tag{5.40}$$

By Strichartz inequality followed by Hölder and Bernstein, we obtain

$$\begin{aligned}
\|u\|_{L_{t,x}^{\frac{2(d+2)}{d}}} &\lesssim \|e^{i(t-t_j)\Delta}u(t_j)\|_{L_{t,x}^{\frac{2(d+2)}{d}}} + \|u\|_{L_{t,x}^{\frac{2(d+2)}{d}}}^{\frac{d+4}{d}}\\
&\lesssim \|u_{\geq N_0}(t_j)\|_{L_x^2} + \|e^{i(t-t_j)\Delta}u_{\leq N_0}(t_j)\|_{L_{t,x}^{\frac{2(d+2)}{d}}} + \|u\|_{L_{t,x}^{\frac{2(d+2)}{d}}}^{\frac{d+4}{d}}\\
&\lesssim \|u_{\geq N_0}(t_j)\|_{L_x^2} + |I_j|^{\frac{d}{2(d+2)}} N_0^{\frac{d}{d+2}}\|u(t_j)\|_{L_x^2} + \|u\|_{L_{t,x}^{\frac{2(d+2)}{d}}}^{\frac{d+4}{d}},
\end{aligned}$$

where all spacetime norms are taken on the slab $I_j \times \mathbb{R}^d$. Choosing N_0 as a large multiple of $N(t_j)$ and using Definition 5.1, one can make the first term as small as one wishes. Subsequently, choosing η sufficiently small depending on $M(u)$ and invoking (5.40), one may also render the second term arbitrarily small. Thus, by the usual bootstrap argument we obtain

$$\int_{I_j}\int_{\mathbb{R}^d} |u(t,x)|^{\frac{2(d+2)}{d}}\,dx\,dt \le 1.$$

Using the bound on the number of intervals I_j, this leads to (5.38).

Now we prove

$$\int_J\int_{\mathbb{R}^d} |u(t,x)|^{\frac{2(d+2)}{d}}\,dx\,dt \gtrsim_u \int_J N(t)^2\,dt. \tag{5.41}$$

Using Definition 5.1 and choosing η sufficiently small depending on $M(u)$, we can guarantee that

$$\int_{|x-x(t)|\le C(\eta)N(t)^{-1}} |u(t,x)|^2\,dx \gtrsim_u 1 \tag{5.42}$$

for all $t\in J$. On the other hand, a simple application of Hölder's inequality yields

$$\int_{\mathbb{R}^d} |u(t,x)|^{\frac{2(d+2)}{d}}\,dx \gtrsim_u \Big(\int_{|x-x(t)|\le C(\eta)N(t)^{-1}} |u(t,x)|^2\Big)^{\frac{d+2}{d}} N(t)^2.$$

Thus, using (5.42) and integrating over J we derive (5.41). □

COROLLARY 5.22 (Maximal-lifespan almost periodic solutions blow up). *Let u be a maximal-lifespan solution to the mass- or energy-critical NLS that is almost periodic modulo symmetries. Then u blows up both forward and backward in time.*

PROOF. In the case of a finite endpoint, this amounts to the definition of maximal-lifespan; see Corollary 3.5. Indeed, the assumption of almost-periodicity is redundant in this case.

In the case of an infinite endpoint, we see that by Corollary 5.19, $N(t)\gtrsim_u \langle t-t_0\rangle^{-1/2}$. Thus by Lemma 5.21, the spacetime norm diverges, which is the definition of blowup. □

We end this subsection with a result concerning the behaviour of almost periodic solutions at the endpoints of their maximal lifespan.

PROPOSITION 5.23 (Asymptotic orthogonality to free evolutions, [**96**]). *Let $u: I\times\mathbb{R}^d\to\mathbb{C}$ be a maximal-lifespan solution to (1.4) that is almost periodic modulo symmetries. Then $e^{-it\Delta}u(t)$ converges weakly to zero in $L^2_x(\mathbb{R}^d)$ as $t\to\sup I$ or $t\to\inf I$. In particular, we have the 'reduced' Duhamel formulae*

$$\begin{aligned} u(t) &= i\lim_{T\to\sup I}\int_t^T e^{i(t-t')\Delta}F(u(t'))\,dt' \\ &= -i\lim_{T\to\inf I}\int_T^t e^{i(t-t')\Delta}F(u(t'))\,dt', \end{aligned} \tag{5.43}$$

where the limits are to be understood in the weak L^2_x topology. In the energy-critical case, the same formulae hold in the weak $\dot H^1_x$ topology.

PROOF. Let us just prove the claim as $t\to\sup I$, since the reverse claim is similar.

Assume first that $\sup I<\infty$. Then by Corollary 5.19,

$$\lim_{t\to\sup I} N(t)=\infty.$$

By Definition 5.1, this implies that $u(t)$ converges weakly to zero as $t \to \sup I$. As $\sup I < \infty$ and the map $t \mapsto e^{it\Delta}$ is continuous in the strong operator topology on L_x^2, we see that $e^{-it\Delta}u(t)$ converges weakly to zero, as desired.

Now suppose instead that $\sup I = \infty$. It suffices to show that

$$\lim_{t\to\infty} \langle u(t), e^{it\Delta}\phi\rangle_{L_x^2(\mathbb{R}^d)} = 0$$

for all test functions $\phi \in C_c^\infty(\mathbb{R}^d)$. Let $\eta > 0$ be a small parameter; using Hölder's inequality and Definition 5.1, we estimate

$$\begin{aligned}
&\Big|\langle u(t), e^{it\Delta}\phi\rangle_{L_x^2(\mathbb{R}^d)}\Big|^2 \\
&\lesssim \Big|\int_{|x-x(t)|\leq C(\eta)/N(t)} u(t,x)\overline{e^{it\Delta}\phi(x)}\,dx\Big|^2 + \Big|\int_{|x-x(t)|\geq C(\eta)/N(t)} u(t,x)\overline{e^{it\Delta}\phi(x)}\,dx\Big|^2 \\
&\lesssim \int_{|x-x(t)|\leq C(\eta)/N(t)} |e^{it\Delta}\phi(x)|^2\,dx + \eta\|\phi\|_{L_x^2}^2.
\end{aligned}$$

The claim now follows from Lemma 4.12, Corollary 5.19, and an easy change of variables. □

5.4. Further refinements: the enemies. The purpose of this subsection is to construct more refined counterexamples than those provided by Theorems 5.2 and 5.12, should the global well-posedness and scattering conjectures fail. These theorems provide little information about the behaviour of $N(t)$ over the lifespan I of the solution. In this subsection we strengthen those results by showing that the failure of Conjecture 1.4 or 1.5 implies the existence of at least one of three types of almost periodic solutions u for which $N(t)$ and I have very particular properties.

We would like to point out that elementary scaling arguments show that one may assume that $N(t)$ is either bounded from above or from below at least on half of its maximal lifespan; see for example, [**97**, Theorem 3.3] or [**38, 57**]. However, several recent results seem to require finer control on the nature of the blowup as one approaches either endpoint of the interval I.

We start with the mass-critical equation.

THEOREM 5.24 (Three enemies: the mass-critical NLS, [**43**]). *Fix μ, d and suppose that Conjecture 1.4 fails for this choice of μ and d. Then there exists a maximal-lifespan solution u to (1.4), which is almost periodic modulo symmetries, blows up both forward and backward in time, and in the focusing case also obeys $M(u) < M(Q)$.*

We can also ensure that the lifespan I and the frequency scale function $N(t)$ match one of the following three scenarios:

I. *(Soliton-like solution) We have $I = \mathbb{R}$ and*

$$N(t) = 1 \quad \text{for all} \quad t \in \mathbb{R}.$$

II. *(Double high-to-low frequency cascade) We have $I = \mathbb{R}$,*

$$\liminf_{t\to-\infty} N(t) = \liminf_{t\to+\infty} N(t) = 0, \quad \text{and} \quad \sup_{t\in\mathbb{R}} N(t) < \infty.$$

III. *(Self-similar solution) We have $I = (0, +\infty)$ and*

$$N(t) = t^{-1/2} \quad \text{for all} \quad t \in I.$$

PROOF. Fix μ and d. Invoking Theorem 5.2, we can find a solution v with maximal lifespan J, which is almost periodic modulo symmetries and blows up both forward and backward in time; also, in the focusing case we have $M(v) < M(Q)$.

Let $N_v(t)$ be the frequency scale function associated to v as in Definition 5.1, and let $C : \mathbb{R}^+ \to \mathbb{R}^+$ be its compactness modulus function. The solution v partially satisfies the conclusions of Theorem 5.24, but we are not necessarily in one of the three scenarios listed there. To extract a solution u with these additional properties, we will have to perform some further manipulations primarily based on the scaling and time-translation symmetries.

For any $T \geq 0$, define the quantity

$$\operatorname{osc}(T) := \inf_{t_0 \in J} \frac{\sup\{N_v(t) : t \in J \text{ and } |t - t_0| \leq TN_v(t_0)^{-2}\}}{\inf\{N_v(t) : t \in J \text{ and } |t - t_0| \leq TN_v(t_0)^{-2}\}}. \tag{5.44}$$

Roughly speaking, this measures the least possible oscillation one can find in N_v on time intervals of normalised duration T. This quantity is clearly non-decreasing in T. If $\operatorname{osc}(T)$ is bounded, we will be able to extract a soliton-like solution; this is

Case I: $\lim_{T\to\infty} \operatorname{osc}(T) < \infty$.

In this case, we have arbitrarily long periods of stability for N_v. More precisely, we can find a finite number $A = A_v$, a sequence t_n of times in J, and a sequence $T_n \to \infty$ such that

$$\frac{\sup\{N_v(t) : t \in J \text{ and } |t - t_n| \leq T_n N_v(t_n)^{-2}\}}{\inf\{N_v(t) : t \in J \text{ and } |t - t_n| \leq T_n N_v(t_n)^{-2}\}} < A$$

for all n. Note that this, together with Lemma 5.18, implies that

$$[t_n - T_n N_v(t_n)^{-2}, t_n + T_n N_v(t_n)^{-2}] \subset J$$

and

$$N_v(t) \sim_v N_v(t_n)$$

for all t in this interval.

Now define the normalisations $v^{[t_n]}$ of v at times t_n as in (5.32). Then $v^{[t_n]}$ is a maximal-lifespan normalised solution with lifespan

$$J_n := \{s \in \mathbb{R} : t_n + N_v(t_n)^{-2} s \in J\} \supset [-T_n, T_n]$$

and mass $M(v)$. It is almost periodic modulo scaling with frequency scale function

$$N_{v^{[t_n]}}(s) := \frac{N_v\big(t_n + N_v(t_n)^{-2} s\big)}{N_v(t_n)}$$

and compactness modulus function C. In particular, we see that

$$N_{v^{[t_n]}}(s) \sim_v 1 \tag{5.45}$$

for all $s \in [-T_n, T_n]$.

We now apply Lemma 5.17 and conclude (passing to a subsequence if necessary) that $v^{[t_n]}$ converge locally uniformly to a maximal-lifespan solution u with mass $M(v)$ defined on an open interval I containing 0 and which is almost periodic modulo symmetries. As $T_n \to \infty$, Lemma 5.15 and (5.45) imply that the frequency scale function $N : I \to \mathbb{R}^+$ of u satisfies

$$0 < \inf_{t \in I} N(t) \leq \sup_{t \in I} N(t) < \infty.$$

In particular, by Corollary 5.19, $I = \mathbb{R}$. By modifying C by a bounded factor we may now normalise $N \equiv 1$. We have thus constructed a soliton-like solution in the sense of Theorem 5.24.

When $\operatorname{osc}(T)$ is unbounded, we must seek a solution belonging to one of the remaining two scenarios. To distinguish between them, we introduce the quantity

$$a(t_0) := \frac{\inf_{t\in J: t\leq t_0} N_v(t) + \inf_{t\in J: t\geq t_0} N_v(t)}{N_v(t_0)}$$

for every $t_0 \in J$. This measures the extent to which $N_v(t)$ decays to zero on both sides of t_0. Clearly, this quantity takes values in the interval $[0, 2]$.

Case II: $\lim_{T\to\infty} \operatorname{osc}(T) = \infty$ and $\inf_{t_0\in J} a(t_0) = 0$.

In this case, there are no long periods of stability but there are times about which there are arbitrarily large cascades from high to low frequencies in both future and past directions. This will allow us to extract a solution with a double high-to-low frequency cascade as defined in Theorem 5.24.

As $\inf_{t_0\in J} a(t_0) = 0$, there exists a sequence of times $t_n \in J$ such that $a(t_n) \to 0$ as $n \to \infty$. By the definition of a, we can also find times $t_n^- < t_n < t_n^+$ with $t_n^-, t_n^+ \in J$ such that

$$\frac{N_v(t_n^-)}{N_v(t_n)} \to 0 \quad \text{and} \quad \frac{N_v(t_n^+)}{N_v(t_n)} \to 0.$$

Choose $t_n^- < t_n' < t_n^+$ so that

$$N_v(t_n') \sim \sup_{t_n^-\leq t\leq t_n^+} N_v(t);$$

then,

$$\frac{N_v(t_n^-)}{N_v(t_n')} \to 0 \quad \text{and} \quad \frac{N_v(t_n^+)}{N_v(t_n')} \to 0.$$

We define the rescaled and translated times $s_n^- < 0 < s_n^+$ by

$$s_n^{\pm} := N_v(t_n')^2(t_n^{\pm} - t_n')$$

and the normalisations $v^{[t_n']}$ at times t_n' by (5.32). These are normalised maximal-lifespan solutions with lifespans containing $[s_n^-, s_n^+]$, which are almost periodic modulo G with frequency scale functions

$$N_{v^{[t_n']}}(s) := \frac{N_v\big(t_n' + N_v(t_n')^{-2}s\big)}{N_v(t_n')}. \tag{5.46}$$

By the way we chose t_n', we see that

$$N_{v^{[t_n']}}(s) \lesssim 1 \tag{5.47}$$

for all $s_n^- \leq s \leq s_n^+$. Moreover,

$$N_{v^{[t_n']}}(s_n^{\pm}) \to 0 \quad \text{as} \quad n\to\infty \tag{5.48}$$

for either choice of sign.

We now apply Lemma 5.17 and conclude (passing to a subsequence if necessary) that $v^{[t_n']}$ converge locally uniformly to a maximal-lifespan solution u of mass $M(v)$ defined on an open interval I containing 0, which is almost periodic modulo symmetries.

Let N be a frequency scale function for u. From Lemma 5.20 we see that $N(t)$ is bounded from below on any compact set $K \subset I$. From this and Lemma 5.15 (and

Lemma 5.13), we see that $N_{v^{[t'_n]}}(t)$ is also bounded from below, uniformly in $t \in K$, for all sufficiently large n (depending on K). As a consequence of this and (5.48), we see that s_n^- and s_n^+ cannot have any limit points in K; thus $K \subset [s_n^-, s_n^+]$ for all sufficiently large n. Therefore, $s_n^\pm$ converge to the endpoints of I. Combining this with Lemma 5.15 and (5.47), we conclude that

$$\sup_{t \in I} N(t) < \infty. \tag{5.49}$$

Corollary 5.19 now implies that I has no finite endpoints, that is, $I = \mathbb{R}$.

In order to prove that u is a double high-to-low frequency cascade, we merely need to show that

$$\liminf_{t \to +\infty} N(t) = \liminf_{t \to -\infty} N(t) = 0. \tag{5.50}$$

By time reversal symmetry, it suffices to establish that $\liminf_{t \to +\infty} N(t) = 0$. Suppose that this is not the case. Then, using (5.49) we may deduce

$$N(t) \sim_u 1$$

for all $t \geq 0$. We conclude from Lemma 5.15 that for every $m \geq 1$, there exists an n_m such that

$$N_{v^{[t'_{n_m}]}}(t) \sim_u 1$$

for all $0 \leq t \leq m$. But by (5.44) and (5.46) this implies that

$$\operatorname{osc}(\varepsilon m) \lesssim_u 1$$

for all m and some $\varepsilon = \varepsilon(u) > 0$ independent of m. Note that ε is chosen as a lower bound on the quantities $N(t''_{n_m})^2 / N(t'_{n_m})^2$ where $t''_{n_m} = t'_{n_m} + \frac{m}{2} N(t'_{n_m})^{-2}$. This contradicts the hypothesis $\lim_{T \to \infty} \operatorname{osc}(T) = \infty$ and so settles Case II.

Case III: $\lim_{T \to \infty} \operatorname{osc}(T) = \infty$ and $\inf_{t_0 \in J} a(t_0) > 0$.

In this case, there are no long periods of stability and no double cascades from high to low frequencies; we will be able to extract a self-similar solution in the sense of Theorem 5.24.

Let $\varepsilon = \varepsilon(v) > 0$ be such that $\inf_{t_0 \in J} a(t_0) \geq 2\varepsilon$. We call a time t_0 *future-focusing* if

$$N_v(t) \geq \varepsilon N_v(t_0) \text{ for all } t \in J \text{ with } t \geq t_0 \tag{5.51}$$

and *past-focusing* if

$$N_v(t) \geq \varepsilon N_v(t_0) \text{ for all } t \in J \text{ with } t \leq t_0. \tag{5.52}$$

From the choice of ε we see that every time $t_0 \in J$ is either future-focusing or past-focusing, or possibly both.

We will now show that either all sufficiently late times are future-focusing or that all sufficiently early times are past-focusing. If this were false, there would be a future-focusing time t_0 and a sequence of past-focusing times t_n that converge to $\sup J$. For sufficiently large n, we have $t_n \geq t_0$. By (5.51) and (5.52) we then see that

$$N_v(t_n) \sim_v N_v(t_0)$$

for all such n. For any $t_0 < t < t_n$, we know that t is either past-focusing or future-focusing; thus we have either $N_v(t_0) \geq \varepsilon N_v(t)$ or $N_v(t_n) \geq \varepsilon N_v(t)$. Also, since t_0 is future-focusing, $N_v(t) \geq \varepsilon N_v(t_0)$. We conclude that

$$N_v(t) \sim_v N_v(t_0)$$

for all $t_0 < t < t_n$; since $t_n \to \sup J$, this claim in fact holds for all $t_0 < t < \sup J$. In particular, from Corollary 5.19 we see that v does not blow up forward in finite time, that is, $\sup J = \infty$. The function N_v is now bounded above and below on the interval $(t_0, +\infty)$, which implies that $\lim_{T\to\infty} \operatorname{osc}(T) < \infty$, a contradiction. This proves the assertion at the beginning of the paragraph.

We may now assume that future-focusing occurs for all sufficiently late times; more precisely, we can find $t_0 \in J$ such that all times $t \geq t_0$ are future-focusing. The case when all sufficiently early times are past-focusing reduces to this via time-reversal symmetry.

We will now recursively construct a new sequence of times t_n. More precisely, we will explain how to choose t_{n+1} from t_n.

As $\lim_{T\to\infty} \operatorname{osc}(T) = \infty$, we have $\operatorname{osc}(B) \geq 2/\varepsilon$ for some sufficiently large $B = B(v) > 0$. Given $J \ni t_n > t_0$ set $A = 2B\varepsilon^{-2}$ and $t_n' = t_n + \frac{1}{2}AN_v(t_n)^{-2}$. As $t_n > t_0$, it is future-focusing and so $N_v(t_n') \geq \varepsilon N_v(t_n)$. From this, we see that

$$\{t : |t - t_n'| \leq BN_v(t_n')^{-2}\} \subseteq [t_n, t_n + AN_v(t_n)^{-2}]$$

and thus, by the definition of B and the fact that all $t \geq t_n$ are future-focusing,

$$\sup_{t\in J\cap[t_n, t_n + AN_v(t_n)^{-2}]} N_v(t) \geq 2N_v(t_n). \tag{5.53}$$

Using this and Lemma 5.18, we see that for every $t_n \in J$ with $t_n \geq t_0$ there exists a time $t_{n+1} \in J$ obeying

$$t_n < t_{n+1} \leq t_n + AN(t_n)^{-2} \tag{5.54}$$

such that

$$2N_v(t_n) \leq N_v(t_{n+1}) \lesssim_v N_v(t_n) \tag{5.55}$$

and

$$N_v(t) \sim_v N_v(t_n) \quad \text{for all } t_n \leq t \leq t_{n+1}. \tag{5.56}$$

From (5.55) we have

$$N_v(t_n) \geq 2^n N_v(t_0)$$

for all $n \geq 0$, which by (5.54) implies

$$t_{n+1} \leq t_n + O_v(2^{-2n} N_v(t_0)^{-2}).$$

Thus t_n converge to a limit and $N_v(t_n)$ to infinity. In view of Lemma 5.20, this implies that $\sup J$ is finite and $\lim_{n\to\infty} t_n = \sup J$.

Let $n \geq 0$. By (5.55),

$$N_v(t_{n+m}) \geq 2^m N_v(t_n)$$

for all $m \geq 0$ and so, using (5.54) we obtain

$$0 < t_{n+m+1} - t_{n+m} \lesssim_v 2^{-2m} N_v(t_n)^{-2}.$$

Summing this series in m, we conclude that

$$\sup J - t_n \lesssim_v N_v(t_n)^{-2}.$$

Combining this with Corollary 5.19, we obtain

$$\sup J - t_n \sim_v N_v(t_n)^{-2}.$$

In particular, we have

$$\sup J - t_{n+1} \sim_v \sup J - t_n \sim_v N_v(t_n)^{-2}.$$

Applying (5.55) and (5.56) shows

$$\sup J - t \sim_v N_v(t)^{-2}$$

for all $t_n \leq t \leq t_{n+1}$. Since t_n converge to $\sup J$, we conclude that

$$\sup J - t \sim_v N_v(t)^{-2}$$

for all $t_0 \leq t < \sup J$.

As we have the freedom to modify $N(t)$ by a bounded function (modifying C appropriately), we may normalise

$$N_v(t) = (\sup J - t)^{-1/2}$$

for all $t_0 \leq t < \sup J$. It is now not difficult to extract our sought-after self-similar solution by suitably rescaling the interval $(t_0, \sup J)$ as follows.

Consider the normalisations $v^{[t_n]}$ of v at times t_n (cf. (5.32)). These are maximal-lifespan normalised solutions of mass $M(v)$, whose lifespans include the interval

$$\Big(-\frac{\sup J - t_0}{\sup J - t_n}, 1\Big),$$

and which are almost periodic modulo scaling with compactness modulus function C and frequency scale functions

$$N_{v^{[t_n]}}(s) = (1 - s)^{-1/2} \tag{5.57}$$

for all $-\frac{\sup J - t_0}{\sup J - t_n} < s < 1$. We now apply Lemma 5.17 and conclude (passing to a subsequence if necessary) that $v^{[t_n]}$ converge locally uniformly to a maximal-lifespan solution u of mass $M(v)$ defined on an open interval I containing $(-\infty, 1)$, which is almost periodic modulo symmetries.

By Lemma 5.15 and (5.57), we see that u has a frequency scale function N obeying

$$N(s) \sim_v (1 - s)^{-1/2}$$

for all $s \in (-\infty, 1)$. By modifying N (and C) by a bounded factor, we may normalise

$$N(s) = (1 - s)^{-1/2}.$$

From this, Lemma 5.18, and Corollary 5.19 we see that we must have $I = (-\infty, 1)$. Applying a time translation (by -1) followed by a time reversal, we obtain our sought-after self-similar solution.

This finishes the proof of Theorem 5.24. □

Finally, we identify the enemies in the energy-critical setting. The precise statement we present is not as ambitious as the one for the mass-critical NLS, but it has proven sufficient to resolve the global well-posedness and scattering conjecture in high dimensions.

THEOREM 5.25 (Three enemies: the energy-critical NLS, [**44**]). *Fix μ and $d \geq 3$ and suppose that Conjecture* 1.5 *fails for this choice of μ and d. Then there exists a minimal kinetic energy, maximal-lifespan solution u to* (1.6), *which is almost periodic modulo symmetries, $\|u\|_{L_{t,x}^{2(d+2)/(d-2)}(I \times \mathbb{R}^d)} = \infty$, and in the focusing case also obeys $\sup_{t \in I} \|\nabla u(t)\|_2 < \|\nabla W\|_2$.*

We can also ensure that the lifespan I and the frequency scale function $N : I \to \mathbb{R}^+$ match one of the following three scenarios:

I. *(Finite-time blowup) We have that either $|\inf I| < \infty$ or $\sup I < \infty$.*

II. *(Soliton-like solution)* *We have* $I = \mathbb{R}$ *and*

$$N(t) = 1 \quad \text{for all} \quad t \in \mathbb{R}.$$

III. *(Low-to-high frequency cascade)* *We have* $I = \mathbb{R}$,

$$\inf_{t \in \mathbb{R}} N(t) \geq 1, \quad \text{and} \quad \limsup_{t \to +\infty} N(t) = \infty.$$

PROOF. Exercise: adapt the proof of Theorem 5.24 to cover this case. □

6. Quantifying the compactness

In this section we continue our study of minimal blowup solutions, particularly, the study of the enemies described in Theorems 5.24 and 5.25. As we have seen in Section 5, one of properties that these minimal blowup solutions enjoy is that their orbit is precompact (modulo symmetries) in L_x^2 (in the mass-critical case) or in $\dot{H}_x^1$ (in the energy-critical case). We will now show that these minimal counterexamples to the global well-posedness and scattering conjectures enjoy additional regularity and decay, properties which one should regard as a strengthening of the precompactness of their profiles, indeed, as a way to quantify this (pre)compactness.

The *goal* is to show that solutions corresponding to the three scenarios described in Theorem 5.24 belong to $L_t^\infty H_x^1$ (or even $L_t^\infty H_x^{1+\varepsilon}$ for some $\varepsilon = \varepsilon(d) > 0$) throughout their lifespan, while solutions corresponding to the three scenarios described in Theorem 5.25 belong to $L_t^\infty L_x^2$ (or even $L_t^\infty \dot{H}_x^{-\varepsilon}$ for some $\varepsilon = \varepsilon(d) > 0$). As we will see in Section 8, this additional regularity and decay is sufficient to preclude the enemies to the global well-posedness and scattering conjectures. To give just a quick example of how this works, let us notice that in order to preclude the self-similar solution described in Theorem 5.24, it suffices to prove that such a solution belongs to $L_t^\infty H_x^1$, since then it is global (see Weinstein [**105**] for the focusing case); this contradicts the fact that a self-similar solution blows up at $t = 0$.

The goal described in the paragraph above is by no means easily achievable; indeed, most of the effort and innovation in proving the global well-posedness and scattering conjectures concentrate in attaining this goal. In the mass-critical case, additional regularity for the enemies described in Theorem 5.24 was so far only proved in dimensions $d \geq 2$ under the additional assumption of spherical symmetry on the initial data; see [**43, 46**] and also [**97**]. Removing the spherical symmetry assumption even in the defocusing case (when one has the advantage of using Morawetz-type inequalities) has proven quite difficult and is still an open problem.

In the energy-critical case, the goal was achieved in dimensions $d \geq 5$ in [**44**], thus resolving the global well-posedness and scattering conjecture in this case. In lower dimensions $d = 3, 4$, the conjecture was only proved under the additional assumption of spherical symmetry on the initial data; see [**38**]. Unlike in the mass-critical case, for the energy-critical NLS this assumption is sufficiently strong that one does not need to achieve the goal in order to rule out the enemies. Indeed, in these low dimensions, the goal described above is presumably too ambitious since even the ground state W does not belong to L_x^2 in this case. Removing the spherical symmetry assumption for $d = 3, 4$ remains quite a challenge.

In the mass-critical case, we will only revisit the proof of additional regularity for the self-similar solution (cf. Theorem 5.24) and only in the spherically symmetric case, as it appears in [**43, 46**]. We will, however, present the complete argument for the energy-critical NLS in dimensions $d \geq 5$, following [**44**].

6.1. Additional regularity: the self-similar scenario.

THEOREM 6.1 (Regularity in the self-similar case, [**43, 46**]). *Let $d \geq 2$ and let u be a spherically symmetric solution to* (1.4) *that is almost periodic modulo scaling and self-similar in the sense of Theorem* 5.24. *Then $u(t) \in H^s_x(\mathbb{R}^d)$ for all $t \in (0,\infty)$ and all $0 \leq s < 1 + \frac{4}{d}$.*

COROLLARY 6.2 (Absence of self-similar solutions). *For $d \geq 2$ there are no spherically symmetric solutions to* (1.4) *that are self-similar in the sense of Theorem* 5.24.

PROOF. By Theorem 6.1, any such solution would obey $u(t) \in H^1_x(\mathbb{R}^d)$ for all $t \in (0,\infty)$. Then, by the H^1_x global well-posedness theory (see Corollary 4.3 in the focusing case), there exists a global solution with initial data $u(t_0)$ at some time $t_0 \in (0,\infty)$. On the other hand, self-similar solutions blow up at time $t = 0$. These two facts (combined with the uniqueness statement in Corollary 3.5) yield a contradiction. □

The remainder of this subsection is devoted to proving Theorem 6.1.

Let u be as in Theorem 6.1. For any $A > 0$, we define

$$\begin{aligned} \mathcal{M}(A) &:= \sup_{T>0} \|u_{>AT^{-1/2}}(T)\|_{L^2_x(\mathbb{R}^d)} \\ \mathcal{S}(A) &:= \sup_{T>0} \|u_{>AT^{-1/2}}\|_{L^{2(d+2)/d}_{t,x}([T,2T]\times\mathbb{R}^d)} \\ \mathcal{N}(A) &:= \sup_{T>0} \|P_{>AT^{-1/2}}F(u)\|_{L^{2(d+2)/(d+4)}_{t,x}([T,2T]\times\mathbb{R}^d)}. \end{aligned} \tag{6.1}$$

The notation chosen indicates the quantity being measured, namely, the mass, the symmetric Strichartz norm, and the nonlinearity in the adjoint Strichartz norm, respectively. As u is self-similar, $N(t)$ is comparable to $T^{-1/2}$ for t in the interval $[T, 2T]$. Thus, the Littlewood-Paley projections are adapted to the natural frequency scale on each dyadic time interval.

To prove Theorem 6.1 it suffices to show that for every $0 < s < 1 + \frac{4}{d}$ we have

$$\mathcal{M}(A) \lesssim_{s,u} A^{-s}, \tag{6.2}$$

whenever A is sufficiently large depending on u and s. To establish this, we need a variety of estimates linking $\mathcal{M}$, $\mathcal{S}$, and $\mathcal{N}$. From mass conservation, Lemma 5.21, self-similarity, and Hölder's inequality, we see that

$$\mathcal{M}(A) + \mathcal{S}(A) + \mathcal{N}(A) \lesssim_u 1 \tag{6.3}$$

for all $A > 0$. From the Strichartz inequality (Theorem 3.2), we also see that

$$\mathcal{S}(A) \lesssim \mathcal{M}(A) + \mathcal{N}(A) \tag{6.4}$$

for all $A > 0$. One more application of Strichartz inequality combined with Lemma 5.21 and self-similarity shows

$$\|u\|_{L^2_t L_x^{\frac{2d}{d-2}}([T,2T]\times\mathbb{R}^d)} \lesssim_u 1. \tag{6.5}$$

Next, we obtain a deeper connection between these quantities.

LEMMA 6.3 (Nonlinear estimate). *Let* $\eta > 0$ *and* $0 < s < 1+\frac{4}{d}$. *For all* $A > 100$ *and* $0 < \beta \leq 1$, *we have*

$$\mathcal{N}(A) \lesssim_u \sum_{N \leq \eta A^\beta} \big(\tfrac{N}{A}\big)^s \mathcal{S}(N) + \big[\mathcal{S}(\eta A^{\frac{\beta}{2(d-1)}}) + \mathcal{S}(\eta A^\beta)\big]^{\frac{4}{d}} \mathcal{S}(\eta A^\beta) \tag{6.6}$$
$$+ A^{-\frac{2\beta}{d^2}} \big[\mathcal{M}(\eta A^\beta) + \mathcal{N}(\eta A^\beta)\big].$$

PROOF. Fix $\eta > 0$ and $0 < s < 1 + \frac{4}{d}$. It suffices to bound

$$\big\|P_{>AT^{-\frac12}} F(u)\big\|_{L_{t,x}^{\frac{2(d+2)}{d+4}}([T,2T]\times\mathbb{R}^d)}$$

by the right-hand side of (6.6) for fixed $T > 0$, $A > 100$, and $0 < \beta \leq 1$.

To achieve this, we decompose

$$F(u) = F(u_{\leq \eta A^\beta T^{-\frac12}}) + O\big(|u_{\leq \eta A^\alpha T^{-\frac12}}|^{\frac4d} |u_{>\eta A^\beta T^{-\frac12}}|\big) \tag{6.7}$$
$$+ O\big(|u_{\eta A^\alpha T^{-\frac12} < \cdot \leq \eta A^\beta T^{-\frac12}}|^{\frac4d} |u_{>\eta A^\beta T^{-\frac12}}|\big) + O\big(|u_{>\eta A^\beta T^{-\frac12}}|^{1+\frac4d}\big),$$

where $\alpha = \frac{\beta}{2(d-1)}$. To estimate the contribution from the last two terms in the expansion above, we discard the projection onto high frequencies and then use Hölder's inequality and (6.1):

$$\big\||u_{\eta A^\alpha T^{-\frac12} < \cdot \leq \eta A^\beta T^{-\frac12}}|^{\frac4d} u_{>\eta A^\beta T^{-\frac12}}\big\|_{L_{t,x}^{\frac{2(d+2)}{d+4}}([T,2T]\times\mathbb{R}^d)} \lesssim \mathcal{S}(\eta A^\alpha)^{\frac4d} \mathcal{S}(\eta A^\beta)$$
$$\big\||u_{>\eta A^\beta T^{-\frac12}}|^{1+\frac4d}\big\|_{L_{t,x}^{\frac{2(d+2)}{d+4}}([T,2T]\times\mathbb{R}^d)} \lesssim \mathcal{S}(\eta A^\beta)^{1+\frac4d}.$$

To estimate the contribution coming from second term on the right-hand side of (6.7), we discard the projection onto high frequencies and then use Hölder's inequality, Lemma A.6, Corollary 4.19, and (6.4):

$$\big\|P_{>AT^{-\frac12}} O\big(|u_{\leq \eta A^\alpha T^{-\frac12}}|^{\frac4d} |u_{>\eta A^\beta T^{-\frac12}}|\big)\big\|_{L_{t,x}^{\frac{2(d+2)}{d+4}}([T,2T]\times\mathbb{R}^d)}$$
$$\lesssim \big\|u_{\leq \eta A^\alpha T^{-\frac12}} u_{>\eta A^\beta T^{-\frac12}}\big\|_{L^2_{t,x}([T,2T]\times\mathbb{R}^d)}^{\frac{8}{d^2}} \big\|u_{>\eta A^\beta T^{-\frac12}}\big\|_{L_{t,x}^{\frac{2(d+2)}{d}}([T,2T]\times\mathbb{R}^d)}^{1-\frac{8}{d^2}}$$
$$\times \big\|u_{\leq \eta A^\alpha T^{-\frac12}}\big\|_{L^2_{t,x}([T,2T]\times\mathbb{R}^d)}^{\frac4d - \frac{8}{d^2}}$$
$$\lesssim_u \big[(\eta A^\beta T^{-\frac12})^{-\frac12} (\eta A^\alpha T^{-\frac12})^{\frac{d-1}{2}}\big]^{\frac{8}{d^2}} \big[\mathcal{M}(\eta A^\beta) + \mathcal{N}(\eta A^\beta)\big]^{\frac{8}{d^2}} \mathcal{S}(\eta A^\beta)^{1-\frac{8}{d^2}} T^{\frac2d - \frac{4}{d^2}}$$
$$\lesssim_u A^{-\frac{2\beta}{d^2}} \big[\mathcal{M}(\eta A^\beta) + \mathcal{N}(\eta A^\beta)\big].$$

We now turn to the first term on the right-hand side of (6.7). By Lemma A.6 and Corollary A.14 combined with (6.3), we estimate

$$\|P_{>AT^{-\frac12}} F(u_{\leq \eta A^\beta T^{-\frac12}})\|_{L_{t,x}^{\frac{2(d+2)}{d+4}}([T,2T]\times\mathbb{R}^d)}$$
$$\lesssim (AT^{-\frac12})^{-s} \big\||\nabla|^s F(u_{\leq \eta A^\beta T^{-\frac12}})\big\|_{L_{t,x}^{\frac{2(d+2)}{d+4}}([T,2T]\times\mathbb{R}^d)}$$
$$\lesssim_u (AT^{-\frac12})^{-s} \big\||\nabla|^s u_{\leq \eta A^\beta T^{-\frac12}}\big\|_{L_{t,x}^{\frac{2(d+2)}{d}}([T,2T]\times\mathbb{R}^d)}$$
$$\lesssim_u \sum_{N \leq \eta A^\beta} \big(\tfrac{N}{A}\big)^s \mathcal{S}(N),$$

which is acceptable. This finishes the proof of the lemma. □

We have some decay as $A \to \infty$:

LEMMA 6.4 (Qualitative decay). *We have*

$$\lim_{A\to\infty} \mathcal{M}(A) = \lim_{A\to\infty} \mathcal{S}(A) = \lim_{A\to\infty} \mathcal{N}(A) = 0. \tag{6.8}$$

PROOF. The vanishing of the first limit follows from Definition 5.1, (6.1), and self-similarity. By interpolation, (6.1), and (6.5),

$$\mathcal{S}(A) \lesssim \mathcal{M}(A)^{\frac{2}{d+2}} \|u_{\geq AT^{-\frac{1}{2}}}\|^{\frac{d}{d+2}}_{L_t^2 L_x^{\frac{2d}{d-2}}([T,2T]\times\mathbb{R}^d)} \lesssim_u \mathcal{M}(A)^{\frac{2}{d+2}}.$$

Thus, as the first limit in (6.8) vanishes, we obtain that the second limit vanishes. The vanishing of the third limit follows from that of the second and Lemma 6.3. □

We have now gathered enough tools to prove some regularity, albeit in the symmetric Strichartz space. As such, the next result is the crux of this subsection.

PROPOSITION 6.5 (Quantitative decay estimate). *Let $0 < \eta < 1$ and $0 < s < 1 + \frac{4}{d}$. If η is sufficiently small depending on u and s, and A is sufficiently large depending on u, s, and η,*

$$\mathcal{S}(A) \leq \sum_{N\leq\eta A} \left(\tfrac{N}{A}\right)^s \mathcal{S}(N) + A^{-\frac{1}{d^2}}. \tag{6.9}$$

In particular, by Lemma A.15,

$$\mathcal{S}(A) \lesssim_u A^{-\frac{1}{d^2}}, \tag{6.10}$$

for all $A > 0$.

PROOF. Fix $\eta \in (0,1)$ and $0 < s < 1+\frac{4}{d}$. To establish (6.9), it suffices to show

$$\big\|u_{>AT^{-1/2}}\big\|_{L_{t,x}^{\frac{2(d+2)}{d}}([T,2T]\times\mathbb{R}^d)} \lesssim_{u,\varepsilon} \sum_{N\leq\eta A} \left(\tfrac{N}{A}\right)^{s+\varepsilon} \mathcal{S}(N) + A^{-\frac{3}{2d^2}} \tag{6.11}$$

for all $T > 0$ and some small $\varepsilon = \varepsilon(d,s) > 0$, since then (6.9) follows by requiring η to be small and A to be large, both depending upon u.

Fix $T > 0$. By writing the Duhamel formula (3.12) beginning at $\frac{T}{2}$ and then using the Strichartz inequality, we obtain

$$\begin{aligned}\big\|u_{>AT^{-1/2}}\big\|_{L_{t,x}^{\frac{2(d+2)}{d}}([T,2T]\times\mathbb{R}^d)} &\lesssim \big\|P_{>AT^{-1/2}} e^{i(t-\frac{T}{2})\Delta} u(\tfrac{T}{2})\big\|_{L_{t,x}^{\frac{2(d+2)}{d}}([T,2T]\times\mathbb{R}^d)} \\ &\quad + \big\|P_{>AT^{-1/2}} F(u)\big\|_{L_{t,x}^{\frac{2(d+2)}{d+4}}([\frac{T}{2},2T]\times\mathbb{R}^d)}.\end{aligned}$$

Consider the second term. By (6.1), we have

$$\big\|P_{>AT^{-1/2}} F(u)\big\|_{L_{t,x}^{\frac{2(d+2)}{d+4}}([\frac{T}{2},2T]\times\mathbb{R}^d)} \lesssim \mathcal{N}(A/2).$$

Using Lemma 6.3 (with $\beta = 1$ and s replaced by $s+\varepsilon$ for some $0 < \varepsilon < 1 + \frac{4}{d} - s$) combined with Lemma 6.4 (choosing A sufficiently large depending on u, s, and η), and (6.3), we derive

$$\big\|P_{>AT^{-1/2}} F(u)\big\|_{L_{t,x}^{\frac{2(d+2)}{d+4}}([\frac{T}{2},2T]\times\mathbb{R}^d)} \lesssim_{u,\varepsilon} \text{RHS}(6.11).$$

Thus, the second term is acceptable.

We now consider the first term. It suffices to show

$$\big\|P_{>AT^{-1/2}}e^{i(t-\frac{T}{2})\Delta}u(\tfrac{T}{2})\big\|_{L_{t,x}^{\frac{2(d+2)}{d}}([T,2T]\times\mathbb{R}^d)} \lesssim_u A^{-\frac{3}{2d^2}}, \tag{6.12}$$

which we will deduce by first proving two estimates at a single frequency scale, interpolating between them, and then summing.

From Theorem 4.29 and mass conservation, we have

$$\big\|P_{BT^{-1/2}}e^{i(t-\frac{T}{2})\Delta}u(\tfrac{T}{2})\big\|_{L^q_{t,x}([T,2T]\times\mathbb{R}^d)} \lesssim_{u,q} (BT^{-1/2})^{\frac{d}{2}-\frac{d+2}{q}} \tag{6.13}$$

for all $\frac{4d+2}{2d-1} < q \leq \frac{2(d+2)}{d}$ and $B > 0$. This is our first estimate.

Using the Duhamel formula (3.12), we write

$$P_{BT^{-1/2}}e^{i(t-\frac{T}{2})\Delta}u(\tfrac{T}{2}) = P_{BT^{-1/2}}e^{i(t-\delta)\Delta}u(\delta) - i\int_\delta^{\frac{T}{2}} P_{BT^{-1/2}}e^{i(t-t')\Delta}F(u(t'))\,dt'$$

for any $\delta > 0$. By self-similarity, the former term converges strongly to zero in L^2_x as $\delta \to 0$. Convergence to zero in $L_x^{2d/(d-2)}$ then follows from Lemma A.6. Thus, using Hölder's inequality followed by the dispersive estimate (3.2), and then (6.5), we estimate

$$\begin{aligned}
\big\|P_{BT^{-1/2}}&e^{i(t-\frac{T}{2})\Delta}u(\tfrac{T}{2})\big\|_{L_{t,x}^{\frac{2d}{d-2}}([T,2T]\times\mathbb{R}^d)} \\
&\lesssim T^{\frac{d-2}{2d}}\bigg\|\int_0^{\frac{T}{2}}\frac{1}{t-t'}\|F(u(t'))\|_{L_x^{\frac{2d}{d+2}}}\,dt'\bigg\|_{L_t^\infty([T,2T])} \\
&\lesssim T^{-\frac{d+2}{2d}}\|F(u)\|_{L_t^1L_x^{\frac{2d}{d+2}}((0,\frac{T}{2}]\times\mathbb{R}^d)} \\
&\lesssim T^{-\frac{d+2}{2d}}\sum_{0<\tau\leq\frac{T}{4}}\|F(u)\|_{L_t^1L_x^{\frac{2d}{d+2}}([\tau,2\tau]\times\mathbb{R}^d)} \\
&\lesssim T^{-\frac{d+2}{2d}}\sum_{0<\tau\leq\frac{T}{4}}\tau^{1/2}\|u\|_{L_t^2L_x^{\frac{2d}{d-2}}([\tau,2\tau]\times\mathbb{R}^d)}\|u\|^{\frac{4}{d}}_{L_t^\infty L_x^2([\tau,2\tau]\times\mathbb{R}^d)} \\
&\lesssim_u T^{-1/d}.
\end{aligned}$$

Interpolating between the estimate just proved and the $q = \frac{2d(d+2)(4d-3)}{4d^3-3d^2+12}$ case of (6.13), we obtain

$$\big\|P_{BT^{-1/2}}e^{i(t-\frac{T}{2})\Delta}u(\tfrac{T}{2})\big\|_{L_{t,x}^{\frac{2(d+2)}{d}}([T,2T]\times\mathbb{R}^d)} \lesssim_u B^{-\frac{3}{2d^2}}.$$

Summing this over dyadic $B \geq A$ yields (6.12) and hence (6.11). □

COROLLARY 6.6. *For any $A > 0$ we have*

$$\mathcal{M}(A) + \mathcal{S}(A) + \mathcal{N}(A) \lesssim_u A^{-1/d^2}.$$

PROOF. The bound on $\mathcal{S}$ was proved in the previous proposition. The bound on $\mathcal{N}$ follows from this, Lemma 6.3 with $\beta = 1$, and (6.3).

We now turn to the bound on $\mathcal{M}$. By Proposition 5.23 and weak lower semicontinuity of the norm,

$$\|P_{>AT^{-1/2}}u(T)\|_2 \leq \sum_{k=0}^\infty\bigg\|\int_{2^kT}^{2^{k+1}T}e^{i(T-t')\Delta}P_{>AT^{-1/2}}F(u(t'))\,dt'\bigg\|_2. \tag{6.14}$$

Intuitively, the reason for using the Duhamel formula forward in time is that the solution becomes smoother as $N(t) \to 0$.

Combining (6.14) with Strichartz inequality and (6.1), we get

$$\mathcal{M}(A) = \sup_{T>0} \|P_{>AT^{-1/2}} u(T)\|_2 \lesssim \sum_{k=0}^{\infty} \mathcal{N}(2^{k/2}A). \tag{6.15}$$

The desired bound on $\mathcal{M}$ now follows from that on $\mathcal{N}$. □

PROOF OF THEOREM 6.1. Let $0 < s < 1 + \frac{4}{d}$. Combining Lemma 6.3 (with $\beta = 1 - \frac{1}{2d^2}$), (6.4), and (6.15), we deduce that if

$$\mathcal{S}(A) + \mathcal{M}(A) + \mathcal{N}(A) \lesssim_u A^{-\sigma}$$

for some $0 < \sigma < s$, then

$$\mathcal{S}(A) + \mathcal{M}(A) + \mathcal{N}(A) \lesssim_u A^{-\sigma}\Big(A^{-\frac{s-\sigma}{2d^2}} + A^{-\frac{(d+1)(3d-2)\sigma}{2d^3(d-1)}} + A^{-\frac{3-\sigma}{2d^2} - \frac{d^2-2}{2d^4}}\Big).$$

More precisely, Lemma 6.3 provides the bound on $\mathcal{N}(A)$, then (6.15) gives the bound on $\mathcal{M}(A)$ and then finally (6.4) gives the bound on $\mathcal{S}(A)$.

Iterating this statement shows that $u(t) \in H^s_x(\mathbb{R}^d)$ for all $0 < s < 1 + \frac{4}{d}$. Note that Corollary 6.6 allows us to begin the iteration with $\sigma = d^{-2}$. □

6.2. Additional decay: the finite-time blowup case. We consider now the energy-critical NLS. The purpose of the next two subsections is to prove that solutions corresponding to the three scenarios described in Theorem 5.25 obey additional decay, in particular, they belong to $L^\infty_t L^2_x$ or better (at least in dimensions $d \geq 5$).

We start with the finite-time blowup scenario and show that in this case, the solution has finite mass; indeed, we will show that the solution must have zero mass, and hence derive a contradiction to the fact that it is, after all, a blowup solution. In this particular case, we do not need to restrict to dimensions $d \geq 5$. The argument is essentially taken from [**38**].

THEOREM 6.7 (No finite-time blowup). *Let $d \geq 3$. Then there are no maximal-lifespan solutions $u : I \times \mathbb{R}^d \to \mathbb{C}$ to* (1.6) *that are almost periodic modulo symmetries, obey*

$$S_I(u) = \infty, \tag{6.16}$$

and

$$\sup_{t \in I} \|\nabla u(t)\|_2 < \infty, \tag{6.17}$$

and are such that either $|\inf I| < \infty$ or $\sup I < \infty$.

PROOF. Suppose for a contradiction that there existed such a solution u. Without loss of generality, we may assume $\sup I < \infty$. By Corollary 5.19, we must have

$$\liminf_{t \nearrow \sup I} N(t) = \infty. \tag{6.18}$$

We now show that (6.18) implies

$$\limsup_{t \nearrow \sup I} \int_{|x| \leq R} |u(t,x)|^2 \, dx = 0 \quad \text{for all } R > 0. \tag{6.19}$$

Indeed, let $0 < \eta < 1$ and $t \in I$. By Hölder's inequality, Sobolev embedding, and (6.17),

$$\begin{aligned}\int_{|x|\leq R} |u(t,x)|^2\,dx &\leq \int_{|x-x(t)|\leq \eta R} |u(t,x)|^2\,dx + \int_{\substack{|x|\leq R\\ |x-x(t)|>\eta R}} |u(t,x)|^2\,dx\\ &\lesssim \eta^2 R^2 \|u(t)\|_{\frac{2d}{d-2}}^2 + R^2\Big(\int_{|x-x(t)|>\eta R} |u(t,x)|^{\frac{2d}{d-2}}\,dx\Big)^{\frac{d-2}{d}}\\ &\lesssim \eta^2 R^2 + R^2\Big(\int_{|x-x(t)|>\eta R} |u(t,x)|^{\frac{2d}{d-2}}\,dx\Big)^{\frac{d-2}{d}}.\end{aligned}$$

Letting $\eta \to 0$, we can make the first term on the right-hand side of the inequality above as small as we wish. On the other hand, by (6.18) and Definition 5.11, we see that

$$\limsup_{t\nearrow \sup I} \int_{|x-x(t)|>\eta R} |u(t,x)|^{\frac{2d}{d-2}}\,dx = 0.$$

This proves (6.19).

The next step is to prove that (6.19) implies the solution u is identically zero, thus contradicting (6.16). For $t \in I$ define

$$M_R(t) := \int_{\mathbb{R}^d} \phi\big(\tfrac{|x|}{R}\big)|u(t,x)|^2\,dx,$$

where ϕ is a smooth, radial function, such that

$$\phi(r) = \begin{cases} 1 & \text{for } r \leq 1\\ 0 & \text{for } r \geq 2.\end{cases}$$

By (6.19),

$$\limsup_{t\nearrow \sup I} M_R(t) = 0 \quad \text{for all } R > 0. \tag{6.20}$$

On the other hand, a simple computation involving Hardy's inequality and (6.17) shows

$$|\partial_t M_R(t)| \lesssim \|\nabla u(t)\|_2 \Big\|\frac{u(t)}{|x|}\Big\|_2 \lesssim \|\nabla u(t)\|_2^2 \lesssim_u 1.$$

Thus, by the Fundamental Theorem of Calculus,

$$M_R(t_1) = M_R(t_2) - \int_{t_1}^{t_2} \partial_t M_R(t)\,dt \lesssim_u M_R(t_2) + |t_2 - t_1|$$

for all $t_1, t_2 \in I$ and $R > 0$. Letting $t_2 \nearrow \sup I$ and invoking (6.20), we deduce

$$M_R(t_1) \lesssim_u |\sup I - t_1|.$$

Now letting $R \to \infty$ we obtain $u(t_1) \in L_x^2(\mathbb{R}^d)$. Finally, letting $t_1 \nearrow \sup I$ and using the conservation of mass, we conclude $u \equiv 0$, contradicting (6.16).

This concludes the proof of Theorem 6.7. □

6.3. Additional decay: the global case. In this subsection we prove

THEOREM 6.8 (Negative regularity in the global case, [**44**]). *Let $d \geq 5$ and let u be a global solution to (1.6) that is almost periodic modulo symmetries. Suppose also that*

$$\sup_{t\in\mathbb{R}} \|\nabla u(t)\|_{L_x^2} < \infty \tag{6.21}$$

and

$$\inf_{t\in\mathbb{R}} N(t) \geq 1. \tag{6.22}$$

Then $u \in L_t^\infty \dot{H}_x^{-\varepsilon}(\mathbb{R}\times\mathbb{R}^d)$ for some $\varepsilon = \varepsilon(d) > 0$. In particular, $u \in L_t^\infty L_x^2$.

The proof of Theorem 6.8 is achieved in two steps: First, we 'break' scaling in a Lebesque space; more precisely, we prove that our solution lives in $L_t^\infty L_x^p$ for some $2 < p < \frac{2d}{d-2}$. Next, we use a double Duhamel trick to upgrade this to $u \in L_t^\infty \dot{H}_x^{1-s}$ for some $s = s(p,d) > 0$. Iterating the second step finitely many times, we derive Theorem 6.8.

The double Duhamel trick was used in [**91**] for a similar purpose; however, in that paper, the breach of scaling comes directly from the subcritical nature of the nonlinearity. An earlier related instance of this trick can be found in [**20**, §14].

Let u be a solution to (1.6) that obeys the hypotheses of Theorem 6.8. Let $\eta > 0$ be a small constant to be chosen later. Then by the almost periodicity modulo symmetries combined with (6.22), there exists $N_0 = N_0(\eta)$ such that

$$\|\nabla u_{\leq N_0}\|_{L_t^\infty L_x^2(\mathbb{R}\times\mathbb{R}^d)} \leq \eta. \tag{6.23}$$

We turn now to our first step, that is, breaking scaling in a Lebesgue space. To this end, we define

$$A(N) := \begin{cases} N^{-\frac{2}{d-2}} \sup_{t\in\mathbb{R}} \|u_N(t)\|_{L_x^{\frac{2(d-2)}{d-4}}} & \text{for } \ d \geq 6 \\ N^{-\frac{1}{2}} \sup_{t\in\mathbb{R}} \|u_N(t)\|_{L_x^5} & \text{for } \ d = 5. \end{cases}$$

for frequencies $N \leq 10N_0$. Note that by Bernstein's inequality combined with Sobolev embedding and (6.21),

$$A(N) \lesssim \|u_N\|_{L_t^\infty L_x^{\frac{2d}{d-2}}} \lesssim \|\nabla u\|_{L_t^\infty L_x^2} < \infty.$$

We next prove a recurrence formula for $A(N)$.

LEMMA 6.9 (Recurrence). *For all $N \leq 10N_0$,*

$$A(N) \lesssim_u \left(\tfrac{N}{N_0}\right)^\alpha + \eta^{\frac{4}{d-2}} \sum_{\frac{N}{10}\leq N_1 \leq N_0} \left(\tfrac{N}{N_1}\right)^\alpha A(N_1) + \eta^{\frac{4}{d-2}} \sum_{N_1 < \frac{N}{10}} \left(\tfrac{N_1}{N}\right)^\alpha A(N_1),$$

where $\alpha := \min\{\frac{2}{d-2}, \frac{1}{2}\}$.

PROOF. We first give the proof in dimensions $d \geq 6$. Once this is completed, we will explain the changes necessary to treat $d = 5$.

Fix $N \leq 10N_0$. By time-translation symmetry, it suffices to prove

$$N^{-\frac{2}{d-2}} \|u_N(0)\|_{L_x^{\frac{2(d-2)}{d-4}}} \lesssim_u \left(\tfrac{N}{N_0}\right)^{\frac{2}{d-2}} + \eta^{\frac{4}{d-2}} \sum_{\frac{N}{10}\leq N_1\leq N_0} \left(\tfrac{N}{N_1}\right)^{\frac{2}{d-2}} A(N_1)$$

$$+\eta^{\frac{4}{d-2}} \sum_{N_1<\frac{N}{10}} \left(\tfrac{N_1}{N}\right)^{\frac{2}{d-2}} A(N_1). \tag{6.24}$$

Using the Duhamel formula (5.43) into the future followed by the triangle inequality, Bernstein, and the dispersive inequality, we estimate

$$\begin{aligned}
N^{-\frac{2}{d-2}}\|u_N(0)\|_{L_x^{\frac{2(d-2)}{d-4}}} &\leq N^{-\frac{2}{d-2}}\Big\|\int_0^{N^{-2}} e^{-it\Delta}P_N F(u(t))\,dt\Big\|_{L_x^{\frac{2(d-2)}{d-4}}}\\
&\quad + N^{-\frac{2}{d-2}}\int_{N^{-2}}^{\infty}\big\|e^{-it\Delta}P_N F(u(t))\big\|_{L_x^{\frac{2(d-2)}{d-4}}}\,dt\\
&\lesssim N\Big\|\int_0^{N^{-2}} e^{-it\Delta}P_N F(u(t))\,dt\Big\|_{L_x^2}\\
&\quad + N^{-\frac{2}{d-2}}\|P_N F(u)\|_{L_t^\infty L_x^{\frac{2(d-2)}{d}}}\int_{N^{-2}}^{\infty} t^{-\frac{d}{d-2}}\,dt\\
&\lesssim N^{-1}\|P_N F(u)\|_{L_t^\infty L_x^2} + N^{\frac{2}{d-2}}\|P_N F(u)\|_{L_t^\infty L_x^{\frac{2(d-2)}{d}}}\\
&\lesssim N^{\frac{2}{d-2}}\|P_N F(u)\|_{L_t^\infty L_x^{\frac{2(d-2)}{d}}}.
\end{aligned} \tag{6.25}$$

Using the Fundamental Theorem of Calculus, we decompose

$$\begin{aligned}
F(u) &= O(|u_{>N_0}||u_{\leq N_0}|^{\frac{4}{d-2}}) + O(|u_{>N_0}|^{\frac{d+2}{d-2}}) + F(u_{\frac{N}{10}\leq\cdot\leq N_0})\\
&\quad + u_{<\frac{N}{10}}\int_0^1 F_z\big(u_{\frac{N}{10}\leq\cdot\leq N_0}+\theta u_{<\frac{N}{10}}\big)\,d\theta\\
&\quad + \overline{u_{<\frac{N}{10}}}\int_0^1 F_{\bar z}\big(u_{\frac{N}{10}\leq\cdot\leq N_0}+\theta u_{<\frac{N}{10}}\big)\,d\theta.
\end{aligned} \tag{6.26}$$

The contribution to the right-hand side of (6.25) coming from terms that contain at least one copy of $u_{>N_0}$ can be estimated in the following manner: Using Hölder, Bernstein, and (6.21),

$$\begin{aligned}
N^{\frac{2}{d-2}}\|P_N O(|u_{>N_0}||u|^{\frac{4}{d-2}})\|_{L_t^\infty L_x^{\frac{2(d-2)}{d}}} &\lesssim N^{\frac{2}{d-2}}\|u_{>N_0}\|_{L_t^\infty L_x^{\frac{2d(d-2)}{d^2-4d+8}}}\|u\|^{\frac{4}{d-2}}_{L_t^\infty L_x^{\frac{2d}{d-2}}}\\
&\lesssim_u N^{\frac{2}{d-2}}N_0^{-\frac{2}{d-2}}.
\end{aligned} \tag{6.27}$$

Thus, this contribution is acceptable.

Next we turn to the contribution to the right-hand side of (6.25) coming from the last two terms in (6.26); it suffices to consider the first of them since similar arguments can be used to deal with the second.

Lemma A.13 yields

$$\big\|P_{>\frac{N}{10}}F_z(u)\big\|_{L_t^\infty L_x^{\frac{d-2}{2}}} \lesssim N^{-\frac{4}{d-2}}\|\nabla u\|^{\frac{4}{d-2}}_{L_t^\infty L_x^2}.$$

Thus, by Hölder's inequality and (6.23),

$$\begin{aligned}
&N^{\frac{2}{d-2}}\Big\|P_N\Big(u_{<\frac{N}{10}}\int_0^1 F_z\big(u_{\frac{N}{10}\leq\cdot\leq N_0}+\theta u_{<\frac{N}{10}}\big)\,d\theta\Big)\Big\|_{L_t^\infty L_x^{\frac{2(d-2)}{d}}}\\
&\quad\lesssim N^{\frac{2}{d-2}}\|u_{<\frac{N}{10}}\|_{L_t^\infty L_x^{\frac{2(d-2)}{d-4}}}\Big\|P_{>\frac{N}{10}}\Big(\int_0^1 F_z\big(u_{\frac{N}{10}\leq\cdot\leq N_0}+\theta u_{<\frac{N}{10}}\big)\,d\theta\Big)\Big\|_{L_t^\infty L_x^{\frac{d-2}{2}}}
\end{aligned}$$

$$\lesssim N^{-\frac{2}{d-2}} \|u_{<\frac{N}{10}}\|_{L_t^\infty L_x^{\frac{2(d-2)}{d-4}}} \|\nabla u_{\leq N_0}\|_{L_t^\infty L_x^2}^{\frac{4}{d-2}}$$

$$\lesssim \eta^{\frac{4}{d-2}} \sum_{N_1 < \frac{N}{10}} \left(\tfrac{N_1}{N}\right)^{\frac{2}{d-2}} A(N_1). \tag{6.28}$$

Hence, the contribution coming from the last two terms in (6.26) is acceptable.

We are left to estimate the contribution of $F(u_{\frac{N}{10}\leq\cdot\leq N_0})$ to the right-hand side of (6.25). We need only show

$$\|F(u_{\frac{N}{10}\leq\cdot\leq N_0})\|_{L_t^\infty L_x^{\frac{2(d-2)}{d}}} \lesssim \eta^{\frac{4}{d-2}} \sum_{\frac{N}{10}\leq N_1\leq N_0} N_1^{-\frac{2}{d-2}} A(N_1). \tag{6.29}$$

As $d \geq 6$, we have $\frac{4}{d-2} \leq 1$. Using the triangle inequality, Bernstein, (6.23), and Hölder, we estimate as follows:

$$\begin{aligned}
&\|F(u_{\frac{N}{10}\leq\cdot\leq N_0})\|_{L_t^\infty L_x^{\frac{2(d-2)}{d}}} \\
&\quad\lesssim \sum_{\frac{N}{10}\leq N_1\leq N_0} \big\| u_{N_1} |u_{\frac{N}{10}\leq\cdot\leq N_0}|^{\frac{4}{d-2}} \big\|_{L_t^\infty L_x^{\frac{2(d-2)}{d}}} \\
&\quad\lesssim \sum_{\frac{N}{10}\leq N_1, N_2\leq N_0} \big\| u_{N_1} |u_{N_2}|^{\frac{4}{d-2}} \big\|_{L_t^\infty L_x^{\frac{2(d-2)}{d}}} \\
&\quad\lesssim \sum_{\frac{N}{10}\leq N_1\leq N_2\leq N_0} \|u_{N_1}\|_{L_t^\infty L_x^{\frac{2(d-2)}{d-4}}} \|u_{N_2}\|_{L_t^\infty L_x^2}^{\frac{4}{d-2}} \\
&\qquad + \sum_{\frac{N}{10}\leq N_2\leq N_1\leq N_0} \|u_{N_1}\|_{L_t^\infty L_x^2}^{\frac{4}{d-2}} \|u_{N_1}\|_{L_t^\infty L_x^{\frac{2(d-2)}{d-4}}}^{\frac{d-6}{d-2}} \|u_{N_2}\|_{L_t^\infty L_x^{\frac{2(d-2)}{d-4}}}^{\frac{4}{d-2}} \\
&\quad\lesssim \sum_{\frac{N}{10}\leq N_1\leq N_2\leq N_0} \|u_{N_1}\|_{L_t^\infty L_x^{\frac{2(d-2)}{d-4}}} \eta^{\frac{4}{d-2}} N_2^{-\frac{4}{d-2}} \\
&\qquad + \sum_{\frac{N}{10}\leq N_2\leq N_1\leq N_0} \eta^{\frac{4}{d-2}} N_1^{-\frac{4}{d-2}} \|u_{N_1}\|_{L_t^\infty L_x^{\frac{2(d-2)}{d-4}}}^{\frac{d-6}{d-2}} \|u_{N_2}\|_{L_t^\infty L_x^{\frac{2(d-2)}{d-4}}}^{\frac{4}{d-2}} \\
&\quad\lesssim \eta^{\frac{4}{d-2}} \sum_{\frac{N}{10}\leq N_1\leq N_0} N_1^{-\frac{2}{d-2}} A(N_1) \\
&\qquad + \eta^{\frac{4}{d-2}} \sum_{\frac{N}{10}\leq N_2\leq N_1\leq N_0} \left(\tfrac{N_2}{N_1}\right)^{\frac{16}{(d-2)^2}} \big(N_1^{-\frac{2}{d-2}} A(N_1)\big)^{\frac{d-6}{d-2}} \big(N_2^{-\frac{2}{d-2}} A(N_2)\big)^{\frac{4}{d-2}} \\
&\quad\lesssim \eta^{\frac{4}{d-2}} \sum_{\frac{N}{10}\leq N_1\leq N_0} N_1^{-\frac{2}{d-2}} A(N_1).
\end{aligned}$$

This proves (6.29) and so completes the proof of the lemma in dimensions $d \geq 6$.

Consider now $d = 5$. Arguing as for (6.25), we have

$$N^{-\frac{1}{2}} \|u_N(0)\|_{L_x^5} \lesssim N^{\frac{1}{2}} \|P_N F(u)\|_{L_t^\infty L_x^{\frac{5}{4}}},$$

which we estimate by decomposing the nonlinearity as in (6.26). The analogue of (6.27) in this case is

$$N^{\frac{1}{2}} \|P_N O(|u_{>N_0}||u|^{\frac{4}{d-2}})\|_{L_t^\infty L_x^{\frac{5}{4}}} \lesssim N^{\frac{1}{2}} \|u_{>N_0}\|_{L_t^\infty L_x^{\frac{5}{2}}} \|u\|_{L_t^\infty L_x^{\frac{10}{3}}}^{\frac{4}{3}} \lesssim_u N^{\frac{1}{2}} N_0^{-\frac{1}{2}}.$$

Using Bernstein and Lemma A.11 together with (6.23), we replace (6.28) by

$$
\begin{aligned}
N^{\frac{1}{2}}\Big\|P_N\Big(u_{<\frac{N}{10}}\int_0^1 F_z\big(u_{\frac{N}{10}\le\cdot\le N_0}+\theta u_{<\frac{N}{10}}\big)\,d\theta\Big)\Big\|_{L_t^\infty L_x^{\frac{5}{4}}}
&\lesssim N^{\frac{1}{2}}\|u_{<\frac{N}{10}}\|_{L_t^\infty L_x^5}\Big\|P_{>\frac{N}{10}}\Big(\int_0^1 F_z\big(u_{\frac{N}{10}\le\cdot\le N_0}+\theta u_{<\frac{N}{10}}\big)\,d\theta\Big)\Big\|_{L_t^\infty L_x^{\frac{5}{3}}}\\
&\lesssim N^{-\frac{1}{2}}\|u_{<\frac{N}{10}}\|_{L_t^\infty L_x^5}\|\nabla u_{\le N_0}\|_{L_t^\infty L_x^2}\|u_{\le N_0}\|^{\frac{1}{3}}_{L_t^\infty L_x^{\frac{10}{3}}}\\
&\lesssim \eta^{\frac{4}{3}}\sum_{N_1<\frac{N}{10}}\big(\tfrac{N_1}{N}\big)^{\frac{1}{2}}A(N_1).
\end{aligned}
$$

Finally, arguing as for (6.29), we estimate

$$
\begin{aligned}
&\|F(u_{\frac{N}{10}\le\cdot\le N_0})\|_{L_t^\infty L_x^{\frac{5}{4}}}\\
&\quad\lesssim \sum_{\frac{N}{10}\le N_1,N_2\le N_0}\big\|u_{N_1}u_{N_2}|u_{\frac{N}{10}\le\cdot\le N_0}|^{\frac{1}{3}}\big\|_{L_t^\infty L_x^{\frac{5}{4}}}\\
&\quad\lesssim \sum_{\frac{N}{10}\le N_1\le N_2,N_3\le N_0}\|u_{N_1}\|_{L_t^\infty L_x^5}\|u_{N_2}\|_{L_t^\infty L_x^{\frac{20}{9}}}\|u_{N_3}\|^{\frac{1}{3}}_{L_t^\infty L_x^{\frac{20}{9}}}\\
&\qquad+\sum_{\frac{N}{10}\le N_3\le N_1\le N_2\le N_0}\|u_{N_1}\|^{\frac{2}{3}}_{L_t^\infty L_x^5}\|u_{N_1}\|^{\frac{1}{3}}_{L_t^\infty L_x^{\frac{20}{9}}}\|u_{N_2}\|_{L_t^\infty L_x^{\frac{20}{9}}}\|u_{N_3}\|^{\frac{1}{3}}_{L_t^\infty L_x^5}\\
&\quad\lesssim \sum_{\frac{N}{10}\le N_1\le N_2,N_3\le N_0}\|u_{N_1}\|_{L_t^\infty L_x^5}\eta N_2^{-\frac{3}{4}}\eta^{\frac{1}{3}}N_3^{-\frac{1}{4}}\\
&\qquad+\sum_{\frac{N}{10}\le N_3\le N_1\le N_2\le N_0}\|u_{N_1}\|^{\frac{2}{3}}_{L_t^\infty L_x^5}\eta^{\frac{1}{3}}N_1^{-\frac{1}{4}}\eta N_2^{-\frac{3}{4}}\|u_{N_3}\|^{\frac{1}{3}}_{L_t^\infty L_x^5}\\
&\quad\lesssim \eta^{\frac{4}{3}}\sum_{\frac{N}{10}\le N_1\le N_0}N_1^{-\frac{1}{2}}A(N_1)\\
&\qquad+\eta^{\frac{4}{3}}\sum_{\frac{N}{10}\le N_3\le N_1\le N_0}\big(\tfrac{N_3}{N_1}\big)^{\frac{1}{3}}\big(N_1^{-\frac{1}{2}}A(N_1)\big)^{\frac{2}{3}}\big(N_3^{-\frac{1}{2}}A(N_3)\big)^{\frac{1}{3}}\\
&\quad\lesssim \eta^{\frac{4}{3}}\sum_{\frac{N}{10}\le N_1\le N_0}N_1^{-\frac{1}{2}}A(N_1).
\end{aligned}
$$

Putting everything together completes the proof of the lemma in the case $d=5$. □

This lemma leads very quickly to our first goal:

PROPOSITION 6.10 (L_x^p breach of scaling). *Let u be as in Theorem 6.8. Then*

$$u\in L_t^\infty L_x^p \quad \text{for} \quad \tfrac{2(d+1)}{d-1}\le p<\tfrac{2d}{d-2}. \tag{6.30}$$

In particular, by Hölder's inequality,

$$\nabla F(u)\in L_t^\infty L_x^r \quad \text{for} \quad \tfrac{2(d-2)(d+1)}{d^2+3d-6}\le r<\tfrac{2d}{d+4}. \tag{6.31}$$

REMARK. As will be seen in the proof, p and r can be allowed to be smaller; however, the statement given will suffice for our purposes.

PROOF. We only present the details for $d \geq 6$. The treatment of $d = 5$ is completely analogous.

Combining Lemma 6.9 with Lemma A.15, we deduce

$$\|u_N\|_{L_t^\infty L_x^{\frac{2(d-2)}{d-4}}} \lesssim_u N^{\frac{4}{d-2}-} \quad \text{for all} \quad N \leq 10N_0. \tag{6.32}$$

In applying Lemma A.15, we set $N = 10 \cdot 2^{-k} N_0$, $x_k = A(10 \cdot 2^{-k} N_0)$, and take η sufficiently small.

By interpolation followed by (6.32), Bernstein, and (6.21),

$$\begin{aligned}
\|u_N\|_{L_t^\infty L_x^p} &\leq \|u_N\|_{L_t^\infty L_x^{\frac{2(d-2)}{d-4}}}^{(d-2)(\frac{1}{2}-\frac{1}{p})} \|u_N\|_{L_t^\infty L_x^2}^{\frac{d-2}{p}-\frac{d-4}{2}} \\
&\lesssim_u N^{\frac{2(p-2)}{p}-} N^{\frac{d-4}{2}-\frac{d-2}{p}} \\
&\lesssim_u N^{\frac{1}{d+1}-}
\end{aligned}$$

for all $N \leq 10N_0$. Thus, using Bernstein together with (6.21), we obtain

$$\|u\|_{L_t^\infty L_x^p} \leq \|u_{\leq N_0}\|_{L_t^\infty L_x^p} + \|u_{>N_0}\|_{L_t^\infty L_x^p} \lesssim_u \sum_{N \leq N_0} N^{\frac{1}{d+1}-} + \sum_{N > N_0} N^{\frac{d-2}{2}-\frac{d}{p}} \lesssim_u 1,$$

which completes the proof of the proposition. □

REMARK. With a few modifications, the argument used in dimension five can be adapted to dimensions three and four. However, while we may extend Proposition 6.10 in this way, $u(t,x) = W(x)$ provides an explicit counterexample to Theorem 6.8 in these dimensions. At a technical level, the obstruction is that the strongest dispersive estimate available is $|t|^{-d/2}$, which is insufficient to perform both integrals in the double Duhamel trick below when $d \leq 4$.

The second step is to use the double Duhamel trick to upgrade (6.30) to 'honest' negative regularity (i.e., in Sobolev sense). This will be achieved by repeated application of the following

PROPOSITION 6.11 (Some negative regularity). *Let $d \geq 5$ and let u be as in Theorem 6.8. Assume further that $|\nabla|^s F(u) \in L_t^\infty L_x^r$ for some $\frac{2(d-2)(d+1)}{d^2+3d-6} \leq r < \frac{2d}{d+4}$ and some $0 \leq s \leq 1$. Then there exists $s_0 = s_0(r,d) > 0$ such that $u \in L_t^\infty \dot{H}_x^{s-s_0+}$.*

PROOF. The proposition will follow once we establish

$$\big\||\nabla|^s u_N\big\|_{L_t^\infty L_x^2} \lesssim_u N^{s_0} \quad \text{for all} \quad N > 0 \quad \text{and} \quad s_0 := \tfrac{d}{r} - \tfrac{d+4}{2} > 0. \tag{6.33}$$

Indeed, by Bernstein combined with this and (6.21),

$$\begin{aligned}
\big\||\nabla|^{s-s_0+} u\big\|_{L_t^\infty L_x^2} &\leq \big\||\nabla|^{s-s_0+} u_{\leq 1}\big\|_{L_t^\infty L_x^2} + \big\||\nabla|^{s-s_0+} u_{>1}\big\|_{L_t^\infty L_x^2} \\
&\lesssim_u \sum_{N \leq 1} N^{0+} + \sum_{N>1} N^{(s-s_0+)-1} \\
&\lesssim_u 1.
\end{aligned}$$

Thus, we are left to prove (6.33). By time-translation symmetry, it suffices to prove

$$\big\||\nabla|^s u_N(0)\big\|_{L_x^2} \lesssim_u N^{s_0} \quad \text{for all} \quad N > 0 \quad \text{and} \quad s_0 := \tfrac{d}{r} - \tfrac{d+4}{2} > 0. \tag{6.34}$$

Using the Duhamel formula (5.43) both in the future and in the past, we write

$$\begin{aligned}
&\big\||\nabla|^s u_N(0)\big\|_{L_x^2}^2\\
&\quad = \lim_{T\to\infty}\lim_{T'\to-\infty}\Big\langle i\int_0^T e^{-it\Delta}P_N|\nabla|^s F(u(t))\,dt, -i\int_{T'}^0 e^{-i\tau\Delta}P_N|\nabla|^s F(u(\tau))\,d\tau\Big\rangle\\
&\quad \le \int_0^\infty\int_{-\infty}^0 \Big|\big\langle P_N|\nabla|^s F(u(t)), e^{i(t-\tau)\Delta}P_N|\nabla|^s F(u(\tau))\big\rangle\Big|\,dt\,d\tau.
\end{aligned}$$

We estimate the term inside the integrals in two ways. On one hand, using Hölder and the dispersive estimate,

$$\begin{aligned}
&\Big|\big\langle P_N|\nabla|^s F(u(t)), e^{i(t-\tau)\Delta}P_N|\nabla|^s F(u(\tau))\big\rangle\Big|\\
&\qquad\qquad \lesssim \big\|P_N|\nabla|^s F(u(t))\big\|_{L_x^r}\big\|e^{i(t-\tau)\Delta}P_N|\nabla|^s F(u(\tau))\big\|_{L_x^{r'}}\\
&\qquad\qquad \lesssim |t-\tau|^{\frac{d}{2}-\frac{d}{r}}\big\||\nabla|^s F(u)\big\|_{L_t^\infty L_x^r}^2.
\end{aligned}$$

On the other hand, using Bernstein,

$$\begin{aligned}
&\Big|\big\langle P_N|\nabla|^s F(u(t)), e^{i(t-\tau)\Delta}P_N|\nabla|^s F(u(\tau))\big\rangle\Big|\\
&\qquad\qquad \lesssim \big\|P_N|\nabla|^s F(u(t))\big\|_{L_x^2}\big\|e^{i(t-\tau)\Delta}P_N|\nabla|^s F(u(\tau))\big\|_{L_x^2}\\
&\qquad\qquad \lesssim N^{2(\frac{d}{r}-\frac{d}{2})}\big\||\nabla|^s F(u)\big\|_{L_t^\infty L_x^r}^2.
\end{aligned}$$

Thus,

$$\begin{aligned}
\big\||\nabla|^s u_N(0)\big\|_{L_x^2}^2 &\lesssim \big\||\nabla|^s F(u)\big\|_{L_t^\infty L_x^r}^2\int_0^\infty\int_{-\infty}^0 \min\{|t-\tau|^{-1}, N^2\}^{\frac{d}{r}-\frac{d}{2}}\,dt\,d\tau\\
&\lesssim N^{2s_0}\big\||\nabla|^s F(u)\big\|_{L_t^\infty L_x^r}^2.
\end{aligned}$$

To obtain the last inequality we used the fact that $\frac{d}{r}-\frac{d}{2}>2$ since $r<\frac{2d}{d+4}$. Thus (6.34) holds, which finishes the proof of the proposition. □

Proof of Theorem 6.8. Proposition 6.10 allows us to apply Proposition 6.11 with $s=1$. We conclude that $u\in L_t^\infty \dot H_x^{1-s_0+}$ for some $s_0=s_0(r,d)>0$. Combining this with the fractional chain rule Lemma A.11 and (6.30), we deduce that $|\nabla|^{1-s_0+}F(u)\in L_t^\infty L_x^r$ for some $\frac{2(d-2)(d+1)}{d^2+3d-6}\le r<\frac{2d}{d+4}$. We are thus in the position to apply Proposition 6.11 again and obtain $u\in L_t^\infty \dot H_x^{1-2s_0+}$. Iterating this procedure finitely many times, we derive $u\in L_t^\infty \dot H_x^{-\varepsilon}$ for any $0<\varepsilon<s_0$.

This completes the proof of Theorem 6.8. □

6.4. Compactness in other topologies. In this subsection we show that solutions to the mass-critical NLS (or energy-critical NLS), which are solitons in the sense of Theorem 5.24 (or Theorem 5.25) and which enjoy sufficient additional regularity (or decay), have orbits that are not only precompact in L_x^2 (or $\dot H_x^1$) but also in $\dot H_x^1$ (or L_x^2). Combining the two gives precompactness in H_x^1.

Lemma 6.12 (H_x^1 compactness for the mass-critical NLS). *Let $d\ge 1$ and let u be a soliton in the sense of Theorem 5.24. Assume further that $u\in L_t^\infty H_x^{1+\varepsilon}$ for some $\varepsilon=\varepsilon(d)>0$. Then for every $\eta>0$ there exists $C(\eta)>0$ such that*

$$\sup_{t\in\mathbb{R}}\int_{|x-x(t)|\ge C(\eta)} |\nabla u(t,x)|^2\,dx \lesssim_u \eta.$$

REMARK. The hypotheses of Lemma 6.12 are known to be satisfied in dimensions $d \geq 2$ for spherically symmetric initial data; see [**43, 46**].

PROOF. The entire argument takes place at a fixed t; in particular, we may assume $x(t) = 0$.

First we control the contribution from the high frequencies. As $u \in L_t^\infty H_x^{1+\varepsilon}$ for some $\varepsilon > 0$, then for any $R > 0$,

$$\big\|\nabla u_{>N}(t)\big\|_{L_x^2(|x|\geq R)} \leq \big\|\nabla u_{>N}(t)\big\|_{L_x^2} \lesssim N^{-\varepsilon}\big\||\nabla|^{1+\varepsilon}u\big\|_{L_t^\infty L_x^2} \lesssim_u N^{-\varepsilon}.$$

This can be made smaller than η by choosing $N = N(\eta)$ sufficiently large.

We now turn to the contribution coming from the low frequencies. A simple application of Schur's test reveals the following: For any $m \geq 0$,

$$\big\|\chi_{|x|\geq 2R}\nabla P_{\leq N}\chi_{|x|\leq R}\big\|_{L_x^2\to L_x^2} \lesssim_m N\langle RN\rangle^{-m}$$

uniformly in $R, N > 0$. Thus, by Bernstein's inequality,

$$\begin{aligned}
&\big\|\nabla u_{\leq N}(t)\big\|_{L_x^2(|x|\geq R)} \\
&\qquad \leq \big\|\chi_{|x|\geq R}\nabla P_{\leq N}\chi_{|x|\leq R/2}u(t)\big\|_{L_x^2} + \big\|\chi_{|x|\geq R}\nabla P_{\leq N}\chi_{|x|\geq R/2}u(t)\big\|_{L_x^2} \\
&\qquad \lesssim_u N\langle RN\rangle^{-100} + N\|u(t)\|_{L_x^2(|x|\geq R/2)}.
\end{aligned}$$

Choosing R sufficiently large (depending on N and η), we can ensure that the contribution of the low frequencies is less than η.

Combining the estimates for high and low frequencies yields the claim. □

We now turn our attention to the energy-critical NLS.

LEMMA 6.13 (H_x^1 compactness for the energy-critical NLS). *Let $d \geq 3$ and let u be a soliton in the sense of Theorem 5.25 that belongs to $L_t^\infty \dot H_x^{-\varepsilon}$ for some $\varepsilon = \varepsilon(d) > 0$. Then for every $\eta > 0$ there exists $C(\eta) > 0$ such that*

$$\sup_{t\in\mathbb{R}} \int_{|x-x(t)|\geq C(\eta)} |u(t,x)|^2\,dx \lesssim_u \eta.$$

REMARK. By Theorem 6.8, the hypotheses of this lemma are satisfied in dimensions $d \geq 5$.

PROOF. The entire argument takes place at a fixed t; in particular, we may assume $x(t) = 0$.

First we control the contribution from the low frequencies: by hypothesis,

$$\big\|u_{<N}(t)\big\|_{L_x^2(|x|\geq R)} \leq \big\|u_{<N}(t)\big\|_{L_x^2} \lesssim N^{\varepsilon}\big\||\nabla|^{-\varepsilon}u\big\|_{L_t^\infty L_x^2} \lesssim_u N^{\varepsilon}.$$

This can be made smaller than η by choosing $N = N(\eta)$ small enough.

We now turn to the contribution from the high frequencies. A simple application of Schur's test reveals the following: For any $m \geq 0$,

$$\big\|\chi_{|x|\geq 2R}\Delta^{-1}\nabla P_{\geq N}\chi_{|x|\leq R}\big\|_{L_x^2\to L_x^2} \lesssim_m N^{-1}\langle RN\rangle^{-m}$$

uniformly in $R, N > 0$. On the other hand, by Bernstein,

$$\big\|\chi_{|x|\geq 2R}\Delta^{-1}\nabla P_{\geq N}\chi_{|x|\geq R}\big\|_{L_x^2\to L_x^2} \lesssim N^{-1}.$$

Together, these lead quickly to

$$\int_{|x|\geq 2R} |u_{\geq N}(t,x)|^2\,dx \lesssim N^{-2}\langle RN\rangle^{-100}\|\nabla u(t)\|_{L_x^2}^2 + N^{-2}\int_{|x|\geq R} |\nabla u(t,x)|^2\,dx.$$

By choosing R large enough, we can render the first term smaller than η; the same is true of the second summand by virtue of $\dot H^1_x$-compactness:

$$\sup_{t\in\mathbb{R}} \int_{|x-x(t)|\geq C(\eta_1)} |\nabla u(t,x)|^2\,dx \leq \eta_1.$$

The lemma follows by combining our estimates for $u_{<N}$ and $u_{\geq N}$. □

7. Monotonicity formulae

The goal of this section it to introduce certain monotonicity formulae for the (non)linear Schrödinger equation. These have proved to be very powerful tools in the analysis of NLS; indeed, they have become *sine qua non* both for proving well-posedness and for describing the behaviour of solutions that blow up. Our goal here is just to give a small taste of what is available and how it can be used. Specific application to the mass- and energy- critical problems is discussed in Section 8.

7.1. The classical Virial theorem. Consider a classical mechanical system with n position coordinates, $q_1,\dots,q_n$, and n corresponding momenta, $p_1,\dots,p_n$. The energy is a sum of kinetic and potential terms,

$$H = K + V \quad \text{with} \quad K = \sum \tfrac{1}{2m_j}p_j^2 \quad \text{and} \quad V = V(q_1,\dots,q_n),$$

where m_j denote the mass of the particle associated to the jth coordinate. The basic precursor of all virial-like identities are the following simple calculations:

$$\tfrac{d}{dt}\sum \tfrac12 m_j q_j^2 = \sum m_j\dot q_j q_j = \sum p_j q_j, \tag{7.1}$$

$$\tfrac{d}{dt}\sum p_j q_j = \sum p_j\dot q_j + \dot p_j q_j = \sum \tfrac{1}{m_j}p_j^2 - \tfrac{\partial V}{\partial q_j}q_j. \tag{7.2}$$

THEOREM 7.1 (The Virial Theorem of Clausius, [**17**]). *If V is a homogeneous function of degree k, then the time averages of kinetic and potential energies are related by $\langle K\rangle = \frac{k}{2}\langle V\rangle$ along any orbit that remains inside a compact set in phase space. More precisely,*

$$\frac{1}{2T}\int_{-T}^{T}\Big[\sum \tfrac{1}{2m_j}p_j^2(t) - \tfrac{k}{2}V(q_1(t),\dots,q_n(t))\Big]\,dt = O(\tfrac1T) \tag{7.3}$$

as $T\to\infty$.

PROOF. The result follows quickly from (7.2) together with

$$\sum \tfrac{\partial V}{\partial q_j}q_j = kV,$$

which is a consequence of the homogeneity of V. □

REMARK. The quantity $\sum \dot p_j q_j$ (or rather, its time average) is known as the *virial.* The name was coined by Clausius and derives from the Latin for 'force'. A more famous notion (and name) due to Clausius is 'entropy'. His nomenclature for kinetic energy, 'vis viva', and potential energy, 'ergal', however, did not catch on.

EXAMPLE 7.1. For gravitational attraction, the potential energy is homogeneous of degree -1. Thus, for the eight major planets (whose orbits are approximately circular), the virial theorem gives a relation between the orbital radius r and the orbital velocity v of the form $v^2 = GM/r$, where M is the solar mass and G is the gravitational constant. As the orbital period is given by $T = 2\pi r/v$, we obtain Kepler's third law: T^2/r^3 is the same for all the major planets. Indeed, we find that this constant is $4\pi^2/GM = 3.0\times 10^{-19}s^2m^{-3}$, which agrees with astronomical data.

EXAMPLE 7.2 (Weighing things in space). Through a telescope, one may approximately measure lengths and speeds (Doppler effect). Now consider applying the virial identity to some form of self-gravitating ensemble of similar objects (e.g., stars or galaxies). The potential energy is quadratic in the mass, while the kinetic energy is linear in the mass. Given the typical distances involved and the typical speeds involved, one can quickly pop out a crude estimate for the total mass.

7.2. Some Lyapunov functions. In the field of ordinary differential equations, functions that are monotone in time (under the flow) are traditionally referred to as Lyapunov functions, in honour of the important work of A. M. Lyapunov on stability. Our applications of monotonicity formulae are perhaps better described as instability. The following two examples convey something of the spirit of this.

EXAMPLE 7.3. Consider a particle in $\mathbb{R}^3\setminus\{0\}$ moving in the presence of a repulsive potential $V(q)$, for example, $V(q)=|q|^{-1}$. The word repulsive is meant in the technical sense that $q\cdot\nabla V(q)<0$, which says that the radial component of the force on the particle always points away from the origin. By referring to (7.2), we see that $\sum p_j q_j$ is strictly increasing (in time) along any trajectory of the system. We immediately see that there can be no periodic orbits; indeed, any orbit must escape to (spatial) infinity as $t\to\pm\infty$.

EXAMPLE 7.4. If we choose $m_j\equiv 1$ and $V(q)=-|q|^{-2}$, then (7.1) and (7.2) become

$$\tfrac{d^2}{dt^2}\,\tfrac{1}{2}|q|^2 = 2H(p,q).$$

If the initial energy is negative, then $|q(t)|^2$ is a concave function of time. It is also non-negative. Thus we see that the particle falls into the origin in finite time.

In this section, we will discuss Lyapunov functionals for the flow

$$iu_t = -\Delta u + Vu + \mu|u|^p u. \tag{7.4}$$

We need only consider as potential Lyapunov functionals those which are odd under time reversal; even functionals, at least, cannot be monotone. Probably the simplest example is the quadratic form associated to a self-adjoint differential operator of first order:

$$\begin{aligned} F(u) &:= \tfrac{1}{i}\int_{\mathbb{R}^d} \bar u(x)\big[a_j(x)\partial_j + \partial_j a_j(x)\big]u(x)\,dx \\ &= 2\int_{\mathbb{R}^d} a_j(x)\,\mathrm{Im}\big(\bar u(x)\partial_j u(x)\big)\,dx, \end{aligned} \tag{7.5}$$

where a_j are real-valued functions on $\mathbb{R}^d$ and (both here and below) the repeated index j is summed over $1\le j\le d$. As we will only consider cases where $F(u)$ has spherical symmetry, we are guaranteed that there is a function $a(x)$ so that $a_j(x)=\partial_j a(x)$. This restriction has the happy consequence that we may use subscripts to denote partial derivatives, which we shall do from now on. A more scientific consequence is the first part of the following:

LEMMA 7.2 (Morawetz/Virial identity). *Under the flow* (7.4),

$$F(u) = \frac{d}{dt}\int_{\mathbb{R}^d} a(x)|u(t,x)|^2\,dx \tag{7.6}$$

$$\frac{d}{dt}F(u(t)) = \int_{\mathbb{R}^d} -a_{jjkk}|u|^2 + 4a_{jk}\bar{u}_j u_k + \mu\tfrac{2p}{p+2}a_{jj}|u|^{p+2} - 2a_j V_j |u|^2. \tag{7.7}$$

Here (as always in this subsection) subscripts indicate partial derivatives and repeated indices are summed.

We will discuss three applications in approximately historical order. Our first relates to the spectral and scattering theory of the linear Schrödinger equation and can be viewed as a quantum version of Example 7.3. Earlier still, identities analogous to (7.7) played an important role in the problem of obstacle scattering for the linear wave equation. Identities of this type are commonly known as Morawetz identities in honour of her pioneering work in this direction; see [**53**] for the link to scattering theory and [**60**] for an early retrospective.

Before discussing the linear Schrödinger equation, we first wish to present some completely abstract results about Lyapunov functions in quantum mechanics. The Putnam of the first theorem is *not* that of the competition; the name of the second theorem was coined in [**70**] and reflects the initials of Ruelle, Amrein, Georgescu, and Enss, rather than any ill-feeling.

THEOREM 7.3 (Putnam–Kato Theorem, [**36, 69**]). *Let H and A be bounded self-adjoint operators on a Hilbert space. If $C := i[H, A]$ is positive definite, then H has purely absolutely continuous spectrum.*

REMARK. Under certain technical assumptions, one may allow H and/or A to be unbounded; indeed, in the PDE context, this is the most common situation. However, our goal here is simply to give a taste of what may be expected.

PROOF. As A is bounded, we can quickly see that $\langle e^{-itH}\phi, Ce^{-itH}\phi\rangle$ belongs to $L^1_t(\mathbb{R})$ for all vectors ϕ. Thus, for all vectors ϕ in the range of $\sqrt{C}$, which is dense in the Hilbert space, we have $\langle \phi, e^{-itH}\phi\rangle \in L^2_t(\mathbb{R})$. The result now follows from the fact that only absolutely continuous measures can have square integrable Fourier transforms (cf. Parseval's Theorem). □

THEOREM 7.4 (RAGE Theorem). *Let H be a self-adjoint operator with purely absolutely continuous spectrum and let C be a bounded self-adjoint operator with $C(H-i)^{-1}$ compact. Then*

$$\langle e^{-itH}\phi, Ce^{-itH}\phi\rangle \to 0, \quad \text{as } t \to \pm\infty,$$

for all ϕ in the Hilbert space. If H has purely continuous spectrum, then

$$\frac{1}{2T}\int_{-T}^{T} \langle e^{-itH}\phi, Ce^{-itH}\phi\rangle\, dt \to 0, \quad \text{as } T \to \pm\infty.$$

PROOF. The results follow (respectively) from the Riemann–Lebesgue lemma and Wiener's lemma,

$$\frac{1}{2T}\int_{-T}^{T} \left| \int e^{-i\omega t}\, d\mu(\omega)\right|^2 dt \longrightarrow \sum_{\omega\in\mathbb{R}} \left|\mu(\{\omega\})\right|^2 \quad \text{as } T \to \infty,$$

after first applying the spectral theorem. □

The connection of Theorem 7.3 to Lyapunov functions is clear. We have included Theorem 7.4 to convey the fact that Theorem 7.3 guarantees that all trajectories escape to infinity in a fairly strong sense; indeed one may deduce the following from the RAGE Theorem:

EXERCISE. Suppose H is a self-adjoint operator and ϕ a vector in the associated Hilbert space. Show that the orbit $\{e^{-itH}\phi : t \in \mathbb{R}\}$ is pre-compact if and only if ϕ is a linear combination of eigenvectors of H, that is, if and only if the spectral measure associated to (H, ϕ) is of pure-point type.

Finally, we turn to our long-promised application to the linear Schrödinger equation. What we present is a special case of results contained in two early papers of R. Lavine, [**51, 52**]. This material is also discussed at some length in [**71**, §XIII.7]. Note that our particular statement has been chosen to simplify the exposition and in no way represents the limit of the method.

THEOREM 7.5. *Suppose $d \geq 3$ and $V : \mathbb{R}^d \to \mathbb{R}$ obeys $|V(x)| \lesssim \langle x\rangle^{-1-\varepsilon}$ and is repulsive in the sense that $x \cdot \nabla V \leq 0$ as a distribution. Then $H := -\Delta + V$ has purely absolutely continuous spectrum. Moreover, the limits $\lim_{t\to\pm\infty} e^{-it\Delta}e^{-itH}$ and $\lim_{t\to\pm\infty} e^{itH}e^{it\Delta}$ exist in the strong topology and define unitary operators.*

PROOF. We will prove absolute continuity by adapting the argument used to prove Theorem 7.3. For the scattering results, see the references given above.

Set $a(x) = \langle x\rangle$. For $\phi \in C_c^\infty(\mathbb{R}^d)$, let $u(t) := e^{-itH}\phi$. Then by (7.7),

$$\frac{d}{dt}F(u(t)) \geq \int_{\mathbb{R}^d} |u(t,x)|^2[-\Delta\Delta a](x)\,dx \gtrsim \int_{\mathbb{R}^d} |u(t,x)|^2\langle x\rangle^{-7}\,dx. \tag{7.8}$$

Note that the missing terms have the right sign for the following reasons: a is convex, so a_{jk} is a positive definite matrix; μ is zero since we consider the linear equation; the potential is assumed repulsive.

Now, mass/energy conservation guarantee that $u \in L_t^\infty H_x^1$, which then implies that $F(u)$ is bounded. Integrating (7.8) in time and using $\phi \in L^2(\langle x\rangle^7\,dx)$, we may deduce that $\langle \phi, e^{-itH}\phi\rangle \in L^2(dt)$. This proves that the spectral measure associated to (H, ϕ) is absolutely continuous (via Parseval's theorem) for a dense set of $\phi \in L_x^2(\mathbb{R}^d)$. Thus, we may conclude that H has purely absolutely continuous spectrum. □

Before turning to the nonlinear Schrödinger equation, we wish to draw the readers attention to two further developments connected to the material just described. The first is Mourre's method, which extends and refines the ideas behind the proof of Theorem 7.5. This is surveyed in [**22**, Ch. 4]. Chapter 5 of that book describes the Enss method in scattering theory. The idea here is that because of the RAGE Theorem, any part of the solution not described by bound states must travel far from the (spatial) origin. Once far away, the wave packet will continue to move outward since the potential is very weak out there. Parts of the argument in [**43**] can be viewed as an NLS incarnation of the Enss approach.

Our first NLS application of the Morawetz/Virial identity is an analogue of Example 7.4 and shows that for certain initial data, the solution of NLS must blow up in finite time. This is the well-known concavity argument; see, for instance, [**31, 102**]:

THEOREM 7.6 (Finite-time blow up). *Consider*

$$iu_t = -\Delta u - |u|^p u \quad \textit{with} \quad \tfrac{4}{d} \leq p \leq \tfrac{4}{d-2}. \tag{7.9}$$

Initial data $u_0 \in \Sigma := \{f \in H_x^1(\mathbb{R}^d) : |x|f \in L_x^2(\mathbb{R}^d)\}$ with negative energy (that is, $E(u_0) < 0$) lead to solutions which blow up in finite time in both the past and future.

REMARK. Such negative energy initial data do exist. Indeed, if $f \in \Sigma$ is non-zero then $u_0 = \lambda f$ will have negative energy for λ sufficiently large, because the kinetic and potential energies contain different powers of u_0. By the same reasoning, $E(u_0) > 0$ for small initial data.

PROOF. By the local theory discussed in Section 3, the H^1_x norm will remain finite (though not necessarily uniformly bounded!) for as long as the solution exists. Choosing $a(x) = |x|^2$ in (7.6) gives

$$\frac{d}{dt} \int_{\mathbb{R}^d} |x|^2 |u(t,x)|^2 \, dx = 4 \int_{\mathbb{R}^d} \operatorname{Im}\big(\bar{u}(x)\, x \cdot \nabla u(x)\big)\, dx = O(\|\nabla u\|_{L^2_x} \|xu\|_{L^2_x}), \tag{7.10}$$

which shows that the the second moment will also remain finite throughout the lifespan of the solution. More importantly, (7.7) from Lemma 7.2 shows that

$$\frac{d^2}{dt^2} \int_{\mathbb{R}^d} |x|^2 |u(t,x)|^2 \, dx = \int_{\mathbb{R}^d} 8|\nabla u(t,x)|^2 - \tfrac{4pd}{p+2} |u(t,x)|^{p+2} \, dx \tag{7.11}$$

$$= 16E(u_0) - \int_{\mathbb{R}^d} \tfrac{4(pd-4)}{p+2} |u(t,x)|^{p+2} \, dx. \tag{7.12}$$

Thus (using the conservation and negativity of energy) we see that a manifestly positive quantity is trapped beneath an inverted parabola, at least on the lifespan of the solution. This guarantees that the lifespan must be finite in both time directions. □

There are two natural directions to try to extend Theorem 7.6. The first is to weaken the hypothesis $u_0 \in \Sigma$; indeed, it certainly seems reasonable to imagine that the result still holds for negative energy data $u_0 \in H^1_x$. At present this is only known under the additional assumption that u_0 is spherically symmetric; see [**65**] where this is proved for $4/d \leq p < \min\{4, 4/(d-2)\}$ and $d \geq 2$. Secondly, one might hope to take advantage of the second term on the right-hand side of (7.12) to prove finite-time blowup for certain positive energy initial data. This is indeed possible:

EXERCISE ([**38**, Remark 3.14]). Use Theorem 4.4 to prove the following in the energy-critical case: if $E(u_0) < E(W)$ then RHS(7.11) cannot change sign. In particular, if $u_0 \in \Sigma$, $E(u_0) < E(W)$, and RHS(7.11) is negative for u_0, then the solution will blow up in finite time.

Combining this with the argument in [**65**], one may show that if $u_0 \in H^1_{\text{rad}}$, $E(u_0) < E(W)$, and RHS(7.11) is negative for u_0, then the solution will blow up in finite time; for complete details see [**44**]. Analogous arguments in the subcritical case can be found in [**34**].

The first application of Lemma 7.2 to the scattering problem for NLS appears to be [**55**], although the authors freely acknowledge their debt to earlier work on the nonlinear Klein–Gordon equation, [**60, 61**]. This innovation led to considerable developments in the scattering theory for the energy-subcritical (but mass-supercritical) defocusing problem, particularly at the hands of Ginibre and Velo; see [**29**], for example, and the references therein.

The Morawetz identity also played a very important role in the first treatment of the large-data energy-critical problem [**7**]; this was for spherically symmetric data:

PROPOSITION 7.7 (Morawetz à la Bourgain, [**7**]). *Let u be a spherically symmetric solution to the defocusing energy-critical NLS on a spacetime slab $I \times \mathbb{R}^d$. Then, for any $K \geq 1$, we have*

$$\int_I \int_{|x| \leq K|I|^{\frac{1}{2}}} \frac{|u(t,x)|^{\frac{2d}{d-2}}}{|x|}\, dx\, dt \lesssim K|I|^{\frac{1}{2}} E(u). \tag{7.13}$$

In particular, for this NLS there are no solitons or low-to-high cascades, in the sense of Theorem 5.25.

PROOF. The inequality (7.13) follows (with a little work) from Lemma 7.2 with $a(x) := R\psi(\frac{x}{R})$, provided we take $R = K|I|^{1/2}$ and choose $\psi(x)$ to be a spherically symmetric nondecreasing (in radius) function obeying

$$\psi(x) = \begin{cases} |x| & \text{if } |x| \leq 1 \\ \frac{3}{2} & \text{if } |x| \geq 2, \end{cases}$$

which is smooth except at the origin.

We now turn our attention to the second assertion. By Lemma 5.18, we may partition $\mathbb{R}$ into intervals I_j so that for some $t_j \in I_j$ we have $|I_j| \sim_u N(t_j)^{-2}$ and $N(t) \sim_u N(t_j)$ for all $t \in I_j$. Let I be the union of some contiguous sub-collection of the intervals I_j. Then, using almost periodicity, (7.13) implies

$$\int_I N(t)\, dt \lesssim_u |I|^{\frac{1}{2}} E(u). \tag{7.14}$$

This shows that $N(t)$ must go to zero rather quickly; it is certainly inconsistent with the scenarios mentioned in the proposition. □

Bourgain's argument [**7**] was simplified and extended in [**89**], which also obtains a much better spacetime bound. See also [**45**], which incorporates some further simplifications made possible by Lemma A.12.

The papers just referenced do not discuss almost periodic solutions, nor did the extraction of the three enemies (Theorem 5.25) exist at that time. It was however known that solutions with large Strichartz norm must regularly contain bubbles of energy concentration; the natural analogue of $N(t)$ is the reciprocal of the characteristic length scale of these bubbles. Following [**89**], the Morawetz inequality was used roughly as follows: by making the most of (7.14), it is shown that there must be a cascade of bubbles of rapidly changing size in a comparatively small amount of time. This is then contradicted using the almost conservation of mass in finite regions.

With the exception of Theorem 7.6, the applications of Lemma 7.2 that we have discussed so far have discarded the kinetic term $a_{jk}u_j\bar{u}_k$. Indeed, as long as a is a convex function, it will have a favourable sign. By choosing a slightly more convex a, one may exhibit a weighted version of the kinetic energy. This non-linear analogue of local smoothing (cf. Proposition 4.14) has proved valuable in the treatment of the mass-critical NLS, at least, for spherically symmetric data; see [**97**].

EXERCISE (See [**90**, p. 87]). Let u be a solution of (7.4) in three or more dimensions with $V \equiv 0$ and $\mu \geq 0$. By using Lemma 7.2 with $a(x) = \langle x\rangle - \varepsilon\langle x\rangle^{1-\varepsilon}$, show that

$$\int_I \int_{\mathbb{R}^d} |\nabla u(t,x)|^2 \langle x\rangle^{-1-\varepsilon}\, dx\, dt \lesssim \|u\|_{L_t^\infty L_x^2} \|\nabla u\|_{L_t^\infty L_x^2}.$$

In fact, (a further exercise) the right-hand side can be upgraded to $\||\nabla|^{1/2}u\|_{\infty,2}^2$.

The restriction to dimensions three and higher stems from the lack of a good choice for a in one and two dimensions, that is, of a convex a with a_k bounded and $-\Delta\Delta a$ positive.

7.3. Interaction Morawetz. The weight appearing in (7.13) is strongly tied to the case of spherically symmetric data. In [**19**], a variant of the Morawetz identity was introduced that is better adapted to the treatment of general (not spherically symmetric) data. This is the topic of this subsection.

One of the early applications of the new monotonicity formula was to the proof of global well-posedness and scattering for the three dimensional energy-critical defocusing nonlinear Schrödinger equation, [**20**]. This argument was subsequently adapted to four dimensions, [**75**], and then to dimensions five and higher, [**103, 104**].

In the papers just mentioned, it was necessary to introduce a frequency cutoff; this means that one needs to consider solutions to an inhomogeneous NLS:

$$iu_t = -\Delta u + |u|^p u + F, \tag{7.15}$$

where F is some function of space and time. Note that we limit ourselves to the defocusing case, since this is where the interaction Morawetz identity has proved most useful.

Beginning with (7.15), a few elementary computations reveal

$$\partial_t |u|^2 = -2\operatorname{Im}(u_k\bar u)_k + 2\operatorname{Im}(F\bar u) \tag{7.16}$$

$$\partial_t 2\operatorname{Im}(u_k\bar u) = \Delta\big(|u|^2\big)_k - 4\operatorname{Re}\big(\bar u_k u_j\big)_j - \tfrac{2p}{p+2}\big(|u|^{p+2}\big)_k + 2\operatorname{Re}\big(u_k\bar F - F_k\bar u\big). \tag{7.17}$$

As in the previous subsection, subscripts denote spatial derivatives and repeated indices are summed.

Proposition 7.8 (Interaction Morawetz, [**19**]). *If u obeys* (7.15) *and*

$$M(t) := 2\iint_{\mathbb{R}^d\times\mathbb{R}^d} |u(y)|^2 a_k(x-y)\operatorname{Im}\{u_k(x)\bar u(x)\}\, dx\, dy, \tag{7.18}$$

for some even convex function $a:\mathbb{R}^d\to\mathbb{R}$, then

$$\begin{aligned}\partial_t M(t) \geq \iint_{\mathbb{R}^d\times\mathbb{R}^d} \Big\{ &- a_{kkjj}(x-y)\,|u(y)|^2|u(x)|^2 + \tfrac{2p}{p+2} a_{kk}(x-y)|u(x)|^{p+2}|u(y)|^2 \\ &+ 2a_k(x-y)|u(y)|^2\operatorname{Re}\big[u_k(x)\bar F(x) - F_k(x)\bar u(x)\big] \\ &+ 4a_k(x-y)(\operatorname{Im} F(y)\bar u(y))(\operatorname{Im} u_k(x)\bar u(x))\Big\}\, dx\, dy.\end{aligned} \tag{7.19}$$

Proof. Patient computation shows that with the addition of one term, (7.19) would become an equality. In this way, one sees that the claim is equivalent to

$$4\iint_{\mathbb{R}^d\times\mathbb{R}^d} a_{jk}(x-y)\big[|u(y)|^2\bar u_j(x)u_k(x) - (\operatorname{Im}\bar u(y)u_j(y))(\operatorname{Im}\bar u(x)u_k(x))\big]\, dx\, dy \geq 0,$$

which is what we will explain here.

Fix x and y. As a is convex, the matrix $a_{jk}(x-y)$ is positive semi-definite. Now suppose e is one of the eigenvectors of this matrix. By elementary considerations,

$$\begin{aligned}\big|e_k e_j(\operatorname{Im}\bar u(y)u_j(y))(\operatorname{Im}\bar u(x)u_k(x))\big| &\leq |u(y)|\,|e\cdot\nabla u(y)|\,|u(x)|\,|e\cdot\nabla u(x)|\\ &\leq \tfrac12|u(x)|^2|e\cdot\nabla u(y)|^2+\tfrac12|u(y)|^2|e\cdot\nabla u(x)|^2.\end{aligned}$$

Writing out $a_{jk}(x-y)$ in terms of its eigenvalues and vectors, this shows that the integrand is indeed non-negative, at least, after symmetrization under $x\leftrightarrow y$. □

EXERCISE (See [**19**]). Show that for $d=3$ and $a(x)=|x|$, Lemma 7.8 implies

$$\int_{\mathbb{R}}\int_{\mathbb{R}^3}|u(t,x)|^4\,dx\,dt \lesssim \|u\|^3_{L^\infty_t L^2_x}\|\nabla u\|_{L^\infty_t L^2_x}$$

for solutions of (7.15) with $F\equiv 0$.

In dimensions $d\geq 4$, there is an analogous result although the left-hand side takes a much less simple form. Nevertheless, it allows one to deduce the following:

PROPOSITION 7.9. *For $d\geq 3$ and $F\equiv 0$, any solution to* (7.15) *obeys*

$$\|u\|_{L^{d+1}_t L_x^{\frac{2(d+1)}{d-1}}(I\times\mathbb{R}^d)} \lesssim \|u\|_{L^\infty_t H^{\frac12}_x(I\times\mathbb{R}^d)}.$$

As noted above, this is in [**19**] when $d=3$. For $d\geq 4$, the result appears as [**95**, Proposition 5.1]; see also [**103**, §5], which uses the same ideas. One application of this lemma given in [**95**] is a simplified proof of scattering for defocusing inter-critical NLS. The original proof by Ginibre and Velo, [**29**], used the standard (Lin–Strauss) Morawetz identity.

As noted at the end of the previous section, there are some difficulties in using the standard Morawetz estimate in one and two dimensions. Some of these difficulties can be alleviated by switching to the interaction Morawetz estimate. See for instance [**67**]. There is also a four-particle interaction Morawetz that has proved effective in the one-dimensional setting:

PROPOSITION 7.10 ([**18**, Proposition 3.1]). *Let u be a solution to a defocusing NLS in one space dimension, then*

$$\int_I\int_{\mathbb{R}}|u(t,x)|^8\,dx\,dt \lesssim \|u\|^2_{L^\infty_t \dot H^{1/2}_x(I\times\mathbb{R})}\|u_0\|^6_{L^\infty_t L^2_x(I\times\mathbb{R})}. \tag{7.20}$$

For a recent review of interaction Morawetz inequalities and their application to the scattering problem for inter-critical NLS see [**30**].

8. Nihilism

In this section we use conservation laws and monotonicity formulae to preclude the global enemies described in Theorems 5.24 and 5.25, provided that these enemies obey additional regularity/decay. More precisely, we show how to dispense with soliton and frequency cascade solutions that belong to $L^\infty_t H^{1+\varepsilon}_x$ for some $\varepsilon>0$ in the mass-critical case or to $L^\infty_t\dot H^{-\varepsilon}_x$ in the energy-critical case. Recall that in the mass-critical case, the spherically symmetric soliton and cascade were shown to enjoy such additional regularity in [**43, 46**] for $d\geq 2$. For the energy-critical NLS, Theorem 6.8 established the decay needed in dimensions $d\geq 5$.

We remind the reader that enemies which are not global, that is, the self-similar solution (in the mass-critical case) or the finite-time blowup solution (in the energy-critical case) can be precluded via more direct techniques. In the former case it is sufficient to prove $u(t) \in H_x^1$ for some $t \in (0, \infty)$, since then the global theory for H_x^1 initial data leads to a contradiction. Theorem 6.1 establishes this for spherically symmetric initial data and $d \geq 2$.

For the energy-critical NLS, finite-time blowup solutions (as described in Theorem 5.25) were precluded in Theorem 6.7 for all dimensions $d \geq 3$.

8.1. Frequency cascade solutions. We first turn our attention to high-to-low frequency cascade solutions of the mass-critical NLS (cf. Theorem 5.24). We will show that no such solutions may belong to $L_t^\infty H_x^{1+\varepsilon}$ for some $\varepsilon > 0$. We would like to point out that regularity above H_x^1 is needed for the argument we present below.

THEOREM 8.1 (Absence of mass-critical cascades). *Let $d \geq 1$. There are no non-zero global solutions to* (1.4) *which are double high-to-low frequency cascades in the sense of Theorem* 5.24 *and which obey $u \in L_t^\infty H_x^{1+\varepsilon}$ for some $\varepsilon = \varepsilon(d) > 0$.*

PROOF. Suppose to the contrary that there is such a solution u. Using a Galilean transformation, we may set its momentum equal to zero, that is,

$$\int_{\mathbb{R}^d} \xi |\hat{u}(t, \xi)|^2 \, d\xi = 0.$$

Note that u remains in $L_t^\infty H_x^{1+\varepsilon}$.

By hypothesis $u \in L_t^\infty H_x^1$ and so the energy

$$E(u) = E(u(t)) = \int_{\mathbb{R}^d} \tfrac{1}{2} |\nabla u(t, x)|^2 + \mu \tfrac{d}{2(d+2)} |u(t, x)|^{\frac{2(d+2)}{d}} \, dx$$

is finite and conserved. Moreover, as $M(u) < M(Q)$ in the focusing case, the sharp Gagliardo-Nirenberg inequality gives

$$\|\nabla u(t)\|_{L_x^2(\mathbb{R}^d)}^2 \sim_u E(u) \sim_u 1 \tag{8.1}$$

for all $t \in \mathbb{R}$. We will now reach a contradiction by proving that $\|\nabla u(t)\|_2 \to 0$ along any sequence where $N(t) \to 0$. The existence of two such time sequences is guaranteed by the fact that u is a double high-to-low frequency cascade.

Let $\eta > 0$ be arbitrary. By Definition 5.1, we can find $C(\eta) > 0$ such that

$$\int_{|\xi - \xi(t)| \geq C(\eta) N(t)} |\hat{u}(t, \xi)|^2 \, d\xi \leq \eta^2$$

for all t. Meanwhile, by hypothesis, $u \in L_t^\infty H_x^{1+\varepsilon}(\mathbb{R} \times \mathbb{R}^d)$ for some $\varepsilon > 0$. Thus,

$$\int_{\mathbb{R}^d} |\xi|^{2+2\varepsilon} |\hat{u}(t, \xi)|^2 \, d\xi \lesssim_u 1$$

for all t. Therefore, combining the two estimates gives

$$\int_{|\xi - \xi(t)| \geq C(\eta) N(t)} |\xi|^2 |\hat{u}(t, \xi)|^2 \, d\xi \lesssim_u \eta^{\frac{2\varepsilon}{1+\varepsilon}}.$$

On the other hand, from mass conservation and Plancherel's theorem we have

$$\int_{|\xi - \xi(t)| \leq C(\eta) N(t)} |\xi|^2 |\hat{u}(t, \xi)|^2 \, d\xi \lesssim_u \big[C(\eta) N(t) + |\xi(t)| \big]^2.$$

Summing these last two bounds and using Plancherel's theorem again, we obtain

$$\|\nabla u(t)\|_{L_x^2(\mathbb{R}^d)} \lesssim_u \eta^{\frac{\varepsilon}{1+\varepsilon}} + C(\eta) N(t) + |\xi(t)|$$

for all t. As u is a double high-to-low frequency cascade, there exists a sequence of times $t_n \to \infty$ such that $N(t_n) \to 0$. As $\eta > 0$ is arbitrary, it remains to prove that $|\xi(t_n)| \to 0$ as $n \to \infty$ in order to deduce $\|\nabla u(t_n)\|_2 \to 0$, which would contradict (8.1), thus concluding the proof of the theorem.

To see that $|\xi(t_n)| \to 0$ as $n \to \infty$ we use mass conservation, the uniform $H_x^{1/2+\varepsilon}$ bound for some $\varepsilon > 0$, and the fact that $N(t_n) \to 0$, together with the vanishing of the total momentum of u. □

We now turn our attention to the energy-critical NLS and preclude low-to-high frequency cascade solutions belonging to $L_t^\infty \dot H_x^{-\varepsilon}$ for some $\varepsilon > 0$.

THEOREM 8.2 (Absence of energy-critical cascades). *Let $d \geq 3$. There are no non-zero global solutions to* (1.6) *that are low-to-high frequency cascades in the sense of Theorem* 5.25 *and that belong to $L_t^\infty \dot H_x^{-\varepsilon}$ for some $\varepsilon > 0$.*

PROOF. Suppose for a contradiction that there existed such a solution u. Then by hypothesis, $u \in L_t^\infty L_x^2$; thus, by the conservation of mass,

$$0 < M(u) = M(u(t)) = \int_{\mathbb{R}^d} |u(t,x)|^2\, dx < \infty \quad \text{for all} \quad t \in \mathbb{R}. \tag{8.2}$$

Let $\eta > 0$ be a small constant. By almost periodicity modulo symmetries, there exists $c(\eta) > 0$ such that

$$\int_{|\xi| \leq c(\eta) N(t)} |\xi|^2 |\hat u(t,\xi)|^2\, d\xi \leq \eta^2$$

for all $t \in \mathbb{R}$. On the other hand, as $u \in L_t^\infty \dot H_x^{-\varepsilon}$ for some $\varepsilon > 0$,

$$\int_{|\xi| \leq c(\eta) N(t)} |\xi|^{-2\varepsilon} |\hat u(t,\xi)|^2\, d\xi \lesssim_u 1$$

for all $t \in \mathbb{R}$. Hence, by Hölder's inequality,

$$\int_{|\xi| \leq c(\eta) N(t)} |\hat u(t,\xi)|^2\, d\xi \lesssim_u \eta^{\frac{2\varepsilon}{1+\varepsilon}} \quad \text{for all} \quad t \in \mathbb{R}. \tag{8.3}$$

Meanwhile, by elementary considerations and recalling that u has uniformly bounded kinetic energy,

$$\int_{|\xi| \geq c(\eta) N(t)} |\hat u(t,\xi)|^2\, d\xi \leq [c(\eta)N(t)]^{-2} \int_{\mathbb{R}^d} |\xi|^2 |\hat u(t,\xi)|^2\, d\xi \lesssim_u [c(\eta)N(t)]^{-2}. \tag{8.4}$$

Collecting (8.3) and (8.4) and using Plancherel's theorem, we obtain

$$0 \leq M(u) \lesssim_u c(\eta)^{-2} N(t)^{-2} + \eta^{\frac{2\varepsilon}{1+\varepsilon}}$$

for all $t \in \mathbb{R}$. As u is a low-to-high cascade, there is a sequence of times $t_n \to \infty$ so that $N(t_n) \to \infty$. As $\eta > 0$ is arbitrary, we conclude $M(u) = 0$ and hence u is identically zero. This contradicts (8.2). □

8.2. Fall of the soliton solutions. We now turn our attention to soliton-like solutions to the mass- and energy-critical NLS as described in Theorem 5.24 and 5.25 and preclude those which obey additional regularity/decay. In the defocusing case, this can be achieved using the interaction Morawetz inequality given in Proposition 7.9. We leave the precise details to the reader, noting only that the assumed regularity/decay allow one to bound the right-hand side.

In order to treat the focusing problem, we need to rely on the virial identity, which is much more closely wedded to $x = 0$. This requires us to control the motion of $x(t)$, which we do next using an argument from [**23**]. This step can be skipped over in the case of spherically symmetric initial data, since then one may take $x(t) \equiv 0$.

LEMMA 8.3 (Control over $x(t)$). *Suppose there is an $L_t^\infty H_x^1$ soliton-like solution to the mass-critical NLS in the sense of Theorem 5.24. Then there exists a solution u with all these properties that additionally obeys*

$$|x(t)| = o(t) \quad \text{as} \quad t \to \infty.$$

Similarly, if u is a is a minimal kinetic energy soliton-like solution to the energy-critical NLS in the sense of Theorem 5.25 that belongs to $L_t^\infty \dot{H}_x^{-\varepsilon}$ for some $\varepsilon > 0$, then the same conclusion holds.

PROOF. We will prove the claim for soliton-like solutions to the energy-critical NLS and leave the mass-critical case as an exercise.

We argue by contradiction. Suppose there exist $\delta > 0$ and a sequence $t_n \to \infty$ such that

$$|x(t_n)| > \delta t_n \quad \text{for all} \quad n \geq 1. \tag{8.5}$$

By spatial-translation symmetry, we may assume $x(0) = 0$.

Let $\eta > 0$ be a small constant to be chosen later. By the almost periodicity of u and Lemma 6.13, there exists $C(\eta) > 0$ such that

$$\sup_{t\in\mathbb{R}} \int_{|x-x(t)|>C(\eta)} \big(|\nabla u(t,x)|^2 + |u(t,x)|^2\big)\, dx \leq \eta. \tag{8.6}$$

Define

$$T_n := \inf\{t \in [0, t_n] : |x(t)| = |x(t_n)|\} \leq t_n \text{ and } R_n := C(\eta) + \sup_{t\in[0,T_n]} |x(t)|. \tag{8.7}$$

Now let ϕ be a smooth, radial function such that

$$\phi(r) = \begin{cases} 1 & \text{for } r \leq 1 \\ 0 & \text{for } r \geq 2, \end{cases}$$

and define the truncated 'position'

$$X_R(t) := \int_{\mathbb{R}^d} x\phi\big(\tfrac{|x|}{R}\big)|u(t,x)|^2\, dx.$$

By hypothesis, $u \in L_t^\infty L_x^2$; together with (8.6) this implies

$$\begin{aligned} |X_{R_n}(0)| &\leq \Big|\int_{|x|\leq C(\eta)} x\phi\big(\tfrac{|x|}{R_n}\big)|u(0,x)|^2\, dx\Big| + \Big|\int_{|x|\geq C(\eta)} x\phi\big(\tfrac{|x|}{R_n}\big)|u(0,x)|^2\, dx\Big| \\ &\leq C(\eta)M(u) + 2\eta R_n. \end{aligned}$$

On the other hand, by the triangle inequality combined with (8.6) and (8.7),

$$\begin{aligned} |X_{R_n}(T_n)| \geq |x(T_n)|M(u) - |x(T_n)|\Big|\int_{\mathbb{R}^d} \Big[1 - \phi\big(\tfrac{|x|}{R_n}\big)\Big]|u(T_n,x)|^2\, dx\Big| \\ - \Big|\int_{|x-x(T_n)|\leq C(\eta)} [x - x(T_n)]\phi\big(\tfrac{|x|}{R_n}\big)|u(T_n,x)|^2\, dx\Big| \end{aligned}$$

$$- \Big| \int_{|x-x(T_n)| \geq C(\eta)} [x - x(T_n)] \phi\big(\tfrac{|x|}{R_n}\big) |u(T_n, x)|^2 \, dx \Big|$$
$$\geq |x(T_n)| [M(u) - \eta] - C(\eta) M(u) - \eta [2R_n + |x(T_n)|]$$
$$\geq |x(T_n)| [M(u) - 4\eta] - 3C(\eta) M(u).$$

Thus, taking $\eta > 0$ sufficiently small (depending on $M(u)$),

$$\big| X_{R_n}(T_n) - X_{R_n}(0) \big| \gtrsim_{M(u)} |x(T_n)| - C(\eta).$$

A simple computation establishes

$$\partial_t X_R(t) = 2 \operatorname{Im} \int_{\mathbb{R}^d} \phi\big(\tfrac{|x|}{R}\big) \nabla u(t,x) \overline{u(t,x)} \, dx$$
$$+ 2 \operatorname{Im} \int_{\mathbb{R}^d} \frac{x}{|x| R} \phi'\big(\tfrac{|x|}{R}\big) \, x \cdot \nabla u(t,x) \overline{u(t,x)} \, dx.$$

As a minimal kinetic energy blowup solution must have zero momentum (see Corollary 2.4), using Cauchy-Schwarz and (8.6) we obtain

$$\big| \partial_t X_{R_n}(t) \big| \leq \Big| 2 \operatorname{Im} \int_{\mathbb{R}^d} \Big[1 - \phi\big(\tfrac{|x|}{R_n}\big) \Big] \nabla u(t,x) \overline{u(t,x)} \, dx \Big|$$
$$+ \Big| 2 \operatorname{Im} \int_{\mathbb{R}^d} \frac{x}{|x| R} \phi'\big(\tfrac{|x|}{R_n}\big) \, x \cdot \nabla u(t,x) \overline{u(t,x)} \, dx \Big|$$
$$\leq 6\eta$$

for all $t \in [0, T_n]$.

Thus, by the Fundamental Theorem of Calculus,

$$|x(T_n)| - C(\eta) \lesssim_{M(u)} \eta T_n.$$

Recalling that $|x(T_n)| = |x(t_n)| > \delta t_n \geq \delta T_n$ and letting $n \to \infty$ we derive a contradiction. □

We are finally in a position to preclude our last enemies.

THEOREM 8.4 (No solitons). *There are no solutions to the mass-critical NLS that are solitons in the sense of Theorem 5.24 and that belong to $L_t^\infty H_x^{1+\varepsilon}$ for some $\varepsilon > 0$. Similarly, there are no solutions to the energy-critical NLS that are solitons in the sense of Theorem 5.25 and that belong to $L_t^\infty \dot{H}_x^{-\varepsilon}$ for some $\varepsilon > 0$.*

PROOF. We only prove the claim for the mass-critical NLS and leave the energy-critical case as exercise. Suppose for a contradiction that there existed such a solution u.

Let $\eta > 0$ be a small constant to be specified later. Then, by Definition 5.1 and Lemma 6.12 there exists $C(\eta) > 0$ such that

$$\sup_{t \in \mathbb{R}} \int_{|x - x(t)| > C(\eta)} \big(|u(t,x)|^2 + |\nabla u(t,x)|^2 \big) \, dx \leq \eta. \tag{8.8}$$

Moreover, by Lemma 8.3, $|x(t)| = o(t)$ as $t \to \infty$. Thus, there exists $T_0 = T_0(\eta) \in \mathbb{R}$ such that

$$|x(t)| \leq \eta t \quad \text{for all} \quad t \geq T_0. \tag{8.9}$$

Now let ϕ be a smooth, radial function such that

$$\phi(r) = \begin{cases} r & \text{for } r \leq 1 \\ 0 & \text{for } r \geq 2, \end{cases}$$

and define

$$V_R(t) := \int_{\mathbb{R}^d} a(x)|u(t,x)|^2\,dx,$$

where $a(x) := R^2\phi(\frac{|x|^2}{R^2})$ for some $R > 0$.

Differentiating V_R with respect to the time variable, we find

$$\partial_t V_R(t) = 4\,\mathrm{Im}\int_{\mathbb{R}^d} \phi'\bigl(\tfrac{|x|^2}{R^2}\bigr)\overline{u(t,x)}\; x\cdot\nabla u(t,x)\,dx.$$

as in (7.6). By hypothesis $u \in L_t^\infty H_x^1$ and so we obtain

$$|\partial_t V_R(t)| \lesssim R\|\nabla u(t)\|_2\|u(t)\|_2 \lesssim_u R \tag{8.10}$$

for all $t \in \mathbb{R}$ and $R > 0$.

Further, using (7.7) for our specific choice of a, we find

$$\begin{aligned}\partial_{tt}V_R(t) = 16E(u) &+ O\biggl(\frac{1}{R^2}\int_{|x|\geq R}|u(t,x)|^2\,dx\biggr)\\ &+ O\biggl(\int_{|x|\geq R}\Bigl[|\nabla u(t,x)|^2 + |u(t,x)|^{\frac{2(d+2)}{d}}\Bigr]\,dx\biggr).\end{aligned}$$

Recall that in the focusing case, $M(u) < M(Q)$. As a consequence, the sharp Gagliardo–Nirenberg inequality implies that the energy is a positive quantity in the focusing case as well as in the defocusing case. Indeed,

$$E(u) \gtrsim_u \int_{\mathbb{R}^d} |\nabla u(t,x)|^2\,dx > 0.$$

Thus, choosing $\eta > 0$ sufficiently small and $R := C(\eta) + \sup_{T_0\leq t\leq T_1}|x(t)|$ and invoking (8.8), we obtain

$$\partial_{tt}V_R(t) \geq 8E(u) > 0. \tag{8.11}$$

Using the Fundamental Theorem of Calculus on the interval $[T_0, T_1]$ together with (8.10) and (8.11), we obtain

$$(T_1 - T_0)E(u) \lesssim_u R \lesssim_u C(\eta) + \sup_{T_0\leq t\leq T_1}|x(t)|$$

for all $T_1 \geq T_0$. Invoking (8.9) and taking η sufficiently small and then T_1 sufficiently large, we derive a contradiction to $E(u) > 0$. □

Appendix A. Background material

A.1. Compactness in L^p. Recall that a family of continuous functions on a compact set $K \subset \mathbb{R}^d$ is precompact in $C^0(K)$ if and only if it is uniformly bounded and equicontinuous. This is the Arzelà–Ascoli theorem. The natural generalization to L^p spaces is due to M. Riesz [**72**] and reads as follows:

PROPOSITION A.1. *Fix $1 \leq p < \infty$. A family of functions $\mathcal{F} \subset L^p(\mathbb{R}^d)$ is precompact in this topology if and only if it obeys the following three conditions:*
(i) *There exists $A > 0$ so that $\|f\|_p \leq A$ for all $f \in \mathcal{F}$.*
(ii) *For any $\varepsilon > 0$ there exists $\delta > 0$ so that $\int_{\mathbb{R}^d}|f(x) - f(x+y)|^p\,dx < \varepsilon$ for all $f \in \mathcal{F}$ and all $|y| < \delta$.*
(iii) *For any $\varepsilon > 0$ there exists R so that $\int_{|x|\geq R}|f|^p\,dx < \varepsilon$ for all $f \in \mathcal{F}$.*

REMARK. By analogy to the case of continuous functions (or of measures) it is natural to refer to the three conditions as uniform boundedness, equicontinuity, and tightness, respectively.

PROOF. If $\mathcal{F}$ is precompact, it may be covered by balls of radius $\frac{1}{2}\varepsilon$ around a finite collection of functions, $\{f_j\}$. As any single function obeys (i)–(iii), these properties can be extended to the whole family by approximation by an f_j.

We now turn to sufficiency. Given $\varepsilon > 0$, our job is to show that there are finitely many functions $\{f_j\}$ so that the ε-balls centered at these points cover $\mathcal{F}$. We will find these points via the usual Arzelà–Ascoli theorem, which requires us to approximate $\mathcal{F}$ by a family of continuous functions of compact support. Let $\phi : \mathbb{R}^d \to [0, \infty)$ be a smooth function supported by $\{|x| \leq 1\}$ with $\phi(x) = 1$ in a neighbourhood of $x = 0$ and $\int_{\mathbb{R}^d} \phi(x)\, dx = 1$. Given $R > 0$ we define

$$f_R(x) := \phi\big(\tfrac{x}{R}\big) \int_{\mathbb{R}^d} R^d \phi\big(R(x-y)\big) f(y)\, dy$$

and write $\mathcal{F}_R := \{f_R : f \in \mathcal{F}\}$. Employing the three conditions, we see that it is possible to choose R so large that $\|f - f_R\|_p < \frac{1}{2}\varepsilon$ for all $f \in \mathcal{F}$. We also see that $\mathcal{F}_R$ is a uniformly bounded family of equicontinuous functions on the compact set $\{|x| \leq R\}$. Thus, $\mathcal{F}_R$ is precompact and we may find a finite family $\{f_j\} \subseteq C^0(\{|x| \leq R\})$ so that $\mathcal{F}_R$ is covered by the L^p-balls of radius $\frac{1}{2}\varepsilon$ around these points. By construction, the ε-balls around these points cover $\mathcal{F}$. □

In the L^2 case it is natural to replace (ii) by a condition on the Fourier transform:

COROLLARY A.2. *A family of functions is precompact in $L^2(\mathbb{R}^d)$ if and only if it obeys the following two conditions:*

(i) *There exists $A > 0$ so that $\|f\| \leq A$ for all $f \in \mathcal{F}$.*

(ii) *For all $\varepsilon > 0$ there exists $R > 0$ so that $\int_{|x|\geq R} |f(x)|^2\, dx + \int_{|\xi|\geq R} |\hat{f}(\xi)|^2\, d\xi < \varepsilon$ for all $f \in \mathcal{F}$.*

PROOF. Necessity follows as before. Regarding the sufficiency of these conditions, we note that

$$\int_{\mathbb{R}^d} |f(x+y) - f(x)|^2\, dx \sim \int_{\mathbb{R}^d} |e^{i\xi y} - 1|^2 |\hat{f}(\xi)|^2\, d\xi,$$

which allows us to rely on the preceding proposition. □

As well as being useful in the treatment of NLS with spherically symmetric data, the following allows one to obtain tightness in the proof of Lemma A.4.

LEMMA A.3 (Weighted radial Sobolev embedding). *Let $f \in H^1_x(\mathbb{R}^d)$ be spherically symmetric. Suppose $\omega : [0, \infty) \to [0, 1]$ obeys $0 \leq \omega(r) \leq C\omega(\rho)$ whenever $r < \rho$. Then*

$$\big| |x|^{\frac{d-1}{2}} \omega(|x|) f(x) \big|^2 \lesssim_d C^2 \|f\|_{L^2_x(\mathbb{R}^d)} \|\omega^2 \nabla f\|_{L^2_x(\mathbb{R}^d)}$$

for all $x \in \mathbb{R}^d$.

PROOF. It suffices to establish the claim for spherically symmetric Schwartz functions f, which we write as functions of radius alone. Let $r \geq 0$. By the

Fundamental Theorem of Calculus and the Cauchy–Schwarz inequality,

$$\begin{aligned} r^{d-1}\omega(r)^2|f(r)|^2 &= 2r^{d-1}\omega(r)^2 \operatorname{Re}\int_r^\infty \bar{f}(\rho)f'(\rho)\,d\rho \\ &\le 2C^2\int_r^\infty \rho^{d-1}\omega(\rho)^2|f(\rho)|\,|f'(\rho)|\,d\rho \\ &\le 2C^2\Big(\int_r^\infty \rho^{d-1}|f(\rho)|^2\,d\rho\Big)^{\frac12}\Big(\int_r^\infty \rho^{d-1}\omega(\rho)^4|f'(\rho)|^2\,d\rho\Big)^{\frac12} \\ &\le 2C^2\|f\|_{L^2(\rho^{d-1}d\rho)}\|\omega^2 f'\|_{L^2(\rho^{d-1}d\rho)}, \end{aligned}$$

from which the claim follows. □

LEMMA A.4 (Compactness in spherically symmetric Gagliardo–Nirenberg). *The embedding* $H^1_{rad}(\mathbb{R}^d) \hookrightarrow L^p(\mathbb{R}^d)$ *is compact for* $d \ge 2$ *and* $2 < p < \frac{2d}{d-2}$.

PROOF. Exercise. □

Our last lemma for this subsection is not strictly a compactness statement; however, it is very helpful to us in some places where we rely on weak-$*$ compactness. Recall that under weak-$*$ limits, the norm may jump down (i.e., the norm is weak-$*$ lower semicontinuous). The question is, by how much? As we have seen in Subsection 4.2, this has a very satisfactory answer in Hilbert space (cf. (4.22)), but less so in other L^p spaces.

In our applications, regularity allows us to upgrade weak-$*$ convergence to almost everywhere convergence. The lower semicontinuity of the norm under this notion of convergence is essentially Fatou's lemma. The following quantitative version of this is due to Brézis and Lieb [**10**] (see also [**54**, Theorem 1.9]):

LEMMA A.5 (Refined Fatou). *Suppose* $\{f_n\} \subseteq L^p_x(\mathbb{R}^d)$ *with* $\limsup \|f_n\|_p < \infty$. *If* $f_n \to f$ *almost everywhere, then*

$$\int_{\mathbb{R}^d} \Big| |f_n|^p - |f_n - f|^p - |f|^p \Big|\,dx \to 0.$$

In particular, $\|f_n\|_p^p - \|f_n - f\|_p^p \to \|f\|_p^p$.

A.2. Littlewood–Paley theory. Let $\varphi(\xi)$ be a radial bump function supported in the ball $\{\xi \in \mathbb{R}^d : |\xi| \le \frac{11}{10}\}$ and equal to 1 on the ball $\{\xi \in \mathbb{R}^d : |\xi| \le 1\}$. For each number $N > 0$, we define the Fourier multipliers

$$\begin{aligned} \widehat{P_{\le N} f}(\xi) &:= \varphi(\xi/N)\hat{f}(\xi) \\ \widehat{P_{> N} f}(\xi) &:= (1 - \varphi(\xi/N))\hat{f}(\xi) \\ \widehat{P_N f}(\xi) &:= (\varphi(\xi/N) - \varphi(2\xi/N))\hat{f}(\xi) \end{aligned}$$

and similarly $P_{<N}$ and $P_{\ge N}$. We also define

$$P_{M<\cdot\le N} := P_{\le N} - P_{\le M} = \sum_{M<N'\le N} P_{N'}$$

whenever $M < N$. We will usually use these multipliers when M and N are *dyadic numbers* (that is, of the form 2^n for some integer n); in particular, all summations over N or M are understood to be over dyadic numbers. Nevertheless, it will occasionally be convenient to allow M and N to not be a power of 2.

Like all Fourier multipliers, the Littlewood-Paley operators commute with the propagator $e^{it\Delta}$, as well as with differential operators such as $i\partial_t + \Delta$. We will use basic properties of these operators many many times, including

LEMMA A.6 (Bernstein estimates). *For* $1 \le p \le q \le \infty$,

$$\||\nabla|^{\pm s} P_N f\|_{L_x^p(\mathbb{R}^d)} \sim N^{\pm s} \|P_N f\|_{L_x^p(\mathbb{R}^d)},$$

$$\|P_{\le N} f\|_{L_x^q(\mathbb{R}^d)} \lesssim N^{\frac{d}{p}-\frac{d}{q}} \|P_{\le N} f\|_{L_x^p(\mathbb{R}^d)},$$

$$\|P_N f\|_{L_x^q(\mathbb{R}^d)} \lesssim N^{\frac{d}{p}-\frac{d}{q}} \|P_N f\|_{L_x^p(\mathbb{R}^d)}.$$

LEMMA A.7 (Square function estimates). *Given a Schwartz function f, let*

$$S(f)(x) := \Big(\sum |P_N f(x)|^2\Big)^{1/2},$$

which is known as the Littlewood–Paley square function. For $1 < p < \infty$,

$$\|S(f)\|_{L_x^p} \sim \|f\|_{L_x^p}.$$

Our next estimate is a weak form of square function estimate that does not require the same amount of sparseness of the Fourier supports. We first saw this estimate as [**93**, Lemma 6.1]. While it is formulated there for rectangles, we prefer to state it for parallepipeds. It makes the proof no more involved, but reduces the amount of arithmetic required when we actually use it.

DEFINITION A.8. A *parallelepiped* in $\mathbb{R}^d$ is a set of the form

$$R = \{Ax + c : x \in [-\tfrac{1}{2}, \tfrac{1}{2}]^d\},$$

where $A \in GL_d(\mathbb{R})$ and $c \in \mathbb{R}^d$. The variable $c = c(R)$ denotes the center of R. Given $\alpha \in (0, \infty)$, we write αR or α*-dilate of* R to refer to the parallelpiped formed from R by replacing A by αA.

Let us adopt a uniform notion of smoothed Fourier restriction operator to a parallelepiped, since we will need it in the proof below. Given $\alpha > 1$, fix a non-negative $\psi \in C_c^\infty(\mathbb{R}^d)$ with

$$\psi(x) = 1 \quad \text{for all } x \in [-\tfrac{1}{2}, \tfrac{1}{2}]^d \quad \text{and} \quad \operatorname{supp}(\psi) \subseteq [-\tfrac{\alpha}{2}, \tfrac{\alpha}{2}]^d.$$

With this fixed, we define P_R by

$$[P_R f]\hat{}(\xi) = \psi\big(A^{-1}(\xi - c)\big)\hat{f}(\xi),$$

or equivalently, by

$$P_R f = K_R * f \quad \text{where} \quad K_R(x) = |\det(A)|\, e^{ix\cdot c}\, \hat{\psi}(A^T x). \tag{A.1}$$

Here A and c are the matrix and vector used to define R. In particular, we note that

$$\int_{\mathbb{R}^d} |K_R(x)|\, dx \lesssim 1 \quad \text{uniformly in } R.$$

LEMMA A.9. *Let $\{R_k\}$ be a family of parallelpipeds in $\mathbb{R}^d$ obeying*

$$\sup_\xi \sum \chi_{\alpha R_k}(\xi) \lesssim 1$$

for some $\alpha > 1$. Fix $1 \le p \le 2$. Then

$$\Big\|\sum P_{R_k} f_k\Big\|_{L_x^p(\mathbb{R}^d)}^p \lesssim \sum \|f_k\|_{L_x^p(\mathbb{R}^d)}^p$$

for any $\{f_k\} \subseteq L_x^p(\mathbb{R}^d)$.

PROOF. When $p = 2$, the result follows from Plancherel's Theorem; when $p = 1$, it follows from the triangle inequality. The remaining cases can then be obtained by interpolation. □

REMARK. The case $2 < p \leq \infty$ is also discussed in [**93**]; in this case, the estimate reads

$$\big\|\sum P_{R_k} f_k\big\|_{L_x^p(\mathbb{R}^d)}^{p'} \lesssim \sum \|f_k\|_{L_x^p(\mathbb{R}^d)}^{p'} \tag{A.2}$$

and the proof is essentially the same. For such p, one can actually recover the full square function estimate; see [**35, 74**].

A.3. Fractional calculus.

LEMMA A.10 (Product rule, [**16**]). *Let $s \in (0,1]$ and $1 < r, p_1, p_2, q_1, q_2 < \infty$ such that $\frac{1}{r} = \frac{1}{p_i} + \frac{1}{q_i}$ for $i = 1, 2$. Then,*

$$\big\||\nabla|^s(fg)\big\|_r \lesssim \|f\|_{p_1}\big\||\nabla|^s g\big\|_{q_1} + \big\||\nabla|^s f\big\|_{p_2}\|g\|_{q_2}.$$

We will also need the following fractional chain rule from [**16**]. For a textbook treatment, see [**98**, §2.4].

LEMMA A.11 (Fractional chain rule, [**16**]). *Suppose $G \in C^1(\mathbb{C})$, $s \in (0,1]$, and $1 < p, p_1, p_2 < \infty$ are such that $\frac{1}{p} = \frac{1}{p_1} + \frac{1}{p_2}$. Then,*

$$\||\nabla|^s G(u)\|_p \lesssim \|G'(u)\|_{p_1}\||\nabla|^s u\|_{p_2}.$$

When the function G is no longer C^1, but merely Hölder continuous, we have the following chain rule:

LEMMA A.12 (Fractional chain rule for a Hölder continuous function, [**104**]). *Let G be a Hölder continuous function of order $0 < \alpha < 1$. Then, for every $0 < s < \alpha$, $1 < p < \infty$, and $\frac{s}{\alpha} < \sigma < 1$ we have*

$$\big\||\nabla|^s G(u)\big\|_p \lesssim \big\||u|^{\alpha - \frac{s}{\sigma}}\big\|_{p_1}\big\||\nabla|^\sigma u\big\|_{\frac{s}{\sigma}p_2}^{\frac{s}{\sigma}}, \tag{A.3}$$

provided $\frac{1}{p} = \frac{1}{p_1} + \frac{1}{p_2}$ and $(1 - \frac{s}{\alpha\sigma})p_1 > 1$.

The next result is formally similar to the preceding lemma; however, the proof is much simpler. It is used in the proof of Lemma 6.9.

LEMMA A.13 (Nonlinear Bernstein). *Let $G : \mathbb{C} \to \mathbb{C}$ be Hölder continuous of order $0 < \alpha \leq 1$. Then*

$$\|P_N G(u)\|_{L_x^{p/\alpha}(\mathbb{R}^d)} \lesssim N^{-\alpha}\|\nabla u\|_{L_x^p(\mathbb{R}^d)}^\alpha$$

for any $1 \leq p < \infty$ and $u \in \dot{W}^{1,p}(\mathbb{R}^d)$.

PROOF. Given $h \in \mathbb{R}^d$, the Fundamental Theorem of Calculus implies

$$u(x+h) - u(x) = \int_0^1 h \cdot \nabla u(x + \theta h)\, d\theta \tag{A.4}$$

and thus,

$$\big\|G(u(x+h)) - G(u(x))\big\|_{L_x^{p/\alpha}(\mathbb{R}^d)} \lesssim |h|^\alpha \|\nabla u\|_{L_x^p(\mathbb{R}^d)}^\alpha.$$

Now let k denote the convolution kernel of the Littlewood-Paley projection P_1, so that

$$[P_N f](x) = \int_{\mathbb{R}^d} N^d k(N(x-y)) f(y)\, dy$$

$$= \int_{\mathbb{R}^d} N^d k(-Nh)[f(x+h) - f(x)]\, dh.$$

Note that in obtaining the second identity, we used the fact that $\int_{\mathbb{R}^d} k(x)\, dx = 0$. Combining this with (A.4) and using the triangle inequality, we obtain

$$\begin{aligned}\|P_N G(u)\|_{L_x^{p/\alpha}(\mathbb{R}^d)} &\lesssim \|\nabla u\|_{L_x^p(\mathbb{R}^d)}^\alpha \int_{\mathbb{R}^d} |h|^\alpha N^d |k(-Nh)|\, dh \\ &\lesssim N^{-\alpha} \|\nabla u\|_{L_x^p(\mathbb{R}^d)}^\alpha,\end{aligned}$$

which proves the lemma. □

Lastly, we record a particular consequence of Lemma A.12 that is used for Lemma 6.3.

COROLLARY A.14. *Let $0 \le s < 1 + \frac{4}{d}$ and $F(u) = |u|^{4/d} u$. Then, on any spacetime slab $I \times \mathbb{R}^d$ we have*

$$\big\| |\nabla|^s F(u) \big\|_{L_{t,x}^{\frac{2(d+2)}{d+4}}} \lesssim \big\| |\nabla|^s u \big\|_{L_{t,x}^{\frac{2(d+2)}{d}}} \|u\|_{L_{t,x}^{\frac{2(d+2)}{d}}}^{\frac{4}{d}}.$$

PROOF. Fix a compact interval I. Throughout the proof, all spacetime estimates will be on $I \times \mathbb{R}^d$.

For $0 < s \le 1$, the claim is an easy consequence of Lemma A.11. It remains to address the case $1 < s < 1 + \frac{4}{d}$. We will only give details for $d \ge 5$; the main ideas carry over to lower dimensions.

Using the chain rule and the fractional product rule, we estimate as follows:

$$\begin{aligned}\big\| |\nabla|^s F(u) \big\|_{L_{t,x}^{\frac{2(d+2)}{d+4}}} &\lesssim \big\| |\nabla|^{s-1} \big(F_z(u) \nabla u + F_{\bar z}(u) \nabla \bar u \big) \big\|_{L_{t,x}^{\frac{2(d+2)}{d+4}}} \\ &\lesssim \big\| |\nabla|^s u \big\|_{L_{t,x}^{\frac{2(d+2)}{d}}} \|u\|_{L_{t,x}^{\frac{2(d+2)}{d}}}^{\frac{4}{d}} \\ &\quad + \|\nabla u\|_{L_{t,x}^{\frac{2(d+2)}{d}}} \Big[\big\| |\nabla|^{s-1} F_z(u) \big\|_{L_{t,x}^{\frac{d+2}{2}}} + \big\| |\nabla|^{s-1} F_{\bar z}(u) \big\|_{L_{t,x}^{\frac{d+2}{2}}} \Big].\end{aligned}$$

The claim will follow from this, once we establish

$$\big\| |\nabla|^{s-1} F_z(u) \big\|_{L_{t,x}^{\frac{d+2}{2}}} + \big\| |\nabla|^{s-1} F_{\bar z}(u) \big\|_{L_{t,x}^{\frac{d+2}{2}}} \lesssim \big\| |\nabla|^\sigma u \big\|_{L_{t,x}^{\frac{2(d+2)}{d}}}^{\frac{s-1}{\sigma}} \|u\|_{L_{t,x}^{\frac{2(d+2)}{d}}}^{\frac{4}{d} - \frac{s-1}{\sigma}} \tag{A.5}$$

for some $\frac{d(s-1)}{4} < \sigma < 1$. Indeed, by interpolation,

$$\big\| |\nabla|^\sigma u \big\|_{L_{t,x}^{\frac{2(d+2)}{d}}} \lesssim \big\| |\nabla|^s u \big\|_{L_{t,x}^{\frac{2(d+2)}{d}}}^{\frac{\sigma}{s}} \|u\|_{L_{t,x}^{\frac{2(d+2)}{d}}}^{1-\frac{\sigma}{s}}$$

and

$$\|\nabla u\|_{L_{t,x}^{\frac{2(d+2)}{d}}} \lesssim \big\| |\nabla|^s u \big\|_{L_{t,x}^{\frac{2(d+2)}{d}}}^{\frac{1}{s}} \|u\|_{L_{t,x}^{\frac{2(d+2)}{d}}}^{1-\frac{1}{s}}.$$

To derive (A.5), we merely observe that F_z and $F_{\bar z}$ are Hölder continuous functions of order $\frac{4}{d}$ and then apply Lemma A.12 (with $\alpha := \frac{4}{d}$ and $s := s - 1$). □

A.4. A Gronwall inequality. Our last technical tool is the most elementary. It is a form of Gronwall's inequality that involves both the past and the future, 'acausal' in the terminology of [**90**]. It is used in Section 6.

LEMMA A.15. *Fix $\gamma > 0$. Given $0 < \eta < \frac{1}{2}(1 - 2^{-\gamma})$ and $\{b_k\} \in \ell^\infty(\mathbb{Z}^+)$, let $x_k \in \ell^\infty(\mathbb{Z}^+)$ be a non-negative sequence obeying*

$$x_k \leq b_k + \eta \sum_{l=0}^{\infty} 2^{-\gamma|k-l|} x_l \qquad \text{for all } k \geq 0. \tag{A.6}$$

Then

$$x_k \lesssim \sum_{l=0}^{k} r^{|k-l|} b_l \qquad \text{for all } k \geq 0 \tag{A.7}$$

for some $r = r(\eta) \in (2^{-\gamma}, 1)$. Moreover, $r \downarrow 2^{-\gamma}$ as $\eta \downarrow 0$.

PROOF. Our proof follows a well-travelled path. By decreasing entries in b_k we can achieve equality in (A.6); since this also reduces the righthand side of (A.7), it suffices to prove the lemma in this case. Note that since $x_k \in \ell^\infty$, b_k will remain a bounded sequence.

Let A denote the doubly infinite matrix with entries $A_{k,l} = 2^{-\gamma|k-l|}$ and let P denote the natural projection from $\ell^2(\mathbb{Z})$ onto $\ell^2(\mathbb{Z}^+)$. Our goal is to show that (A.7) holds for any solution of

$$(1 - \eta PAP^*)x = b. \tag{A.8}$$

First we observe that since

$$\|A\| = \sum_{k\in\mathbb{Z}} 2^{-\gamma|k|} = \frac{1 + 2^{-\gamma}}{1 - 2^{-\gamma}},$$

ηA is a contraction on ℓ^∞. Thus, we may write

$$x = \sum_{p=0}^{\infty} (\eta PAP^*)^p b \leq \sum_{p=0}^{\infty} P(\eta A)^p P^* b = P(1 - \eta A)^{-1} P^* b,$$

where the inequality is meant entry-wise. The justification for this inequality is simply that the matrix A has non-negative entries. We will complete the proof of (A.7) by computing the entries of $(1 - \eta A)^{-1}$. This is easily done via Fourier methods: Let

$$a(z) := \sum_{k\in\mathbb{Z}} 2^{-\gamma|k|} z^k = 1 + \frac{2^{-\gamma} z}{1 - 2^{-\gamma} z} + \frac{2^{-\gamma} z^{-1}}{1 - 2^{-\gamma} z^{-1}}$$

and

$$\begin{aligned} f(z) := \frac{1}{1 - \eta a(z)} &= \frac{(z - 2^{\gamma})(z - 2^{-\gamma})}{z^2 - (2^{-\gamma} + 2^{\gamma} - \eta 2^{\gamma} + \eta 2^{-\gamma})z + 1} \\ &= 1 + \frac{(1 - r2^{-\gamma})(r2^{\gamma} - 1)}{(1 - r^2)} \Big[1 + \frac{rz}{1 - rz} + \frac{rz^{-1}}{1 - rz^{-1}}\Big], \end{aligned}$$

where $r \in (0,1)$ and $1/r$ are the roots of $z^2 - (2^{-\gamma} + 2^{\gamma} - \eta 2^{\gamma} + \eta 2^{-\gamma})z + 1 = 0$. From this formula, we can immediately read off the Fourier coefficients of f, which give us the matrix elements of $(1 - \eta A)^{-1}$. In particular, they are $O(r^{|k-l|})$. □

References

[1] V. I. Arnold, V. V. Kozlov, and A. I. Neishtadt, *Mathematical aspects of classical and celestial mechanics.* Translated from the 1985 Russian original by A. Iacob. Reprint of the original English edition. Springer-Verlag, Berlin, 1997. MR1656199

[2] T. Aubin, *Problémes isopérimétriques et espaces de Sobolev.* J. Diff. Geom. **11** (1976), 573–598. MR0448404

[3] H. Bahouri and P. Gérard, *High frequency approximation of solutions to critical nonlinear wave equations.* Amer. J. Math. **121** (1999), 131–175. MR1705001

[4] P. Begout and A. Vargas, *Mass concentration phenomena for the L^2-critical nonlinear Schrödinger equation.* Trans. Amer. Math. Soc. **359** (2007), 5257–5282. MR2327030

[5] G. A. Bliss, *An integral inequality.* J. London Math. Soc. **5** (1930), 40–46.

[6] J. Bourgain, *Refinements of Strichartz inequality and applications to 2d-NLS with critical nonlinearity.* Int. Math. Res. Not. (1998), 253–283. MR1616917

[7] J. Bourgain, *Global wellposedness of defocusing critical nonlinear Schrödinger equation in the radial case.* J. Amer. Math. Soc. **12** (1999), 145–171. MR1626257

[8] J. Bourgain, *Global solutions of nonlinear Schrödinger equations.* American Mathematical Society Colloquium Publications, **46**. American Mathematical Society, Providence, RI, 1999. MR1691575

[9] H. Brézis and J.-M. Coron, *Convergence of solutions of H-systems or how to blow bubbles.* Arch. Rational Mech. Anal. **89** (1985), 21–56. MR0784102

[10] H. Brézis and E. Lieb, *A relation between pointwise convergence of functions and convergence of functionals.* Proc. Amer. Math. Soc. **88** (1983), 486–490. MR0699419

[11] J. E. Brothers and W. P. Ziemer, *Minimal rearrangements of Sobolev functions.* J. Reine Angew. Math. **384** (1988), 153–179. MR0929981

[12] R. Carles and S. Keraani, *On the role of quadratic oscillations in nonlinear Schrödinger equation II. The L^2-critical case.* Trans. Amer. Math. Soc. **359** (2007), 33–62. MR2247881

[13] T. Cazenave and F. B. Weissler, *Some remarks on the nonlinear Schrödinger equation in the critical case.* In "Nonlinear Semigroups, Partial Differential Equations and Attractors." Lecture Notes in Math. **1394** (1989), 18–29. MR1021011

[14] T. Cazenave and F. B. Weissler, *The Cauchy problem for the critical nonlinear Schrödinger equation in H^s.* Nonlinear Anal. **14** (1990), 807–836. MR1055532

[15] T. Cazenave, *Semilinear Schrödinger equations.* Courant Lecture Notes in Mathematics, **10**. American Mathematical Society, 2003. MR2002047

[16] M. Christ and M. Weinstein, *Dispersion of small amplitude solutions of the generalized Korteweg-de Vries equation.* J. Funct. Anal. **100** (1991), 87–109. MR1124294

[17] R. Clausius, *On a mechanical theorem applicable to heat.* Philosophical Magazine, Ser. 4 **40** (1870), 122-127.

[18] J. Colliander, J. Holmer, M. Visan, and X. Zhang, *Global existence and scattering for rough solutions to generalized nonlinear Schrödinger equations on $\mathbb{R}$.* Commun. Pure Appl. Anal. **7** (2008), 467–489. MR2379437

[19] J. Colliander, M. Keel, G. Staffilani, H. Takaoka, and T. Tao, *Global existence and scattering for rough solutions of a nonlinear Schrödinger equation on $\mathbb{R}^3$.* Comm. Pure Appl. Math. **57** (2004), 987–1014. MR2053757

[20] J. Colliander, M. Keel, G. Staffilani, H. Takaoka, and T. Tao, *Global well-posedness and scattering for the energy-critical nonlinear Schrödinger equation in $\mathbb{R}^3$.* Ann. Math. **167** (2008), 767–865.

[21] P. Constantin and J.-C. Saut, *Local smoothing properties of dispersive equations.* J. Amer. Math. Soc. **1** (1988), 413–439. MR0928265

[22] H. L. Cycon, R. G. Froese, W. Kirsch, and B. Simon, *Schrödinger operators with application to quantum mechanics and global geometry.* Texts and Monographs in Physics. Springer-Verlag, Berlin, 1987. MR0883643

[23] T. Duyckaerts, J. Holmer, and S. Roudenko, *Scattering for the non-radial 3D cubic nonlinear Schrödinger equation.* Math. Res. Lett. **15** (2008), 1233–1250. MR2470397

[24] C. Fefferman, *Inequalities for strongly singular convolution operators.* Acta Math. **124** (1970), 9–36. MR0257819

[25] D. Foschi, *Inhomogeneous Strichartz estimates.* J. Hyperbolic Differ. Equ. **2** (2005), 1–24. MR2134950

[26] P. Gérard, *Description du défaut de compacité de l'injection de Sobolev.* ESAIM Control Optim. Calc. Var. **3** (1998), 213–233. MR1632171

[27] P. Gérard, Y. Meyer, and F. Oru, *Inégalités de Sobolev précisées.* Séminaire É.D.P. (1996–1997), Exp. No. IV, 11pp. MR1482810

[28] J. Ginibre and G. Velo, *Smoothing properties and retarded estimates for some dispersive evolution equations.* Comm. Math. Phys. **144** (1992), 163–188. MR1151250

[29] J. Ginibre and G. Velo, *Scattering theory in the energy space for a class of nonlinear Schrödinger equations.* J. Math. Pures Appl. **64** (1985), 363–401. MR0839728

[30] J. Ginibre and G. Velo, *Quadratic Morawetz inequalities and asymptotic completeness in the energy space for nonlinear Schrödinger and Hartree equations.* Quart. Appl. Math. **68** (2010), 113–134. MR2598884

[31] R. T. Glassey, *On the blowing up of solutions to the Cauchy problem for nonlinear Schrödinger equations.* J. Math. Phys. **18** (1977), 1794–1797. MR0460850

[32] G. Grillakis, *On nonlinear Schrödinger equations.* Comm. PDE **25** (2000), 1827–1844. MR1778782

[33] T. Hmidi and S. Keraani, *Blowup theory for the critical nonlinear Schrödinger equations revisited.* Int. Math. Res. Not. **2005**, 2815–2828. MR2180464

[34] J. Holmer and S. Roudenko, *On blow-up solutions to the 3D cubic nonlinear Schrödinger equation.* Appl. Math. Res. Express. AMRX (2007), 31 pp. MR2354447

[35] J.-L. Journé, *Calderón-Zygmund operators on product spaces.* Rev. Mat. Iberoamericana **1** (1985), 55–91. MR0836284

[36] T. Kato, *Smooth operators and commutators.* Studia Math. **31** (1968), 535–546. MR0234314

[37] M. Keel and T. Tao, *Endpoint Strichartz estimates.* Amer. J. Math. **120** (1998), 955–980. MR1646048

[38] C. E. Kenig and F. Merle, *Global well-posedness, scattering and blow up for the energy-critical, focusing, nonlinear Schrödinger equation in the radial case.* Invent. Math. **166** (2006), 645–675. MR2257393

[39] C. E. Kenig and F. Merle, *Global well-posedness, scattering and blow-up for the energy critical focusing non-linear wave equation.* Acta Math. **201** (2008), 147–212. MR2461508

[40] C. E. Kenig and F. Merle, *Scattering for $H^{1/2}$ bounded solutions to the cubic, defocusing NLS in 3 dimensions.* Trans. Amer. Math. Soc. **362** (2010), 1937–1962. MR2574882

[41] S. Keraani, *On the defect of compactness for the Strichartz estimates for the Schrödinger equations.* J. Diff. Eq. **175** (2001), 353–392. MR1855973

[42] S. Keraani, *On the blow-up phenomenon of the critical nonlinear Schrödinger equation.* J. Funct. Anal. **235** (2006), 171–192. MR2216444

[43] R. Killip, T. Tao, and M. Visan, *The cubic nonlinear Schrödinger equation in two dimensions with radial data.* J. Eur. Math. Soc. (JEMS) **11** (2009), 1203–1258. MR2557134

[44] R. Killip and M. Visan, *The focusing energy-critical nonlinear Schrödinger equation in dimensions five and higher.* Amer. J. Math. **132** (2010), 361–424. MR2654778

[45] R. Killip, M. Visan, and X. Zhang, *Energy-critical NLS with quadratic potentials.* Comm. Partial Differential Equations **34** (2009), 1531–1565. MR2581982

[46] R. Killip, M. Visan, and X. Zhang, *The mass-critical nonlinear Schrödinger equation with radial data in dimensions three and higher.* Anal. PDE **1** (2008), 229–266. MR2472890

[47] R. Killip, M. Visan, and X. Zhang, *The focusing energy-critical nonlinear Schrödinger equation with radial data.* Unpublished manuscript, Sept. 2007.

[48] L. D. Kudryavtsev and S. M. Nikol'skiĭ, *Spaces of differentiable functions of several variables and imbedding theorems.* In "Analysis. III. Spaces of differentiable functions." Encyclopaedia of Mathematical Sciences, **26**. Springer-Verlag, Berlin, 1991. MR1094115

[49] M. K. Kwong, *Uniqueness of positive solutions of $\Delta u - u + u^p = 0$ in $\mathbb{R}^n$.* Arch. Rational Mech. Anal. **105** (1989), 243–266. MR0969899

[50] L. D. Landau and E. M. Lifshitz, *Course of theoretical physics. Vol. 1. Mechanics.* Third edition. Pergamon Press, Oxford-New York-Toronto, 1976.

[51] R. B. Lavine, *Absolute continuity of Hamiltonian operators with repulsive potential.* Proc. Amer. Math. Soc. **22** (1969), 55–60. MR0247529

[52] R. B. Lavine, *Commutators and scattering theory. I. Repulsive interactions.* Comm. Math. Phys. **20** (1971), 301–323. MR0293945

[53] P. D. Lax and R. S. Phillips, *Scattering theory.* Second edition. With appendices by Cathleen S. Morawetz and Georg Schmidt. Pure and Applied Mathematics, **26**. Academic Press, Inc., Boston, MA, 1989. MR1037774

[54] E. H. Lieb and M. Loss, *Analysis.* Second edition. Graduate Studies in Mathematics, **14**. American Mathematical Society, Providence, RI, 2001. MR1817225

[55] J. E. Lin and W. A. Strauss, *Decay and scattering of solutions of a nonlinear Schrödinger equation.* J. Funct. Anal. **30** (1978), 245–263. MR0515228

[56] P.-L. Lions, *The concentration-compactness principle in the calculus of variations. The limit case. I.* Rev. Mat. Iberoamericana **1** (1985), 145–201. MR0834360

[57] F. Merle, *Existence of blow-up solutions in the energy space for the critical generalized KdV equation.* J. Amer. Math. Soc. **14** (2001), 555–578. MR1824989

[58] F. Merle and L. Vega, *Compactness at blow-up time for L^2 solutions of the critical nonlinear Schrödinger equation in 2D.* Int. Math. Res. Not. **8** (1998), 399–425. MR1628235

[59] S. J. Montgomery-Smith, *Time decay for the bounded mean oscillation of solutions of the Schrödinger and wave equation.* Duke Math J. **19** (1998), 393–408. MR1600602

[60] C. S. Morawetz, *Notes on time decay and scattering for some hyperbolic problems.* Regional Conference Series in Applied Mathematics, No. 19. Society for Industrial and Applied Mathematics, Philadelphia, PA, 1975. MR0492919

[61] C. S. Morawetz and W. A. Strauss, *Decay and scattering of solutions of a nonlinear relativistic wave equation.* Comm. Pure Appl. Math. **25** (1972), 1–31. MR0303097

[62] A. Moyua, A. Vargas, and L. Vega, *Restriction theorems and maximal operators related to oscillatory integrals in $\mathbb{R}^3$.* Duke Math. J. **96** (1999), 547–574. MR1671214

[63] B. V. Sz. Nagy, *Über Integralgleichungen zwischen einer Funktion und ihrer Ableitung.* Acta Sci. Math. (Szeged) **10** (1941), 64–74.

[64] K. Nakanishi, *Scattering theory for nonlinear Klein-Gordon equation with Sobolev critical power.* Internat. Math. Res. Notices **1** (1999), 31–60. MR1666973

[65] T. Ogawa and Y. Tsutsumi, *Blow-up of H^1 solution for the nonlinear Schrödinger equation.* J. Diff. Eq. **92** (1991), 317–330. MR1120908

[66] T. Ozawa and Y. Tsutsumi, *Space-time estimates for null gauge forms and nonlinear Schrödinger equations.* Differential Integral Equations **11** (1998), 201–222. MR1741843

[67] F. Planchon and L. Vega, *Bilinear virial identities and applications.* Ann. Sci. Éc. Norm. Supér. (4) **42** (2009), 261–290. MR2518079

[68] S. I. Pohožaev, *On the eigenfunctions of the equation $\Delta u + \lambda f(u) = 0$.* Dokl. Akad. Nauk SSSR **165** (1965), 36–39. English translation, Soviet Math. Dokl. **6** (1965), 1408–1411. MR0192184

[69] C. R. Putnam, *Commutation properties of Hilbert space operators and related topics.* Ergebnisse der Mathematik und ihrer Grenzgebiete, Band **36**. Springer-Verlag, New York 1967. MR0217618

[70] M. Reed and B. Simon, *Methods of modern mathematical physics. III. Scattering theory.* Academic Press, New York-London, 1979. MR0529429

[71] M. Reed and B. Simon, *Methods of modern mathematical physics. IV. Analysis of operators.* Academic Press, New York-London, 1978. MR0493421

[72] M. Riesz, *Sur les ensembles compacts de fonctions sommable.* Acta Sci. Math. (Szeged) **6** (1933), 136–142.

[73] G. Rosen, *Minimum value for c in the Sobolev inequality $\|\phi^3\| \leq c\|\nabla\phi\|^3$.* SIAM J. Appl. Math. **21** (1971). 30–32. MR0289739

[74] J. L. Rubio de Francia, *A Littlewood-Paley inequality for arbitrary intervals.* Rev. Mat. Iberoamericana **1** (1985), 1–14. MR0850681

[75] E. Ryckman and M. Visan, *Global well-posedness and scattering for the defocusing energy-critical nonlinear Schrödinger equation in $\mathbb{R}^{1+4}$.* Amer. J. Math. **129** (2007), 1–60. MR2288737

[76] J. Sacks and K. Uhlenbeck, *The existence of minimal immersions of 2-spheres.* Ann. of Math. **113** (1981), 1–24. MR0604040

[77] S. Shao, *Sharp linear and bilinear restriction estimates for the paraboloid in the cylindrically symmetric case.* Preprint `arXiv:0706.3759`.

[78] J. Shatah and M. Struwe, *Geometric wave equations.* Courant Lecture Notes in Mathematics, **2**. Courant Institute of Mathematical Sciences, New York, NY; American Mathematical Society, Providence, RI, 1998. MR1674843

[79] P. Sjölin, *Regularity of solutions to the Schrödinger equation.* Duke Math. J. **55** (1987), 699–715. MR0904948

[80] E. M. Stein, *Some problems in harmonic analysis.* In "Harmonic analysis in Euclidean spaces (Proc. Sympos. Pure Math., Williams Coll., Williamstown, Mass., 1978), Part 1." Proc. Sympos. Pure Math., XXXV, Part 1, Amer. Math. Soc., Providence, R.I., 1979. MR0545235

[81] E. M. Stein, *Harmonic analysis: real-variable methods, orthogonality, and oscillatory integrals.* Princeton Mathematical Series, **43**. Princeton University Press, Princeton, NJ, 1993. MR1232192

[82] A. Stefanov, *Strichartz estimates for the Schrödinger equation with radial data,* Proc. Amer. Math. Soc. **129** (2001), 1395–1401. MR1814165

[83] R. S. Strichartz, *Restriction of Fourier transform to quadratic surfaces and deay of solutions of wave equations.* Duke Math. J. **44** (1977), 705–714. MR0512086

[84] M. Struwe, *A global compactness result for elliptic boundary value problems involving limiting nonlinearities.* Math. Z. **187** (1984), 511–517. MR0760051

[85] M. Struwe, *Large H-surfaces via the mountain-pass-lemma.* Math. Ann. **270** (1985), 441–459.

[86] G. Talenti, *Best constant in Sobolev inequality.* Ann. Mat. Pura. Appl. **110** (1976), 353–372. MR0463908

[87] T. Tao, *Spherically averaged endpoint Strichartz estimates for the two-dimensional Schrödinger equation,* Comm. PDE **25** (2000), 1471–1485. MR1765155

[88] T. Tao, *A sharp bilinear restrictions estimate for paraboloids.* Geom. Funct. Anal. **13** (2003), 1359–1384. MR2033842

[89] T. Tao, *Global well-posedness and scattering for the higher-dimensional energy-critical non-linear Schrödinger equation for radial data.* New York J. of Math. **11** (2005), 57–80. MR2154347

[90] T. Tao, *Nonlinear dispersive equations. Local and global analysis.* CBMS Regional Conference Series in Mathematics, **106**. American Mathematical Society, Providence, RI, 2006. MR2233925

[91] T. Tao, *A (concentration-)compact attractor for high-dimensional non-linear Schrödinger equations.* Dyn. Partial Differ. Equ. **4** (2007), 1–53. MR2304091

[92] T. Tao, *A pseudoconformal compactification of the nonlinear Schrödinger equation and applications.* New York J. Math. **15** (2009), 265–282. MR2530148

[93] T. Tao, A. Vargas, and L. Vega, *A bilinear approach to the restriction and Kakeya conjectures.* J. Amer. Math. Soc. **11** (1998), 967–1000. MR1625056

[94] T. Tao and M. Visan, *Stability of energy-critical nonlinear Schrödinger equations in high dimensions.* Electron. J. Diff. Eqns. **118** (2005), 1–28. MR2174550

[95] T. Tao, M. Visan, and X. Zhang, *The nonlinear Schrödinger equation with combined power-type nonlinearities.* Comm. Partial Differential Equations **32** (2007), 1281–1343. MR2354495

[96] T. Tao, M. Visan, and X. Zhang, *Minimal-mass blowup solutions of the mass-critical NLS.* Forum Math. **20** (2008), 881–919. MR2445122

[97] T. Tao, M. Visan, and X. Zhang, *Global well-posedness and scattering for the mass-critical nonlinear Schrödinger equation for radial data in high dimensions.* Duke Math. J. **140** (2007), 165–202. MR2355070

[98] M. E. Taylor, *Tools for PDE.* Mathematical Surveys and Monographs, **81**. American Mathematical Society, Providence, RI, 2000. MR1766415

[99] H. Triebel, *The structure of functions.* Monographs in Mathematics, **97**. Birkhäuser Verlag, Basel, 2001. MR1851996

[100] L. Vega, *Schrödinger equations: pointwise convergence to the initial data.* Proc. Amer. Math. Soc. **102** (1988), 874–878. MR0934859

[101] M. C. Vilela, *Inhomogeneous Strichartz estimates for the Schrödinger equation.* Trans. Amer. Math. Soc. **359** (2007), 2123–2136. MR2276614

[102] S. N. Vlasov, V. A. Petrishchev, and V. I. Talanov, *Averaged description of wave beams in linear and nonlinear media (the method of moments).* Radiophys. Quantum Electron. **14** (1971), 1062–1070.

[103] M. Visan, *The defocusing energy-critical nonlinear Schrödinger equation in dimensions five and higher.* Ph.D. Thesis, UCLA, 2006. MR2709575

[104] M. Visan, *The defocusing energy-critical nonlinear Schrödinger equation in higher dimensions.* Duke Math. J. **138** (2007), 281–374. MR2318286

[105] M. Weinstein, *Nonlinear Schrödinger equations and sharp interpolation estimates.* Comm. Math. Phys. **87** (1983), 567–576. MR0691044

[106] H. Wente,*Large solutions to the volume constrained Plateau problem.* Arch. Rational Mech. Anal. **75** (1980/81), 59–77. MR0592104

[107] T. Wolff, *A sharp bilinear cone restriction estimate.* Ann. of Math. **153** (2001), 661–698. MR1836285

[108] X. Zhang, *On the Cauchy problem of 3-D energy-critical Schrödinger equations with subcritical perturbations.* J. Differential Equations **230** (2006), 422–445. MR2271498

[109] A. Zygmund, *On Fourier coefficients and transforms of functions of two variables.* Studia Math. **50** (1974), 189–201. MR0387950

Department of Mathematics, University of California, Los Angeles, Los Angeles, California 90095

E-mail address: killip@math.ucla.edu

Department of Mathematics, University of California, Los Angeles, Los Angeles, California 90095

E-mail address: visan@math.ucla.edu

Clay Mathematics Proceedings
Volume **17**, 2013

Foreword to 'Geometry and Analysis in Many-Body Scattering'

András Vasy

The original of these notes was given as a lecture series at the Université de Nantes in 2002 and then was published in 2003 in a volume for the MSRI inverse problems program. The lectures given at the Clay Mathematics Institute summer school in Zürich in 2008 were based on the same notes, though a slightly different emphasis was made and some updates were given. The purpose of this foreword is to briefly describe the developments that took place between 2002 and 2008, as well as developments that took place since 2008.

The motivation for the propagation phenomena in N-body scattering described in Section 3, as explained in Section 2, is wave propagation on domains or manifolds with smooth boundary, corresponding to 3-body scattering, or on domains or manifolds with corners, corresponding to 4- or more body scattering. At the time of the original lectures, the situation on manifolds with smooth boundaries was well understood, but the analogous results on manifolds with corners only existed in the analytic setting thanks to Lebeau's work [**18**]. By 2008 propagation of singularities has also been proved for manifold with corners by the author in [**31**], see also [**32**].

In a different direction, the study of radial points, discussed in Section 3 and corresponding to the propagation set of Sigal and Soffer [**28**], has expanded to include the papers [**16, 17**] of Herbst and Skibsted, and in a more microlocal setting the papers [**14, 15**] of Hassell, Melrose and the author, in which isolated radial points were analyzed (the motivation being scattering by symbolic potentials of order 0 with non-degenerate critical points on the sphere at infinity), as well as other papers in these settings, but what is more directly relevant, the techniques of [**14, 15**] were used later to study Lagrangian sets of radial points in the work of the author and later of Haber [**33**], [**13**] and [**12**], with some applications in papers mentioned below. While these are not yet in the N-body setting, it is expected that eventually a similarly detailed picture will be available there as well.

Threshold behavior in long-range two-body type setting was studied by Dereziński, Fournais, Skibsted and Wang [**8, 6, 29**] as well as Bony and Häfner, and

---

2010 *Mathematics Subject Classification.* Primary 35P25, 47A40, 81U10.

The author gratefully acknowledges partial support from the NSF under grant number DMS-1068742, and is grateful to the Clay Mathematics Institute for the hospitality at the summer school in Zürich where the original lecture were given in 2008.

Bouclet [**3, 2, 4**], and in a more geometric sitting by Wang, and then Hassell, Guillarmou and Sikora, Wunsch and the author, and later Sher [**36, 10, 11, 9, 34, 27**], as well as Metcalfe and Tataru [**30, 26**] and Wunsch and the author [**35**] from a more evolution equation point of view. While this is significant progress, it has taken place in a geometric two-body type setting, though the papers focusing on evolutionary problems do have some 3-body aspects. Thus, much remains to be understood for thresholds.

The first lectures in Zürich also discussed evolution equations, in particular the Schrödinger, the Klein-Gordon and the wave equations, and the role the radial compactification of $\mathbb{R}^n \times \mathbb{R}_t$ plays (see also here the remark after Theorem 3.1). Without an energy localization away from the threshold 0 of the spatial operator (on $\mathbb{R}^n$), even if the Hamiltonian (on $L^2(\mathbb{R}^n)$) is positive, the wave equation is rather ill-behaved from the scattering perspective espoused in these lecture notes since the whole zero section of the scattering cotangent bundle over the boundary consists of radial points; in fact the Hamilton vector field vanishes there (since the principal symbol vanishes quadratically) so the zero section in fact consists of degenerate radial points (unlike the non-degenerate ones playing a role elsewhere in scattering theory). The recent work with Baskin and Wunsch [**1**], building on [**33**], provides a new point of view in asymptotically Minkowski-type spaces, though currently not yet in a truly N-body type setting, via Melrose's b-pseudodifferential framework, see [**25**]; here the radial points in an appropriate sense also play a crucial role. This also relates to the Schwarzschild and Kerr black hole analysis; the latter problems were explained by Dafermos and Rodnianski in Zürich in 2008; see their lecture notes for details and references. See also [**33**] for some explanation of the connection, though the latter paper concentrates on Kerr-de Sitter and Minkowski type settings, not the combination of these two; this paper also contains numerous references to other works.

Although Faddeev's original approach to 3-body scattering, espoused in [**7**], was not described in the original lectures, since then quite intriguing connections have emerged between wave propagation and Faddeev's method. While not explained in these terms, Faddeev's work relates to the so-called diffractive improvement when waves scatter from cone points, edges or corners. Starting with the work of Cheeger and Taylor [**5**] on product cones, this diffractive improvement was extended to all of the just mentioned cases in works of Melrose, Wunsch and the author [**24, 22, 23**]. It would be interesting to make this connection with Faddeev's work more explicit.

Finally, the papers of Mazzeo and the author on the analysis of the Laplacian on symmetric spaces, described in the final section of the lecture notes, were published in [**21, 19, 20**].

References

[1] D. Baskin, A. Vasy, and J. Wunsch. Asymptotics of radiation fields in asymptotically Minkowski space. *Preprint, arxiv:1212.5141*, 2012.

[2] Jean-François Bony and Dietrich Häfner. Local energy decay for several evolution equations on asymptotically euclidean manifolds. *arxiv:1008.2357*, 2010.

[3] Jean-François Bony and Dietrich Häfner. Low frequency resolvent estimates for long range perturbations of the Euclidean Laplacian. *Math. Res. Lett.*, 17(2):301–306, 2010.

[4] Jean-Marc Bouclet. Low frequency estimates and local energy decay for asymptotically Euclidean Laplacians. *Comm. Partial Differential Equations*, 36(7):1239–1286, 2011.

[5] J. Cheeger and M. Taylor. Diffraction by conical singularities, I, II. *Comm. Pure Applied Math.*, 35:275–331, 487–529, 1982.

[6] J. Dereziński and E. Skibsted. Quantum scattering at low energies. *J. Funct. Anal.*, 257(6):1828–1920, 2009.
[7] L. D. Faddeev. *Mathematical aspects of the three-body problem in the quantum scattering theory.* Steklov Institute, 1963.
[8] S. Fournais and E. Skibsted. Zero energy asymptotics of the resolvent for a class of slowly decaying potentials. *Math. Z.*, 248(3):593–633, 2004.
[9] C. Guillarmou, A. Hassell, and A. Sikora. Resolvent at low energy III: The spectral measure. *Transactions of the AMS*, to appear; arxiv:1009.3084.
[10] Colin Guillarmou and Andrew Hassell. Resolvent at low energy and Riesz transform for Schrödinger operators on asymptotically conic manifolds. I. *Math. Ann.*, 341(4):859–896, 2008.
[11] Colin Guillarmou and Andrew Hassell. Resolvent at low energy and Riesz transform for Schrödinger operators on asymptotically conic manifolds. II. *Ann. Inst. Fourier (Grenoble)*, 59(4):1553–1610, 2009.
[12] N. Haber. A normal form around a Lagrangian submanifold of radial points. *Preprint, arvix:1210.1275*, 2012.
[13] N. Haber and A. Vasy. Propagation of singularities around a Lagrangian submanifold of radial points. *Preprint, arxiv:1110.1419*, 2011.
[14] A. Hassell, R. B. Melrose, and A. Vasy. Spectral and scattering theory for symbolic potentials of order zero. *Advances in Mathematics*, 181:1–87, 2004.
[15] A. Hassell, R. B. Melrose, and A. Vasy. Microlocal propagation near radial points and scattering for symbolic potentials of order zero. *Analysis and PDE*, 1:127–196, 2008.
[16] Ira Herbst and Erik Skibsted. Quantum scattering for potentials independent of $|x|$: asymptotic completeness for high and low energies. *Comm. Partial Differential Equations*, 29(3-4):547–610, 2004.
[17] Ira Herbst and Erik Skibsted. Absence of quantum states corresponding to unstable classical channels. *Ann. Henri Poincaré*, 9(3):509–552, 2008.
[18] G. Lebeau. Propagation des ondes dans les variétés à coins. *Ann. Scient. Éc. Norm. Sup.*, 30:429–497, 1997.
[19] R. Mazzeo and A. Vasy. Analytic continuation of the resolvent of the Laplacian on SL(3)/SO(3). *American Journal of Mathematics*, 126:821–844, 2004.
[20] R. Mazzeo and A. Vasy. Analytic continuation of the resolvent of the Laplacian on symmetric spaces of noncompact type. *J. Func. Anal.*, 228:311–368, 2005.
[21] Rafe Mazzeo and András Vasy. Scattering theory on SL(3)/SO(3): connections with quantum 3-body scattering. *Proc. Lond. Math. Soc. (3)*, 94(3):545–593, 2007.
[22] R. B. Melrose, A. Vasy, and J. Wunsch. Propagation of singularities for the wave equation on manifolds with edges. *Duke Math. J.*, 144:109–193, 2008.
[23] R. B. Melrose, A. Vasy, and J. Wunsch. Diffraction of singularities for the wave equation on manifolds with corners. *Astérisque*, to appear; arXiv:0903.3208.
[24] Richard Melrose and Jared Wunsch. Propagation of singularities for the wave equation on conic manifolds. *Invent. Math.*, 156(2):235–299, 2004.
[25] Richard B. Melrose. *The Atiyah-Patodi-Singer index theorem*, volume 4 of *Research Notes in Mathematics*. A K Peters Ltd., Wellesley, MA, 1993.
[26] Jason Metcalfe and Daniel Tataru. Global parametrices and dispersive estimates for variable coefficient wave equations. *Math. Ann.*, 353(4):1183–1237, 2012.
[27] D. Sher. The heat kernel on an asymptotically conic manifold. *Preprint, arxiv:1208.1808*, 2012.
[28] I. M. Sigal and A. Soffer. N-particle scattering problem: asymptotic completeness for short range systems. *Ann. Math.*, 125:35–108, 1987.
[29] Erik Skibsted and Xue Ping Wang. Two-body threshold spectral analysis, the critical case. *J. Funct. Anal.*, 260(6):1766–1794, 2011.
[30] Daniel Tataru. Local decay of waves on asymptotically flat stationary spacetimes. *Preprint, arxiv:0910.5290*, 2009.
[31] A. Vasy. Propagation of singularities for the wave equation on manifolds with corners. *Annals of Mathematics*, 168:749–812, 2008.
[32] A. Vasy. Correction to "Propagation of singularities for the wave equation on manifolds with corners". *Ann. Math.*, to appear.

[33] A. Vasy. Microlocal analysis of asymptotically hyperbolic and Kerr-de Sitter spaces. *Inventiones Math.*, to appear; arxiv:1012.4391. With an appendix by S. Dyatlov.
[34] A. Vasy and J. Wunsch. Positive commutators at the bottom of the spectrum. *J. Func. Anal*, 259:503–523, 2010.
[35] A. Vasy and J. Wunsch. Morawetz estimates for the wave equation at low frequency. *Math. Annalen*, To appear.
[36] Xue Ping Wang. Threshold resonance in geometric scattering. *Mat. Contemp.*, 26:135–164, 2004.

Department of Mathematics, Stanford University, California 94305-2125
E-mail address: andras@math.stanford.edu

Clay Mathematics Proceedings
Volume **17**, 2013

Geometry and Analysis in Many-Body Scattering

András Vasy

Abstract. The present notes are intended to explain in relatively non-technical terms recent results in many-body scattering and related topics. Thus, many results in the many-body setting should be understood as new results on the propagation of singularities, here understood as lack of decay of wave functions at infinity, with much in common with real principal type propagation, i.e. wave phenomena. Classical mechanics plays the role that geometric optics has in the study of the wave equation, but even at this point quantum phenomena emerge. Propagation of singularities has immediate applications to the structure of scattering matrices and to inverse scattering; topics that are addressed here. The final section studies a problem very closely related to many-body scattering, namely scattering on higher rank non-compact symmetric spaces.

1. Introduction

The present notes are an effort to explain in relatively non-technical terms recent results in many-body scattering and related topics. Thus, many results in the many-body setting should be understood as new results on the propagation of singularities, here understood as lack of decay of wave functions at infinity, with much in common with real principal type propagation, i.e. wave phenomena. Motivated by this, I first briefly describe propagation of singularities for the wave equation. This is a remarkable relationship between geometric optics (the particle view of light) and the solutions of the wave equation (the wave view).

Next, in Section 3, I explain the geometry of many-body scattering, which includes both that of the configuration space and phase space. This geometry is closely related to classical mechanics, playing the role of geometric optics, but even at this point quantum phenomena emerge. This leads to the analytic results, namely the propagation of singularities connecting classical and quantum mechanics.

Much as for the wave equation, such a result has immediate applications, including the description of the scattering matrices and of the scattering phase. Slightly

2010 *Mathematics Subject Classification.* Primary 35P25, 47A40, 81U10.

This article originally appeared in Inside Out: Inverse Problems and Applications (Gunther Uhlmann, Ed.), pp. 333-380, Mathematical Sciences Research Institute Publications, 47, Cambridge University Press, Cambridge, 2003. Reprinted with permission from Cambridge University Press.

This work is partially supported by NSF grant #DMS-0201092, a Fellowship from the Alfred P. Sloan Foundation, and the Université de Nantes, where these lectures were originally given.

stronger versions can even lead to inverse results, a topic covered in the following section.

After so explaining the results, in Sections 5-6, I will try to at least give a flavor of how they are proved. This uses a many-body pseudo-differential algebra and positive commutator estimates, so these are discussed. We remark that these techniques are closely related to the proofs of the propagation of singularities for the wave equation, but there are significant differences as well, mostly arising from bound states of particles, which have no analogues for the wave equation. The pseudo-differential algebra itself is very interesting from the viewpoint of non-commutative geometry: there is a hierarchy of operator valued symbols at infinity.

Asymptotic completeness was the main focus of work in many-body scattering for a long period. In Section 7, I briefly explain how it relates to the microlocal estimates.

There is another area that is very closely related to many-body scattering, namely scattering on higher rank non-compact symmetric spaces. Here, in Section 8, we only discuss rank two, which corresponds to three-body scattering, since this is the only part that has been properly written up, but it is expected that very soon these results will extend to all higher rank spaces.

I hope that the notes will make many of these results more accessible, the connections more transparent, and explain the motivation behind them. Many-body scattering has a long history, and here I can only talk about the most recent developments. An excellent overview of results known in the early 1990s can be found in Hiroshi Isozaki's lecture notes [**30**]. Indeed, in some sense, the current notes continue where [**30**] left off. I introduce a fully microlocal picture, motivated by the geometric approach of Richard Melrose [**42**], and emphasize the results these give, but the basic spectral and scattering results follow from a simpler 'partial' microlocalization, which is one of the subjects of [**30**].

The notes were originally prepared for a mini-course at the Université de Nantes at the invitation of Professor Xue-Ping Wang, whose hospitality I gratefully acknowledge. The analytic continuation of the resolvent on symmetric spaces is a more recent development, but it was fueled by a discussion during the visit of Rafe Mazzeo, my collaborator, to Nantes. Over an espresso, Gilles Carrón mentioned that the existence of the analytic continuation was not known, something that was hard to believe, but we immediately realized that our methods should yield such a continuation rather directly. I also thank Gunther Uhlmann for urging me to write up these notes: without him, they may never have been written up, and Rafe Mazzeo for a careful reading of the manuscript.

2. Geometric optics and the wave equation

According to the rules of geometric optics, light propagates in straight lines, and reflects/refracts from surfaces according to Snell's/Descartes' law. That is to say, considering light as a stream of billiard balls, the energy as well as the tangential component of the momentum (tangential to the surface hit) is conserved upon hitting the surface.

But light satisfies the wave equation, i.e. if $u = u(x,t)$ is the electromagnetic field on $\Omega_x \times \mathbb{R}_t$, $\Omega \subset \mathbb{R}^n$, then $Pu = 0$ where P is the wave operator $c^2\Delta - D_t^2$, and a boundary condition also holds (say, Dirichlet), if Ω is not the whole space.

(Here $D_t = \frac{1}{i}\frac{\partial}{\partial t}$ and $\Delta = \sum_j D_{x_j}^2$ is the *positive* Laplacian.) How are these two viewpoints related?

One can phrase the connection in different ways. The most usual one in physics is that the billiard ball picture is accurate in the high frequency, i.e. low wave length, limit. That is to say, for high frequency light, geometric optics is accurate up to a 'small' error. A slightly different way of looking at this, which however does not involve approximations, is that the location of singularities of the solution of the wave equation is *exactly* predicted by geometric optics. Here singularities are understood as lack of smoothness, or possibly lack of analyticity.

Indeed, it is convenient at this point to generalize the setting somewhat. So let (Ω, g) be a Riemannian manifold with corners, $P = c^2\Delta_g - D_t^2$, $c > 0$. The speed of light, c, may be absorbed in the metric g, of course, we keep the notation in analogy with the usual wave equation.

For simplicity of notation in this paragraph we assume that $M_z = \Omega_x \times \mathbb{R}_t$ is boundaryless; in general, the same definitions hold in the interior of M. Thus, we associate a homogeneous real function on T^*M to P, namely its principal symbol: $p = c^2|\xi|_g^2 - \tau^2$, where we write $\zeta = (\xi, \tau)$ as the dual variable of $z = (x, t)$. Now T^*M is a symplectic manifold with symplectic form $\omega = \sum d\zeta_j \wedge dz_j$. Thus, p gives rise to a vector field H_p, called the Hamilton vector field, by requiring that $\omega(V, H_p) = Vp$ for any vector field V on T^*M. Hence H_p is a smooth vector field on T^*M explicitly given by

$$H_p = \frac{\partial p}{\partial \zeta}\frac{\partial}{\partial z} - \frac{\partial p}{\partial z}\frac{\partial}{\partial \zeta}.$$

Note that p is constant along the integral curves of H_p since taking $V = H_p$, $0 = \omega(H_p, H_p) = H_p p$. Null bicharacteristics are the integral curves of H_p inside its characteristic set $\Sigma = p^{-1}(\{0\})$. Thus, if $\gamma : I \to T^*M$ is a null bicharacteristic (here I is an interval), and $z(s) = z(\gamma(s))$, $\zeta(s) = \zeta(\gamma(s))$, then these solve the ODE's $\frac{dz}{ds} = \frac{\partial p}{\partial \zeta}$, $\frac{d\zeta}{ds} = -\frac{\partial p}{\partial z}$. Hence, when $M = \Omega \times \mathbb{R}$, $\Omega \subset \mathbb{R}^n$, $p = c^2|\xi|^2 - \tau^2$ as above, we deduce that ξ and τ are constant along the integral curves of H_p, hence their projection to M consists of straight line segments. More generally, the projection of null-bicharacteristics to Ω are geodesics of g.

There is an appropriate extension of this at boundary surfaces and even at corners, called generalized broken bicharacteristics, see [**43, 36**], which I will not explain in full generality, though I remark that many-body scattering, discussed in the next section in detail, is rather similar. However, a somewhat typical example is that of broken bicharacteristics. These are piecewise bicharacteristics, i.e. there is a sequence s_j, j in a subset of integers, such that for each j, $\gamma|_{(s_j, s_{j+1})}$ is a null bicharacteristic in the sense described above, the projection $z \circ \gamma$ of γ to M is continuous, and $\gamma(s_j+) - \gamma(s_j-)$ is conormal to the smallest dimensional boundary face containing $z(\gamma(s_j))$. Thus, the tangent vectors to $z\circ\gamma|_{(s_j,s_{j+1})}$ and $z\circ\gamma|_{(s_{j-1},s_j)}$ differ by a vector normal to the smallest boundary face containing $z(\gamma(s_j))$. This expresses that the normal component of the momentum may change, while the tangential component is conserved, when a light ray hits a boundary.

Now one can describe the singularities of u using null bicharacteristics. Let o be the zero section of T^*M. The location of the singularities is described by an object

$$\mathrm{WF}(u) \subset T^*M \setminus o = \{(z, \zeta) :\ \zeta \neq 0\}$$

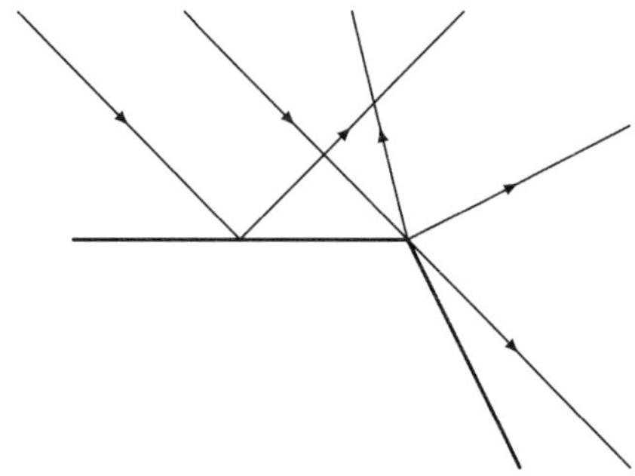

FIGURE 1. Projection of broken bicharacteristics to Ω. When rays hit the boundary hypersurfaces, the tangential component of the momentum and the kinetic energy are conserved, but the normal component may change. At the corner, there is no tangential component (though there would be if the time variable were not projected out), so the only constraint is the conservation of kinetic energy.

that is conic in ζ, i.e. $(z,\zeta)\in \mathrm{WF}(u)$ if and only if $(z,r\zeta)\in\mathrm{WF}(u)$ for every $r>0$. $\mathrm{WF}(u)$ is called the wave front set of u, and it describes where (in z) and in which codirection ζ is the distribution u not $\mathcal{C}^\infty$. More precisely, the definition of $\mathrm{WF}(u)$ is that $(z_0,\zeta_0)\notin\mathrm{WF}(u)$ if and only if there exists $\phi\in\mathcal{C}^\infty_c(M)$, $\phi(z_0)\neq 0$ such that the Fourier transform $\mathcal{F}(\phi u)$ of ϕu is rapidly decreasing in an open cone around ζ_0. Here we assume that M is boundaryless; otherwise we need to require that ϕ is supported in the interior of M. Again, there is a natural definition at ∂M which we do not give here. (There are more natural versions of this definition using pseudo-differential operators that I will describe later.) As an example, consider the step function: write $z=(z_1,z'')$, $u(z)=1$ if $z_1>0$, $u(z)=0$ if $z_1<0$. Then

$$\mathrm{WF}(u)=N^*\{z_1=0\}\setminus o=\{(0,z'',\zeta_1,0):\ \zeta_1\neq 0\},$$

the conormal bundle of the hypersurface $z_1=0$, with its zero section removed. The same statement holds, with $=$ possibly replaced by $\subset$, if we take any $\mathcal{C}^\infty$ function u_0 on M, and then define $u=u_0$ in $z_1>0$ and $u=0$ in $z_1<0$. Informally, one might say that u is singular in z_1 at $z_1=0$, but it depends smoothly on z''. The wave front set thus pinpoints not only the locations z of singularities (lack of smoothness) in M, but it refines it by also giving the frequencies (or rather direction of frequencies) at which these appear at z.

The theorem we are after is the following. In early versions it goes back to Lax [**35**], its boundaryless version is due to Hörmander [**28**], the smooth boundary versions are due to Melrose, Sjöstrand, Taylor and Ivrii [**33, 43, 44, 61**], and the corner version in the analytic category is due to Lebeau [**36**] (the $\mathcal{C}^\infty$ version is still not known in the corner setting) while a different extension, to conic points, is due to Melrose and Wunsch [**46**].

THEOREM 2.1. *Suppose $Pu\in\mathcal{C}^\infty(M)$, and if $\partial M\neq 0$ then $u|_{\partial M}=0$. Then $\mathrm{WF}(u)\subset\Sigma=p^{-1}(\{0\})$ (microlocal elliptic regularity). Moreover, $\mathrm{WF}(u)$ is a union of maximally extended generalized broken bicharacteristics inside Σ (propagation of singularities).*

This theorem states that if a point $(z,\zeta) \in T^*M \setminus o$ is in $\mathrm{WF}(u)$ and u solves $Pu \in \mathcal{C}^\infty(M)$, and satisfies a boundary condition if appropriate, then there is at least one maximally extended generalized broken bicharacteristic through (z,ζ) that is completely contained in $\mathrm{WF}(u)$. Of course, in the absence of boundaries, and often even in their presence, there is a unique maximally extended generalized broken bicharacteristic through (z,ζ), so the statement is that this bicharacteristic is completely in $\mathrm{WF}(u)$. However, as soon as codimension two or higher corners appear, there is no hope for such uniqueness, and this theorem is the optimal statement.

At least in the nicest settings (no boundaries, or non-degeneracy assumption at the boundaries which are assumed to be smooth), this theorem can be improved significantly to predict not only the location, but also the amplitude of the singularities of u.

3. Propagation in many-body scattering

There is an analogous setup for scattering. Now we want to understand how interacting particles behave. Again, there is a classical mechanical setup (the analogue of geometric optics) and a quantum mechanical setup (the analogue of the wave equation). To focus on the most relevant points, I formulate the problem in a time-independent fashion, though it is easy to reformulate everything in a time dependent way. We only do this in a remark following Theorem 3.1.

Thus, we want to understand tempered distributional solutions u of $(H-\lambda)u = 0$; here $\lambda \in \mathbb{R}$ is the energy, and H is the Hamiltonian, i.e. the analogue of $H-\lambda$ is P above. Namely, if we have N particles, each of which is d-dimensional with positions $x_1, \dots, x_N \in \mathbb{R}^d$, mass $m_1, \dots, m_N$, and the interaction between particle i and j is given by a potential V_{ij} (which is a function on $\mathbb{R}^d$), then the Hamiltonian describing this system is

$$H = \sum_{i=1}^N \frac{1}{2m_i}\Delta_{x_i} + \sum_{i<j} V_{ij}(x_i - x_j) = \Delta + V,$$

which is an operator on (functions on) $\mathbb{R}^n = \mathbb{R}^{Nd}$. Planck's constant $\hbar$ is here taken to be 1; it could be absorbed in the x_i by a simple rescaling.

Now H is elliptic in the standard sense, namely its principal symbol is $\sum \frac{1}{2m_i}|\xi|^2$, which never vanishes outside the zero section o. Note that the potential is lower order than Δ in the standard sense, so it is not part of the principal symbol. So, by the previous theorem,

$$(H-\lambda)u = 0 \Rightarrow \mathrm{WF}(u) = \emptyset \Rightarrow u \in \mathcal{C}^\infty(\mathbb{R}^n).$$

So the only possibility of interesting behavior for u is at infinity, and this is exactly what we want to understand.

The main feature of many-body problems is that even if V_{ij} decays at infinity on $\mathbb{R}^d$, it does *not* decay at infinity in $\mathbb{R}^n$ since it is a constant along $X_{ij} = \{x_i = x_j\}$, as well as along its translates X'_{ij}, X''_{ij}, so it does not decay if we go to infinity, say, along X_{ij}; see Figure 2. The X_{ij} are called collision planes (as are their intersections) since at X_{ij} particles i and j are at the same place.

In the two-body problem one actually has $H = \Delta_{x_1,x_2} + V_{12}(x_1 - x_2)$, i.e. V_{12} still does not decay at infinity, e.g. if one keeps $x_1 = x_2$ but lets $x_1 \to \infty$. However, one can easily remove the center of mass by performing a Fourier transform along

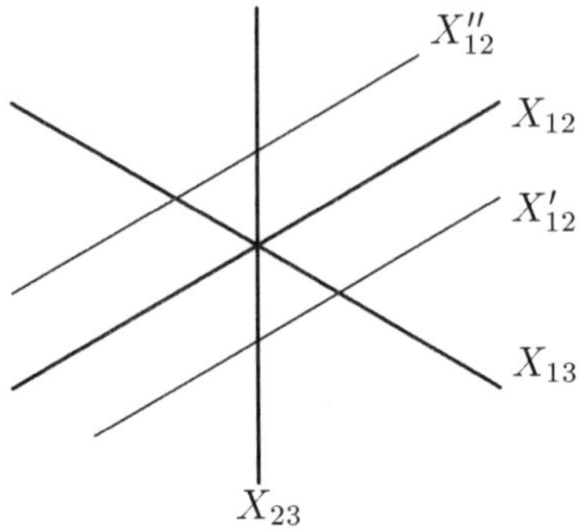

FIGURE 2. Collision planes X_{12}, X_{13} and X_{23} and translates X'_{12} and X''_{12} of X_{12}. V_{12} is constant along X_{12}, is a (typically different) constant along X'_{12}, etc., so it does not decay at infinity unless it is identically zero.

X_{12}. This conjugates $H - \lambda$ to $H^{12} + |\xi_{12}|^2 - \lambda$, where ξ_{12} is the variable on X^*_{12}, and $H^{12} = \Delta_{X^{12}} + V_{12}$, X^{12} being the orthocomplement of X_{12}. Thus, one reduces the study of $H - \lambda$ to that of a Hamiltonian on X^{12}, namely $H^{12} - \lambda'$, $\lambda' = \lambda - |\xi_{12}|^2$ being a shifted spectral parameter. Now V_{12} decays at infinity (we are working on X^{12}!), so H^{12} can be considered as a perturbation of $\Delta_{X^{12}}$, hence its analysis is rather simple. Notice that the point spectrum of H^{12} gives rise to a branch of the continuous spectrum of H: this is a phenomenon that is very typical in many-body scattering. The center of mass can also be removed in any actual many-body problem, but one still obtains a Hamiltonian with non-decaying potentials as before.

One can still talk about classical mechanics, just as for the wave equation, using bicharacteristics. These are deterministic – if V is smooth enough (we usually assume that V is $\mathcal{C}^\infty$). But much like for corners, there is a compressed description of dynamics near infinity. This is somewhat more complicated than for the wave equation, but only because particles can be bound together. Thus, even the 'classical' description is partly quantum. These two facts, the presence of collision planes and the bound states, are the two crucial features of many-body scattering.

The compressed dynamics in the absence of bound states looks just like in the wave equation setting. One should think of this as a good description when a classical trajectory is uniformly near infinity.

More precisely, it is convenient to introduce Agmon's generalization of the many-body problem, which amounts to using the vector space structure of $\mathbb{R}^n$ as the setting. One can also give geometric generalizations (in the sense of differential geometry) that arose from the work of Melrose [**41, 42**], and I will do this later.

So we work on the vector space $X_0 = \mathbb{R}^n$, equipped with the Euclidean metric. We are also given a finite collection $\mathcal{X} = \{X_a : a \in I\}$ of linear subspaces X_a of $\mathbb{R}^n$, called the collision planes. We assume that $\mathcal{X}$ is closed under intersections, and $X_0 = \mathbb{R}^n \in \mathcal{X}$, $X_1 = \{0\} \in \mathcal{X}$. We let $X^a = X_a^\perp$ be the orthocomplement of X_a in $\mathbb{R}^n$, so $\mathbb{R}^n = X_a \oplus X^a$. (Agmon's generalization is thus that the X_a do not have to come from intersections of the planes $X_{ij} = \{x_i = x_j\}$.) We write the corresponding coordinates as (x_a, x^a), and denote the orthogonal projection to X^a

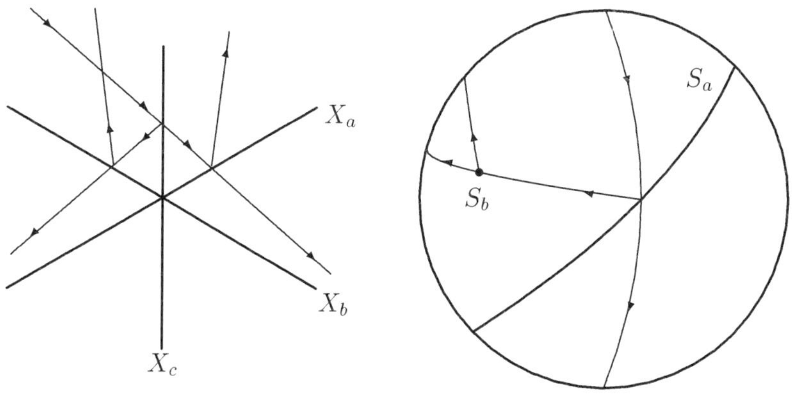

FIGURE 3. On the left, broken geodesics in $\mathbb{R}^n \setminus \{0\}$, $n = 2$, broken at the collision planes X_a, X_b and X_c. On the right, the projection of broken geodesics in $\mathbb{R}^n \setminus \{0\}$, $n = 3$, emanating from the north pole, to the unit sphere S_0, better understood as the sphere at infinity. The C_a, C_b are the intersection of the collision planes X_a, X_b with S_0; $\dim X_a = 2$, $\dim X_b = 1$.

by π^a. A many-body Hamiltonian in potential scattering is an operator of the form

$$H = \Delta + \sum_a (\pi^a)^* V_a,$$

where V_a is a real valued function on X^a in a certain class, for example V_a is a symbol on X^a of negative order: $V_a \in S^{-\rho}(X^a)$, $\rho > 0$. We also assume that $V_0 = 0$ for normalization; note that $X^0 = \{0\}$, so V_0 would simply play the role of the spectral parameter. We sometimes drop the pull-back notation from now on and write $H = \Delta + \sum_a V_a$.

Another useful piece of terminology is the following. We say that V_a is short range if $V_a \in S^{-\rho}(X^a)$ for some $\rho > 1$. We say that V_a is long-range if $V_a \in S^{-\rho}(X^a)$ for some $\rho \in (0, 1]$. The Coulomb potential is thus 'marginally long-range', at least if we ignore its singularity at 0 (which is not a serious problem anyway). Whether V_a is short- or long-range does not make any difference for the propagation phenomena we discuss in this section. However, it does make a major difference for the precise behavior of generalized eigenfunctions at the 'radial sets' which we discuss later. This also shows up in the related issue of asymptotic completeness.

Yet another notation we use on occasion is that of a k-cluster. Physically, a cluster describes particles that are close (or collide), and a k-cluster means that there are k clusters of particles, inside each of which the particles are close to each other. So in N-particle scattering, the N-cluster describes N asymptotically free particles (none is close to any other), hence we say that the collision plane $X_0 = \mathbb{R}^n$ is the N-cluster. On the other hand, if $X_a \neq \{0\}$ is such that $X_b \subsetneq X_a$ implies that $X_b = \{0\}$, then X_a, or rather a, is a 2-cluster. E.g. given five particles, a 2-cluster is where $x_1 = x_2$ and $x_3 = x_4 = x_5$, i.e. the particles 1 and 2, resp. 3, 4 and 5, are close to each other. In general, a k-cluster X_a can be defined by the length of nested chains of collision planes inside X_a.

One need not assume that all interactions between the particles are via potentials. Indeed, V_a may be allowed to be any first order differential operator on the vector space X^a with symbolic coefficients of negative order. Also, one may generalize the metric g in an analogous fashion, as discussed later, which in effect allows V_a to be second order provided that H remains elliptic. To simplify the notation, and due to the traditions, we mostly talk as if V_a were potentials, but the generalization to such higher order perturbations requires only occasional and minor modifications, which will be pointed out.

The subsystem Hamiltonians are defined by

$$H^a = \Delta_{X^a} + \sum_{X_a \subset X_b} V_b.$$

Note that $X_a \subset X_b$ if and only if $X^a \supset X^b$, so above V_b is really the pull-back of V_b from X^b to to X^a by the orthogonal projection. Thus, H^a is an operator on (functions on) X^a, and indeed it is a many-body Hamiltonian.

We also let

$$X_{a,\text{sing}} = \bigcup \{X_b : X_b \subsetneq X_a\}, \ X_{a,\text{reg}} = X_a \setminus X_{a,\text{sing}},$$

be the singular, respectively the regular, part of X_a. Thus, if X_c is a collision plane and X_a is not a subset of X_c, then $X_a \cap X_c$ is a proper subset of X_a, and is a collision plane (since $\mathcal{X}$ is closed under intersections), so $X_a \cap X_c \subset X_{a,\text{sing}}$. Correspondingly, V_c decays at $X_{a,\text{reg}}$, so

$$H_a = \Delta_{X_a} + H^a,$$

which is an operator on (functions on) $X_0 = \mathbb{R}^n$, has the property that $H - H_a$ is a function that decreases at $X_{a,\text{reg}}$. So H_a should be thought of as a good approximation of H at $X_{a,\text{reg}}$. Note that $X_{a,\text{sing}}$ is a finite union of codimension ≥ 1 submanifolds of X_a, so $X_{a,\text{reg}}$ is in particular an open dense subset of X_a. Also, note that Δ_{X_a} plays a role analogous to the kinetic energy of the center of mass in the two-body setting, but now this description only valid locally, at $X_{a,\text{reg}}$.

Having thus described the configuration space $X = X_0 = \mathbb{R}^n$, the next step is to describe the phase space, as was done first in [**65**] and [**66**]. The main goal in the process is to obtain a space on which broken bicharacteristics behave well. We remind the reader that we are concerned with singularities at infinity, hence with bicharacteristics that are uniformly close to infinity. Later we give a compactified description, but here for simplicity we give its homogeneous version, much as for the wave equation where bicharacteristics were integral curves of the homogeneous principal symbol. So we start with T^*X, but we wish to compress it at X_a in such a way that at $X_{a,\text{reg}}$, $T^*_{X_{a,\text{reg}}}X$ is replaced by $T^*_{X_{a,\text{reg}}}X_a = T^*X_{a,\text{reg}}$. For broken bicharacteristics this has the effect that only the X_a-tangential component of the momentum is preserved at $X_{a,\text{reg}}$. So we define the compressed cotangent bundle as

$$\dot{T}^*X = \cup_{a \neq 1} T^*X_{a,\text{reg}}.$$

Note that this is at first just a set, equipped with a projection $\dot{T}^*X \to X \setminus \{0\}$ induced by the bundle projections $T^*X_{a,\text{reg}} \to X_{a,\text{reg}}$. There is also a natural $\mathbb{R}^+$-action on $\dot{T}^*X$ via dilation in the configuration variables:

$$\mathbb{R}^+_r \times T^*X_{a,\text{reg}} \ni (r, x_a, \xi_a) \mapsto (rx_a, \xi_a) \in T^*X_{a,\text{reg}}. \tag{3.1}$$

We topologize $\dot{T}^*X$ via the projection

$$\pi : T^*_{X\setminus\{0\}}X \to \dot{T}^*X,$$

whose restriction to $T^*_{X_{a,\mathrm{reg}}}X$ is the pull-back of one-forms by the inclusion map $X_{a,\mathrm{reg}} \hookrightarrow X$. Thus, writing (ξ_a, ξ^a) as the momenta dual to (x_a, x^a), π projects out the normal component of the momentum, ξ^a. The topology is then the weakest topology that makes π continuous, i.e. a set C in $\dot{T}^*X$ is closed if and only if $\pi^{-1}(C)$ is closed.

We can now describe the contribution of the bound states to the characteristic sets. As mentioned above, this is one of the most interesting features of many-body scattering that has no analogue for the wave equation. These are conic subsets of $\dot{T}^*X$ (conic with respect to the $\mathbb{R}^+_r$-action in (3.1)). The characteristic sets describe where certain operators are not elliptic, i.e. invertible, at infinity, in a precise sense described in the subsequent sections. They correspond to the 'energy shell', i.e. being on the characteristic set at energy λ means that the particles have total energy λ. We let

$$\mathrm{Char}_0(\lambda) = \{(x, \xi) \in T^*X : \; g(\xi) = \lambda\}$$

be the free characteristic variety, with g being the metric function on T^*X, and more generally we set

$$\mathrm{Char}_a(\lambda) = \{(x_a, \xi_a) \in T^*X_a : \; \lambda - g_a(\xi_a) \in \mathrm{spec}_{pp}(H^a)\} \subset T^*X_a.$$

Notice that $\lambda = g_a(\xi_a) + \epsilon_\alpha$, $\epsilon_\alpha \in \mathrm{spec}_{pp}(H^a)$, corresponds to the splitting of the total energy λ to the kinetic energy of the cluster, $g_a(\xi_a)$, plus the energy of the bound state, ϵ_α. Thus, $\mathrm{Char}_a(\lambda)$ describes that particles may exist in a bound state of H^a, of energy ϵ_α, along X_a, with kinetic energy $g_a(\xi_a) = \lambda - \epsilon_\alpha$. Moreover, H^0 is the zero operator on $X^0 = \{0\}$, so if $a = 0$, these two definitions are consistent. If $X_a \subset X_b$, the pull-back of one-forms gives a projection $\pi_{ba} : T^*_{X_a}X_b \to T^*X_a$. Let

$$\dot{\mathrm{C}}\mathrm{har}(\lambda) = \cup\, \dot{\mathrm{C}}\mathrm{har}_a(\lambda) \subset \dot{T}^*X,$$
$$\dot{\mathrm{C}}\mathrm{har}_a(\lambda) = \cup_{X_b \supset X_a} \pi_{ba}(\mathrm{Char}_b(\lambda)) \cap T^*X_{a,\mathrm{reg}}.$$

In order to understand $\dot{\mathrm{C}}\mathrm{har}(\lambda)$ it is important to keep in mind several results on the structure of the eigenvalues of the subsystems. So let

$$\Lambda_a = \cup_{b : X^b \subsetneq X^a} \mathrm{spec}_{pp}(H^b)$$

be the set of thresholds of H^a. Fundamental results of Perry, Sigal and Simon [**53**] and of Froese and Herbst [**16**] show that Λ_a is closed, countable, and the countable set $\mathrm{spec}_{pp}(H^a)$ can only accummulate at Λ_a, so

$$\Lambda'_a = \Lambda_a \cup \mathrm{spec}_{pp}(H^a) = \cup_{b : X^b \subset X^a} \mathrm{spec}_{pp}(H^b)$$

is also closed. Hence, $\dot{\mathrm{C}}\mathrm{har}(\lambda)$ is a closed subset of $\dot{T}^*X$. In fact, the quotient $\dot{\Sigma}(\lambda)$ of $\dot{\mathrm{C}}\mathrm{har}(\lambda)$ by the $\mathbb{R}^+$ action (which can be realized by restricting the various bundles to the unit sphere, $S_0 = \{x \in X_0 : \; |x| = 1\}$) is compact, and indeed it is metrizable, see [**65**]. Since compact topological spaces have better properties than non-compact ones, it is quite natural to work with $\dot{\Sigma}(\lambda)$, although we do not follow this route in this section. We also remark that it is much better to think of $\dot{\Sigma}(\lambda)$ lying at the sphere at infinity, rather than at S_0, since it is the dynamics at infinity that is described here. We will take up this approach in later sections.

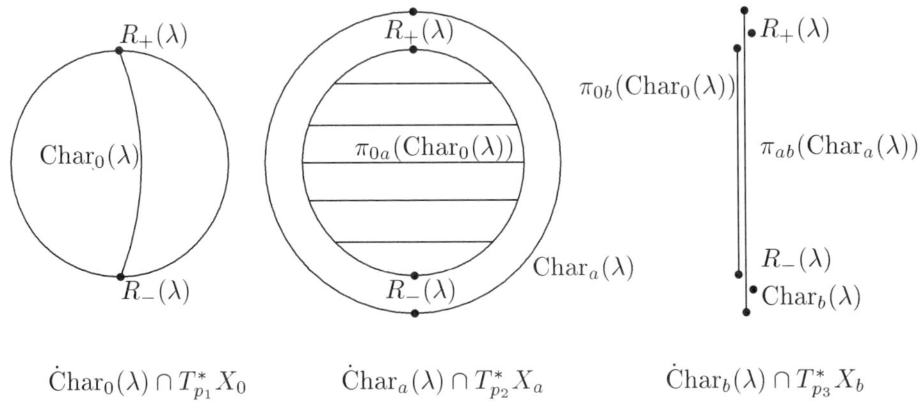

FIGURE 4. The characteristic set of many-body Hamiltonians. Here H is a 4-body Hamiltonian, a is a 3-cluster, b is a 2-cluster, $p_1 \in X_{0,\mathrm{reg}}$, $p_2 \in X_{a,\mathrm{reg}}$, $p_3 \in X_{b,\mathrm{reg}}$. The solid dots are the radial sets, defined below.

We also recall another result of Froese and Herbst [**15**], namely that eigenfunctions ψ_α of H^a with eigenvalue ϵ_α decay exponentially on X^a, at a specified rate, if $\epsilon_\alpha \notin \Lambda_a$. This generalizes to higher order perturbations, but requires a somewhat different approach, see [**67**]. In fact, this is the only place where second order perturbations behave differently from first or zeroth order ones. For the latter, there can be no positive energy bound states, while for the former this has been only proved for small metric perturbations in [**67**], and it is not clear whether it holds more generally, especially for trapping perturbations. (Note that if H^a, $a \neq 1$, is trapping then H is trapping at infinity!)

A generalized broken bicharacteristic (at energy λ) is then a continuous map $\gamma : I \to \dot{\mathrm{C}}\mathrm{har}(\lambda)$, I an interval, such that a Hamilton vector field condition holds. To see what this is, we consider a subset of continuous functions on $\dot{T}^*X$, namely the class of π-invariant $\mathcal{C}^\infty$ functions on T^*X. π-invariance means that if $\zeta, \zeta' \in T^*X$ and $\pi(\zeta) = \pi(\zeta')$ then $f(\zeta) = f(\zeta')$. If f is π-invariant then it induces a function f_π on $\dot{T}^*X$ by $f_\pi(q) = f(\zeta)$ if $q = \pi(\zeta)$. Moreover, if f is smooth (or indeed just continuous) then f_π is continuous by the definition of the topology on $\dot{T}^*X$.

Now, if $\tilde{\gamma}$ is a curve in a manifold, one way to put that it is an integral curve of a vector field V is that for all smooth functions f, $\frac{d}{ds}(f \circ \tilde{\gamma})|_{s=s_0} = (Vf)(\tilde{\gamma}(s_0))$. If f is a smooth π-invariant function on T^*X, then f defines a $\mathcal{C}^\infty$ function on T^*X_a for all a, so $H_{g_a} f$ makes sense. Here H_{g_a} is the Hamilton vector field of the metric function g_a on T^*X_a, so explicitly, $H_{g_a} = 2\xi_a \cdot \partial_{x_a}$. Now we would like to say that along a generalized broken bicharacteristic γ, $\frac{d}{ds}(f_\pi \circ \gamma)|_{s=s_0}$ should be given by $H_{g_b} f$ for some b and some ζ with $\pi(\zeta) = \gamma(s_0)$. The problem is that there are many such points ζ and clusters b, so this statement does not make any sense. However, we may replace the derivative by the lim inf of the difference quotients, i.e. by $D_\pm h(s_0) = \liminf_{s \to s_0} \frac{h(s) - h(s_0)}{s - s_0}$, and demand an inequality instead of the equality. That is, we may demand that $D_\pm(f_\pi \circ \gamma)(s_0)$ may not be less than the worst possible scenario as we run over all such b and ζ. Thus, the condition for a continuous map $\gamma : I \to \dot{\mathrm{C}}\mathrm{har}(\lambda)$ to be a generalized broken bicharacteristic is then that for any $s_0 \in I$, if $\gamma(s_0) \in T^*X_{a,\mathrm{reg}}$ then

$$D_\pm(f_\pi \circ \gamma)(s_0) \geq \inf\{(H_{g_b} f)(\zeta) :\ \zeta \in \mathrm{Char}_b(\lambda),\ \pi(\zeta) = \gamma(s_0),\ X_a \subset X_b\}.$$

If the set of bound states is discrete, then such a curve γ is piecewise an integral curve of the Hamilton vector field of g_b inside $\mathrm{Char}_b(\lambda)$, where b may of course vary. In particular, if there are no bound states in any proper subsystem, the picture is very similar to wave propagation: the definition can be reduced to the analogue of Lebeau's [**36**].

The structure of the generalized broken bicharacteristics, including the above claims, depends on having a large supply of π-invariant functions. But these exist, since the pull-backs of all functions on X to T^*X is π-invariant, so one can localize in X using smooth cutoffs. Moreover, near $X_{a,\mathrm{reg}}$, each component of ξ_a is π-invariant, as is $\xi^a \cdot x^a$. Note that the generalized broken bicharacteristics depend on V, but only via the characteristic set $\dot{\mathrm{C}}\mathrm{har}(\lambda)$, i.e. only via the bound states of the subsystem Hamiltonians.

There is also a wave front set associated to many-body scattering which measures the microlocal decay of tempered distributions at infinity. For a tempered distribution u, $\mathrm{WF}_{\mathrm{sc}}(u)$ is a closed conic subset of $\dot{T}^*X$. Apart from u, it depends on $\mathcal{X}$, since $\dot{T}^*X$ depends on $\mathcal{X}$, but we suppress this in the notation, and write

$$\mathrm{WF}_{\mathrm{sc}}(u) = \mathrm{WF}_{\mathrm{sc},\mathcal{X}}(u).$$

Its definition is slightly complicated, and I only refer to [**65**] for the general definition, which uses the structure of the pseudo-differential algebra, in particular the operator-valued nature of symbols at infinity. However, for generalized eigenfunctions of many-body Hamiltonians it is simple. Namely, suppose that $(H-\lambda)u \in \mathcal{S}(X)$, where $\mathcal{S}(X)$ is the space of Schwartz functions. For $\bar{x} = \bar{x}_a \in X_{a,\mathrm{reg}}$ and $\bar{\xi}_a \in X_a^*$ we say that $(\bar{x}_a, \bar{\xi}_a) \notin \mathrm{WF}_{\mathrm{sc}}(u)$ if there exists $\phi \in \mathcal{C}_c^\infty(X_a^*)$ such that $\phi(\bar{\xi}_a) \neq 0$ and $\mathcal{F}^{-1}\phi\mathcal{F}u$ is rapidly decreasing in an open cone in X around $\bar{x}_a$. Two examples are:

$$\mathrm{WF}_{\mathrm{sc}}(e^{ix\cdot\xi_0}) = \pi(\{(x,\xi_0) :\ x \neq 0\}),\ \xi_0 \in X_0^*,$$
$$\mathrm{WF}_{\mathrm{sc}}(e^{i\alpha|x|}) = \pi(\{(x, \frac{\alpha x}{|x|}) :\ x \neq 0\}),\ \alpha \in \mathbb{R}.$$

More generally, if v is a symbol of any order on X_0, say $v \in S^k(X_0)$, and $\phi \in \mathcal{C}^\infty(X_0)$ is homogeneous degree 1 for $|x| > 1$, then

$$\mathrm{WF}_{\mathrm{sc}}(e^{i\phi(x)}v(x)) \subset \pi(\mathrm{graph}\, d\phi) = \pi(\{(x,(d\phi)(x)) : |x| > 1\}).$$

The condition $|x| > 1$ is due to the requirement of the homogeneity of ϕ only for $|x| > 1$; technically we should add a subset of $|x| \leq 1$ to the right hand side to make it conic. The theorem on the propagation of singularities is the following.

THEOREM 3.1. *Suppose that $\lambda \in \mathbb{R}$ and H is a many-body Hamiltonian. If $u \in \mathcal{S}'(\mathbb{R}^n)$ and $(H-\lambda)u \in \mathcal{S}(\mathbb{R}^n)$ then $\mathrm{WF}_{sc}(u) \subset \dot{\mathrm{C}}\mathrm{har}(\lambda)$ (microlocal elliptic regularity). Moreover, $\mathrm{WF}_{sc}(u)$ is a union of maximally extended generalized broken bicharacteristics of $H-\lambda$ (propagation of singularities).*

REMARK. In the time dependent version, one considers tempered distributional solutions of $(D_t+H)u = 0$ on $X_0 \times \mathbb{R}_t$. Then D_t+H still has the structure of a many-body Hamiltonian, with $D_t+\Delta$ in place of Δ, with collision planes given by $X_a \times \mathbb{R}$, with $\{0\}$ added for the sake of completeness. Thus, t is always a variable along the collision planes, so in particular, its dual variable τ, is π-invariant. Moreover, $\mathrm{Char}_a(\lambda)$ is replaced by

$$\mathrm{Char}_a = \{(x_a, t, \xi_a, \tau) \in T^*(X_a \times \mathbb{R}) : \ -\tau - g_a(\xi_a) \in \mathrm{spec}_{pp}(H^a)\} \subset T^* X_a,$$

so effectively $-\tau$ plays the role of the energy λ. Generalized broken bicharacteristics can be defined as before with H_{g_b} replaced by $H_{\tau+g_b} = \partial_t + 2\xi_b \cdot \partial_{x_b}$. The main additional issue is that they can only be expected to give a good description of propagation at finite energies since $D_t + H$ is not elliptic in the usual sense. So the analogue of Theorem 3.1 is that if $u \in \mathcal{S}'(\mathbb{R}^n \times \mathbb{R})$, $(D_t + H)u = 0$, and $u = \psi(H)u$ for some $\psi \in \mathcal{C}_c^\infty(\mathbb{R})$, then $\mathrm{WF}_{\mathrm{sc}}(u)$ is a subset of the characteristic set, and in fact $\mathrm{WF}_{\mathrm{sc}}(u)$ is a union of maximally extended generalized broken bicharacteristics of $D_t + H$ inside it. The proof of this statement only requires simple modifications of the proof of the theorem.

The interpretation of the theorem is much analogous to that for the wave equation. However, there is a difference which also occurs in the traditional microlocal setting for more general operators (i.e. for operators other than the wave operator), see [**21**]. Namely, the orbits of the $\mathbb{R}^+$-action may be bicharacteristics, and then the statement of the theorem is empty at the points lying on these orbits since the wave front set is *a priori* conic. This happens for $(x_a, \xi_a) \in T^* X_{a,\mathrm{reg}} \subset \dot{T}^* X$ if and only if there exists some cluster b with $X_b \supset X_a$, and $\zeta \in \mathrm{Char}_b(\lambda)$ such that H_{g_b} at ζ is tangent to the orbits of the $\mathbb{R}^+$-action. This happens, in turn, if and only if ξ_a is parallel to x_a and $\lambda - |\xi_a|^2 \in \mathrm{spec}_{pp}(H^b)$. Such points are called radial points, and their collection is denoted by

$$\mathcal{R}(\lambda) = \bigcup_{a \neq 1} \{(x_a, \xi_a) \in T^* X_{a,\mathrm{reg}} : \exists c \in \mathbb{R}, \ \xi_a = c x_a, \\ \exists b, \ X_b \supset X_a, \ \lambda - |\xi_a|^2 \in \mathrm{spec}_{pp}(H^b)\}.$$

As we discuss in Section 7, $\mathcal{R}(\lambda)$ plays an important role in asymptotic completeness. In many-body scattering it appeared in the work of Sigal and Soffer [**55**] and was called 'propagation set' because in the time-dependent picture this is where particles end up as time goes to infinity. (In the stationary semiclassical picture, this is where non-trapped classical trajectories starting in a compact region end up.) It is thus unfortunate, in terms of terminology, that this is also the region where there is no real principal type propagation.

REMARK. In the time-dependent problem, the set of radial points is

$$\mathcal{R} = \bigcup_a \{(x_a, t, \xi_a, \tau) \in T^*(X_{a,\mathrm{reg}} \times \mathbb{R}) : x_a = 2t\xi_a, \\ \exists b, \ X_b \supset X_a, \ -\tau - |\xi_a|^2 \in \mathrm{spec}_{pp}(H^b)\}.$$

In terms of radial points, the difference between threshold energies $\lambda \in \Lambda = \Lambda_0$ and non-threshold energies is that if $\lambda \in \Lambda$, then there are *constant* generalized broken bicharacteristics, i.e. bicharacteristics whose image is a single point. Namely, if $x_a \in X_{a,\mathrm{reg}}$ and $\lambda \in \mathrm{spec}_{pp}(H^a)$, then

$$(x_a, 0) \in \mathrm{Char}_a(\lambda) \cap T^* X_{a,\mathrm{reg}} \subset \dot{\mathrm{C}}\mathrm{har}(\lambda),$$

and $H_{g_a} = 2\xi_a \cdot \partial_{x_a}$ vanishes there, so $(x_a, 0)$ is indeed the image of a constant bicharacteristic. While this does not make any difference for the propagation of singularities, it does for the related limiting absorption principle, which in this generality is due to Perry, Sigal and Simon [**53**].

THEOREM 3.2. *If $\lambda \notin \Lambda$, then the limits $R(\lambda \pm i0) = (H - (\lambda \pm i0))^{-1}$ exists as bounded operators between L^2_s and H^2_{-s} for $s > \frac{1}{2}$. Here H^m_l is the weighted Sobolev space $\langle x \rangle^{-l} H^m(\mathbb{R}^n)$, $L^2_s = H^0_s$.*

In fact, the proofs of the limiting absorption principle and the propagation of singularities are related. Indeed, the statement on propagation of singularities can be strengthened for $R(\lambda + i0)f$, $f \in \mathcal{S}(\mathbb{R}^n)$, by saying that $\mathrm{WF}_{\mathrm{sc}}(R(\lambda + i0)f)$ is not only a union of maximally extended generalized broken bicharacteristics, as follows from Theorem 3.1, but in fact it is a union of generalized broken bicharacteristics $\gamma : \mathbb{R}_s \to \dot{T}^* X$ which go to

$$\mathcal{R}_+(\lambda) = \mathcal{R}(\lambda) \cap \bigcup_{a \neq 1} \{(x_a, \xi_a) \in T^* X_{a,\mathrm{reg}} : \ x_a \cdot \xi_a > 0\}$$

as $s \to -\infty$. That is, the singularities at $\mathcal{R}_+(\lambda)$ (where the statement of Theorem 3.1 is empty) can only leave $\mathcal{R}_+(\lambda)$ in the *forward* direction. The limiting absorption principle is thus strengthened to:

THEOREM 3.3. *If $\lambda \notin \Lambda$, then for $f \in \mathcal{S}(\mathbb{R}^n)$, $\mathrm{WF}_{sc}(R(\lambda + i0)f)$ is a subset of the image of $\mathcal{R}_+(\lambda)$ under the* forward *generalized broken bicharacteristic relation. A similar statement holds for $R(\lambda - i0)f$ with $\mathcal{R}_+(\lambda)$ replaced by*

$$\mathcal{R}_-(\lambda) = \mathcal{R}(\lambda) \cap \bigcup_{a \neq 1} \{(x_a, \xi_a) \in T^* X_{a,reg} : \ x_a \cdot \xi_a < 0\},$$

and the forward relation by the backward relation.

In fact, if $u \in \mathcal{S}'(\mathbb{R}^n)$ and $\mathrm{WF}_{sc}(u)$ is disjoint from the image of $\mathcal{R}_-(\lambda)$ under the backward generalized broken bicharacteristic relation, then $R(\lambda + i0)u$ is defined by duality and $\mathrm{WF}_{sc}(R(\lambda + i0)u)$ is a subset of the image of $\mathcal{R}_+(\lambda) \cup \mathrm{WF}_{sc}(u)$ under the forward relation.

REMARK. $\lambda \notin \Lambda$ can be also characterized by $\mathcal{R}(\lambda) = \mathcal{R}_+(\lambda) \cup \mathcal{R}_-(\lambda)$, i.e. that $x_a \cdot \xi_a$ never vanishes on $\mathcal{R}(\lambda) \cap T^* X_{a,\mathrm{reg}}$ for any a.

In the time-dependent setting, $x_a = 2t\xi_a$ on $\mathcal{R}$, so $x \cdot \xi_a = 2t$. So $\mathcal{R}_+$, defined in $\mathcal{R}$ by $x_a \cdot \xi_a > 0$, is the subset of $\mathcal{R}$ where $t > 0$. Hence the 'outgoing' terminology for $R(\lambda + i0)$ and 'incoming' for $R(\lambda - i0)$. In fact, the solution of $(D_t + H)u = 0$ with $u|_{t=0} = \phi$, $\phi \in \mathcal{S}(X_0)$, say, is $u(.,t) = e^{-iHt}\phi$. The time-dependent propagation of singularities shows that $\mathrm{WF}_{\mathrm{sc}}(u)$ is a subset of the union of the image of $\mathcal{R}_+$ under the forward broken bicharacteristic relation and the image of $\mathcal{R}_-$ under the backward bicharacteristic relation. Using the spectral measure and Stone's theorem,

$$u(.,t) = \frac{1}{2\pi i} \int_{\mathbb{R}} e^{-i\lambda t} (R(\lambda + i0) - R(\lambda - i0))\phi \, d\lambda$$

Fixing some $\psi \in \mathcal{C}^\infty_c(\mathbb{R})$, for ϕ in the range of $\psi(H)$ we thus deduce that in $t > 0$, $\mathrm{WF}_{\mathrm{sc}}(u)$ arises from the $R(\lambda + i0)$ term, and in $t < 0$ from the $R(\lambda - i0)$ term. So the time-dependent and stationary settings are very close: the only difference is that in the latter, λ is a parameter, while in the former, it is a variable, $\lambda = -\tau$.

Again, one can make more precise propagation statements in some circumstances, such as three-body scattering, where the precise nature of the singularities can be analyzed, see [**22, 63**]. Here we only state the stronger implication for the structure of the scattering matrices, which we proceed to analyze.

4. Scattering matrices

Physically, the scattering matrices relate incoming and outgoing data in an experiment. In the time independent framework (where $-\lambda$ is the dual variable of time), for short-range potentials an incoming wave of energy λ in channel α (a channel is the choice of a cluster a and an L^2-eigenfunction ψ_α of H^a of energy ϵ_α) takes the following form in $|x| > 1$:

$$u_{\alpha,-} = e^{-i\sqrt{\lambda-\epsilon_\alpha}|x_a|}|x_a|^{-\frac{\dim X_a - 1}{2}} g_{\alpha,-}(\frac{x_a}{|x_a|})\psi_\alpha(x^a) + u'_-$$

Similarly, an outgoing wave has the form

$$u_{\alpha,+} = e^{i\sqrt{\lambda-\epsilon_\alpha}|x_a|}|x_a|^{-\frac{\dim X_a - 1}{2}} g_{\alpha,+}(\frac{x_a}{|x_a|})\psi_\alpha(x^a) + u'_+$$

i.e. the sign of the phase has changed. Here $g_{\alpha,\pm}$ may be taken e.g. L^2 functions on S_a, the unit sphere in X_a, or ideally, at least one of them may be taken $\mathcal{C}^\infty$. In either case, $u'_\pm$ are 'lower order terms', namely they must be in $L^2_{-1/2}$. (Note that $\langle x_a\rangle^{-\frac{\dim X_a - 1}{2}} \in L^2_s(X_a)$ for $s < -1/2$ but not for $s = -1/2$.) In fact, for $g_{\alpha,\pm} \in \mathcal{C}^\infty_c(S_{a,\mathrm{reg}})$ we may take them to be of the form $e^{-i\sqrt{\lambda-\epsilon_\alpha}|x_a|}|x_a|^{-\frac{\dim X_a + 1}{2}} v$ where v is a 0th order symbol, with $S_{a,\mathrm{reg}}$ denoting $X_{a,\mathrm{reg}} \cap S_0$.

One can now produce tempered distribution with given incoming, or alternatively of given outgoing, asymptotics. A typical example is of the form

$$P_{\alpha,+}(\lambda)g_{\alpha,-} = u_{\alpha,-} - (H - (\lambda + i0))^{-1}((H - \lambda)u_{\alpha,-}); \tag{4.1}$$

here the lower order terms may be dropped from $u_{\alpha,-}$ without affecting $u = P_{\alpha,+}(\lambda)g_{\alpha,-}$ and $g_{\alpha,-}$ can be specified to be any smooth function on S_a. In general, even if the incoming data are in a single channel α, as in (4.1), the corresponding generalized eigenfunction u of H will have outgoing waves in all channels. The S-matrix $S_{\alpha\beta}(\lambda)$ picks out the component in channel β by projection in a certain sense, see [**64**]. Thus, $S_{\alpha\beta}(\lambda)$ maps functions on S_a, the unit sphere in X_a, to functions on S_b, by

$$S_{\alpha\beta}(\lambda)g_{\alpha,-} = g_{\beta,+}$$

for u as in (4.1). For example, the free-to-free (i.e. N-cluster to N-cluster in N-body scattering) S-matrix $S_{00}(\lambda)$ maps functions on S_0, the unit sphere in $\mathbb{R}^n$, to functions on S_0, more precisely $S_{00}(\lambda) : L^2(S_0) \to L^2(S_0)$ is bounded.

More precisely, let T_+ be a pseudodifferential operator that is identically 1 on the outgoing radial set and identically 0 on the incoming radial set; see the paragraph of (5.6) for a precise statement. Then

$$S_{\alpha\beta}(\lambda) = \frac{1}{2i\sqrt{\lambda - \epsilon_\beta}}((H-\lambda)T_+P_{\beta,-}(\lambda))^* P_{\alpha,+}(\lambda), \tag{4.2}$$

i.e. for any $g \in \mathcal{C}^\infty(S_{a,\mathrm{reg}})$, $h \in \mathcal{C}^\infty(S_{b,\mathrm{reg}})$,

$$\langle h, S_{\alpha\beta}(\lambda)g\rangle = \langle (H-\lambda)T_+P_{\beta,-}(\lambda)h, \frac{1}{2i\sqrt{\lambda-\epsilon_\beta}}P_{\alpha,+}(\lambda)g\rangle.$$

This is equivalent to the usual wave operator definition in the time-dependent setting, see [**64**]. An immediate consequence of the propagation of singularities and the definition of the scattering matrices is the following:

THEOREM 4.1. *The wave front relation of $S_{\alpha\beta}(\lambda)$ is given by the broken bicharacteristic relation. In particular, if no proper subsystem of H has bound states, the wave front relation of $S_{00}(\lambda)$ is given by the broken geodesic flow on S_0 at distance π.*

While typically broken bicharacteristics can be continued in many ways when they hit a collision plane, it is important to keep in mind that under suitable assumptions (which rule out geometric complications) the broken bicharacteristic relation is Lagrangian, hence its dimension is the same as if there were no collision planes. The reason is that only a low dimensional family of broken bicharacteristics hits any specified collision plane, with the dimension of the possible continuations of each of these these bicharacteristics compensating to yield the correct dimension for Lagrangian submanifolds.

The significant improvement in the three-body case, as shown by Hassell and the author [**63, 22**], is that one can pinpoint not only the location of the singularities, but also their precise form. This theorem was motivated by the geometric result of Melrose and Zworski [**45**], showing that the scattering matrix on asymptotically Euclidean manifolds is a Fourier integral operator.

THEOREM 4.2. *Suppose that H is a three-body Hamiltonian and the V_a are Schwartz on X^a for all a. Then $S_{00}(\lambda)$ is a finite sum of Fourier integral operators (FIOs) associated to the broken geodesic relation on S_0 to distance π. Its canonical relation corresponds to the various collision patterns. The principal symbol of the term corresponding to a single collision at X_a is given by, and in turn determines, the 2-body S-matrix of H^a at energies $\lambda' \in (0, \lambda)$.*

REMARK. This result presumably extends to short range symbolic potentials, using the same methods, though it is technically more complicated to write down the argument in that case, and it has not been done. In fact, it should also extend to the N-body problem, provided that there are no bound states in any proper subsystem. Some assumption on the bound states is necessary, for otherwise the generalize broken bicharacteristic relation can become fairly complicated, see [**66**]. The reason why one does not need any assumption on bound states in three-body scattering is that for any 2-cluster a, $\mathrm{Char}_a(\lambda) \cap \pi_{0a}(\mathrm{Char}_0(\lambda))$ is either empty (if 0 is not an eigenvalue of H^a) or consists of the boundary of $\pi_{0a}(\mathrm{Char}_0(\lambda))$. In the former case there is no interaction (modulo smoothing terms) between the 0-cluster and the a-cluster dynamics, while in the latter case in the only place they interact, the two dynamics give the same propagation.

It should also be noted that the normalization of $S_{\alpha\beta}(\lambda)$ is not the standard one in many-body scattering (which is based on wave operators), but rather follows the geometric conventions [**41**]. The difference is that in the wave operator approach, free motion is factored out, so the free scattering matrix is the identity operator. On the other hand, in the geometric approach we describe the asymptotics of generalized eigenfunctions, or alternatively of the Schrödinger equation. Since free particles move to infinity in the opposite direction from which they came, it is reasonable that the two should differ by (a constant multiple of) pull-back by the antipodal map, and this is indeed the case, see [**64**]. The distance π propagation along (not broken!) geodesics on the sphere indeed takes particles to the antipodal point.

An immediate corollary, when combined with two-body results (e.g. analyticity of the S-matrix in λ' and the Born approximation) is the following inverse result.

COROLLARY. *If the* V_a *decay exponentially and* $\dim X_a \geq 2$ *for all* a *then* $S_{00}(\lambda)$ *for a single value of* λ *determines all interactions.*

This result is analogous to the recovery of cracks in a material by directing sound waves at it and observing the singularities of the reflected waves, except the last step which uses two-body results to get the potentials from the two-body S-matrices.

The other extremal scattering matrices are the 2-cluster to 2-cluster ones, and they describe the physically most interesting events. Indeed, it is hard to make more than two particles collide in an accelerator, so the initial state in a physical experiment tends to be a 2-cluster. The following result is due to Skibsted [**59**], and it also follows from the propagation of singularities and the definition of the S-matrices.

THEOREM 4.3. *Let* α *and* β *be two-clusters, and suppose that either* $\epsilon_\alpha \in \operatorname{spec}_d(H^a)$ *and* $\epsilon_\beta \in \operatorname{spec}_d(H^b)$, *or* V_c *is Schwartz for all* c. *Then the two-cluster to two-cluster S-matrix* $S_{\alpha\beta}(\lambda)$ *has* $\mathcal{C}^\infty$ *Schwartz kernel, except if* $\alpha = \beta$ *in which case the Scwartz kernel of* $S_{\alpha\alpha}(\lambda)$ *is conormal to the graph of the antipodal map on* S_a, *corresponding to free motion.*

Thus, principal symbol calculations do not help in this inverse problem. Note that if H is a 3-body Hamiltonian, then $\epsilon_\alpha \in \operatorname{spec}_d(H^a)$ and $\epsilon_\beta \in \operatorname{spec}_d(H^b)$ holds for any non-threshold bound state energies. The new result, in a joint project with Gunther Uhlmann, is the following [**62**].

THEOREM 4.4. *Suppose that* H *is a 3-body Hamiltonian,* a *is a 2-cluster,* α *is a channel of energy* $\epsilon_\alpha < 0$, V_a *is a symbol of negative order (i.e. may be long range). For any* $\mu > \dim X_a$ *there exists* $\delta > 0$ *such that the following holds.*

Suppose that $\sup |(1+|x^b|)^\mu V_b(x^b)| < \delta$ *for all* $b \neq a$. *Suppose also that* $I \subset (\epsilon_\alpha, 0)$ *is a non-empty open set, and let*

$$R = 2\sqrt{\sup I - \epsilon_\alpha}.$$

Then $S_{\alpha'\alpha''}(\lambda)$ *given for all* $\lambda \in I$ *and for all bound states* α', α'' *of* H^a *with* $\epsilon_{\alpha'}, \epsilon_{\alpha''} < \sup I$, *determines the Fourier transform of the effective interaction* $V_{\alpha,\text{eff}}$ *in the ball of radius* R *centered at* 0.

The effective interaction is the interaction that arises if we consider the 3-body problem as a 2-body problem, i.e. if we regard the two particles forming the cluster a as a single particle. Mathematically, this amounts to projecting to the state ψ_α in X^a and obtaing a new Hamiltonian $\Delta_{X_a} + V_{\alpha,\text{eff}}$ on X_a. Thus, the effective interaction is physically relevant. Moreover, there is no hope for recovering anything better than $V_{\alpha,\text{eff}}$ as shown by the high-energy inverse results of Enss and Weder [**11, 13**], Novikov [**49**] and Wang [**70, 71**].

This theorem says that if the unknown interactions are small then the effective interaction can be determined from the knowledge of all S-matrices with incoming and outgoing data in the cluster a in the relevant energy range. In fact, near-forward information suffices as in two-body scattering, where this was observed recently by Novikov [**50**]. Also, if one is willing to take small R and α is the ground state of H^a, it suffices to know $S_{\alpha\alpha}(\lambda)$ to recover $\hat{V}_{\alpha,\text{eff}}$ in a small ball.

In case V_b decay exponentially on X^b for all $b \neq a$, then $V_{\alpha,\text{eff}}$ decays exponentially on X_a, hence its Fourier transform is analytic, so $V_{\alpha,\text{eff}}$ itself can be recovered from these S-matrices.

REMARK. It is clear from the proof in [**62**] that there is a natural extension of this theorem to many-body scattering at low energies.

This result should extend to higher energies, i.e. $\sup I \leq 0$ is *not* expected to be essential. But it is hard to make R greater than $2\sqrt{-\epsilon_\alpha}$ even then. The reason is that our method relies on the construction of exponential solutions following Faddeev [**14**], Calderón [**6**], Sylvester and Uhlmann [**60**] and Novikov and Khenkin [**48**], but in the three-body setting. One thus allows complex momenta $\rho \in \mathbb{C}(X_a)$, the complexification of X_a, and one wants to construct solutions of $(H-\lambda)u = 0$ of the form

$$e^{i\rho\cdot x_a}(\psi_\alpha(x^a) + v),$$

where $v = v_\rho$ is supposed to be 'small' in the sense that it goes to 0 as $\rho \to \infty$ in an appropriate fashion. Note that with $v = 0$ these complex plane waves solve $(H_a - \lambda)u = 0$ with

$$\lambda = \rho\cdot\rho + \epsilon_\alpha; \tag{4.3}$$

this expresses that the total energy λ is the sum of the kinetic energy, $\rho\cdot\rho$, and the potential energy ϵ_α.

To construct u, we need to find v, and its study reduces to that of the conjugated Hamiltonian

$$e^{-i\rho\cdot x_a}(H-\lambda)e^{i\rho\cdot x_a} = H^a + \Delta_{X_a} + 2\rho\cdot D_{X_a} + I_a - \epsilon_\alpha$$

with $\rho \in \mathbb{C}(X_a)$ the complex frequency. Here we used (4.3). Now, I_a is considered as a perturbation (this is the reason for the smallness assumption in the theorem), so we really study the model operator,

$$H^a + \Delta_{X_a} + 2\rho\cdot D_{X_a} - \epsilon_\alpha.$$

Taking the Fourier transform in the X_a variables, one obtains

$$H^a + |\xi_a|^2 + 2\rho\cdot\xi_a - \epsilon_\alpha.$$

Writing $\rho = z\nu + \rho_\perp$ with $|\nu| = 1$, $\rho_\perp\cdot\nu = 0$, ρ, ν real, $z \in \mathbb{C}$, this operator becomes

$$H^a + |\xi_a|^2 + 2\rho_\perp\cdot\xi_a + 2z\nu\cdot\xi_a - \epsilon_\alpha = H^a + (\xi_a + \rho_\perp)^2 + 2z\nu\cdot\xi_a - |\rho_\perp|^2 - \epsilon_\alpha.$$

If ρ is not real, then neither is z, so this operator is invertible if $\nu\cdot\xi_a \neq 0$ since H^a is self-adjoint. On the other hand, if $\nu\cdot\xi_a = 0$, this operator becomes

$$H^a + (\xi_a + \rho_\perp)^2 - |\rho_\perp|^2 - \epsilon_\alpha,$$

i.e. its invertibility properties correspond to the behavior of the boundary values of the resolvent of H^a at the real axis. If $|\rho_\perp|^2 + \epsilon_\alpha < 0$, i.e. if $|\rho_\perp| < \sqrt{-\epsilon_\alpha}$, then the spectral parameter $|\rho_\perp|^2 + \epsilon_\alpha - (\xi_a + \rho_\perp)^2$ is negative, so only the bound states of H^a contribute to the characteristic variety, i.e. the two-cluster a may not break up. On the other hand, if $|\rho| \geq \sqrt{-\epsilon_\alpha}$, such a break-up is possible *even if* $\lambda < 0$, i.e. where the break up may not happen for *real* frequencies. The break-up greatly influences analyticity properties, hence one cannot easily use large $\rho_\perp$. On the other hand, one needs such large $\rho_\perp$ to recover $V_{\alpha,\text{eff}}$ on larger balls, hence the limitation in the theorem. This also suggests that the fixed energy problem would be hard, since then one always needs to let $\rho_\perp \to \infty$ to keep $\rho \cdot \rho = |\rho_\perp|^2 + z^2$ fixed and yet have $\rho \to \infty$.

5. Many-body scattering pseudo-differential operators

I will present the calculus from the compactified point of view. Both the one-step polyhomogeneous (i.e. 'classical') and the non-polyhomogeneous calculus can be described in non-compact terms, i.e. directly on X_0, but this is more complicated and less natural. Indeed, one of the beauties of compactification is that it exactly captures the structure of many-body Hamiltonians. *We warn the reader here that from now on the Euclidean variable is written as z, rather than x in the preceeding sections, for compatibility with previous papers espousing this approach, such as* [**41, 42**].

To see how the compactification should go, recall first that a classical symbol of order 0 on $\mathbb{R}^n_z$ has an asymptotic expansion

$$a(r\omega) \sim \sum_{j=0}^{\infty} r^{-j} a_j(\omega),\ a_j \in \mathcal{C}^\infty(\mathbb{S}^{n-1}),$$

in the polar coordinates (r, ω): $z = r\omega$. The meaning of such an expansion is that, for any k, the difference of a and the sum of the first k terms on the right hand side is a symbol of order $-k$. This expansion is just a Taylor series at $r = \infty$, or rather at '$r^{-1} = 0$'. So we compactify $\mathbb{R}^n$ into a ball $\overline{\mathbb{B}^n}$ by adding points $(0, \omega)$, $\omega \in \mathbb{S}^{n-1}$, and making $(r^{-1}, \omega) = (x, \omega)$ coordinates near these points. The resulting space is called the radial compactification $\overline{\mathbb{R}^n}$ of $\mathbb{R}^n$. Thus, a classical symbol of order 0 is simply a smooth function of $\overline{\mathbb{R}^n}$; the asymptotic expansion at infinity is its Taylor series around the boundary, $x = 0$.

This compactification, whose utility in this context was emphasized by Melrose [**41**], can also be realized as the closed unit upper hemisphere via a modified stereographic projection. So let

$$\mathrm{RC} : \mathbb{R}^n \to \mathbb{S}^n_+,\ \mathrm{RC}(z) = (\frac{1}{\langle z \rangle}, \frac{z}{\langle z \rangle}),\ \langle z \rangle = (1 + |z|^2)^{1/2},\ z \in \mathbb{R}^n.$$

Then n of the $n+1$ variables $(\frac{1}{\langle z \rangle}, \frac{z}{\langle z \rangle})$ give local coordinates on various regions of $\mathbb{S}^n_+$. In particular, in coordinate patches near the equator, which is $\partial\mathbb{S}^n_+$, $\frac{1}{\langle z \rangle}$ (or indeed $x = |z|^{-1}$) and $n-1$ of $\frac{z_j}{\langle z \rangle}$ (or indeed $\omega_j = \frac{z_j}{|z|}$) can be taken as coordinates, showing that $\mathbb{S}^n_+$ can be identified with the radial compactification $\overline{\mathbb{R}^n}$. A slightly modified version of x (it needs to be smoothed at $z = 0$, where '$x = \infty$'), or $\langle z \rangle^{-1}$, can be taken as a boundary defining function. We will usually write x for this, so $x = |z|^{-1}$ for $|z| \geq 1$, say. (A boundary defining function is a non-negative function whose zero set is exactly the boundary, and whose differential does not vanish there.)

How can we adapt this to many-body scattering? Let $\bar{X}_a$ denote the closure of X_a in the compactification $\overline{\mathbb{R}^n}$ of $\mathbb{R}^n$, and let $C_a = \partial\bar{X}_a \subset \partial\mathbb{S}^n_+ = C_0$. The closure of any translate of X_a intersects C_0 in the same submanifold (a sphere) as X_a itself. Indeed, writing the coordinates as (z_a, z^a) on $X_0 = X_a \oplus X^a$, local coordinates near C_a are given by $Z^a = \frac{z^a}{|z|}$, $|z|^{-1}$ and $\dim X_a - 1$ of $\frac{(z_a)_j}{|z|}$. Thus, $Z^a \to 0$ as $x \to 0$ along any translate, since z^a is constant along these. So V_a is not even continuous on $\bar{X}_0$, as it takes different values on the different translates of X_a. However, it *is* a

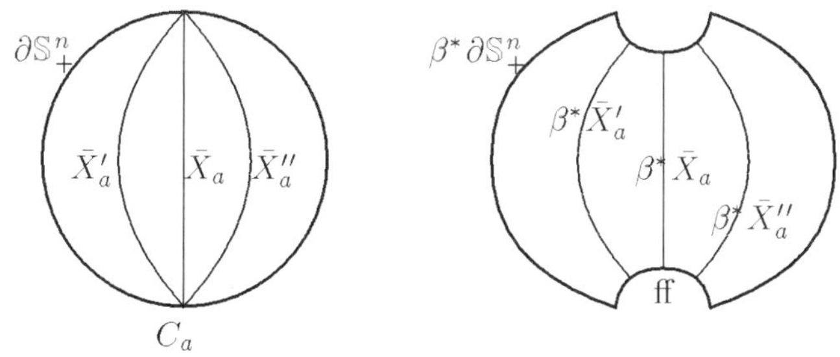

FIGURE 5. Translates of X_a on $[\bar{X}_0; C_a]$.

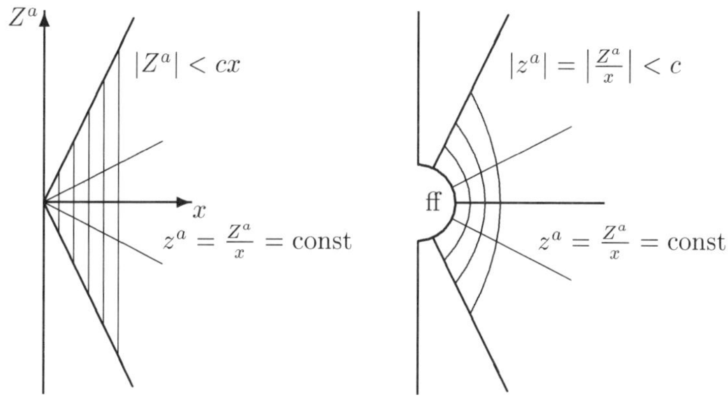

FIGURE 6. The blow up of C_a, given by $Z^a = 0$, $x = 0$.

negative order symbol (in particular continuous with boundary value 0) on $\bar{X}_0 \setminus C_a$, if V_a is such on X^a; see Figure 5.

So the compactification works for V_a, except at C_a. To remedy this, we blow up C_a. This is an invariant way of introducing polar coordinates about it (i.e. projective coordinates in various charts). That is, curves approaching C_a from various normal directions will correspond to different points on the blown-up space $[\bar{X}_0; C_a]$. Since C_a is given by $x = 0$, $Z^a = 0$, in local coordinates, this means concretely that the components of Z^a/x become coordinate functions on the part of $[\bar{X}_0; C_a]$ where this quotient is finite. (For the sake of completeness, a complete set of coordinates in this region is given by x, the components of Z^a as well as the $\dim X_a - 1$ coordinates on the sphere $y_a = \frac{z_a}{|z_a|}$; see Figure 6.) But $Z^a/x = z^a$, so it is now easy to see that for classical symbols V_a on X^a (of negative integer order), V_a is a $\mathcal{C}^\infty$ function on $[\bar{X}_0; C_a]$.

In general, there are many collision planes, and we blow them up recursively, starting with ones of the largest codimension, to get $[\bar{X}_0; \mathcal{C}]$,

$$\mathcal{C} = \{C_a : \ X_a \in \mathcal{X}, \ a \neq 1\}.$$

We refer to [**65**] for details.

There is no reason at all to take $\bar{X}_0$ as the space we start with. Given any compact manifold with boundary, $\bar{X}$, and a cleanly intersecting family of closed embedded submanifolds $\mathcal{C}$ of $\partial\bar{X}$, we can define $[\bar{X}; \mathcal{C}]$ analogously. For instance, one can start with $\bar{X} = \overline{\mathbb{R}^n} \times \mathbb{S}^k$. The space $[\bar{X};\mathcal{C}]$ is equipped with boundary fibrations given by the blow-down maps, see [**38**] for a simpler case where these first appeared explicitly.

Having described the configuration space, we turn to differential operators. X_0 has a nice algebra of differential operators, consisting of operators with symbolic coefficients: $\sum_{|\alpha|\leq m} a_\alpha(z)D_z^\alpha$, $a_\alpha \in S^0(X_0)$. We may require instead that a_α is 'classical', i.e. that $a_\alpha \in \mathcal{C}^\infty(\bar{X}_0)$. The resulting algebras were denoted $\mathrm{Diff}_{\mathrm{scc}}(\bar{X}_0)$ and $\mathrm{Diff}_{\mathrm{sc}}(\bar{X}_0)$ by Melrose; he called them 'scattering differential operators'.

This setup generalizes to the geometric set-up as follows. Let (x,y), $y = (y_1,\ldots,y_{n-1})$, be local coordinates near $\partial\bar{X}_0$. Then the vector fields in $\mathrm{Diff}_{\mathrm{sc}}(\bar{X}_0)$ are linear combinations of x^2D_x and the xD_{y_j} with coefficients in $\mathcal{C}^\infty(\bar{X}_0)$, as can be seen easily by an explicit calculation.

Now, if $\bar{X}$ is a manifold with boundary, $\mathcal{V}_{\mathrm{b}}(\bar{X})$ is the Lie algebra of vector fields tangent to $\partial\bar{X}$, and $\mathcal{V}_{\mathrm{sc}}(\bar{X}) = x\mathcal{V}_{\mathrm{b}}(\bar{X})$, where x is a defining function of $\partial\bar{X}$. $\mathcal{V}_{\mathrm{sc}}(\bar{X})$ is independent of the choice of x. Then $\mathcal{V}_{\mathrm{b}}(\bar{X})$ is spanned by $x\partial_x$ and ∂_y over $\mathcal{C}^\infty(\bar{X})$, so $\mathcal{V}_{\mathrm{sc}}(\bar{X})$ is spanned by $x^2\partial_x$ and $x\partial_y$ over $\mathcal{C}^\infty(\bar{X})$. By definition, these generate $\mathrm{Diff}_{\mathrm{sc}}(\bar{X})$. Also, $\mathcal{V}_{\mathrm{sc}}(\bar{X})$ is the set of all smooth sections of a vector bundle over $\bar{X}$, this is denoted by ${}^{\mathrm{sc}}T\bar{X}$. Its dual bundle is the scattering cotangent bundle, denoted by ${}^{\mathrm{sc}}T^*\bar{X}$. In the Euclidean setting,

$$ {}^{\mathrm{sc}}T\bar{X}_0 = \bar{X}_0 \times X_0, \quad {}^{\mathrm{sc}}T^*\bar{X}_0 = \bar{X}_0 \times X_0^*. $$

The way to generalize this differential operator algebra to one that includes many-body potentials is to allow singular coefficients $a_\alpha \in \mathcal{C}^\infty([\bar{X};\mathcal{C}])$. Thus,

$$ \mathrm{Diff}_{\mathrm{sc}}(\bar{X};\mathcal{C}) = \mathcal{C}^\infty([\bar{X};\mathcal{C}]) \otimes_{\mathcal{C}^\infty(\bar{X})} \mathrm{Diff}_{\mathrm{sc}}(\bar{X}). $$

In particular, if H is a many-body Hamiltonian, with either potential or higher order interactions, then $H \in \mathrm{Diff}^2_{\mathrm{sc}}(\bar{X}_0;\mathcal{C})$.

We let ${}^{\mathrm{sc}}T[\bar{X};\mathcal{C}] = \beta^{*\mathrm{sc}}T\bar{X}$ and ${}^{\mathrm{sc}}T^*[\bar{X};\mathcal{C}] = \beta^{*\mathrm{sc}}T^*\bar{X}$, where $\beta : [\bar{X};\mathcal{C}] \to \bar{X}$ is the blow-down map, and we are pulling back the vector bundles by it. Hence in the Euclidean setting,

$$ {}^{\mathrm{sc}}T[\bar{X}_0;\mathcal{C}] = [\bar{X}_0;\mathcal{C}] \times X_0, \quad {}^{\mathrm{sc}}T^*[\bar{X}_0;\mathcal{C}] = [\bar{X}_0;\mathcal{C}] \times X_0^*. $$

Now it is natural to define pseudo-differential operators using these bundles. Although I restrict the discussion to the Euclidean setting, the construction generalizes to any $\bar{X}$ via localization.

So we consider symbols

$$ a \in \langle z\rangle^{-l}\langle\zeta\rangle^m\mathcal{C}^\infty([\bar{X}_0;\mathcal{C}] \times \bar{X}_0^*), \tag{5.1} $$

$X_0 = \mathbb{R}^n$, where ζ is the dual variable of z, i.e. the variable on X_0^*. Note that this means that a is a classical symbol of order m in ζ. As usual, we define the Schwartz kernel of the left quantization of a by

$$ A = q_L(a) = (2\pi)^{-n}\int_{\mathbb{R}^n} e^{i(z-z')\cdot\zeta}a(z,\zeta)\,d\zeta, \tag{5.2} $$

understood as an oscillatory integral. In particular, for any $f \in \mathcal{S}(\mathbb{R}^n)$,

$$ Af(z) = (2\pi)^{-n}\int_{\mathbb{R}^n}\int_{\mathbb{R}^n} e^{i(z-z')\cdot\zeta}a(z,\zeta)f(z')\,d\zeta\,dz', $$

again understood as an oscillatory integral. We write $A \in \Psi_{\mathrm{sc}}(\bar{X}_0;\mathcal{C})$ for this class of operators.

Note that $\langle z\rangle^l a \in S^m_\infty(X_0;X_0^*)$, Hörmander's uniform symbol space [**27**, Section 18.1], so $A = \langle z\rangle^l \tilde{A}$, $\tilde{A} \in \Psi^m_\infty(X_0)$, the uniform ps.d.o.-algebra arising by quantizing $S^m_\infty(X_0;X_0^*)$ as in (5.2). In particular, since the mapping properties of $\Psi^m_\infty(X_0)$ between weighted Sobolev spaces $H^{r,s}$ are well known, the corresponding properties of A follow. Namely, $A : H^{r,s} \to H^{r-m,s+l}$ for all r, s, where

$$H^{r,s} = \langle z\rangle^{-s} H^r = \langle f \in \mathcal{S}'(\mathbb{R}^n) : \ \langle z\rangle^s f \in H^r\}.$$

Now, $\Psi_{\mathrm{sc}}(\bar{X}_0;\mathcal{C})$ is a $*$-algebra, in particular is closed under composition. Indeed, since $\Psi_{\mathrm{sc}}(\bar{X}_0;\mathcal{C}) \subset \Psi_\infty(X_0)$, and the latter is closed under composition, it suffices to follow the usual proof and make sure that the product is in $\Psi_{\mathrm{sc}}(\bar{X}_0;\mathcal{C})$, rather than merely in $\Psi_\infty(X_0)$. Thus, the key fact is that for any

$$b \in \langle z\rangle^{-l}\langle\zeta\rangle^m \mathcal{C}^\infty([\bar{X}_0;\mathcal{C}]_z \times [\bar{X}_0;\mathcal{C}]_{z'} \times (\bar{X}_0^*)_\zeta)$$

there exists a as in (5.1) such that the induced operators

$$B = (2\pi)^{-n} \int_{\mathbb{R}^n} e^{i(z-z')\cdot\zeta} b(z,z',\zeta)\, d\zeta, \tag{5.3}$$

and A as in (5.2) are the same. The proof of this claim is standard. Indeed, we can expand b in Taylor series in z' around $z = z'$ to finite order k. The finite order terms depend on z' only via $(z'-z)^\alpha$, $|\alpha| \le k$. We rewrite $(z'-z)^\alpha e^{i(z-z')\cdot\zeta}$ as $(-1)^{|\alpha|} D^\alpha_\zeta e^{i(z-z')\cdot\zeta}$, and integrate by parts with respect to ζ. Thus, the α-term is the left quantization of

$$\frac{1}{\alpha!} D^\alpha_{z'} D^\alpha_\zeta b(z,z',\zeta)|_{z'=z}, \tag{5.4}$$

which is of the desired form, i.e. is in $\langle z\rangle^{-l}\langle\zeta\rangle^m \mathcal{C}^\infty([\bar{X}_0;\mathcal{C}] \times \bar{X}_0^*)$. In fact, the weight $\langle\zeta\rangle^m$ can be replaced by $\langle\zeta\rangle^{m-|\alpha|}$ due to the symbolic properties of b in $(X_0^*)_\zeta$, but no corresponding change may be made for the z weight. Similarly, the remainder term is of the form

$$\begin{gathered} K_k(z,z') = (2\pi)^{-n} \int_{\mathbb{R}^n} e^{i(z-z')\cdot\zeta} b_k(z,z',\zeta)\, d\zeta, \\ b_k \in \langle z\rangle^{-l}\langle\zeta\rangle^{m-k-1} \mathcal{C}^\infty([\bar{X}_0;\mathcal{C}]_z \times [\bar{X}_0;\mathcal{C}]_{z'} \times (\bar{X}_0^*)_\zeta). \end{gathered} \tag{5.5}$$

Now we can asymptotically sum the b_α to get a new symbol

$$c \in \langle z\rangle^{-l}\langle\zeta\rangle^m \mathcal{C}^\infty([\bar{X}_0;\mathcal{C}] \times \bar{X}_0^*).$$

Let C be the left quantization of c. Then $B - C$ is of the form (5.5) for all k, with b_k replaced by some b'_k with the same properties. It is then straightforward to show that the Schwartz kernel K' of $B - C$ is $\mathcal{C}^\infty$, decays rapidly with all derivatives as $\langle z - z'\rangle \to \infty$, and more precisely it is of the form

$$K' \in \mathcal{C}^\infty([\bar{X}_0;\mathcal{C}]_z \times (\bar{X}_0)_{z-z'})$$

with infinite order vanishing at the boundary of the second factor. Taking its Fourier transform b' in $z - z'$, K' is thus the left quantization of $a = c + b'$, proving the claim, hence in turn that $\Psi_{\mathrm{sc}}(\bar{X}_0;\mathcal{C})$ is closed under composition.

In the two-body setting, where $\mathcal{C} = \emptyset$, there is a principal symbol at infinity. Namely, if $A \in \Psi^{m,l}_{\mathrm{sc}}(\bar{X})$, $A = q_L(a)$, then $\sigma_{m,l}(A)$ is given by the restriction of $\langle z\rangle^l\langle\zeta\rangle^{-m} a \in \mathcal{C}^\infty(\bar{X}_0 \times \bar{X}_0^*)$ to $\partial(\bar{X}_0 \times \bar{X}_0^*) = (\partial\bar{X}_0 \times \bar{X}_0^*) \cup (\bar{X}_0 \times \partial\bar{X}_0^*)$. Of the two boundary hypersurfaces, the restriction to $\bar{X}_0 \times \partial\bar{X}_0^*$ yields the usual principal symbol, while the restriction to $\partial\bar{X}_0 \times \bar{X}_0^*$ is the principal symbol at infinity. More precisely, if $l = 0$, we can indeed define the part of $\sigma_{m,0}(A)$ at infinity to be the restriction of a to $(\partial\bar{X}_0) \times \bar{X}_0^*$. The principal symbol is multiplicative, i.e. $\sigma_{m+m',l+l'}(AB) = \sigma_{m,l}(A)\sigma_{m',l'}(B)$. Thus, $[A,B] \in \Psi_{\mathrm{sc}}^{m+m'-1,l+l'+1}(\bar{X})$, and its principal symbol is given by the Poisson bracket of their symbols, see Section 6.

Since in the many-body setting we do not gain decay in z in (5.4), we cannot expect to have a commutative principal symbol at infinity, i.e. at $\partial[\bar{X}_0;\mathcal{C}]$. For $C \in \Psi_{\mathrm{sc}}^{m,0}(\bar{X},\mathcal{C})$, $y_a \in C_{a,\mathrm{reg}}$, $\zeta_a \in X_a^*$, we let

$$\hat{C}_a(y_a,\zeta_a) = (2\pi)^{-\dim X^a} \int e^{i(z^a-(z')^a)\cdot\zeta^a} c(y_a,z^a,\zeta)\,d\zeta \in \mathcal{S}'(X^a \times X^a)$$

be the operator valued principal symbol of C at (y_a,ζ_a). Thus, $\hat{C}_a(y_a,\zeta_a)$ is a tempered distribution on $X^a \times X^a$ (denoted by the variables $(z^a,(z^a)')$), and it is in fact a many-body ps.d.o. itself: $\hat{C}_a(y_a,\zeta_) \in \Psi_{\mathrm{sc}}^{m,0}(\bar{X}^a,\mathcal{C}^a)$ corresponding to the collision planes $X^a \cap X_b$, with b satisfying $X_b \supset X_a$. We also call it the indicial operator of C to make it clear we are not talking about the standard principal symbol. We also write $\hat{C}_a(z_a,\zeta_a)$ in the same setting, where we extend $\hat{C}_a(y_a,\zeta_a)$ to be homogeneous degree 0 in z_a. It can be easily seen to satisfy

$$\hat{A}_a\hat{B}_a = \widehat{(AB)}_a,$$

where on the left hand side we compose the operators $\hat{A}_a(z_a,\zeta_a)$ and $\hat{B}_a(z_a,\zeta_a)$. Thus, multiplication of operators is only partially commutative, even to top order. This can be observed already from $[D_{z_a},V_a] = 0$, hence certainly lower order at infinity, while $[D_{z^a},V_a] \in \mathcal{C}^\infty([\bar{X}_0;\mathcal{C}])$ without *any* decay at C_a.

This observation has important implications for the positive commutator estimates that we take up in the next section. Namely, H must commute to leading order with the operators we want to microlocalize with. This means that these operators A must have $\hat{A}_a$ commute with $\hat{H}_a$, and the most reasonable way of achieving this is to have $\hat{A}_a$ be a scalar multiple of $\psi(\hat{H}_a)$, where e.g. $\psi \in \mathcal{C}_c^\infty(\mathbb{R})$. This multiple defines a function on $\dot{T}^*\bar{X}_0$; we want this to arise from a smooth π-invariant function for our estimates. On the other hand, $\psi(\hat{H}_a)$ provides localization at the characteristic set.

Here, however, I would like to talk about pseudo-differential constructions first. Namely, if $\lambda \notin \mathbb{R}$, or indeed if $\lambda \in \mathbb{C} \setminus [\inf\Lambda, +\infty)$ then there exists a parametrix $G(\lambda) \in \Psi_{\mathrm{sc}}^{-2,0}(\bar{X}_0;\mathcal{C})$ for $H - \lambda$, i.e. such that

$$(H-\lambda)G(\lambda) - \mathrm{Id},\ G(\lambda)(H-\lambda) - \mathrm{Id} \in \Psi_{\mathrm{sc}}^{-\infty,\infty}(\bar{X}_0;\mathcal{C}).$$

Then the parametrix identities show that

$$\lambda \in \mathbb{C} \setminus \mathrm{spec}(H) \Rightarrow (H-\lambda)^{-1} \in \Psi_{\mathrm{sc}}^{-2,0}(\bar{X}_0;\mathcal{C}).$$

The parametrix construction proceeds inductively by constructing $(\hat{H}_a - \lambda)^{-1}$ in $\Psi_{\mathrm{sc}}^{-2,0}(\bar{X}_a,\mathcal{C}^a)$ for every $a \neq 1$ and then combining these: there exists a $G_0(\lambda) \in \Psi_{\mathrm{sc}}^{-2,0}(\bar{X}_0;\mathcal{C})$ with specified indicial operators $(\hat{H}_a - \lambda)^{-1}$, hence satisfying

$$(H-\lambda)G_0(\lambda) - \mathrm{Id},\ G_0(\lambda)(H-\lambda) - \mathrm{Id} \in \Psi_{\mathrm{sc}}^{-1,1}(\bar{X}_0;\mathcal{C}).$$

Then the standard Neumann series argument yields $G(\lambda)$.

The Helffer-Sjöstrand argument [**24**] then shows that for any $\phi \in \mathcal{C}_c^\infty(\mathbb{R})$,

$$\phi(H) = \frac{-1}{2\pi i} \int_{\mathbb{C}} \overline{\partial}_\lambda \tilde{\phi}(\lambda)(H-\lambda)^{-1}\,d\lambda \wedge d\bar{\lambda},$$

where $\tilde{\phi}$ is an almost analytic extension of ϕ: $\tilde{\phi} \in \mathcal{C}_c^\infty(\mathbb{C})$, $|\bar{\partial}_\lambda \tilde{\phi}| \leq C_k |\operatorname{Im}\lambda|^k$ for all k. We can control $(H-\lambda)^{-1}$ in $\Psi_{\mathrm{sc}}^{-2,0}(\bar{X}, \mathcal{C})$ as $\lambda \to \mathbb{R}$ with semi-norm estimates $\mathcal{O}(|\operatorname{Im}\lambda|^{-j})$ (j depends on the norm), so we conclude that $\phi(H) \in \Psi_{\mathrm{sc}}^{-\infty,0}(\bar{X}; \mathcal{C})$.

We can now explain the precise specifications on T_+ in (4.2). Namely, we require that on a neighborhood of $\mathcal{R}_+(\lambda)$ in $\dot{T}^*X$, the indicial operators $\widehat{T_+}$ should equal $\widehat{\phi(H)}$ for some $\phi \in \mathcal{C}_c^\infty(\mathbb{R})$ identically 1 near λ, and on a neighborhood of $\mathcal{R}_-(\lambda)$ they should vanish. Explicitly this can be arranged by taking any ϕ as above, and any $\chi \in \mathcal{C}^\infty(\mathbb{R})$ identically 1 on $(\frac{\sqrt{\lambda}}{2}, +\infty)$, identically 0 on $(-\infty, -\frac{\sqrt{\lambda}}{2})$. Then let $T_+ = \phi(H) q_R(\chi(\frac{z\cdot\zeta}{\langle z\rangle}))$, with q_R denoting the 'right quantization' (i.e. where we take $b = \chi(\frac{z'\cdot\zeta}{\langle z'\rangle})$ in (5.3)). Although $q_R(\chi(\frac{z\cdot\zeta}{\langle z\rangle}))$ is not in $\Psi_{\mathrm{sc}}(\bar{X};\mathcal{C})$, due to the non-symbolic behavior of b as $\zeta \to \infty$, T_+ is, namely $T_+ \in \Psi_{\mathrm{sc}}^{-\infty,0}(\bar{X};\mathcal{C})$, since $\phi(H)$ is smoothing: see [**65**]. Moreover,

$$\widehat{T_+}_a(z_a, \zeta_a) = \chi(\frac{z_a \cdot \zeta_a}{|z_a|})\phi(\hat{H}_a(z_a, \zeta_a)), \tag{5.6}$$

hence has the desired properties.

Our construction of $\phi(H)$ in fact shows that if all potentials are in $S^{-\rho}(X_a)$, $\rho > 0$, and $\chi_a \in \mathcal{C}^\infty(\bar{X}_0)$ is supported away from C_b such that $C_b \supset C_a$ does *not* hold, then $\chi_a(\phi(H) - \phi(H_a)) \in \Psi_{\mathrm{sc}}^{-\infty,\rho}(\bar{X};\mathcal{C})$, hence trace class if $\rho > \dim X_0$. In the three-body setting this shows that

$$\phi(H) - \phi(H_0) - \sum_{\#b=2} (\phi(H_b) - \phi(H_0))$$

is trace class. Indeed, near C_a this can be written as

$$(\phi(H) - \phi(H_a)) - \sum_{\#b=2,\ b\neq a} (\phi(H_b) - \phi(H_0)),$$

and now all terms in parantheses are in $\Psi_{\mathrm{sc}}^{-\infty,\rho}(\bar{X};\mathcal{C})$ near C_a. So we conclude, with a proof that shows much more, a result of Buslaev and Merkureev:

$$\sigma(\phi) = \operatorname{tr}((\phi(H) - \phi(H_0)) - \sum_{\#b=2} (\phi(H_b) - \phi(H_0)))$$

defines a distribution $\sigma \in \mathcal{C}^{-\infty}(\mathbb{R})$. Writing $\sigma = \xi'$ defines the spectral shift function, up to a constant, which in turn, in two-body scattering, is the well-known generalization of the eigenvalue counting function on compact manifolds. These statements, as well as the following theorem, which is joint work with Xue-Ping Wang [**68**], generalize to arbitrary many-body Hamiltonians (with short-range interactions as indicated).

THEOREM 5.1. *Suppose H is a three-body Hamiltonian with Schwartz interactions: $V_a \in \mathcal{S}(X^a)$, and that all interactions are pair interactions (i.e. $V_a \neq 0$ implies that a is a 2-cluster.) Then the spectral shift function σ is $\mathcal{C}^\infty$ in $\mathbb{R} \setminus (\Lambda \cup \operatorname{spec}_{pp}(H))$, and it is a classical symbol at infinity (i.e. outside a compact set) with a complete asymptotic expansion:*

$$\sigma(\lambda) \sim \lambda^{\frac{n}{2}-3} \sum_{j=0}^{\infty} c_j \lambda^{-j},\ c_0 = C_0(n) \sum_{a,b:a\neq b} \int_{\mathbb{R}^n} V_a V_b \, dg.$$

Note that σ decays one order faster than in 2-body scattering, and two orders faster than Weyl's law on compact manifolds. This is because $\phi(H_0) + \sum_{\#b=2}(\phi(H_b) - \phi(H_0))$ is, in a high-energy sense, closer to $\phi(H)$ than $\phi(H_0)$ is to $\phi(H)$ in two body scattering. If not all interactions are pair interactions, the order of the leading term changes, namely becomes $\lambda^{\frac{n}{2}-2}$ as in 2-body scattering.

The proof of this theorem relies on the propagation of singularities, applied to the Schwartz kernel of the resolvent, $R(\lambda + i0)$. (In fact, the theorem should generalize to symbolic potentials, but the proof would require a more precise microlocalization than provided by $\mathrm{WF}_{\mathrm{sc}}$.) So we now turn to the positive commutator methods that prove this.

6. Microlocal positive commutator estimates

First I sketch, somewhat vaguely, the idea of positive commutator estimates. So suppose that we want to obtain estimates on the solutions of $Pu = f$, where f is known, and is 'nice', and P is self-adjoint. Suppose that we can construct an operator A which is self-adjoint and is such that

$$i[A, P] = B^*B + E. \tag{6.1}$$

Here B^*B is the positive term, giving the name to the estimate. The point is that we can estimate Bu in terms of Eu. Indeed, at least formally,

$$\langle u, i[A, P]u\rangle = \langle u, B^*Bu\rangle + \langle u, Eu\rangle = \|Bu\|^2 + \langle u, Eu\rangle.$$

On the other hand,

$$\langle u, i[A, P]u\rangle = \langle u, iAPu\rangle - \langle u, iPAu\rangle = \langle u, iAPu\rangle + \langle iAPu, u\rangle = 2\,\mathrm{Re}\langle u, iAPu\rangle.$$

Combining these yields

$$\|Bu\|^2 \leq 2|\,\mathrm{Re}\langle u, iAPu\rangle| + |\langle u, Eu\rangle|. \tag{6.2}$$

This means that Bu can be estimated in terms of Pu, which is known from the PDE, and Eu, on which we need to make assumptions. The typical application is that E is supported in one region of phase space and B in another, in which case we can *propagate* estimates of u.

In fact, one can also apply this estimate if one does not know a priori that $Bu \in L^2$. Namely, an approximation argument gives that if Pu and Eu are in appropriate spaces so that the right hand side of (6.2) makes sense, then $Bu \in L^2$, and (6.2) holds. Considering pseudo-differential operators A of various orders, this means that we obtain microlocal weighted Sobolev estimates for u. Also, typically one has an error term F, i.e. $i[A, P] = B^*B + E + F$, but F is 'lower order' in some sense. Thus, $|\langle u, Fu\rangle|$ is added to the right hand side of (6.2), but being 'lower order' means that $|\langle u, Fu\rangle|$ automatically makes sense, hence is irrelevant when proving that $Bu \in L^2$.

In fact, this method also yields estimates for the resolvent very directly. Since for $t \in \mathbb{R}$, $i[A, P - it] = i[A, P]$, and

$$\begin{aligned}\langle u, i[A, P - it]u\rangle &= \langle u, iA(P - it)u\rangle - \langle u, i(P - it)Au\rangle \\ &= \langle u, iA(P - it)u\rangle + \langle iA(P + it)u, u\rangle = 2\,\mathrm{Re}\langle u, iA(P - it)u\rangle - 2t\langle Au, u\rangle.\end{aligned}$$

Thus, we deduce

$$\|Bu\|^2 + 2t\langle Au, u\rangle \leq 2|\,\mathrm{Re}\langle u, iAPu\rangle| + |\langle u, Eu\rangle|. \tag{6.3}$$

This is in particular an estimate for $\|Bu\|$ provided that $t \geq 0$ and A is positive. Here we may take $u = u_t = (P - it)^{-1}f$, defined for $t > 0$, say, and we find a uniform estimate for Bu_t as $t \to 0$.

The question is thus how one can produce operators A which have a positive commutator with P as above. First, we recall how this happens in the scattering calculus. Namely, if $A \in \Psi_{\mathrm{sc}}^{m,l}(\bar{X})$, $P \in \Psi_{\mathrm{sc}}^{m',l'}(\bar{X})$ then $[A, P] \in \Psi_{\mathrm{sc}}^{m+m'-1,l+l'+1}(\bar{X})$ and

$$\sigma_{m+m'-1,l+l'+1}(i[A, P]) = H_a p = -H_p a, \ a = \sigma_{m,l}(A), \ p = \sigma_{m',l'}(P),$$

where H_a is the Hamilton vector field of a, H_p the Hamilton vector field of p. So modulo lower terms, which I ignore here and which are easy to deal with, we need to arrange that

$$H_p a = -b^2 + e, \tag{6.4}$$

and then take B, E with $\sigma(B) = b$, $\sigma(E) = e$. Indeed, under these assumptions (6.2) shows that $\|Bu\|$ can be estimated in terms of Pu and Eu. That is, u microlocally on $\operatorname{supp} b$ is estimated by u on $\operatorname{supp} e$ (and Pu) in the precise sense described in the next paragraph, so we can *propagate* estimates of u from $\operatorname{supp} e$ to $\operatorname{supp} b$. (Incidentally, this is a good example of the F term: only the principal symbols of E and B are specified. Take any E and B with these principal symbols, $F = i[A, P] - B^*B - E$ is lower order.)

This can be used in a very straightforward manner to obtain bounds on $\mathrm{WF}_{\mathrm{sc}}(u)$. Namely, one works with 'relative wave front sets', relative to $x^s H^r = H^{r,s}$, that is. Thus, for $\bar{X} = \overline{\mathbb{R}^n}$, $(z, \zeta) \notin \mathrm{WF}_{\mathrm{sc}}^{r,s}(u)$ means that there is a cutoff function $\phi \in \mathcal{C}_c^\infty(\mathbb{R}^n)$ with $\phi(\zeta) \neq 0$ such that $\mathcal{F}^{-1}\phi\mathcal{F}u$ is in $H^{r,s}$ in an open cone around z. But this is equivalent to the existence of some $Q \in \Psi_{\mathrm{sc}}^{r,-s}(\bar{X})$ such that $\sigma(Q)(z, \zeta) \neq 0$ and $Qu \in L^2$. Note that $\sigma(Q)(z, \zeta) \neq 0$ means that Q is elliptic at (z, ζ). So if we find $A \in \Psi_{\mathrm{sc}}^{m,l}(\bar{X})$, and consequently $B \in \Psi_{\mathrm{sc}}^{\frac{m-1}{2},\frac{l+1}{2}}(\bar{X})$, $E \in \Psi_{\mathrm{sc}}^{m-1,l+1}(\bar{X})$ as above, then the conclusion is that (if $Pu \in \dot{\mathcal{C}}^\infty(\bar{X})$)

$$\mathrm{WF}_{\mathrm{sc}}^{\frac{m-1}{2},-\frac{l+1}{2}}(u) \cap \operatorname{supp} e = \emptyset \Rightarrow \mathrm{WF}_{\mathrm{sc}}^{\frac{m-1}{2},-\frac{l+1}{2}}(u) \cap \operatorname{supp} b = \emptyset.$$

In scattering theory m is usually irrelevant by standard elliptic regularity. Thus, one iteratively reduces l, proving that $\operatorname{supp} b$ is disjoint from the wave front set with respect to more and more decaying Sobolev spaces. (In fact, b is shrunk slightly during the iterative procedure for technical reasons.)

I now illustrate how to prove the propagation of singularities at $\partial\bar{X}$ for real principal type $P \in \Psi_{\mathrm{sc}}^{m,0}(\bar{X})$. For example, we may take $P = H - \lambda$, $\lambda > 0$, and H is a two-body Hamiltonian. (Note that microlocal elliptic regularity is the consequence of the standard microlocal parametrix construction.) We thus want to prove that if $Pu \in \dot{\mathcal{C}}^\infty(\bar{X})$ (or a microlocal version of it holds), $\bar{\xi} \in {}^{\mathrm{sc}}T^*_{\partial\bar{X}}\bar{X}$ and there is a point on the backward bicharacteristic through $\bar{\xi}$ which is not in $\mathrm{WF}_{\mathrm{sc}}(u)$, then $\bar{\xi} \notin \mathrm{WF}_{\mathrm{sc}}(u)$. In fact, by a simple argument it suffices to prove that there exists a neighborhood U of $\bar{\xi}$ such that if there is a point $\tilde{\xi}$ in U which is also on the backward bicharacteristic through $\bar{\xi}$ and which is not in $\mathrm{WF}_{\mathrm{sc}}(u)$, then $\bar{\xi} \notin \mathrm{WF}_{\mathrm{sc}}(u)$.

The standard proof proceeds via linearization of H_p, see [**28**]. Thus, first note that $x^{-1}H_p$ is a smooth vector field on ${}^{\mathrm{sc}}T^*\bar{X}$ which is tangent to the boundary.

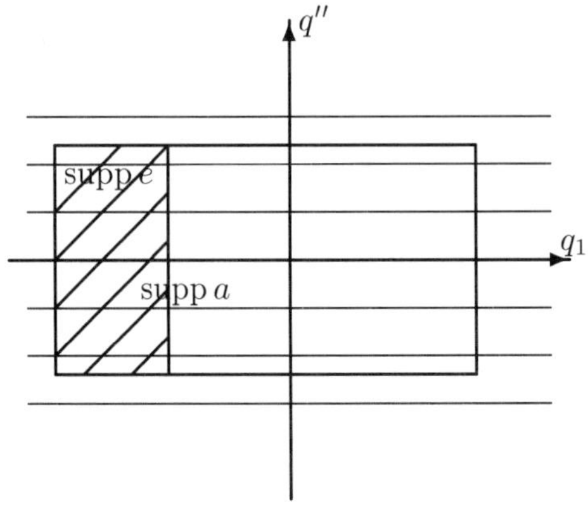

FIGURE 7. supp a superimposed on the linearized Hamilton flow. supp e is the shaded region on the left.

(For example, for Euclidean two-body Hamiltonians,

$$x^{-1}H_p = 2|z|\zeta \cdot \partial_z = 2\sum_j \zeta_j |z|\partial_{z_j},$$

and $|z|\partial_{z_j}$ is a smooth vector field tangent to $\partial\bar{X}$, i.e. it is in $\mathcal{V}_{\mathrm{b}}(\bar{X})$.) Thus, given any point $\bar{\xi} \in {}^{\mathrm{sc}}T^*_{\partial\bar{X}}\bar{X}$ one can introduce local coordinates $(q_1, \ldots, q_{2n-1}) = (q_1, q'')$ on ${}^{\mathrm{sc}}T^*_{\partial\bar{X}}\bar{X}$ centered at $\bar{\xi}$ such that $x^{-1}H_p = \partial_{q_1}$ at ${}^{\mathrm{sc}}T^*_{\partial\bar{X}}\bar{X}$. Thus, bicharacteristics at $\partial\bar{X}$ are curves $q'' =$ constant. Now let $\chi_1 \in \mathcal{C}^\infty_c(\mathbb{R}_{q_1})$ and $\chi_2 \in \mathcal{C}^\infty_c(\mathbb{R}^{2n-2}_{q''})$ be smooth functions supported near 0 with the property that

$$\chi_1' = -b_1^2 + e_1,$$

$b_1, e_1 \in \mathcal{C}^\infty_c(\mathbb{R})$, and $\operatorname{supp} e_1 \subset (-\infty, 0)$. Let

$$a = \chi_1\chi_2^2,\ b = b_1\chi_2,\ e = e_1\chi_2^2.$$

Then (6.4) holds. In fact, we can even allow weights and take

$$a_s = x^s\chi_1\chi_2^2, s \in \mathbb{R},$$

since $(x^{-1}H_p x^s)\chi_1$ can be absorbed in $x^s(x^{-1}H_p\chi_1)$ by choosing χ_1' large compared to χ_1. This gives microlocal weighted L^2 estimates in $x^{-s-1/2}L^2$. The iterative argument, in which we gradually let $s \to -\infty$, then allows one to conclude that

$$\operatorname{supp} e \cap \mathrm{WF}_{\mathrm{sc}}(u) = \emptyset \text{ and } \operatorname{supp} a \cap \mathrm{WF}_{\mathrm{sc}}(Pu) = \emptyset \Rightarrow \{b > 0\} \cap \mathrm{WF}_{\mathrm{sc}}(u) = \emptyset.$$

By choosing supp χ_1 and supp χ_2 appropriately, we may arrange that e is supported near $\tilde{\xi}$ so that $\operatorname{supp} e \cap \mathrm{WF}_{\mathrm{sc}}(u) = \emptyset$, and so that $b(\bar{\xi}) > 0$, as shown below.

There are several directions from here. One can use finer notion of regularity, such as Lagrangian regularity, which would correspond to using χ_2 that vanishes simply on a Lagrangian submanifold, or such as regularity at radial points, which is the subject of a joint paper with Andrew Hassell and Richard Melrose [**23**].

Here I will talk about a rougher version, namely what happens if the bicharacteristic 'flow' is more complicated, e.g. in the presence of boundaries or corners for the wave equation [**43, 36**] or many-body scattering. In fact, here I will not explain the detailed behavior of bicharacteristics, rather just show how to microlocalize positive commutator estimates in a versatile fashion. This method goes back to the work of Melrose and Sjöstrand [**43**].

The main point is that if we cannot put the operator P, or at least its Hamilton vector field H_p in a model form, the previous construction will not work. Indeed, unless $H_p\chi_2 = 0$, $H_p(\chi_1\chi_2)$ will always yield a term $\chi_1 H_p\chi_2$, which cannot be controlled by $(H_p\chi_1)\chi_2$: the problem being near the boundary of $\operatorname{supp}\chi_2$. So instead use a different form of localization. First let $\eta \in \mathcal{C}^\infty({}^{\mathrm{sc}}T^*\bar{X})$ be a function with

$$\eta(\bar{\xi}) = 0,\ H_p\eta(\bar{\xi}) > 0.$$

Thus, η measures propagation along bicharacteristics, e.g. $\eta = q_1$ in the above example would work, but so would many other choices. We will use a function ω to localize near putative bicharacteristics. This statement is deliberately vague; at first we only assume that $\omega \in \mathcal{C}^\infty({}^{\mathrm{sc}}T^*\bar{X})$ is the sum of the squares of $\mathcal{C}^\infty$ functions σ_j, $j = 1, \ldots, l$, with non-zero differentials at $\bar{\xi}$ such that $d\eta$ and $d\sigma_j$, $j = 1, \ldots, l$, span $T_{\bar{\xi}}{}^{\mathrm{sc}}T^*_{\partial\bar{X}}\bar{X}$. Such a function ω is non-negative and it vanishes quadratically at $\bar{\xi}$, i.e. $\omega(\bar{\xi}) = 0$ and $d\omega(\bar{\xi}) = 0$. An example is $\omega = q_2^2 + \ldots + q_{2n-1}^2$ with the notation from before, but again there are many other possible choices. We now consider a family symbols, parameterized by constants $\delta \in (0,1)$, $\epsilon \in (0, \delta]$, of the form

$$a = \chi_0(2 - \frac{\phi}{\epsilon})\chi_1(\frac{\eta + \delta}{\epsilon\delta} + 1),$$

where

$$\phi = \eta + \frac{1}{\epsilon}\omega,$$

$\chi_0(t) = 0$ if $t \leq 0$, $\chi_0(t) = e^{-1/t}$ if $t > 0$, $\chi_1 \in \mathcal{C}^\infty(\mathbb{R})$, $\operatorname{supp}\chi_1 \subset [0, +\infty)$, $\operatorname{supp}\chi_1' \subset [0,1]$. Although we do not do it explicitly here, weights such as x^s can be accommodated for any $s \in \mathbb{R}$, by replacing the factor $\chi_0(2 - \frac{\phi}{\epsilon})$ by $\chi_0(A_0^{-1}(2 - \frac{\phi}{\epsilon}))$ and taking $A_0 > 0$ large.

We analyze the properties of a step by step. First, note that $\phi(\bar{\xi}) = 0$, $H_p\phi(\bar{\xi}) = H_p\eta(\bar{\xi}) > 0$, and $\chi_1(\frac{\eta+\delta}{\epsilon\delta} + 1)$ is identically 1 near $\bar{\xi}$, so $H_p a(\bar{\xi}) < 0$. Thus, $H_p a$ has the correct sign, and is in particular non-zero, at $\bar{\xi}$.

Next,

$$\xi \in \operatorname{supp} a \Rightarrow \phi(\xi) \leq 2\epsilon \text{ and } \eta(\xi) \geq -\delta - \epsilon\delta.$$

Since $\epsilon < 1$, we deduce that in fact $\eta = \eta(\xi) \geq -2\delta$. But $\omega \geq 0$, so $\phi = \phi(\xi) \leq 2\epsilon$ implies that $\eta = \phi - \epsilon^{-1}\omega \leq \phi \leq 2\epsilon$. Hence, $\omega = \omega(\xi) = \epsilon(\phi - \eta) \leq 4\epsilon\delta$. Since ω vanishes quadratically at $\bar{\xi}$, it is useful to rewrite the estimate as $\omega^{1/2} \leq 2(\epsilon\delta)^{1/2}$. Combining these, we have seen that on $\operatorname{supp} a$,

$$-\delta - \epsilon\delta \leq \eta \leq 2\epsilon \text{ and } \omega^{1/2} \leq 2(\epsilon\delta)^{1/2}. \tag{6.5}$$

Moreover, on $\operatorname{supp} a \cap \operatorname{supp}\chi_1'$,

$$-\delta - \epsilon\delta \leq \eta \leq -\delta \text{ and } \omega^{1/2} \leq 2(\epsilon\delta)^{1/2}.$$

Note that given any neighborhood U of $\bar{\xi}$, we can thus make a supported in U by choosing ϵ and δ sufficiently small. Below we illustrate the parabola shaped region given by $\operatorname{supp} a$.

Note that as $\epsilon \to 0$, but δ fixed, the parabola becomes very sharply localized at $\omega = 0$. In particular, for very small $\epsilon > 0$ we obtain a picture quite analogous to letting $\operatorname{supp}\chi_2 \to \{0\}$ in Figure 7.

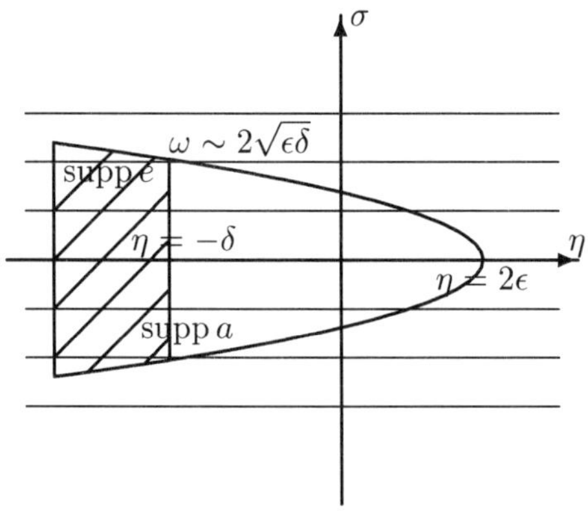

FIGURE 8. supp a in (η, σ) coordinates. supp e is again the shaded region on the left.

So we have shown that a is supported near $\bar{\xi}$. We define

$$e = \chi_0(2 - \frac{\phi}{\epsilon}) H_p \chi_1(\frac{\eta + \delta}{\epsilon\delta} + 1),$$

so the crucial question is whether $H_p\phi \geq 0$ on supp a. Note that choosing $\delta_0 \in (0,1)$ and $\epsilon_0 \in (0, \delta_0)$ sufficiently small, one has $H_p\eta \geq c_0 > 0$ where $|\eta| \leq 2\delta_0$, $\omega^{1/2} \leq 2(\epsilon_0\delta_0)^{1/2}$. So $H_p\phi \geq 0$ on supp a, provided that $|H_p\omega| \leq \frac{c_0}{2}\epsilon$ there.

But being a sum of squares of functions with non-zero differentials, $H_p\omega$ vanishes at $\omega = 0$ and satisfies $|H_p\omega| \leq C\omega^{1/2}$. Due to (6.5), we deduce that $|H_p\omega| \leq 2C(\epsilon\delta)^{1/2}$. So $|H_p\omega| \leq \frac{c_0}{2}\epsilon$ holds if $\frac{c_0}{2}\epsilon \geq 2C(\epsilon\delta)^{1/2}$, i.e. if $\epsilon \geq C'\delta$ for some constant $C' > 0$ independent of ϵ, δ. Note that this constraint on ϵ, i.e. that it cannot be too small, gives very rough localization: the width of the parabola at $\eta = -\delta$ is roughly $\omega^{1/2} \sim \delta$, i.e. it is very wide, and in particular insufficient to prove the propagation of singularities along the bicharacteristics. The reason is simple: our localizing function ω has no relation to H_p, so we cannot expect a more precise estimate. Note, however, that the estimate is still non-trivial! Indeed, it shows that singularities propagate in the sense that $\bar{\xi}$ cannot be an isolated point of $\mathrm{WF}_{\mathrm{sc}}(u)$. (We required $\epsilon \in (0, \delta]$ beforehand, but in fact we could have dealt with $\epsilon \leq \mu\delta$, even if $\mu > 1$, if we localized slightly differently.)

We need to adapt ω to H_p to get a better estimate. If we linearize H_p as above, and take $\omega = q_2^2 + \ldots + q_{2n-1}^2$, then $H_p\omega = 0$ and any $\epsilon > 0$ works. Thus, in this case, we can prove propagation of singularities much like by the previous, simpler, construction.

However, we do not need such a strong relationship to H_p. Suppose instead that we merely get ω 'right' at $\bar{\xi}$, in the sense that $\omega = \sum \sigma_j^2$ and $H_p\sigma_j(\bar{\xi}) = 0$. Then $|H_p\sigma_j| \leq C_0(\omega^{1/2} + |\eta|)$, so $|H_p\omega| \leq C\omega^{1/2}(\omega^{1/2} + |\eta|)$. Using (6.5), we deduce that $|H_p\omega| \leq \frac{c_0}{2}\epsilon$ provided that $\frac{c_0}{2}\epsilon \geq C''(\epsilon\delta)^{1/2}\delta$, i.e. that $\epsilon \geq C'\delta^3$ for some constant C' independent of ϵ, δ. Now the size of the parabola at $\eta = -\delta$ is roughly $\omega^{1/2} \sim \delta^2$, i.e. we have localized along a single direction, namely the direction of H_p at $\bar{\xi}$. By a relatively simple argument, one can piece together such estimates (i.e. where the direction is correct 'to first order') and deduce the propagation of singularities. We emphasize that the lower bound for ϵ is natural. Indeed, with q_j as above, we may

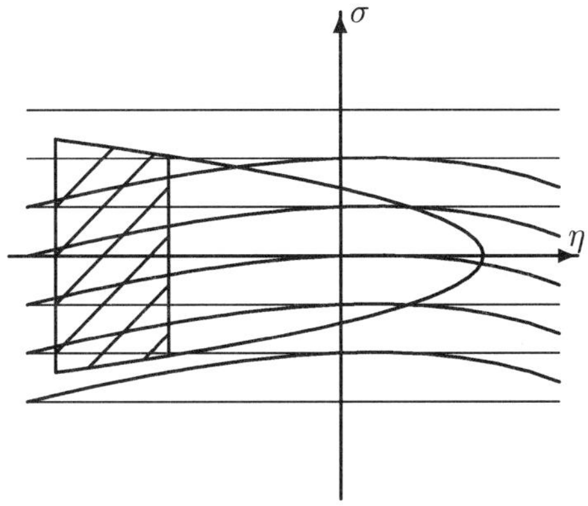

FIGURE 9. Bicharacteristics and supp a. The labels from Figure 8 have been removed to make the picture less cluttered. The straight horizontal lines are the σ = constant lines, while the nearby parabolae are the bicharacteristics.

take σ_j e.g. to be $\sigma_j = q_j + q_1^2$, $j \geq 2$. The bicharacteristics are q_j = constant, but we are localizing near σ_j = constant, and at $\eta = -\delta$ these differ by δ^2. So any localization better than $\omega^{1/2} \sim \delta^2$ would in fact contradict the propagation of singularities!

The microlocal positive commutator estimates in many-body scattering arise by this method. In particular, one can take $\eta = \frac{z \cdot \zeta}{|z|}$, which is the radial component of the momentum. The function ω needs to be π-invariant, so if $\bar{\xi} \in {}^{\mathrm{sc}}T^*\bar{X}_a$, it involves functions on ${}^{\mathrm{sc}}T^*\bar{X}_a$ as well as $Z^a = \frac{z^a}{|z|}$ and η. The only additional argument needed is to show that the commutator is indeed positive, which has to be understood in an operator sense. Thus, the key point is that the commutator of $H - \lambda$ with a quantization B of η is positive, modulo compact terms, when localized at λ in the spectrum of H and microlocalized away from the radial set $\mathcal{R}(\lambda)$. Note that, as usual, there is nothing to prove at $\mathcal{R}(\lambda)$, since each point in it is the image of a maximally extended generalized broken bicharacteristic.

This positivity can be proved directly by showing that the indicial operators of the commutator are positive away from $\mathcal{R}(\lambda)$, which follows from an iterative argument. However, it also reduces to the Mourre estimate, involving the generator of dilations $A = \frac{1}{2}(\cdot D_z + D_z \cdot z)$, which has principal symbol at $\zeta \cdot z$. The Mourre estimate states the following. Suppose that $\lambda \notin \Lambda$ and $\epsilon > 0$. Then there is a $\delta > 0$ such that if $\phi \in \mathcal{C}_c^\infty(\mathbb{R})$ is supported in $(\lambda - \delta, \lambda + \delta)$ then there exists $K' \in \Psi_{\mathrm{sc}}^{-\infty,1}(\bar{X}, \mathcal{C})$ such that

$$\phi(H) i[A, H]\phi(H)) \geq 2(d(\lambda) - \epsilon)\phi(H)^2 + K', \tag{6.6}$$

where

$$d(\lambda) = \inf\{\lambda - \lambda' : \lambda' \leq \lambda,\ \lambda' \in \Lambda\} \geq 0$$

is the distance of λ to the next threshold below it if $\lambda \geq \inf \Lambda$, and $d(\lambda)$ an arbitrary positive number if $\lambda < d(\lambda)$. Since $d(\lambda) > 0$ if $\lambda \notin \Lambda$, (6.6) is indeed a positive commutator estimate, which does not even have a 'negative' term E, unlike (6.1). The Mourre estimate, originating in [**47**], has been well understood since the work of Perry, Sigal and Simon [**53**] and Froese and Herbst [**16**]. Here I just outline the argument in the simplest case, namely if no proper subsystem has any L^2-eigenvalues.

In this simplest case, the argument of [**16**] (see also [**8**] and [**67**]) proceeds as follows. In order to prove (6.6), one only needs to show that for all b, the corresponding indicial operators satisfy the corresponding inequality, i.e. that

$$\phi(\hat{H}_b)i\widehat{[A,H]}_b\phi(\hat{H}_b) \geq 2(d(\lambda)-\epsilon)\phi(\hat{H}_b)^2. \tag{6.7}$$

(This means that the operators on the two sides, which are families of operators on X^b, depending on $(y_b,\zeta_b)\in {}^{\mathrm{sc}}T^*_{C_b}\bar{X}$, satisfy the inequality for all $(y_b,\zeta_b)\in {}^{\mathrm{sc}}T^*_{C_b}\bar{X}$.) It is convenient to assume that ϕ is identically 1 near λ; if (6.7) holds for such ϕ, it holds for any ϕ_0 with slightly smaller support, as follows by multiplication by $\phi_0(\hat{H}_b)$ from the left and right.

Note that for $b=0$ the estimate certainly holds: it comes from the Poisson bracket formula in the scattering calculus, or from a direct computation yielding $i\widehat{[A,H]}_0 = 2\Delta_{g_0}$. Hence, if the the localizing factor $\phi(\hat{H}_0)=\phi(|\zeta|^2)$ is supported in $(\lambda-\delta,\lambda+\delta)$ and $\lambda>0$, then (6.7) holds even with $d(\lambda)-\epsilon$ replaced with $\lambda-\delta$. Note that $\lambda\geq d(\lambda)$, if $\lambda>0$, since 0 is a threshold of H. On the other hand, if $\lambda<0$, both sides of (6.7) vanish for ϕ supported near λ, so the inequality holds trivially.

In general, we may assume inductively that at all clusters c with $C_c\subsetneq C_b$, i.e. $X^b\subsetneq X^c$, (6.7) has been proved with ϕ replaced by a cutoff $\tilde{\phi}$ and ϵ replaced by ϵ', i.e. we may assume that for all $\epsilon'>0$ there exists $\delta'>0$ such that for all c with $C_c\subsetneq C_b$, and for all $\tilde{\phi}\in\mathcal{C}^\infty_c(\mathbb{R};[0,1])$ supported in $(\lambda-\delta',\lambda+\delta')$,

$$\tilde{\phi}(\hat{H}_c)i\widehat{[A,H]}_c\tilde{\phi}(\hat{H}_c) \geq 2(d(\lambda)-\epsilon')\tilde{\phi}(\hat{H}_c)^2. \tag{6.8}$$

But these are exactly the indicial operators of $\tilde{\phi}(\hat{H}_b)i\widehat{[A,H]}_b\tilde{\phi}(\hat{H}_b)$, so, as discussed in [**65**, Proposition 8.2], (6.7) implies that

$$\tilde{\phi}(\hat{H}_b)i\widehat{[A,H]}_b\tilde{\phi}(\hat{H}_b) \geq 2(d(\lambda)-\epsilon')\tilde{\phi}(\hat{H}_b)^2 + K_b,\ K_b\in\Psi^{-\infty,1}_{\mathrm{sc}}(X^b,\mathcal{C}^b). \tag{6.9}$$

This implication relies on a square root construction in the many-body calculus, which is particularly simple in this case.

Now, we first multiply (6.9) through by $\phi(H)$ from both the left and the right. Recall that we use coordinates (z_b,z^b) on $X_b\oplus X^b$ and (ζ_b,ζ^b) are the dual coordinates. We remark that $\hat{H}_b=|\zeta_b|^2+H^b$, so if $\lambda-|\zeta_b|^2$ is not an eigenvalue of H^b, then as $\operatorname{supp}\phi\to\{\lambda\}$, $\phi(H^b+|\zeta_b|^2)\to 0$ strongly, so as K_b is compact, $\phi(H^b+|\zeta_b|^2)K_b\to 0$ in norm; in particular it can be made to have norm smaller than $\epsilon'-\epsilon>0$. After multiplication from both sides by $\phi_1(\hat{H}_b)$, with ϕ_1 having even smaller support, (6.7) follows (with ϕ_1 in place of ϕ), with the size of $\operatorname{supp}\phi_1$ a priori depending on ζ_b. However, $i\phi_1(\hat{H}_b)\widehat{[A,H]}_b\phi_1(\hat{H}_b)$ is continuous in ζ_b with values in bounded operators on $L^2(X^b)$, so if (6.7) holds at one value of ζ_b, then it holds nearby. Moreover, for large $|\zeta_b|$ both sides vanish as $\hat{H}_b=H^b+|\zeta_b|^2$, with H^b bounded below, so the estimate is in fact uniform if we slightly increase $\epsilon>0$.

In general, the proof requires to treat the range of E, the spectral projection of H^b to $\{\lambda\}$, separately. Roughly, the positivity estimate on the range of E comes from the virial theorem, $iE[z^bD_{z^b},H^b]E=0$, which is formally clear, and is easy to prove. Thus,

$$iE[A, H_b]E = iE[z^b D_{z^b}, H^b]E + iE[z_b D_{z_b}, \Delta_{X_b}]E = i[z_b D_{z_b}, \Delta_{X_b}]E,$$

and the commutator $i\phi(H_b)[z_b D_{z_b}, \Delta_{X_b}]\phi(H_b)$ is easily computed to be positive. Of course, there are also cross-terms that need to be considered, but they can be estimated by Cauchy-Schwartz estimates, see [**16**] or [**67**].

I refer to [**66**] and [**65**] for the detailed arguments proving propagation of singularities in the many-body setting, and to [**68**, Appendix] for weaker estimates with simplified proofs.

7. Asymptotic completeness

Asymptotic completeness (AC) is an L^2-based statement describing the long-term behavior of solutions of the Schrödinger equation. In the short-range setting it says that for any $\phi \in L^2(X_0)$ in the range of $\mathrm{Id} - E_{\mathrm{pp}}$, E_{pp} being the projection onto the bound states of H (i.e. onto its L^2-eigenfunctions), there exist $\phi_\alpha \in L^2(X_a)$ such that

$$\|e^{-iHt}\phi - \sum_\alpha e^{-iH_a t}(\phi_\alpha \otimes \psi_\alpha)\| \to 0 \text{ as } t \to +\infty.$$

In the long-range setting, $e^{-iH_a t}$ must be somewhat modified. After the ground-breaking work of Enss [**12, 10**], AC was first proved by Sigal and Soffer [**55**] in the short-range setting (see Graf's paper [**20**] for a different proof), and later by Dereziński [**7**], and also by Sigal and Soffer [**56, 57**], in the long-range setting. In the short range case the main ingredient is equivalent to certain estimates of the resolvent at the radial sets in a sense that I now describe. In the long-range setting, as already in two-body scattering, additional constructive steps are needed. The estimates, in a different language, appeared first in the work of Sigal and Soffer [**55**]. I hope that the following discussion makes it clearer how they relate to the propagation of singularities.

While asymptotic completeness gives a complete long-term L^2-description of solutions of the Schrödinger equation, the question remains whether an analogous description exists on other spaces, such as weighted L^2-spaces. For example, if ϕ is Schwartz, are the ϕ_α Schwartz? Or dually, starting with a tempered distribution ϕ, are there tempered distributions α such that the convergence holds, as $t \to +\infty$, in a suitable sense? A different point of view is the parameterization of generalized eigenfunctions of H using the Poisson operators $P_{\alpha,+}(\lambda)$, and the analogues of these questions can be asked there as well. The answer is affirmative in the two-body setting (even in the geometric setting, see [**41, 45**]). However, as indicated by the related issue of the mapping properties of the scattering matrices, discussed at the end of this section, it is unlikely that the same holds in the many-body setting. One can then ask weaker question, e.g. whether it holds in weighted spaces L^2_s, s near 0. Or, one may ask whether one can give a precise description of the map $\phi \mapsto \phi_\alpha$ e.g. as some sort of Fourier integral operator.

As a starting point of relating the propagation of singularities to AC, we note that the propagation of singularities is proved by showing its 'relative' versions, i.e. that for any l, $\mathrm{WF}^{*,l}_{\mathrm{sc}}(u)$ is also a union of maximally extended generalized broken bicharacteristics. When considering the resolvent, first recall that for $f \in \dot{C}^\infty(\bar{X})$, $R(\lambda + i0)f \in H^{\infty,l}$ for all $l < -1/2$, so we only need to find $\mathrm{WF}^{*,l}_{\mathrm{sc}}(R(\lambda + i0)f)$ for $l \geq -1/2$. Theorem 3.3 is also valid for $\mathrm{WF}^{*,l}_{\mathrm{sc}}$, i.e. the following holds.

THEOREM 7.1. *If $\lambda \notin \Lambda$, then for $f \in \mathcal{S}(\mathbb{R}^n)$, $l \geq -1/2$, $\mathrm{WF}_{sc}(R(\lambda+i0)f)$ is a subset of the image of $\mathcal{R}_+(\lambda)$ under the* forward *generalized broken bicharacteristic relation.*

This result allows $u = R(\lambda + i0)f$ *not* to lie in $H^{*,-1/2}$ on the image of $\mathcal{R}_+(\lambda)$ under the forward generalized broken bicharacteristic relation. This is a small set, but it is important to know whether $\mathrm{WF}_{\mathrm{sc}}^{*,l}(u)$ may indeed intersect the forward image of $\mathcal{R}_+(\lambda)$. Of course, we cannot expect an improvement at $\mathcal{R}_+(\lambda)$, as shown already by the example of the free Euclidean Laplacian. The crucial improvement is the following estimate, due to Sigal and Soffer [**55**].

THEOREM 7.2. *If $\lambda \notin \Lambda$, then for $f \in \mathcal{S}(\mathbb{R}^n)$, $\mathrm{WF}_{sc}^{*,-1/2}(R(\lambda+i0)f) \subset \mathcal{R}_+(\lambda)$.*

REMARK. This theorem also has a time-dependent analogue. If u is a solution of the Schrödinger equation $(D_t + H)u = 0$ with $u|_{t=0} \in \mathcal{S}(X_0)$ then on the one hand $u \in H^{\infty,l}(X_0 \times \mathbb{R})$ for $l < -1/2$, on the other hand $\mathrm{WF}_{\mathrm{sc}}^{*,-1/2}(u) \subset \mathcal{R}$.

In fact, this theorem can be improved along the lines of the distributional statement in Theorem 3.3:

COROLLARY 7.3. *Suppose that $\lambda \notin \Lambda$, $f \in H^{*,1/2}$ and $\mathrm{WF}_{sc}^{*,1/2+\epsilon}(f) \cap \mathcal{R}_-(\lambda) = \emptyset$ for some $\epsilon > 0$. Then $R(\lambda + i0)f = \lim_{t\to 0} R(\lambda + it)f$ exists in $H^{*,-1/2-\epsilon'}(\bar{X})$, $\epsilon' > 0$, and $\mathrm{WF}_{sc}^{*,-1/2}(R(\lambda + i0)f) \subset \mathcal{R}_+(\lambda)$.*

Theorem 7.2 can be proved rather simply. The main issue is how to obtain a positive commutator at the radial point. Away from the radial sets arbitrary weights can be accommodated by suitable construction, as pointed out in the previous section. At radial points only the weights can give positive commutators. Now, one has to use weights x^{-2l-1} to obtain estimates for $\mathrm{WF}_{\mathrm{sc}}^{*,l}$, and these weights will give a commutator whose sign depends on that of $-2l-1$, hence on whether $l > -1/2$, $l < -1/2$ or $l = -1/2$. It turns out that the sign is correct for (6.3) to be of use if $l < -1/2$; this yields the limiting absorption principle. The sign is wrong if $l > -1/2$, so no results can be expected then. In the borderline case $l = -1/2$, the weight vanishes. The way to obtain a positive commutator is thus to consider operators A which are microlocally (a multiple of) the identity near $\mathcal{R}_+(\lambda)$. The commutator then vanishes microlocally near $\mathcal{R}_+(\lambda)$, which is reasonable since no estimate on $\mathrm{WF}_{\mathrm{sc}}^{*,-1/2}(u)$ can be expected there.

It is then straightfoward to construct A so that (6.3) can be used to prove Theorem 7.2. Indeed, it suffices to show that on $\mathrm{WF}_{\mathrm{sc}}^{*,-1/2}(u)$, $\eta = \frac{z\cdot\zeta}{|z|}$ must satisfy $\lambda - \eta^2 \in \Lambda$, for then the full statement of the theorem follows by the propagation of singularities for $\mathrm{WF}_{\mathrm{sc}}^{*,-1/2}(u)$. So we proceed to prove this simpler result, namely that if $\lambda - \bar{\eta}^2 \notin \Lambda$ then for any point ξ, $\eta(\xi) = \bar{\eta}$ implies $\xi \notin \mathrm{WF}_{\mathrm{sc}}^{*,-1/2}(u)$.

To do so, we let $a = \chi(\eta)$ where $\chi \in \mathcal{C}_c^\infty(\mathbb{R})$, $\chi \geq 0$, is chosen so that $\chi \equiv 1$ on $[0, \bar{\eta} - \delta]$ for some $\delta > 0$, $\chi' \leq 0$ on $(0, \infty)$, $\chi'(\bar{\eta}) < 0$, and $t \in \operatorname{supp} \chi'$ implies that $\lambda - t^2 \notin \Lambda$. This can be arranged as Λ is closed. We can further make sure that $\sqrt{-\chi'}$ is $\mathcal{C}^\infty$ on $(0, \infty)$. Then the positive commutator methods outlined show the commutator of A, a quantization of a, with $H - \lambda$ is positive, in the region $\eta > 0$, yielding the estimate that proves the theorem. We remark that partial microlocalization, using functions of η, hes been used extensively in many-body scattering, especially by Gérard, Isozaki and Skibsted [**18, 19**] and Wang [**69**], to obtain partially microlocal statements such as radiation conditions and uniqueness statements [**32, 31**], and indeed to prove the smoothness of 2-cluster to 2-cluster scattering matrices [**59**].

It turns out that there is an even simpler way of proving Theorem 7.2, or indeed a stronger statement, which is due to Yafaev [**73**]. His estimate states that in a neighborhood of $C_{a,\mathrm{reg}}$, where we write y_a for the coordinates $z_a/|z_a|$ along $C_{a,\mathrm{reg}}$, $xD_{y_a}R(\lambda+i0)f$ is in $H^{*,-1/2}(\bar{X})$. Since the principal symbol of xD_{y_a} is invertible on $(T^*X_{a,\mathrm{reg}} \cap \dot{\mathrm{C}}\mathrm{har}(\lambda)) \setminus \mathcal{R}(\lambda)$, this result implies Theorem 7.2. Yafaev's proof relies on a simple and explicit commutator calculation, which allows one to deal with various error terms that one may, a priori, expect. However, exactly because of its explicit nature, it is presumably hard to generalize to more geometric settings, while the argument we sketched does not face this difficulty.

As discussed by Yafaev [**73**] in the usual time-dependent version, short-range asymptotic clustering, hence asymptotic completeness, are relatively easy consequences of Corollary 7.3, and we refer to [**73**] and [**8**] for more details. However, it is worth pointing out that the reason why Coulomb-type potentials (i.e. those in S^{-1}) are not 'short-range' is that the Hamilton vector field in some subsystem vanishes at radial points. This degeneracy makes even the subprincipal term important in describing the precise behavior of generalized eigenfunctions microlocally near this point.

Before turning to scattering theory on symmetric spaces, we note the implications of Theorem 7.2 for the scattering matrices. Previously, $S_{\alpha\beta}(\lambda)$ was only defined as a map $S_{\alpha\beta}(\lambda) : \mathcal{C}_c^\infty(S_{a,\mathrm{reg}}) \to \mathcal{C}^{-\infty}(S_{b,\mathrm{reg}})$. Indeed, part of the broken bicharacteristic relation connects $\mathcal{R}_+(\lambda)$ with its image, and this can a priori give a singularity in the kernel of $S_{\alpha\beta}(\lambda)$ of the kind that does not even allow one to conclude that $S_{\alpha\beta}(\lambda) : \mathcal{C}_c^\infty(S_{a,\mathrm{reg}}) \to \mathcal{C}^\infty(S_{b,\mathrm{reg}})$. The pairing formula, (4.2), combined with Theorem 7.2, show that in fact

$$S_{\alpha\beta}(\lambda) : L^2(S_a) \to L^2(S_b). \tag{7.1}$$

It is an interesting question whether this can be improved if we restrict $S_{\alpha\beta}(\lambda)$ to $\mathcal{C}_c^\infty(S_{a,\mathrm{reg}})$. Namely, except in special cases such as N-clusters and two-clusters, the best known result is the trivial consequence of (7.1):

$$S_{\alpha\beta}(\lambda) : \mathcal{C}_c^\infty(S_{a,\mathrm{reg}}) \to L^2(S_b).$$

(In the case of N-clusters and 2-clusters, the geometry of generalized broken bicharacteristics gives $S_{\alpha\beta}(\lambda) : \mathcal{C}_c^\infty(S_{a,\mathrm{reg}}) \to \mathcal{C}^\infty(S_{b,\mathrm{reg}})$.) The putative improvement would have to be connected to an improvement of Theorem 7.2, namely to the existence of *some* $l > -1/2$ such that $\mathrm{WF}_{\mathrm{sc}}^{*,l}(R(\lambda+i0)f) \subset \mathcal{R}_+(\lambda)$. It would also be connected to a better understanding of $R(\lambda \pm i0)$ at the thresholds, in which direction Wang's paper [**72**] is the only one I am aware of.

8. Scattering on higher rank symmetric spaces

In this section I discuss $\mathrm{SL}(N,\mathbb{R})/\mathrm{SO}(N,\mathbb{R})$, indeed, mostly $\mathrm{SL}(3,\mathbb{R})/\mathrm{SO}(3,\mathbb{R})$. The books [**25**], [**34**] and [**9**] are good general references. $N = 2$ yields the hyperbolic plane $\mathbb{H}^2$, which is a rank one symmetric space on which many aspects of analysis, such as the asymptotic behavior of the resolvent kernel and the analytic continuation of the resolvent are well understood. Indeed, these have been described on asymptotically hyperbolic spaces by Mazzeo and Melrose [**37**] and Perry [**51, 52**].

Higher rank symmetric spaces, such as $\mathrm{SL}(N)/\mathrm{SO}(N)$, $N \geq 3$, are much less understood. For example, using results of Harish-Chandra, and Trombi and Varadarajan (see [**25**]), Anker and Ji only recently obtained the leading order behavior of the Green's function [**2, 3, 4**]. Also, while spherical functions, which are most analogous to partial plane-partial spherical waves in the Euclidean setting, have been analyzed by Harish-Chandra, Trombi and Varadarajan, and in particular their analytic continuation is understood, the same cannot be said about the Green's function. The analysis of spherical functions relies on perturbation series expansions, much like in the proof of the Cauchy-Kovalevskaya theorem, and it does not work well at the walls of the Weyl chambers. Here I only illustrate some recent joint results with Rafe Mazzeo [**40, 39**], that illuminate the connections with many-body scattering, and in particular give rather direct results for the resolvent.

First I describe the space $\mathrm{SL}(3)/\mathrm{SO}(3)$. The polar decomposition states that any $C \in \mathrm{SL}(3)$ can be written uniquely as $C = VR$, $V = (CC^t)^{1/2}$ is positive definite and has determinant 1, $R \in \mathrm{SO}(3)$. Thus, $\mathrm{SL}(3)/\mathrm{SO}(3)$ can be identified with the set M of positive definite matrices of determinant 1; this is a five-dimensional real analytic manifold. The Killing form provides a Riemannian metric g. The associated Laplacian $\Delta = \Delta_g$ gives a self-adjoint unbounded operator on $L^2(M, dg)$, with spectrum $[\lambda_0, +\infty)$, $\lambda_0 = \frac{1}{3}$. Let $R(\lambda) = (\Delta - \lambda)^{-1}$ be the resolvent of Δ_g, $\lambda \notin [\lambda_0, +\infty)$.

Fix a point $o \in M$, which we may as well assume is the image of the identity matrix I in the identification above. The stabilizer subgroup K_o (in the natural $\mathrm{SL}(3)$ action on M) is isomorphic to $\mathrm{SO}(3)$. The Green function $G_o(\lambda)$ with pole at o and at eigenvalue λ is, by definition $R(\lambda)\delta_o$. It is standard that G_o lies in the space of K_o-invariant distributions on M. It is thus natural to study Δ on K_o-invariant functions.

Perhaps the most interesting property is the analytic continuation of the resolvent, which I state before indicating how it, and other results, relate to many-body scattering.

Fix the branch of the square root function $\sqrt{\ }$ on $\mathbb{C} \setminus [0, +\infty)$ which has negative imaginary part when $w \in \mathbb{C} \setminus [0, +\infty)$. Let S denote that part of the Riemann surface for $\lambda \mapsto \sqrt{\lambda - \lambda_0}$ where we continue from $\lambda - \lambda_0 \notin [0, +\infty)$ and allow $\arg(\lambda - \lambda_0)$ to change by any amount less than π. In other words, starting in the region $\operatorname{Im} \sqrt{\lambda - \lambda_0} < 0$, we continue across either of the rays where $\operatorname{Im} \sqrt{\lambda - \lambda_0} = 0$ and $\operatorname{Re} \sqrt{\lambda - \lambda_0} > 0$, respectively < 0, allowing the argument of $\sqrt{\lambda - \lambda_0}$ to change by any amount less than $\pi/2$ (so that only the positive imaginary axis is not reached).

THEOREM 8.1. *With all notation as above, the Green function $G_o(\lambda)$ continues meromorphically to S as a distribution. Similarly, as an operator between appropriate spaces of K_o-invariant functions, the resolvent $R(\lambda)$ itself has a meromorphic continuation in this region, with all poles of finite rank.*

Having stated the theorem, I indicate how it relates to many-body scattering. To do so, fix the point o – we may as well take it to be the identity matrix I. Now, M is a perfectly nice real analytic manifold and Δ is an elliptic operator on it in the usual sense, so the only question is its behavior at infinity. In order to describe this, we remark that any matrix $A \in M$ can be diagonalized, i.e. written as $A = O\Lambda O^t$, with $O \in \mathrm{SO}(3)$ and Λ diagonal, with entries given by the eigenvalues of A. If $\mathfrak{a}$ is the set of diagonal matrices of trace 0, then $\Lambda \in \exp(\mathfrak{a})$. If all eigenvalues of A are distinct, then Λ is determined except for the ordering of the eigenvalues, and there are only finitely many possibilities for O as well. However, at the walls, which are defined to be the places where two eigenvalues coincide, there is much more indeterminacy. For example, if two eigenvalues coincide, only their joint eigenspace is well-defined. Correspondingly, we may replace O by $O'O$ for any $O' \in \mathrm{SO}(3)$ preserving the eigenspace decomposition and still obtain the desired diagonalization.

This is closely reflected in the structure of the Laplacian at infinity. In fact, it turns out that on SO(3)-invariant functions, Δ is essentially a three-body Hamiltonian on $\mathfrak{a}$ with first order interactions and with collision 'planes' given by the walls (they are lines), see e.g. [**25**, Chapter II, Proposition 3.9]. So rather than particles, eigenvalues scatter in this case! Consequently, many-body results can be adapted to this setting.

We indicate how this is done. The most succint way of describing the geometry of M at infinity is to compactify it to a manifold $\bar{M}$ with codimension two corners. It has two boundary hypersurfaces, $H_\sharp$ and $H^\sharp$, which are perhaps easiest to describe in terms of a natural system of local coordinates derived from the matrix representation of elements in M. As above, we write $A \in M$ as $A = O\Lambda O^t$, with $O \in \mathrm{SO}(3)$ and Λ diagonal. The ordering of the diagonal entries of Λ is undetermined, but in the region where no two of them are equal, we denote them as $0 < \lambda_1 < \lambda_2 < \lambda_3$ (but recall also that $\lambda_1\lambda_2\lambda_3 = 1$). In this region the ratios

$$\mu = \frac{\lambda_1}{\lambda_2}, \qquad \nu = \frac{\lambda_2}{\lambda_3}$$

are independent functions, and near the submanifold $\exp(\mathfrak{a})$ in M we can complete them to a full coordinate system by adding the above-diagonal entries c_{12}, c_{13}, c_{23} in the skew-symmetric matrix $T = \log O$. On M we have $\mu, \nu > 0$, and locally the compactification consists of replacing $(\mu, \nu) \in (0,1) \times (0,1)$ by $(\mu, \nu) \in [0,1) \times [0,1)$. Then $H^\sharp = \{\mu = 0\}$ and $H_\sharp = \{\nu = 0\}$, and this coordinate system gives the $\mathcal{C}^\infty$ structure near the corner $H_\sharp \cap H^\sharp$.

On the other hand, in a neighborhood of the interior of $H_\sharp$, for example, we obtain the compactification and its $\mathcal{C}^\infty$ structure as follows. Write the eigenvalues of $A \in M$, i.e. the diagonal entries of Λ in the decomposition for A above, as λ_1, λ_2 and λ_3. Suppose that A lies in a small neighbourhood $\mathcal{U}$ where

$$c < \lambda_1/\lambda_2 < c^{-1}, \quad \lambda_3 > 1/c,$$

for some fixed $c \in (0,1)$. Recall also that $\lambda_3 = 1/\lambda_1\lambda_2$. These inequalities imply that $\lambda_1 = (\lambda_1/\lambda_2)^{1/2}\lambda_3^{-1/2} < 1$ and $\lambda_2 = (\lambda_2/\lambda_1)^{1/2}\lambda_3^{-1/2} < 1$, and $\lambda_3 > 1$ in $\mathcal{U}$. Hence there is a well-defined decomposition $\mathbb{R}^3 = E_{12} \oplus E_3$ for any $A \in \mathcal{U}$, where E_{12} is the sum of the first two eigenspaces and E_3 is the eigenspace corresponding to λ_3, regardless of whether or not λ_1 and λ_2 coincide. We could write equivalently $A = OCO^t$, where C is block-diagonal, preserving the splitting $\mathbb{R}^2 \oplus \mathbb{R}$ of $\mathbb{R}^3$. The ambiguity in this factorization is that C can be conjugated by an element of O(2) (acting in the upper left block), and O(2) can be included in the top left corner of SO(3) (the bottom right entry being set equal to ± 1 appropriately). Let C' denote the upper-left block of C; the bottom right entry of C is just λ_3, and so $\lambda_3 \det C' = 1$. In other words, $C' = \lambda_3^{-1/2} C''$ where C'' is positive definite and symmetric with determinant 1, hence represents an element of $\mathrm{SL}(2)/\mathrm{SO}(2) \equiv \mathbb{H}^2$. Hence for an appropriate neighbourhood $\mathcal{V}$ of I in $\mathrm{SL}(2)/\mathrm{SO}(2)$, the neighbourhood

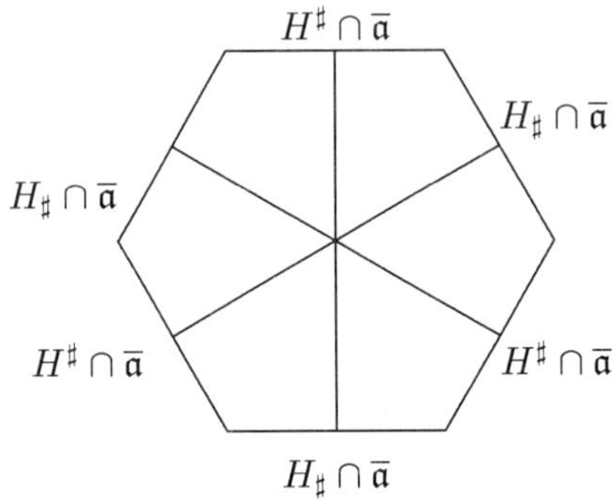

FIGURE 10. The closure of $\mathfrak{a}$, or rather $\exp(\mathfrak{a})$, in the compactification $\bar{M}$ of M. The lines in the interior are the Weyl chamber walls, playing the role of collision planes in many-body scattering. The side faces $H^\sharp \cap \bar{\mathfrak{a}}$ and $H_\sharp \cap \bar{\mathfrak{a}}$ correspond to the front faces on Figure 5. The main face on Figure 5 would only show up if we did a logarithmic blow-up of all boundary hypersurfaces of $\bar{M}$ and then blew up the corner.

$\mathcal{U}$ is identified with $(\mathcal{V} \times \mathrm{SO}(3))/\,\mathrm{O}(2) \times (0, c^{3/2})$, where the variable on the last factor is $s = \lambda_3^{-3/2}$. The compactification then simply replaces $(0, c^{3/2})_s$ by $[0, c^{3/2})$. Note that although the action of O(2) on $\mathcal{V}$ has a fixed point (namely I), its action on SO(3), and hence on the product, is free. The neighbourhood $\mathcal{V}$ can be chosen larger when λ_3 is larger, and the limiting 'cross-section' $\lambda_3 = \text{const}$ has the form $(\mathbb{H}^2 \times \mathrm{SO}(3))/\mathrm{O}(2)$. This space is a fibre bundle over $\mathrm{SO}(3)/\,\mathrm{O}(2)$ $(= \mathbb{R}P^2)$ with fibre $\mathbb{H}^2$. Notice that the Weyl chamber wall corresponds to the origin (i.e. the point fixed by the SO(2) action) in $\mathbb{H}^2$. I refer to [**40**] for a more thorough description of $\bar{M}$.

On each boundary hypersurface of M, it is now easy to describe model operators for Δ acting on SO(3)-invariant functions. For instance, at $H_\sharp$ this model can be considered as an operator $L_\sharp$ on $\mathbb{R}_s \times \mathbb{H}^2$, acting on SO(2)-invariant functions. Explicitly,

$$L_\sharp = \frac{1}{4}(sD_s)^2 + i\frac{1}{2}(sD_s) + \frac{1}{3}\Delta_{\mathbb{H}^2}.$$

This is tensor product type, so its resolvent can be obtained from an integral of the resolvents of $\frac{1}{4}(sD_s)^2 + i\frac{1}{2}(sD_s)$ and $\frac{1}{3}\Delta_{\mathbb{H}^2}$. (Note that I am ignoring the weights of the L^2 spaces on which we are working, hence the appearance of the perhaps strange first order terms.)

This framework allows one to develop the elliptic theory, for example to analyze $(\Delta - \lambda)^{-1}$ for $\lambda \in \mathbb{C} \setminus [\lambda_0, +\infty)$. In particular, one can construct a parametrix for Δ on $\bar{M}$ that has a smoothing error. Since this error has no decay at infinity, it is not compact. However, the error can be improved by pasting together the resolvents of $L_\sharp$ and $L^\sharp$, and applying the result to the error term to remove it modulo a decaying, hence compact, new error term. One of the consequences is then the description of the asymptotic behavior of the Green's function, see [**40**].

The proof of Theorem 8.1 relies on the method of complex scaling. This is an extension of dilation analyticity, and was originally developed in the setting of

2-body scattering by Aguilar-Combes [**1**] and generalized to the many-body setting by Balslev-Combes [**5**] and later further generalized by Gérard [**17**]. We refer to [**26**] and [**54**, Volume 4] for an exposition, and to the paper [**58**] of Sjöstrand and Zworski for a slightly different point of view.

The point of complex scaling is to rotate the essential spectrum of the operator being studied, in this case the Laplacian. To give the reader a rough idea how this works, consider the hyperbolic space $\mathbb{H}^2 = \mathrm{SL}(2,\mathbb{R})/\mathrm{SO}(2,\mathbb{R})$, which may be identified with the set of two-by-two positive definite matrices A of determinant 1. In terms of geodesic normal coordinates (r,ω) about $o = I$, the Laplacian is given by

$$\Delta_{\mathbb{H}^2} = D_r^2 - i(\coth r)D_r + (\sinh r)^{-2}D_\omega^2.$$

Now consider the diffeomorphism $\Phi_\theta : A \mapsto A^w$, $w = e^\theta$, on $\mathbb{H}^2$, $\theta \in \mathbb{R}$. This corresponds to dilation along the geodesics through o, since these have the form $\gamma_A : s \mapsto A^{cs}$, $c > 0$. Thus, in geodesic normal coordinates, $\Phi_\theta : (r,\omega) \mapsto (e^\theta r, \omega)$. Φ_θ defines a group of unitary operators on $L^2(\mathbb{H}^2)$ via

$$(U_\theta f)(A) = (\det D_A\Phi_\theta)^{1/2}(\Phi_\theta^* f)(A),\ J = \det D_A\Phi_\theta = w\frac{\sinh wr}{\sinh r},\ w = e^\theta.$$

Now, for θ real, consider the scaled Laplacian

$$\begin{aligned}(\Delta_{\mathbb{H}^2})_\theta &= U_\theta \Delta_{\mathbb{H}^2} U_\theta^{-1} = J^{1/2}\Phi_\theta^*\Delta_{\mathbb{H}^2}\Phi_{-\theta}^* J^{-1/2}\\ &= J^{1/2}(w^{-2}D_r^2 - iw^{-1}\coth(wr)D_r + (\sinh(wr))^{-2}D_\omega^2)J^{-1/2}\end{aligned}$$

This is an operator on $\mathbb{H}^2$, with coefficients which extend analytically in the strip $|\operatorname{Im}\theta| < \frac{\pi}{2}$. The square root is continued from the standard branch near $w > 0$. (The singularity of the coefficients at $r = 0$ is only an artifact of the polar coordinate representation.) Note that $(\Delta_{\mathbb{H}^2})_\theta$ and $(\Delta_{\mathbb{H}^2})_{\theta'}$ are unitary equivalent if $\operatorname{Im}\theta = \operatorname{Im}\theta'$ because of the group properties of U_θ. The scaled operator, $(\Delta_{\mathbb{H}^2})_\theta$, is not elliptic on all of $\mathbb{H}^2$ when $0 < |\operatorname{Im}\theta| < \frac{\pi}{2}$ because for r large enough, $w^2\sinh(wr)^{-2}$ can lie in $\mathbb{R}^-$. However, it is elliptic in some uniform neighbourhood of o in $\mathbb{H}^2$, and its radial part

$$(\Delta_{\mathbb{H}^2})_{\theta,\mathrm{rad}} = J^{1/2}(w^{-2}D_r^2 - iw^{-1}\coth(wr)D_r)J^{-1/2},$$

which corresponds to its action on SO(2)-invariant functions, is elliptic on the entire half-line $r > 0$. The model operator for $(\Delta_{\mathbb{H}^2})_{\theta,\mathrm{rad}} - \lambda$ at infinity,

$$\begin{aligned}&e^{(w-1)r/2}(w^{-2}D_r^2 - iw^{-1}D_r - \lambda)e^{-(w-1)r/2}\\ &= e^{(w-1)r/2}((w^{-1}D_r - \frac{i}{2})^2 - (\lambda - \frac{1}{4}))e^{-(w-1)r/2},\end{aligned}$$

is also invertible on the model space at infinity, $L^2(\mathbb{R}; e^r\,dr)$, since this is equivalent to the invertibility of

$$e^{wr/2}((w^{-1}D_r - \frac{i}{2})^2 - (\lambda - \frac{1}{4}))e^{-wr/2} = w^{-2}D_r^2 - (\lambda - \frac{1}{4})$$

on $L^2(\mathbb{R}; dr)$. Thus, a parametrix with compact remainder can be constructed for $(\Delta_{\mathbb{H}^2})_{\theta,\mathrm{rad}}$, and this show that its essential spectrum lies in $\frac{1}{4} + e^{-2i\operatorname{Im}\theta}[0,+\infty)$. Hence $((\Delta_{\mathbb{H}^2})_{\theta,\mathrm{rad}} - \lambda)^{-1}$ is meromorphic outside this set. In fact, it is well known that there are no poles in this entire strip (although there are an infinite number on $\arg\sqrt{\lambda - \lambda_0} = \pi/2$).

Combining this with some more standard technical facts, we are in a position to apply the theory of Aguilar-Balslev-Combes to prove that $((\Delta_{\mathbb{H}^2})_{\mathrm{rad}} - \lambda)^{-1}$, and hence $(\Delta_{\mathbb{H}^2} - \lambda)^{-1}$, has an analytic continuation in λ across $(\frac{1}{4}, +\infty)$. This is done by noting that for SO(2)-invariant functions $f, g \in L^2(\mathbb{H}^2)$ and $\theta \in \mathbb{R}$,

$$\langle f, ((\Delta_{\mathbb{H}^2})_{\mathrm{rad}} - \lambda)^{-1} g\rangle = \langle U_{\bar{\theta}} f, ((\Delta_{\mathbb{H}^2})_{\theta,\mathrm{rad}} - \lambda)^{-1} U_\theta g\rangle$$

by the unitarity of U_θ. Now if f, g lie in a smaller (dense) class of functions such that $U_\theta f$ and $U_\theta g$ continue analytically from $\theta \in \mathbb{R}$, then the meromorphic continuation in λ of the right hand side is obtained by first making θ complex with imaginary part of the appropriate sign, and then allowing λ to cross the continuous spectrum of $\Delta_{\mathbb{H}^2}$ without encountering the essential spectrum of $(\Delta_{\mathbb{H}^2})_{\theta,\mathrm{rad}}$. Hence the left hand side continues meromorphically also. With some additional care, one can even allow g to be the delta distribution at o, yielding the meromorphic continuation of the Green's function.

The argument on the higher rank symmetric space $M = \mathrm{SL}(3)/\mathrm{SO}(3)$ is similar. We still use the same scaling $\Phi_\theta : A \mapsto A^w$, $w = e^\theta$, on M, $\theta \in \mathbb{R}$. Again, the first concern is that, allowing θ to become complex, the scaled operator Δ_θ is not elliptic. However, it is elliptic near $o = \mathrm{Id}$, and the scaled models for it near the walls, such as $(L_\sharp)_\theta$, remain elliptic at the walls. After all, for the latter, this is just the ellipticity of $(\Delta_{\mathbb{H}^2})_\theta$ near the origin, which we have already observed. This again allows the elliptic parametrix construction to proceed, supplying the results we needed in order to reach the framework of complex scaling. This in turn finishes the proof of Theorem 8.1.

References

[1] J. Aguilar and J. M. Combes. A class of analytic perturbations for one-body Schrödinger operators. *Comm. Math. Phys.*, 22:269–279, 1971.

[2] J.-P. Anker. La forme exacte de l'estimation fondamentale de Harish-Chandra. *C. R. Acad. Sci. Paris*, Sér. I, 305:371–374, 1987.

[3] J.-P. Anker and L. Ji. Comportement exact du noyau de la chaleur et de la fonction de Green sur les espaces symétriques non-compacts. *C. R. Acad. Sci. Paris*, Sér. I, 326:153–156, 1998.

[4] J.-P. Anker and L. Ji. Heat kernel and Green function estimates on noncompact symmetric spaces. *Geom. Funct. Anal.*, 9(6):1035–1091, 1999.

[5] E. Balslev and J. M. Combes. Spectral properties of many body Schrödinger operators with dilation analytic potentials. *Commun. Math. Phys.*, 22:280–294, 1971.

[6] Alberto-P. Calderón. On an inverse boundary value problem. In *Seminar on Numerical Analysis and its Applications to Continuum Physics (Rio de Janeiro, 1980)*, pages 65–73. Soc. Brasil. Mat., Rio de Janeiro, 1980.

[7] J. Dereziński. Asymptotic completeness of long-range N-body quantum systems. *Ann. Math.*, 138:427–476, 1993.

[8] J. Dereziński and C. Gérard. *Scattering theory of classical and quantum N-particle systems.* Springer, 1997.

[9] P. Eberlein. *Geometry of nonpositively curved manifolds.* University of Chicago Press, 1996.

[10] V. Enss. *Quantum scattering theory two- and three-body systems with potentials of short and long range*, pages 39–176. Lecture notes in mathematics. Springer-Verlag, 1985.

[11] V. Enss and R. Weder. The geometrical approach to multidimensional inverse scattering. *J. Math. Phys.*, 36:3902–3921, 1995.

[12] Volker Enss. Asymptotic completeness for quantum mechanical potential scattering. I. Short range potentials. *Comm. Math. Phys.*, 61(3):285–291, 1978.

[13] Volker Enss and Ricardo Weder. Inverse two-cluster scattering. *Inverse Problems*, 12(4):409–418, 1996.

[14] L. D. Faddeev. The inverse problem in the quantum theory of scattering. II. In *Current problems in mathematics, Vol. 3 (Russian)*, pages 93–180, 259. (loose errata). Akad. Nauk SSSR Vsesojuz. Inst. Naučn. i Tehn. Informacii, Moscow, 1974.

[15] R. G. Froese and I. Herbst. Exponential bounds and absence of positive eigenvalues of N-body Schrödinger operators. *Commun. Math. Phys.*, 87:429–447, 1982.

[16] R. G. Froese and I. Herbst. A new proof of the Mourre estimate. *Duke Math. J.*, 49:1075–1085, 1982.

[17] C. Gérard. Distortion analyticity for N-particle Hamiltonians. *Helv. Phys. Acta*, 66(2):216–225, 1993.

[18] C. Gérard, H. Isozaki, and E. Skibsted. *Commutator algebra and resolvent estimates*, volume 23 of *Advanced studies in pure mathematics*, pages 69–82. 1994.

[19] C. Gérard, H. Isozaki, and E. Skibsted. N-body resolvent estimates. *J. Math. Soc. Japan*, 48:135–160, 1996.

[20] G. M. Graf. Asymptotic completeness for N-body short range systems: a new proof. *Commun. Math. Phys.*, 132:73–101, 1990.

[21] Victor Guillemin and David Schaeffer. On a certain class of Fuchsian partial differential equations. *Duke Math. J.*, 44(1):157–199, 1977.

[22] A. Hassell. Distorted plane waves for the 3 body Schrödinger operator. *Geom. Funct. Anal.*, 10:1–50, 2000.

[23] A. Hassell, R. B. Melrose, and A. Vasy. Spectral and scattering theory for symbolic potentials of order zero. *Advances in Mathematics*, 181:1–87, 2004.

[24] B. Helffer and J. Sjöstrand. Équation de Schrödinger avec champ magnétique et équation de Harper. In *Schrödinger operators (Sønderborg, 1988)*, pages 118–197. Springer, Lecture Notes in Physics, No. 345, Berlin, 1989.

[25] S. Helgason. *Groups and geometric analysis.* Academic Press, 1984.

[26] P. D. Hislop and I. M. Sigal. *Introduction to spectral theory.* Springer-Verlag, 1996.

[27] L. Hörmander. *The analysis of linear partial differential operators,* vol. 1-4. Springer-Verlag, 1983.

[28] Lars Hörmander. On the existence and the regularity of solutions of linear pseudo-differential equations. *Enseignement Math. (2)*, 17:99–163, 1971.

[29] M. Ikawa, editor. *Spectral and scattering theory.* Marcel Dekker, 1994.

[30] H. Isozaki. On N-body Schrödinger operators. *Proc. Indian Acad. Sci. Math. Sci.*, 104:667–703, 1993.

[31] H. Isozaki. A generalization of the radiation condition of Sommerfeld for N-body Schrödinger operators. *Duke Math. J.*, 74:557–584, 1994.

[32] H. Isozaki. *A uniqueness theorem for the N-body Schrödinger equation and its applications.* In Ikawa [**29**], 1994.

[33] V. Ja. Ivriĭ. Wave fronts of solutions of boundary value problems for a class of symmetric hyperbolic systems. *Sibirsk. Mat. Zh.*, 21(4):62–71, 236, 1980.

[34] J. Jost. *Riemannian geometry and geometric analysis.* Springer, 1998.

[35] Peter D. Lax. Asymptotic solutions of oscillatory initial value problems. *Duke Math. J.*, 24:627–646, 1957.

[36] G. Lebeau. Propagation des ondes dans les variétés à coins. *Ann. Scient. Éc. Norm. Sup.*, 30:429–497, 1997.

[37] R. Mazzeo and R. B. Melrose. Meromorphic extension of the resolvent on complete spaces with asymptotically constant negative curvature. *J. Func. Anal.*, 75:260–310, 1987.

[38] R. Mazzeo and R. B. Melrose. Pseudodifferential operators on manifolds with fibred boundaries. *Asian Journal of Mathematics*, 2, 1998.

[39] R. Mazzeo and A. Vasy. Analytic continuation of the resolvent of the Laplacian on SL(3)/SO(3). *American Journal of Mathematics*, 126:821–844, 2004.

[40] Rafe Mazzeo and András Vasy. Scattering theory on SL(3)/SO(3): connections with quantum 3-body scattering. *Proc. Lond. Math. Soc. (3)*, 94(3):545–593, 2007.

[41] R. B. Melrose. *Spectral and scattering theory for the Laplacian on asymptotically Euclidian spaces.* In Ikawa [**29**], 1994.

[42] R. B. Melrose. *Geometric scattering theory.* Cambridge University Press, 1995.

[43] R. B. Melrose and J. Sjöstrand. Singularities of boundary value problems. I. *Comm. Pure Appl. Math*, 31:593–617, 1978.

[44] R. B. Melrose and J. Sjöstrand. Singularities of boundary value problems. II. *Comm. Pure Appl. Math*, 35:129–168, 1982.

[45] R. B. Melrose and M. Zworski. Scattering metrics and geodesic flow at infinity. *Inventiones Mathematicae*, 124:389–436, 1996.

[46] Richard Melrose and Jared Wunsch. Propagation of singularities for the wave equation on conic manifolds. *Invent. Math.*, 156(2):235–299, 2004.

[47] E. Mourre. Absence of singular continuous spectrum of certain self-adjoint operators. *Commun. Math. Phys.*, 78:391–408, 1981.

[48] R. G. Novikov and G. M. Khenkin. The $\overline{\partial}$-equation in the multidimensional inverse scattering problem. *Uspekhi Mat. Nauk*, 42(3(255)):93–152, 255, 1987.

[49] Roman G. Novikov. On inverse scattering for the N-body Schrödinger equation. *J. Funct. Anal.*, 159(2):492–536, 1998.

[50] Roman G. Novikov. On determination of the fourier transform of a potential from the scattering amplitude. *Inverse Problems*, 17:1243–1251, 2001.

[51] P. Perry. The Laplace operator on a hyperbolic manifold. I. Spectral and scattering theory. *J. Funct. Anal.*, 75:161–187, 1987.

[52] P. Perry. The Laplace operator on a hyperbolic manifold. II. Eisenstein series and the scattering matrix. *J. Reine. Angew. Math.*, 398:67–91, 1989.

[53] P. Perry, I. M. Sigal, and B. Simon. Spectral analysis of N-body Schrödinger operators. *Ann. Math.*, 114:519–567, 1981.

[54] M. Reed and B. Simon. *Methods of modern mathematical physics*. Academic Press, 1979.

[55] I. M. Sigal and A. Soffer. N-particle scattering problem: asymptotic completeness for short range systems. *Ann. Math.*, 125:35–108, 1987.

[56] I. M. Sigal and A. Soffer. Long-range many-body scattering. *Inventiones Math.*, 99:115–143, 1990.

[57] I. M. Sigal and A. Soffer. Asymptotic completeness of N-particle long-range scattering. *J. Amer. Math. Soc.*, 7:307–334, 1994.

[58] Johannes Sjöstrand and Maciej Zworski. Complex scaling and the distribution of scattering poles. *J. Amer. Math. Soc.*, 4(4):729–769, 1991.

[59] E. Skibsted. Smoothness of N-body scattering amplitudes. *Reviews in Math. Phys.*, 4:619–658, 1992.

[60] J. Sylvester and G. Uhlmann. A global uniqueness theorem for an inverse boundary value problem. *Ann. of Math.*, 125:153–169, 1987.

[61] M. Taylor. Grazing rays and reflection of singularities of solutions to wave equations. *Comm. Pure Appl. Math.*, 29:1–38, 1976.

[62] Gunther Uhlmann and András Vasy. Low-energy inverse problems in three-body scattering. *Inverse Problems*, 18(3):719–736, 2002.

[63] A. Vasy. Structure of the resolvent for three-body potentials. *Duke Math. J.*, 90:379–434, 1997.

[64] A. Vasy. Scattering matrices in many-body scattering. *Commun. Math. Phys.*, 200:105–124, 1999.

[65] A. Vasy. Propagation of singularities in many-body scattering. *Ann. Sci. École Norm. Sup. (4)*, 34:313–402, 2001.

[66] A. Vasy. Propagation of singularities in many-body scattering in the presence of bound states. *J. Func. Anal.*, 184:177–272, 2001.

[67] A. Vasy. Exponential decay of eigenfunctions in many-body type scattering with second order perturbations. *J. Func. Anal.*, 209:468–492, 2004.

[68] A. Vasy and X. P. Wang. Smoothness and high energy asymptotics of the spectral shift function in many-body scattering. *Comm. PDEs*, 27:2139–2186, 2002.

[69] X. P. Wang. Microlocal estimates for N-body Schrödinger operators. *J. Fac. Sci. Univ. Tokyo Sect. IA, Math.*, 40:337–385, 1993.

[70] X. P. Wang. High energy asymptotics for N-body scattering matrices with arbitrary channels. *Ann. Inst. H. Poincaré Phys. Théor.*, 65(1):81–108, 1996.

[71] X. P. Wang. An inverse problem related to channel scattering operators. *Asymptot. Anal.*, 18(1-2):147–164, 1998.

[72] X. P. Wang. Spectral analysis of N-body Schrödinger operators near a threshold. *Preprint*, 2001.

[73] D. Yafaev. Radiation conditions and scattering theory for N-particle Hamiltonians. *Commun. Math. Phys.*, 154:523–554, 1993.

Department of Mathematics, Stanford University, Stanford, California 94305-2125
E-mail address: `andras@math.stanford.edu`

Clay Mathematics Proceedings
Volume **17**, 2013

Wave Maps with and without Symmetries

Michael Struwe

Introduction

Many of the results on wave maps seem highly technical and require deep results from harmonic analysis for a complete understanding. In these three lectures we present direct approaches to certain global aspects of the wave map problem, with powerful conclusions.

The Cauchy problem for wave maps

In this first lecture we recall the approach presented in [**20**] for showing global existence and uniqueness for the Cauchy problem for wave maps from the $(1+m)$-dimensional Minkowski space, $m \geq 4$, to any complete Riemannian manifold with bounded curvature, provided the initial data are small in the critical norm.

1.1. Wave maps. Let (N,h) be a complete Riemannian manifold of dimension k with $\partial N = \emptyset$. We denote space-time coordinates on $\mathbb{R}^{m+1}$ as $(t,x) = (x^\alpha), 0 \leq \alpha \leq m$. A wave map $u\colon \mathbb{R}^{m+1} \to N$ is a solution to the equation

$$D^\alpha \partial_\alpha u = 0, \tag{1}$$

where $\partial_\alpha = \frac{\partial}{\partial x^\alpha}$ and where we raise and lower indices with the Minkowski metric $(\eta_{\alpha\beta}) = \mathrm{diag}(-1,1,\dots,1)$. We tacitly sum over repeated indices. Moreover, D is the covariant pull-back derivative in the bundle u^*TN.

The equivalent extrinsic form of equation (1) reveals that this is a quasilinear wave equation. Recall that the Nash embedding theorem permits to regard N as a submanifold of some Euclidean $\mathbb{R}^n$. Letting $u = (u^1,\dots,u^n)\colon \mathbb{R}^{m+1} \to N \hookrightarrow \mathbb{R}^n$ be the corresponding extrinsic representation of our wave map u, equation (1) then takes the form

$$\Box u^i = -\partial^\alpha \partial_\alpha u^i = u^i_{tt} - \Delta u^i = B^i_{jk}(u)\partial_\alpha u^j \partial^\alpha u^k, 1 \leq i \leq n, \tag{2}$$

where $B(p)\colon T_pN \times T_pN \to (T_pN)^\perp$ is the second fundamental form of $N \subset \mathbb{R}^n$ at any $p \in N$. This extrinsic form of the wave map equation (1) will be very useful in the sequel.

Note that equation (2) geometrically can be interpreted simply as saying that $\Box u \perp T_uN$, which immediately gives the intrinsic form (1). Moreover, in the case

2010 *Mathematics Subject Classification.* Primary 35L70, 35L71.

when $N = S^k \hookrightarrow \mathbb{R}^{k+1}$ equation (2) takes the form $\Box u = \lambda u$ for some scalar function λ. Taking account of the fact that $|u|^2 \equiv 1$, we compute

$$\lambda = \Box u \cdot u = -\partial^\alpha(\partial_\alpha u \cdot u) + \partial_\alpha u \partial^\alpha u = \partial_\alpha u \partial^\alpha u = |\nabla u|^2 - |u_t|^2$$

and thus find the equation

$$\Box u = u_{tt} - \Delta u = (|\nabla u|^2 - |u_t|^2)u \tag{3}$$

for a wave map $u \colon \mathbb{R}^{m+1} \to S^k \hookrightarrow \mathbb{R}^{k+1}$.

We study the Cauchy problem for wave maps with initial data

$$(u, u_t)_{|t=0} = (u_0, u_1) \in \dot{H}^{\frac{m}{2}} \times \dot{H}^{\frac{m}{2}-1}(\mathbb{R}^m; TN), \tag{4}$$

where $\dot{H}^s$ for any s denotes the homogenous Sobolev space. Note that from any solution u to equation (1) or (2), we can obtain further solutions by scaling $u^R(t,x) = u(Rt, Rx)$. In view of the invariance

$$||(u, u_t)_{|t=0}||_{\dot{H}^{\frac{m}{2}} \times \dot{H}^{\frac{m}{2}-1}(\mathbb{R}^m;TN)} = ||(u^R, u_t^R)_{|t=0}||_{\dot{H}^{\frac{m}{2}} \times \dot{H}^{\frac{m}{2}-1}(\mathbb{R}^m;TN)} \tag{5}$$

the $\dot{H}^{\frac{m}{2}} \times \dot{H}^{\frac{m}{2}-1}$-regularity is critical.

With $L^{(2m,2)}(\mathbb{R}^m) \hookrightarrow L^{2m}(\mathbb{R}^m)$ denoting the Lorentz space, the main result from [**20**] may now be stated, as follows.

THEOREM 1.1. *Suppose N is complete, without boundary and has bounded curvature in the sense that the curvature operator R and the second fundamental form B and all their derivatives are bounded, and let $m \geq 4$. Then there is a constant $\varepsilon_0 > 0$ such that for any $(u_0, u_1) \in H^{\frac{m}{2}} \times H^{\frac{m}{2}-1}(\mathbb{R}^m; TN)$ satisfying*

$$||u_0||_{\dot{H}^{\frac{m}{2}}} + ||u_1||_{\dot{H}^{\frac{m}{2}-1}} < \varepsilon_0$$

there exists a unique global solution $u \in C^0(\mathbb{R}; H^{\frac{m}{2}}) \cap C^1(\mathbb{R}; H^{\frac{m}{2}-1})$ of (1), (4) satisfying

$$\sup_t ||du(t)||_{\dot{H}^{\frac{m}{2}-1}} + \int_{\mathbb{R}} ||du(t)||^2_{L^{(2m,2)}(\mathbb{R}^m)}\, dt \leq C\varepsilon_0 \tag{6}$$

and preserving any higher regularity of the data.

For $N = S^k$, global wellposedness of the Cauchy problem (1), (4) for initial data having small energy in the critical norm was first shown by Tao [**26**], [**27**], initially only for $m \geq 5$ and finally for all $m \geq 2$. For $m \geq 5$, by a variant of Tao's method, Klainerman-Rodnianski [**10**] were able to extend his results to general targets, independently and almost simultaneously with our work [**20**] with Shatah. Similar results are due to Nahmod - Stefanov - Uhlenbeck [**16**]. In the low-dimensional cases $2 \leq m \leq 3$ for wave maps $u \colon \mathbb{R}^{m+1} \to H^2$ to hyperbolic space H^2, the analogue of Theorem 1.1 was obtained by Krieger [**12**], [**13**]. Finally, Tataru [**30**] established well-posedness of the Cauchy problem for (1), (4) for initial data of small critical energy in the low-dimensional cases $2 \leq m \leq 3$ also for general targets. Previous work of Tataru [**28**], [**29**] already had shown the problem to be wellposed for initial data of small energy in a critical Besov space.

Whereas the methods of Tao, Klainerman-Rodnianski, Tataru, and many others working on this problem strongly rely on Littlewood-Paley theory and a sophisticated analysis of the interaction between different frequency components of a solution, the approach in [**20**] requires no microlocalization. It proceeds in physical space and is very direct, using as a tool essentially only the Strichartz estimate and its recent subtle improvement by Keel and Tao [**9**].

Terence Tao, and independently also Sergiu Klainerman and Igor Rodnianski pointed out that estimates similar to the crucial $L^1_t L^\infty_x$-estimate in Lemma 1.2 below can also be obtained from bilinear estimates for the wave equation obtained by Klainerman-Tataru [**11**]. Tristan Rivière has brought to our attention further applications of Lorentz spaces in gauge theory related to our use of Lorentz spaces here.

1.2. Uniqueness and higher regularity. The condition (6) easily yields uniqueness when we consider the extrinsic form (2) of the wave map system. Indeed, let u and v be solutions to (2) of class $H^{\frac{m}{2}}$ with $u, v \in C^0(\mathbb{R}; H^{\frac{m}{2}}) \cap C^1(\mathbb{R}; H^{\frac{m}{2}-1})$, and suppose that

$$u_{|t=0} = v_{|t=0},\ u_{t|t=0} = v_{t|t=0}.$$

Moreover, we assume (6), that is, in particular,

$$||du||^2_{L^2_t L^{2m}_x} = \int_{\mathbb{R}} ||du(t)||^2_{L^{2m}(\mathbb{R}^m)}\, dt < \infty,$$

and similarly for v. Then $w = u - v$ satisfies

$$w_{tt} - \Delta w = [B(u) - B(v)](\partial_\alpha u, \partial^\alpha u) + B(v)(\partial_\alpha u + \partial_\alpha v, \partial^\alpha w).$$

Multiplying by w_t, we obtain

$$\frac{1}{2}\frac{d}{dt}||dw(t)||^2_{L^2} = I(t) + II(t),$$

where by Sobolev's embedding $\dot{H}^1(\mathbb{R}^m) \hookrightarrow L^{\frac{2m}{m-2}}(\mathbb{R}^m)$ we can estimate

$$\begin{aligned} I(t) &= \int_{\mathbb{R}^m} \langle [B(u) - B(v)](\partial_\alpha u, \partial^\alpha u), w_t\rangle\, dx \le C \int_{\mathbb{R}^m} |du|^2 |w||dw|\, dx \\ &\le C||du||^2_{L^{2m}} ||w||_{L^{\frac{2m}{m-2}}} ||dw||_{L^2} \le C||du||^2_{L^{2m}} ||dw||^2_{L^2}. \end{aligned}$$

In order to bound the term $II(t)$, we note that orthogonality $\langle B(u)(\cdot,\cdot), u_t\rangle = 0 = \langle B(v)(\cdot,\cdot), v_t\rangle$ implies

$$\begin{aligned} |\langle B(v)(\partial_\alpha u, \partial^\alpha w), w_t\rangle| &= |\langle B(v)(\partial_\alpha u, \partial^\alpha w), u_t\rangle| \\ &= |\langle [B(v) - B(u)](\partial_\alpha u, \partial^\alpha w), u_t\rangle| \le C|du|^2 |w||dw|, \end{aligned}$$

and similarly for the term involving $\partial_\alpha v$.

Thus also this term can be bounded

$$II(t) \le C(||du||^2_{L^{2m}} + ||dv||^2_{L^{2m}})||dw||^2_{L^2},$$

yielding the inequality

$$\frac{d}{dt}||dw||^2_{L^2} \le C(||du||^2_{L^{2m}} + ||dv||^2_{L^{2m}})||dw||^2_{L^2}.$$

Hence we obtain the uniform estimate

$$||dw||^2_{L^\infty_t L^2_x} \le ||dw(0)||^2_{L^2} \cdot \exp(C(||du||^2_{L^2_t L^{2m}_x} + ||dv||^2_{L^2_t L^{2m}_x})).$$

Since $dw(0) = 0$, uniqueness follows.

Higher regularity estimates for (smooth) solutions u of (2) satisfying (6) for sufficiently small $\varepsilon > 0$ can be obtained in similar fashion by differentiating the intrinsic form of the wave map equation covariantly in spatial directions and using standard energy estimates; see [**20**] for details.

1.3. Moving frames and Gauge condition. Our approach requires the construction of a suitable frame for the pull-back bundle u^*TN, as pioneered by Christodoulou-Tahvildar-Zadeh [**2**] and Hélein [**7**]. With no loss of generality, we may assume that TN is parallelizable, that is, there exist smooth vector fields $\overline{e}_1, \ldots, \overline{e}_k$ such that at each $p \in N$ the collection $\overline{e}_1(p), \ldots, \overline{e}_k(p)$ is an orthonormal basis for T_pN; see [**2**], [**7**]. Given a (smooth) map $u \colon \mathbb{R}^{m+1} \to N$ then the vector fields $\overline{e}_a \circ u, 1 \leq a \leq k$, yield a smooth orthonormal frame for the pull-back bundle u^*TN. Moreover, we may freely rotate this frame at any point $z = (t, x) \in \mathbb{R}^{m+1}$ with a matrix $(R_a^b) = (R_a^b(z)) \in SO(k)$, thus obtaining the frame

$$e_a = R_a^b \overline{e}_b \circ u, 1 \leq a \leq k.$$

Expressing du as

$$du = q^a e_a \tag{7}$$

with an $\mathbb{R}^k$-valued 1-form $q = q_\alpha\, dx^\alpha$, then we have

$$|du|^2 = |q|^2 = \sum_{\alpha=0}^{m} |q_\alpha|^2 .$$

In particular, for $1 \leq p \leq \infty$ the L^p-norm of du is well-defined, independently of the choice of "gauge" (R_a^b), and coincides with the L^p-norm of du in the extrinsic representation of u as a map $u \colon \mathbb{R}^{m+1} \to N \subset \mathbb{R}^m$. Later we will see that if the gauge R is suitably chosen, and if $\varepsilon_0 > 0$ is sufficiently small, also the norms of the derivatives of du and the derivatives of q agree up to a multiplicative constant.

Letting $D = (D_\alpha)_{0 \leq \alpha \leq m}$ be the pull-back covariant derivative, we have

$$De_a = A_a^b e_b, 1 \leq a \leq k, \tag{8}$$

for some matrix-valued 1-form $A = A_\alpha dx^\alpha$. Fix a pair of space-time indices $0 \leq \alpha, \beta \leq m$. The curvature of D enters in the commutation relation

$$\begin{aligned} D_\alpha D_\beta e_a - D_\beta D_\alpha e_a &= D_\alpha(A_{a,\beta}^b e_b) - D_\beta(A_{a,\alpha}^b e_b) \\ &= (\partial_\alpha A_{a,\beta}^c - \partial_\beta A_{a,\alpha}^c + A_{b,\alpha}^c A_{a,\beta}^b - A_{b,\beta}^c A_{a,\alpha}^b) e_c = F_{a,\alpha\beta}^c e_c, \end{aligned}$$

or

$$\partial_\alpha A_\beta - \partial_\beta A_\alpha + [A_\alpha, A_\beta] = F_{\alpha\beta} = R(\partial_\alpha u, \partial_\beta u) \tag{9}$$

for short. (The comma separates the form subscript from the vector subscript and does not indicate a differential.)

Following Hélein [**7**] we choose the columb gauge

$$\sum_{i=1}^{m} \partial_i A_i = 0. \tag{10}$$

This results in the equation

$$\Delta A_\beta + \partial_i [A_i, A_\beta] = \partial_i F_{i\beta} = \partial_i (R(\partial_i u, \partial_\beta u)), 0 \leq \beta \leq m, \tag{11}$$

where we tacitly sum over $1 \leq i \leq m$. Given $u \colon \mathbb{R}^{m+1} \to N$ with du having sufficiently small L^m-norm, this equation admits a unique solution A which for any fixed time we may represent as

$$A_\beta = G_i * ([A_i, A_\beta] - F_{i\beta}), \tag{12}$$

where

$$G(x) = \frac{c}{|x|^{m-2}}$$

is the fundamental solution to the Laplace operator on $\mathbb{R}^m$ and $G_i = -\partial_i G$.

Indeed, from (11) and elliptic regularity theory we have the a-priori estimate

$$\begin{aligned}||A||_{L^m} &\le C||A||_{\dot W^{1,\frac{m}{2}}} \le C||[A,A]||_{L^{\frac{m}{2}}} + C||F||_{L^{\frac{m}{2}}}\\ &\le C||A||^2_{L^m} + C||R||_{L^\infty}||du||^2_{L^m};\end{aligned}$$

confer [**5**], Section 4.3. For sufficiently small $||A||_{L^m}$ we may absorb the first term on the right on the left hand side of this equation to obtain at any fixed time the estimate with constants C independent of t

$$(13)\qquad ||A||_{L^m} \le C||A||_{\dot W^{1,\frac{m}{2}}} \le C||du||^2_{L^m} \le C||du||^2_{\dot H^{\frac{m}{2}-1}} \le C\varepsilon_0.$$

For later use we derive further estimates for the connection 1-form A and the curvature F, assuming that $\varepsilon_0 > 0$ is sufficiently small. For the sake of exposition, we indicate these estimates only in the case when $m = 4$ and refer to [**20**] for the general case. For $1 \le s \le \infty$ again denote as $L^{(p,s)}(\mathbb{R}^m)$ the Lorentz space.

LEMMA 1.2. *Let $m = 4$, and fix $r = 8/5$.*

(i) For any time t there holds

$$\|\nabla^2 A\|_{L^r} + \|\nabla\partial_0 A\|_{L^r} \le C\|\nabla F\|_{L^r} \le C\|du\|_{L^8}\|du\|_{\dot H^1}\,.$$

(ii) For any time t we have

$$\|A\|_{L^\infty} \le C\|du\|^2_{L^{(8,2)}}\,.$$

Proof. (i) To estimate $\nabla^2 A$, observe that equation (11) implies

$$(14)\qquad \|\nabla^2 A\|_{L^r} \le C\|\nabla[A,A]\|_{L^r} + C\|\nabla F\|_{L^r}\,.$$

By Hölder's inequality and Sobolev's embedding we can estimate

$$\|\nabla[A,A]\|_{L^r} \le 2\|\nabla A\|_{L^{r_1}}\|A\|_{L^m} \le C\|\nabla^2 A\|_{L^r}\|A\|_{L^m}\,,$$

where

$$\frac{1}{r_1} = \frac{1}{r} - \frac{1}{m} = \frac{3}{8}.$$

From (13) and (14) then, for sufficiently small $\varepsilon_0 > 0$ we obtain

$$\|\nabla^2 A\|_{L^r} \le C\|\nabla F\|_{L^r}.$$

The term ∇F only involves terms of the form $R(\nabla\partial_\alpha u, \partial_\beta u)$ and $\nabla R(\partial_\alpha u, \partial_\beta u)$ and therefore may be estimated

$$|\nabla F| \le C(|\nabla du||du| + |du|^3).$$

Letting $q = 8 = 2m$, so that $1/r = 5/8 = 1/q + 1/2$, upon estimating

$$\|\nabla F\|_{L^r} \le C(\|\nabla du\|_{L^2}\|du\|_{L^q} + \|du\|^2_{L^4}\|du\|_{L^q})\,,$$

from Sobolev's embedding $\|du\|_{L^4} \le C\|du\|_{\dot H^1} \le C$ we conclude that

$$\|\nabla^2 A\|_{L^r} \le C\|\nabla F\|_{L^r} \le C\|du\|_{L^8}\|du\|_{\dot H^1}.$$

To estimate $\nabla\partial_0 A$ we note that the equations

$$\partial_0 A_i = \partial_i A_0 + [A_i, A_0] + F_{0i}$$

and

$$\Delta\partial_0 A_0 + \partial_i\partial_0[A_i, A_0] = \partial_i\partial_0 F_{i0}\,,$$

from (11) make exchanging of time derivative by spatial derivative possible and thus imply the desired estimate.

(ii) By the Sobolev embedding into Lorentz spaces and i), we have

$$\|A\|_{L^{(8,2)}} \le C\|A\|_{L^{(8,\frac{8}{5})}} \le \|A\|_{\dot{W}^{2,\frac{8}{5}}} \le C\|du\|_{L^8}\,.$$

Therefore, and since for any $m \geq 4$ we have $G_i \in L^{(\frac{m}{m-1},\infty)}$, the dual of $L^{(m,1)}$, using the representation of A given by (12) we obtain

$$\|A\|_{L^\infty} \le C(\|[A,A]\|_{L^{(4,1)}} + \|F\|_{L^{(4,1)}}) \le C(\|A\|^2_{L^{(8,2)}} + \|du\|^2_{L^{(8,2)}}) \le C\|du\|^2_{L^{(8,2)}}\,,$$

as claimed. □

1.4. Equivalence of Norms. Estimate (13) implies the equivalence of the extrinsic H^ℓ-norm of du and the H^ℓ-norm of q for any ℓ, provided $\varepsilon_0 > 0$ is sufficiently small. To see this consider a vector field W in u^*TN whose coordinates in the frame $\{e_a\}$ are given by

$$W = Q^a e_a = Qe$$

with

$$\|W\|_{L^2} = \|Q\|_{L^2}\,.$$

The extrinsic partial derivative of W can be computed from the covariant derivative and the second fundamental form B as

$$D_k W = \partial_k W + B(u)(\partial_k u, W) = (\partial_k Q + AQ)e\,;$$

that is,

$$\partial_k W = (\partial_k Q + AQ)e - B(u)(\partial_k u, Qe).$$

Therefore from (12), Sobolev embedding, and boundedness of the second fundamental form B we obtain

$$\begin{aligned}\big|\, \|\partial W\|_{L^2} - \|\partial Q\|_{L^2}\big| &\le C(\|AQ\|_{L^2} + \|duQ\|_{L^2}) \\ &\le C(\|A\|_{L^m} + \|du\|_{L^m})\|\partial Q\|_{L^2} \le C\varepsilon_0\|\partial Q\|_{L^2})\,.\end{aligned}$$

By linearity of the map $Q \mapsto W$ and interpolation we conclude the equivalence of the H^s-norms of Q and W for all $0 \le s \le 1$. The same argument establishes the equivalence of the covariant and extrinsic H^s-norms of W for $0 \le s \le 1$. By applying this argument iteratively to $W = \nabla^\ell du$ for $\ell = 0, 1, \ldots$, we then obtain the equivalence of the H^s-norm of du and H^s-norm of q for any $s \geq 0$, provided $\varepsilon_0 > 0$ is sufficiently small.

1.5. A priori bounds. In order to obtain the a-priori bounds from which we may derive existence, we represent a local smooth solution u of (1), (4) in terms of the 1-form q given by (7), where the frame (e_a) is in Coulomb gauge.

From (8) then we have the equations

$$0 = D_\alpha \partial_\beta u - D_\beta \partial_\alpha u = (D_\alpha q_\beta - D_\beta q_\alpha)e,$$

where we denote

$$D_\alpha q_\beta = (\partial_\alpha + A_\alpha) q_\beta; \tag{15}$$

in components, this is

$$D_\alpha(q^a_\beta e_a) = (\partial_\alpha q^c_\beta + A^c_{a,\alpha} q^a_\beta) e_c.$$

Again the comma separates the form subscript from the vector subscript and does not indicate a differential.

That is, we have

$$D_\alpha q_\beta - D_\beta q_\alpha = 0. \tag{16}$$

Moreover, the wave map equation (1) yields the equation

$$D^\alpha q_\alpha = 0. \tag{17}$$

Differentiating (17) with respect to x^β and using (9), (16), we derive the covariant wave equation

$$0 = D_\beta D^\alpha q_\alpha = D^\alpha D_\beta q_\alpha + F^\alpha_\beta q_\alpha = D^\alpha D_\alpha q_\beta + F^\alpha_\beta q_\alpha.$$

Expanding this identity using (15), we obtain

$$(\partial_t^2 - \Delta) q_\beta = 2A^\alpha \partial_\alpha q_\beta + (\partial^\alpha A_\alpha) q_\beta + A^\alpha A_\alpha q_\beta + F^\alpha_\beta q_\alpha =: h_\beta. \tag{18}$$

We can estimate q in terms of the initial data and h by using the Strichartz estimate for the linear wave equation

$$\Box v = h, v_{|_{t=0}} = f, v_{t|_{t=0}} = g. \tag{19}$$

Again denoting as $\dot{H}^\gamma = (\sqrt{-\Delta})^{-\gamma} L^2(\mathbb{R}^m)$ the homogeneous Sobolev space, and as $L^{(p,r)}(\mathbb{R}^m)$ the Lorentz space, from Keel-Tao [9], Corollary 1.3, if $h = 0$ for any $T > 0$ we have

$$\begin{aligned} ||v||_{L^2([0,T];L^{\frac{2(m-1)}{m-3}}(\mathbb{R}^m))} &+ ||v||_{C^0([0,T];\dot{H}^\gamma(\mathbb{R}^m))} + ||v_t||_{C^0([0,T];\dot{H}^{\gamma-1}(\mathbb{R}^m))} \\ &\leq C(||f||_{\dot{H}^\gamma(\mathbb{R}^m)} + ||g||_{\dot{H}^{\gamma-1}(\mathbb{R}^m)}). \end{aligned}$$

where $\gamma = \frac{m+1}{2(m-1)}$. If $m = 4$, we have $\gamma = \frac{5}{6}$ and the preceding becomes

$$\begin{aligned} ||v||_{L^2([0,T];L^6(\mathbb{R}^4))} &+ ||v||_{C^0([0,T];\dot{H}^{5/6}(\mathbb{R}^4))} + ||v_t||_{C^0([0,T];\dot{H}^{-1/6}(\mathbb{R}^4))} \\ &\leq C(||f||_{\dot{H}^{5/6}(\mathbb{R}^4)} + ||g||_{\dot{H}^{-1/6}(\mathbb{R}^4)}). \end{aligned} \tag{20}$$

By real interpolation between this estimate and the analogous estimate for derivatives of v, and using the embedding (in the notation of [9])

$$(L_t^2 L_x^6, L_t^2 \dot{W}_x^{1,6})_{\frac{1}{6},2} \hookrightarrow L_t^2 L_x^{(8,2)},$$

we obtain

$$||v||_{L_t^2 L_x^{(8,2)}} + ||dv||_{C^0([0,T];L^2)} \leq C(||f||_{\dot{H}^1} + ||g||_{L^2}). \tag{21}$$

By Duhamel's principle, for general h it then follows that

$$||v||_{L_t^2 L_x^{(8,2)}} + ||dv||_{C_t^0 L_x^2} \leq C(||f||_{\dot{H}^1} + ||g||_{L^2} + ||h||_{L_t^1 L_x^2}). \tag{22}$$

(The crucial gain of the Lorentz exponent by real interpolation was already observed by Keel and Tao [9] but was omitted in the final statement of their theorem.)

We will apply estimate (22) to equation (18) on any time interval $[0, T]$ such that $||du||_{\dot{H}^1}$ remains sufficiently small, uniformly for $0 < t < T$. Also using the equivalence of the H^s-norms of du and q for $s \leq 1$ on any such time interval, we obtain

$$\begin{aligned} ||du||_{C_t^0 \dot{H}_x^1} + ||du||_{L_t^2 L_x^{(8,2)}} &\leq C(||dq||_{C_t^0 L_x^2} + ||q||_{L_t^2 L_x^{(8,2)}}) \\ &\leq C(||dq(0)||_{L^2} + ||h||_{L_t^1 L_x^2}) \leq C(||du(0)||_{\dot{H}^1} + ||h||_{L_t^1 L_x^2}) \\ &\leq C(||u_0||_{\dot{H}^2} + ||u_1||_{\dot{H}^1} + ||h||_{L_t^1 L_x^2}). \end{aligned}$$

To estimate the various terms in h we observe that by Lemma 1.2 at any time t with $r_1 = 8/3$ we have

$$\begin{aligned}||h||_{L^2} &\le 2\|A\partial q\|_{L^2} + \|\partial A q\|_{L^2} + \|A^2 q\|_{L^2} + \|Fq\|_{L^2} \\ &\le 2\|A\|_{L^\infty}\|q\|_{\dot H^1} + \big(\|\nabla A\|_{L^{r_1}} + \|A^2\|_{L^{r_1}} + \|F\|_{L^{r_1}}\big)\|q\|_{L^8}\,.\end{aligned}$$

But Lemma 1.2 with $r = 8/5$ implies

$$\begin{aligned}\|\nabla A\|_{L^{r_1}} + \|A^2\|_{L^{r_1}} + \|F\|_{L^{r_1}} &\le C(\|\nabla^2 A\|_{L^r} + \|\nabla(A^2)\|_{L^r} + \|\nabla F\|_{L^r}) \\ &\le C\|du\|_{L^8}\|du\|_{\dot H^1}.\end{aligned}$$

Here we also used Sobolev's embedding and (13) to bound

$$\|\nabla(A^2)\|_{L^r} \le C\|\nabla A\|_{L^{r_1}}\|A\|_{L^4} \le C\|\nabla^2 A\|_{L^r}.$$

From Lemma 1.2 we then obtain

$$||h||_{L^2} \le C\|q\|_{L^8}\|du\|_{L^8}\|du\|_{\dot H^1} + 2\|A\|_{L^\infty}\|q\|_{\dot H^1} \le C\|du\|^2_{L^{(8,2)}}\|du\|_{\dot H^1}\,.$$

Using these estimates, we can bound h by

$$||h||_{L^1_t L^2_x} \le C||du||^2_{L^2_t L^{(8,2)}_x}||du||_{L^\infty_t \dot H^1_x}$$

and we conclude that

$$||du||_{L^\infty_t \dot H^1} + ||du||_{L^2_t L^{(8,2)}_x} \le C(||u_0||_{\dot H^2} + ||u_1||_{\dot H^1} + ||du||^2_{L^2_t L^{(8,2)}_x}||du||_{L^\infty_t \dot H^1_x})\,.$$

A global priori bound on $||du||_{L^\infty_t \dot H^1_x} + ||du||_{L^2_t L^8_x}$ thus follows, provided $||u_0||_{\dot H^2} + ||u_1||_{\dot H^1}$ is sufficiently small.

1.6. Existence. Recall $C^\infty \times C^\infty(\mathbb{R}^m; TN)$ is dense in $H^{\frac{m}{2}} \times H^{\frac{m}{2}-1}(\mathbb{R}^m; TN)$. We can thus find smooth data $(u_0^{(k)}, u_1^{(k)}) \to (u_0, u_1)$ in $H^{\frac{m}{2}} \times H^{\frac{m}{2}-1}(\mathbb{R}^m; TN)$. The local solutions $u^{(k)}$ to the Cauchy problem for (1) with data $(u_0^{(k)}, u_1^{(k)})$ by our a-priori bounds and regularity results for sufficiently small energy

$$||u_0||^2_{\dot H^{\frac{m}{2}}} + ||u_1||^2_{\dot H^{\frac{m}{2}-1}} < \varepsilon_0$$

then may be extended as smooth solutions to (1), (4) for all time and will satisfy the uniform estimates

$$||du^{(k)}||_{C^0_t \dot H^{\frac{m}{2}-1}_x} + ||du^{(k)}||_{L^2_t L^{(2m,2)}_x} \le C(||u_0^{(k)}||_{\dot H^{\frac{m}{2}}} + ||u_1^{(k)}||_{\dot H^{\frac{m}{2}-1}}) < C\varepsilon_0$$

for sufficiently large k.

Hence as $k \to \infty$ a subsequence $u^{(k)} \rightharpoonup u$ weakly in $H^{\frac{m}{2}}_{loc}(\mathbb{R}^{m+1})$, where

$$||du||_{C^0_t \dot H^{\frac{m}{2}-1}} + ||du||_{L^2_t L^{(2m,2)}_x} \le C(||u_0||_{\dot H^{\frac{m}{2}}} + ||u_1||_{\dot H^{\frac{m}{2}-1}})\,.$$

Since $\frac{m}{2} \ge 2$, by Rellich's theorem for a further subsequence $du^{(k)} \to du$ converges pointwise almost everywhere, and u solves (1), (4), as claimed.

Wave maps with symmetries I

The H^1-energy is the only known conserved quantity for the wave map system. The case when $m = 2$ therefore is particularly interesting, because in this dimension the H^1-energy is critical and one may hope to obtain also global results and a characterization of singularities. Indeed, this is possible in the case of symmetry.

In this second lecture, we study co-rotational wave maps from (1+2)-dimensional Minkowski space into a target surface of revolution. In the third lecture, finally, we investigate rotationally symmetric wave maps on $\mathbb{R}^{1+2}$.

2.1. Corotational wave maps. Let N be a surface of revolution with metric

$$ds^2 = d\rho^2 + g^2(\rho)d\theta^2,$$

where $\theta \in S^1$ and with $g \in C^\infty(\mathbb{R})$ satisfying $g(0) = 0$, $g'(0) = 1$. Moreover, we assume that g is odd and either

$$g(\rho) > 0 \text{ for all } \rho > 0 \tag{23}$$

with

$$\int_0^\infty |g(\rho)|\, d\rho = \infty, \tag{24}$$

or, if N is compact, that g has a first zero $\rho_1 > 0$ where $g'(\rho_1) = -1$, and that g is periodic with period $2\rho_1$. Note that in this second case assumption (24) is trivially satisfied. The case (23) corresponds to non-compact surfaces; condition (24) is a technical assumption needed to rule out that N contains a "sphere at infinity".

We regard (ρ, θ) as polar coordinates on N. Letting (r, ϕ) be the usual polar coordinates on $\mathbb{R}^2$, we then consider equivariant wave maps $u\colon \mathbb{R} \times \mathbb{R}^2 \to N$ given by

$$\rho = h(t, r), \theta = \phi.$$

The equation (2) for a wave map $u = (u^1, \ldots, u^n)\colon \mathbb{R}^{2+1} \to N \hookrightarrow \mathbb{R}^n$, that is

$$\Box u^i = B^i_{jk}(u)\partial_\alpha u^j \partial^\beta u^k,\ 1 \leq i \leq n, \tag{25}$$

in this co-rotational case simplifies to the nonlinear scalar equation

$$\Box h + \frac{f(h)}{r^2} = 0, \tag{26}$$

where

$$\Box h = h_{tt} - \Delta h = h_{tt} - \frac{1}{r}\left(rh_r\right)_r = h_{tt} - h_{rr} - \frac{h_r}{r}$$

and with $f(h) = g(h)g'(h)$. If $N = S^2$, for example, we have $g(h) = sin(h)$ and $f(h) = \frac{1}{2}sin(2h)$

In [**21**], Shatah and Tahvildar-Zadeh showed that the initial value problem for (25) with smooth equivariant data

$$(u, u_t)_{|t=0} = (u_0, u_1) \tag{27}$$

of finite energy admits a unique smooth solution for small time, which may be extended for all time if the target surface N is geodesically convex.

The latter condition is equivalent to the assumption $g'(\rho) \geq 0$ for all $\rho > 0$. This condition was later weakened by Grillakis [**4**] who showed that it suffices to assume

$$(g(\rho)\rho)' = g(\rho) + g'(\rho)\rho > 0 \text{ for } \rho > 0\,.$$

Note that this hypothesis, in particular, implies conditions (23) and (24).

In [**23**] we improve these results and show that conditions (23) and (24) already suffice for proving global well-posedness of the Cauchy problem for (26). In fact, we show that for general target surfaces N satisfying (24) the appearance of a singularity in (26) is related to the existence of a non-constant harmonic map $\overline{u}\colon S^2 \to N$, thereby confirming a long-standing conjecture about wave maps in this special, co-rotational case. But if N also satisfies (23), any co-rotational harmonic map $\overline{u}\colon S^2 \to N$ is constant, and global well-posedness follows.

On the other hand, when $N = S^2$ on the basis of numerical work of Bizon et al. [**1**] and Isenberg-Liebling [**8**] it had been conjectured that for suitable initial data equivariant wave maps $u\colon \mathbb{R} \times \mathbb{R}^2 \to S^2$ indeed may develop singularities in finite time. In a penetrating analysis, Krieger-Schlag-Tataru [**14**] and Rodnianski-Sterbenz [**17**] recently were able to confirm this conjecture also theoretically and give a rigorous proof of blow-up.

2.2. Results. By the results of Shatah-Tahvildar-Zadeh [**21**] singularities of co-rotational maps may be detected by measuring their energy

$$E(u(t), R) = \frac{1}{2}\int_{B_R(0)} |Du(t)|^2\, dx,$$

with $|Du|^2 = |u_t|^2 + |\nabla u|^2$. In terms of $h = h(t)$ we have

$$E(u(t), R) = \pi \int_0^R \left(|Dh|^2 + \frac{g^2(h)}{r^2}\right) rdr.$$

We also let

$$E(u(t)) = \lim_{R\to\infty} E(u(t), R).$$

By [**21**] there exists a number $\varepsilon_0 = \varepsilon_0(N) > 0$ such that the Cauchy problem for co-rotational wave maps for smooth data with energy $E(u(0)) < \varepsilon_0$ admits a global smooth solution; confer also [**19**], Theorem 8.1. By finite speed of propagation, similarly we obtain well-posedness of the Cauchy problem for time $t \le R$, provided $E(u(0), R) < \varepsilon_0$.

Conversely, let $u\colon [0, t_0[\times\mathbb{R}^2 \to N$ be a smooth co-rotational wave map. Then $z_0 = (t_0, 0)$ is a (first) singularity and t_0 is the blow-up time of u if and only if there holds

$$\inf_{0\le t<t_0} E(u(t), t_0 - t) \ge \varepsilon_0 > 0. \tag{28}$$

In fact, for any map u satisfying (28) the space-time gradient Du cannot be bounded near the origin $(0, 0)$. On the other hand, negating condition (28) we can find a time $t < t_0$ such that

$$E(u(t), R) < \varepsilon_0$$

for some $R > t_0 - t$ and the results quoted above will allow us to extend u smoothly as a solution to (25) on a neighborhood of $z_0 = (t_0, 0)$. Observe that, by symmetry, u can only blow up at the origin.

We can now state our main result.

THEOREM 2.1. *Let u be a smooth co-rotational solution to* (25) *blowing up at time t_0. Then there exist sequences $R_i \downarrow 0, t_i \uparrow t_0 (i \to \infty)$ such that*

$$u_i(t, x) = u(t_i + R_i t, R_i x) \to u_\infty(t, x)$$

strongly in $H^1_{loc}(]-1,1[\times\mathbb{R}^2)$*, where* u_∞ *is a non-constant, time-independent solution of* (25) *giving rise to a non-constant, smooth co-rotational harmonic map* $\overline{u}\colon S^2\to N$.

As a consequence, for target manifolds that do not admit non-constant co-rotational harmonic spheres we obtain global existence of smooth solutions to the Cauchy problem (25), (27) for smooth co-rotational data. In particular, we can improve Grillakis' result as follows.

THEOREM 2.2. *Suppose* N *is a surface of revolution with metric* $ds^2 = d\rho^2 + g^2(\rho)d\theta^2$ *satisfying* (23) *and* (24). *Then for any smooth co-rotational data the Cauchy problem* (25), (27) *admits a unique global smooth solution.*

As we shall see in Lecture 3, similar results also hold true in the case of radially symmetric wave maps $u = u(t,r)$ from $\mathbb{R}^{1+2}$ to an arbitrary closed target manifold; confer [**24**], [**25**].

2.3. Notation. Let $u\colon [0,t_0[\times\mathbb{R}^2\to N$ be a smooth co-rotational wave map blowing up at time t_0 and let $h = h(t,r)$ be the associated solution of (26).

For convenience we shift and reverse time and then scale our space-time coordinate $z = (t,x)$ so that in our new coordinates u is an equivariant solution to (25) on $]0,1]\times\mathbb{R}^2$ blowing up at the origin.

Letting

$$K^T = \{z=(t,x); 0\le |x|\le t\le T\}$$

be the forward light cone with vertex at the origin, truncated at height T, with lateral boundary

$$M^T = \{(t,x)\in K^T; |x| = t\},$$

we also introduce the flux

$$\mathrm{Flux}(u,T) = \frac{1}{2}\int_{M^T} |D^{||}u|^2\,do = \pi\int_0^T \left(|h_t+h_r|^2 + \frac{g^2(h)}{r^2}\right)\Big|_{t=r} r\,dr.$$

Here, $|D^{||}u|^2$ denotes the energy of all derivatives in directions tangent to M^T.

2.4. Basic estimates. We recall the energy bounds and decay estimates for (25) from [**21**]; these can also be found in [**19**], Chapter 8.1. Since $B(u)(v,w)\perp T_uN$ from (25) we obtain the conservation law

$$0 = \Box u\cdot u_t = \frac{\partial}{\partial t}e - div\,m \tag{29}$$

for the densities

$$e = \frac{1}{2}(|\nabla u|^2 + |u_t|^2),\ \ m = \nabla u\cdot u_t$$

of energy and momentum. Observe that $|m|\le e$. Integrating (29) over a truncated cone $K^{T_0}\setminus K^T$ for $0<T\le T_0\le 1$ we then find the identity

$$\int_{\{T\}\times B_T(0)} e\,dx + \frac{1}{2}\int_{M^{T_0}\setminus M^T} |D^{||}u|^2\,do = \int_{\{T_0\}\times B_{T_0}(0)} e\,dx\,.$$

From this we deduce the *energy inequality*

$$E(u(t),R)\le E(u(t+\tau),R+|\tau|). \tag{30}$$

for any $t,\tau,R>0$. (Of course, in the present case we only consider values such that $0<t,t+\tau\le 1$.)

Moreover, we conclude that

$$\lim_{T\downarrow 0}\int_{\{T\}\times B_T(0)} e\,dx$$

exists and we have *decay of the flux*

$$\text{Flux}(u,T)\to 0 \text{ as } T\downarrow 0. \tag{31}$$

Condition (24) together with the energy inequality implies the *uniform bounds*

$$\sup_{r<R}|h(t,r)|\le C(E(u(t),R)) \text{ for any } R>0 \tag{32}$$

for the function h associated with u, where $C(s)\to 0$ as $s\to 0$. Indeed, let

$$G(s)) = \int_0^s |g(\rho)|\,d\rho.$$

Since (24) implies that $G(s)\to\infty$ as $s\to\infty$ it then suffices to estimate

$$\begin{aligned} G(|h(t,R)|) &= \int_0^R (G(|h(t,r)|)_r\,dr \le \int_0^R |g(h(t,r))||h_r(t,r)|\,dr \\ &\le \frac{1}{2}\int_0^R (|h_r|^2 + \frac{g^2(h(t,r))}{r^2})rdr \le CE(u(t),R)\,. \end{aligned}$$

Moreover we have *exterior energy decay:* For any $0<\lambda\le 1$ as $t\to 0$ there holds

$$E(u(t),t) - E(u(t),\lambda t)\to 0. \tag{33}$$

An immediate consequence of (33) is the *decay of time derivatives:* Suppose that N satisfies (24). Then

$$\frac{1}{T}\int_{K^T} |u_t|^2\,dz \to 0 \text{ as } T\to 0. \tag{34}$$

These estimates seem particular to the rotationally symmetric setting. The (lengthy) proof of (33) and the derivation of (34) are given in the appendix.

Finally, as is also well-known, in view of the uniform energy bounds (30) above, we have uniform Hölder continuity away from $x=0$.

Lemma 2.3. *For any $r_0>0$, any (t,r) and (s,q) with $2r_0\le q\le s<t\le 1, 2r_0\le r\le t$ there holds*

$$|h(t,r)-h(s,q)|^2 \le C(|r-q|+|t-s|) \tag{35}$$

with a constant C depending only on the energy $E(u(1),1)$ and r_0.

Proof. Given $r_0>0$, for any t and $r_0\le r'<r\le t\le 1$ by Hölder's inequality and (30) we have

$$|h(t,r)-h(t,r')|^2 \le \left(\int_{r'}^r |h_r|\,dr''\right)^2 \le \frac{r-r'}{r'}\cdot\int_{r'}^r |h_r|^2\,r''\,dr'' \le C\frac{r-r'}{r_0},$$

while for any $s<t$ and $r_0\le r'\le s$ we find

$$|h(s,r')-h(t,r')|^2 \le \left(\int_s^t |h_t(t',r')|\,dt'\right)^2 \le \frac{t-s}{r_0}\int_s^t |h_t(t',r')|^2\,r'\,dt'.$$

Combining these inequalities, for any (t,r) and (s,q) with $2r_0 \le q \le s < t \le 1, 2r_0 \le r \le t$ and any r' with $r_0 \le r' \le r_1 := \inf\{q,r\}$ we find

$$|h(t,r) - h(s,q)|^2 \le C\frac{r - r' + q - r'}{r_0} + 2\frac{t-s}{r_0}\int_s^t |h_t(t',r')|^2\, r'\, dt'.$$

Taking the average with respect to $r' \in [r_1 - \min\{r_0, |r-q| + |t-s|\}, r_1]$, we obtain the claim. □

2.5. Proofs of Theorems 2.1 and 2.2. Fix a number $\varepsilon_1 = \varepsilon_1(N) > 0$ to be determined below. For $0 < t \le 1$ then choose $R = R(t) > 0$ so that

$$\varepsilon_1 \le E(u(t), 6R(t)) \le 2\varepsilon_1. \tag{36}$$

Applying the energy inequality (30), for any $|\tau| \le 5R$ we have

$$E(u(t+\tau), R) \le E(u(t), 6R) \le 2\varepsilon_1 \tag{37}$$

and similarly

$$\varepsilon_1 \le E(u(t+\tau), 6R + |\tau|) \le E(u(t+\tau), 11R). \tag{38}$$

We will choose ε_1 so that $2\varepsilon_1 < \varepsilon_0$. Then, in particular, from (28) and (36) we deduce the inequality

$$6R(t) < t \tag{39}$$

for all t. In fact, we obtain a much stronger result.

LEMMA 2.4. *$R(t)/t \to 0$ as $t \to 0$.*

Proof. Suppose by contradiction that for some sequence $t_i \downarrow 0$ $(i \to \infty)$ with associated radii $R_i = R(t_i)$ there holds $6R_i \ge \lambda t_i$ for some constant $\lambda > 0$. Then from (28) and (36) we deduce that

$$0 < \varepsilon_0 - 2\varepsilon_1 \le E(u(t_i), t_i) - E(u(t_i), 6R_i) \le E(u(t_i), t_i) - E(u(t_i), \lambda t_i),$$

contradicting (33) for large $i \in \mathbb{N}$. □

The following lemma is the main new technical ingredient in our work [**23**].

Consider the intervals $\Lambda_{R(t)}(t) =]t - R(t), t + R(t)[$, $0 < t \le 1$. By Vitali's theorem we can find a countable subfamily of disjoint intervals $\Lambda_i = \Lambda_{R(t_i)}(t_i), i \in \mathbb{N}$, such that $]0,1] \subset \cup_{i=1}^{\infty} \Lambda_i^*$, where $\Lambda_i^* = \Lambda_{5R(t_i)}(t_i)$. Observe that (39) implies

$$\inf \Lambda_i^* = t_i - 5R(t_i) > R(t_i) =: R_i \tag{40}$$

for each i. For any $\tau > 0$ the interval $[\tau, 1]$ is covered by finitely many intervals Λ_i^* which, however, fail to cover $]0,1]$ completely in view of (40). Therefore, we may assume that $t_i \to 0$ as $i \to \infty$.

LEMMA 2.5. *With the above notations there holds*

$$\liminf_{i\to\infty} \frac{1}{R_i}\int_{\Lambda_i}\int_{B_t(0)} |u_t|^2\, dx\, dt = 0.$$

Proof. Negating the assertion, we can find a number $\delta > 0$ and an index $i_0 \in \mathbb{N}$ such that

$$\int_{\Lambda_i}\int_{B_t(0)} |u_t|^2\, dx\, dt \ge \delta R_i \quad \text{for } i \ge i_0. \tag{41}$$

Given $0 < T < \inf \cup_{i<i_0} \Lambda_i^*$, let $I_0 = \{i\,;\, \inf \Lambda_i^* < T\} \subset \{i_0, i_0+1, \dots\}$. Observe that

$$]0, T[\subset \cup_{i \in I_0} \Lambda_i^* .$$

By (40) we have

$$R_i < \inf \Lambda_i^* = t_i - 5R_i < T$$

and therefore

$$t_i + R_i < T + 6R_i < 7T$$

for all $i \in I_0$. It follows that

$$\cup_{i \in I_0} \Lambda_i \subset]0, 7T]. \tag{42}$$

By choice of I_0, our assumption (41), and in view of (42) we now obtain that

$$\begin{aligned} \delta T &\le \delta \sum_{i \in I_0} \operatorname{diam} \Lambda_i^* = 10\,\delta \sum_{i \in I_0} R_i \le 10 \sum_{i \in I_0} \int_{\Lambda_i} \int_{B_t(0)} |u_t|^2 \,dx\,dt \\ &= 10 \int_{\cup_{i \in I_0} \Lambda_i} \int_{B_t(0)} |u_t|^2 \,dx\,dt \le 10 \int_{K^{7T}} |u_t|^2 \,dz, \end{aligned} \tag{43}$$

where we also used the fact that the intervals Λ_i are disjoint. But for small $T > 0$ this contradicts (34), thus proving the lemma. □

Proof of Theorem 2.1. i) Letting

$$u_i(t, x) = u(t_i + R_i t, R_i x), i \in \mathbb{N},$$

from Lemma 2.5 for a suitable subsequence we obtain

$$\int_{-1}^{1} \int_{B_{r_i}(0)} |\partial_t u_i|^2 \,dx\,dt \to 0 \text{ as } i \to \infty, \tag{44}$$

where $r_i = t_i/R_i - 1 \to \infty$ as $i \to \infty$ on account of Lemma 2.4. Relabelling, we may assume that (44) holds true for the original sequence (u_i).

Moreover, the energy inequality (30) implies the uniform bound

$$E(u_i(t), r_i) \le E(u(1), 1) =: E_0 \tag{45}$$

for all $i \in \mathbb{N}$ and $|t| \le 1$.

Hence we may extract a further subsequence such that $u_i \rightharpoonup u_\infty$ weakly in H^1_{loc} and locally uniformly away from $x = 0$ on $[-1, 1] \times \mathbb{R}^2$ as $i \to \infty$, and similarly for the associated functions h_i. Their limit h_∞ then is associated with u_∞ and is a time-independent solution of (26) away from $x = 0$. It follows that $u_\infty(t, x) = \overline{u}(x)$ is a time-independent solution of (25) on $]-1, 1[\times(\mathbb{R}^2 \setminus \{0\})$; that is, $\overline{u}\colon \mathbb{R}^2 \setminus \{0\} \to N$ is a smooth, co-rotational harmonic map with finite energy

$$E(\overline{u}) = \int_{\mathbb{R}^2} |\nabla u|^2 \,dx \le \liminf_{i \to \infty} \sup_{|t| \le 1} E(u_i(t), r_i) \le E_0.$$

By [**18**] then $\overline{u}$ extends to a smooth harmonic map $\overline{u}\colon \mathbb{R}^2 \to N$. Since $\mathbb{R}^2$ is conformal to $S^2 \setminus \{p_0\}$ by stereographic projection from any point $p_0 \in S^2$ and since the composition of a harmonic map with a conformal transformation again yields a harmonic map with the same energy, we may thus regard $\overline{u}$ as a harmonic map from $S^2 \setminus \{p_0\}$ to N. Finally, recalling that $E(\overline{u}) < \infty$ and again using [**18**], we see that the map $\overline{u}$ extends to a smooth equivariant harmonic map $\overline{u}\colon S^2 \to N$.

ii) To show that $\overline{u}$ is non-constant we now establish strong convergence

$$u_i \to u_\infty \text{ in } H^1_{loc}(]-1, 1[\times\mathbb{R}^2)$$

as $i \to \infty$. Recalling (37), we have

$$E(u_i(t), 1) \le 2\varepsilon_1, \quad E(u_\infty(t), 1) \le 2\varepsilon_1$$

uniformly in i and $|t| \le 1$. Hence, from (32) for sufficiently small $\varepsilon_1 > 0$ the images of $B_1(0)$ under $u_i(t)$ or u_∞ are all contained in a fixed coordinate system around the center of symmetry $O \in N$. In addition, we can achieve that

$$\sup_{|t|,|x|\le 1} |B(u_i)||u_i - u_\infty| \le \frac{1}{4} \tag{46}$$

uniformly in $i \in \mathbb{N}$, provided $\varepsilon_1 > 0$ is chosen sufficiently small.

For any $\varphi \in C_0^\infty(]-1,1[\times\mathbb{R}^2)$ with $0 \le \varphi \le 1$ then, upon multiplying the equation (25) for u_i by $(u_i - u_\infty)\varphi$ and integrating by parts we obtain

$$\int_{\mathbb{R}^{1+2}} |D(u_i - u_\infty)|^2 \varphi \, dz \le \int_{\mathbb{R}^{1+2}} |B(u_i)||Du_i|^2 |u_i - u_\infty| \varphi \, dz + I, \tag{47}$$

with error

$$\begin{aligned} |I| \le C \int_{\mathbb{R}^{1+2}} (|\partial_t u_i|^2 \varphi + |Du_i||u_i - u_\infty||D\varphi|) \, dz \\ + \sum_\alpha |\int_{\mathbb{R}^{1+2}} \partial_\alpha u_\infty \partial_\alpha (u_i - u_\infty) \varphi \, dz| \to 0 \text{ as } i \to \infty \end{aligned}$$

in view of (44) and since $u_i \to u_\infty$ strongly in L^2_{loc} by Rellich's theorem.

Now we estimate

$$|Du_i|^2 \le 2|D(u_i - u_\infty)|^2 + 2|Du_\infty|^2$$

and observe that

$$\int_{\mathbb{R}^{1+2}} |Du_\infty|^2 |u_i - u_\infty| \varphi \, dz \to 0$$

as $i \to \infty$ by bounded almost everywhere convergence $u_i \to u_\infty$ and Lebesgue's theorem on dominated convergence. Also recalling (46), we thus may absorb the first term on the right of (47) on the left to obtain that

$$\int_{\mathbb{R}^{1+2}} |D(u_i - u_\infty)|^2 \varphi \, dz \to 0$$

as $i \to \infty$. Since φ as above is arbitrary, this yields the desired convergence $u_i \to u_\infty$ in $H^1_{loc}(]-1,1[\times\mathbb{R}^2)$.

But, recalling (38), we also have the uniform lower bound

$$\varepsilon_1 \le E(u_i(t), 11)$$

for all $i \in \mathbb{N}$ and $|t| \le 1$ and we conclude that $u_\infty \not\equiv const$, as claimed. Therefore, also $\overline{u}\colon S^2 \to N$ is non-constant, and the proof of Theorem 2.1 is complete. □

Proof of Theorem 2.2. In view of Theorem 2.1 it suffices to show that any co-rotational harmonic map $\overline{u}\colon S^2 \to N$ with finite energy is constant. Let $\overline{u}$ be such a map, viewed as a map $\overline{u}\colon \mathbb{R}^2 \to N$. Also consider the associated distance function $\rho = \overline{h}(r)$, a time-independent solution of (26). The image $\overline{u}(S^2)$ being compact there exists $r_0 > 0$ such that

$$|\overline{h}(r_0)| = \max_{r>0} |\overline{h}(r)|.$$

Hence $\overline{h}_r(r_0) = 0$ and therefore $\overline{u}_r(x) = 0$ for any $x \in \partial B_{r_0}(0)$.

Since any harmonic map $\overline{u}\colon \mathbb{R}^2 \to N$ with finite energy is conformal, the vanishing of $\overline{u}_r$ implies that also $\overline{u}_\phi$ vanishes along $\partial B_{r_0}(0)$, and we conclude that $\overline{u} \equiv const$ on $\partial B_{r_0}(0)$. Equivariance of $\overline{u}$ then implies that $g(\overline{h}(r_0)) = 0$ and hence $\overline{h}(r_0) = 0$ on account of (23). But then $\overline{h} \equiv 0$ by choice of r_0, and $\overline{u} \equiv const \equiv O$, as desired. □

Wave maps with symmetries II

In this final lecture we show that the Cauchy problem for radially symmetric wave maps $u(t,x) = u(t,|x|)$ from the $(1+2)$-dimensional Minkowski space to an arbitrary smooth, compact Riemannian manifold without boundary is globally well-posed for arbitrary smooth, radially symmetric data.

3.1. The result. Again let N be a smooth, compact Riemmanian k-manifold without boundary, isometrically embedded in $\mathbb{R}^n$. Given smooth, radially symmetric data $(u_0, u_1) = (u_0(|x|), u_1(|x|))\colon \mathbb{R}^2 \to TN$, by a result of Christodoulou-Tahvildar-Zadeh **[2]** there is a unique smooth solution $u = (u^1, \dots, u^n) = u(t,|x|)$ for small time to the Cauchy problem for the equation

$$\Box u = u_{tt} - \Delta u = B(u)(\partial_\alpha u, \partial^\alpha u) \perp T_u N, \tag{48}$$

with initial data

$$(u, u_t)_{|t=0} = (u_0, u_1). \tag{49}$$

Here B again denotes the second fundamental form of N.

As shown by Christodoulou-Tahvildar-Zadeh **[2]**, the solution may be extended globally, if the energy of u is small or if the range of u is contained in a convex part of the target N. Either condition, however, turns out to be unnecessary. In fact, by using the blow-up analysis from **[23]** that we presented in the second lecture, in **[24]**, **[25]** we showed that the local solution may be extended globally for any target manifold.

THEOREM 3.1. *Let $N \subset \mathbb{R}^n$ be a smooth, compact Riemannian manifold without boundary. Then for any radially symmetric data $(u_0, u_1) = (u_0(|x|), u_1(|x|)) \in C^\infty(\mathbb{R}^2; TN)$ there exists a unique, smooth solution $u = u(t,|x|)$ to the Cauchy problem* (48), (49), *defined for all time.*

The regularity requirements on the data may be relaxed; we consider smooth data mainly for ease of exposition.

Summarizing the ideas of the proof, as in the co-rotational symmetric setting of **[23]** that we described in the second lecture, again we argue indirectly. Thus, we suppose that the local solution u to (48), (49) becomes singular in finite time. As before we then obtain a sequence of rescaled solutions u_l on the region $]-1,1[\times\mathbb{R}^2$ with energy bounds and such that $\partial_t u_l \to 0$ in $L^2_{loc}(]-1,1[\times\mathbb{R}^2)$. Finally, rephrasing the wave map equation intrinsically as described in the first lecture, and imposing the exponential gauge, we establish energy decay. But this contradicts the blow-up criterion of Christodoulou and Tahvildar-Zadeh **[2]** and completes the proof.

I would like to thank Jalal Shatah for suggesting the use of the exponential gauge.

3.2. Basic estimates. Let $u = u(t, |x|)\colon [0, t_0[\times\mathbb{R}^2 \to N \subset \mathbb{R}^n$ be a smooth radially symmetric wave map blowing up at time t_0. Necessarily, blow-up occurs at $x = 0$. As before, upon shifting and reversing time and then scaling our space-time coordinates suitably, we may assume that u is a smooth radial solution to (48) on $]0, 1] \times \mathbb{R}^2$ blowing up at the origin. Again let

$$K^T = \{z = (t, x); 0 \le |x| \le t \le T\}$$

be the truncated forward light cone from the origin with lateral boundary

$$M^T = \{(t, x) \in K^T; |x| = t\}.$$

Denoting as

$$e = \frac{1}{2}|Du|^2 = \frac{1}{2}(|u_t|^2 + |u_r|^2), \quad f = \frac{1}{2}|D^{||}u|^2 = \frac{1}{2}|u_t + u_r|^2$$

the energy and flux density of u, and letting

$$E(u, R) = \int_{B_R(0)} e\,dx, \quad \mathrm{Flux}(u, T) = \int_{M^T} f\,do$$

be the local energy and the flux through M^T, then from [**2**], [**21**] we have the following results just as in the co-rotational setting. The identity (29) again leads to the *energy inequality:* For any $t, \tau, R > 0$ there holds

(50) $$E(u(t), R) \le E(u(t + \tau), R + |\tau|).$$

Again, we only consider values such that $0 < t, t + \tau \le 1$. Together with [**2**] this yields the *blow-up criterion:* There exists $\varepsilon_0 = \varepsilon_0(N) > 0$ such that

(51) $$E(u(t), t) \ge \varepsilon_0 \text{ for all } 0 < t \le 1\,.$$

Moreover, we have *flux decay:*

(52) $$Flux(u, T) \to 0 \text{ as } T \to 0.$$

As shown in the Appendix, similar to (33) and (34) we also have *exterior energy decay* and *decay of time derivatives:* For any $0 < \lambda \le 1$ as $t \to 0$ there holds

(53) $$E(u(t), t) - E(u(t), \lambda t) \to 0,$$

and

(54) $$\frac{1}{T}\int_{K^T} |u_t|^2\,dz \to 0 \text{ as } T \to 0.$$

Moreover, as shown in Lemma 2.3, the function u is locally uniformly Hölder continuous on $]0, 1] \times B_1(0)$ away from $x = 0$.

Fix a number $0 < \varepsilon_1 = \varepsilon_1(N) < \varepsilon_0/2$ as determined below. For $0 < t \le 1$ we again choose $R = R(t)$ so that

(55) $$\varepsilon_1 \le E(u(t), 6R) \le 2\varepsilon_1.$$

Then from (50) for any $|\tau| \le 5R$ we have

(56) $$E(u(t + \tau), R) \le E(u(t), 6R) \le 2\varepsilon_1 < \varepsilon_0$$

and similarly

(57) $$\varepsilon_1 \le E(u(t + \tau), 6R + |\tau|) \le E(u(t + \tau), 11R).$$

In particular, combining (51) and (55) we deduce the inequality

(58) $$6R(t) \le t$$

for all t. In fact, from (51), (53), and (55) as in Lemma 2.4 we even obtain that

$$R(t)/t \to 0 \text{ as } t \downarrow 0. \tag{59}$$

As in Lemma 2.5 we consider the intervals $\Lambda_{R(t)}(t) =]t - R(t), t + R(t)[, 0 < t \leq 1$. An application of Vitali's covering theorem and (54) then yields a sequence $t_l \to 0$ with corresponding radii $R_l = R(t_l)$ such that

$$\frac{1}{R_l}\int_{\Lambda_l}\left(\int_{B_t(0)} |u_t|^2\,dx\right)dt \to 0$$

as $l \to \infty$, where $\Lambda_l = \Lambda_{R_l}(t_l), l \in \mathbb{N}$. Rescale, letting

$$u_l(t,x) = u(t_l + R_l t, R_l x), l \in \mathbb{N}.$$

Observe that u_l solves (48) on $[-1,1] \times \mathbb{R}^2$ with

$$\int_{-1}^{1}\left(\int_{D_l(t)} |\partial_t u_l|^2\,dx\right)dt \to 0 \text{ as } l \to \infty, \tag{60}$$

where

$$D_l(t) = \{x; R_l|x| \leq t_l + R_l t\}$$

exhausts $\mathbb{R}^2$ as $l \to \infty$ uniformly in $|t| \leq 1$ on account of (59).

Moreover, from (50), (51), (56), and (57) we have the uniform energy estimates

$$\frac{1}{2}E(u_l(t),1) \leq \varepsilon_1 \leq E(u_l(t),11) \tag{61}$$

and

$$\varepsilon_0 \leq \frac{1}{2}\int_{D_l(t)} |Du_l|^2\,dx = E(u(t_l + R_l t), t_l + R_l t) \leq E(u(1),1) =: E_0, \tag{62}$$

uniformly for $|t| \leq 1$ and sufficiently large $l \in \mathbb{N}$. Hence, we may assume that $u_l \rightharpoonup u_\infty$ weakly in $H^1_{loc}(]-1,1[\times\mathbb{R}^2)$ and locally uniformly away from $x = 0$, where $u_\infty(t,x) = u_\infty(|x|)$ is a time-independent radial map $u_\infty \colon \mathbb{R}^2 \to N$ with finite energy $E(u_\infty) \leq E_0$.

LEMMA 3.2. *We have $u_\infty \equiv const$, and $Du_l \to 0$ in $L^2_{loc}(]-1,1[\times(\mathbb{R}^2 \setminus \{0\})$ as $l \to \infty$.*

Proof. We claim that u_∞ is smooth and harmonic. Indeed, fix any function $\varphi \in C_0^\infty(]-1,1[\times\mathbb{R}^2)$ vanishing near $x = 0$. Upon multiplying (48) by $(u_l - u_\infty)\varphi$ and integrating by parts, we then have

$$\int_{\mathbb{R}^{1+2}} |D(u_l - u_\infty)|^2\varphi\,dz = \int_{\mathbb{R}^{1+2}} \langle B(u_l)(\partial_\alpha u_l, \partial^\alpha u_l), u_l - u_\infty\rangle\varphi\,dz + I,$$

where

$$\begin{aligned}|I| \leq 2\int_{\mathbb{R}^{1+2}} |\partial_t u_l|^2\varphi\,dz &+ \int_{\mathbb{R}^{1+2}} |Du_l||u_l - u_\infty||D\varphi|\,dz \\ &+ \left|\int_{\mathbb{R}^{1+2}} Du_\infty \cdot D(u_l - u_\infty)\varphi\,dz\right| \to 0\end{aligned}$$

as $l \to \infty$. Observing that $(u_l - u_\infty)\varphi \to 0$ uniformly, moreover, we have

$$\int_{\mathbb{R}^{1+2}} \langle B(u_l)(\partial_\alpha u_l, \partial^\alpha u_l), u_l - u_\infty\rangle\varphi\,dz \to 0$$

as $l \to \infty$, and $u_l \to u_\infty$ strongly in $H^1_{loc}(]-1,1[\times\mathbb{R}^2 \setminus \{0\})$. Thus, we may pass to the distribution limit in equation (48) for u_l and find that u_∞ is weakly harmonic on $\mathbb{R}^2 \setminus \{0\}$. Since u_∞ has finite energy, by results of [**18**] then u_∞ is smooth and extends to a smooth, radially symmetric harmonic map $u_\infty \colon \mathbb{R}^2 \to N$.

Next recall that a harmonic map $u_\infty \colon \mathbb{R}^2 \to N$ with finite energy is conformal; in particular, there holds $|\partial_r u_\infty| = \frac{1}{r}|\partial_\phi u_\infty| \equiv 0$, and u_∞ must be constant. □

Finally we note the following estimate similar to [**2**], Lemma 4.

LEMMA 3.3. *For any $\psi = \psi(t) \in C_0^\infty(]-1,1[)$ there holds*

$$\int_{-1}^{1}\int_{B_1(0)} |\partial_t u_l|^2 \psi| \log|x|| \, dx\, dt = \int_{-1}^{1}\int_{B_1(0)} e(u_l)\psi \, dx\, dt + o(1),$$

where $o(1) \to 0$ as $l \to \infty$.

Proof. In radial coordinates $r = |x|$, equation (48) for $u = u_l$ may be written in the form

$$u_{tt} - \frac{1}{r}\,\partial_r(r u_r) \perp T_u N. \tag{63}$$

Multiplying by $u_r \psi r^2 \log r$, we obtain

$$\begin{aligned} 0 = \frac{d}{dt}\left(\langle u_t, u_r\rangle \psi r^2 \log r\right) - \frac{d}{dr}\left(\frac{|u_t|^2 + |u_r|^2}{2}\psi r^2 \log r\right) \\ + |u_t|^2 \psi r \log r - \langle u_t, u_r\rangle \psi_t r^2 \log r + e(u) r\psi. \end{aligned}$$

Upon integrating this identity over the domain $0 < r < 1, |t| < 1$ and observing that the boundary terms vanish, we find

$$\int_{-1}^{1}\int_{0}^{1} |u_t|^2 \psi r \log r \, dr\, dt + \int_{-1}^{1}\int_{0}^{1} e(u) r\psi \, dr\, dt = \int_{-1}^{1}\int_{0}^{1} \langle u_t, u_r\rangle \psi_t r^2 \log r \, dr\, dt.$$

In view of (60), (62), and Hölder's inequality the last term may be estimated

$$\begin{aligned} \left|\int_{-1}^{1}\int_{0}^{1} \langle u_t, u_r\rangle \psi_t r^2 \log r \, dr\, dt\right|^2 &= \left|\frac{1}{2\pi}\int_{-1}^{1}\int_{B_1(0)} \langle u_t, u_r\rangle \psi_t r \log r \, dx\, dt\right|^2 \\ &\leq C\int_{-1}^{1}\int_{B_1(0)} |u_t|^2 \, dx\, dt \cdot \int_{-1}^{1}\int_{B_1(0)} |u_r|^2 \, dx\, dt \to 0 \text{ as } l \to \infty, \end{aligned}$$

proving the claim. □

3.3. Intrinsic setting. Recalling the set-up from our first lecture, in terms of the pull-back covariant derivative D in u^*TN we may write equation (63) as

$$D_t u_t - \frac{1}{r} D_r(r u_r) = 0. \tag{64}$$

Again we may assume that TN is parallelizable and we let $\overline{e}_1, \ldots, \overline{e}_k$ be a smooth orthonormal frame field such that at any point $p \in N$ the vectors $\overline{e}_1(p), \ldots, \overline{e}_k(p)$ form an orthonormal basis for T_pN. From $(\overline{e}_i)_{1\leq i\leq k}$ we then obtain a frame $e_i = R_i^j(\overline{e}_j \circ u), 1 \leq i \leq k$, for the pull-back bundle, where $R = R(t,r) = (R_i^j)$ is a smooth map from $\mathbb{R}^{1+2}$ into $SO(k)$.

Denoting

$$De_i = A_i^j e_j$$

with a matrix-valued connection 1-form $A = A_0\,dt + A_1\,dr$, we compute the curvature F of D via the commutation relation (9), or, more concisely,

$$dA + \frac{1}{2}[A, A] = F.$$

Moreover, we now impose the "exponential gauge" condition $A_1 = 0$. This yields the relation

$$*dA = -\partial_r A_0 = F_{01}.$$

If we normalize $A_0(t, 1) = 0$ for all t, from this relation we obtain

$$A_0 = \int_r^1 F_{01}\,ds.$$

Observing that

$$|F_{01}| \le C|du|^2, \tag{65}$$

from (61) we then deduce the estimate

$$|A_0| \le a_0 := \int_r^1 |F_{01}|\,ds \le C\int_r^1 |du|^2\,ds \le C\varepsilon_1 r^{-1}.$$

Note that in the exponential gauge for any fixed time t the frame field $e = e(t, r)$ is obtained by parallel transport along the curve $\gamma(r) = u(t, r)$ from the frame $e(t, 1)$ at $r = 1$.

Expressing du as

$$du = u_t\,dt + u_r\,dr = q^i e_i,$$

where $q = q_0\,dt + q_1\,dr$ is a vector-valued 1-form with coefficients $q = (q^i)_{1\le i\le k}$, and using the notation

$$D_\alpha\partial_\beta u = D_\alpha(q^i_\beta e_i) = (\partial_\alpha q^j_\beta + A^j_{i\alpha} q^i_\beta)e_j =: (D_\alpha q_\beta)^j e_j$$

from our first lecture, we then may write equation (64) in the form

$$D_t q_0 - \frac{1}{r}D_r(rq_1) = \partial_t q_0 + A_0 q_0 - \frac{1}{r}\partial_r(rq_1) = 0. \tag{66}$$

Moreover, we have the commutation relation $D_r q_0 = D_t q_1$; that is,

$$\partial_r q_0 = \partial_t q_1 + A_0 q_1. \tag{67}$$

Finally there holds

$$|q_0| = |u_t|, |q_1| = |u_r|. \tag{68}$$

3.4. Proof of Theorem 3.1. By using Lemma 3.3 we show that (60) for sufficiently small $\varepsilon_1 > 0$ leads to a contradiction with (61).

Fix a cut-off function $0 \le \varphi = \varphi(r) \le 1$ in $C_0^\infty([0, 1[)$ such that $\varphi(r) = 1$ for $r \le 1/2$. Also fix $0 \le \psi = \psi(t) \le 1$ in $C_0^\infty(]-1, 1[)$ such that $\psi(t) = 1$ for $|t| \le 1/2$. For $u = u_l$ with associated 1-forms q, let

$$Q = Q_l = \int_r^1 q_1\varphi\,ds.$$

Note that by Hölder's inequality and (61) for $0 < r < 1$ we can estimate

$$|Q|^2 \le \Big(\int_r^1 |q|\,ds\Big)^2 \le \int_r^1 s|q|^2\,ds \cdot \int_r^1 \frac{ds}{s} \le C\varepsilon_1 \log(\frac{1}{r}). \tag{69}$$

We will also use the bound

$$\Big(\int_0^r s|q|\varphi\,ds\Big)^2 \le \int_0^r s|q|^2\,ds\cdot\int_0^r s\,ds \le C\varepsilon_1 r^2 \tag{70}$$

resulting from (61). Similarly, we have

$$\Big(\int_0^r s|q_0||\log s|^{1/2}\varphi\,ds\Big)^2 \le \frac{r^2}{2}\int_0^1 s|q_0|^2|\log s|\,ds,$$

which in view of (61), (65), and Lemma 3.3 allows to estimate

$$\begin{aligned}\int_{-1}^1\int_0^1\Big(\int_0^r s|q_0||\log s|^{1/2}\varphi\,ds\Big)|F_{01}|\psi\,dr\,dt\\ \le \int_{-1}^1\Big(\int_0^1 s|q_0|^2|\log s|\,ds\Big)^{1/2}\Big(\int_0^1 r|F_{01}|\,dr\Big)\psi\,dt\\ \le C\varepsilon_1\Big(\int_{-1}^1\int_0^1 s|q_0|^2|\log s|\psi\,ds\,dt\Big)^{1/2} \le C\varepsilon_1^{3/2}.\end{aligned}\tag{71}$$

Also note that Lemma 3.2 implies

$$\int_{-1}^1\int_0^1 r|\log r|^{1/2}|q|\psi\,dr\,dt \le C\Big(\int_{-1}^1\int_{B_1(0)} r|\log r||Du|^2\psi\,dx\,dt\Big)^{1/2}\to 0 \tag{72}$$

as $l\to\infty$.

Using the function $Q\varphi\psi r$ as a multiplier, from (67) then we obtain

$$\begin{aligned}\int_{-1}^1\int_0^1 \partial_t q_0 Q\varphi\psi r\,dr\,dt &= -\int_{-1}^1\int_0^1 q_0\Big(\int_r^1 \partial_t q_1\varphi\,ds\Big)\varphi\psi r\,dr\,dt + I\\ &= \int_{-1}^1\int_0^1 |q_0|^2\varphi^2\psi r\,dr\,dt + \int_{-1}^1\int_0^1 q_0\Big(\int_r^1 A_0 q_1\varphi\,ds\Big)\varphi\psi r\,dr\,dt + II,\end{aligned}$$

where, in view of (69), and (72),

$$|I| = |\int_{-1}^1\int_0^1 q_0 Q\varphi\psi_t r\,dr\,dt| \le C\int_{-1}^1\int_0^1 r|q_0||\log r|^{1/2}|\psi_t|\,dr\,dt\to 0$$

as $l\to\infty$. Similarly,

$$\begin{aligned}|II| &\le |I| + |\int_{-1}^1\int_0^1 q_0\Big(\int_r^1 q_0\partial_r\varphi\,ds\Big)\varphi\psi r\,dr\,dt|\\ &\le |I| + C\int_{-1}^1\int_0^1 r|q_0||\log r|^{1/2}\psi\,dr\,dt\to 0.\end{aligned}$$

On the other hand, noting that

$$\frac{1}{r}\partial_r(rq_1)rQ = \partial_r(rq_1Q) + r|q_1|^2\varphi,$$

we obtain

$$\int_{-1}^1\int_0^1 \frac{1}{r}\partial_r(rq_1)rQ\varphi\psi\,dr\,dt = \int_{-1}^1\int_0^1 r|q_1|^2\varphi^2\psi\,dr\,dt + III,$$

where, by (69) and (72),

$$|III| \le \int_{-1}^1\int_0^1 r|q_1||Q||\varphi_r|\psi\,dr\,dt \le C\int_{-1}^1\int_0^1 r|\log r|^{1/2}|q_1|\psi\,dr\,dt\to 0$$

as $l \to \infty$. Thus, from (66) we deduce the identity

$$\begin{aligned}\int_{-1}^{1}\int_{0}^{1} & r(|q_1|^2 - |q_0|^2)\varphi^2\psi\,dr\,dt + o(1)\\ &= \int_{-1}^{1}\int_{0}^{1} q_0\Big(\int_r^1 A_0 q_1\varphi\,ds\Big)\varphi\psi r\,dr\,dt + \int_{-1}^{1}\int_{0}^{1} A_0 q_0\Big(\int_r^1 q_1\varphi\,ds\Big)\varphi\psi r\,dr\,dt,\end{aligned}$$

where $o(1) \to 0$ as $l \to \infty$. Using (70), (65) and repeated integration by parts, we find

$$\begin{aligned}\int_{-1}^{1}\int_{0}^{1} q_0\Big(\int_r^1 A_0 q_1\varphi\,ds\Big)\varphi\psi r\,dr\,dt &= \int_{-1}^{1}\int_{0}^{1}\Big(\int_0^r q_0\varphi s\,ds\Big)A_0 q_1\varphi\psi\,dr\,dt\\ &\le C\varepsilon_1^{1/2}\int_{-1}^{1}\int_{0}^{1} r a_0|q_1|\varphi\psi\,dr\,dt = C\varepsilon_1^{1/2}\int_{-1}^{1}\int_{0}^{1} r|q_1|\varphi\Big(\int_r^1 |F_{01}|\,ds\Big)\psi\,dr\,dt\\ &= C\varepsilon_1^{1/2}\int_{-1}^{1}\int_{0}^{1}\Big(\int_0^r s|q_1|\varphi\,ds\Big)|F_{01}|\psi\,dr\,dt \le C\varepsilon_1\int_{-1}^{1}\int_{0}^{1} r|F_{01}|\psi\,dr\,dt\\ &\le C\varepsilon_1\int_{-1}^{1}\int_{0}^{1} r|du|^2\psi\,dr\,dt \le C\varepsilon_1^2.\end{aligned}$$

Similarly, we estimate, now using (69) and (71),

$$\begin{aligned}\int_{-1}^{1}\int_{0}^{1} A_0 q_0\Big(\int_r^1 q_1\varphi\,ds\Big)\varphi\psi r\,dr\,dt &\le C\varepsilon_1^{1/2}\int_{-1}^{1}\int_{0}^{1} a_0|q_0||\log r|^{1/2}\varphi\psi r\,dr\,dt\\ &= C\varepsilon_1^{1/2}\int_{-1}^{1}\int_{0}^{1} r|q_0||\log r|^{1/2}\varphi\Big(\int_r^1 |F_{01}|\,ds\Big)\psi\,dr\,dt\\ &= C\varepsilon_1^{1/2}\int_{-1}^{1}\int_{0}^{1}\Big(\int_0^r s|q_0||\log s|^{1/2}\varphi\,ds\Big)|F_{01}|\psi\,dr\,dt \le C\varepsilon_1^2.\end{aligned}$$

But then from (61), Lemma 3.2, and (60), with error $o(1) \to 0$ as $l \to \infty$ we obtain

$$\begin{aligned}\varepsilon_1 &\le \frac{1}{2}\int_{-1}^{1}\int_{B_{11}(0)} |Du|^2\psi\,dx\,dt \le \pi\int_{-1}^{1}\int_{0}^{1} r|q|^2\varphi^2\psi\,dr\,dt + o(1)\\ &\le \pi\int_{-1}^{1}\int_{0}^{1} r(|q_1|^2 - |q_0|^2)\varphi^2\psi\,dr\,dt + o(1) \le C\varepsilon_1^2 + o(1),\end{aligned}$$

which is impossible for sufficiently small $\varepsilon_1 > 0$ and large l. The proof of Theorem 3.1 is complete.

Appendix A: Exterior energy decay

In this Appendix we recall the proof of the following lemma which is fundamental for the treatment of the equivariant and rotationally symmetric case.

LEMMA 4.1. *Let u be a radially symmetric solution of* (48) *or a co-rotational wave map on $K = K^1$ which is smooth away from the origin. Then for any $0 < \lambda \le 1$ as $t \to 0$ there holds*

$$E(u(t), t) - E(u(t), \lambda t) \to 0.$$

Proof. We follow the presentation in [**19**]. Therefore in the following we change time t to $-t$.

With the notation

$$e = \frac{1}{2}(|u_r|^2 + |u_t|^2),\ m = u_r \cdot u_t,\ l = \frac{1}{2}(|u_r|^2 - |u_t|^2) \tag{73}$$

for a radially symmetric solution u of (48) we compute

$$\frac{\partial}{\partial t}(rm) - \frac{\partial}{\partial r}(re) = ru_r \cdot (u_{tt} - \frac{1}{r}(ru_r)_r) + l = l, \tag{74}$$

thereby observing the geometric interpretation (63) of (48) and the fact that $u_r \in T_u N$. Moreover, recalling the equation (29) we have

$$\frac{\partial}{\partial t}(re) - \frac{\partial}{\partial r}(rm) = 0. \tag{75}$$

Similarly, for a co-rotational wave map u with associated function h solving (26) we let

$$\begin{aligned} e &= \frac{1}{2}(|u_r|^2 + |u_t|^2) = \frac{1}{2}(|h_r|^2 + |h_t|^2 + \frac{g^2(h)}{r^2}),\ m = h_r \cdot h_t, \\ L &= \frac{1}{2}(|h_r|^2 + \frac{g^2(h)}{r^2} - |h_t|^2) - \frac{2}{r}f(h)h_r \end{aligned} \tag{76}$$

and we compute

$$\partial_t(re) - \partial_r(rm) = 0,\ \partial_t(rm) - \partial_r(re) = L. \tag{77}$$

Changing coordinates to

$$\eta = t + r\,, \qquad \xi = t - r\,, \tag{78}$$

and introducing

$$\mathcal{A}^2 = r(e+m),\ \mathcal{B}^2 = r(e-m)\,,$$

identities (74), (75) turn into

$$\begin{aligned} \partial_\xi \mathcal{A}^2 &= \frac{l}{2}\,, \\ \partial_\eta \mathcal{B}^2 &= -\frac{l}{2}\,, \end{aligned}$$

where

$$r^2 l^2 \le \mathcal{A}^2 \mathcal{B}^2\,.$$

Likewise, (77) can be written as

$$\begin{aligned} \partial_\xi \mathcal{A}^2 &= \frac{L}{2}\,, \\ \partial_\eta \mathcal{B}^2 &= -\frac{L}{2}\,, \end{aligned}$$

where now, with $F = g^2/2$, and using the fact that $|h| \le C(E_0)$ by (32) to bound $f^2(h) \le CF(h)$,

$$\begin{aligned} L^2 &\le \frac{3}{4}\left(h_t^2 - h_r^2\right)^2 + \frac{12}{r^2}h_r^2 f^2(h) + \frac{3}{r^4}F^2(h) \\ &\le C\left[\frac{1}{4}(h_t^2 - h_r^2)^2 + \frac{1}{r^2}(h_t^2 + h_r^2)F(h) + \frac{1}{r^4}F^2(h)\right] \\ &= \frac{C}{r^2}\mathcal{A}^2\mathcal{B}^2\,. \end{aligned}$$

Thus in both cases we get the inequalities

$$|\partial_\xi \mathcal{A}| \le \frac{C}{r}\mathcal{B}, \quad |\partial_\eta \mathcal{B}| \le \frac{C}{r}\mathcal{A}. \tag{79}$$

Upon integrating (79) on a rectangle $\Gamma = [\eta, 0] \times [\xi_0, \xi]$, as shown in Figure 1, we obtain

$$\mathcal{A}(\eta,\xi) \le \mathcal{A}(\eta,\xi_0) + C\int_{\xi_0}^{\xi} \frac{\mathcal{B}(0,\xi')}{\eta-\xi'}\,d\xi' + C^2\int_{\xi_0}^{\xi}\int_{\eta}^{0} \frac{\mathcal{A}(\eta',\xi')}{(\eta-\xi')(\eta'-\xi')}\,d\eta'\,d\xi'.$$

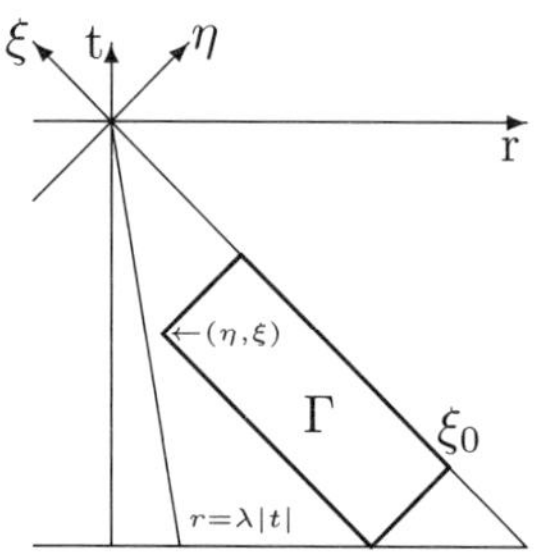

FIGURE 1. Domain of integration Γ.

First we estimate the second term on the right.

$$\begin{aligned}\int_{\xi_0}^{\xi} \frac{\mathcal{B}(0,\xi')}{\eta-\xi'}\,d\xi' &\le \left(\int_{\xi_0}^{\xi} \mathcal{B}^2(0,\xi')\,d\xi'\right)^{1/2}\left(\int_{\xi_0}^{\xi}\frac{d\xi'}{(\eta-\xi')^2}\right)^{1/2}\\ &= (\mathrm{Flux}(\xi_0)-\mathrm{Flux}(\xi))^{1/2}\sqrt{\frac{1}{\eta-\xi}-\frac{1}{\eta-\xi_0}}\\ &\le C\sqrt{\frac{\mathrm{Flux}(\xi_0)}{|\eta-\xi|}}.\end{aligned}$$

Letting

$$a(\eta,\xi) = \sup_{\eta\le\eta'\le 0}\sqrt{\eta'-\xi}\,\mathcal{A}(\eta',\xi), \tag{80}$$

the third term may be bounded

(81)

$$\begin{aligned}\int_{\xi_0}^{\xi}\int_{\eta}^{0}\frac{\mathcal{A}(\eta',\xi')}{(\eta-\xi')(\eta'-\xi')}\,d\eta'\,d\xi' &\le \int_{\xi_0}^{\xi}\int_{\eta}^{0}\frac{a(\eta,\xi')}{(\eta-\xi')(\eta'-\xi')^{3/2}}\,d\eta'\,d\xi'\\ &\le \int_{\xi_0}^{\xi}\frac{a(\eta,\xi')}{\eta-\xi'}\left(\frac{1}{\sqrt{\eta-\xi'}}-\frac{1}{\sqrt{-\xi'}}\right)d\xi' \le \int_{\xi_0}^{\xi} a(\eta,\xi')\frac{\eta}{\xi'(\eta-\xi')^{3/2}}\,d\xi'.\end{aligned}$$

Also observing that

$$\sup_{\eta\le\eta'\le 0}\sqrt{\eta'-\xi}\,\mathcal{A}(\eta',\xi_0) \le \sup_{\eta\le\eta'\le 0}\frac{\sqrt{\eta'-\xi}}{\sqrt{\eta'-\xi_0}}a(\eta,\xi_0) = \frac{\sqrt{-\xi}}{\sqrt{-\xi_0}}\,a(\eta,\xi_0) \tag{82}$$

with constants C_1, C_2 we then obtain

$$a(\eta,\xi) \le \frac{\sqrt{-\xi}}{\sqrt{-\xi_0}}\, a(\eta,\xi_0) + C_1\sqrt{\mathrm{Flux}(\xi_0)} + C_2\int_{\xi_0}^{\xi} a(\eta,\xi')\frac{\eta}{\xi'(\eta-\xi')}\, d\xi' .$$

Setting

$$\rho(\xi') = \frac{\eta}{\xi'(\eta-\xi')}, \tag{83}$$

and letting

$$F(\xi) = \int_{\xi_0}^{\xi} a(\eta,\xi')\rho(\xi')\, d\xi',\ G(\xi) = \frac{\sqrt{-\xi}}{\sqrt{-\xi_0}}\, a(\eta,\xi_0) + C_1\sqrt{\mathrm{Flux}(\xi_0)}\, , \tag{84}$$

for any fixed η we then find the differential inequality

$$F' \le G\rho + C_2 F\rho \text{ in } [\xi_0, \lambda'\eta]\, , \tag{85}$$

where $\lambda' = (1+\lambda)/(1-\lambda) > 1$. Applying Gronwall's lemma we obtain

$$F(\xi) \le \int_{\xi_0}^{\xi} G(\xi')\rho(\xi')e^{C_2\int_{\xi'}^{\xi}\rho(\xi'')d\xi''}\, d\xi' . \tag{86}$$

But for $\xi_0 \le \xi' \le \xi = \lambda'\eta$ we have

$$\int_{\xi'}^{\xi}\rho(\xi'')d\xi'' = \int_{\xi'}^{\xi}\frac{\eta}{\xi''(\eta-\xi'')}\, d\xi'' = \log\frac{\xi(\eta-\xi')}{\xi'(\eta-\xi)} = \log\frac{\xi(\xi-\lambda'\xi')}{\xi'(\xi-\lambda'\xi)} \le \log\frac{\lambda'}{\lambda'-1} .$$

Hence we can estimate

$$\begin{aligned} a(\eta,\xi) &\le G + C_2 F \\ &\le \frac{\sqrt{-\xi}}{\sqrt{-\xi_0}}\, a(\eta,\xi_0) + C_1\sqrt{\mathrm{Flux}(\xi_0)} \\ &\quad + C_3\int_{\xi_0}^{\xi}\left(\frac{\sqrt{-\xi'}}{\sqrt{-\xi_0}}\, a(\eta,\xi_0) + C_1\sqrt{\mathrm{Flux}(\xi_0)}\right)\frac{\eta}{\xi'(\eta-\xi')}\, d\xi' , \end{aligned} \tag{87}$$

where $C_3 = e^{C_2\log\frac{\lambda'}{\lambda'-1}}$. We also know that

$$a(\eta,\xi_0) \le \sup_{\eta\le\eta'\le 0}\sqrt{\eta'-\xi_0}\sup_{\eta\le\eta'\le 0}\mathcal{A}(\eta',\xi_0) \le C(\xi_0)\sqrt{-\xi_0}\, ,$$

because u is assumed to be regular away from the origin, implying that $\mathcal{A}$ is bounded by a constant depending on ξ_0. Now, given $\epsilon > 0$, we can fix $\xi_0 < 0$ small enough such that $C_1\sqrt{\mathrm{Flux}(\xi_0)} < \epsilon$. Then,

$$\begin{aligned} a(\xi/\lambda',\xi) &\le C(\xi_0)\sqrt{-\xi} + \epsilon + C(\xi_0)\int_{\xi_0}^{\xi}\frac{\xi/\lambda'}{\sqrt{-\xi'}(\xi/\lambda'-\xi')}\, d\xi' + C\epsilon \\ &\le C(\xi_0)\sqrt{-\xi} + C\epsilon \le C\epsilon \end{aligned}$$

for $\xi < 0$ small enough. Therefore,

$$\mathcal{A}(\eta,\xi) \le \frac{a(\xi/\lambda',\xi)}{\sqrt{\eta-\xi}} \le \frac{C\epsilon}{\sqrt{\eta-\xi}}$$

for (η,ξ) small enough inside $K^{\lambda}_{\mathrm{ext}}$. Hence,

$$\int_{\eta}^{0}\mathcal{A}^2(\eta',\xi)d\eta' \le C\epsilon^2\int_{\xi/\lambda'}^{0}\frac{d\eta'}{\eta'-\xi} = C\epsilon^2\log\frac{1}{(\lambda'-1)} = C\epsilon^2 .$$

Finally, if we integrate the energy identity (75)) on the triangle Δ (as shown in

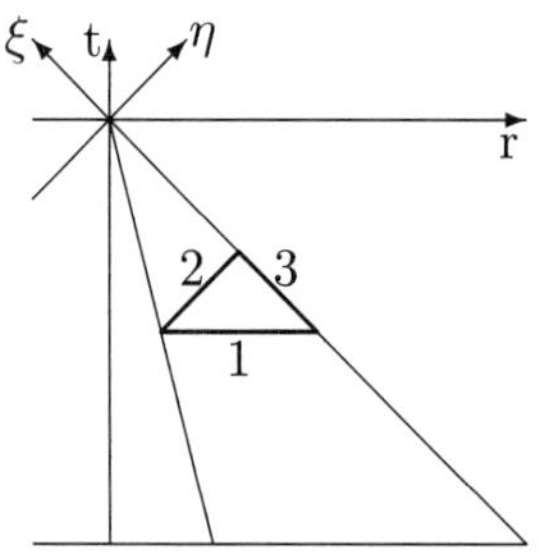

FIGURE 2. Triangular region Δ.

Figure 2 with vertices at (η,ξ), $(0,\xi)$, and $(0,\eta+\xi)$, with $\eta=\xi/\lambda'$ as before), we obtain

$$0=-\int_{\lambda|t|}^{|t|} e(r,t)r\,dr-\int_{\eta}^{0} r(e+m)d\eta'+\int_{\xi+\eta}^{\xi} r(e-m)d\xi'=\mathrm{I}+\mathrm{II}+\mathrm{III}\,.$$

As $t\to 0$ we proved that $\mathrm{II}\to 0$; moreover, $\mathrm{III}\to 0$ because it is the flux, and therefore $\mathrm{I}\to 0$. □

As consequence we obtain the decay of time derivatives.

COROLLARY 4.2. *Let u be a radially symmetric solution of* (48) *or a co-rotational wave map on $K=K^1$ which is smooth away from the origin. In the latter case also suppose that N satisfies* (24). *Then*

$$\frac{1}{T}\int_{K^T}|u_t|^2\,dz\to 0 \text{ as } T\to 0.$$

Proof. Again we change time t to $-t$. Multiply the identity (74), (77), respectively, by r and integrate on the truncated cone

$$K_T^{-\epsilon}\cong\{(t,r);t\le-\epsilon,0\le r\le -t\le -T\},$$

and let $\epsilon\to 0$ to obtain

$$\left|\iint_{K_T^0} u_t^2 r\,dr\,dt-\int_0^{|T|}(u_tu_r)\big|_{t=T}r^2\,dr\right|\le C|T|\,\mathrm{Flux}(T)\,.$$

Therefore, for any $\lambda\in]0,1[$ we have

$$\begin{aligned}\frac{1}{|T|}\int_T^0\int_0^{-t}u_t^2r\,dr\,dt&\le\frac{1}{|T|}\int_0^{|T|}|(u_tu_r)\big|_{t=T}|r^2\,dr+C\,\mathrm{Flux}(T)\\&\le\frac{C}{|T|}\int_0^{|T|}e(T,r)r^2\,dr+C\,\mathrm{Flux}(T)\\&\le\frac{C}{|T|}\left(\int_0^{\lambda|T|}e(T,r)r^2\,dr+\int_{\lambda|T|}^{|T|}e(T,r)r^2dr\right)+C\,\mathrm{Flux}(T)\\&\le C(\lambda E_0+E_{\mathrm{ext}}^{\lambda}(T)+\mathrm{Flux}(T))\,.\end{aligned}$$

Given $\epsilon>0$ we then may choose $\lambda>0$ such that the first term on the right is less then $\epsilon/3$. By Lemma 4.1 and by decay of the flux the second and third terms also will be less than $\epsilon/3$ for T sufficiently close to 0. □

References

[1] P. Bizo'n, T. Chmaj, Z. Tabor: *Formation of singularities for equivariant* $(2+1)$-*dimensional wave maps into the* 2-*sphere*, Nonlinearity 14 (2001), no. 5, 1041–1053.

[2] D. Christodoulou, A.S. Tahvildar-Zadeh: *On the regularity of spherically symmetric wave maps.* Comm. Pure Appl. Math. 46 (1993), no. 7, 1041–1091.

[3] R. Côte: *Instability of nonconstant harmonic maps for the (1 + 2)-dimensional equivariant wave map system.* Int. Math. Res. Notices, 2005 (2005), 3525-3549.

[4] M. Grillakis: *Classical solutions for the equivariant wave map in 1+2 dimensions* , preprint, 1991

[5] M. Giaquinta: *Introduction to regularity theory for nonlinear elliptic systems*, Lect. in Math. ETH Zürich, Birkhäuser, 1993

[6] F. Hélein: *Régularité des applications faiblement harmoniques entre une surface et une varieté riemannienne.* C. R. Acad. Sci. Paris Sér. I Math. 312 (1991), no. 8, 591–596.

[7] F. Hélein: *Harmonic maps, conservation laws and moving frames.* Translated from the 1996 French original. With a foreword by James Eells. Second edition. Cambridge Tracts in Mathematics, 150. Cambridge University Press, Cambridge, 2002.

[8] J. Isenberg, S. Liebling: *Singularity formation in* $2+1$ *wave maps*, J. Math. Phys. 43 (2002), no. 1, 678–683.

[9] M. Keel, T. Tao: *Endpoint Strichartz estimates*, Amer. Math. J. 120 (1998), 955–980.

[10] S. Klainerman, I. Rodnianski: *On the global regularity of wave maps in the critical Sobolev norm.* Internat. Math. Res. Notices 2001, no. 13, 655–677.

[11] S. Klainerman, D. Tataru: *On the optimal local regularity for Yang-Mills equations in* R^{4+1}. J. Amer. Math. Soc. 12 (1999), no. 1, 93–116.

[12] J. Krieger: *Global regularity of wave maps from* $\mathbf{R}^{2+1}$ *to* H^2. *Small energy.* Comm. Math. Phys. 250 (2004), no. 3, 507–580.

[13] J. Krieger: *Global regularity of wave maps from* R^{3+1} *to surfaces.* Comm. Math. Phys. 238 (2003), no. 1-2, 333–366.

[14] J. Krieger, W. Schlag, D. Tataru: *Renormalization and blow up for charge one equivariant critical wave maps.* Invent. Math. 171 (2008), no. 3, 543–615.

[15] S. Müller, M. Struwe: *Global existence of wave maps in* $1+2$ *dimensions with finite energy data*, Topological methods in nonlinear analysis 7 (1996), 245-259.

[16] A. Nahmod, A. Stefanov, K. Uhlenbeck: *On the well-posedness of the wave map problem in high dimensions.* Comm. Anal. Geom. 11 (2003), no. 1, 49–83.

[17] I. Rodnianski, J. Sterbenz: *On the formation of singularities in the critical* $O(3)$ σ-*model.* Preprint (2006)

[18] Sacks, K. Uhlenbeck: *The existence of minimal immersions of* 2-*spheres.* Ann. of Math. (2) 113 (1981), no. 1, 1–24.

[19] J. Shatah, M. Struwe: *Geometric wave equations*, Courant Lecture Notes 2, New York University (1998), 2nd edition: AMS (2000).

[20] J. Shatah, M. Struwe: *The Cauchy problem for wave maps.* Int. Math. Res. Not. 2002, no. 11, 555–571.

[21] J. Shatah, A.S. Tahvildar-Zadeh: *Regularity of harmonic maps from the Minkowski space into rotationally symmetric manifolds*, Comm. Pure Appl. Math. 45 (1992), 947-971.

[22] J. Shatah, A.S. Tahvildar-Zadeh: *On the Cauchy problem for equivariant wave maps.* Comm. Pure Appl. Math. 47 (1994), no. 5, 719–754.

[23] M. Struwe: *Equivariant wave maps in two space dimensions*, Comm. Pure Appl. Math. 56 (2003), no. 7, 815–823.

[24] M. Struwe: *Radially symmetric wave maps from* $(1+2)$-*dimensional Minkowski space to the sphere.* Math. Z. 242 (2002), no. 3, 407–414.

[25] M. Struwe: *Radially symmetric wave maps from* $(1+2)$-*dimensional Minkowski space to general targets.* Calc. Var. Partial Differential Equations 16 (2003), no. 4, 431–437.

[26] T. Tao: *Global regularity of wave maps. I. Small critical Sobolev norm in high dimension.* Internat. Math. Res. Notices 2001, no. 6, 299–328.

[27] T. Tao: *Global regularity of wave maps. II. Small energy in two dimensions.* Comm. Math. Phys. 224 (2001), no. 2, 443–544.

[28] D. Tataru: *Local and global results for wave maps. I.* Comm. Partial Differential Equations 23 (1998), no. 9-10, 1781–1793.

[29] D. Tataru: *On global existence and scattering for the wave maps equation.* Amer. J. Math. 123 (2001), no. 1, 37–77.
[30] D. Tataru: *Rough solutions for the wave maps equation.* Amer. J. Math. 127 (2005), no. 2, 293–377.

Mathematik, ETH-Zentrum, CH-8092 Zürich
E-mail address: michael.struwe@math.ethz.ch

Clay Mathematics Proceedings
Volume **17**, 2013

Derivation of Effective Evolution Equations from Microscopic Quantum Dynamics

Benjamin Schlein

Contents

1. Introduction

A quantum mechanical system of N particles in d dimensions can be described by a complex valued wave function $\psi_N \in L^2(\mathbb{R}^{dN}, \mathrm{d}x_1 \dots \mathrm{d}x_N)$. The variables $x_1, \dots, x_N \in \mathbb{R}^d$ represent the position of the N particles. Physically, the absolute value squared of $\psi_N(x_1, x_2, \dots, x_N)$ is interpreted as the probability density for finding particle one at x_1, particle two at x_2, and so on. Because of this probabilistic interpretation, we will always consider wave functions ψ_N with L^2-norm equal to one.

In Nature there exist two different types of particles; bosons and fermions. Bosonic systems are described by wave functions which are symmetric with respect to permutations, in the sense that

$$\psi_N(x_{\pi 1}, x_{\pi 2}, \dots, x_{\pi N}) = \psi_N(x_1, \dots, x_N) \tag{1.1}$$

for every permutation $\pi \in S_N$. Fermionic systems, on the other hand, are described by antisymmetric wave functions satisfying

$$\psi_N(x_{\pi 1}, x_{\pi 2}, \dots, x_{\pi N}) = \sigma_\pi \psi_N(x_1, \dots, x_N) \qquad \text{for all } \pi \in S_N,$$

where σ_π is the sign of the permutation π; $\sigma_\pi = +1$ if π is even (in the sense that it can be written as the composition of an even number of transpositions) and $\sigma_\pi = -1$ if it is odd. In these notes we are only going to consider bosonic systems; the wave function ψ_N will always be taken from the Hilbert space $L^2_s(\mathbb{R}^{dN})$, the subspace of $L^2(\mathbb{R}^{dN})$ consisting of all functions satisfying (1.1).

2010 *Mathematics Subject Classification.* Primary 35Q55, 81Q15, 81T18, 81V70.

Observables of the N-particle system are self adjoint operators over $L^2_s(\mathbb{R}^{dN})$. The expected value of an observable A in a state described by the wave function ψ_N is given by the inner product

$$\langle \psi_N, A\psi_N \rangle = \int \mathrm{d}x_1 \dots \mathrm{d}x_N \, \overline{\psi}_N(x_1,\dots,x_N)\,(A\psi_N)(x_1,\dots,x_N).$$

The multiplication operator x_j is the observable measuring the position of the j-th particle. The differential operator $p_j = -i\nabla_j$ is the observable measuring the momentum of the j-th particle (p_j is called the momentum operator of the j-th particle).

The time evolution of an N-particle wave function $\psi_N \in L^2_s(\mathbb{R}^{dN})$ is governed by the Schrödinger equation

$$i\partial_t \psi_{N,t} = H_N \psi_{N,t}\,. \tag{1.2}$$

Here, and in the rest of these notes, the subscript t indicates the time dependence of the wave function; all time-derivatives will be explicitly written as ∂_t. On the right hand side of (1.2), H_N is a self-adjoint operator acting on the Hilbert space $L^2_s(\mathbb{R}^{dN})$, usually known as the Hamilton operator (or Hamiltonian) of the system. We will consider only time-independent Hamilton operators with two body interactions, which have the form

$$H_N = \sum_{j=1}^N \left(-\Delta_{x_j} + V_{\text{ext}}(x_j)\right) + \lambda \sum_{i<j}^N V(x_i - x_j)\,.$$

The first part of the Hamiltonian is a sum of one-body operators (operators acting on one particle only); the sum of the Laplacians is the kinetic part of the Hamiltonian. The function V_{ext} describes an external potential which acts in the same way on all N particles. For example, V_{ext} may describe a confining potential which traps the particles in a certain region. The second part of the Hamiltonian, given by a sum over all pairs of particles, describes the interactions among the particles ($\lambda \in \mathbb{R}$ is a coupling constant). The Hamilton operator is the observable associated with the energy of the system. In other words, the expectation

$$\langle \psi_N, H_N\psi_N \rangle = \int \mathrm{d}x_1 \dots \mathrm{d}x_N \, \overline{\psi}_N(x_1,\dots,x_N)(H_N\psi_N)(x_1,\dots,x_N)$$

gives the energy of the system in the state described by the wave function ψ_N.

Note that the Schrödinger equation (1.2) is linear and, since H_N is a self-adjoint operator, it preserves the L^2-norm of the wave function (moreover, since H_N is invariant with respect to permutations, it also preserves the permutation symmetry of the wave function). In fact, the solution to (1.2), with initial condition $\psi_{N,t=0} = \psi_N$, can be written by means of the unitary group generated by H_N as

$$\psi_{N,t} = e^{-iH_N t}\psi_N \qquad \text{for all } t \in \mathbb{R}\,. \tag{1.3}$$

The global well-posedness of (1.2) is not an issue here. The study of (1.2) is focused, therefore, on other questions concerning the qualitative and quantitative behavior of the solution $\psi_{N,t}$. Despite the linearity of the equation, these questions are usually quite hard to answer, because in physically interesting situation the number of particles N is very large; it varies between $N \simeq 10^3$ for very dilute Bose-Einstein samples, up to values of the order $N \simeq 10^{30}$ in stars. For such huge values of N, it is of course impossible to compute the solution (1.3) explicitly; numerical methods are completely useless as well (unless the interaction among the particles is switched off).

Fortunately, also from the point of view of physics, it is not so important to know the precise solution to (1.2); it is much more important, for physicists performing experiments, to have information about the macroscopic properties of the system, which describe the typical behavior of the particles, and result from averaging over a large number of particles. Restricting the attention to macroscopic quantities simplifies the study of the solution $\psi_{N,t}$, but it still does not make it accessible to mathematical analysis. To further simplify matters, we are going to let the number of particles N tend to infinity. The macroscopic properties of the system, computed in the limiting regime $N \to \infty$, are then expected to be a good approximation for the macroscopic properties observed in experiments, where the number of particles N is very large, but finite (in some cases it is possible to obtain explicit bounds on the difference between the limiting behavior as $N \to \infty$ and the behavior for large but finite N; see Section 4.2).

To consider the limit of large N, we are going to make use of the marginal or reduced density matrices associated with an N particle wave function $\psi_N \in L^2_s(\mathbb{R}^{dN})$. First of all, we define the density matrix $\gamma_N = |\psi_N\rangle\langle\psi_N|$ associated with ψ_N as the orthogonal projection onto ψ_N; we use here and henceforth the notation $|\psi\rangle\langle\psi|$ to indicate the orthogonal projection onto ψ (Dirac bracket notation). Note that, expressed through the density matrix γ_N, the expectation $\langle\psi_N, A\psi_N\rangle$ of the observable A can be written as $\operatorname{Tr} A\gamma_N$. The kernel of the operator γ_N is then given by

$$\gamma_N(\mathbf{x};\mathbf{x}') = \psi(\mathbf{x})\overline{\psi}(\mathbf{x}'),$$

where we introduced the notation $\mathbf{x} = (x_1, \ldots, x_N), \mathbf{x}' = (x'_1, \ldots, x'_N) \in \mathbb{R}^{dN}$. Note that the L^2-normalization of ψ_N implies that $\operatorname{Tr}\gamma_N = 1$. For $k = 1, \ldots, N$, we define the k-particle marginal density $\gamma_N^{(k)}$ associated with ψ_N as the partial trace of γ_N over the degrees of freedom of the last $(N-k)$ particles:

$$\gamma_N^{(k)} = \operatorname{Tr}_{k+1,k+2,\ldots,N} |\psi_N\rangle\langle\psi_N|$$

where $\operatorname{Tr}_{k+1,\ldots,N}$ denotes the partial trace over the particle $k+1, k+2, \ldots, N$. In other words, $\gamma_N^{(k)}$ is defined as the non-negative trace class operator on $L^2_s(\mathbb{R}^{dk})$ with kernel given by

$$\begin{aligned} \gamma_N^{(k)}(\mathbf{x}_k;\mathbf{x}'_k) &= \int \mathrm{d}\mathbf{x}_{N-k}\, \gamma_N(\mathbf{x}_k, \mathbf{x}_{N-k}; \mathbf{x}'_k, \mathbf{x}_{N-k}) \\ &= \int \mathrm{d}\mathbf{x}_{N-k}\, \overline{\psi}_N(\mathbf{x}_k, \mathbf{x}_{N-k})\, \psi_N(\mathbf{x}'_k, \mathbf{x}_{N-k}) . \end{aligned} \tag{1.4}$$

The last equation can be considered as the definition of partial trace. Here we used the notation $\mathbf{x}_k = (x_1, \ldots, x_k), \mathbf{x}'_k = (x'_1, \ldots, x'_k) \in \mathbb{R}^{dk}$ as well as $\mathbf{x}_{N-k} = (x_{k+1}, \ldots, x_N) \in \mathbb{R}^{d(N-k)}$. By definition, $\operatorname{Tr}\gamma_N^{(k)} = 1$ for all N and for all $k = 1, \ldots, N$ (note that, in the physics literature, one normally uses a different normalization of the reduced density matrices). For fixed $k < N$, the k-particle density matrix does not contain the full information about the state described by ψ_N. Knowledge of the k-particle marginal $\gamma_N^{(k)}$, however, is sufficient to compute the expectation of every k-particle observable in the state described by the wave function ψ_N. In fact, if $A^{(k)}$ denotes an arbitrary bounded operator on $L^2(\mathbb{R}^{dk})$, and if $A^{(k)} \otimes 1^{(N-k)}$ denotes the operator on $L^2(\mathbb{R}^{dN})$ which acts as $A^{(k)}$ on the first k particles, and as the identity on the last $(N-k)$ particles, we have

$$\left\langle \psi_N, \left(A^{(k)} \otimes 1^{(N-k)}\right) \psi_N \right\rangle = \text{Tr}\, \left(A^{(k)} \otimes 1^{(N-k)}\right) \gamma_N = \text{Tr}\, A^{(k)} \gamma_N^{(k)}. \tag{1.5}$$

Thus, $\gamma_N^{(k)}$ is sufficient to compute the expectation of arbitrary observables which depend non-trivially on at most k particles (because of the permutation symmetry, it is not important on which particles it acts, just that it acts at most on k particles).

Marginal densities play an important role in the analysis of the $N \to \infty$ limit because, in contrast to the wave function ψ_N and to the density matrix γ_N, the k-particle marginal $\gamma_N^{(k)}$ can have, for every fixed $k \in \mathbb{N}$, a well-defined limit as $N \to \infty$ (because, if we fix $k \in \mathbb{N}$, $\{\gamma_N^{(k)}\}$ defines a sequence of operators all acting on the same space $L^2(\mathbb{R}^{dk})$).

In these notes we are going to study macroscopic properties of the dynamics of bosonic N-particle systems, in the limit $N \to \infty$. We are interested in the time-evolution of marginal densities $\gamma_{N,t}^{(k)}$ associated with the solution $\psi_{N,t}$ to the Schrödinger equation (1.2), for fixed k, and as $N \to \infty$. Unfortunately, it is not so simple to describe the time-dependence of $\gamma_{N,t}^{(k)}$ in the limit of large N; in fact it is in general impossible to obtain closed equations for the evolution of the limiting k-particle density $\gamma_{\infty,t}^{(k)} = \lim_{N\to\infty} \gamma_{N,t}^{(k)}$ (in general it is not even clear that this limit exists). Nevertheless, there are some physically interesting situations for which it is indeed possible to prove the existence of $\gamma_{\infty,t}^{(k)}$ and to derive closed equations to describe its dynamics. In Section 2, we are going to study the time evolution of factorized initial wave functions of the form $\psi_N(\mathbf{x}) = \prod_{j=1}^N \varphi(x_j)$ in the so-called mean field limit. We will show that in this case, for every fixed $k \in \mathbb{N}$, the k-particle marginal associated with the solution to the Schrödinger equation $\psi_{N,t}$ converges, as $N \to \infty$, to the limiting k-particle density $\gamma_{\infty,t}^{(k)} = |\varphi_t\rangle\langle\varphi_t|^{\otimes k}$, where φ_t is the solution of a certain one-particle nonlinear Schrödinger equation, known as the Hartree equation. In Section 3, we are going to study the time-evolution of Bose-Einstein condensates, in the so-called Gross-Pitaevskii limit. As we will see, although it describes a very different physical situation, the Gross-Pitaevskii scaling can be formally interpreted as a mean-field limit, with a very singular interaction potentials. Also in this case we will prove that the time evolution of the marginal densities can be described through a one-particle nonlinear Schrödinger equation (known as the time-dependent Gross-Pitaevskii equation); however, because of the singularity of the interaction potential, the analysis in this case is going to be much more involved. In Section 4, we will come back to the study of mean field models, and we will discuss how to prove quantitative estimates on the rate of convergence towards the Hartree dynamics.

Notation. Throughout these notes, we will make use of the notation $\mathbf{x} = (x_1, \dots, x_N), \mathbf{x}' = (x_1, \dots x'_N) \in \mathbb{R}^{dN}$, and for $k = 1, \dots, N$, $\mathbf{x}_k = (x_1, \dots, x_k), \mathbf{x}'_k = (x'_1, \dots, x'_k) \in \mathbb{R}^{dk}$, and $\mathbf{x}_{N-k} = (x_{k+1}, \dots, x_N) \in \mathbb{R}^{d(N-k)}$ (starting from Section 3, we will fix $d = 3$). We will also use the shorthand notation $\nabla_j = \nabla_{x_j}$ and $\Delta_j = \Delta_{x_j}$.

Acknowledgements. These notes were completed while the author was at the Institute of Mathematics of the Ludwig-Maximilian University of Munich, Germany, supported by a Sofja Kovalevskaja Award of the Humboldt Foundation.

2. Derivation of the Hartree Equation in the Mean Field Limit

2.1. Mean Field Systems. A mean-field system is described by an N-body Hamilton operator of the form

$$H_N = \sum_{j=1}^{N} (-\Delta_j + V_{\text{ext}}(x_j)) + \frac{1}{N} \sum_{i<j}^{N} V(x_i - x_j) \tag{2.1}$$

acting on the Hilbert space $L^2_s(\mathbb{R}^{dN})$, the subspace of $L^2(\mathbb{R}^{dN})$ consisting of permutation symmetric functions. In these notes we will only discuss bosonic systems, which are described by symmetric wave functions. Note, however, that the mean field limit for fermionic system has also been considered in the literature; see, for example, [**26, 31, 7**]. In (2.1) and henceforth, we use the notation $\Delta_j = \Delta_{x_j}$ (similarly, we will use the notation $\nabla_j = \nabla_{x_j}$). The mean-field character of the Hamiltonian is expressed by the factor $1/N$ in front of the interaction; this factor guarantees that the kinetic and potential energies are typically both of order N.

We are interested in the solution $\psi_{N,t} = e^{-iH_N t}\psi_N$ of the Schrödinger equation (1.2) with Hamiltonian H_N given by (2.1) and with factorized initial data

$$\psi_N(\mathbf{x}) = \prod_{j=1}^{N} \varphi(x_j)\,, \tag{2.2}$$

for some $\varphi \in L^2(\mathbb{R}^d)$. The physical motivation for studying the evolution of factorized wave functions is that states close to the ground state of H_N (the eigenvector associated with the lowest eigenvalue), which are the most accessible and thus the most interesting states, can be approximately described by wave functions like (2.2) (the results which we are going to discuss in this section do not require strict factorization as in (2.2); instead condensation of the initial wave function in the sense of (3.1) would be sufficient).

Because of the interaction among the particles, the factorization (2.2) is not preserved by the time evolution; in other words, the evolved wave function $\psi_{N,t}$ is not given by the product of one-particle wave functions, if $t \neq 0$. However, due to the mean-field character of the interaction each particle interacts very weakly (the strength of the interaction is of the order $1/N$) with all other $(N-1)$ particles (at least in the initial state, every particle is described by the same one-particle orbital; every particles therefore "sees" all other particles). For this reason, we may expect that, in the limit of large N, the total interaction potential experienced by a typical particle in the system can be effectively replaced by an averaged, mean-field, potential, and therefore that factorization is approximately, and in an appropriate sense, preserved by the time evolution. In other words, we may expect that, in a sense to be made precise,

$$\psi_{N,t}(x_1, \ldots, x_N) \simeq \prod_{j=1}^{N} \varphi_t(x_j) \qquad \text{as } N \to \infty \tag{2.3}$$

for an evolved one-particle orbital φ_t. Assuming (2.3), it is simple to derive a self-consistent equation for the time-evolution of the one-particle orbital φ_t. In fact, (2.3) states that, for every fixed time t, the N particles are independently distributed in space with density $|\varphi_t(x)|^2$. If this is true, the total potential experienced, for example, by the first particle can be approximated by

$$\frac{1}{N}\sum_{j\geq 2} V(x_1-x_j) \simeq \frac{1}{N}\sum_{j\geq 2}\int \mathrm{d}y\, V(x_1-y)|\varphi_t(y)|^2 = \frac{N-1}{N}(V*|\varphi_t|^2) \simeq (V*|\varphi_t|^2)$$

as $N \to \infty$. It follows that, if (2.3) holds true, the one-particle orbital φ_t must satisfy the self-consistent equation

$$i\partial_t\varphi_t = -\Delta\varphi_t + (V*|\varphi_t|^2)\varphi_t \tag{2.4}$$

with initial data $\varphi_{t=0} = \varphi$ given by (2.2). Equation (2.4) is known as the Hartree equation; it is an example of a cubic nonlinear Schrödinger equation on $\mathbb{R}^d$. Starting from the linear Schrödinger equation (1.2) on $\mathbb{R}^{dN}$, we obtain, for the evolution of factorized wave functions, a nonlinear Schrödinger equation on $\mathbb{R}^d$; the nonlinearity in the Hartree equation is a consequence of the many-body effects in the linear dynamics.

The convergence of $\psi_{N,t}$ to the factorized wave function on the r.h.s. of (2.3) as $N \to \infty$ cannot hold in the L^2-sense; we cannot expect, in other words, that $\|\psi_{N,t} - \varphi_t^{\otimes N}\| \to 0$ as $N \to \infty$ (we use here the notation $\varphi^{\otimes N}(\mathbf{x}) = \prod_{j=1}^N \varphi(x_j)$). Instead, (2.3) has to be understood as convergence of marginal densities. Recall that, for $k = 1, \dots, N$, the k-particle marginal $\gamma_{N,t}^{(k)}$ associated with $\psi_{N,t}$ is defined as the non-negative trace class operator on $L^2(\mathbb{R}^{dk})$ with kernel given by

$$\begin{aligned}\gamma_{N,t}^{(k)}(\mathbf{x}_k;\mathbf{x}'_k) &= \int \mathrm{d}\mathbf{x}_{N-k}\, \gamma_{N,t}(\mathbf{x}_k,\mathbf{x}_{N-k};\mathbf{x}'_k,\mathbf{x}_{N-k})\\ &= \int \mathrm{d}\mathbf{x}_{N-k}\, \overline{\psi}_{N,t}(\mathbf{x}_k,\mathbf{x}_{N-k})\,\psi_{N,t}(\mathbf{x}'_k,\mathbf{x}_{N-k})\,.\end{aligned} \tag{2.5}$$

It turns out that (2.3) holds in the sense that, for every fixed $k \in \mathbb{N}$, the k-particle marginal density associated with the left hand side converges, as $N \to \infty$, to the k-particle marginal density associated with the right hand side (which is independent of N, if $N \geq k$). In other words, assuming (2.2), one can show that, for a large class of interaction potentials, and for every fixed $t \in \mathbb{R}$ and $k \in \mathbb{N}$,

$$\mathrm{Tr}\,\left|\gamma_{N,t}^{(k)} - |\varphi_t\rangle\langle\varphi_t|^{\otimes k}\right| \to 0 \qquad \text{as } N \to \infty\,, \tag{2.6}$$

where φ_t is the solution to the Hartree equation (2.4) with initial data $\varphi_{t=0} = \varphi$. It is important here that the time $t \in \mathbb{R}$ and the integer $k \geq 1$ are fixed; the convergence is not uniform in these two parameters. From (1.5), we observe that (2.6) implies (and is actually equivalent to the condition) that, for every fixed $t \in \mathbb{R}$ and $k \in \mathbb{N}$, and for every fixed compact operator $J^{(k)}$ on $L^2_s(\mathbb{R}^{dk})$,

$$\left\langle \psi_{N,t}, \left(J^{(k)} \otimes 1^{(N-k)}\right)\psi_{N,t}\right\rangle \to \langle \varphi_t^{\otimes k}, J^{(k)}\varphi_t^{\otimes k}\rangle \tag{2.7}$$

as $N \to \infty$. This means that, if we are interested in the expectation of observables which depend non-trivially on a fixed number k of particles, then we can approximate, as $N \to \infty$, the true solution $\psi_{N,t}$ to the N-body Schrödinger equation by the product of N copies of the solution φ_t to the Hartree equation (2.4). This approximation, however, is in general not valid if we are interested in the expectation of observables depending on a macroscopic (that is, proportional to N) number of particles.

The first rigorous results establishing a relation between the many body Schrödinger evolution and the nonlinear Hartree dynamics were obtained by Hepp in [**21**] (for smooth interaction potentials) and then generalized by Ginibre and Velo to singular potentials in [**20**]. These works were inspired by techniques used in quantum field theory. We will discuss this method in Section 4, where we present a recent proof of (2.6), obtained in collaboration with I. Rodnianski in [**28**], which provides a quantitative control of the rate of convergence and makes use of the original idea of Hepp.

The first proof of the convergence (2.6) was obtained by Spohn in [**30**], for bounded potentials. The method introduced by Spohn was then extended to singular potentials. In [**16**], Erdős and Yau proved (2.6) for a Coulomb potential $V(x) = \pm 1/|x|$; partial results for the Coulomb potential were also obtained by Bardos, Golse and Mauser in [**5**] (note that recently a new proof of (2.6) for the case of a Coulomb interaction has been proposed by Fröhlich, Knowles, and Schwarz in [**19**]). In [**9**], a joint work with A. Elgart, we considered again the Coulomb potential, but this time assuming a relativistic dispersion for the bosons. Recently, in a series of papers [**11, 12, 13, 14, 15**] in collaboration with L. Erdős and H.-T. Yau (and also in [**8**], a collaboration with A. Elgart, L. Erdős and H.-T. Yau) the strategy of [**30**] was applied to systems with an N-dependent interaction potential, which converges, in the limit $N \to \infty$, to a delta-function. Note that in the one-dimensional setting, potentials converging to a delta-interaction have been considered by Adami, Golse and Teta in [**2**] (making use of previous results obtained by the same authors in collaboration with Bardos in [**1**]) . We will discuss these systems in Section 3.

Recently, a different approach to the proof of (2.6) has been proposed by Fröhlich, Schwarz and Graffi in [**17**]. For smooth potentials, they can consider the mean-field limit uniformly in Planck's constant $\hbar$ (up to errors exponentially small in time); this allows them to combine the semiclassical limit and the mean field limit. It is also interesting to remark that the mean-field limit (2.6) can be interpreted as a Egorov-type theorem; this was observed by Fröhlich, Knowles, and Pizzo in [**18**].

2.2. Derivation of the Hartree Equation for Bounded Potentials. We consider, in this section, the dynamics generated by the mean field Hamiltonian (2.1) under the assumption that the interaction potential is a bounded operator. We will assume, in other words, that $V \in L^\infty(\mathbb{R}^d)$ (recall that the operator norm of the multiplication operator $V(x_i - x_j)$ is given by the L^∞-norm of the function V). To simplify a little bit the analysis we will also assume the external potential V_{ext} in the Hamiltonian (2.1) to vanish; the techniques discussed here can however be easily extended to $V_{\text{ext}} \neq 0$.

THEOREM 2.1 (Spohn, [**30**]). *Suppose that*

$$H_N = \sum_{j=1}^N -\Delta_j + \frac{1}{N} \sum_{i<j} V(x_i - x_j)$$

with $V \in L^\infty(\mathbb{R}^d)$. Let $\psi_N = \varphi^{\otimes N} \in L^2(\mathbb{R}^{dN})$ for some $\varphi \in L^2(\mathbb{R}^d)$ with $\|\varphi\| = 1$. Let $\psi_{N,t} = e^{-iH_N t}\psi_N$, and denote by $\gamma^{(k)}_{N,t}$ the k-particle marginal density associated with $\psi_{N,t}$. Then, for every fixed $t \in \mathbb{R}$, and for every fixed $k \geq 1$, we have

$$\text{Tr} \left| \gamma^{(k)}_{N,t} - |\varphi_t\rangle\langle\varphi_t|^{\otimes k} \right| \to 0 \tag{2.8}$$

as $N \to \infty$. Here φ_t denotes the solution to the Hartree equation

$$i\partial_t \varphi_t = -\Delta\varphi_t + (V * |\varphi_t|^2)\varphi_t \tag{2.9}$$

with initial data $\varphi_{t=0} = \varphi$.

PROOF. The proof is based on the study of the time evolution of the marginal densities $\gamma^{(k)}_{N,t}$ in the limit $N \to \infty$. From (1.2), it is simple to show that the dynamics of the marginals is governed by a hierarchy of N coupled equation, commonly known as the BBGKY hierarchy:

$$\begin{aligned} i\partial_t \gamma^{(k)}_{N,t} = &\sum_{j=1}^k \left[-\Delta_j, \gamma^{(k)}_{N,t}\right] + \frac{1}{N}\sum_{i<j}^k \left[V(x_i - x_j), \gamma^{(k)}_{N,t}\right] \\ &+ \frac{(N-k)}{N} \sum_{j=1}^k \text{Tr}_{k+1} \left[V(x_j - x_{k+1}), \gamma^{(k+1)}_{N,t}\right]. \end{aligned} \tag{2.10}$$

We use here the convention that $\gamma^{(k)}_{N,t} = 0$ if $k > N$. Moreover $[A, B] = AB - BA$ denotes the commutator of the two operators A and B. The symbol Tr_{k+1} denotes the partial trace over the $(k+1)$-th particle; the kernel of the k-particle operator $\text{Tr}_{k+1}\,[V(x_j - x_{k+1}), \gamma^{(k+1)}_{N,t}]$ is given by

$$\begin{aligned} &\left(\text{Tr}_{k+1} \left[V(x_j - x_{k+1}), \gamma^{(k+1)}_{N,t}\right]\right)(\mathbf{x}_k; \mathbf{x}'_k) \\ &\quad = \int \mathrm{d}x_{k+1} \left(V(x_j - x_{k+1}) - V(x'_j - x_{k+1})\right) \gamma^{(k+1)}(\mathbf{x}_k, x_{k+1}; \mathbf{x}'_k, x_{k+1}). \end{aligned} \tag{2.11}$$

Rewriting the BBGKY hierarchy (2.10) in integral form, we find

$$\begin{aligned} \gamma^{(k)}_{N,t} = &\,\mathcal{U}^{(k)}(t)\gamma^{(k)} + \frac{1}{N}\int_0^t \mathrm{d}s\, \mathcal{U}^{(k)}(t-s) A^{(k)} \gamma^{(k)}_{N,s} \\ &+ \left(1 - \frac{k}{N}\right) \int_0^t \mathrm{d}s\, \mathcal{U}^{(k)}(t-s) B^{(k)} \gamma^{(k+1)}_{N,s} \end{aligned} \tag{2.12}$$

where $\mathcal{U}^{(k)}(t)$ denotes the free evolution of k particles, defined by

$$\mathcal{U}^{(k)}(t)\gamma^{(k)} = e^{it\sum_{j=1}^k \Delta_j} \gamma^{(k)} e^{-it\sum_{j=1}^k \Delta_j} \tag{2.13}$$

and the maps $A^{(k)}$ and $B^{(k)}$ are defined by

$$A^{(k)}\gamma^{(k)} = -i\sum_{j=1}^k \left[V(x_i - x_j), \gamma^{(k)}\right] \tag{2.14}$$

and, respectively, by

$$B^{(k)}\gamma^{(k+1)} = -i\sum_{j=1}^k \text{Tr}_{k+1} \left[V(x_j - x_{k+1}), \gamma^{(k+1)}\right]. \tag{2.15}$$

Note that $B^{(k)}$ maps $(k+1)$-particle operators into k-particle operators (while $A^{(k)}$ maps k-particle operators into k-particle operators). Since we are interested in the limit $N \to \infty$ with fixed $k \geq 1$, it is clear that the second term on the r.h.s. of (2.12), as well as the contribution proportional to k/N to the third term on the r.h.s. of (2.12) should be considered as small perturbations. Iterating the integral equation (2.12) for n times, and stopping the iteration every time we hit a perturbation, we obtain the Duhamel type series

(2.16)
$$\begin{aligned}
\gamma_{N,t}^{(k)} =& \mathcal{U}^{(k)}(t)\gamma_0^{(k)} \\
&+ \sum_{m=1}^{n-1} \int_0^t \mathrm{d}s_1 \ldots \int_0^{s_{m-1}} \mathrm{d}s_m \, \mathcal{U}^{(k)}(t-s_1)B^{(k)}\mathcal{U}^{(k+1)}(s_1-s_2)\ldots \\
&\qquad \ldots \times B^{(k+m-1)}\mathcal{U}^{(k+m)}(s_m)\gamma_0^{(k+m)} \\
&+ \int_0^t \mathrm{d}s_1 \ldots \int_0^{s_{n-1}} \mathrm{d}s_n \, \mathcal{U}^{(k)}(t-s_1)B^{(k)}\mathcal{U}^{(k+1)}(s_1-s_2)\ldots B^{(k+n-1)}\gamma_{N,s_n}^{(k+n)} \\
&+ \frac{1}{N}\sum_{m=1}^{N} \int_0^t \mathrm{d}s_1 \ldots \int_0^{s_{m-1}} \mathrm{d}s_m \mathcal{U}^{(k)}(t-s_1)B^{(k)}\ldots \\
&\qquad \ldots \times \mathcal{U}^{(k+m-1)}(s_{m-1}-s_m)A^{(k+m-1)}\gamma_{N,s_m}^{(k+m-1)} \\
&- \sum_{m=1}^{n} \frac{k+m-1}{N} \int_0^t \mathrm{d}s_1 \ldots \int_0^{s_{m-1}} \mathrm{d}s_m \, \mathcal{U}^{(k)}(t-s_1)B^{(k)}\ldots \\
&\qquad \ldots \times B^{(k+m-1)}\gamma_{N,s_m}^{(k+m)} .
\end{aligned}$$

To show (2.8), we need to compare $\gamma_{N,t}^{(k)}$ with $\gamma_{\infty,t}^{(k)} = |\varphi_t\rangle\langle\varphi_t|^{\otimes k}$, where φ_t is the solution to the Hartree equation (2.9). It is simple to check that the family $\{\gamma_{\infty,t}^{(k)}\}_{k\geq 1}$ solves the infinite hierarchy (written directly in integral form)

$$\gamma_{\infty,t}^{(k)} = \mathcal{U}^{(k)}(t)\gamma_0^{(k)} + \int_0^t \mathrm{d}s \, \mathcal{U}^{(k)}(t-s)B^{(k)}\gamma_{\infty,s}^{(k+1)} \tag{2.17}$$

which leads, after iteration, to the expansion

$$\begin{aligned}
\gamma_{\infty,t}^{(k)} =& \mathcal{U}^{(k)}(t)\gamma_0^{(k)} \\
&+ \sum_{m=1}^{n-1} \int_0^t \mathrm{d}s_1 \ldots \int_0^{s_{m-1}} \mathrm{d}s_m \, \mathcal{U}^{(k)}(t-s_1)B^{(k)}\mathcal{U}^{(k+1)}(s_1-s_2)\ldots \\
&\qquad \ldots \times B^{(k+m-1)}\mathcal{U}^{(k+m)}(s_m)\gamma_0^{(k+m)} \\
&+ \int_0^t \mathrm{d}s_1 \ldots \int_0^{s_{n-1}} \mathrm{d}s_n \, \mathcal{U}^{(k)}(t-s_1)B^{(k)}\mathcal{U}^{(k+1)}(s_1-s_2)\ldots \\
&\qquad \ldots \times B^{(k+n-1)}\gamma_{\infty,s_n}^{(k+n)} .
\end{aligned} \tag{2.18}$$

The difference between $\gamma^{(k)}_{N,t}$ and $\gamma^{(k)}_{\infty,t}$ can thus be bounded by

$$
\begin{aligned}
&\operatorname{Tr}\left|\gamma^{(k)}_{N,t}-\gamma^{(k)}_{\infty,t}\right| \\
&\quad\leq \int_0^t ds_1 \ldots \int_0^{s_{n-1}} ds_n \operatorname{Tr}\Big|\mathcal{U}^{(k)}(t-s_1)B^{(k)}\mathcal{U}^{(k+1)}(s_1-s_2)\ldots \\
&\qquad\qquad \ldots \times B^{(k+n-1)}\left(\gamma^{(k+n)}_{N,s_n}-\gamma^{(k+n)}_{\infty,s_n}\right)\Big| \\
&\quad + \frac{1}{N}\sum_{m=1}^{N}\int_0^t ds_1\ldots\int_0^{s_{m-1}} ds_m \operatorname{Tr}\Big|\mathcal{U}^{(k)}(t-s_1)B^{(k)}\ldots \\
&\qquad\qquad \ldots\times \mathcal{U}^{(k+m-1)}(s_{m-1}-s_m)A^{(k+m-1)}\gamma^{(k+m-1)}_{N,s_m}\Big| \\
&\quad + \sum_{m=1}^{n}\frac{k+m-1}{N}\int_0^t ds_1\ldots\int_0^{s_{m-1}} ds_m \operatorname{Tr}\Big|\mathcal{U}^{(k)}(t-s_1)B^{(k)}\ldots \\
&\qquad\qquad \ldots\times B^{(k+m-1)}\gamma^{(k+m)}_{N,s_m}\Big|.
\end{aligned}
\tag{2.19}
$$

Next we observe that, since $\mathcal{U}^{(k)}(t)$ is a unitary operator,

$$\operatorname{Tr}\left|\mathcal{U}^{(k)}(t)\gamma^{(k)}\right| = \operatorname{Tr}\left|\gamma^{(k)}\right|. \tag{2.20}$$

Moreover, since V is a bounded potential, we have

$$\operatorname{Tr}\left|A^{(k)}\gamma^{(k)}\right| \leq k^2\|V\|\operatorname{Tr}\left|\gamma^{(k)}\right| \tag{2.21}$$

and

$$\operatorname{Tr}\left|B^{(k)}\gamma^{(k+1)}\right| \leq 2k\|V\|\operatorname{Tr}\left|\gamma^{(k+1)}\right| \tag{2.22}$$

where we used the fact that

$$\operatorname{Tr}\left|\operatorname{Tr}_{k+1}\gamma^{(k+1)}\right| \leq \operatorname{Tr}\left|\gamma^{(k+1)}\right|. \tag{2.23}$$

(Here the trace on the r.h.s. is a trace over $(k+1)$ particles.) Applying these bounds iteratively to the terms on the r.h.s. of (2.19), and using the a-priori information $\operatorname{Tr}\left|\gamma^{(k+n)}_{N,t}\right| = \operatorname{Tr}\gamma^{(k+n)}_{N,t} = 1$ (and analogously for $\gamma^{(k+n)}_{\infty,t}$), we obtain

$$
\begin{aligned}
\operatorname{Tr}\left|\gamma^{(k)}_{N,t}-\gamma^{(k)}_{\infty,t}\right| &\leq 2\,(2\|V\|t)^n\frac{(k+n-1)!}{(k-1)!n!} \\
&\qquad + \frac{2}{N}\sum_{m=1}^{n}(2\|V\|t)^m(k+m-1)\frac{(k+m-1)!}{m!(k-1)!} \\
&\leq 2^k\,(4\|V\|t)^n + \frac{k\,2^{k+1}}{N}\sum_{m=1}^{N}(4\|V\|t)^m.
\end{aligned}
\tag{2.24}
$$

If $0<t\leq t_0$, with $t_0 = 1/(8\|V\|)$, it follows that

$$\operatorname{Tr}\left|\gamma^{(k)}_{N,t}-\gamma^{(k)}_{\infty,t}\right| \leq \frac{2^k}{2^n}+\frac{k2^{k+1}}{N}.$$

Since the l.h.s. is independent of the order n of the expansion, it follows that

$$\text{Tr}\left|\gamma^{(k)}_{N,t} - \gamma^{(k)}_{\infty,t}\right| \leq \frac{k2^{k+1}}{N} \tag{2.25}$$

and thus that

$$\text{Tr}\left|\gamma^{(k)}_{N,t} - \gamma^{(k)}_{\infty,t}\right| \to 0 \qquad \text{as } N \to \infty \tag{2.26}$$

for all $0 \leq t \leq t_0$ and for all $k \geq 1$. Next, set

$$t_1 := \sup\left\{t > 0 : \lim_{N\to\infty} \text{Tr}\left|\gamma^{(k)}_{N,s} - \gamma^{(k)}_{\infty,s}\right| = 0 \quad \text{for all fixed } 0 \leq s \leq t \text{ and } k \geq 1\right\}.$$

From (2.26), it follows that $t_1 \geq t_0$. We show that $t_1 = \infty$ by contradiction. Suppose that $t_1 < \infty$. Then, if $t_2 = t_1 - (t_0/2)$, we have, by definition,

$$\lim_{N\to\infty} \text{Tr}\left|\gamma^{(k)}_{N,t_2} - \gamma^{(k)}_{\infty,t_2}\right| = 0 \qquad \text{for all } k \geq 1. \tag{2.27}$$

Starting from (2.27), we are going to prove that

$$\lim_{N\to\infty} \text{Tr}\left|\gamma^{(k)}_{N,t} - \gamma^{(k)}_{\infty,t}\right| = 0 \tag{2.28}$$

for all $k \geq 1$ and for all $0 \leq t \leq t_1 + (t_0/2)$; this contradicts the definition of t_1. To show (2.28), we expand $\gamma^{(k)}_{N,t}$ and $\gamma^{(k)}_{\infty,t}$ in Duhamel series similar to (2.16) and (2.18), but starting at time $t_2 = t_1 - (t_0/2)$. Analogously to (2.19), we obtain, for $t = t_2 + \tau$,

(2.29)

$$\begin{aligned}
&\text{Tr}\left|\gamma^{(k)}_{N,t} - \gamma^{(k)}_{\infty,t}\right| \\
&\leq \text{Tr}\left|\mathcal{U}^{(k)}(\tau)\left(\gamma^{(k)}_{N,t_2} - \gamma^{(k)}_{\infty,t_2}\right)\right| \\
&\quad + \sum_{m=1}^{n-1}\int_0^\tau ds_1 \ldots \int_0^{s_{m-1}} ds_m \,\text{Tr}\,\Big|\mathcal{U}^{(k)}(\tau - s_1)B^{(k)}\mathcal{U}^{(k+1)}(s_1 - s_2)\ldots \\
&\qquad\qquad \ldots \times B^{(k+m-1)}\mathcal{U}^{(k+m)}(s_m)\left(\gamma^{(k+n)}_{N,t_2} - \gamma^{(k+n)}_{\infty,t_2}\right)\Big| \\
&\quad + \int_0^\tau ds_1 \ldots \int_0^{s_{n-1}} ds_n \,\text{Tr}\,\left|\mathcal{U}^{(k)}(\tau - s_1)B^{(k)}\ldots B^{(k+n-1)}\left(\gamma^{(k+n)}_{N,s_n} - \gamma^{(k+n)}_{\infty,s_n}\right)\right| \\
&\quad + \frac{1}{N}\sum_{m=1}^{N}\int_0^\tau ds_1 \ldots \int_0^{s_{m-1}} ds_m \text{Tr}\,\Big|\mathcal{U}^{(k)}(\tau - s_1)B^{(k)}\ldots \\
&\qquad\qquad \ldots \times \mathcal{U}^{(k+m-1)}(s_{m-1} - s_m)A^{(k+m-1)}\gamma^{(k+m-1)}_{N,s_m}\Big| \\
&\quad + \sum_{m=1}^{n}\frac{k+m-1}{N}\int_0^\tau ds_1 \ldots \int_0^{s_{m-1}} ds_m \,\text{Tr}\,\Big|\mathcal{U}^{(k)}(\tau - s_1)B^{(k)}\ldots \\
&\qquad\qquad \ldots \times B^{(k+m-1)}\gamma^{(k+m)}_{N,s_m}\Big|.
\end{aligned}$$

With respect to (2.19), we have one more term on the r.h.s. of the last equation, due to the fact that at time $t = t_2$ the densities do not coincide (while they do at time $t = 0$). Analogously to (2.24) we find

$$\text{Tr}\left|\gamma^{(k)}_{N,t} - \gamma^{(k)}_{\infty,t}\right| \leq 2^k \sum_{m=0}^{n-1}\frac{1}{2^m}\,\text{Tr}\,\left|\left(\gamma^{(k+m)}_{N,t_2} - \gamma^{(k+m)}_{\infty,t_2}\right)\right| + \frac{2^k}{2^n} + \frac{k2^{k+1}}{N},$$

if $t_1 - (t_0/2) \leq t \leq t_1 + (t_0/2)$ (that is, if $0 \leq \tau \leq t_0$). Choosing first $n > 0$ sufficiently large (to make the second term on the r.h.s. smaller than $\varepsilon/3$), and then $N > 0$ sufficiently large (this guarantees that the third term, and, by (2.27), also the first term, are smaller than $\varepsilon/3$), the quantity on the l.h.s. can be made smaller than any $\varepsilon > 0$ (for arbitrary $k \geq 1$ and $t_1 - (t_0/2) \leq t \leq t_1 + (t_0/2)$). This shows (2.28) and completes the proof of the theorem. □

2.3. Another Proof of Theorem 2.1. From the proof of Theorem 2.1 presented above, we notice that the expansion of the BBGKY hierarchy in (2.16) is much more involved than the corresponding expansion (2.18) of the infinite hierarchy (2.17). It turns out that it is possible to avoid the expansion of the BBGKY hierarchy making use of a simple compactness argument; this will be especially important when dealing with singular potentials. In the following we explain the main steps of this alternative proof to Theorem 2.1. Then, in the next section, we will illustrate how to extend it to potentials with a Coulomb singularity.

The idea, which was first presented in [**5, 4, 16**], consists in characterizing the limit of the densities $\gamma^{(k)}_{N,t}$ as the unique solution to the infinite hierarchy of equations (2.17); combined with the compactness, this information provides a proof of Theorem 2.1. More precisely, the proof is divided into three main steps. First of all, one shows the compactness of the sequence $\{\gamma^{(k)}_{N,t}\}_{k\geq 1}$ with respect to an appropriate weak topology. Then, one proves that an arbitrary limit point $\{\gamma^{(k)}_{\infty,t}\}_{k\geq 1}$ of the sequence $\{\gamma^{(k)}_{N,t}\}_{k=1}^N$ is a solution to the infinite hierarchy (2.17) (one proves, in other words, the convergence to the infinite hierarchy). Finally, one shows the uniqueness of the solution to the infinite hierarchy (2.17). Since it is simple to verify that the factorized family $\{\gamma^{(k)}_{\infty,t}\}_{k\geq 1}$, with $\gamma^{(k)}_t = |\varphi_t\rangle\langle\varphi_t|^{\otimes k}$ for all $k \geq 1$, is a solution to the infinite hierarchy, it follows immediately that $\gamma^{(k)}_{N,t} \to |\varphi_t\rangle\langle\varphi_t|^{\otimes k}$ as $N \to \infty$ (at first only in the weak topology with respect to which we have compactness; since the limit is an orthogonal rank one projection, it is however simple to check that weak convergence implies strong convergence, in the sense (2.8)). Next, we discuss these three main steps (compactness, convergence, and uniqueness) in some more details.

Compactness: Let $\mathcal{L}^1_k \equiv \mathcal{L}^1(L^2(\mathbb{R}^{dk}))$ denote the space of trace class operators on $L^2(\mathbb{R}^{dk})$, equipped with the trace norm

$$\|A\|_1 = \mathrm{Tr}\,|A| = \mathrm{Tr}\,(A^*A)^{1/2} \qquad \text{for all } A \in \mathcal{L}^1_k\,.$$

Moreover, let $\mathcal{K}_k \equiv \mathcal{K}(L^2(\mathbb{R}^{dk}))$ be the space of compact operators on $L^2(\mathbb{R}^{dk})$, equipped with the operator norm. Then $\mathcal{L}^1_k$ and $\mathcal{K}_k$ are Banach spaces and $\mathcal{L}^1_k = \mathcal{K}^*_k$ (see, for example, [**27**][Theorem VI.26]). By definition, the k-particle marginal density $\gamma^{(k)}_{N,t}$ is a non-negative operator in $\mathcal{L}^1_k$, with

$$\|\gamma^{(k)}_{N,t}\|_1 = \mathrm{Tr}\,|\gamma^{(k)}_{N,t}| = \mathrm{Tr}\,\gamma^{(k)}_{N,t} = 1$$

for all $N \geq k$. For fixed $t \in \mathbb{R}$ and $k \geq 1$, it follows from the Banach-Alaouglu Theorem that the sequence $\{\gamma^{(k)}_{N,t}\}_{N\geq k}$ is compact with respect to the weak* topology of $\mathcal{L}^1_k$.

Since we want to identify limit points of the sequence $\gamma^{(k)}_{N,t}$ as solutions to the system of integral equations (2.17), compactness for fixed $t \in \mathbb{R}$ is not enough. To make sure that there are subsequences of $\gamma_{N,t}^{(k)}$ which converge for all times in a certain interval, we use the fact that, since $\mathcal{K}_k$ is separable, the weak* topology on the unit ball of $\mathcal{L}_k^1$ is metrizable. It is possible, in other words, to introduce a metric η_k on $\mathcal{L}_k^1$ such that a uniformly bounded sequence $\{A_n\}_{n\in\mathbb{N}} \in \mathcal{L}_k^1$ converges to $A \in \mathcal{L}_k^1$ as $n \to \infty$ with respect to the weak* topology of $\mathcal{L}_k^1$ if and only if $\eta_k(A_n, A) \to 0$ (see [**29**][Theorem 3.16], for the explicit construction of the metric η_k). For arbitrary $T > 0$ let $C([0,T], \mathcal{L}_1^k)$ be the space of functions of $t \in [0,T]$ with values in $\mathcal{L}_k^1$ which are continuous with respect to the metric η_k; on $C([0,T], \mathcal{L}_k^1)$ we can define the metric

$$\widehat{\eta}_k(\gamma^{(k)}(\cdot), \bar{\gamma}^{(k)}(\cdot)) := \sup_{t\in[0,T]} \eta_k(\gamma^{(k)}(t), \bar{\gamma}^{(k)}(t)) \,. \tag{2.30}$$

Finally, we denote by τ_{prod} the topology on the space $\bigoplus_{k\geq 1} C([0,T], \mathcal{L}_k^1)$ given by the product of the topologies generated by the metrics $\widehat{\eta}_k$ on $C([0,T], \mathcal{L}_k^1)$.

The metric structure introduced on the space $\bigoplus_{k\geq 1} C([0,T], \mathcal{L}_k^1)$ allows us to invoke the Arzela-Ascoli Theorem to prove the compactness of the sequence $\Gamma_{N,t} = \{\gamma_{N,t}^{(k)}\}_{k=1}^N$. We obtain the following proposition (for the detailed proof, see, for example, [**13**, Section 6]).

PROPOSITION 2.2. *Fix $T > 0$. Then $\Gamma_{N,t} = \{\gamma_{N,t}^{(k)}\}_{k=1}^N \in \bigoplus_{k\geq 1} C([0,T], \mathcal{L}_k^1)$ is a compact sequence with respect to the product topology τ_{prod} defined above. For any limit point $\Gamma_{\infty,t} = \{\gamma_{\infty,t}^{(k)}\}_{k\geq 1}$, $\gamma_{\infty,t}^{(k)}$ is symmetric w.r.t. permutations, non-negative and such that*

$$Tr\, \gamma_{\infty,t}^{(k)} \leq 1 \tag{2.31}$$

for every $k \geq 1$.

Remark. Convergence of $\Gamma_{N,t} = \{\gamma_{N,t}^{(k)}\}_{k=1}^N$ to $\Gamma_{\infty,t} = \{\gamma_{\infty,t}^{(k)}\}_{k\geq 1}$ with respect to the topology τ_{prod} is equivalent to the statement that, for every fixed $k \geq 1$, and for every fixed compact operator $J^{(k)} \in \mathcal{K}_k$,

$$\text{Tr} J^{(k)} \left(\gamma_{N,t}^{(k)} - \gamma_{\infty,t}^{(k)} \right) \to 0 \tag{2.32}$$

as $N \to \infty$, uniformly in t for $t \in [0,T]$. Compactness of $\Gamma_{N,t}$ with respect to the topology τ_{prod} means therefore that for every sequence $\{M_j\}_{j\in\mathbb{N}}$ there exists a subsequence $\{N_j\}_{j\in\mathbb{N}} \subset \{M_j\}_{j\in\mathbb{N}}$ and a limit point $\Gamma_{\infty,t}$ such that $\Gamma_{N_j,t} \to \Gamma_{\infty,t}$ in the sense (2.32).

Convergence: The second main step consists in characterizing the limit points of the (compact) sequence $\Gamma_{N,t} = \{\gamma_{N,t}^{(k)}\}_{k\geq 1}$ as solutions to the infinite hierarchy of equations (2.17).

PROPOSITION 2.3. *Suppose that $V \in L^\infty(\mathbb{R}^d)$ such that $V(x) \to 0$ as $|x| \to \infty$. Assume moreover that $\Gamma_{\infty,t} = \{\gamma_{\infty,t}^{(k)}\}_{k\geq 1} \in \bigoplus_{k\geq 1} C([0,T], \mathcal{L}_k^1)$ is a limit point of the sequence $\Gamma_{N,t} = \{\gamma_{N,t}^{(k)}\}_{k=1}^N$ with respect to the product topology τ_{prod}. Then $\gamma_{\infty,0}^{(k)} = |\varphi\rangle\langle\varphi|^{\otimes k}$ and*

$$\gamma_{\infty,t}^{(k)} = \mathcal{U}^{(k)}(t)\gamma_{0,\infty}^{(k)} + \int_0^t \mathrm{d}s\, \mathcal{U}^{(k)}(t-s) B^{(k)} \gamma_{\infty,s}^{(k+1)} \tag{2.33}$$

for all $k \geq 1$. Here $\mathcal{U}^{(k)}(t)$, and $B^{(k)}$ are defined as in (2.13) and, respectively, in (2.15).

Note that in Proposition 2.3 we assume the potential to vanish at infinity. This condition, which was not required in Section 2.2, is not essential but it simplifies the proof and it is also satisfied for the singular potentials (like the Coulomb potential) that we are going to study in the next sections.

PROOF. Passing to a subsequence we can assume that $\Gamma_{N,t} \to \Gamma_{\infty,t}$ as $N \to \infty$, with respect to the product topology τ_{prod}; this implies immediately that $\gamma_{\infty,0} = |\varphi\rangle\langle\varphi|^{\otimes k}$. To prove (2.33), on the other hand, it is enough to show that for every fixed $k \geq 1$, and for every fixed $J^{(k)}$ from a dense subset of $\mathcal{K}_k$,

$$\text{Tr } J^{(k)}\gamma^{(k)}_{\infty,t} = \text{Tr} J^{(k)}\mathcal{U}^{(k)}(t)\gamma^{(k)}_{\infty,0} + \int_0^t \mathrm{d}s\, \mathcal{U}^{(k)}(t-s) \text{Tr } J^{(k)} B^{(k)} \gamma^{(k+1)}_{\infty,s} \,. \tag{2.34}$$

To demonstrate (2.34), we start from the BBGKY hierarchy (2.12) which leads to the relations

$$\begin{aligned} \text{Tr } J^{(k)}\gamma^{(k)}_{N,t} = \; & \text{Tr } J^{(k)}\mathcal{U}^{(k)}(t)\gamma^{(k)}_{N,0} \\ & + \frac{1}{N}\sum_{j=1}^k \int_0^t \mathrm{d}s \, \text{Tr} J^{(k)}\mathcal{U}^{(k)}(t-s)\left[V(x_i - x_j), \gamma^{(k)}_{N,s}\right] \\ & + \frac{N-k}{N}\int_0^t \mathrm{d}s \, \text{Tr} J^{(k)}\mathcal{U}^{(k)}(t-s) B^{(k)}\gamma^{(k+1)}_{N,s} \,. \end{aligned} \tag{2.35}$$

Since, by assumption, the l.h.s. and the first term on the r.h.s. of the last equation converge, as $N \to \infty$, to the l.h.s. and, respectively, to the first term on the r.h.s. of (2.34) (for every compact operator $J^{(k)}$), (2.33) follows if we can prove that

$$\frac{1}{N}\sum_{j=1}^k \int_0^t \mathrm{d}s \, \text{Tr} J^{(k)}\mathcal{U}^{(k)}(t-s)\left[V(x_i - x_j), \gamma^{(k)}_{N,s}\right] \to 0 \tag{2.36}$$

and that

$$\frac{N-k}{N}\int_0^t \mathrm{d}s \, \text{Tr} J^{(k)}\mathcal{U}^{(k)}(t-s) B^{(k)}\gamma^{(k+1)}_{N,s} \to \int_0^t \mathrm{d}s \text{Tr} J^{(k)}\mathcal{U}^{(k)}(t-s) B^{(k)}\gamma^{(k+1)}_{\infty,s} \tag{2.37}$$

as $N \to \infty$. Eq. (2.36) follows because

$$\left|\text{Tr} J^{(k)}\mathcal{U}^{(k)}(t-s)\left[V(x_i - x_j), \gamma^{(k)}_{N,s}\right]\right| \leq 2\|J^{(k)}\|\|V\|\text{Tr }\left|\gamma^{(k)}_{N,s}\right| \leq 2\|J^{(k)}\|\|V\|$$

is finite, uniformly in N. To prove Eq. (2.37) one can use a similar argument, combined with the observation that

$$\begin{aligned} &\text{Tr } J^{(k)}\mathcal{U}^{(k)}(t-s)B^{(k)}\left(\gamma^{(k+1)}_{N,s} - \gamma^{(k+1)}_{\infty,s}\right) \\ &\quad = k\text{Tr }\left[\left(\mathcal{U}^{(k)}(s-t)J^{(k)}\right)V(x_1 - x_{k+1}) - V(x_1 - x_{k+1})\left(\mathcal{U}^{(k)}(s-t)J^{(k)}\right)\right] \\ &\qquad\qquad \times \left(\gamma^{(k+1)}_{N,s} - \gamma^{(k+1)}_{\infty,s}\right) \to 0 \end{aligned}$$

as $N \to \infty$. This does not follow directly from the assumption that $\Gamma_{N,t} \to \Gamma_{\infty,t}$ with respect to the topology τ_{prod} because the operators $(\mathcal{U}^{(k)}(s-t)J^{(k)})V(x_1 - x_{k+1})$ and $V(x_1 - x_{k+1})(\mathcal{U}^{(k)}(s-t)J^{(k)})$ are not compact on $L^2(\mathbb{R}^{d(k+1)})$. Instead we have to apply an approximation argument, cutting off high momenta in the

x_{k+1}-variable, and using the fact that, by energy conservation, $\text{Tr}\,\nabla^*_{k+1}\gamma^{(k+1)}_{N,t}\nabla_{k+1}$ is bounded, uniformly in N and in t (and that, therefore, $\text{Tr}\,\nabla^*_{k+1}\gamma^{(k+1)}_{\infty,t}\nabla_{k+1}$ is bounded as well). Note that, because of the assumption that $V(x) \to 0$ as $|x| \to \infty$, we only need a cutoff in momentum, and no cutoff in position space is necessary. The details of this approximation argument can be found, for example, in Eq. (7.35) and Eq. (7.36) in the proof of Theorem 7.1 in [**13**] (after replacing δ_β through the bounded potential V). □

Uniqueness: to conclude the proof of Theorem 2.1, we still have to prove the uniqueness of the solution to the infinite hierarchy (2.33).

PROPOSITION 2.4. *Fix* $\Gamma_{\infty,0} = \{\gamma^{(k)}_{\infty,0}\}_{k\geq 1} \in \bigoplus_{k\geq 1}\mathcal{L}^1_k$. *Then there exists at most one solution* $\Gamma_{\infty,t} = \{\gamma^{(k)}_{\infty,t}\}_{k\geq 1} \in \bigoplus_{k\geq 1} C([0,T],\mathcal{L}^k_1)$ *to the infinite hierarchy (2.33) such that* $\gamma^{(k)}_{\infty,t=0} = \gamma^{(k)}_{\infty,0}$ *and* $\text{Tr}|\gamma^{(k)}_{\infty,t}| \leq 1$ *for all* $k \geq 1$ *and all* $t \in [0,T]$.

PROOF. Suppose that $\{\gamma^{(k)}_{\infty,1,t}\}_{k\geq 1}$ and $\{\gamma^{(k)}_{\infty,2,t}\}_{k\geq 1}$ are two solutions of (2.33) with the same initial data $\{\gamma^{(k)}_{\infty,0}\}_{k\geq 1}$, such that $\text{Tr}\,|\gamma^{(k)}_{\infty,i,t}| \leq 1$, for all $k \geq 1$, $t \in [0,T]$, and for $i = 1,2$. Then we can expand $\gamma^{(k)}_{\infty,1,t}$ and $\gamma^{(k)}_{\infty,2,t}$ in the Duhamel series (2.19). It follows that

$$\begin{aligned}\text{Tr}\left|\gamma^{(k)}_{\infty,1,t} - \gamma^{(k)}_{\infty,2,t}\right| \leq \int_0^t \mathrm{d}s_1 \ldots \int_0^{s_{n-1}} \mathrm{d}s_n \\ \times \text{Tr}\left|\mathcal{U}^{(k)}(t-s_1)B^{(k)}\ldots B^{(k+n-1)}\left(\gamma^{(k+n)}_{\infty,1,s_n} - \gamma^{(k+n)}_{\infty,2,s_n}\right)\right|.\end{aligned}$$

Applying recursively the bounds (2.20) and (2.22), we obtain

$$\text{Tr}\left|\gamma^{(k)}_{\infty,1,t} - \gamma^{(k)}_{\infty,2,t}\right| \leq \frac{(k+n-1)!}{(k-1)!n!}(2\|V\|t)^n \leq 2^k\,(4\|V\|t)^n$$

and thus, for $0 < t < 1/8\|V\|$,

$$\text{Tr}\left|\gamma^{(k)}_{\infty,1,t} - \gamma^{(k)}_{\infty,2,t}\right| \leq 2^{k-n}.$$

Since the l.h.s. is independent of $n \geq 1$, it has to vanish. This proves uniqueness for short time. Iterating the same argument, we obtain uniqueness for all times. □

2.4. Derivation of the Hartree Equation for a Coulomb Potential. The arguments presented in Section 2.2 and in Section 2.3 required the interaction potential V to be bounded. Unfortunately, several systems of physical interest are described by unbounded potential. For example, in a non-relativistic approximation, a system of gravitating bosons (a boson star) can be described by the Hamiltonian

$$H_N = \sum_{j=1}^N -\Delta_j - \frac{\lambda}{N}\sum_{i<j}^N \frac{1}{|x_i - x_j|} \tag{2.38}$$

with a singular Coulomb interaction among the particles. The factor of $1/N$ in front of the potential energy can be justified, when describing gravitating particles, by the smallness of the gravitational constant. As in the case of bounded potential, we are interested in the dynamics generated by the Hamiltonian (2.38) on factorized initial N-particle wave functions. We specialize here in the physically most interesting case of particles moving in three dimensions; however, the theorem remains valid in all dimensions $d \geq 2$.

THEOREM 2.5 (Erdős-Yau, [**16**]). *Let $\psi_N = \varphi^{\otimes N}$ for some $\varphi \in H^1(\mathbb{R}^3)$ and let $\psi_{N,t} = e^{-iH_N t}\psi_N$ where the Hamiltonian H_N is defined as in (2.38). Then, for arbitrary $k \geq 1$ and $t \in \mathbb{R}$, we have*

$$\text{Tr} \left| \gamma^{(k)}_{N,t} - |\varphi_t\rangle\langle\varphi_t|^{\otimes k} \right| \to 0 \tag{2.39}$$

as $N \to \infty$. Here φ_t is the solution to the nonlinear Hartree equation

$$i\partial_t \varphi_t = -\Delta\varphi_t - \lambda \left(\frac{1}{|\cdot|} * |\varphi_t|^2 \right) \varphi_t$$

with initial data $\varphi_{t=0} = \varphi$.

Remark. Although, physically, the value of the constant λ is positive (corresponding to the Coulomb attraction among gravitating particles), the theorem remains valid also for negative values of λ (corresponding to repulsive Coulomb interaction).

The general strategy used in [**16**] to prove Theorem 2.5 is the same as the one outlined in Section 2.3. First one proves the compactness of the sequence of marginal $\{\gamma^{(k)}_{N,t}\}_{k=1}^N$ with respect to an appropriate weak topology (the product topology τ_{prod} introduced after (2.30)), then one shows that an arbitrary limit point $\{\gamma^{(k)}_{\infty,t}\}_{k\geq 1}$ of the sequence $\{\gamma^{(k)}_{N,t}\}_{k=1}^N$ is a solution to the infinite hierarchy of equations

$$\gamma^{(k)}_t = \mathcal{U}^{(k)}(t)\gamma^{(k)} + \int_0^t ds\, \mathcal{U}^{(k)}(t-s) B^{(k)} \gamma^{(k+1)}_s \tag{2.40}$$

where $\mathcal{U}^{(k)}$ is the free evolution defined in (2.13), and the collision map $B^{(k)}$ is now given by

$$B^{(k)}\gamma^{(k+1)} = -i\lambda \sum_{j=1}^k \text{Tr}_{k+1} \left[\frac{1}{|x_j - x_{k+1}|}, \gamma^{(k+1)} \right]. \tag{2.41}$$

Finally, one proves the uniqueness of the solution to (2.40). Although the proof of the compactness and of the convergence also require several changes with respect to what we discussed in Section 2.3, the main difficulty one has to face when the bounded potential is replaced by the Coulomb interaction is the proof of the uniqueness of the solution to the infinite hierarchy. The key idea introduced by Erdős and Yau in [**16**] was to restrict the class of densities for which uniqueness must be proven. In Proposition 2.4, uniqueness is proved in the class of densities with $\text{Tr}\,|\gamma^{(k)}_t| \leq 1$ for all $k \geq 1$, and all $t \in [0,T]$ (but the same argument works under the weaker assumption $\text{Tr}\,|\gamma^{(k)}_t| \leq C^k$, for some constant $C < \infty$). Following [**16**], in the case of a Coulomb potential we are only going to show the uniqueness of (2.40) in the class of densities $\Gamma_t = \{\gamma^{(k)}_t\}_{k\geq 1}$ satisfying the a-priori bound

$$\text{Tr} \left| (1-\Delta_1)^{1/2} \dots (1-\Delta_k)^{1/2} \gamma^{(k)}_t (1-\Delta_k)^{1/2} \dots (1-\Delta_1)^{1/2} \right| \leq C^k \tag{2.42}$$

for all $k \geq 1$ and for all $t \in [0,T]$. Note that, for non-negative densities $\gamma^{(k)}_t \geq 0$ (in the sense of operators, that is, in the sense that $\langle \psi^{(k)}, \gamma^{(k)}_t \psi^{(k)} \rangle \geq 0$ for all

$\psi^{(k)} \in L^2(\mathbb{R}^{3k})$) we have

$$\operatorname{Tr} \left|(1-\Delta_1)^{1/2} \dots (1-\Delta_k)^{1/2} \gamma_t^{(k)} (1-\Delta_k)^{1/2} \dots (1-\Delta_1)^{1/2}\right| \\ = \operatorname{Tr} (1-\Delta_1) \dots (1-\Delta_k) \gamma_t^{(k)} .$$

There is, of course, a price to pay in order to restrict the proof of the uniqueness to this class of densities. In fact, to apply this uniqueness result to the proof of Theorem 2.5, one has to show that an arbitrary limit point $\Gamma_{\infty,t} = \{\gamma_{\infty,t}^{(k)}\}_{k\geq 1}$ of the sequence of densities $\Gamma_{N,t} = \{\gamma_{N,t}^{(k)}\}_{k=1}^N$ associated with $\psi_{N,t}$ satisfies the a-priori bound (2.42). Due to the Coulomb singularity, this is actually not so simple and requires an additional approximation argument.

Approximation of the Coulomb singularity: For a fixed $\varepsilon > 0$ we define the regularized Hamiltonian

$$\widetilde{H}_N = \sum_{j=1}^N -\Delta_j - \frac{\lambda}{N} \sum_{i<j}^N \frac{1}{|x_i - x_j| + \varepsilon N^{-1}} . \tag{2.43}$$

Moreover, for a fixed sufficiently small $\delta > 0$, we introduce the regularized initial data

$$\widetilde{\psi}_N = \frac{\chi(\delta \widetilde{H}_N/N)\psi_N}{\|\chi(\delta \widetilde{H}_N/N)\psi_N\|} \qquad \text{(recall that } \psi_N = \varphi^{\otimes N}) \tag{2.44}$$

where $\chi \in C_0^\infty(\mathbb{R})$ is a monotone decreasing function such that $\chi(s) = 1$ for all $s \leq 1$ and $\chi(s) = 0$ for all $s \geq 2$. We consider then the regularized evolution of the regularized initial wave function

$$\widetilde{\psi}_{N,t} = e^{-i\widetilde{H}_N t} \widetilde{\psi}_N .$$

The advantage of working with the regularized wave function $\widetilde{\psi}_{N,t}$ instead of $\psi_{N,t}$ is that it satisfies the following strong a-priori bounds.

PROPOSITION 2.6. *Let $\widetilde{\psi}_{N,t} = e^{-i\widetilde{H}_N t}\widetilde{\psi}_N$, for some fixed $\varepsilon, \delta > 0$. Then there exists a constant $C > 0$ (depending on ε, δ) and, for all $k \geq 1$, there exists $N_0 = N_0(k) > k$ such that*

$$\langle \widetilde{\psi}_{N,t}, (1-\Delta_1) \dots (1-\Delta_k)\, \widetilde{\psi}_{N,t} \rangle \leq C^k \tag{2.45}$$

for all $N \geq N_0$.

Remark. Expressed in terms of the k-particle marginal $\widetilde{\gamma}_{N,t}^{(k)}$ associated with $\widetilde{\psi}_{N,t}$, the bound (2.45) reads

$$\operatorname{Tr}\, (1-\Delta_1) \dots (1-\Delta_k)\, \widetilde{\gamma}_{N,t}^{(k)} \leq C^k . \tag{2.46}$$

We will show Proposition 2.6 below, making use of Proposition 2.7; we will see there that the regularization of the Coulomb singularity and of the initial wave function both play an important role. For the solution $\psi_{N,t}$ of the original Schrödinger equation with the original factorized initial data $\psi_N = \varphi^{\otimes N}$ it is not known whether bounds like (2.45) hold true.

In order for the regularized wave function $\widetilde{\psi}_{N,t}$ to be useful, one needs to prove that it approximates, in an appropriate sense, the wave function $\psi_{N,t}$. To compare the two N-particle wave function, we introduce a third wave function $\widehat{\psi}_{N,t} = e^{-i\widetilde{H}_N t}\psi_N$, and we use the triangle inequality

$$\|\psi_{N,t} - \widetilde{\psi}_{N,t}\| \leq \|\psi_{N,t} - \widehat{\psi}_{N,t}\| + \|\widehat{\psi}_{N,t} - \widetilde{\psi}_{N,t}\|\,. \tag{2.47}$$

The second term is actually independent of time because of the unitarity of the evolution. Using the definition of the regularized initial data $\widetilde{\psi}_N$, one can prove that

$$\|\widehat{\psi}_{N,t} - \widetilde{\psi}_{N,t}\| = \|\psi_N - \widetilde{\psi}_N\| \leq C\delta^{1/2}$$

uniformly in N. To control the first term on the r.h.s. of (2.47), we observe that

$$\frac{\mathrm{d}}{\mathrm{d}t}\left\|\psi_{N,t} - \widehat{\psi}_{N,t}\right\|^2 = 2\mathrm{Im}\left\langle\left(H_N - \widetilde{H}_N\right)\widehat{\psi}_{N,t}, \psi_{N,t} - \widehat{\psi}_{N,t}\right\rangle$$

and thus that

$$\left|\frac{\mathrm{d}}{\mathrm{d}t}\left\|\psi_{N,t} - \widehat{\psi}_{N,t}\right\|^2\right| \leq 2\left\|\left(H_N - \widetilde{H}_N\right)\widehat{\psi}_{N,t}\right\|\left\|\psi_{N,t} - \widehat{\psi}_{N,t}\right\|\,. \tag{2.48}$$

We have

$$\|(H_N - \widetilde{H}_N)\widehat{\psi}_{N,t}\| = \left\|\frac{\varepsilon}{N^2}\sum_{i<j}^N \frac{1}{|x_i - x_j|\left(|x_i - x_j| + \varepsilon N^{-1}\right)}\widehat{\psi}_{N,t}\right\|\,.$$

Using the permutation symmetry of $\widehat{\psi}_{N,t}$, it follows that

$$\begin{aligned}
\|(H_N - &\widetilde{H}_N)\widehat{\psi}_{N,t}\|^2 \\
&\leq \varepsilon^2\Big\langle\widehat{\psi}_{N,t}, \frac{1}{|x_1 - x_2|\left(|x_1 - x_2| + \varepsilon N^{-1}\right)} \\
&\qquad\qquad \times \frac{1}{|x_3 - x_4|\left(|x_3 - x_4| + \varepsilon N^{-1}\right)}\widehat{\psi}_{N,t}\Big\rangle \\
&\quad + \varepsilon^2 N^{-1}\Big\langle\widehat{\psi}_{N,t}, \frac{1}{|x_1 - x_2|\left(|x_1 - x_2| + \varepsilon N^{-1}\right)} \\
&\qquad\qquad \times \frac{1}{|x_2 - x_3|\left(|x_2 - x_3| + \varepsilon N^{-1}\right)}\widehat{\psi}_{N,t}\Big\rangle \\
&\quad + \varepsilon^2 N^{-2}\Big\langle\widehat{\psi}_{N,t}, \frac{1}{|x_1 - x_2|^2\left(|x_1 - x_2| + \varepsilon N^{-1}\right)^2}\widehat{\psi}_{N,t}\Big\rangle
\end{aligned}$$

and thus

$$\begin{aligned}
\|(H_N - \widetilde{H}_N)\widehat{\psi}_{N,t}\|^2 \leq\; & \varepsilon^2\Big\langle\widehat{\psi}_{N,t}, \frac{1}{|x_1 - x_2|^2|x_3 - x_4|^2}\widehat{\psi}_{N,t}\Big\rangle \\
& + \varepsilon^2 N^{-1}\Big\langle\widehat{\psi}_{N,t}, \frac{1}{|x_1 - x_2|^2|x_2 - x_3|^2}\widehat{\psi}_{N,t}\Big\rangle \\
& + \varepsilon^{1/2} N^{-1/2}\Big\langle\widehat{\psi}_{N,t}, \frac{1}{|x_1 - x_2|^{5/2}}\widehat{\psi}_{N,t}\Big\rangle\,.
\end{aligned}$$

Applying Hardy inequalities in the form

$$\frac{1}{|x_i - x_j|^{\alpha}} \leq C(1 - \Delta_i)^{\beta/2}(1 - \Delta_j)^{\gamma/2} \tag{2.49}$$

for every $0 \leq \alpha < 3$, if $\beta + \gamma \geq \alpha$ (see **[9]**[Lemma 9.1] for a proof) we find that

$$\|(H_N - \widetilde{H}_N)\widehat{\psi}_{N,t}\|^2 \leq \varepsilon^2(1+N^{-1})\langle \widehat{\psi}_{N,t}, (1-\Delta_1)(1-\Delta_3)\widehat{\psi}_{N,t}\rangle + \varepsilon^{1/2}N^{-1/2}\langle \widehat{\psi}_{N,t}, (1-\Delta_1)(1-\Delta_2)\widehat{\psi}_{N,t}\rangle$$

and thus, from (2.46),

$$\|(H_N - \widetilde{H}_N)\widehat{\psi}_{N,t}\| \leq C\varepsilon^{1/4}$$

uniformly in N. From (2.48), applying Gronwall's Lemma, it follows that

$$\|\psi_{N,t} - \widehat{\psi}_{N,t}\| \leq C\varepsilon^{1/4}t\,.$$

From (2.47), we obtain that

$$\|\psi_{N,t} - \widetilde{\psi}_{N,t}\| \leq C\left(\varepsilon^{1/4}t + \delta^{1/2}\right)$$

and thus

$$\text{Tr}\left|\gamma^{(k)}_{N,t} - \widetilde{\gamma}^{(k)}_{N,t}\right| \leq C\left(\varepsilon^{1/4}t + \delta^{1/2}\right) \tag{2.50}$$

for every $k \in \mathbb{N}$, uniformly in $N \geq k$. Because of (2.50), it suffices to prove (2.39) with $\gamma^{(k)}_{N,t}$ (the k-particle marginal associated with $\psi_{N,t}$) replaced by $\widetilde{\gamma}^{(k)}_{N,t}$ (the k-particle marginal associated with the regularized wave function $\widetilde{\psi}_{N,t}$) for fixed $\varepsilon, \delta > 0$; at the end (2.39) follows by letting $\varepsilon, \delta \to 0$. Note that in **[16]** a slightly different approximation of the initial data was used; the details of the approximation presented above can be found (for a different model) in **[12]**[Section 5].

Energy estimates: To prove the a-priori bounds of Proposition 2.6, one can use so called energy estimates; these are estimates that compare the expectation of high powers of the Hamiltonian with corresponding powers of the kinetic energy.

PROPOSITION 2.7. *Suppose that $\widetilde{H}_N$ is defined as in (2.43) with $\lambda > 0$ (the case $\lambda < 0$ is simpler). Then there exist constants $C_1 > 1$ and $C_2 > 0$ and, for every $k \geq 1$, there exists an $N_0 = N_0(k) \in \mathbb{N}$ such that*

$$\langle \psi_N, (\widetilde{H}_N + C_1 N)^k\, \psi_N\rangle \geq C_2^k N^k\, \langle \psi_N, (-\Delta_1 + C_1)\dots(-\Delta_k + C_1)\psi_N\rangle \tag{2.51}$$

for every $\psi_N \in L^2_s(\mathbb{R}^{3N})$ (symmetric with respect to permutations) and for every $N > N_0$.

PROOF. Using the operator inequality

$$\frac{1}{|x_i - x_j|} \leq \frac{\pi}{4}\,|\nabla_j|$$

we can find a constant $C_1 > 1$ (depending on the coupling constant λ) such that

$$\frac{\lambda}{|x_i - x_j|} \leq \frac{1}{2}\,(-\Delta_j + C_1) = \frac{1}{2}S_j^2, \tag{2.52}$$

where we defined $S_j = (-\Delta_j + C_1)^{1/2}$. Note also that, for every $0 < \alpha < 3$ there exists a constant $C_\alpha < \infty$ such that

$$\frac{1}{|x_i - x_j|^\alpha} \leq C_\alpha S_j^\alpha\,. \tag{2.53}$$

We are going to prove (2.51) for C_1 fixed as in (2.52), and for an arbitrary $0 < C_2 < 1/2$. The proof is by a two-step induction over $k \geq 0$. For $k = 0$, the claim is trivial. For $k = 1$, it follows from (2.52) because, as an operator inequality on the permutation symmetric space $L^2_s(\mathbb{R}^{3N})$, we have

$$(\widetilde{H}_N + C_1 N) \geq N S_1^2 - \frac{N}{2} \frac{\lambda}{|x_1 - x_2|} \geq C_2 N S_1^2 \,. \tag{2.54}$$

Next we assume that (2.51) holds for all $k \leq n$ and we prove it for $k = n+2$, for an arbitrary $n \in \mathbb{N}$. To this end, we observe that, because of the induction assumption,

$$\begin{aligned}(\widetilde{H}_N + C_1 N)^{n+2} &= (\widetilde{H}_N + C_1 N)(\widetilde{H}_N + C_1 N)^n (\widetilde{H}_N + C_1 N) \\ &\geq C_2^n N^n (\widetilde{H}_N + C_1 N) S_1^2 \dots S_n^2 (\widetilde{H}_N + C_1 N),\end{aligned}$$

for all $N \geq N_0(n)$. Writing

$$(\widetilde{H}_N + C_1 N) = \sum_{j \geq n+1} S_j^2 + h_N, \quad \text{with} \quad h_N = \sum_{j=1}^n S_j^2 - \frac{\lambda}{N} \sum_{i<j}^N \frac{1}{|x_i - x_j| + \varepsilon N^{-1}}$$

it follows that

$$\begin{aligned}(\widetilde{H}_N + C_1 N)^{n+2} \geq\ & C_2^n N^n (N-n)(N-n-1) S_1^2 \dots S_{n+2}^2 \\ & + C_2^n N^n (N-n) S_1^4 S_2^2 \dots S_{n+1}^2 \\ & + C_2^n N^n (N-n) \left(S_1^2 \dots S_{n+1}^2 \, h_N + \text{h.c.} \right) .\end{aligned} \tag{2.55}$$

The first two terms are positive. As for the third term, by the definition of h_N, we find that

$$\begin{aligned}&\left(S_1^2 \dots S_{n+1}^2 \, h_N + \text{h.c.} \right) \\ &\quad \geq - \frac{(N-n)(N-n-1)}{2N} \left(S_1^2 \dots S_{n+1}^2 \frac{\lambda}{|x_{n+2} - x_{n+3}| + \varepsilon N^{-1}} + \text{h.c.} \right) \\ &\quad\quad - \frac{(N-n)n}{N} \left(S_1^2 \dots S_{n+1}^2 \frac{\lambda}{|x_1 - x_{n+2}| + \varepsilon N^{-1}} + \text{h.c.} \right) \\ &\quad\quad - \frac{n(n-1)}{2N} \left(S_1^2 \dots S_{n+1}^2 \frac{\lambda}{|x_1 - x_2| + \varepsilon N^{-1}} + \text{h.c.} \right) .\end{aligned} \tag{2.56}$$

The first term on the r.h.s. of (2.56) can be bounded by

$$\begin{aligned}&\left(S_1^2 \dots S_{n+1}^2 \frac{\lambda}{|x_{n+2} - x_{n+3}| + \varepsilon N^{-1}} + \text{h.c.} \right) \\ &\qquad \leq 2 S_1 \dots S_{n+1} \frac{\lambda}{|x_{n+2} - x_{n+3}|} S_{n+1} \dots S_1 \leq S_1^2 \dots S_{n+2}^2 \,.\end{aligned} \tag{2.57}$$

As for the second term on the r.h.s. of (2.56), we remark that

$$\begin{aligned}
&\left(S_1^2\dots S_{n+1}^2 \frac{\lambda}{|x_1-x_{n+2}|+\varepsilon N^{-1}} + \text{h.c}\right) \\
&= S_{n+1}\dots S_2 \left((-\Delta_1 + C_1)\frac{\lambda}{|x_1-x_{n+2}|+\varepsilon N^{-1}} + \text{h.c.}\right) S_2 \dots S_{n+1} \\
&\leq 2C_1\, S_{n+1}\dots S_2 \frac{\lambda}{|x_1-x_{n+2}|+\varepsilon N^{-1}} S_2\dots S_{n+1} \\
&\quad + 2S_{n+1}\dots S_2\, \nabla_1^* \frac{\lambda}{|x_1-x_{n+2}|+\varepsilon N^{-1}} \nabla_1 S_2 \dots S_{n+1} \\
&\quad + \lambda\, S_{n+1}\dots S_2 \left(\nabla_1^* \frac{(x_1-x_{n+2})}{|x_1-x_{n+2}|\left(|x_1-x_{n+2}|+\varepsilon N^{-1}\right)^2} + \text{h.c.}\right) S_2\dots S_{n+1}\,.
\end{aligned}$$

Applying a Schwarz inequality in the last term, we conclude that there exists a constant $D>0$ (depending on λ) such that

$$\left(S_1^2\dots S_{n+1}^2 \frac{\lambda}{|x_1-x_{n+2}|+\varepsilon N^{-1}} + \text{h.c.}\right) \leq D\, S_1^2 \dots S_{n+2}^2\,. \tag{2.58}$$

Similarly, using the operator inequalities (2.53), the last term on the r.h.s. of (2.56) can be bounded by
(2.59)

$$\left(S_1^2\dots S_{n+1}^2 \frac{1}{|x_1-x_2|+\varepsilon N^{-1}} + \text{h.c.}\right) \leq D\varepsilon^{-1} N\, S_1^2 \dots S_{n+1}^2 + D S_1^4 S_2^2 \dots S_{n+1}^2$$

for all $0<\varepsilon<1$ and for a constant D depending only on λ (it is at this point that the condition $\varepsilon>0$ is needed). Inserting (2.57), (2.58), and (2.59) in the r.h.s. of (2.56), and the resulting bound in the r.h.s. of (2.55), we obtain that there exists $N_0>0$ (depending on n) such that

$$(\widetilde{H}_N + C_1 N)^{n+2} \geq C_2^{n+2} S_1^2 \dots S_{n+2}^2$$

for all $N>N_0$. Note that the value of N_0 also depends on the parameter $\varepsilon>0$. □

Using the result of Proposition 2.7, it is simple to complete the proof of the a-priori bounds for $\widetilde{\psi}_{N,t} = e^{-i\widetilde{H}_N t}\widetilde{\psi}_N$ (recall the definition of the regularized initial data $\widetilde{\psi}_N$ in (2.44)).

PROOF OF PROPOSITION 2.6. From (2.51), and since $C_1>1$, we have

$$\begin{aligned}
\langle \widetilde{\psi}_{N,t}, (1-\Delta_1)\dots(1-\Delta_k)\widetilde{\psi}_{N,t}\rangle &\leq \langle \widetilde{\psi}_{N,t}, (C_1-\Delta_1)\dots(C_1-\Delta_k)\widetilde{\psi}_{N,t}\rangle \\
&\leq \frac{1}{C_2^k N^k}\langle \widetilde{\psi}_{N,t}, (\widetilde{H}_N + C_1 N)^k \widetilde{\psi}_{N,t}\rangle \\
&= \frac{1}{C_2^k N^k}\langle \widetilde{\psi}_{N}, (\widetilde{H}_N + C_1 N)^k \widetilde{\psi}_{N}\rangle
\end{aligned}$$

where in the last line we used the fact that the expectation of any power of $\widetilde{H}_N$ is preserved by the time-evolution. From the definition (2.44) of $\widetilde{\psi}_N$, we immediately obtain (2.45). □

Since the a-priori bounds for $\widetilde{\gamma}^{(k)}_{N,t}$ obtained in Proposition 2.6 hold uniformly in N, they can also be used to derive a-priori bounds on the limit points $\{\gamma^{(k)}_{\infty,t}\}_{k\geq 1}$ of the sequence $\{\gamma^{(k)}_{N,t}\}_{k=1}^N$.

Corollary 2.8. *Suppose that* $\Gamma_{\infty,t} = \{\gamma^{(k)}_{\infty,t}\}_{k\geq 1} \in \bigoplus_{k\geq 1} C([0,T],\mathcal{L}^1_k)$ *is a limit point of the sequence* $\widetilde{\Gamma}_{N,t} = \{\widetilde{\gamma}^{(k)}_{N,t}\}_{k=1}^N$ *with respect to the product topology* τ_{prod} *defined after (2.30). Then* $\gamma^{(k)}_{\infty,t} \geq 0$ *and there exists a constant* C *such that*

$$\text{Tr}\,(1-\Delta_1)\dots(1-\Delta_k)\gamma^{(k)}_{\infty,t} \leq C^k \tag{2.60}$$

for all $k \geq 1$.

Uniqueness: The bounds of Corollary 2.8 are crucial; from (2.60) it follows that it is enough to show the uniqueness of the infinite hierarchy (2.33) in the class of densities satisfying (2.60), a much simpler task than proving uniqueness for all densities with $\text{Tr}\,|\gamma^{(k)}_t| \leq C^k$.

Theorem 2.9. *Fix* $\{\gamma^{(k)}\}_{k\geq 1} \in \bigoplus_{k\geq 1} \mathcal{L}^1_k$. *Then there exists at most one solution* $\{\gamma^{(k)}_t\}_{k\geq 1} \in \bigoplus_{k\geq 1} C([0,T],\mathcal{L}^1_k)$ *to the infinite hierarchy (2.40), such that*

$$\text{Tr}\,\left|(1-\Delta_1)^{1/2}\dots(1-\Delta_k)^{1/2}\gamma^{(k)}_t(1-\Delta_k)^{1/2}\dots(1-\Delta_1)^{1/2}\right| \leq C^k \tag{2.61}$$

for all $k \geq 1$, *and all* $t \in [0,T]$.

Proof. We define the norm

$$\|\gamma^{(k)}\|_{\mathcal{H}_k} = \text{Tr}\,\left|(1-\Delta_1)^{1/2}\dots(1-\Delta_k)^{1/2}\gamma^{(k)}(1-\Delta_k)^{1/2}\dots(1-\Delta_1)^{1/2}\right|$$

and we observe that there exists a constant $C > 0$ with (recall the definition (2.41) for the collision map $B^{(k)}$)

$$\|B^{(k)}\gamma^{(k+1)}\|_{\mathcal{H}_k} \leq Ck\|\gamma^{(k+1)}\|_{\mathcal{H}_{k+1}}\,. \tag{2.62}$$

To prove (2.62), we write

$$\begin{aligned}\|B^{(k)}\gamma^{(k+1)}\|_{\mathcal{H}_k} \leq &\sum_{j=1}^k \text{Tr}\,\left|S_1\dots S_k\left(\text{Tr}_{k+1}\frac{1}{|x_j-x_{k+1}|}\gamma^{(k+1)}\right)S_k\dots S_1\right| \\ &+\sum_{j=1}^k \text{Tr}\,\left|S_1\dots S_k\left(\text{Tr}_{k+1}\,\gamma^{(k+1)}\frac{1}{|x_j-x_{k+1}|}\right)S_k\dots S_1\right|\,.\end{aligned}$$

All terms can be handled similarly. We show how to bound the summand with $j = 1$ on the first line.

$$\begin{aligned}
\operatorname{Tr}&\left|S_1\dots S_k\left(\operatorname{Tr}_{k+1}\frac{1}{|x_1-x_{k+1}|}\gamma^{(k+1)}\right)S_k\dots S_1\right| \\
&= \operatorname{Tr}\left|S_1\dots S_k\left(\operatorname{Tr}_{k+1}S_{k+1}^{-1}\frac{1}{|x_1-x_{k+1}|}S_{k+1}^{-1}S_{k+1}\gamma^{(k+1)}S_{k+1}\right)S_k\dots S_1\right| \\
&\leq \operatorname{Tr}\left|S_1S_{k+1}^{-1}\frac{1}{|x_j-x_{k+1}|}S_{k+1}^{-1}S_1^{-1}S_1\dots S_kS_{k+1}\gamma^{(k+1)}S_{k+1}S_k\dots S_1\right| \\
&\leq \|S_1S_{k+1}^{-1}\frac{1}{|x_1-x_{k+1}|}S_{k+1}^{-1}S_1^{-1}\|\,\|\gamma^{(k+1)}\|_{\mathcal{H}_{k+1}} \\
&\leq C\|\gamma^{(k+1)}\|_{\mathcal{H}_{k+1}}
\end{aligned}$$

where in the second line we used the cyclicity of the partial trace, in the third line we used (2.23) and, in the last line, we used the bound

$$\|S_1S_{k+1}^{-1}\frac{1}{|x_1-x_{k+1}|}S_{k+1}^{-1}S_1^{-1}\| < \infty\,. \tag{2.63}$$

To prove (2.63) we write, assuming for example that $k = 1$,

$$\begin{aligned}
S_1&S_2^{-1}\frac{1}{|x_1-x_2|}S_2^{-1}S_1^{-1} \\
&= S_1^{-1}S_2^{-1}(1-\Delta_1)\frac{1}{|x_1-x_2|}S_2^{-1}S_1^{-1} \\
&= S_1^{-1}S_2^{-1}\frac{1}{|x_1-x_2|}S_2^{-1}S_1^{-1} \\
&\quad + S_1^{-1}S_2^{-1}\nabla_1^*\frac{1}{|x_1-x_2|}\nabla_1S_2^{-1}S_1^{-1} + S_1^{-1}S_2^{-1}\nabla_1^*\frac{(x_1-x_2)}{|x_1-x_2|^3}S_1^{-1}S_2^{-1},
\end{aligned}$$

and we use the norm-estimates $\|\nabla_1S_1^{-1}\| < \infty$ and $\|S_2^{-1}|x_1-x_2|^{-\alpha}S_2^{-1}\| < \infty$ for all $0 \leq \alpha \leq 2$ (by (2.49)).

Suppose now that $\{\gamma_{i,t}^{(k)}\}_{k\geq 1}$, for $i = 1, 2$ are two solutions to the infinite hierarchy (2.40). Using (2.18), we can expand both $\gamma_{1,t}^{(k)}$ and $\gamma_{2,t}^{(k)}$ in a Duhamel series. From (2.62), and from the fact that $\|\mathcal{U}^{(k)}\gamma^{(k)}\|_{\mathcal{H}_k} = \|\gamma^{(k)}\|_{\mathcal{H}_k}$, we obtain that

$$\begin{aligned}
\left\|\gamma_{1,t}^{(k)}-\gamma_{2,t}^{(k)}\right\|_{\mathcal{H}_k} &\leq C^n\frac{(k+n)!}{k!}\int_0^t ds_1\dots\int_0^{s_{n-1}}ds_n\,\|\gamma_{1,s_n}^{(k+n)}-\gamma_{2,s_n}^{(k+n)}\|_{\mathcal{H}_{k+n}} \\
&\leq C^k(Ct)^n
\end{aligned}$$

for any n. Here we used the a-priori bounds (2.61). For $t \leq 1/(2C)$, the l.h.s. must vanish. This shows uniqueness for short time, and thus, by iteration, for all times. □

3. Dynamics of Bose-Einstein Condensates: the Gross-Pitaevskii Equation

Dilute Bose gases at very low temperature are characterized by the macroscopic occupancy of a single one-particle state; a non-vanishing fraction of the total number of particles N is described by the same one-particle orbital. Although this phenomenon, known as Bose-Einstein condensation, has been predicted in the early days of quantum mechanics, the first experimental evidence for its existence was only obtained in 1995, in experiments performed by groups led by Cornell and Wieman at the University of Colorado at Boulder and by Ketterle at MIT (see [**3, 6**]). In these important experiments, atomic gases were initially trapped by magnetic fields and cooled down at very low temperatures. Then the magnetic traps were switched off and the consequent time evolution of the gas was observed; for sufficiently small temperatures, it was observed that the gas coherently moves as a single particle, a clear sign for the existence of condensation.

To describe these experiments from a theoretical point of view, we have, first of all, to give a precise definition of Bose-Einstein condensation. It is simple to understand the meaning of condensation if one considers factorized wave functions, given by the (symmetrization of the) product of one-particle orbitals. In this case, to decide whether we have condensation, we only have to count the number of particles occupying every orbital; if there is a single orbital with macroscopic occupancy the wave function exhibits Bose-Einstein condensation, otherwise it does not. In particular, wave functions of the form $\psi_N(\mathbf{x}) = \prod_{j=1}^N \varphi(x_j)$, for some $\varphi \in L^2(\mathbb{R}^3)$ (we consider in this section three dimensional systems only), exhibit Bose-Einstein condensation; since in these examples all particles occupy the same one-particle orbital, we say that ψ_N exhibits complete Bose-Einstein condensation in the state φ.

Although factorized wave functions were used as initial data in Theorem 2.1 and Theorem 2.5, they are, from a physical point of view, not very satisfactory, because they do not allow for any correlation among the particles. Since we would like to consider systems of interacting particles, the complete absence of correlations is not a realistic assumption. For this reason, we want to give a definition of Bose-Einstein condensation, in particular of complete Bose-Einstein condensation, that applies also to wave functions which are not factorized. To this end, we will make use of the one-particle density $\gamma_N^{(1)}$ associated with an N-particle wave function ψ_N. By definition (see (1.4)), the one-particle density is a non-negative trace class operator on $L^2(\mathbb{R}^3)$ with trace equal to one. It is simple to verify that the eigenvalues of $\gamma_N^{(1)}$ (which are all non-negative and sum up to one) can be interpreted as probabilities for finding particles in the state described by the corresponding eigenvector (a one-particle orbital). This observation justifies the following definition of Bose-Einstein condensation. We will say that a sequence $\{\psi_N\}_{N\in\mathbb{N}}$ with $\psi_N \in L^2_s(\mathbb{R}^{3N})$ exhibits complete Bose-Einstein condensation in the one-particle state with orbital $\varphi \in L^2(\mathbb{R}^3)$ if

$$\text{Tr} \left| \gamma_N^{(1)} - |\varphi\rangle\langle\varphi| \right| \to 0 \tag{3.1}$$

as $N \to \infty$. In particular, complete Bose-Einstein condensation implies that the largest eigenvalue of $\gamma_N^{(1)}$ converges to one, as $N \to \infty$. More generally, we say that a sequence $\{\psi_N\}_{N\in\mathbb{N}}$ exhibits (not necessarily complete) Bose-Einstein condensation if the largest eigenvalue of $\gamma_N^{(1)}$ remains strictly positive in the limit $N \to \infty$. Note that condensation is not a property of a single N-particle wave function ψ_N, but it is a property characterizing a sequence $\{\psi_N\}_{N\in\mathbb{N}}$ in the limit $N \to \infty$.

It is in general very difficult to verify that Bose-Einstein condensation occurs in physically interesting wave functions of interacting systems. There exists, however, a class of interacting systems for which complete condensation of the ground state has been recently established.

In [**24**], Lieb, Yngvason, and Seiringer considered a trapped Bose gas consisting of N three-dimensional particles described by the Hamiltonian

$$H_N^{\text{trap}} = \sum_{j=1}^{N} (-\Delta_j + V_{\text{ext}}(x_j)) + \sum_{i<j}^{N} V_N(x_i - x_j), \tag{3.2}$$

where V_{ext} is an external confining potential with $\lim_{|x|\to\infty} V_{\text{ext}}(x) = \infty$, and $V_N(x) = N^2 V(Nx)$, where V is pointwise positive, spherically symmetric, and rapidly decaying (for simplicity, V can be thought of as being compactly supported). Note that the potential V_N scales with N so that its scattering length is of the order $1/N$ (Gross-Pitaevskii scaling). The scattering length of V is a physical quantity measuring the effective range of the potential; two particles interacting through V see each others, when they are far apart, as hard spheres with radius given by the scattering length of V. More precisely, if f denotes the spherical symmetric solution to the zero-energy scattering equation

$$\left(-\Delta + \frac{1}{2}V(x)\right) f = 0 \qquad \text{with boundary condition } f(x) \to 1 \quad \text{as } |x| \to \infty, \tag{3.3}$$

the scattering length of V is defined by

$$a_0 = \lim_{|x|\to\infty} |x| - |x| f(x)\,.$$

This limit can be proven to exist if V decays sufficiently fast at infinity. Another equivalent characterization of the scattering length is given by

$$8\pi a_0 = \int \mathrm{d}x\, V(x) f(x)\,. \tag{3.4}$$

It is simple to verify that, if f solves (3.3), the rescaled function $f_N(x) = f(Nx)$ solves the zero energy scattering equation with rescaled potential V_N, that is

$$\left(-\Delta + \frac{1}{2}V_N\right) f_N = 0 \qquad \text{with} \qquad f_N(x) \to 1 \quad \text{as } |x| \to \infty\,. \tag{3.5}$$

This implies immediately that the scattering length of V_N is given by $a = a_0/N$, where a_0 is the scattering length of the unscaled potential V. Note that, for $|x| \gg a$, $f_N(x) \simeq 1 - a/|x|$. For $|x| < a$, f_N remains bounded; for practical purposes, we can think of this function as $f_N(x) \simeq 1 - a/(|x| + a)$.

Letting $N \to \infty$, Lieb, Yngvason, and Seiringer showed that the ground state energy $E(N)$ of (3.2) divided by the number of particle N converges to

$$\lim_{N\to\infty} \frac{E(N)}{N} = \min_{\varphi \in L^2(\mathbb{R}^3):\, \|\varphi\|=1} \mathcal{E}_{\text{GP}}(\varphi)$$

where $\mathcal{E}_{\text{GP}}$ is the Gross-Pitaevskii energy functional

$$\mathcal{E}_{\text{GP}}(\varphi) = \int \mathrm{d}x\, \left(|\nabla\varphi(x)|^2 + V_{\text{ext}}(x)|\varphi(x)|^2 + 4\pi a_0 |\varphi(x)|^4\right)\,. \tag{3.6}$$

Later, in [**22**], Lieb and Seiringer also proved that the ground state of the Hamiltonian (3.2) exhibits complete Bose-Einstein condensation into the minimizer of the Gross-Pitaevskii energy functional $\mathcal{E}_{\text{GP}}$. More precisely they showed that, if

ψ_N is the ground state wave function of the Hamiltonian (3.2) and if $\gamma_N^{(1)}$ denotes the corresponding one-particle marginal, then

$$\gamma_N^{(1)} \to |\phi_{\rm GP}\rangle\langle\phi_{\rm GP}| \qquad \text{as } N \to \infty\,, \tag{3.7}$$

where $\phi_{\rm GP} \in L^2(\mathbb{R}^3)$ is the minimizer of the Gross-Pitaevskii energy functional (3.6).

To describe the experiments mentioned above, it is important to understand the time-evolution of the Bose-Einstein condensate after removing the external traps. We define therefore the translation invariant Hamiltonian

$$H_N = \sum_{j=1}^N -\Delta_j + \sum_{i<j}^N V_N(x_i - x_j) \tag{3.8}$$

and we consider solutions to the N-particle Schrödinger equation

$$i\partial_t \psi_{N,t} = H_N \psi_{N,t} \quad \Rightarrow \quad \psi_{N,t} = e^{-iH_N t}\psi_N \tag{3.9}$$

with initial data ψ_N exhibiting complete Bose-Einstein condensation. In a series of joint articles with L. Erdős and H.-T. Yau, see [**12, 13, 14, 15**], we prove that, for every fixed time $t \in \mathbb{R}$, the evolved N-particle wave function $\psi_{N,t}$ still exhibits complete Bose-Einstein condensation. Moreover we show that the time evolution of the condensate wave function evolves according to the one-particle time-dependent Gross-Pitaevskii equation associated with the energy functional $\mathcal{E}_{\rm GP}$. Our main result is the following theorem.

THEOREM 3.1. *Suppose that $V \geq 0$ is spherically symmetric and $V(x) \leq C\langle x\rangle^{-\sigma}$, for some $\sigma > 5$, and for all $x \in \mathbb{R}^3$. Assume that the family $\{\psi_N\}_{N\in\mathbb{N}}$ with $\psi_N \in L^2_s(\mathbb{R}^{3N})$ and $\|\psi_N\| = 1$ for all N, has finite energy per particle, that is*

$$\langle \psi_N, H_N \psi_N\rangle \leq CN \tag{3.10}$$

for all $N \in \mathbb{N}$, and that it exhibits complete Bose-Einstein condensation in the sense that

$$\text{Tr} \left| \gamma_N^{(1)} - |\varphi\rangle\langle\varphi| \right| \to 0 \tag{3.11}$$

as $N \to \infty$ for some $\varphi \in L^2(\mathbb{R}^3)$. Then, for every $k \geq 1$ and $t \in \mathbb{R}$, we have

$$\text{Tr} \left| \gamma_{N,t}^{(k)} - |\varphi_t\rangle\langle\varphi_t|^{\otimes k} \right| \to 0 \tag{3.12}$$

as $N \to \infty$. Here φ_t is the solution of the nonlinear Gross-Pitaevskii equation

$$i\partial_t \varphi_t = -\Delta\varphi_t + 8\pi a_0 |\varphi_t|^2 \varphi_t \tag{3.13}$$

with initial data $\varphi_{t=0} = \varphi$.

Making use of an approximation of the initial N-particle wave function (similarly to (2.44)), it is possible to replace the assumption (3.10) with the much more stringent condition

$$\langle \psi_N, H_N^k \psi_N\rangle \leq C^k N^k \tag{3.14}$$

for all $k \in \mathbb{N}$. In the following we will illustrate the main ideas involved in the proof of Theorem 3.1, assuming the initial wave function ψ_N to satisfy (3.14).

3.1. Comparison with Mean-Field Systems. The formal relation of the Hamiltonian (3.8) with the mean-field Hamiltonian (2.1) considered in Section 2 is evident; (3.8) can in fact be rewritten as

$$H_N = \sum_{j=1}^{N} -\Delta_j + \frac{1}{N} \sum_{i<j} v_N(x_i - x_j) \tag{3.15}$$

with $v_N(x) = N^3 V(Nx)$. Since we are considering three dimensional systems, v_N converges to a delta-function in the limit of large N; at least formally, as $N \to \infty$, we have $v_N(x) \to b_0 \delta(x)$, where $b_0 = \int V(x) \mathrm{d}x$. In other words, the Hamiltonian (3.8) in the Gross-Pitaevskii scaling can be formally interpreted as a mean field Hamiltonian with an N-dependent potential which converges, as $N \to \infty$, to a δ-function. Despite the formal similarity, it should be stressed that the physics described by the Gross-Pitaevskii Hamiltonian is completely different from the physics described by the mean field Hamiltonian (2.1). In a mean-field system, each particle typically interacts with all other particles through a very weak potential. The Gross-Pitaevskii Hamiltonian (3.8), on the other hand, describes a very dilute gas, where interactions are very rare and at the same time very strong. Although the physics described by (2.1) and (3.8) are completely different, due to the formal similarity of the two models, we may try to apply the strategy discussed in Section 2.3 to prove Theorem 3.1. In other words, we may try to prove Theorem 3.1 by showing the compactness of the sequence $\Gamma_{N,t} = \{\gamma_{N,t}^{(k)}\}_{k=1}^N$ with respect to an appropriate weak topology (it is going to be the same topology introduced in Section 2.3), the convergence to an infinite hierarchy similar to (2.33), and the uniqueness of the solution to the infinite hierarchy. It turns out that it is indeed possible to extend the general strategy introduced in Section 2.3 to prove Theorem 3.1; however, as we will see, many important modifications of the arguments used for bounded or for Coulomb potential are required. We discuss next the main changes.

First of all, the simple observation that, formally, $v_N(x) \to b_0 \delta(x)$ as $N \to \infty$ may lead to the conclusion that the evolution of the condensate wave function φ_t should be described by the nonlinear Hartree equation (2.9) with V replaced by $b_0 \delta$, that is by the equation

$$i\partial_t \varphi_t = -\Delta \varphi_t + b_0 (\delta * |\varphi_t|^2) \varphi_t = -\Delta \varphi_t + b_0 |\varphi_t|^2 \varphi_t \,. \tag{3.16}$$

Comparing with (3.13), we note that (3.16) is characterized by a different coupling constant in front of the nonlinearity. The emergence of the scattering length in the Gross-Pitaevskii equation (3.13) is the consequence of a subtle interplay between the N-dependent interaction potential and the short scale correlation structure developed by the solution of the N-particle Schrödinger equation $\psi_{N,t}$ (as we will see, the correlation structure varies on lengths of the order $1/N$, the same lengthscale characterizing the interaction potential). This remark implies that, in order to prove the convergence to the infinite hierarchy (where the coupling constant $8\pi a_0$ already appears), we will need to identify the singular correlation structure of $\psi_{N,t}$ (which is then inherited by the marginal densities $\gamma_{N,t}^{(k)}$). This is one of the main difficulties in the proof of the convergence, which was completely absent in the analysis of mean-field systems presented in Section 2; we will discuss it in more details in Section 3.2 and in Section 3.3.

The presence of the correlation structure in $\psi_{N,t}$ also affects the proof of a-priori bounds

$$\text{Tr}\,(1-\Delta_1)\dots(1-\Delta_k)\gamma^{(k)}_{\infty,t} \leq C^k \tag{3.17}$$

for all $k \geq 1$, $t \in \mathbb{R}$, for the limit points $\Gamma_{\infty,t} = \{\gamma^{(k)}_{\infty,t}\}_{k\geq 1}$ of the marginal densities $\Gamma_{N,t} = \{\gamma^{(k)}_{N,t}\}_{k=1}^N$ associated with $\psi_{N,t}$. As in the case of a Coulomb potential discussed in Section 2.4, these a-priori bounds play a fundamental role because they allow us to restrict the proof of the uniqueness to a smaller class of densities. In Section 2.4, we derived the a-priori bounds for $\gamma^{(k)}_{\infty,t}$ making use of the estimates (2.46) which hold uniformly in N, for N sufficiently large (in turns, the bounds (2.46) were obtained through the energy estimates of Proposition 2.7). In the present setting, however, bounds of the form

$$\text{Tr}\,(1-\Delta_1)\dots(1-\Delta_k)\gamma^{(k)}_{N,t} \leq C^k \tag{3.18}$$

cannot hold uniformly in N, because of the short scale correlation structure developed by the solution of the Schrödinger equation. Remember in fact that the short scale structure varies on the length scale $1/N$; therefore, when we take derivatives of $\gamma^{(k)}_{N,t}$ as in (3.18), we cannot expect to obtain bounds uniform in N (unless we take only one derivative, because of energy conservation). Although the marginals $\gamma^{(k)}_{N,t}$, for large but finite N, do not satisfy the strong estimates (3.18), it turns out that one can still prove the a-priori bound (3.17) on the limit point $\gamma^{(k)}_{\infty,t}$. This is indeed possible because in the weak limit $N \to \infty$, the singular short scale correlation structure characterizing the marginal densities $\gamma^{(k)}_{N,t}$ disappears, producing limit points $\gamma^{(k)}_{\infty,t}$ which are much more regular than the densities $\gamma^{(k)}_{N,t}$. Because of the absence of estimates of the form (3.18) for $\gamma^{(k)}_{N,t}$, the proof of the a-priori bounds for the limit points $\gamma^{(k)}_{\infty,t}$ requires completely new ideas with respect to what has been discussed in Section 2.4; we briefly discuss the most important ones in Section 3.4.

Finally, the singularity of the interaction potential strongly affects the proof of the uniqueness of the solution to the infinite hierarchy. In Section 2.4, the main idea to prove the uniqueness of the infinite hierarchy was to expand the solution in a Duhamel series and to control all Coulomb potentials appearing in the expansion through Laplacians acting on appropriate variables and at the end to control the expectation of the Laplacians through the a-priori bounds (2.46) on the densities $\gamma^{(k)}_{\infty,t}$. In this argument, it was very important that the Coulomb potential can be controlled by the kinetic energy, in the sense of the operator inequality

$$\frac{1}{|x|} \leq C(1-\Delta)\,. \tag{3.19}$$

In the present setting, the Coulomb potential has to be replaced by a δ-function. In three dimensions, the δ-potential cannot be controlled by the kinetic energy. In other words, the bound

$$\delta(x) \leq C(1-\Delta)^{\alpha}$$

is not true for $\alpha = 1$; it only holds if $\alpha > 3/2$ (in three dimensions, the L^∞ norm of a function can be controlled by the H^α-norm, only if $\alpha > 3/2$). This observation implies that the a-priori bounds (3.17) are not sufficient to conclude the proof of the uniqueness of the infinite hierarchy with delta-interaction (while similar bounds were enough to prove the uniqueness of the infinite hierarchy with Coulomb potential). Since it does not seem possible to improve the a-priori bounds to gain control of higher derivatives (one would need more than 3/2 derivatives per particle), we need new techniques to prove the uniqueness of the infinite hierarchy. We will briefly discuss these new methods in Section 3.5.

3.2. Convergence to the Infinite Hierarchy. The goal of this section is to discuss the main ideas used to prove the next proposition which identifies limit points of the sequence $\Gamma_{N,t} = \{\gamma^{(k)}_{N,t}\}_{k=1}^N$ as solutions to a certain infinite hierarchy of equations (this proposition replaces Proposition 2.3, which was stated for mean-field systems with bounded interaction potential).

PROPOSITION 3.2. *Suppose that $V \geq 0$, with $V(x) \leq C\langle x\rangle^{-\sigma}$, for some $\sigma > 5$, and for all $x \in \mathbb{R}^3$. Assume that the sequence ψ_N satisfies (3.11) and the additional assumption (3.14). Fix $T > 0$ and let $\Gamma_{\infty,t} = \{\gamma^{(k)}_{\infty,t}\}_{k\geq 1} \in \bigoplus_{k\geq 1} C([0,T], \mathcal{L}^1_k)$ be a limit point of $\Gamma_{N,t} = \{\gamma^{(k)}_{N,t}\}_{k=1}^N$ (with respect to the product topology τ_{prod} defined in Section 2.3). Then $\Gamma_{\infty,t}$ is a solution to the infinite hierarchy*

(3.20)
$$\gamma^{(k)}_{\infty,t} = \mathcal{U}^{(k)}(t)\gamma^{(k)}_{\infty,0} - 8\pi a_0 i \sum_{j=1}^k \int_0^t \mathrm{d}s\, \mathcal{U}^{(k)}(t-s)\, Tr_{k+1}\left[\delta(x_j - x_{k+1}), \gamma^{(k+1)}_{\infty,s}\right]$$

with initial data $\gamma^{(k)}_{\infty,0} = |\varphi\rangle\langle\varphi|^{\otimes k}$ (see (2.13) for the definition of $\mathcal{U}^{(k)}$).

The detailed proof of this proposition can be found in [**15**, Theorem 8.1] (for small interaction potential, see also [**13**, Theorem 7.1]).

To prove the proposition, we start by studying the time-evolution of the marginal densities $\gamma^{(k)}_{N,t}$, which is governed by the BBGKY hierarchy. In integral form, the BBGKY hierarchy is given by

$$\begin{aligned} \gamma^{(k)}_{N,t} &= \mathcal{U}^{(k)}(t)\gamma^{(k)}_{N,0} - i\sum_{i<j}\int_0^t \mathrm{d}s\, \mathcal{U}^{(k)}(t-s)\left[V_N(x_i - x_j), \gamma^{(k)}_{N,s}\right] \\ &\quad - i(N-k)\sum_{j=1}^k \int_0^t \mathrm{d}s\, \mathcal{U}^{(k)}(t-s)\,\mathrm{Tr}_{k+1}\left[V_N(x_j - x_{k+1}), \gamma^{(k+1)}_{N,s}\right]. \end{aligned} \tag{3.21}$$

Assuming (by passing to an appropriate subsequence) that $\Gamma_{N,t} \to \Gamma_{\infty,t}$ as $N \to \infty$ with respect to the product topology τ_{prod} introduced in Section 2.3, it is simple to prove that the l.h.s. and the first term on the r.h.s. of (3.21) converge, as $N \to \infty$, to the l.h.s. and, respectively, to the first term on the r.h.s. of (3.20). The second term on the r.h.s. of (3.21), on the other hand, can be proven to vanish in the limit $N \to \infty$ (at least formally, this follows by the observation that the second term is smaller by a factor of N w.r.t. the third term). The fact that the second term on the r.h.s. of (3.21) is negligible in the limit $N \to \infty$ (compared with the third term) corresponds to the physical intuition that the interactions among the first k particles affect their time-evolution less than their interaction with the other $(N-k)$ particles.

To conclude the proof of Proposition 3.2, we only need to show that the third term on the r.h.s. of (3.21) converges, as $N \to \infty$, to the last term on the r.h.s.

of (3.20). As already remarked in Section 3.1, this convergence relies critically on the correlation structure characterizing the $(k+1)$-particle density $\gamma_{N,t}^{(k+1)}$. A naive approach, based on the observation that $(N-k)V_N(x_j - x_{k+1}) \simeq N^3 V(N(x_j - x_{k+1})) \simeq b_0 \delta(x_j - x_{k+1})$ for large N, fails to explain the coupling constant in front of the last term on the r.h.s. of (3.20). The emergence of the scattering length can only be understood by taking into account the correlation structure of $\gamma_{N,t}^{(k+1)}$. Assuming for a moment that the correlations can be described, in good approximation, by the solution f_N to the zero-energy scattering equation (3.5), we can expect that, for large N,

$$\gamma_{N,t}^{(k+1)}(\mathbf{x}_{k+1}; \mathbf{x}'_{k+1}) \simeq f_N(x_j - x_{k+1}) \gamma_{\infty,t}^{(k+1)}(\mathbf{x}_{k+1}; \mathbf{x}'_{k+1}) \tag{3.22}$$

in the region where $x_j - x_{k+1}$ is of the order $1/N$ (and all other variables are at larger distances). Assuming some regularity of the limit point $\gamma_{\infty,t}^{(k+1)}$, and using (3.4), the approximation (3.22) immediately leads to

$$\begin{aligned} &\left(\operatorname{Tr}_{k+1}(N-k)V_N(x_j - x_{k+1})\gamma_{N,t}^{(k+1)}\right)(\mathbf{x}_k; \mathbf{x}'_k) \\ &\quad \simeq \int \mathrm{d}x_{k+1}\, N^3 V(N(x_j - x_{k+1})) f(N(x_j - x_{k+1})) \gamma_{\infty,t}^{(k+1)}(\mathbf{x}_k, x_{k+1}; \mathbf{x}'_k, x_{k+1}) \\ &\quad = \int \mathrm{d}y\, V(y) f(y) \gamma_{\infty,t}^{(k+1)}\left(\mathbf{x}_k, x_j + \frac{y}{N}; \mathbf{x}'_k, x_j + \frac{y}{N}\right) \\ &\quad \simeq \left(\int \mathrm{d}y\, V(y) f(y)\right) \gamma_{\infty,t}^{(k+1)}(\mathbf{x}_k, x_j; \mathbf{x}'_k, x_j) \\ &\quad = 8\pi a_0 \int \mathrm{d}x_{k+1}\, \delta(x_j - x_{k+1}) \gamma_{\infty,t}^{(k+1)}(\mathbf{x}_k, x_{k+1}; \mathbf{x}'_k, x_{k+1}) \end{aligned} \tag{3.23}$$

and thus explains the emergence of the scattering length on the r.h.s. of (3.20) (note that the third term on r.h.s. of (3.21) is a commutator and thus produces two summands; in (3.23) we only consider one of these terms, the other can be handled analogously). This heuristic argument shows that in order to prove Proposition 3.2 we need to identify the short scale structure of the marginal densities and prove that it can be described by the function f_N as in (3.22). To this end we are going to use energy estimates. In [**13**] and [**15**], we developed two different approaches to this problem. The first approach is simpler, but it only works for sufficiently small interaction potentials. The second approach is a little bit more involved, but it can be used for all potentials satisfying the assumptions of Theorem 3.1. In the following we will focus on the first, simpler, approach; in the next subsection, we present the main ideas of the second approach.

To measure the strength of the interaction potential V, we define the dimensionless constant

$$\rho = \sup_{x \in \mathbb{R}^3} |x|^2 V(x) + \int \frac{\mathrm{d}x}{|x|} V(x). \tag{3.24}$$

PROPOSITION 3.3. *Assume that the potential V satisfies the conditions of Theorem 3.1, and suppose that $\rho > 0$ is sufficiently small. Then there exists $C > 0$ such that*

$$\langle \psi, H_N^2 \psi \rangle \geq C N^2 \int \mathrm{d}\mathbf{x} \left| \nabla_i \nabla_j \frac{\psi(\mathbf{x})}{f_N(x_i - x_j)} \right|^2 \tag{3.25}$$

for all $i \neq j$ and for all $\psi \in L^2_s(\mathbb{R}^{3N}, \mathrm{d}\mathbf{x})$.

This energy estimate, combined with the assumption (3.14) on the initial wave function ψ_N, leads to the following a-priori bounds on the solution $\psi_{N,t} = e^{-iH_N t}\psi_N$ of the Schrödinger equation (3.9).

COROLLARY 3.4. *Assume that V satisfies the conditions of Theorem 3.1, and suppose that $\rho > 0$ is sufficiently small. Suppose that ψ_N satisfies (3.10) and (3.14). Then we have*

$$\int \mathrm{d}\mathbf{x} \left| \nabla_i \nabla_j \frac{\psi_{N,t}(\mathbf{x})}{f_N(x_i - x_j)} \right|^2 \leq C \tag{3.26}$$

for all $i \neq j$, uniformly in $N \in \mathbb{N}$ and in $t \in \mathbb{R}$. Therefore, if $\gamma^{(k)}_{N,t}$ denotes the k-particle marginal associated with $\psi_{N,t}$, we have, for every $1 \leq i, j \leq k$ with $i \neq j$,

$$\mathrm{Tr}\, (1 - \Delta_i)(1 - \Delta_j) \frac{1}{f_N(x_i - x_j)} \gamma^{(k)}_{N,t} \frac{1}{f_N(x_i - x_j)} \leq C$$

uniformly in $N \in \mathbb{N}$ and in $t \in \mathbb{R}$.

PROOF. Using (3.25), the conservation of the energy along the time evolution, and the assumption (3.14) on the initial wave function ψ_N, we find

$$\int \mathrm{d}\mathbf{x} \left| \nabla_i \nabla_j \frac{\psi_{N,t}(\mathbf{x})}{f_N(x_i - x_j)} \right|^2 \leq CN^{-2} \langle \psi_{N,t}, H_N^2 \psi_{N,t} \rangle = CN^{-2} \langle \psi_N, H_N^2 \psi_N \rangle \leq C.$$

□

Remark that the a-priori bounds (3.26) cannot hold true if we do not divide the solution $\psi_{N,t}$ of the Schrödinger equation by $f_N(x_i - x_j)$. In fact, using that $f_N(x) \simeq 1 - a_0/(N|x| + 1)$, it is simple to check that

$$\int \mathrm{d}x \, |\nabla^2 f_N(x)|^2 \simeq N \, .$$

This implies that, if we replace $\psi_{N,t}(\mathbf{x})/f_N(x_i - x_j)$ by $\psi_N(\mathbf{x})$ the integral in (3.26) would be of order N. Only after removing the singular factor $f_N(x_i - x_j)$ from $\psi_{N,t}(\mathbf{x})$ we can obtain useful bounds on the regular part of the wave function (regular in the variable $(x_i - x_j)$). These a-priori bounds allow us to identify the correlation structure of the wave function $\psi_{N,t}$ and to show that, when x_i and x_j are close to each other, $\psi_{N,t}(\mathbf{x})$ can be approximated by the time independent correlation factor $f_N(x_i - x_j)$, which varies on the length scale $1/N$, multiplied with a regular part (which only varies on scales of order one). In other words, the bounds (3.26) establish a strong separation of scales for the solution $\psi_{N,t}$ of the N-particle Schrödinger equation, and for its marginal densities; on length scales of order $1/N$, $\psi_{N,t}$ is characterized by a singular, time independent, short scale correlation structure described by the the solution f_N to the zero-energy scattering equation. On scales of order one, on the other hand, the wave function $\psi_{N,t}$ is regular, and, as it follows from Theorem 3.1, it can be approximated, in an appropriate sense, by products of the solution to the time-dependent Gross-Pitaevskii equation. Remark that although the short-scale correlation structure is time independent, it still affects, in a non-trivial way, the time-evolution on length scales of order one (because it produces the scattering length in the Gross-Pitaevskii equation).

PROOF OF PROPOSITION 3.3. We decompose the Hamiltonian (3.8) as

$$H_N = \sum_{j=1}^N h_j \qquad \text{with} \qquad h_j = -\Delta_j + \frac{1}{2}\sum_{i\neq j} V_N(x_i - x_j)\,.$$

For an arbitrary permutation symmetric wave function ψ and for any fixed $i \neq j$, we have

$$\langle \psi, H_N^2 \psi\rangle = N\langle \psi, h_i^2\psi\rangle + N(N-1)\langle\psi, h_i h_j \psi\rangle \geq N(N-1)\langle \psi, h_i h_j \psi\rangle\,.$$

Using the positivity of the potential, we find

$$\langle \psi, H_N^2\psi\rangle \geq N(N-1)\left\langle \psi, \left(-\Delta_i + \frac{1}{2}V_N(x_i - x_j)\right)\left(-\Delta_j + \frac{1}{2}V_N(x_i-x_j)\right)\psi\right\rangle\,. \tag{3.27}$$

Next, we define $\phi(\mathbf{x})$ by $\psi(\mathbf{x}) = f_N(x_i - x_j)\,\phi(\mathbf{x})$ (ϕ is well defined because $f_N(x) > 0$ for all $x \in \mathbb{R}^3$); note that the definition of the function ϕ depends on the choice of i,j. Then

$$\begin{aligned}\frac{1}{f_N(x_i-x_j)}&\Delta_i\,(f_N(x_i-x_j)\phi(\mathbf{x})) \\ &= \Delta_i\phi(\mathbf{x}) + \frac{(\Delta f_N)(x_i-x_j)}{f_N(x_i-x_j)}\phi(\mathbf{x}) + 2\frac{\nabla f_N(x_i-x_j)}{f_N(x_i-x_j)}\nabla_i\phi(\mathbf{x})\,.\end{aligned}$$

From (3.3) it follows that

$$\frac{1}{f_N(x_i-x_j)}\left(-\Delta_i + \frac{1}{2}V_N(x_i-x_j)\right) f_N(x_i-x_j)\phi(\mathbf{x}) = L_i\phi(\mathbf{x})$$

and analogously

$$\frac{1}{f_N(x_i-x_j)}\left(-\Delta_j + \frac{1}{2}V_N(x_i-x_j)\right) f_N(x_i-x_j)\phi(\mathbf{x}) = L_j\phi(\mathbf{x})$$

where we defined

$$L_\ell = -\Delta_\ell + 2\frac{\nabla_\ell f_N(x_i-x_j)}{f_N(x_i-x_j)}\nabla_\ell, \qquad \text{for} \qquad \ell = i,j\,.$$

Remark that, for $\ell = i,j$, the operator L_ℓ satisfies

$$\begin{aligned}\int \mathrm{d}\mathbf{x}\, f_N^2(x_i-x_j)\, L_\ell\overline{\phi}(\mathbf{x})\,\psi(\mathbf{x}) &= \int \mathrm{d}\mathbf{x}\, f_N^2(x_i-x_j)\,\overline{\phi}(\mathbf{x})\, L_\ell\,\psi(\mathbf{x}) \\ &= \int \mathrm{d}\mathbf{x}\, f_N^2(x_i-x_j)\,\nabla_\ell\overline{\phi}(\mathbf{x})\,\nabla_\ell\,\psi(\mathbf{x})\,.\end{aligned}$$

Therefore, from (3.27), we obtain

(3.28)

$$\begin{aligned}\langle\psi, H_N^2\psi\rangle &\geq N(N-1)\int \mathrm{d}\mathbf{x}\, f_N^2(x_i-x_j)\, L_i\,\overline{\phi}(\mathbf{x})\, L_j\,\phi(\mathbf{x})\\ &= N(N-1)\int \mathrm{d}\mathbf{x}\, f_N^2(x_i-x_j)\, \nabla_i\overline{\phi}(\mathbf{x})\, \nabla_i L_j\,\phi(\mathbf{x})\\ &= N(N-1)\int \mathrm{d}\mathbf{x}\, f_N^2(x_i-x_j)\, \nabla_i\overline{\phi}(\mathbf{x})\, L_j\,\nabla_i\phi(\mathbf{x})\\ &\quad + N(N-1)\int \mathrm{d}\mathbf{x}\, f_N^2(x_i-x_j)\, \nabla_i\overline{\phi}(\mathbf{x})\, [\nabla_i, L_j]\phi(\mathbf{x})\\ &= N(N-1)\int \mathrm{d}\mathbf{x}\, f_N^2(x_i-x_j)\, |\nabla_j\nabla_i\phi(\mathbf{x})|^2\\ &\quad + N(N-1)\int \mathrm{d}\mathbf{x}\, f_N^2(x_i-x_j)\, \left(\nabla_i\frac{\nabla f_N(x_i-x_j)}{f_N(x_i-x_j)}\right)\, \nabla_i\overline{\phi}(\mathbf{x})\,\nabla_j\phi(\mathbf{x})\,.\end{aligned}$$

To control the second term on the right hand side of the last equation we use bounds on the function f_N, which can be derived from the zero energy scattering equation (3.3):

$$1 - C\rho \leq f_N(x) \leq 1, \quad |\nabla f_N(x)| \leq C\frac{\rho}{|x|}, \quad |\nabla^2 f_N(x)| \leq C\frac{\rho}{|x|^2} \tag{3.29}$$

for constants C independent of N and of the potential V (recall the definition of the dimensionless constant ρ from (3.24)). Therefore, for $\rho < 1$,

$$\begin{aligned}\Big|\int \mathrm{d}\mathbf{x}\, f_N^2(x_i-x_j)\, &\left(\nabla_i\frac{\nabla f_N(x_i-x_j)}{f_N(x_i-x_j)}\right)\, \nabla_i\overline{\phi}(\mathbf{x})\,\nabla_j\phi(\mathbf{x})\Big|\\ &\leq C\rho\int \mathrm{d}\mathbf{x}\,\frac{1}{|x_i-x_j|^2}\,|\nabla_i\phi(\mathbf{x})|\,|\nabla_j\phi(\mathbf{x})|\\ &\leq C\rho\int \mathrm{d}\mathbf{x}\,\frac{1}{|x_i-x_j|^2}\,\left(|\nabla_i\phi(\mathbf{x})|^2 + |\nabla_j\phi(\mathbf{x})|^2\right)\\ &\leq C\rho\int \mathrm{d}\mathbf{x}\,|\nabla_i\nabla_j\phi(\mathbf{x})|^2\end{aligned}$$

where we used Hardy inequality. Thus, from (3.28), and using again the first bound in (3.29), we obtain

$$\langle\psi, H_N^2\psi\rangle \geq N(N-1)(1-C\rho)\int \mathrm{d}\mathbf{x}\,|\nabla_i\nabla_j\phi(\mathbf{x})|^2$$

which implies (3.25). □

Equipped with the a-priori bounds of Corollary 3.4, we can now come back to the problem of proving the convergence of the last term on the r.h.s. of (3.21) to the last term on the r.h.s. of (2.6). For simplicity, we consider the case $k = 1$, and we only discuss the term with the interaction potential on the left of the density (the commutator also has a term with the interaction on the right of the density, which can be handled analogously). After multiplying with a smooth one-particle observable $J^{(1)}$ (a compact operator on $L^2(\mathbb{R}^3)$, with sufficiently smooth kernel), we need to prove that

$$\mathrm{Tr}\, \left(\mathcal{U}^{(1)}(s-t)J^{(1)}\right)\left(N^3V(N(x_1-x_2))\gamma^{(2)}_{N,t} - 8\pi a_0\delta(x_1-x_2)\gamma^{(2)}_{\infty,t}\right) \to 0$$

as $N\to\infty$. To this end we decompose the difference in several terms. We use the notation $J^{(1)}_t = \mathcal{U}^{(1)}(t)J^{(1)}$, and, for a bounded function $h(x)\geq 0$ with $\int \mathrm{d}x\, h(x) = 1$, we define $h_\alpha(x) = \alpha^{-3}h(\alpha^{-1}x)$ for all $\alpha>0$. Then we have

(3.30)

$$\begin{aligned}
&\mathrm{Tr}\, \left(\mathcal{U}^{(1)}(s-t)J^{(1)}\right)\left(N^3V(N(x_1-x_2))\gamma^{(2)}_{N,t} - 8\pi a_0\delta(x_1-x_2)\gamma^{(2)}_{\infty,t}\right)\\
&= \mathrm{Tr}\, J^{(1)}_{s-t}\, N^3V(N(x_1-x_2))f(N(x_1-x_2))\\
&\qquad\qquad \times \frac{1}{f(N(x_1-x_2))}\gamma^{(2)}_{N,t}\frac{1}{f(N(x_1-x_2))}(f(N(x_1-x_2))-1)\\
&\quad + \mathrm{Tr}\, J^{(1)}_{s-t}\left(N^3V(N(x_1-x_2))f(N(x_1-x_2)) - 8\pi a_0\delta(x_1-x_2)\right)\\
&\qquad\qquad \times \frac{1}{f(N(x_1-x_2))}\gamma^{(2)}_{N,t}\frac{1}{f(N(x_1-x_2))}\\
&\quad + 8\pi a_0\, \mathrm{Tr}\, J^{(1)}_{s-t}\left(\delta(x_1-x_2)-h_\alpha(x_1-x_2)\right)\frac{1}{f(N(x_1-x_2))}\gamma^{(2)}_{N,t}\frac{1}{f(N(x_1-x_2))}\\
&\quad + 8\pi a_0\, \mathrm{Tr}\, J^{(1)}_{s-t}h_\alpha(x_1-x_2)\left(\frac{1}{f(N(x_1-x_2))}\gamma^{(2)}_{N,t}\frac{1}{f(N(x_1-x_2))} - \gamma^{(2)}_{N,t}\right)\\
&\quad + 8\pi a_0\, \mathrm{Tr}\, J^{(1)}_{s-t}h_\alpha(x_1-x_2)\left(\gamma^{(2)}_{N,t}-\gamma^{(2)}_{\infty,t}\right)\\
&\quad + 8\pi a_0\, \mathrm{Tr}\, J^{(1)}_{s-t}\left(h_\alpha(x_1-x_2)-\delta(x_1-x_2)\right)\gamma^{(2)}_{\infty,t}\,.
\end{aligned}$$

The idea here is that in order to compare the N-dependent potential $N^3V(N(x_1-x_2))$ with the limiting δ-potential, we have to test it against a regular density (using an appropriate Poincaré inequality). For this reason, we first regularize the density $\gamma^{(2)}_{N,t}$ in the variable (x_1-x_2) dividing it by the correlation function $f_N(x_1-x_2)$ on the left and the right (first term on the r.h.s. of the last equation). Using the regularity of $f_N^{-1}(x_1-x_2)\gamma^{(2)}_{N,t}f_N^{-1}(x_1-x_2)$ from Corollary 3.4, we can then compare, in the regime of large N, the interaction potential with the delta-function (second term on the r.h.s.). At this point we are still not done, because, in order to remove the regularizing factors $f_N^{-1}(x_1-x_2)$ (fourth term on the r.h.s. of (3.30)) and in order to replace the density $\gamma^{(2)}_{N,t}$ by its limit point $\gamma^{(2)}_{\infty,t}$ (fifth term on the r.h.s. of (3.30)), we need to test the density against a compact observable. For this reason, in the third term on the r.h.s. of (3.30), we replace the δ-function (which is of course not bounded) by the function h_α which approximate the delta-function on the length scale α; it is important here that α is now decoupled from N. In the last term, after removing all the N dependence, we go back to the δ-potential using the regularity of the limiting density $\gamma^{(2)}_{\infty,t}$.

To control the first and fourth term on the r.h.s. of (3.30), we use the fact that $1-f_N(x_1-x_2)\simeq 1/(N|x_1-x_2|+1)$ varies on a length scale of order $1/N$. It follows that the first term converges to zero as $N\to\infty$, as well as the fourth term, for every fixed $\alpha>0$. To estimate the second, the third and the last term, we make use of appropriate Poincaré inequalities, combined with the result of Corollary 3.4

and, for the last term, of Proposition 3.7 (we present an example of a Poincaré inequality, which can be used to estimate these terms in Appendix A). It follows that the second term converges to zero as $N \to \infty$, and that the third and the fifth terms converge to zero as $\alpha \to 0$, uniformly in N. Finally, the fifth term on the r.h.s. of (3.30) converges to zero as $N \to \infty$, for every fixed α; this follows from thc assumption that $\gamma^{(2)}_{N,t} \to \gamma^{(2)}_{\infty,t}$ as $N \to \infty$ with respect to the weak* topology (some additional work has to be done here, because the operator $J^{(1)}_{s-t} h_\alpha(x_1 - x_2)$ is not compact). Therefore, if we first fix $\alpha > 0$ and let $N \to \infty$ and then we let $\alpha \to 0$ all terms on the r.h.s. of (3.30) converge to zero; this concludes the proof of Proposition 3.2.

3.3. Convergence for Large Interaction Potentials. As pointed out in Section 3.2, the energy estimate given in Proposition 3.3, which was a crucial ingredient for the proof of Proposition 3.2, only holds for sufficiently small potentials (for sufficiently small values of the parameter ρ defined in (3.24)). For large potentials, we need a different approach. The new technique, developed in [**15**], is based on the use of the wave operator associated with the one-particle Hamiltonian $\mathfrak{h}_N = -\Delta + (1/2)V_N$, defined through the strong limit

$$W_N = s - \lim_{t\to\infty} e^{i\mathfrak{h}_N t} e^{i\Delta t} . \tag{3.31}$$

Under the assumptions of Theorem 3.1 on the potential V, it is simple to show that the limit (3.31) exists, that the wave operator W_N is complete, in the sense that

$$W_N^{-1} = W_N^* = s - \lim_{t\to\infty} e^{-i\Delta t} e^{-i\mathfrak{h}_N t} ,$$

and that it satisfies the intertwining relation

$$W_N^* \, \mathfrak{h} \, W_N = -\Delta . \tag{3.32}$$

It is also important to observe that the wave operator W_N is related by simple scaling to the wave operator W associated with the one-particle Hamiltonian $\mathfrak{h} = -\Delta + (1/2)V$ (and defined analogously to (3.31)). In fact, if $W_N(x; x')$ and $W(x; x')$ denote the kernels of W_N and, respectively, of W, we have

$$W_N(x; x') = N^3 W(Nx; Nx') \qquad \text{and} \qquad W_N^*(x; x') = N^3 W^*(Nx; Nx') .$$

In particular this implies that the norm of W_N, as an operator from $L^p(\mathbb{R}^3)$ to $L^p(\mathbb{R}^3)$, for arbitrary $1 \le p \le \infty$, is independent of N. From the work of Yajima, see [**32, 33**], we know that, under the conditions on V assumed in Theorem 3.1, W is a bounded operator from $L^p(\mathbb{R}^3)$ to $L^p(\mathbb{R}^3)$, for all $1 \le p \le \infty$. Therefore

$$\|W_N\|_{L^p \to L^p} = \|W\|_{L^p\to L^p} < \infty \qquad \text{for all } 1 \le p \le \infty .$$

In the following we will denote by $W_{N,(i,j)}$ the wave operator W_N acting only on the relative variable $x_j - x_i$. In other words, the action of $W_{N,(i,j)}$ on a N-particle wave function $\psi_N \in L^2(\mathbb{R}^{3N})$ is given by

$$\begin{aligned} &\left(W_{N,(i,j)} \psi_N\right)(\mathbf{x}) \\ &= \int \mathrm{d}v \, W_N(x_j - x_i; v) \, \psi_N \left(x_1, \dots, \frac{x_i + x_j}{2} + \frac{v}{2}, \dots, \frac{x_i + x_j}{2} - \frac{v}{2}, \dots, x_N\right) \end{aligned} \tag{3.33}$$

if $j < i$ (the formula for $i > j$ is similar). Similarly, we define $W^*_{N,(i,j)}$. Using the wave operator we have the following energy estimate, which replaces Proposition 3.3, and whose proof can be found in [**15**, Proposition 5.2].

PROPOSITION 3.5. *Suppose $V \geq 0$, $V \in L^1(\mathbb{R}^3) \cap L^2(\mathbb{R}^3)$ and $V(x) = V(-x)$ for all $x \in \mathbb{R}^3$. Then we have, for every $i \neq j$,*

$$\langle \psi_N, H_N^2 \psi_N \rangle \geq CN^2 \int \mathrm{d}\mathbf{x} \, \left| (\nabla_i \cdot \nabla_j) \, W^*_{N,(i,j)} \psi_N \right|^2 . \tag{3.34}$$

From Proposition 3.5, we obtain immediately an a-priori bound on $\psi_{N,t}$ and on its marginal densities.

COROLLARY 3.6. *Assume that V satisfies the conditions of Theorem 3.1. Suppose that ψ_N satisfies (3.10) and (3.14). Then we have, for all $i \neq j$,*

$$\int \mathrm{d}\mathbf{x} \, \left| (\nabla_i \cdot \nabla_j) \, W^*_{N,(i,j)} \psi_{N,t}(\mathbf{x}) \right|^2 \leq C \tag{3.35}$$

uniformly in $N \in \mathbb{N}$ and $t \in \mathbb{R}$. Therefore, if $\gamma^{(k)}_{N,t}$ denote the k-particle marginal associated with $\psi_{N,t}$, we have, for every $1 \leq i, j \leq k$ with $i \neq j$,

$$\mathrm{Tr} \, \left((\nabla_i \cdot \nabla_j)^2 - \Delta_i - \Delta_j + 1 \right) W^*_{N,(i,j)} \gamma^{(k)}_{N,t} W_{N,(i,j)} \leq C$$

uniformly in $N \in \mathbb{N}$ and in $t \in \mathbb{R}$.

The philosophy of the bounds (3.35) and (3.26) is the same; first we have to regularize the wave function $\psi_{N,t}$, and then we can prove useful bounds on its derivatives. There are however important differences. In (3.26) we regularized $\psi_{N,t}$ in position space, by factoring out the short scale correlation structure $f_N(x_i - x_j)$. In (3.35), instead, we regularize $\psi_{N,t}$ applying the wave operator $W^*_{N,(i,j)}$. Another important difference is that (3.35) is weaker than (3.26); in fact, (3.35) only gives a control on the combination $\sum_{\alpha=1}^3 \partial_{x_{i,\alpha}} \partial_{x_{j,\alpha}}$, while (3.26) controls $\partial_{x_{i,\alpha}} \partial_{x_{j,\beta}}$ for all $1 \leq \alpha, \beta \leq 3$. The weakness of the bound (3.35) makes the proof of the convergence more difficult. In particular we have to establish new Poincaré inequalities, which only require control of the inner product $\nabla_i \cdot \nabla_j$. It turns out that the weaker control provided by (3.35) is still enough to conclude the proof of convergence to the infinite hierarchy (Proposition 3.2). For more details, see [**15**, Section 8].

3.4. A-Priori Estimates on Limit Points $\Gamma_{\infty,t}$. In this section we present some of the arguments involved in the proof of the a-priori bounds (3.17).

PROPOSITION 3.7. *Assume that V satisfies the conditions of Theorem 3.1. Suppose ψ_N satisfies (3.10) and (3.14). Let $\Gamma_{\infty,t} = \{\gamma^{(k)}_{\infty,t}\}_{k\geq 1} \in \bigoplus_{k\geq 1} C([0,T], \mathcal{L}^1_k)$ be a limit point of the sequence $\Gamma_{N,t} = \{\gamma^{(k)}_{N,t}\}_{k=1}^N$ with respect to the product topology τ_{prod} defined in Section 2.3. Then $\gamma^{(k)}_{\infty,t} \geq 0$ and there exists a constant C such that*

$$\mathrm{Tr} \, (1 - \Delta_1) \dots (1 - \Delta_k) \gamma^{(k)}_{\infty,t} \leq C^k \tag{3.36}$$

for all $k \geq 1$ and $t \in [0,T]$.

The main difficulty in proving Proposition 3.7 is the fact that the estimate (3.36) does not hold true if we replace $\gamma^{(k)}_{\infty,t}$ with the marginal density $\gamma^{(k)}_{N,t}$. More precisely,

$$\mathrm{Tr} \, (1 - \Delta_1) \dots (1 - \Delta_k) \gamma^{(k)}_{N,t} \leq C^k \tag{3.37}$$

cannot hold true with a constant C independent of N. In fact, for finite N and $k > 1$, the k-particle density $\gamma_{N,t}^{(k)}$ still contains the singular short scale correlation structure. For example, when particle one and particle two are very close to each other (at distances of order $1/N$), we can expect the two-particle density to be approximately given by

$$\gamma_{N,t}^{(2)}(\mathbf{x}_2, \mathbf{x}_2') \simeq \text{const}\, f_N(x_1 - x_2) f_N(x_1' - x_2')$$

(the constant part takes into account factors which vary on larger scales). It is then simple to check that

$$\text{Tr}\,(1 - \Delta_1)(1 - \Delta_2)\gamma_{N,t}^{(2)} \simeq N\,.$$

Only after taking the weak limit $N \to \infty$, the short scale correlation structure disappears (because it varies on a length scale of order $1/N$), and one can hope to prove bounds like (3.36).

To overcome this problem, we cutoff the wave function $\psi_{N,t}$ when two or more particles come at distances smaller than some intermediate length scale ℓ, with $N^{-1} \ll \ell \ll 1$ (more precisely, the cutoff will be effective only when one or more particles come close to one of the variable x_j over which we want to take derivatives). For fixed $j = 1, \ldots, N$, we define $\theta_j \in C^\infty(\mathbb{R}^{3N})$ such that

$$\theta_j(\mathbf{x}) \simeq \begin{cases} 1 & \text{if } \;|x_i - x_j| \gg \ell \quad \text{for all } i \neq j \\ 0 & \text{if there exists } i \neq j \quad \text{with} \quad |x_i - x_j| \lesssim \ell \end{cases}.$$

It is important, for our analysis, that θ_j controls its derivatives (in the sense that, for example, $|\nabla_i \theta_j| \leq C\ell^{-1}\theta_j^{1/2}$); for this reason we cannot use standard compactly supported cutoffs. Instead we have to construct appropriate functions which decay exponentially when particles come close together (the prototype of such function is $\theta(x) = \exp[-\ell^{-\varepsilon} \exp(-\sqrt{(x/\ell)^2 + 1})])$. Making use of the functions $\theta_j(\mathbf{x})$, we prove the following higher order energy estimates.

PROPOSITION 3.8. *Choose $\ell \ll 1$ such that $N\ell^2 \gg 1$. Then there exist constants C_1 and C_2 such that, for any $\psi \in L_s^2(\mathbb{R}^{3N})$,*

$$\langle \psi, (H_N + C_1 N)^k \psi \rangle \geq C_2 N^k \int \mathrm{d}\mathbf{x}\, \theta_1(\mathbf{x}) \ldots \theta_{k-1}(\mathbf{x})\, |\nabla_1 \ldots \nabla_k \psi(\mathbf{x})|^2\,. \tag{3.38}$$

The meaning of the bound (3.38) is clear. The L^2-norm of the k-th derivative $\nabla_1 \ldots \nabla_k \psi$ can be controlled by the expectation of the k-th power of the energy per particle, if we restrict the integration domain to regions where the first $(k-1)$ particles are "isolated" (in the sense that there is no particle at distances smaller than ℓ from $x_1, x_2, \ldots, x_{k-1}$).

Note that we can allow one "free derivative"; in (3.38) we take the derivative over x_k although there is no cutoff $\theta_k(\mathbf{x})$. The reason is that the correlation structure becomes singular, in the L^2-sense, only when we derive it twice (if one uses the zero energy solution f_N introduced in (3.3) to describe the correlations, this can be seen by observing that $|\nabla f_N(x)| \leq 1/|x|$, which is locally square integrable). Remark that the condition $N\ell^2 \gg 1$ is necessary to control the error due to the localization of the kinetic energy on distances of order ℓ. The proof of Proposition 3.8 is based on induction over k; for details see Section 7 in [**15**].

From the estimates (3.38), using the preservation of the expectation of H_N^k along the time evolution and the condition (3.14), we obtain the following bounds for the solution $\psi_{N,t} = e^{-iH_N t}\psi_N$ of the Schrödinger equation (3.9).

$$\int d\mathbf{x}\, \theta_1(\mathbf{x}) \dots \theta_{k-1}(\mathbf{x})\; |\nabla_1 \dots \nabla_k \psi_{N,t}(\mathbf{x})|^2 \leq C^k$$

uniformly in N and t, and for all $k \geq 1$. Translating these bounds in the language of the density matrix $\gamma_{N,t}$, we obtain

$$\text{Tr}\; \theta_1 \dots \theta_{k-1} \nabla_1 \dots \nabla_k \gamma_{N,t} \nabla_1^* \dots \nabla_k^* \leq C^k \,. \tag{3.39}$$

The idea now is to use the freedom in the choice of the cutoff length ℓ. If we fix the position of all particles but x_j, it is clear that the cutoff θ_j is effective at most in a volume of the order $N\ell^3$. If we choose ℓ such that $N\ell^3 \to 0$ as $N \to \infty$ (which is of course compatible with the condition that $N\ell^2 \gg 1$), we can expect that, in the limit of large N, the cutoff becomes negligible. This approach yields in fact the desired results; starting from (3.39), and choosing ℓ such that $N\ell^3 \ll 1$, we can complete the proof of Proposition 3.7 (see Proposition 6.3 in [**13**] for more details).

3.5. Uniqueness of the Solution to the Infinite Hierarchy. To complete the proof of Theorem 3.1 we have to prove the uniqueness of the solution to the infinite hierarchy (3.20) in the class of densities satisfying the a-priori bounds (3.36). Remark that the uniqueness of the infinite hierarchy (3.20), in a different class of densities, was recently proven by Klainerman and Machedon in [**25**]. The proof proposed by Klainerman and Machedon is simpler than the proof of Proposition 3.9 which we discuss below. Unfortunately, the result of [**25**] cannot be applied to the proof of Theorem 3.1, because it is not yet clear whether limit points of the sequence of marginal densities $\Gamma_{N,t} = \{\gamma_{N,t}^{(k)}\}_{k=1}^N$ fit into the class of densities for which uniqueness is proven.

Proposition 3.9. *Fix $T > 0$ and $\Gamma = \{\gamma^{(k)}\}_{k\geq 1} \in \bigoplus_{k\geq 1} \mathcal{L}_k^1$. Then there exists at most one solution $\Gamma_t = \{\gamma_t^{(k)}\}_{k\geq 1} \in \bigoplus C([0,T], \mathcal{L}_k)$ of the infinite hierarchy (3.20) with $\Gamma_{t=0} = \Gamma$, such that $\gamma_t^{(k)} \geq 0$ is symmetric with respect to permutations, and*

$$Tr (1-\Delta_1) \dots (1-\Delta_k)\, \gamma_t^{(k)} \leq C^k \tag{3.40}$$

for all $k \geq 1$ and all $t \in [0,T]$.

In this section we briefly explain some of the main steps involved in the proof of Proposition 3.9; the details can be found in [**12**][Section 9].

To shorten the notation, we write the infinite hierarchy (3.20) in the form

$$\gamma_t = \mathcal{U}^{(k)}(t)\gamma_0 + \int_0^t ds\, \mathcal{U}^{(k)}(t-s)\, B^{(k)} \gamma_s^{(k+1)} \,, \tag{3.41}$$

where $\mathcal{U}^{(k)}(t)$ denotes the free evolution of k particles

$$\mathcal{U}^{(k)}(t)\gamma^{(k)} = e^{it\sum_{j=1}^k \Delta_j} \gamma^{(k)} e^{-it\sum_{j=1}^k \Delta_j}$$

and the collision operator $B^{(k)}$ maps $(k+1)$-particle operators into k-particle operators according to

$$B^{(k)}\gamma^{(k+1)} = -8i\pi a_0 \sum_{j=1}^{k} \mathrm{Tr}_{k+1}\left[\delta(x_j - x_{k+1}), \gamma^{(k+1)}\right]. \tag{3.42}$$

The map $B^{(k)}$ is defined as in Section 2; in particular the kernel of $B^{(k)}\gamma^{(k+1)}$ is given by the expression on the r.h.s. of (2.11), with $V(x)$ replaced by $8\pi a_0\delta(x)$.

Iterating (3.41) n times we obtain the Duhamel type series

$$\gamma_t^{(k)} = \mathcal{U}^{(k)}(t)\gamma_0^{(k)} + \sum_{m=1}^{n-1} \xi_{m,t}^{(k)} + \eta_{n,t}^{(k)} \tag{3.43}$$

with

(3.44)

$$\begin{aligned}\xi_{m,t}^{(k)} = & \int_0^t \mathrm{d}s_1 \ldots \int_0^{s_{m-1}} \mathrm{d}s_m\, \mathcal{U}^{(k)}(t-s_1)B^{(k)}\mathcal{U}^{(k+1)}(s_1-s_2)B^{(k+1)}\ldots \\ & \times B^{(k+m-1)}\mathcal{U}^{(k+m)}(s_m)\gamma_0^{(k+m)} \\ = & \sum_{j_1=1}^{k}\sum_{j_2=1}^{k+1}\cdots\sum_{j_m=1}^{k+m}\int_0^t \mathrm{d}s_1\ldots\int_0^{s_{m-1}} \mathrm{d}s_m\, \mathcal{U}^{(k)}(t-s_1) \\ & \times \mathrm{Tr}_{k+1}\Big[\delta(x_{j_1}-x_{k+1}), \mathcal{U}^{(k+1)}(s_1-s_2)\mathrm{Tr}_{k+2}\Big[\delta(x_{j_2}-x_{k+2}),\ldots \\ & \times \mathrm{Tr}_{k+m}\left[\delta(x_{j_m}-x_{k+m}), \mathcal{U}^{(k+m)}(s_m)\gamma_0^{(k+m)}\right]\ldots\Big]\Big]\end{aligned}$$

and the error term

(3.45)

$$\begin{aligned}\eta_{n,t}^{(k)} = \int_0^t \mathrm{d}s_1 \int_0^{s_1}\mathrm{d}s_2\ldots\int_0^{s_{n-1}}\mathrm{d}s_n\, \mathcal{U}^{(k)}(t-s_1)B^{(k)}\mathcal{U}^{(k+1)}(s_1-s_2)B^{(k+1)}\ldots \\ \ldots B^{(k+n-1)}\gamma_{s_n}^{(k+m)}.\end{aligned}$$

Note that the error term (3.45) has exactly the same form as the terms in (3.44), with the only difference that the last free evolution is replaced by the full evolution $\gamma_{s_n}^{(k+m)}$.

To prove the uniqueness of the infinite hierarchy, it is enough to prove that the fully expanded terms (3.44) are well-defined and that the error term (3.45) converges to zero as $n \to \infty$ (in some norm, or even after testing it against a sufficiently large class of smooth observables). The main problem here is that the delta function in the collision operator $B^{(k)}$ cannot be controlled by the kinetic energy (in the sense that, in three dimensions, the operator inequality $\delta(x) \leq C(1-\Delta)$ does not hold true). For this reason, the a-priori estimates (3.40) are not sufficient to show that (3.45) converges to zero, as $n \to \infty$. Instead, we have to make use of the smoothing effects of the free evolutions $\mathcal{U}^{(k+j)}(s_j - s_{j+1})$ in (3.45) (in a similar way, Stricharzt estimates are used to prove the well-posedness of nonlinear Schrödinger equations). To this end, we rewrite each term in the series (3.43) as a sum of contributions associated with certain Feynman graphs, and then we prove the convergence of the Duhamel expansion by controlling each contribution separately.

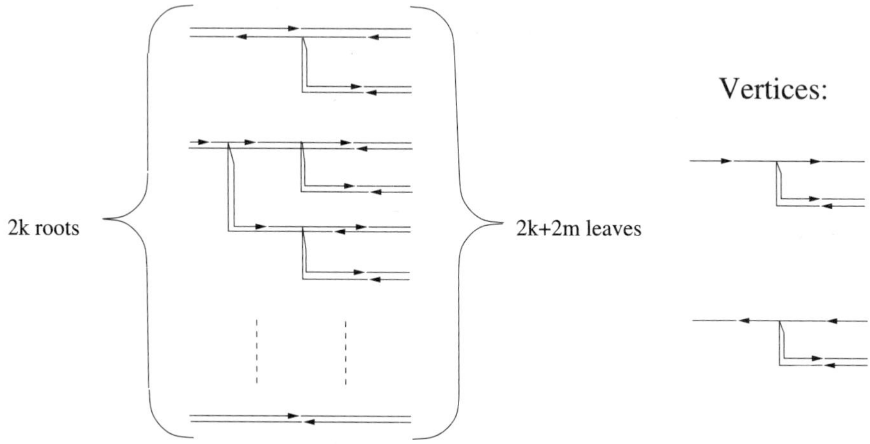

FIGURE 1. A Feynman graph in $\mathcal{F}_{m,k}$ and its two types of vertices

The details of the diagrammatic expansion can be found in [**12**, Section 9]. Here we only sketch the main ideas. We start by considering the term $\xi_{m,t}^{(k)}$ in (3.44). After multiplying it with a compact k-particle observable $J^{(k)}$ and taking the trace, we expand the result as

$$\operatorname{Tr} J^{(k)} \xi_{m,t}^{(k)} = \sum_{\Lambda \in \mathcal{F}_{m,k}} K_{\Lambda,t} \tag{3.46}$$

where $K_{\Lambda,t}$ is the contribution associated with the Feynman graph Λ. Here $\mathcal{F}_{m,k}$ denotes the set of all graphs consisting of $2k$ disjoint, paired, oriented, and rooted trees with m vertices. An example of a graph in $\mathcal{F}_{m,k}$ is drawn in Figure 1. Each vertex has one of the two forms drawn in Figure 1, with one "father"-edge on the left (closer to the root of the tree) and three "son"-edges on the right. One of the son edge is marked (the one drawn on the same level as the father edge; the other two son edges are drawn below). Graphs in $\mathcal{F}_{m,k}$ have $2k+3m$ edges, $2k$ roots (the edges on the very left), and $2k+2m$ leaves (the edges on the very right). It is possible to show that the number of different graphs in $\mathcal{F}_{m,k}$ is bounded by 2^{4m+k}.

The particular form of the graphs in $\mathcal{F}_{m,k}$ is due to the quantum mechanical nature of the expansion; the presence of a commutator in the collision operator (3.42) implies that, for every $B^{(k+j)}$ in (3.44), we can choose whether to write the interaction on the left or on the right of the density. When we draw the corresponding vertex in a graph in $\mathcal{F}_{m,k}$, we have to choose whether to attach it on the incoming or on the outgoing edge.

Graphs in $\mathcal{F}_{m,k}$ are characterized by a natural partial ordering among the vertices ($v \prec v'$ if the vertex v is on the path from v' to the roots); there is, however, no total ordering. The absence of total ordering among the vertices is the consequence of a rearrangement of the summands on the r.h.s. of (3.44); by removing the order between times associated with non-ordered vertices we substantially reduce the number of terms in the expansion. In fact, while (3.44) contains $(m+k)!/k!$ summands, in (3.46) we are only summing over at most 2^{4m+k} contributions. The price we have to pay is that the apparent gain of a factor $1/m!$ due to the ordering of the time integrals in (3.44) is lost in the new expansion (3.46). However, since we want to use the time integrations to smooth out singularities it seems quite difficult to make use of this factor $1/m!$. In fact, we find that the expansion (3.46) is better suited for analyzing the cumulative space-time smoothing effects of the multiple free evolutions than (3.44).

Because of the pairing of the $2k$ trees, there is a natural pairing between the $2k$ roots of the graph. Moreover, it is also possible to define a natural pairing of the leaves of the graph (this is evident in Figure 1); two leaves ℓ_1 and ℓ_2 are paired if there exists an edge e_1 on the path from ℓ_1 back to the roots, and an edge e_2 on the path from ℓ_2 to the roots, such that e_1 and e_2 are the two unmarked son-edges of the same vertex (or, in case there is no unmarked sons in the path from ℓ_1 and ℓ_2 to the roots, if the two roots connected to ℓ_1 and ℓ_2 are paired).

For $\Lambda \in \mathcal{F}_{m,k}$, we denote by $E(\Lambda)$, $V(\Lambda)$, $R(\Lambda)$ and $L(\Lambda)$ the set of all edges, vertices, roots and, respectively, leaves in the graph Λ. For every edge $e \in E(\Lambda)$, we introduce a three-dimensional momentum variable p_e and a one-dimensional frequency variable α_e. Then, denoting by $\widehat{\gamma}_0^{(k+m)}$ and by $\widehat{J}^{(k)}$ the kernels of the density $\gamma_0^{(k+m)}$ and of the observable $J^{(k)}$ in Fourier space, the contribution $K_{\Lambda,t}$ in (3.46) is given by

(3.47)

$$K_{\Lambda,t} = \int \prod_{e\in E(\Lambda)} \frac{\mathrm{d}p_e \mathrm{d}\alpha_e}{\alpha_e - p_e^2 + i\tau_e\mu_e} \prod_{v\in V(\Lambda)} \delta\left(\sum_{e\in v} \pm\alpha_e\right) \delta\left(\sum_{e\in v} \pm p_e\right)$$
$$\times \exp\left(-it \sum_{e\in R(\Lambda)} \tau_e(\alpha_e + i\tau_e\mu_e)\right) \widehat{J}^{(k)}\left(\{p_e\}_{e\in R(\Lambda)}\right) \widehat{\gamma}_0^{(k+m)}\left(\{p_e\}_{e\in L(\Lambda)}\right).$$

Here $\tau_e = \pm 1$, according to the orientation of the edge e. We observe from (3.47) that the momenta of the roots of Λ are the variables of the kernel of $J^{(k)}$, while the momenta of the leaves of Λ are the variables of the kernel of $\gamma_0^{(k+m)}$ (this also explain why roots and leaves of Λ need to be paired).

The denominators $(\alpha_e - p_e^2 + i\tau_e\mu_e)^{-1}$ are called propagators; they correspond to the free evolutions in the expansion (3.44) and they enter the expression (3.47) through the formula

$$e^{itp^2} = \int_{-\infty}^{\infty} \mathrm{d}\alpha \, \frac{e^{it(\alpha+i\mu)}}{\alpha - p^2 + i\mu}$$

(here and in (3.47) the measure $\mathrm{d}\alpha$ is defined by $\mathrm{d}\alpha = \mathrm{d}'\alpha/(2\pi i)$ where $\mathrm{d}'\alpha$ is the Lebesgue measure on $\mathbb{R}$).

The regularization factors μ_e in (3.47) have to be chosen such that $\mu_{\text{father}} = \sum_{e=\text{ son}} \mu_e$ at every vertex. The delta-functions in (3.47) express momentum and frequency conservation (the sum over $e \in v$ denotes the sum over all edges adjacent to the vertex v; here $\pm\alpha_e = \alpha_e$ if the edge points towards the vertex, while $\pm\alpha_e = -\alpha_e$ if the edge points out of the vertex, and analogously for $\pm p_e$).

An analogous expansion can be obtained for the error term $\eta_{n,t}^{(k)}$ in (3.45). The problem now is to analyze the integral (3.47) (and the corresponding integral for the error term). Through an appropriate choice of the regularization factors μ_e one can extract the time dependence of $K_{\Lambda,t}$ and show that

$$\begin{aligned}|K_{\Lambda,t}| \leq C^{k+m}\, t^{m/4} \int \prod_{e\in E(\Gamma)} \frac{\mathrm{d}\alpha_e \mathrm{d}p_e}{\langle \alpha_e - p_e^2\rangle} \prod_{v\in V(\Gamma)} \delta\left(\sum_{e\in v} \pm\alpha_e\right) \delta\left(\sum_{e\in v} \pm p_e\right) \\ \times \left| \widehat{J}^{(k)}\left(\{p_e\}_{e\in R(\Gamma)}\right)\right| \left| \widehat{\gamma}_0^{(k+m)}\left(\{p_e\}_{e\in L(\Gamma)}\right)\right|\end{aligned} \tag{3.48}$$

where we introduced the notation $\langle x\rangle = (1+x^2)^{1/2}$.

Because of the singularity of the interaction at zero, we may be faced here with an ultraviolet problem; we have to show that all integrations in (3.48) are finite in the regime of large momenta and large frequency. Because of (3.40), we know that the kernel $\widehat{\gamma}_0^{(k+m)}(\{p_e\}_{e\in L(\Lambda)})$ in (3.48) provides decay in the momenta of the leaves. From (3.40) we have, in momentum space,

$$\int \mathrm{d}p_1 \ldots \mathrm{d}p_n\, (p_1^2+1)\ldots(p_n^2+1)\, \widehat{\gamma}_0^{(n)}(p_1,\ldots,p_n;p_1,\ldots,p_n) \leq C^n$$

for all $n\geq 1$. Heuristically, this suggests that

$$|\widehat{\gamma}_0^{(k+m)}(\{p_e\}_{e\in L(\Lambda)})| \lesssim \prod_{e\in L(\Lambda)} \langle p_e\rangle^{-5/2}, \tag{3.49}$$

where $\langle p\rangle = (1+p^2)^{1/2}$. Using this decay in the momenta of the leaves and the decay of the propagators $\langle \alpha_e - p_e^2\rangle^{-1}, e\in E(\Lambda)$, we can prove the finiteness of all the momentum and frequency integrals in (3.47). On the heuristic level, this can be seen using a simple power counting argument. Fix $\kappa \gg 1$, and cutoff all momenta $|p_e|\geq\kappa$ and all frequencies $|\alpha_e|\geq \kappa^2$. Each p_e-integral scales then as κ^3, and each α_e-integral scales as κ^2. Since we have $2k+3m$ edges in Λ, we have $2k+3m$ momentum- and frequency integrations. However, because of the m delta functions (due to momentum and frequency conservation), we effectively only have to perform $2k+2m$ momentum- and frequency-integrations. Therefore the whole integral in (3.47) carries a volume factor of the order $\kappa^{5(2k+2m)} = \kappa^{10k+10m}$. Now, since there are $2k+2m$ leaves in the graph Λ, the estimate (3.49) guarantees a decay of the order $\kappa^{-5/2(2k+2m)} = \kappa^{-5k-5m}$. The $2k+3m$ propagators, on the other hand, provide a decay of the order $\kappa^{-2(2k+3m)} = \kappa^{-4k-6m}$. Choosing the observable $J^{(k)}$ so that $\widehat{J}^{(k)}$ decays sufficiently fast at infinity, we can also gain an additional decay κ^{-6k}. Since

$$\kappa^{10k+10m}\cdot\kappa^{-5k-5m-4k-6m-6k} = \kappa^{-m-5k} \ll 1$$

for $\kappa\gg 1$, we can expect (3.47) to converge in the large momentum and large frequency regime. Remark the importance of the decay provided by the free evolution (through the propagators); without making use of it, we would not be able to prove the uniqueness of the infinite hierarchy.

This heuristic argument is clearly far from rigorous. To obtain a rigorous proof, we use an integration scheme dictated by the structure of the graph Λ; we start by integrating the momenta and the frequency of the leaves (for which (3.49) provides sufficient decay). The point here is that when we perform the integrations over the momenta of the leaves we have to propagate the decay to the next edges on the left. We move iteratively from the right to the left of the graph, until we reach the roots; at every step we integrate the frequencies and momenta of the son edges

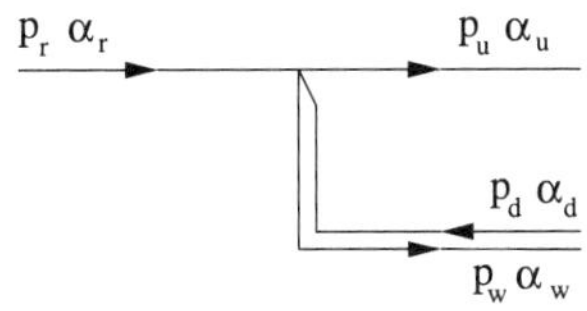

FIGURE 2. Integration scheme: a typical vertex

of a fixed vertex and as a result we obtain decay in the momentum of the father edge. When we reach the roots, we use the decay of the kernel $\widehat{J}^{(k)}$ to complete the integration scheme. In a typical step, we consider a vertex as the one drawn in Figure 2 and we assume to have decay in the momenta of the three son-edges, in the form $|p_e|^{-\lambda}$, $e = u, d, w$ (for some $2 < \lambda < 5/2$). Then we integrate over the frequencies $\alpha_u, \alpha_d, \alpha_w$ and the momenta p_u, p_d, p_w of the son-edges and as a result we obtain a decaying factor $|p_r|^{-\lambda}$ in the momentum of the father edge. In other words, we prove bounds of the form
(3.50)
$$\int \frac{d\alpha_u d\alpha_d d\alpha_w dp_u dp_d dp_w}{|p_u|^\lambda |p_d|^\lambda |p_w|^\lambda} \frac{\delta(\alpha_r = \alpha_u + \alpha_d - \alpha_w)\delta(p_r = p_u + p_d - p_w)}{\langle \alpha_u - p_u^2 \rangle \langle \alpha_d - p_d^2 \rangle \langle \alpha_w - p_w^2 \rangle} \leq \frac{\text{const}}{|p_r|^\lambda}.$$
Power counting implies that (3.50) can only be correct if $\lambda > 2$. On the other hand, to start the integration scheme we need $\lambda < 5/2$ (from (3.49) this is the decay in the momenta of the leaves, obtained from the a-priori estimates). It turns out that, choosing $\lambda = 2 + \varepsilon$ for a sufficiently small $\varepsilon > 0$, (3.50) can be made precise, and the integration scheme can be completed.

After integrating all the frequency and momentum variables, from (3.48) we obtain that
$$|K_{\Lambda,t}| \leq C^{k+m}\, t^{m/4}$$
for every $\Lambda \in \mathcal{F}_{m,k}$. Since the number of diagrams in $\mathcal{F}_{m,k}$ is bounded by C^{k+m}, it follows immediately that
$$\left|\text{Tr}\ J^{(k)}\, \xi_{m,t}^{(k)}\right| \leq C^{k+m} t^{m/4}.$$
Note that, from (3.44), one may expect $\xi_{m,t}^{(k)}$ to be proportional to t^m. The reason why we only get a bound proportional to $t^{m/4}$ is that we effectively use part of the time integration to control the singularity of the potentials.

The only property of $\gamma_0^{(k+m)}$ used in the analysis of (3.47) is the estimate (3.40), which provides the necessary decay in the momenta of the leaves. Since the a-priori bound (3.40) hold uniformly in time, we can use a similar argument to bound the contribution arising from the error term $\eta_{n,t}^{(k)}$ in (3.45) (as explained above, also $\eta_{n,t}^{(k)}$ can be expanded analogously to (3.46), with contributions associated to Feynman graphs similar to (3.47); the difference, of course, is that these contributions will depend on $\gamma_s^{(k+n)}$ for all $s \in [0, t]$, while (3.47) only depends on the initial data). We get
(3.51)
$$\left|\text{Tr}\ J^{(k)}\, \eta_{n,t}^{(k)}\right| \leq C^{k+n}\, t^{n/4}.$$

This bound immediately implies the uniqueness. In fact, given two solutions $\Gamma_{1,t} = \{\gamma_{1,t}^{(k)}\}_{k\geq 1}$ and $\Gamma_{2,t} = \{\gamma_{2,t}^{(k)}\}_{k\geq 1}$ of the infinite hierarchy (3.41), both satisfying the a-priori bounds (3.40) and with the same initial data, we can expand both in a Duhamel series of order n as in (3.43). If we fix $k \geq 1$, and consider the difference between $\gamma_{1,t}^{(k)}$ and $\gamma_{2,t}^{(k)}$, all terms (3.44) cancel out because they only depend on the initial data. Therefore, from (3.51), we immediately obtain that, for arbitrary (sufficiently smooth) compact k-particle operators $J^{(k)}$,

$$\left| \text{Tr} J^{(k)} \left(\gamma_{1,t}^{(k)} - \gamma_{2,t}^{(k)} \right) \right| \leq 2\, C^{k+n}\, t^{n/4}\,.$$

Since it is independent of n, the left side has to vanish for all $t < 1/(2C)^4$. This proves uniqueness for short times. But then, since the a-priori bounds hold uniformly in time, the argument can be repeated to prove uniqueness for all times.

3.6. Other Microscopic Models Leading to the Nonlinear Schrödinger Equation. As discussed in Section 3.1, the strategy used to prove Theorem 3.1 is dictated by the formal similarity with the mean-field systems discussed in Section 2; from (3.15) the Hamiltonian characterizing dilute Bose gases in the Gross-Pitaevskii scaling can be formally interpreted as a mean field Hamiltonian with an N-dependent potential converging to a delta-function as $N \to \infty$ (the physics described by the two models is however completely different). The choice of the N-dependent potential $V_N(x) = N^2 V(Nx)$ in the Gross-Pitaevskii scaling is, of course, not the only choice for which the formal identification with a mean-field model is possible. For arbitrary $\beta > 0$, we can for example define the N-particle Hamiltonian

$$H_{N,\beta} = \sum_{j=1}^{N} -\Delta_j + \frac{1}{N} \sum_{i<j} N^{3\beta} V(N^{\beta}(x_i - x_j))$$

acting on the Hilbert space $L^2_s(\mathbb{R}^{3N})$. The Hamiltonian (3.8) is recovered by choosing $\beta = 1$. For $0 < \beta < 1$, the potential $N^{3\beta} V(N^{\beta} x)$ still converges to a delta-function as $N \to \infty$, but the convergence is slower. This fact has important consequences for the macroscopic dynamics; it turns out, in fact, that for $0 < \beta < 1$ the correlation structure developed by the evolved wave function $\psi_{N,\beta,t} = e^{-iH_{N,\beta}t}$ varies on much shorter length scales compared with the length scale $N^{-\beta}$ characterizing the potential. Therefore, for $0 < \beta < 1$, the time evolution of the condensate wave function is still governed by a cubic nonlinear Schrödinger equation; this time, however, the coupling constant in front of the nonlinearity is given by $b_0 = \int V$ instead of $8\pi a_0$ (recall that the emergence of the scattering length in the Gross-Pitaevskii equation was a consequence of the interplay between the correlation structure in the many body wave function and the interaction potential; since, for $0 < \beta < 1$, the potential and the correlation structure vary on different length scales, this interplay is suppressed). The following theorem can be proven using the techniques developed in [**15**] (the statement for $0 < \beta < 1/2$ was proven in [**12**]; in [**13**], the whole range $0 < \beta \leq 1$ was covered, but only for sufficiently small potentials).

THEOREM 3.10. *Suppose that $V \geq 0$ satisfies the same assumption as in Theorem 3.1, and assume that $0 < \beta \leq 1$. Let $\psi_N(\mathbf{x}) = \prod_{j=1}^N \varphi(x_j)$, for some*

$\varphi \in H^1(\mathbb{R}^3)$ and $\psi_{N,t} = e^{-iH_{\beta,N}t}\psi_N$ with the mean-field Hamiltonian

$$H_{\beta,N} = \sum_{j=1}^{N} -\Delta_j + \frac{1}{N}\sum_{i<j}^{N} N^{3\beta}V(N^\beta(x_i - x_j)) .$$

Then, for every fixed $k \geq 1$ and $t \in \mathbb{R}$, we have

$$\gamma_{N,t}^{(k)} \to |\varphi_t\rangle\langle\varphi_t|^{\otimes k}$$

as $N \to \infty$, where φ_t is the solution to the nonlinear Schrödinger equation

$$i\partial_t\varphi_t = -\Delta\varphi_t + \sigma|\varphi_t|^2\varphi_t$$

with initial data $\varphi_{t=0} = \varphi$ and with

$$\sigma = \begin{cases} 8\pi a_0 & \text{if } \beta = 1 \\ b_0 & \text{if } 0 < \beta < 1 \end{cases} .$$

4. Rate of Convergence towards Mean-Field Dynamics

From the results of Section 2, we obtain that, for every fixed $t \in \mathbb{R}$, and for every fixed $k \in \mathbb{N}$,

$$\gamma_{N,t}^{(k)} \to |\varphi_t\rangle\langle\varphi_t|^{\otimes k}$$

where $\gamma_{N,t}^{(k)}$ is the k-particle density associated with the solution $\psi_{N,t}$ of the N-particle Schrödinger equation, and φ_t is the solution of the Hartree equation. From Theorem 2.1 and Theorem 2.5, we do not get any information about the rate of convergence of $\gamma_{N,t}^{(k)}$ to $|\varphi_t\rangle\langle\varphi_t|^{\otimes k}$. We only know that the difference $\gamma_{N,t}^{(k)} - |\varphi_t\rangle\langle\varphi_t|^{\otimes k}$ converges to zero, but we do not know how fast. Also in Section 3, we do not obtain any information about the rate of convergence of the N-particle Schrödinger evolution towards the Gross-Pitaevskii equation. This is not only a question of academic interest; in order to apply these results to physically relevant situations, bounds on the error are essential.

For bounded potential, (2.25) implies (specializing to the case $k = 1$) that

$$\text{Tr}\left|\gamma_{N,t}^{(1)} - |\varphi_t\rangle\langle\varphi_t|\right| \leq CN^{-1} \tag{4.1}$$

for sufficiently short times $0 \leq t \leq t_0 = 1/(8\|V\|)$. Iterating the argument leading to (4.1) to obtain estimates valid for larger times, it is possible to derive bounds of the form

$$\text{Tr}\left|\gamma_{N,t}^{(1)} - |\varphi_t\rangle\langle\varphi_t|\right| \leq \frac{C}{N^{2^{-t}}} . \tag{4.2}$$

Although (4.2) shows that, for every fixed $t > 0$, the difference $\gamma_{N,t}^{(1)} - |\varphi_t\rangle\langle\varphi_t|$ converges to zero, it is not very useful in applications because it deteriorates too fast; one would like to find bounds on the difference $\gamma_{N,t}^{(1)} - |\varphi_t\rangle\langle\varphi_t|$ which are of the same order in N for every fixed time.

In [**28**], a joint work with I. Rodnianski, we obtain such bounds for mean-field systems with potential having at most a Coulomb singularity; the problem of obtaining error estimates for the Gross-Pitaevskii dynamics is still open. To prove such bounds, we do not make use of the BBGKY hierarchy. Instead, we use an approach, introduced by Hepp in [**21**] and extended by Ginibre and Velo in [**20**], based on embedding the N-body Schrödinger system into the second quantized Fock-space representation and on the use of coherent states as initial data (coherent states do not have a fixed number of particles; this is what makes the use of a Fock-space representation necessary). The Hartree dynamics emerges as the main component of the evolution of coherent states in the mean field limit. To obtain bounds on the difference $\gamma^{(1)}_{N,t} - |\varphi_t\rangle\langle\varphi_t|$ for initial data describing coherent states, it is therefore enough to study the fluctuations around the Hartree dynamics, and to prove that, in an appropriate sense, they are small. Since factorized N-particle wave functions can be written as appropriate linear combinations of coherent states, the estimates for coherent initial data can be translated into bounds for factorized initial data. Using these techniques, we prove in [**28**] the following theorem. We focus in this section on three dimensional systems, which are the most interesting from the point of view of physics; most of the results however can be extended to dimension $d \neq 3$.

THEOREM 4.1. *Suppose that there exists $D > 0$ such that the operator inequality*

$$V^2(x) \leq D\,(1-\Delta_x) \tag{4.3}$$

holds true. Let

$$\psi_N(\mathbf{x}) = \prod_{j=1}^N \varphi(x_j),$$

for some $\varphi \in H^1(\mathbb{R}^3)$ with $\|\varphi\| = 1$. Denote by $\psi_{N,t} = e^{-iH_N t}\psi_N$ the solution to the Schrödinger equation (1.2) with initial data $\psi_{N,0} = \psi_N$, and let $\gamma^{(1)}_{N,t}$ be the one-particle density associated with $\psi_{N,t}$. Then there exist constants C, K, depending only on the H^1 norm of φ and on the constant D on the r.h.s. of (4.3) such that

$$\text{Tr}\,\left|\gamma^{(1)}_{N,t} - |\varphi_t\rangle\langle\varphi_t|\right| \leq \frac{C}{N^{1/2}}\, e^{K|t|}\,, \tag{4.4}$$

for every $t \in \mathbb{R}$ and every $N \in \mathbb{N}$. Here φ_t is the solution to the nonlinear Hartree equation

$$i\partial_t\varphi_t = -\Delta\varphi_t + (V * |\varphi_t|^2)\varphi_t \tag{4.5}$$

with initial data $\varphi_{t=0} = \varphi$.

Remarks. Condition (4.3) is in particular satisfied by bounded potentials and by potentials with an attractive or repulsive Coulomb singularity. Theorem 4.1 implies therefore Theorem 2.1 and Theorem 2.5. Note, moreover, that the decay of the order $N^{-1/2}$ on the r.h.s. of (4.4) is not expected to be optimal. In fact, for initially coherent states we obtain in Theorem 4.4 the expected decay of the order $1/N$ for every fixed time $t \in \mathbb{R}$; unfortunately, when factorized initial data are expressed as a superposition of coherent states, part of the decay is lost (note, however, that for a certain class of bounded potential a decay of the order N^{-1} for factorized initial data has recently been established in [**10**]).

4.1. Fock-Space Representation. We define the bosonic Fock space over $L^2(\mathbb{R}^3, \mathrm{d}x)$ as the Hilbert space

$$\mathcal{F} = \bigoplus_{n\geq 0} L^2(\mathbb{R}^3, \mathrm{d}x)^{\otimes_s n} = \mathbb{C} \oplus \bigoplus_{n\geq 1} L^2_s(\mathbb{R}^{3n}, \mathrm{d}x_1 \ldots \mathrm{d}x_n)\,,$$

where we put $L^2(\mathbb{R}^3)^{\otimes_s 0} = \mathbb{C}$. Vectors in $\mathcal{F}$ are sequences $\psi = \{\psi^{(n)}\}_{n\geq 0}$ of n-particle wave functions $\psi^{(n)} \in L^2_s(\mathbb{R}^{3n})$ with $\sum_{n\geq 0}\|\psi^{(n)}\|^2 < \infty$. The scalar product on $\mathcal{F}$ is defined by

$$\begin{aligned}\langle\psi_1,\psi_2\rangle &= \sum_{n\geq 0}\langle\psi_1^{(n)},\psi_2^{(n)}\rangle_{L^2(\mathbb{R}^{3n})}\\ &= \overline{\psi_1^{(0)}}\psi_2^{(0)} + \sum_{n\geq 1}\int \mathrm{d}x_1\ldots\mathrm{d}x_n\,\overline{\psi_1^{(n)}}(x_1,\ldots,x_n)\psi_2^{(n)}(x_1,\ldots,x_n)\,.\end{aligned}$$

An N particle state with wave function ψ_N is described on $\mathcal{F}$ by the sequence $\{\psi^{(n)}\}_{n\geq 0}$ where $\psi^{(n)} = 0$ for all $n\neq N$ and $\psi^{(N)} = \psi_N$. The vector $\{1,0,0,\ldots\}\in\mathcal{F}$ is called the vacuum, and will be denoted by Ω.

On $\mathcal{F}$, we define the number of particles operator $\mathcal{N}$, by $(\mathcal{N}\psi)^{(n)} = n\psi^{(n)}$. Eigenvectors of $\mathcal{N}$ are vectors of the form $\{0,\ldots,0,\psi^{(m)},0,\ldots\}$ with a fixed number of particles m. For $f\in L^2(\mathbb{R}^3)$ we also define the creation operator $a^*(f)$ and the annihilation operator $a(f)$ on $\mathcal{F}$ by

$$\begin{aligned}\left(a^*(f)\psi\right)^{(n)}(x_1,\ldots,x_n) &= \frac{1}{\sqrt{n}}\sum_{j=1}^{n} f(x_j)\psi^{(n-1)}(x_1,\ldots,x_{j-1},x_{j+1},\ldots,x_n)\\ \left(a(f)\psi\right)^{(n)}(x_1,\ldots,x_n) &= \sqrt{n+1}\int\mathrm{d}x\;\overline{f(x)}\,\psi^{(n+1)}(x,x_1,\ldots,x_n)\,.\end{aligned}$$

The operators $a^*(f)$ and $a(f)$ are unbounded, densely defined and closed; $a^*(f)$ creates a particle with wave function f, $a(f)$ annihilates it. It is simple to check that, for arbitrary $n\geq 1$,

$$\frac{\left(a^*(f)\right)^n}{\sqrt{n!}}\,\Omega = \{0,\ldots,0,f^{\otimes n},0,\ldots\}\,.$$

The creation operator $a^*(f)$ is the adjoint of the annihilation operator $a(f)$ (note that by definition $a(f)$ is anti-linear in f), and they satisfy the canonical commutation relations

$$[a(f),a^*(g)] = \langle f,g\rangle_{L^2(\mathbb{R}^3)},\qquad [a(f),a(g)] = [a^*(f),a^*(g)] = 0\,. \tag{4.6}$$

For every $f\in L^2(\mathbb{R}^3)$, we introduce the self adjoint operator

$$\phi(f) = a^*(f) + a(f)\,.$$

We will also make use of operator valued distributions a_x^* and a_x ($x\in\mathbb{R}^3$), defined so that

$$\begin{aligned}a^*(f) &= \int\mathrm{d}x\,f(x)\,a_x^*\\ a(f) &= \int\mathrm{d}x\,\overline{f(x)}\,a_x\end{aligned}$$

for every $f\in L^2(\mathbb{R}^3)$. The canonical commutation relations take the form

$$[a_x,a_y^*] = \delta(x-y)\qquad [a_x,a_y] = [a_x^*,a_y^*] = 0\,.$$

The number of particle operator, expressed through the distributions a_x, a_x^*, is given by

$$\mathcal{N} = \int\mathrm{d}x\,a_x^*a_x\,.$$

The following lemma provides some useful bounds to control creation and annihilation operators in terms of the number of particle operator $\mathcal{N}$.

LEMMA 4.2. *Let $f \in L^2(\mathbb{R}^3)$. Then*

$$\|a(f)\psi\| \leq \|f\| \, \|\mathcal{N}^{1/2}\psi\|$$
$$\|a^*(f)\psi\| \leq \|f\| \, \|(\mathcal{N}+1)^{1/2}\psi\|$$
$$\|\phi(f)\psi\| \leq 2\|f\| \|(\mathcal{N}+1)^{1/2}\psi\| .$$

PROOF. The last inequality clearly follows from the first two. To prove the first bound we note that

$$\begin{aligned}\|a(f)\psi\| &\leq \int \mathrm{d}x\, |f(x)|\, \|a_x\psi\| \leq \left(\int \mathrm{d}x\, |f(x)|^2\right)^{1/2} \left(\int \mathrm{d}x\, \|a_x\psi\|^2\right)^{1/2} \\ &= \|f\| \, \|\mathcal{N}^{1/2}\psi\| .\end{aligned}$$

The second estimate follows by the canonical commutation relations (4.6) because

$$\begin{aligned}\|a^*(f)\psi\|^2 &= \langle\psi, a(f)a^*(f)\psi\rangle = \langle\psi, a^*(f)a(f)\psi\rangle + \|f\|^2\|\psi\|^2 \\ &= \|a(f)\psi\|^2 + \|f\|^2\|\psi\|^2 \\ &\leq \|f\|^2\left(\|\mathcal{N}^{1/2}\psi\| + \|\psi\|^2\right) = \|f\|^2\|(\mathcal{N}+1)^{1/2}\psi\|^2 .\end{aligned}$$

□

Given $\psi \in \mathcal{F}$, we define the one-particle density $\gamma_\psi^{(1)}$ associated with ψ as the operator on $L^2(\mathbb{R}^3)$ with kernel given by

$$\gamma_\psi^{(1)}(x;y) = \frac{1}{\langle\psi, \mathcal{N}\psi\rangle} \langle\psi, a_y^* a_x \psi\rangle . \tag{4.7}$$

By definition, $\gamma_\psi^{(1)}$ is a positive trace class operator on $L^2(\mathbb{R}^3)$ with $\operatorname{Tr}\gamma_\psi^{(1)} = 1$. For every N-particle state with wave function $\psi_N \in L_s^2(\mathbb{R}^{3N})$ (described on $\mathcal{F}$ by the sequence $\{0, 0, \dots, \psi_N, 0, 0, \dots\}$) it is simple to see that this definition is equivalent to the definition (1.4).

We define the Hamiltonian $\mathcal{H}_N$ on $\mathcal{F}$ by $(\mathcal{H}_N\psi)^{(n)} = \mathcal{H}_N^{(n)}\psi^{(n)}$, with

$$\mathcal{H}_N^{(n)} = -\sum_{j=1}^n \Delta_j + \frac{1}{N}\sum_{i<j}^n V(x_i - x_j) .$$

Using the distributions a_x, a_x^*, $\mathcal{H}_N$ can be rewritten as

$$\mathcal{H}_N = \int \mathrm{d}x \nabla_x a_x^* \nabla_x a_x + \frac{1}{2N}\int \mathrm{d}x \mathrm{d}y\, V(x-y) a_x^* a_y^* a_y a_x . \tag{4.8}$$

By definition the Hamiltonian $\mathcal{H}_N$ leaves sectors of $\mathcal{F}$ with a fixed number of particles invariant. Moreover, it is clear that on the N-particle sector, $\mathcal{H}_N$ agrees with the Hamiltonian H_N (the subscript N in $\mathcal{H}_N$ is a reminder of the scaling factor $1/N$ in front of the potential energy). We will study the dynamics generated by the operator $\mathcal{H}_N$. In particular we will consider the time evolution of coherent states, which we introduce next.

For $f \in L^2(\mathbb{R}^3)$, we define the Weyl-operator

$$W(f) = \exp\left(a^*(f) - a(f)\right) = \exp\left(\int \mathrm{d}x\, \left(f(x)a_x^* - \overline{f}(x)a_x\right)\right) .$$

Then the coherent state $\psi(f) \in \mathcal{F}$ with one-particle wave function f is defined by

$$\psi(f) = W(f)\Omega .$$

Notice that

$$\psi(f) = W(f)\Omega = e^{-\|f\|^2/2} \sum_{n \geq 0} \frac{(a^*(f))^n}{n!} \Omega = e^{-\|f\|^2/2} \sum_{n \geq 0} \frac{1}{\sqrt{n!}} f^{\otimes n} , \tag{4.9}$$

where $f^{\otimes n}$ indicates the Fock-vector $\{0, \dots, 0, f^{\otimes n}, 0, \dots\}$. This follows from

$$\exp(a^*(f) - a(f)) = e^{-\|f\|^2/2} \exp(a^*(f)) \exp(-a(f))$$

which is a consequence of the fact that the commutator $[a(f), a^*(f)] = \|f\|^2$ commutes with $a(f)$ and $a^*(f)$. From (4.9), it follows that coherent states are superpositions of states with different number of particles (the probability of having n particles in $\psi(f)$ is given by $e^{-\|f\|^2} \|f\|^{2n}/n!$).

In the following lemma we collect some important and well known properties of Weyl operators and coherent states.

LEMMA 4.3. *Let $f, g \in L^2(\mathbb{R}^3)$.*

i) *The Weyl operator satisfy the relations*

$$W(f)W(g) = W(g)W(f)e^{-2i\, Im \langle f,g \rangle} = W(f+g)e^{-i\, Im \langle f,g \rangle} .$$

ii) *$W(f)$ is a unitary operator and*

$$W(f)^* = W(f)^{-1} = W(-f).$$

iii) *For every $x \in \mathbb{R}^3$, we have*

$$W^*(f) a_x W(f) = a_x + f(x), \qquad \text{and} \quad W^*(f) a_x^* W(f) = a_x^* + \overline{f}(x) .$$

iv) *From iii) it follows that coherent states are eigenvectors of annihilation operators*

$$a_x \psi(f) = f(x) \psi(f) \qquad \Rightarrow \qquad a(g)\psi(f) = \langle g, f \rangle_{L^2} \psi(f) .$$

v) *The expectation of the number of particles in the coherent state $\psi(f)$ is given by $\|f\|^2$, that is*

$$\langle \psi(f), \mathcal{N} \psi(f) \rangle = \|f\|^2 .$$

Also the variance of the number of particles in $\psi(f)$ is given by $\|f\|^2$ (the distribution of $\mathcal{N}$ is Poisson), that is

$$\langle \psi(f), \mathcal{N}^2 \psi(f) \rangle - \langle \psi(f), \mathcal{N} \psi(f) \rangle^2 = \|f\|^2 .$$

vi) *Coherent states are normalized but not orthogonal to each other. In fact*

$$\langle \psi(f), \psi(g) \rangle = e^{-\frac{1}{2}\left(\|f\|^2 + \|g\|^2 - 2(f,g) \right)} \quad \Rightarrow \quad |\langle \psi(f), \psi(g) \rangle| = e^{-\frac{1}{2}\|f-g\|^2} .$$

4.2. Time Evolution of Coherent States. Next we study the dynamics of coherent states with expected number of particles N in the limit $N \to \infty$. We choose the initial data

$$\psi(\sqrt{N}\varphi) = W(\sqrt{N}\varphi)\Omega \qquad \text{for } \varphi \in H^1(\mathbb{R}^3) \text{ with } \|\varphi\| = 1$$

and we consider its time evolution $\psi(N,t) = e^{-i\mathcal{H}_N t}\psi(\sqrt{N}\varphi)$ with the Hamiltonian $\mathcal{H}_N$ defined in (4.8).

THEOREM 4.4. *Suppose that there exists $D > 0$ such that the operator inequality*

$$V^2(x) \le D(1-\Delta_x) \tag{4.10}$$

holds true. Let $\Gamma^{(1)}_{N,t}$ be the one-particle marginal associated with the Fock-space vector $\psi(N,t) = e^{-i\mathcal{H}_N t}W(\sqrt{N}\varphi)\Omega$ (as defined in (4.7)). Then there exist constants $C, K > 0$ (only depending on the H^1-norm of φ and on the constant D appearing in (4.10)) such that

$$\mathrm{Tr}\left|\Gamma^{(1)}_{N,t} - |\varphi_t\rangle\langle\varphi_t|\right| \le \frac{C}{N}\, e^{K|t|} \tag{4.11}$$

for all $t \in \mathbb{R}$.

We explain next the main steps in the proof of Theorem 4.4. By (4.7), the kernel of $\Gamma^{(1)}_{N,t}$ is given by

$$\begin{aligned}
\Gamma^{(1)}_{N,t}(x;y) =& \frac{1}{N}\left\langle \Omega, W^*(\sqrt{N}\varphi)e^{i\mathcal{H}_N t}a^*_y a_x e^{-i\mathcal{H}_N t}W(\sqrt{N}\varphi)\Omega\right\rangle \\
&= \varphi_t(x)\overline{\varphi}_t(y) \\
&+ \frac{\overline{\varphi}_t(y)}{\sqrt{N}}\left\langle \Omega, W^*(\sqrt{N}\varphi)e^{i\mathcal{H}_N t}(a_x - \sqrt{N}\varphi_t(x))e^{-i\mathcal{H}_N t}W(\sqrt{N}\varphi)\Omega\right\rangle \\
&+ \frac{\varphi_t(x)}{\sqrt{N}}\left\langle \Omega, W^*(\sqrt{N}\varphi)e^{i\mathcal{H}_N t}(a^*_y - \sqrt{N}\overline{\varphi}_t(y))e^{-i\mathcal{H}_N t}W(\sqrt{N}\varphi)\Omega\right\rangle \\
&+ \frac{1}{N}\Big\langle \Omega, W^*(\sqrt{N}\varphi)e^{i\mathcal{H}_N t}(a^*_y - \sqrt{N}\overline{\varphi}_t(y)) \\
&\qquad\qquad \times (a_x - \sqrt{N}\varphi_t(x))e^{-i\mathcal{H}_N t}W(\sqrt{N}\varphi)\Omega\Big\rangle .
\end{aligned} \tag{4.12}$$

It was observed by Hepp in [**21**] (see also Eqs. (1.17)-(1.28) in [**20**]) that

$$\begin{aligned}
W^*(\sqrt{N}\varphi_s)\, e^{i\mathcal{H}_N(t-s)}(a_x - \sqrt{N}\varphi_t(x))e^{-i\mathcal{H}_N(t-s)}W(\sqrt{N}\varphi_s) \\
= \mathcal{U}_N(t;s)^*\, a_x\, \mathcal{U}_N(t;s) = \mathcal{U}_N(s;t)\, a_x\, \mathcal{U}_N(t;s)
\end{aligned} \tag{4.13}$$

where the unitary evolution $\mathcal{U}_N(t;s)$ is determined by the equation

$$i\partial_t \mathcal{U}_N(t;s) = \mathcal{L}_N(t)\mathcal{U}_N(t;s) \qquad \text{and} \quad \mathcal{U}_N(s;s) = 1 \tag{4.14}$$

with the generator

$$\begin{aligned}\mathcal{L}_N(t) = &\int \mathrm{d}x\, \nabla_x a_x^* \nabla_x a_x + \int \mathrm{d}x\, \left(V * |\varphi_t|^2\right)(x)\, a_x^* a_x \\ &+ \int \mathrm{d}x\mathrm{d}y\, V(x-y)\, \overline{\varphi_t}(x)\varphi_t(y) a_y^* a_x \\ &+ \frac{1}{2}\int \mathrm{d}x\mathrm{d}y\, V(x-y)\, \left(\varphi_t(x)\varphi_t(y) a_x^* a_y^* + \overline{\varphi_t}(x)\overline{\varphi_t}(y) a_x a_y\right) \\ &+ \frac{1}{\sqrt{N}}\int \mathrm{d}x\mathrm{d}y\, V(x-y)\, a_x^* \left(\varphi_t(y) a_y^* + \overline{\varphi_t}(y) a_y\right) a_x \\ &+ \frac{1}{2N}\int \mathrm{d}x\mathrm{d}y\, V(x-y)\, a_x^* a_y^* a_y a_x\,.\end{aligned} \tag{4.15}$$

It follows from (4.12) that

$$\begin{aligned}\Gamma_{N,t}^{(1)}(x,y) - \varphi_t(x)\overline{\varphi}_t(y) = &\frac{1}{N}\left\langle \Omega, \mathcal{U}_N(t;0)^* a_y^* a_x \mathcal{U}_N(t;0)\Omega\right\rangle \\ &+ \frac{\varphi_t(x)}{\sqrt{N}}\left\langle \Omega, \mathcal{U}_N(t;0)^* a_y^* \mathcal{U}_N(t;0)\Omega\right\rangle \\ &+ \frac{\overline{\varphi}_t(y)}{\sqrt{N}}\left\langle \Omega, \mathcal{U}_N(t;0)^* a_x \mathcal{U}_N(t;0)\Omega\right\rangle\,.\end{aligned} \tag{4.16}$$

In order to produce another decaying factor $1/\sqrt{N}$ in the last two term on the r.h.s. of the last equation, we compare the evolution $\mathcal{U}_N(t;0)$ with another evolution $\widetilde{\mathcal{U}}_N(t;0)$ defined through the equation

$$i\partial_t \widetilde{\mathcal{U}}_N(t;s) = \widetilde{\mathcal{L}}_N(t)\widetilde{\mathcal{U}}_N(t;s) \qquad \text{with} \quad \widetilde{\mathcal{U}}_N(s;s) = 1 \tag{4.17}$$

and

$$\begin{aligned}\widetilde{\mathcal{L}}_N(t) = &\int \mathrm{d}x\, \nabla_x a_x^* \nabla_x a_x + \int \mathrm{d}x\, \left(V * |\varphi_t|^2\right)(x)\, a_x^* a_x \\ &+ \int \mathrm{d}x\mathrm{d}y\, V(x-y)\overline{\varphi_t}(x)\varphi_t(y) a_y^* a_x \\ &+ \frac{1}{2}\int \mathrm{d}x\mathrm{d}y\, V(x-y)\, \left(\varphi_t(x)\varphi_t(y) a_x^* a_y^* + \overline{\varphi_t}(x)\overline{\varphi_t}(y) a_x a_y\right) \\ &+ \frac{1}{2N}\int \mathrm{d}x\mathrm{d}y\, V(x-y)\, a_x^* a_y^* a_y a_x\,.\end{aligned}$$

From (4.16) we find

$$\begin{aligned}&\Gamma_{N,t}^{(1)}(x;y) - \varphi_t(x)\overline{\varphi}_t(y) \\ &= \frac{1}{N}\langle \Omega, \mathcal{U}_N(t;0)^* a_y^* a_x \mathcal{U}_N(t;0)\Omega\rangle \\ &\quad + \frac{\varphi_t(x)}{\sqrt{N}}\Big(\Big\langle \Omega, \mathcal{U}_N(t;0)^* a_y^* \left(\mathcal{U}_N(t;0) - \widetilde{\mathcal{U}}_N(t;0)\right)\Omega\Big\rangle \\ &\qquad\qquad + \Big\langle \Omega, \left(\mathcal{U}_N(t;0)^* - \widetilde{\mathcal{U}}_N(t;0)^*\right) a_y^* \widetilde{\mathcal{U}}_N(t;0)\Omega\Big\rangle\Big) \\ &\quad + \frac{\overline{\varphi}_t(y)}{\sqrt{N}}\Big(\Big\langle \Omega, \mathcal{U}_N(t;0)^* a_x \left(\mathcal{U}_N(t;0) - \widetilde{\mathcal{U}}_N(t;0)\right)\Omega\Big\rangle \\ &\qquad\qquad + \Big\langle \Omega, \left(\mathcal{U}_N(t;0)^* - \widetilde{\mathcal{U}}_N(t;0)^*\right) a_x \widetilde{\mathcal{U}}_N(t;0)\Omega\Big\rangle\Big),\end{aligned} \tag{4.18}$$

because

$$\left\langle \Omega, \widetilde{\mathcal{U}}_N(t;0)^* a_y \widetilde{\mathcal{U}}_N(t;0)\Omega \right\rangle = \left\langle \Omega, \widetilde{\mathcal{U}}_N(t;0)^* a_x^* \widetilde{\mathcal{U}}_N(t;0)\Omega \right\rangle = 0\,.$$

This follows from the observation that, although the evolution $\widetilde{\mathcal{U}}_N(t)$ does not preserve the number of particles, it preserves the parity (it commutes with $(-1)^{\mathcal{N}}$). From (4.18), it easily follows that

(4.19)

$$\begin{aligned} \left\| \Gamma^{(1)}_{N,t} - |\varphi_t\rangle\langle\varphi_t| \right\|_{\rm HS} \leq \; & \frac{1}{N} \, \langle \mathcal{U}_N(t;0)\Omega, \mathcal{N}\mathcal{U}_N(t;0)\Omega\rangle \\ & + \frac{2}{\sqrt{N}} \| (\mathcal{U}_N(t;0) - \widetilde{\mathcal{U}}_N(t;0))\Omega\| \, \| (\mathcal{N}+1)^{1/2} \mathcal{U}_N(t;0)\Omega\| \\ & + \frac{2}{\sqrt{N}} \| (\mathcal{U}_N(t;0) - \widetilde{\mathcal{U}}_N(t;0))\Omega\| \, \| (\mathcal{N}+1)^{1/2} \widetilde{\mathcal{U}}_N(t;0)\Omega\| \,. \end{aligned}$$

To bound the r.h.s. of (4.19), we need to compare the dynamics $\mathcal{U}_N(t;0)$ and $\widetilde{\mathcal{U}}_N(t;0)$, and to control the growth of the number of particle $\mathcal{N}$ with respect to the fluctuation dynamics $\mathcal{U}_N(t;0)$ and $\widetilde{\mathcal{U}}_N(t;0)$. We show, first of all, that

$$\langle \widetilde{\mathcal{U}}_N(t;0)\,\Omega, \mathcal{N}\widetilde{\mathcal{U}}_N(t;0)\Omega\rangle \leq C\, e^{K|t|}\,. \tag{4.20}$$

To prove this bound, we compute the time derivative

$$\begin{aligned} & \frac{\rm d}{{\rm d}t} \langle \widetilde{\mathcal{U}}_N(t;0)\,\Omega, (\mathcal{N}+1) \widetilde{\mathcal{U}}_N(t;0)\Omega\rangle \\ & \quad = \langle \widetilde{\mathcal{U}}_N(t;0)\Omega, [\widetilde{\mathcal{L}}_N(t), \mathcal{N}] \widetilde{\mathcal{U}}_N(t;0)\Omega\rangle \\ & \quad = \langle \widetilde{\mathcal{U}}_N(t;0)\Omega, [\widetilde{\mathcal{L}}_N(t), \mathcal{N}] \widetilde{\mathcal{U}}_N(t;0)\Omega\rangle \\ & \quad = 2{\rm Im} \int {\rm d}x{\rm d}y V(x-y)\varphi_t(x)\varphi_t(y) \langle \widetilde{\mathcal{U}}_N(t;0)\Omega, [a_x^* a_y^*, \mathcal{N}] \widetilde{\mathcal{U}}_N(t;0)\Omega\rangle \\ & \quad = 4{\rm Im} \int {\rm d}x{\rm d}y V(x-y)\varphi_t(x)\varphi_t(y) \langle \widetilde{\mathcal{U}}_N(t;0)\Omega, a_x^* a_y^* \widetilde{\mathcal{U}}_N(t;0)\Omega\rangle\,. \end{aligned}$$

Thus, from Lemma 4.2, we obtain

$$\begin{aligned} & \left| \frac{\rm d}{{\rm d}t} \langle \widetilde{\mathcal{U}}_N(t;0)\Omega, (\mathcal{N}+1) \widetilde{\mathcal{U}}_N(t;0)\Omega\rangle \right| \\ & \qquad \leq 4 \int {\rm d}x |\varphi_t(x)| \|a_x \widetilde{\mathcal{U}}_N(t;0)\Omega\| \, \|a^*(V(x-.)\varphi_t)\widetilde{\mathcal{U}}_N(t;0)\Omega\| \\ & \qquad \leq 4 \sup_x \|V(x-.)\varphi_t\| \|(\mathcal{N}+1)^{1/2} \widetilde{\mathcal{U}}_N(t;0)\Omega\|^2 \\ & \qquad \leq C \,\langle \widetilde{\mathcal{U}}_N(t;0)\Omega, (\mathcal{N}+1) \widetilde{\mathcal{U}}_N(t;0)\Omega\rangle\,, \end{aligned} \tag{4.21}$$

where we used the fact that

$$\|V(x-.)\varphi_t\|^2 = \int {\rm d}y\, |V(x-y)|^2 |\varphi_t(y)|^2 \leq C \|\varphi_t\|^2_{H_1} \leq C\|\varphi\|^2_{H^1}$$

because of the assumption (4.10). From (4.21), we obtain (4.20) applying Gronwall's Lemma.

Making use of (4.20) (and of an analogous bound for the growth of the expectation of $\mathcal{N}^4$ w.r.t. the evolution $\widetilde{\mathcal{U}}_N(t;0)$; see **[28]**[Lemma 3.7]), we can derive the bound

$$\left\| \left(\mathcal{U}_N(t;0) - \widetilde{\mathcal{U}}_N(t;0) \right) \Omega \right\| \le \frac{C}{\sqrt{N}} e^{K|t|} \tag{4.22}$$

for the difference between the two time evolutions $\mathcal{U}_N(t;0)$ and $\widetilde{\mathcal{U}}_N(t;0)$ (note that, at least formally, the difference between the two generators $\mathcal{L}_N(t)$ and $\widetilde{\mathcal{L}}_N(t)$ is a term of the order $N^{-1/2}$; this explains the decay in N on the r.h.s. of (4.22)).

In [**21, 20**] the time evolution $\mathcal{U}(t;s)$ was proven to converge, as $N \to \infty$, to a limiting dynamics $\mathcal{U}_\infty(t;s)$ defined by

$$i\partial_t \mathcal{U}_\infty(t;s) = \mathcal{L}_\infty(t)\mathcal{U}_\infty(t;s) \qquad \text{and} \qquad \mathcal{U}_\infty(s;s) = 1$$

with generator

$$\begin{aligned} \mathcal{L}_\infty(t) = & \int \mathrm{d}x \, \nabla_x a_x^* \nabla_x a_x + \int \mathrm{d}x \, \left(V * |\varphi_t|^2 \right)(x) \, a_x^* a_x \\ & + \int \mathrm{d}x \mathrm{d}y \, V(x-y) \overline{\varphi_t}(x) \varphi_t(y) a_y^* a_x \\ & + \frac{1}{2} \int \mathrm{d}x \mathrm{d}y \, V(x-y) \left(\varphi_t(x)\varphi_t(y) a_x^* a_y^* + \overline{\varphi_t}(x)\overline{\varphi_t}(y) a_x a_y \right) . \end{aligned}$$

In this sense, Hepp (in [**21**], for smooth potentials) and Ginibre-Velo (in [**20**], for singular potentials) were able to identify the limiting time evolution of the fluctuations around the Hartree dynamics. Analogously to (4.22), the bound (4.20) can also be used to show that, for a dense set of $\Psi \in \mathcal{F}$, there exists constants C, K such that $\|(\mathcal{U}_N(t;s) - \mathcal{U}_\infty(t;s))\Psi\| \le C N^{-1/2} e^{K|t-s|}$, giving therefore a quantitative control on the convergence established in [**21, 20**].

Note, however, that the convergence of $\mathcal{U}_N(t;s)$ to the dynamics $\mathcal{U}_\infty(t;s)$ is still not enough to obtain estimates on the difference between $\Gamma^{(1)}_{N,t}$ and $|\varphi_t\rangle\langle\varphi_t|$. In fact, to reach this goal, we still need, by (4.19), to control the growth of $\mathcal{N}$ with respect to the time evolution $\mathcal{U}_N(t;s)$. We are going to prove that

$$\langle \mathcal{U}_N(t;0)\,\Omega, \mathcal{N}\,\mathcal{U}_N(t;0)\Omega \rangle \le C\, e^{K|t|} . \tag{4.23}$$

Inserting (4.20), (4.22) and (4.23) on the r.h.s. of (4.19), it follows immediately that

$$\left\| \Gamma^{(1)}_{N,t} - |\varphi_t\rangle\langle\varphi_t| \right\|_{\mathrm{HS}} \le C \frac{e^{K|t|}}{N} , \tag{4.24}$$

which implies the claim (4.11). In fact, since $|\varphi_t\rangle\langle\varphi_t|$ is a rank one projection, the operator $\delta\gamma = \gamma^{(1)}_{N,t} - |\varphi_t\rangle\langle\varphi_t|$ has at most one negative eigenvalue. Noticing that $\mathrm{Tr}\,\delta\gamma = 0$, it follows that $\delta\gamma$ has a negative eigenvalue, and that the negative eigenvalue must equal, in absolute value, the sum of all its positive eigenvalues. Therefore, the trace norm of $\delta\gamma$ is twice as large as the operator norm of $\delta\gamma$. Since the operator norm is always controlled by the Hilbert Schmidt norm, we obtain (4.11) (this nice argument was pointed out to us by R. Seiringer).

The proof of (4.23) is much more involved than the proof of the analogous bound (4.20). This is a consequence of the presence, in the generator $\mathcal{L}_N(t)$, of terms which are cubic in the creation and annihilation operators (these terms are absent from $\widetilde{\mathcal{L}}_N(t)$). Because of these terms, also the commutator $[\mathcal{L}_N(t), \mathcal{N}]$ is cubic in creation and annihilation operators, and thus its expectation (in absolute value) cannot be controlled by the expectation of $\mathcal{N}$. For this reason, to prove (4.23), we have to introduce yet another approximate dynamics $\mathcal{W}_N(t;s)$, defined by

$$i\partial_t \mathcal{W}_N(t;s) = \mathcal{K}_N(t)\mathcal{W}_N(t;s), \qquad \text{with} \quad \mathcal{W}_N(s;s) = 1$$

and with generator

$$\begin{aligned}\mathcal{K}_N(t) = &\int \mathrm{d}x\, \nabla_x a_x^* \nabla_x a_x + \int \mathrm{d}x\, \left(V * |\varphi_t|^2\right)(x)\, a_x^* a_x \\ &+ \int \mathrm{d}x\mathrm{d}y\, V(x-y)\, \overline{\varphi_t}(x)\varphi_t(y) a_y^* a_x \\ &+ \frac{1}{2}\int \mathrm{d}x\mathrm{d}y\, V(x-y)\, \left(\varphi_t(x)\varphi_t(y) a_x^* a_y^* + \overline{\varphi_t}(x)\overline{\varphi_t}(y) a_x a_y\right) \\ &+ \frac{1}{\sqrt{N}}\int \mathrm{d}x\mathrm{d}y\, V(x-y)\, a_x^* \left(\varphi_t(y)\chi(\mathcal{N} < N) a_y^* + \overline{\varphi_t}(y) a_y \chi(\mathcal{N} < N)\right) a_x \\ &+ \frac{1}{2N}\int \mathrm{d}x\mathrm{d}y\, V(x-y)\, a_x^* a_y^* a_y a_x \,.\end{aligned} \tag{4.25}$$

Observe, that the generator $\mathcal{K}_N(t)$ has exactly the same form as the generator $\mathcal{L}_N(t)$; the only difference is the presence of a cutoff in the number of particles $\mathcal{N}$ inserted in the cubic term. Thanks to the cutoff in $\mathcal{N}$ and to the factor $N^{-1/2}$ in front of the cubic term in $\mathcal{K}_N(t)$, we can prove, making use of a Gronwall-type argument, that

$$\langle \mathcal{W}_N(t;s)\,\Omega, \mathcal{N}\,\mathcal{W}_N(t;s)\Omega\rangle \leq C\, e^{K|t-s|}\,. \tag{4.26}$$

Actually, it is simple to see that the last inequality can be improved to

$$\langle \mathcal{W}_N(t;s)\,\Omega, \mathcal{N}^j\,\mathcal{W}_N(t;s)\Omega\rangle \leq C_j\, e^{K_j|t-s|}\,. \tag{4.27}$$

for every $j \in \mathbb{N}$ and for appropriate j-dependent constants C_j and K_j. To obtain (4.23), we still have to compare the dynamics $\mathcal{U}_N(t;s)$ and $\mathcal{W}_N(t;s)$. To this end, we first show weak a-priori bounds of the form

$$\langle \mathcal{U}_N(t;s)\,\psi, \mathcal{N}^j\,\mathcal{U}_N(t;s)\psi\rangle \leq C\,\langle \psi, (\mathcal{N}+N+1)^j\psi\rangle \tag{4.28}$$

for every $\psi \in \mathcal{F}$ and for $j = 1,2,3$ (these bounds hold uniformly in $t,s \in \mathbb{R}$ and can be proven using the very definition of the unitary group $\mathcal{U}_N(t;s)$; see [**28**][Lemma 3.6]). Using (4.28), we find

$$\begin{aligned}&\langle \mathcal{U}_N(t;0)\Omega, \mathcal{N}\left(\mathcal{U}_N(t;0) - \mathcal{W}_N(t;0)\right)\Omega\rangle \\ &\quad = \langle \mathcal{U}_N(t;0)\Omega, \mathcal{N}\mathcal{U}_N(t;0)\left(1 - \mathcal{U}_N(t;0)^*\mathcal{W}_N(t;0)\right)\Omega\rangle \\ &\quad = -i\int_0^t \mathrm{d}s\, \langle \mathcal{U}_N(t;0)\Omega, \mathcal{N}\mathcal{U}_N(t;s)\left(\mathcal{L}_N(s) - \mathcal{K}_N(s)\right)\mathcal{W}_N(s;0)\Omega\rangle \\ &\quad = -\frac{i}{\sqrt{N}}\int_0^t \mathrm{d}s \int \mathrm{d}x\,\mathrm{d}y V(x-y)\langle \mathcal{U}_N(t;0)\Omega, \mathcal{N}\mathcal{U}_N(t;s) a_x^* \\ &\qquad\qquad \times \left(\overline{\varphi}_t(y) a_y \chi(\mathcal{N} > N) + \varphi_t(y)\chi(\mathcal{N} > N) a_y^*\right) a_x \mathcal{W}_N(s;0)\Omega\rangle\end{aligned}$$

Therefore

$$\begin{aligned}
&\langle \mathcal{U}_N(t;0)\Omega, \mathcal{N}\left(\mathcal{U}_N(t;0) - \mathcal{W}_N(t;0)\right)\Omega\rangle \\
&\quad = -\frac{i}{\sqrt{N}} \int_0^t \mathrm{d}s \int \mathrm{d}x \, \langle a_x \mathcal{U}_N(t;s)^* \mathcal{N} \mathcal{U}_N(t;0)\Omega, \\
&\qquad\qquad\qquad \times a(V(x-.)\varphi_t)\chi(\mathcal{N} > N) a_x \mathcal{W}_N(s;0)\Omega\rangle \\
&\qquad - \frac{i}{\sqrt{N}} \int_0^t \mathrm{d}s \int \mathrm{d}x \langle a_x \mathcal{U}_N(t;s)^* \mathcal{N} \mathcal{U}_N(t;0)\Omega, \\
&\qquad\qquad\qquad \times \chi(\mathcal{N} > N) a^*(V(x-.)\varphi_t) a_x \mathcal{W}_N(s;0)\Omega\rangle \,.
\end{aligned}$$

Hence

$$\begin{aligned}
&\left|\langle \mathcal{U}_N(t;0)\Omega, \mathcal{N}\left(\mathcal{U}_N(t;0) - \mathcal{W}_N(t;0)\right)\Omega\rangle\right| \\
&\quad \leq \frac{1}{\sqrt{N}} \int_0^t \mathrm{d}s \int \mathrm{d}x \|a_x \mathcal{U}_N(t;s)^* \mathcal{N} \mathcal{U}_N(t;0)\Omega\| \\
&\qquad\qquad\qquad \times \|a(V(x-.)\varphi_t) a_x \chi(\mathcal{N} > N+1) \mathcal{W}_N(s;0)\Omega\| \\
&\qquad + \frac{1}{\sqrt{N}} \int_0^t \mathrm{d}s \int \mathrm{d}x \|a_x \mathcal{U}_N(t;s)^* \mathcal{N} \mathcal{U}_N(t;0)\Omega\| \\
&\qquad\qquad\qquad \times \|a^*(V(x-.)\varphi_t) a_x \chi(\mathcal{N} > N) \mathcal{W}_N(s;0)\Omega\| \\
&\quad \leq \frac{2\sup_x \|V(x-.)\varphi_t\|}{\sqrt{N}} \int_0^t \mathrm{d}s \|\mathcal{N}^{1/2} \mathcal{U}_N(t;s)^* \mathcal{N} \mathcal{U}_N(t;0)\psi\| \\
&\qquad\qquad\qquad \times \|\mathcal{N}\chi(\mathcal{N} > N) \mathcal{W}_N(s;0)\psi\| \,.
\end{aligned}$$

Therefore, using the inequality $\chi(\mathcal{N} > N) \leq (\mathcal{N}/N)^2$ and applying (4.27) (with $j = 4$) and (4.28) (first with $j = 1$, and then with $j = 3$) we can bound the two norms in the s-integral. It follows that

$$\left|\langle \mathcal{U}_N(t;0)\Omega, \mathcal{N}\left(\mathcal{U}_N(t;0) - \mathcal{W}_N(t;0)\right)\Omega\rangle\right| \leq Ce^{Kt}$$

which, combined with (4.26), implies (4.23).

4.3. Time-evolution of Factorized Initial Data. To prove Theorem 4.1, we express the factorized initial data as a linear combination of coherent states. Using the properties listed in Lemma 4.3, it is simple to check that

$$\{0, 0, \dots, 0, \varphi^{\otimes N}, 0, 0, \dots\} = \frac{(a^*(\varphi))^N}{\sqrt{N!}}\Omega = d_N \int_0^{2\pi} \frac{\mathrm{d}\theta}{2\pi} \, e^{i\theta N} W(e^{-i\theta}\sqrt{N}\varphi)\Omega$$

with the constant

$$d_N = \frac{\sqrt{N!}}{N^{N/2} e^{-N/2}} \simeq N^{1/4} \,.$$

The kernel of the one-particle density $\gamma^{(1)}_{N,t}$ associated with the solution of the Schrödinger equation $\{0, \dots, 0, e^{-iH_N t}\varphi^{\otimes N}, 0, \dots\}$ is thus given by (see (4.7))

$$\begin{aligned}
&\gamma^{(1)}_{N,t}(x;y) \\
&= \frac{1}{N}\left\langle \frac{(a^*(\varphi))^N}{\sqrt{N!}}\Omega, e^{i\mathcal{H}_N t} a_y^* a_x e^{-i\mathcal{H}_N t} \frac{(a^*(\varphi))^N}{\sqrt{N!}}\Omega \right\rangle \\
&= \frac{d_N^2}{N} \int_0^{2\pi} \frac{\mathrm{d}\theta_1}{2\pi} \int_0^{2\pi} \frac{\mathrm{d}\theta_2}{2\pi} \, e^{-i\theta_1 N} e^{i\theta_2 N} \langle W(e^{-i\theta_1}\sqrt{N}\varphi)\Omega, a_y^*(t) a_x(t) W(e^{-i\theta_2}\sqrt{N}\varphi)\Omega\rangle
\end{aligned}$$

where we introduced the notation $a_x(t) = e^{i\mathcal{H}_N t} a_x e^{-i\mathcal{H}_N t}$. Next, we expand

(4.29)

$$\begin{aligned}
\gamma^{(1)}_{N,t}(x;y) = \; & \frac{d_N^2}{N} \int_0^{2\pi} \frac{d\theta_1}{2\pi} \int_0^{2\pi} \frac{d\theta_2}{2\pi} \, e^{-i\theta_1 N} e^{i\theta_2 N} \Big\langle W(e^{-i\theta_1}\sqrt{N}\varphi)\Omega, \\
& \times \left(a_y^*(t) - e^{i\theta_1}\sqrt{N}\overline{\varphi}_t(y)\right) \left(a_x(t) - e^{-i\theta_2}\sqrt{N}\varphi_t(x)\right) W(e^{-i\theta_2}\sqrt{N}\varphi)\Omega \Big\rangle \\
& + \frac{d_N^2 \, \overline{\varphi}_t(y)}{\sqrt{N}} \int_0^{2\pi} \frac{d\theta_1}{2\pi} \int_0^{2\pi} \frac{d\theta_2}{2\pi} \, e^{-i\theta_1 (N-1)} e^{i\theta_2 N} \\
& \times \Big\langle W(e^{-i\theta_1}\sqrt{N}\varphi)\Omega, \left(a_x(t) - e^{-i\theta_2}\sqrt{N}\varphi_t(x)\right) W(e^{-i\theta_2}\sqrt{N}\varphi)\Omega \Big\rangle \\
& + \frac{d_N^2 \, \varphi_t(x)}{\sqrt{N}} \int_0^{2\pi} \frac{d\theta_1}{2\pi} \int_0^{2\pi} \frac{d\theta_2}{2\pi} \, e^{-i\theta_1 N} e^{i\theta_2 (N-1)} \\
& \times \Big\langle W(e^{-i\theta_1}\sqrt{N}\varphi)\Omega, \left(a_y^*(t) - e^{i\theta_1}\sqrt{N}\overline{\varphi}_t(y)\right) W(e^{-i\theta_2}\sqrt{N}\varphi)\Omega \Big\rangle \\
& + d_N^2 \, \varphi_t(x)\overline{\varphi}_t(y) \int_0^{2\pi} \frac{d\theta_1}{2\pi} \int_0^{2\pi} \frac{d\theta_2}{2\pi} \, e^{-i\theta_1 (N-1)} e^{i\theta_2 (N-1)} \\
& \times \Big\langle W(e^{-i\theta_1}\sqrt{N}\varphi)\Omega, \, W(e^{-i\theta_2}\sqrt{N}\varphi)\Omega \Big\rangle \, .
\end{aligned}$$

Since

$$\begin{aligned}
d_N \int_0^{2\pi} & \frac{d\theta}{2\pi} e^{i\theta(N-1)} W(e^{-i\theta}\sqrt{N}\varphi)\Omega \\
& = d_N e^{-N/2} \sum_{j=0}^{\infty} \left(\int_0^{2\pi} \frac{d\theta}{2\pi} e^{i\theta(N-1-j)} \right) N^{j/2} \frac{(a^*(\varphi))^j}{j!} \Omega \\
& = d_N \frac{e^{-N/2} N^{(N-1)/2}}{\sqrt{N-1!}} \frac{(a^*(\varphi))^{N-1}}{N-1!} \Omega = \varphi^{\otimes N-1} \, ,
\end{aligned}$$

we find that

(4.30)

$$\begin{aligned}
& \gamma^{(1)}_{N,t}(x;y) - \varphi_t(x)\overline{\varphi}_t(y) \\
& = \frac{d_N^2}{N} \int_0^{2\pi} \frac{d\theta_1}{2\pi} \int_0^{2\pi} \frac{d\theta_2}{2\pi} \, e^{-i\theta_1 N} e^{i\theta_2 N} \Big\langle W(e^{-i\theta_1}\sqrt{N}\varphi)\Omega, \left(a_y^*(t) - e^{i\theta_1}\sqrt{N}\overline{\varphi}_t(y)\right) \\
& \qquad \times \left(a_x(t) - e^{-i\theta_2}\sqrt{N}\varphi_t(x)\right) W(e^{-i\theta_2}\sqrt{N}\varphi)\Omega \Big\rangle \\
& \quad + \frac{d_N \, \overline{\varphi}_t(y)}{\sqrt{N}} \int_0^{2\pi} \frac{d\theta_2}{2\pi} \, e^{i\theta_2 N} \Big\langle \varphi^{\otimes(N-1)}, \left(a_x(t) - e^{-i\theta_2}\sqrt{N}\varphi_t(x)\right) \\
& \qquad \times W(e^{-i\theta_2}\sqrt{N}\varphi)\Omega \Big\rangle \\
& \quad + \frac{d_N \, \varphi_t(x)}{\sqrt{N}} \int_0^{2\pi} \frac{d\theta_1}{2\pi} \, e^{-i\theta_1 N} \Big\langle W(e^{-i\theta_1}\sqrt{N}\varphi)\Omega, \left(a_y^*(t) - e^{i\theta_1}\sqrt{N}\overline{\varphi}_t(y)\right) \\
& \qquad \times \varphi^{\otimes(N-1)} \Big\rangle
\end{aligned}$$

and thus

$$\begin{aligned}\left|\gamma^{(1)}_{N,t}(x;y) - \varphi_t(x)\overline{\varphi}_t(y)\right| \leq \; & \frac{d_N^2}{N}\int_0^{2\pi}\frac{\mathrm{d}\theta_1}{2\pi}\int_0^{2\pi}\frac{\mathrm{d}\theta_2}{2\pi}\,\|a_y\mathcal{U}_N^{\theta_1}(t;0)\Omega\|\,\|a_x\mathcal{U}_N^{\theta_2}(t;0)\Omega\| \\ & + \frac{d_N\,|\varphi_t(y)|}{\sqrt{N}}\int_0^{2\pi}\frac{\mathrm{d}\theta}{2\pi}\|a_x\mathcal{U}_N^{\theta}(t;0)\Omega\| \\ & + \frac{d_N\,|\varphi_t(x)|}{\sqrt{N}}\int_0^{2\pi}\frac{\mathrm{d}\theta}{2\pi}\|a_y\mathcal{U}_N^{\theta}(t;0)\Omega\|\end{aligned}$$

where $\mathcal{U}_N^\theta(t;s)$ is defined as (4.14), with φ_t replaced by $e^{i\theta}\varphi_t$ (we are using, here, the fact that, although the Hartree equation is nonlinear, $e^{i\theta}\varphi_t$ is always a solution if φ_t is). Since $d_N \simeq N^{1/4}$, it follows from the bound (4.23) for the growth of the expectation of $\mathcal{N}$ with respect to the fluctuation evolution $\mathcal{U}_N^\theta(t;s)$ (the bound clearly holds uniformly in θ) that

$$\left\|\gamma^{(1)}_{N,t} - |\varphi_t\rangle\langle\varphi_t|\right\|_{\mathrm{HS}} \leq \frac{C}{N^{1/4}}e^{Kt}$$

and therefore (using the argument presented after (4.24) that

$$\mathrm{Tr}\left|\gamma^{(1)}_{N,t} - |\varphi_t\rangle\langle\varphi_t|\right| \leq \frac{C}{N^{1/4}}e^{Kt}\,. \tag{4.31}$$

To improve the decay in N on the r.h.s. of (4.31) from $N^{-1/4}$ to $N^{-1/2}$ (as claimed in Theorem 4.1), it is necessary to study the second and third error terms on the r.h.s. of (4.30) more precisely; for the details, see [**28**][Lemma 4.2].

Appendix A. Non-Standard Sobolev- and Poincaré Inequalities

In this section, we collect some non-standard Sobolev- and Poincaré-type inequalities which are very useful when dealing with singular potentials.

LEMMA A.1 (Sobolev-type inequalities). *Let $\psi \in L^2(\mathbb{R}^6, \mathrm{d}x_1\mathrm{d}x_2)$. If $V \in L^{3/2}(\mathbb{R}^3)$, we have*

$$\langle\psi, V(x_1-x_2)\psi\rangle \leq C\|V\|_{3/2}\,\langle\psi,(1-\Delta_1)\psi\rangle\,. \tag{A.1}$$

If $V \in L^1(\mathbb{R}^3)$, then

$$\langle\psi, V(x_1-x_2)\psi\rangle \leq C\|V\|_1\,\langle\psi,(1-\Delta_1)(1-\Delta_2)\psi\rangle \tag{A.2}$$

The first bound follows from a Hölder inequality followed by a standard Sobolev inequality (in the variable x_1, with fixed x_2). The proof of (A.2) can be obtained through the same arguments used in the proof of the next Poincaré-type inequality (see [**16**]).

LEMMA A.2 (Poincaré-type inequality). *Suppose that $h \in L^1(\mathbb{R}^3)$ is a probability density with $\int \mathrm{d}x\,|x|^{1/2}\,h(x) < \infty$. For $\alpha > 0$, let $h_\alpha(x) = \alpha^{-3}h(x/\alpha)$. Then we have, for every $0 \leq \kappa < 1/2$,*

$$\begin{aligned}&|\langle\varphi,(h_\alpha(x_1-x_2)-\delta(x_1-x_2))\,\psi\rangle| \\ &\qquad \leq C\alpha^\kappa\langle\varphi,(1-\Delta_1)(1-\Delta_2)\varphi\rangle^{1/2}\langle\psi,(1-\Delta_1)(1-\Delta_2)\psi\rangle^{1/2}\,.\end{aligned}$$

PROOF. We rewrite the inner product in Fourier space.

$$\Big\langle\varphi,\Big(h_\alpha(x_1-x_2)-\delta(x_1-x_2)\Big)\psi\Big\rangle=\int \mathrm{d}p_1\mathrm{d}p_2\mathrm{d}q_1\mathrm{d}q_2\mathrm{d}x\,\delta(p_1+p_2-q_1-q_2)\times\overline{\widehat{\varphi}}(p_1,p_2)\,\widehat{\psi}(q_1,q_2)\,h(x)\,\Big(e^{i\alpha(p_1-q_1)\cdot x}-1\Big).$$

Using that $|e^{i\alpha(p_1-q_1)\cdot x}-1|\leq\alpha^\kappa|x|^\kappa|p_1-q_1|^\kappa$, we obtain

$$\begin{aligned}&\Big|\langle\varphi,\Big(h_\alpha(x_1-x_2)-\delta(x_1-x_2)\Big)\psi\rangle\Big|\\&\leq\alpha^\kappa\left(\int\mathrm{d}x\,h(x)|x|^\kappa\right)\\&\qquad\times\int\mathrm{d}p_1\mathrm{d}p_2\mathrm{d}q_1\mathrm{d}q_2\,\delta(p_1+p_2-q_1-q_2)\,(|p_1|^\kappa+|q_1|^\kappa)\,|\widehat{\varphi}(p_1,p_2)|\,|\widehat{\psi}(q_1,q_2)|\,.\end{aligned}$$

We show how to control the term proportional to $|p_1|^\kappa$; the other term can be handled similarly.

$$\begin{aligned}&\Big|\langle\varphi,(h_\alpha(x_1-x_2)-\delta(x_1-x_2))\,\psi\rangle\Big|\\&\leq C\alpha^\kappa\int\mathrm{d}p_1\mathrm{d}p_2\mathrm{d}q_1\mathrm{d}q_2\,\delta(p_1+p_2-q_1-q_2)\\&\quad\times\frac{|p_1|^\kappa(1+p_1^2)^{(1-\kappa)/2}(1+p_2^2)^{1/2}}{(1+q_1^2)^{1/2}(1+q_2^2)^{1/2}}|\widehat{\varphi}(p_1,p_2)|\,\frac{(1+q_1^2)^{1/2}(1+q_2^2)^{1/2}}{(1+p_1^2)^{(1-\kappa)/2}(1+p_2^2)^{1/2}}|\widehat{\psi}(q_1,q_2)|\\&\leq C\alpha^\kappa\left(\int\mathrm{d}p_1\mathrm{d}p_2\mathrm{d}q_1\mathrm{d}q_2\,\delta(p_1+p_2-q_1-q_2)\frac{(1+p_1^2)(1+p_2^2)}{(1+q_1^2)(1+q_2^2)}|\widehat{\varphi}(p_1,p_2)|^2\right)^{1/2}\\&\quad\times\left(\int\mathrm{d}p_1\mathrm{d}p_2\mathrm{d}q_1\mathrm{d}q_2\,\delta(p_1+p_2-q_1-q_2)\frac{(1+q_1^2)(1+q_2^2)}{(1+p_1^2)^{1-\kappa}(1+p_2^2)}|\widehat{\psi}(q_1,q_2)|^2\right)^{1/2}\\&\leq C\alpha^{1/2}\langle\varphi,(1-\Delta_1)(1-\Delta_2)\varphi\rangle^{1/2}\langle\psi,(1-\Delta_1)(1-\Delta_2)\psi\rangle^{1/2}\\&\quad\times\left(\sup_p\int\mathrm{d}q\frac{1}{(1+q^2)(1+(p-q)^2)}\right)^{\frac12}\left(\sup_q\int\mathrm{d}p\frac{1}{(1+p^2)(1+(q-p)^2)^{1-\kappa}}\right)^{\frac12}.\end{aligned}$$

The claim follows because

$$\sup_{q\in\mathbb{R}^3}\int\mathrm{d}p\frac{1}{(1+p^2)(1+(q-p)^2)^{1-\kappa}}\leq C \tag{A.3}$$

for all $\kappa<1/2$. To prove (A.3) we consider the three regions $|p|>2|q|$, $|q|/2\leq|p|\leq2|q|$ and $|p|<|q|/2$ separately. Since $|p-q|>|p|/2$ for $|p|>2|q|$, it follows that

$$\begin{aligned}\int_{|p|>2|q|}\frac{\mathrm{d}p}{(1+p^2)(1+(q-p)^2)^{1-\kappa}}&\leq\int_{|p|>2|q|}\frac{\mathrm{d}p}{\left(1+\frac{p^2}{4}\right)^{2-\kappa}}\\&<C\int\frac{\mathrm{d}p}{(1+p^2)^{2-\kappa}}<\infty\end{aligned}$$

for $\kappa<1/2$, uniformly in q. For $|p|<|q|/2$, we use the fact that $|q-p|>|q|/2$, and we obtain

$$\int_{|p|<|q|/2}\frac{\mathrm{d}p}{(1+p^2)(1+(q-p)^2)^{1-\kappa}}\leq\frac{C}{(1+q^2)^{1-\kappa}}\int_{|p|<|q|/2}\frac{\mathrm{d}p}{1+p^2}\leq\frac{C|q|}{(1+q^2)^{1-\kappa}}$$

which is bounded uniformly in q. Finally, in the region $|q|/2 \leq |p| \leq 2|q|$, we use that

$$\int_{|q|/2<|p|<2|q|} \frac{\mathrm{d}p}{(1+p^2)(1+(q-p)^2)^{1-\kappa}} \leq \frac{C}{(1+q^2)} \int_{|p|<3|q|} \frac{\mathrm{d}p}{(1+p^2)^{1-\kappa}} \leq C\frac{|q|^{2\kappa+1}}{1+q^2} < \infty$$

uniformly in $q \in \mathbb{R}^3$, for all $\kappa < 1/2$. □

In the approach developed in [**15**] for the case of large interaction potential we can only prove weaker estimates on the solution $\psi_{N,t}$ of the Schrödinger equation. As discussed in Section 3.3, we can only prove that

$$\langle W^*_{N,(i,j)}\psi_{N,t}, \left((\nabla_i \cdot \nabla_j)^2 - \Delta_i - \Delta_j + 1\right) W^*_{N,(i,j)}\psi_{N,t}\rangle \leq C$$

uniformly in N and t. For this reason, we need estimates which only require the boundedness of the expectation of this particular combination of derivatives. The next lemma gives a Sobolev inequality of this type.

LEMMA A.3. *Suppose $V \in L^1(\mathbb{R}^3)$. Then*

$$|\langle\varphi, V(x_1-x_2)\psi\rangle| \leq C\|V\|_1 \langle\psi, \left((\nabla_1\cdot\nabla_2)^2 - \Delta_1 - \Delta_2 + 1\right)\psi\rangle^{1/2} \times \langle\varphi, \left((\nabla_1\cdot\nabla_2)^2 - \Delta_1 - \Delta_2 + 1\right)\varphi\rangle^{1/2}$$

for every $\psi, \varphi \in L^2(\mathbb{R}^6, \mathrm{d}x_1\mathrm{d}x_2)$.

PROOF. Switching to Fourier space, we find

$$\langle\varphi, V(x_1-x_2)\psi\rangle = \int \mathrm{d}p_1\mathrm{d}p_2\mathrm{d}q_1\mathrm{d}q_2 \times \overline{\widehat{\varphi}(p_1,p_2)}\widehat{\psi}(q_1,q_2)\,\widehat{V}(q_1-p_1)\,\delta(p_1+p_2-q_1-q_2)\,.$$

Therefore, with a weighted Schwarz inequality,

$$\begin{aligned}
&\left|\langle\varphi, V(x_1-x_2)\psi\rangle\right| \\
&\leq \|\widehat{V}\|_\infty \\
&\quad\times\left(\int \mathrm{d}p_1\mathrm{d}p_2\mathrm{d}q_1\mathrm{d}q_2\, \frac{(p_1\cdot p_2)^2+p_1^2+p_2^2+1}{(q_1\cdot q_2)^2+q_1^2+q_2^2+1}\,|\widehat{\varphi}(p_1,p_2)|^2\delta(p_1+p_2-q_1-q_2)\right)^{\frac12} \\
&\quad\times\left(\int \mathrm{d}p_1\mathrm{d}p_2\mathrm{d}q_1\mathrm{d}q_2 \frac{(q_1\cdot q_2)^2+q_1^2+q_2^2+1}{(p_1\cdot p_2)^2+p_1^2+p_2^2+1}|\widehat{\psi}(q_1,q_2)|^2\,\delta(p_1+p_2-q_1-q_2)\right)^{\frac12} \\
&\leq \|V\|_1\left(\sup_p \int \mathrm{d}q\, \frac{1}{(q\cdot(p-q))^2+q^2+(p-q)^2+1}\right) \\
&\quad\times\left\langle\psi, \left((\nabla_1\cdot\nabla_2)^2-\Delta_1-\Delta_2+1\right)\psi\right\rangle^{1/2}\left\langle\varphi, \left((\nabla_1\cdot\nabla_2)^2-\Delta_1-\Delta_2+1\right)\varphi\right\rangle^{1/2}.
\end{aligned}$$

The lemma follows from

$$\sup_{p\in\mathbb{R}^3}\int \mathrm{d}q\, \frac{1}{(q\cdot(p-q))^2+q^2+(p-q)^2+1} < \infty\,. \tag{A.4}$$

To prove (A.4), we write

$$\begin{aligned}
&\int \mathrm{d}q \, \frac{1}{(q\cdot(p-q))^2 + q^2 + (p-q)^2 + 1} \\
&\quad = \int_{|q-\frac{p}{2}|>|p|} \mathrm{d}q \, \frac{1}{\left(\left(q-\frac{p}{2}\right)^2 - \frac{p^2}{4}\right)^2 + q^2 + (p-q)^2 + 1} \\
&\quad\quad + \int_{|q-\frac{p}{2}|<|p|} \mathrm{d}q \, \frac{1}{\left(\left(q-\frac{p}{2}\right)^2 - \frac{p^2}{4}\right)^2 + q^2 + (p-q)^2 + 1}.
\end{aligned} \tag{A.5}$$

The first term on the r.h.s. of the last equation is bounded by

$$\begin{aligned}
&\int_{|q-\frac{p}{2}|>|p|} \mathrm{d}q \, \frac{1}{\left(\left(q-\frac{p}{2}\right)^2 - \frac{p^2}{4}\right)^2 + q^2 + (p-q)^2 + 1} \\
&\quad \leq \int_{|q-\frac{p}{2}|>|p|} \mathrm{d}q \, \frac{1}{\frac{9}{16}\left|q-\frac{p}{2}\right|^4 + 1} \leq \frac{16}{9}\int_{\mathbb{R}^3} \mathrm{d}q \, \frac{1}{|q|^4+1} < \infty,
\end{aligned}$$

uniformly in $p \in \mathbb{R}^3$. As for the second term on the r.h.s. of (A.5), we observe that

$$\begin{aligned}
&\int_{|q-\frac{p}{2}|<|p|} \mathrm{d}q \, \frac{1}{\left(\left(q-\frac{p}{2}\right)^2 - \frac{p^2}{4}\right)^2 + q^2 + (p-q)^2 + 1} \\
&\quad = \int_{|x|<|p|} \mathrm{d}x \, \frac{1}{\left(x^2 - \frac{p^2}{4}\right)^2 + \left(x+\frac{p}{2}\right)^2 + \left(x-\frac{p}{2}\right)^2 + 1} \\
&\quad = 4\pi \int_0^{|p|} \mathrm{d}r \frac{r^2}{\left(r^2 - \frac{|p|^2}{4}\right)^2 + 2r^2 + \frac{|p|^2}{2} + 1} \\
&\quad \leq C|p|^2 \int_{-|p|/2}^{|p|/2} \mathrm{d}r \frac{1}{r^2\left(r+|p|\right)^2 + \left(r+\frac{|p|}{2}\right)^2 + \frac{|p|^2}{4} + 1} \\
&\quad \leq C\int_{-|p|/2}^{|p|/2} \mathrm{d}r \, \frac{1}{r^2+1} \leq C\int_{\mathbb{R}} \mathrm{d}r \, \frac{1}{r^2+1} < \infty,
\end{aligned}$$

uniformly in p. □

References

[1] Adami, R.; Bardos, C.; Golse, F.; Teta, A.: Towards a rigorous derivation of the cubic nonlinear Schrödinger equation in dimension one. *Asymptot. Anal.* **40** (2004), no. 2, 93–108.

[2] Adami, R.; Golse, F.; Teta, A.: Rigorous derivation of the cubic NLS in dimension one. *J. Stat. Phys.* **127** (2007), no. 6, 1193–1220.

[3] Anderson, M.H.; Ensher, J.R.; Matthews, M.R.; Wieman, C.E.; Cornell, E.A.; *Science* (**269**), 198 (1995).

[4] Bardos, C.; Erdős, L.; Golse, F.; Mauser, N.; Yau, H.-T.: Derivation of the Schrödinger-Poisson equation from the quantum N-body problem. *C.R. Math. Acad. Sci. Paris* **334** (2002), no. 6, 515–520.

[5] Bardos, C.; Golse, F.; Mauser, N.: Weak coupling limit of the N-particle Schrödinger equation. *Methods Appl. Anal.* **7** (2000), 275–293.

[6] Davis, K.B.; Mewes, M.-O.; Andrews, M. R.; van Druten, N. J.; Durfee, D. S.; Kurn, D. M.; Ketterle, W.: *Phys. Rev. Lett.* **75**, 3969 (1995).

[7] Elgart, A.; Erdős, L.; Schlein, B.; Yau, H.-T.: Nonlinear Hartree equation as the mean field limit of weakly coupled fermions. *J. Math. Pures Appl. (9)* **83** (2004), no. 10, 1241–1273.

[8] Elgart, A.; Erdős, L.; Schlein, B.; Yau, H.-T.: Gross–Pitaevskii equation as the mean filed limit of weakly coupled bosons. *Arch. Rat. Mech. Anal.* **179** (2006), no. 2, 265–283.

[9] Elgart, A.; Schlein, B.: Mean Field Dynamics of Boson Stars. *Commun. Pure Appl. Math.* **60** (2007), no. 4, 500–545.

[10] Erdős, L.; Schlein, B.: Quantum dynamics with mean field interactions: a new approach. *J. Statist. Phys.* **134** (2009), no. 5, 859–870.

[11] Erdős, L.; Schlein, B.; Yau, H.-T.: Derivation of the Gross-Pitaevskii Hierarchy for the Dynamics of Bose-Einstein Condensate. *Commun. Pure Appl. Math.* **59** (2006), no. 12, 1659–1741.

[12] Erdős, L.; Schlein, B.; Yau, H.-T.: Derivation of the cubic non-linear Schrödinger equation from quantum dynamics of many-body systems. *Invent. Math.* **167** (2007), 515–614.

[13] Erdős, L.; Schlein, B.; Yau, H.-T.: Derivation of the Gross-Pitaevskii Equation for the Dynamics of Bose-Einstein Condensate. *Ann. of Math. (2)* **172** (2010), no. 1, 291–370.

[14] Erdős, L.; Schlein, B.; Yau, H.-T.: Rigorous derivation of the Gross-Pitaevskii equation. *Phys. Rev Lett.* **98** (2007), no. 4, 040404.

[15] Erdős, L.; Schlein, B.; Yau, H.-T.: Rigorous Derivation of the Gross-Pitaevskii Equation with a Large Interaction Potential. *J. Amer. Math. Soc.* **22** (2009), no. 4, 1099-1156.

[16] Erdős, L.; Yau, H.-T.: Derivation of the nonlinear Schrödinger equation from a many body Coulomb system. *Adv. Theor. Math. Phys.* **5** (2001), no. 6, 1169–1205.

[17] Fröhlich, J.; Graffi, S.; Schwarz, S.: Mean-field- and classical limit of many body Schrödinger dynamics for bosons. *Comm. Math. Phys.* **271** (2007), no. 3, 681-697.

[18] Fröhlich, J.; Knowles, A.; Pizzo, A.: Atomism and quantization. *J. Phys. A* **40** (2007), no. 12, 3033-3045.

[19] Fröhlich, J.; Knowles, A.; Schwarz, S.: On the mean-field limit of bosons with Coulomb two-body interaction. *Comm. Math. Phys.* **288** (2009), no. 3, 1023–1059.

[20] Ginibre, J.; Velo, G.: The classical field limit of scattering theory for nonrelativistic many-boson systems. I.-II. *Comm. Math. Phys.* **66** (1979), no. 1, 37–76, and **68** (1979), no. 1, 45–68.

[21] Hepp, K.: The classical limit for quantum mechanical correlation functions. *Comm. Math. Phys.* **35** (1974), 265–277.

[22] Lieb, E.H.; Seiringer, R.: Proof of Bose-Einstein condensation for dilute trapped gases. *Phys. Rev. Lett.* **88** (2002), 170409-1-4.

[23] Lieb, E.H.; Seiringer, R.; Solovej, J.P.; Yngvason, J.: *The mathematics of the Bose gas and its condensation.* Oberwolfach Seminars, **34.** Birkhauser Verlag, Basel, 2005.

[24] Lieb, E.H.; Seiringer, R.; Yngvason, J.: Bosons in a trap: a rigorous derivation of the Gross-Pitaevskii energy functional. *Phys. Rev A* **61** (2000), 043602.

[25] Klainerman, S.; Machedon, M.: On the uniqueness of solutions to the Gross-Pitaevskii hierarchy. *Commun. Math. Phys.* **279** (2008), no. 1, 169–185.

[26] Narnhofer, H.; Sewell, G. L.: Vlasov hydrodynamics of a quantum mechanical model. *Commun. Math. Phys.* **79** (1981), 9–24.

[27] Reed, M.; Simon, B.: *Methods of Modern Mathematical Physics.* Vol.I. Academic Press, 1975.

[28] Rodnianski, I.; Schlein, B.: Quantum fluctuations and rate of convergence towards mean field dynamics. *Comm. Math. Phys.* **291** (2009), no. 1, 31–61.

[29] Rudin, W.: *Functional analysis.* McGraw-Hill Series in Higher Mathematics, McGraw-Hill Book Co., New York, 1973.

[30] Spohn, H.: Kinetic Equations from Hamiltonian Dynamics. *Rev. Mod. Phys.* **52** (1980), no. 3, 569–615.

[31] Spohn, H.: On the Vlasov hierarchy. *Math. Methods Appl. Sci.* **3** (1981), no. 4, 445–455.

[32] Yajima, K.: The $W^{k,p}$-continuity of wave operators for Schrödinger operators. *J. Math. Soc. Japan* **47** (1995), no. 3, 551–581.

[33] Yajima, K.: The $W^{k,p}$-continuity of wave operators for Schrödinger operators. *Proc. Japan Acad. Ser. A Math. Sci.* **69** (1993), no. 4, 94–98.

CURRENT ADDRESS: INSTITUTE OF APPLIED MATHEMATICS, UNIVERSITY OF BONN, ENDENICHER ALLEE 60, 53115 BONN, GERMANY

E-mail address: benjamin.schlein@hcm.uni-bonn.de

Published Titles in This Series

17 **G. Staffilani, Igor Rodnianski, Gigliola Staffilani, and Jared Wunsch, Editors,** Evolution Equations, 2013

16 **Guillermo Cortiñas, Editor,** Topics in Noncommutative Geometry, 2012

15 **David Ellwood, Charles Newman, Vladas Sidoravicius, and Wendelin Werner, Editors,** Probability and Statistical Physics in Two and More Dimensions, 2012

14 **David Ellwood and Emma Previato, Editors,** Grassmannians, Moduli Spaces and Vector Bundles, 2011

13 **James Arthur, James W. Cogdell, Steve Gelbart, David Goldberg, Dinakar Ramakrishnan, and Jiu-Kang Yu, Editors,** On Certain L-Functions, 2011

12 **Alan Carey, David Ellwood, Sylvie Paycha, and Steve Rosenberg, Editors,** Motives, Quantum Field Theory, and Pseudodifferential Operators, 2010

11 **Etienne Blanchard, David Ellwood, Masoud Khalkhali, Matilde Marcolli, Henri Moscovici, and Sorin Popa, Editors,** Quanta of Maths, 2010

10 **Manfred Leopold Einsiedler, David Ellwood, Alex Eskin, Dmitry Kleinbock, Elon Lindenstrauss, Gregory Margulis, Stefano Marmi, and Jean-Christophe Yoccoz, Editors,** Homogeneous Flows, Moduli Spaces and Arithmetic, 2010

9 **Reza Akhtar, Patrick Brosnan, and Roy Joshua, Editors,** The Geometry of Algebraic Cycles, 2010

8 **Henri Darmon, David Ellwood, Brendan Hassett, and Yuri Tschinkel, Editors,** Arithmetic Geometry, 2009

7 **William Duke and Yuri Tschinkel, Editors,** Analytic Number Theory, 2007

6 **Nigel Higson and John Roe, Editors,** Surveys in Noncommutative Geometry, 2006

5 **David Ellwood, Peter S. Ozsváth, András I. Stipsicz, and Zoltán Szabó, Editors,** Floer Homology, Gauge Theory, and Low-Dimensional Topology, 2006

4 **James Arthur, David Ellwood, and Robert Kottwitz, Editors,** Harmonic Analysis, the Trace Formula, and Shimura Varieties, 2005

3 **Michael Douglas, Jerome Gauntlett, and Mark Gross, Editors,** Strings and Geometry, 2004

2 **David Hoffman, Editor,** Global Theory of Minimal Surfaces, 2005

1 **Atish Dabholkar, Sunil Mukhi, and Spenta R. Wadia, Editors,** Strings 2001, 2002